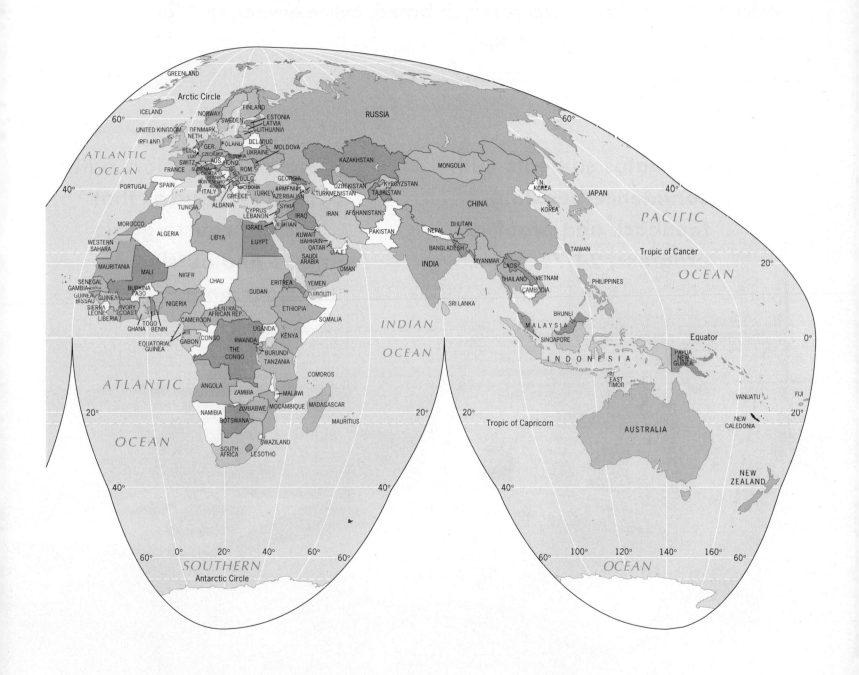

WILEY PLUS

www.wileyplus.com

accessible, affordable, active learning

WileyPLUS is an innovative, research-based, online environment for effective teaching and learning.

WileyPLUS...

...motivates students with confidence-boosting feedback and proof of progress, 24/7.

...supports instructors with reliable resources that reinforce course goals inside and outside of the classroom.

Includes Interactive Textbook & Resources

WileyPLUS... **Learn More.**

www.wile**plus**.com

www.wileyplus.com

ALL THE HELP, RESOURCES, AND PERSONAL SUPPORT YOU AND YOUR STUDENTS NEED!

www.wileyplus.com/resources

2-Minute Tutorials and all of the resources you & your students need to get started.

Student support from an experienced student user.

Collaborate with your colleagues, find a mentor, attend virtual and live events, and view resources.
www.WhereFacultyConnect.com

Pre-loaded, ready-to-use assignments and presentations. Created by subject matter experts.

Technical Support 24/7 FAQs, online chat, and phone support.
www.wileyplus.com/support

Your *WileyPLUS* Account Manager. Personal training and implementation support.

FIFTH EDITION

THE WORLD TODAY

CONCEPTS AND REGIONS IN GEOGRAPHY

To
NLJ

FIFTH EDITION

H. J. de Blij
John A. Hannah Professor of Geography
Michigan State University

Peter O. Muller
Professor, Department of Geography and Regional Studies
University of Miami

Jan Nijman
Professor, Department of Geography and Regional Studies
University of Miami

Antoinette M. G. A. WinklerPrins
Associate Professor, Department of Geography
Michigan State University

THE WORLD TODAY

CONCEPTS AND REGIONS IN GEOGRAPHY

JOHN WILEY & SONS, INC.

VICE PRESIDENT AND EXECUTIVE PUBLISHER	Jay O'Callaghan
EXECUTIVE EDITOR	Ryan Flahive
ASSOCIATE EDITOR	Veronica Armour
PRODUCTION SERVICES MANAGER	Dorothy Sinclair
PRODUCTION EDITOR	Janet Foxman
MARKETING MANAGER	Margaret Barrett
CREATIVE DIRECTOR	Harry Nolan
COVER DESIGN	Maureen Eide
INTERIOR DESIGN	Carole Anson
PHOTO EDITOR	Jennifer MacMillan
ILLUSTRATION EDITOR	Anna Melhorn
EDITORIAL ASSISTANTS	Meredith Leo/Darnell Sessoms
MEDIA EDITOR	Lynn Pearlman
PRODUCTION SERVICES	Furino Production
FRONT COVER PHOTO	© Images & Stories/Alamy
BACK COVER PHOTO	© H. J. de Blij

This book was set in Adobe Garamond Regular by Aptara and printed and bound by Quad/Graphics/Dubuque. The cover was printed by Quad/Graphics/Dubuque.

This book is printed on acid-free paper. ∞

Library of Congress Cataloging-in-Publication Data
The world today : concepts and regions in geography/H. J. de Blij . . . [et al.].—5th ed.
   p. cm.

  Includes bibliographical references.
 ISBN 978-0-470-64638-0 (pbk.)
  1. Geography—Textbooks.   I. De Blij, Harm J.   II. De Blij, Harm J. World Today.

 G128.D394 2010
 910—dc22

                                    2010023278

Main Book ISBN 978-0-470-64638-0
Binder-Ready Version ISBN 978-0-470-91750-3

Printed in the United States of America

10 9 8 7 6 5 4 3 2

PREFACE

This is the Fifth Edition of a book that, since its first appearance in 2003, has undergone several significant changes. Beginning with the Third Edition, its original title became the subtitle because its substantive content became ever more global and current even as its geographic perspective continued to provide the structural framework. Readers told us that they wanted geographic background and explanation for events and circumstances that are changing our increasingly complex world, and a more appropriate title, *The World Today*, signaled our response. The Fourth Edition benefited from the views and insights of co-author Antoinette M.G.A. WinklerPrins of Michigan State University. And a substantial part of this Fifth Edition was prepared by co-author Jan Nijman of the University of Miami.

NEW FORMAT

Those familiar with this book will note that the format has changed markedly. All previous editions were laid out in what is called a "landscape" format in the publishing world, that is, a horizontal shape with dimensions resembling those of a postcard showing a wide view of scenery. The resulting advantage was that maps, especially world maps, could be printed at a relatively large scale without being interrupted by the "gutter"—the place where the pages are bound into the spine. But the disadvantage was a relatively short page, much shorter than in the alternative "portrait" version, when a book has the approximate shape of a postcard held vertically. This new edition introduces a format that combines advantages of "landscape" as well as "portrait"—a wide page, almost as wide as our earlier editions had, but also a tall page, allowing for far greater flexibility of layout and design.

New Structure

Coupled with this change in dimensions is an even more consequential transformation of "*TWT*", as this book has come to be known. As the Table of Contents indicates, a wide-ranging introduction in global perspective is followed by a dozen chapters, each focusing on one of the world's great geographic **realms**. In previous editions, these realms and their constituent parts, geographic **regions**, were discussed separately, but the separation was often informal. We decided to divide each of the major regional chapters into Parts **A** and **B**, with Part **A** dealing with the realm under scrutiny as a whole, and Part **B** focusing on the regional components. This increases *TWT*'s flexibility as a learning guide, because it offers the option to confine study to the realm level by targeting only **A** chapters; but it also facilitates further study at the regional level in particular, targeted realms. As everyone who has taught or taken a course in World Regional Geography knows, the topic is complex and semesters are short. *TWT*'s new structure addresses this challenge.

New Approaches

Every new edition of *TWT* constitutes a combination of retention and innovation, and this Fifth Edition is no exception. A heavier emphasis on matters environmental and demographic marked the Fourth Edition; less stress on details of historical geography marks this revision. While some of our *Field Notes* were retained, others, based on fieldwork by co-author Jan Nijman, are published here for the first time.

As readers of this book will be aware, this edition appears at a time when the debate over globalization and its merits (and demerits) has been heating up, especially following the appearance of a book by a newspaper columnist, Thomas Friedman, entitled *The World is Flat* (New York: Farrar, Straus & Giroux, 2005). For many years, geographers have been debating the causes of the very opposite: the continuing and in some ways growing contrasts between and among haves and have-nots in this un-flat world of wealth and poverty, privilege and penury, fortune and failure. Under the rubric of "core-periphery contrasts," any number of books and research papers asserted that people living in core areas—that is, the rich world, or pockets of wealth amid poverty in poorer societies—were invariably locked in an exploitive relationship. That is, clusters of the rich and powerful would only be able to perpetuate their advantage through exploitation of the poor and weak. Globalization, argue opponents of this view, proves that there is nothing inevitable about this. Look at societies like South Korea, Taiwan, Chile, and Mauritius, and you can see how political and economic decisions can improve the lot of those in the periphery.

As is so often the case in disputes like this, realities on the ground get obscured. Of course the world is not flat, but is it flattening? And, of course, not all core-periphery relationships are exploitive, but are things getting better or worse?

In this edition we try to shed some light on this worldwide issue, which is raised in the Introduction and then forms a unifying thread throughout our book, through maps, text, and, importantly, the Population Table in Appendix B at the end of this volume. If you have never heard of Mauritius, take a look at it in the list of African states and ask yourself—why are the people of this island-state earning six times what the average African earns? Or, given its location and ethnically-mixed population, four times what the average South Asian makes annually? You will find this Table full of surprises. It's a unique feature of this book and we urge you to refer to it when you get into discussions of topics ranging from women's life expectancy (longer than men's—except in a few countries), endemic corruption (you won't believe where Russia ranks), and even the "real" cost of a hamburger around the world.

More than likely you will have seen or heard references in the media to Americans' "geographic illiteracy"—the jokes come up in columns like Dear Abby's, one of whose readers recently reported having sat next to someone in an airplane on the way from Los Angeles to Hawai'i who wanted to know whether it would be better to change money at the Honolulu airport or downtown. But there is nothing amusing about it, because a lack of geographic awareness can lead to poor decision-making and global disadvantage. By taking this course and reading this book, you will travel a long stretch on the information highway to geographic literacy.

The World Today

During the semester in which you read this book, the world is sure to experience some sudden, unexpected events. Consider the two years just past. What did we know about the clockwise-churning "loop current" in the Gulf of Mexico before the oil-spill disaster in the spring of 2010, when suddenly the currents and winds in the Gulf became matters of life or death for wildlife and economic threat or ruin for coastal communities? When we got news of a massive earthquake in Qinghai Province (China) not long before,

why was it that Tibetans were affected when the quake didn't happen in Tibet? When terrorists exploded and derailed two trains in West Bengal State in May 2010, killing more than 150 people, what was their objective in India's east, where Islam is not the key issue but communist insurgency is growing? When North Korea torpedoed a South Korean warship in March 2010 killing 46 sailors and risking war on the peninsula, were we aware that the infamous DMZ boundary that divides the Korean Peninsula extends westward into the Yellow Sea, where it is in dispute? When Israel in June 2010 intercepted a flotilla of boats headed for Gaza from Cyprus with deathly consequences, were those boats in international waters or not? When the Russian delegation at the Copenhagen conference on climate change in December 2009 stayed out of the five-nation draft agreement presented to the participants, why did Russia see global warming in a different light? When Greece faced an economic crisis in April 2010, how did this relate to core-periphery realities in the still-evolving European Union?

For questions like this, **The World Today** will provide context, and the maps may even offer direct answers (as in the case of Figures 9A-5 and 9B-8). But such events punctuate longer-term trends in our fast-changing world ranging from climate change to China's ascendancy on the global stage and from migration patterns and problems to terrorism and its motivations. For all this and much more, **TWT** will—as does geography generally—widen your horizons and deepen your perspectives.

DATA SOURCES

Numerous print and Internet sources were consulted during the updating of the book. For all matters geographical, of course, we consult *The Annals of the Association of American Geographers, The Professional Geographer, The Geographical Review, The Journal of Geography,* and many other academic journals published regularly in North America—plus an array of similar periodicals published in English-speaking countries from Scotland to New Zealand.

All quantitative information was updated to the year of publication and checked rigorously. Hundreds of other modifications were made, many in response to readers' and reviewers' comments. New spellings of geographic names continue, and we pride ourselves in being a reliable source for current and correct usage.

The statistical data that constitute Appendix B are derived from numerous sources. As users of such data are aware, considerable inconsistency marks the reportage by various agencies, and it is often necessary to make informed decisions on contradictory information. For example, some sources still do not reflect the rapidly declining rates of population increase or life expectancies in AIDS-stricken African countries. Others list demographic averages without accounting for differences between males and females in this regard.

In formulating Appendix B we have used among our sources the United Nations, the Population Reference Bureau, the World Bank, the Encyclopaedia Britannica *Books of the Year*, *The Economist* Intelligence Unit, the *Statesman's Year-Book*, and *The New York Times Almanac*. The urban population figures—which also entail major problems of reliability and comparability—are mainly drawn from the most recent database published by the United Nations' Population Division. For cities of less than 750,000, we developed our own estimates from a variety of sources. At any rate, the urban population figures used here are estimates for 2011 and they represent *metropolitan-area totals* unless otherwise specified.

PEDAGOGY

We continue to devise ways to help students learn important geographic concepts and ideas, and to make sense of our complex and rapidly changing world. Continuing special features include the following:

Two-Part (A/B) Chapter Organization. As explained in the third paragraph of this preface above, **A** chapters focus on the featured geographic realm as a whole whereas **B** chapters concentrate on that realm's major regional components. This scheme is followed for Chapters 1–10; Chapters 11 and 12, because they are much shorter, preserve the new structure within a single chapter. The opening Introduction chapter, which treats global perspectives, does not adhere to this divisional scheme.

Concepts, Ideas, and Terms. Each chapter begins with a boxed sequential listing of the key geographic concepts, ideas, and terms that appear in the pages that follow. These are noted by numbers in the margins (e.g., **1**) that correspond to the introduction of each item in the text.

List of Regions. On the opening page of each **B** chapter, a list of the regions within the particular realm provides a preview and helps to organize the chapter.

Major Geographic Qualities. Near the beginning of each **A** chapter (except in the Introduction), we list, in boxed format, the major geographic qualities that best summarize that portion of the Earth's surface.

Points to Ponder. This boxed feature at the end of each chapter offers a few brief statements about things of geographic import happening (or likely to happen) in a realm or region that may give one pause.

From the Field Notes. Many of the photographs in this book were taken by the senior author while doing fieldwork. By linking them to extended captions entitled *From the Field Notes*, valuable insights are provided into how a geographer observes and interprets information in the field. We are also pleased to introduce a number of Field Notes contributed by our third author, who is new to this edition.

Table of Area and Demographic Data for the World's States. This seven-page table in Appendix B displays the most important geographic and population data for each of the world's countries, updated to 2010. Definitions and brief discussions of each data category in the table are found at the end of Appendix B.

Appendices, Glossary, and Index. In addition to the data table that is Appendix B, we feature Appendix A that lists metric and customary units and their conversions. In the text we consistently use metric measurements with British units in parentheses. The extensive Glossary as well as the Index are available at the end of the book. Further material that was included in the final sections of the book in previous editions is available on the book's website.

ANCILLARIES

A broad spectrum of print and electronic ancillaries are available to accompany *The World Today*. Additional information, including prices and ISBNs for ordering, can be obtained by contacting John Wiley & Sons, Inc. Go to: *www.wiley.com/college/deblij*

The Teaching and Learning Package

This Fifth Edition of *The World Today: Concepts and Regions in Geography* is supported by a comprehensive supplements package that includes an extensive selection of print, visual, and electronic materials.

WileyPLUS

WileyPLUS is an innovative, research-based, online environment for effective teaching and learning.

What Do Students Receive with *WileyPLUS*?

A Research-based Design. *WileyPLUS* provides an online environment that integrates relevant resources, including the entire digital textbook, in an easy-to-navigate framework that helps students study more effectively.

- *WileyPLUS* adds structure by organizing textbook content into smaller, more manageable "chunks."
- Related media, examples, and sample practice items reinforce the learning objectives.
- Innovative features such as calendars, visual progress tracking, and self-evaluation tools improve time management and strengthen areas of weakness.

One-on-One Engagement. With *WileyPLUS* for *The World Today, 5th Edition*, students receive 24/7 access to resources that promote positive learning outcomes. Students engage with related examples (in various media) and sample practice items. These include:

- Videos: People & Place videos from the National Geographic Collection and *On Location with Harm de Blij* videos.
- Map Testing: students build mental maps with these automatically-graded map tests. Grades flow automatically to your gradebook.
- GeoDiscoveries Media Library.
- Practice Questions.
- www.ConceptCaching.com: caches linked directly to the ebook provide additional examples of the concepts.

Measurable Outcomes. Throughout each study session, students can assess their progress and gain immediate feedback. *WileyPLUS* provides precise reporting of strengths and weaknesses, as well as individualized quizzes, so that students are confident they are spending their time on the right things. With *WileyPLUS*, students always know the exact outcome of their efforts.

What do Instructors Receive with *WileyPLUS*? *WileyPLUS* provides reliable, customizable resources that reinforce course goals inside and outside of the classroom as well as visibility into individual student progress. Precreated materials and activities help instructors optimize their time:

Customizable Course Plan: *WileyPLUS* comes with a pre-created Course Plan designed by a subject matter expert uniquely for this course. Simple drag-and-drop tools make it easy to assign the Course Plan as-is or modify it to reflect your course syllabus.

Pre-Created Activity Types Include:
- o Questions
- o Readings and resources
- o Presentations
- o Printed Tests
- o Concept Mastery
- o Projects

Course Materials and Assessment Content:
- o Lecture Notes PowerPoint Slides
- o Classroom Response System (Clicker) Questions
- o Image Gallery
- o Instructor's Manual
- o Gradable Reading Assignment Questions (embedded within online text)
- o Google Earth™ Tours and Activities
- o Test Bank

Gradebook: *WileyPLUS* provides instant access to reports on trends in class performance, student use of course materials, and progress towards learning objectives, helping inform decisions and drive classroom discussions.

WileyPLUS. Learn More. www.wileyplus.com.

Powered by proven technology and built on a foundation of cognitive research, *WileyPLUS* has enriched the education of millions of students in over 20 countries around the world.

Resources That Help Teachers Teach

***People and Place* Lecture Launcher Videos.** Working closely with Wiley's Media Team, Erin Fouberg (Northern State University [South Daota]) has created a new collection of video lecture launchers that allow instructors to provide a visual context for key concepts, ideas, and terms that they plan to introduce during their lectures. These videos are available in a DVD format that is optimized for in-class presentation as well as in a streaming format that is available online. Streaming videos will also be made available to students in the context of *Wiley-PLUS* assignments that can be graded online and added to the *WileyPLUS* instructor gradebook.

***WileyPLUS* Online Map Testing.** We know that map tests are important to your course but also recognize that grading these map tests takes time out of your busy schedule. Therefore, we have built secure online tests to support every chapter of *The World Today* that can be customized to ask the questions that you want within a time frame that you set. With *WileyPLUS*, you are able to set all of the parameters for your tests. Moreover, your *WileyPLUS* gradebook will automatically grade every test and provide you with the ability to measure and track each student's grade as well as uncover trends for the entire class. Your students will also have separate Map Quizzes for each chapter that allow them to study for your map tests at their own pace.

Online Geography *On-Location* Videos. Because of their enduring popularity, we have digitized many of the videos from the original VHS series. This rich collection of original and relevant footage was taken during H. J. de Blij's travels. These videos cover a wide range of themes and are integrated throughout each student's *WileyPLUS* eBook.

***Google Earth* Links, Tours, and Activities.** Photos from the *WileyPlus eBook* are linked from the text to their actual location on the Earth using Google Earth. Tours and activities created by professors engage students with geographic concepts addressed in the text. Contributing professors include: Randy Rutberg, Hunter College (New York), and Jeff DeGrave, University of Wisconsin, Eau Claire and James Hayes-Bohanan, Bridgewater State College (MA).

Geodiscoveries Interactions. A collection of activities, animations, and videos that allow students to explore key concepts in greater depth.

ConceptCaching Website *www.conceptcaching.com* ConceptCaching is a pedagogical technique that promotes student spatial awareness by relating specific concepts from their learning to their visual character and GPS coordinates. Photographs and GPS coordinates that demonstrate core concepts in geography are "cached" for viewing by core concept and by region. Additionally, all of the *From the Field Notes* photo boxes in the text are linked to the ConceptCaching website.

Additional Videos and Podcasts. We have created a robust collection of streaming video resources and podcasts to support the Fifth Edition of *The World Today: Concepts and Regions in Geography*. Our streaming videos provide short lecture-launching video clips that can be used to introduce new topics, enhance your presentations, and stimulate classroom discussion.

***The World Today 5e* Instructor's Website** can be found at *www.wiley.com/college/deblij*. This comprehensive website includes numerous resources to help you enhance your current presentations, create new presentations, and employ our pre-made PowerPoint presentations. These resources include:

- **Image Gallery.** We provide online electronic files for the line illustrations and maps in the text, which the instructor can customize for presenting in class (for example, in handouts, overhead transparencies, or PowerPoints).
- A complete collection of **PowerPoint presentations** is available in beautifully rendered, four-color format, which have been resized and edited for maximum effectiveness in large lecture halls. Lecture PowerPoints, prepared by Heidi Lannon, Santa Fe College, include links to videos and animations.
- Revised by Antoinette WinklerPrins, the **Test Bank** for the Fifth Edition of *The World Today* contains over 3000 test items including multiple-choice, true-false, and fill-in questions. It is distributed via the secure Instructor's website as electronic files, which can be saved into all major word processing programs.
- Also revised by co-author Antoinette WinklerPrins, **Clicker Questions** provide instructors with an opportunity for immediate feedback from their students. These clicker questions are available in a number of formats

and also include system instructions and an instructor's manual.
- **Photo and Image Galleries** provide instructors with images from the text for use in their classrooms.

Wiley Faculty Network. This peer-to-peer network of faculty is ready to support your use of online course management tools and discipline-specific software-learning systems in the classroom. The WFN will help you apply innovative classroom techniques, implement software packages, tailor the technology experience to the needs of each individual class, and provide you with virtual training sessions led by faculty for faculty.

Course Management. Online course management assets are available to accompany the Fifth Edition of *The World Today*.

Resources That Help Students Learn

Student Companion Website can be accessed at *www.wiley.com/college/deblij*. This easy-to-use and student-focused website helps reinforce and illustrate key concepts from the text. It also provides interactive media content that helps students prepare for exams and improve their grades. This website provides additional resources that complement the textbook and enhance students' understanding of geography:

- **Map Quizzes** help students master the place names that are crucial to their success in this course. Three game-formatted place-name activities are provided for each chapter.
- **The Interactive Globe** allows the students to explore the Earth in three dimensions. They are able to apply different textures that reveal spatial information about climate, soils, and demographics. Area and demographic data are available by clicking on each country or world realm.
- **Flashcards** offer an excellent way to drill and practice key concepts, ideas, and terms from the text.
- **Base Maps** are available for student study and practice.
- **Virtual Field Trips** and **Photo Gallery** allow students to "visit" many of the locations from their text.
- **Chapter Review Quizzes** provide immediate feedback to true-false, multiple-choice, and other short-answer questions.
- **Audio Pronunciation** is provided for over 2000 key words and place names from the text.
- **Annotated Web Links** allow students to further explore key topics on their own.
- **Area and Demographic Data** are provided in Appendix B for every country and world realm.

Student Study Guide. Written by Justin Scheidt, Ferris State University (Michigan), the *Study Guide* gives students and faculty access to content questions-and-answers, outline maps of each realm, map exercises, and more.

Wiley/National Geographic *College Atlas of the World.* Wiley is proud to offer the *College Atlas of the World* to your students through our exclusive partnership with the National Geographic Society. State-of-the-art cartographic technology plus award-winning design and content make this affordable, compact, yet comprehensive atlas the ultimate resource for every geography student. The past decade has seen an explosion of innovative and sophisticated digital mapping technologies—and the Society has been at the forefront of developing these powerful new tools to create the finest, most functional, and informative atlases available anywhere. This powerful resource can be purchased individually or at a significant discount when packaged with any Wiley textbook.

ACKNOWLEDGMENTS

Over the 40 years since the publication of the First Edition of *Geography: Realms, Regions, and Concepts* (and joined more recently by the appearance of this book), we have been fortunate to receive advice and assistance from literally hundreds of people. One of the rewards associated with the publication of a book of this kind is the steady stream of correspondence and other feedback it generates. Geographers, economists, political scientists, education specialists, and others have written us, often with fascinating enclosures. We make it a point to respond personally to every such letter, and our editors have communicated with many of our correspondents as well. Moreover, we have considered every suggestion made and many who wrote or transmitted their reactions through other channels will see their recommendations in print in this edition.

Student Response

A major part of the correspondence we receive comes from student readers. We would like to take this opportunity to extend our deep appreciation to the several million students around the world who have studied from our books. In particular, we thank the students from more than 130 different colleges across the United States who took the time to send us their opinions. Students told us they found the maps and graphics attractive and functional. We have not only enhanced the map program with exhaustive updating but have added a number of new and updated maps to this Fifth Edition as well as making significant changes in many others. Generally, students have told us that they found the pedagogical devices quite useful. We have kept

the study aids the students cited as effective: a boxed list of each chapter's key concepts, ideas, and terms (numbered for quick reference in both the box and text margins); a box summarizing each realm's major geographic qualities; and an extensive and still-expanding Glossary.

Faculty Feedback

In assembling the Fifth Edition, we are indebted to the following people for advising us on a number of matters:

CHRISTOPHER BADUREK, *Appalachian State University*
JILL (ALICE) BLACK, *Missouri State University*
JEFF DE GRAVE, *University of Wisconsin, Eau Claire*
BRYANT EVANS, *Houston Community College*
WILLIAM FLYNN, *Oklahoma State University*
HARI GARBHARRAN, *Middle Tennessee State University*
TRUMAN HARTSHORN, *Georgia State University*
UWE KACKSTAETTER, *Front Range Community College, Westminster*
CUB KAHN, *Oregon State University*
HEIDI LANNON, *Santa Fe College*
J. MIGUEL KANAI, *University of Miami*
CHRIS MAYDA, *Eastern Michigan University*
ROSANN POLTRONE, *Arapahoe Community College*
JOEL QUAM, *College of DuPage (Illinois)*
RHONDA REAGAN, *Blinn College (Texas)*
PAUL ROLLINSON, *Missouri State University*
JUSTIN SCHEIDT, *Ferris State University (Michigan)*
KATHLEEN SCHROEDER, *Appalachian State University*
DMITRII SIDOROV, *California State University, Long Beach*

In addition, several faculty colleagues from around the world assisted us with earlier editions, and their contributions continue to grace the pages of this book. Among them are:

JAMES P. ALLEN, *California State University, Northridge*
STEPHEN S. BIRDSALL, *University of North Carolina*
J. DOUGLAS EYRE, *University of North Carolina*
FANG YONG-MING, *Shanghai, China*
EDWARD J. FERNALD, *Florida State University*
RAY HENKEL, *Arizona State University*
RICHARD C. JONES, *University of Texas at San Antonio*
GIL LATZ, *Portland State University (Oregon)*
IAN MACLACHLAN, *University of Lethbridge (Alberta)*
MELINDA S. MEADE, *University of North Carolina*
HENRY N. MICHAEL, *late of Temple University (Pennsylvania)*
CLIFTON W. PANNELL, *University of Georgia*
J. R. VICTOR PRESCOTT, *University of Melbourne (Australia)*
JOHN D. STEPHENS, *University of Washington*
CANUTE VANDER MEER, *University of Vermont*

Faculty members from a large number of North American colleges and universities continue to supply us with vital feedback and much-appreciated advice. Our publishers arranged several feedback sessions, and we are most grateful to the following professors for showing us where the text could be strengthened and made more precise:

MARTIN ARFORD, *Saginaw Valley State University (Michigan)*
DONNA ARKOWSKI, *Pikes Peak Community College (Colorado)*
GREG ATKINSON, *Tarleton State University (Texas)*
DENIS BEKAERT, *Middle Tennessee State University*
THOMAS L. BELL, *University of Tennessee*
DONALD J. BERG, *South Dakota State University*
KATHLEEN BRADEN, *Seattle Pacific University*
DAVID COCHRAN, *University of Southern Mississippi*
JOSEPH COOK, *Wayne County Community College (Michigan)*
DEBORAH CORCORAN, *Southwest Missouri State University*
MARCELO CRUZ, *University of Wisconsin at Green Bay*
WILLIAM V. DAVIDSON, *Louisiana State University*
LARRY SCOTT DEANER, *Kansas State University*
JASON DITTMER, *Georgia Southern University*
JAMES DOERNER, *University of Northern Colorado*
STEVEN DRIEVER, *University of Missouri-Kansas City*
ELIZABETH DUDLEY-MURPHY, *University of Utah*
DENNIS EHRHARDT, *University of Louisiana-Lafayette*
WILLIAM FORBES, *Stephen F. Austin State University (Texas)*
BILL FOREMAN, *Oklahoma City Community College*
ERIC FOURNIER, *Samford University (Alabama)*
GARY A. FULLER, *University of Hawai'i*
RANDY GABRYS ALEXSON, *University of Wisconsin-Superior*
WILLIAM GARBARINO, *Community College of Allegheny County (Pennsylvania)*
CHAD GARICK, *Jones County Junior College*
JON GOSS, *University of Hawai'i*
DEBRA GRAHAM, *Messiah College*
RICHARD J. GRANT, *University of Miami*
JASON B. GREENBERG, *Sullivan University*
MARGARET M. GRIPSHOVER, *University of Tennessee*
SARA HARRIS, *Neosho County Community College (Kansas)*
JOHN HAVIR, *Ashland Community & Technical College*
JOHN HICKEY, *Inver Hills Community College (Minnesota)*
SHIRLENA HUANG, *National University of Singapore*
INGRID JOHNSON, *Towson State University (Maryland)*
KRIS JONES, *Saddleback College (California)*
THOMAS KARWOSKI, *Anne Arundel Community College (Maryland)*
ROBERT KERR, *University of Central Oklahoma*
ERIC KEYS, *University of Florida*
JACK KINWORTHY, *Concordia University-Nebraska*
MARTI KLEIN, *Saddleback College (California)*
CHRISTOPHER LAINGEN, *Kansas State University*
UNNA LASSITER, *Stephen F. Austin State University (Texas)*
RICHARD LISICHENKO, *Fort Hays State University (Kansas)*
CATHERINE LOCKWOOD, *Chadron State College (Nebraska)*
GEORGE LONBERGER, *Georgia Perimeter College*
CLAUDE MAJOR, *Stratford Career*
STEVE MATCHAK, *Salem State College (Massachusetts)*

TRINA MEDLEY, *Oklahoma City Community College*
DALTON W. MILLER, JR., *Mississippi State University*
ERNANDO F. MINGHINE, *Wayne State University (Michigan)*
VERONICA MORMINO, *Harper College (Illinois)*
TOM MUELLER, *California University (Pennsylvania)*
IRENE NAESSE, *Orange Coast College (California)*
VALIANT C. NORMAN, *Lexington Community College (Kentucky)*
RICHARD OLMO, *University of Guam*
PAI YUNG-FENG, *New York City*
J. L. PASZTOR, *Delta College (Michigan)*
IWONA PETRUCZYNIK, *Mercyhurst College (Pennsylvania)*
PAUL E. PHILLIPS, *Fort Hays State University (Kansas)*
ROSANN POLTRONE, *Arapahoe Community College (Colorado)*
RINKU ROY CHOWDHURY, *Indiana University*
JEFF POPKE, *East Carolina University*
DAVID PRIVETTE, *Central Piedmont Community College (North Carolina)*
RHONDA REAGAN, *Blinn College (Texas)*
A. L. RYDANT, *Keene State College (New Hampshire)*
NANCY SHIRLEY, *Southern Connecticut State University*
DEAN SINCLAIR, *Northwestern State University (Louisiana)*
RICHARD SLEASE, *Oakland, North Carolina*
SUSAN SLOWEY, *Blinn College (Texas)*
DEAN B. STONE, *Scott Community College (Iowa)*
JAMIE STRICKLAND, *University of North Carolina at Charlotte*
RUTHINE TIDWELL, *Florida Community College at Jacksonville*
IRINA VAKULENKO, *Collin County Community College (Texas)*
CATHY WEIDMAN, *Austin, Texas*
KIRK WHITE, *York College of Pennsylvania*
THOMAS WHITMORE, *University of North Carolina, Chapel Hill*
KEITH YEARMAN, *College of DuPage (Illinois)*
LAURA ZEEMAN, *Red Rocks Community College (Colorado)*
YU ZHO, *Vassar College (New York)*

We also received input from a much wider circle of academic geographers. The list that follows is merely representative of a group of colleagues across North America to whom we are grateful for taking the time to share their thoughts and opinions with us:

MEL AAMODT, *California State University, Stanislaus*
GILLIAN ACHESON, *Southern Illinois University, Edwardsville*
WILLIAM V. ACKERMAN, *Ohio State University*
W. FRANK AINSLEY, *University of North Carolina, Wilmington*
SIAW AKWAWUA, *University of Northern Colorado*
VICTORIA ALAPO, *Metropolitan Community College*
DONALD P. ALBERT, *Sam Houston State University*
TONI ALEXANDER, *Louisiana State University*
R. GABRYS ALEXSON, *University of Wisconsin-Superior*
KHALED MD. ALI, *Geologist, Dhaka, Bangladesh*
NIGEL ALLAN, *University of California, Davis*

JOHN L. ALLEN, *University of Wyoming*

KRISTIN J. ALVAREZ, *Keene State College (New Hampshire)*

DAVID L. ANDERSON, *Louisiana State University, Shreveport*

KEAN ANDERSON, *Freed-Hardeman University*

JEFF ARNOLD, *Southwestern Illinois College*

JERRY R. ASCHERMANN, *Missouri Western State College*

JOSEPH M. ASHLEY, *Montana State University*

PATRICK ASHWOOD, *Hawkeye Community College*

THEODORE P. AUFDEMBERGE, *Concordia College (Michigan)*

JAIME M. AVILA, *Sacramento City College*

EDWARD BABIN, *University of South Carolina-Spartanburg*

ROBERT BAERENT, *Randolph-Macon College*

MARVIN W. BAKER, *University of Oklahoma*

GOURI BANERJEE, *Boston University*

MICHELE BARNABY, *Pittsburg State University (Kansas)*

J. HENRY BARTON, *Thiel College (Pennsylvania)*

STEVEN BASS, *Paradise Valley Community College (Arizona)*

RENE BATAILLE, *Nashville State Technical Community College*

THOMAS F. BAUCOM, *Jacksonville State University (Alabama)*

KLAUS J. BAYR, *Keene State College (New Hampshire)*

DENIS A. BEKAERT, *Middle Tennessee State University*

JAMES BELL, *Linn Benton Community College (Oregon)*

KEITH M. BELL, *Volunteer State Community College (Tennessee)*

WILLIAM H. BERENTSEN, *University of Connecticut*

DONALD J. BERG, *South Dakota State University*

ROYAL BERGLEE, *Morehead State University (Kentucky)*

RIVA BERLEANT-SCHILLER, *University of Connecticut*

LEE LUCAS BERMAN, *Southern Connecticut State University*

RANDY BERTOLAS, *Wayne State College*

THOMAS BITNER, *University of Wisconsin-Marshfield/Wood County*

WARREN BLAND, *California State University, Northridge*

DAVIS BLEVINS, *Huntington College (Alabama)*

HUBERTUS BLOEMER, *Ohio University*

S. BO JUNG, *Bellevue College (Nebraska)*

R. DENISE BLANCHARD-BOEHM, *Texas State University*

MARTHA BONTE, *Clinton Community College (Idaho)*

GEORGE R. BOTJER, *University of Tampa (Florida)*

R. LYNN BRADLEY, *Belleville Area College (Illinois)*

KEN BREHOB, *Elmhurst College (Illinois)*

JAMES A. BREY, *University of Wisconsin-Fox Valley*

ROBERT BRINSON, *Santa Fe Community College (Florida)*

REUBEN H. BROOKS, *Tennessee State University*

PHILLIP BROUGHTON, *Arizona Western College*

LARRY BROWN, *Ohio State University*

LAWRENCE A. BROWN, *Troy State-Dothan (Alabama)*

ROBERT N. BROWN, *Delta State University (Mississippi)*

STANLEY D. BRUNN, *University of Kentucky*

RANDALL L. BUCHMAN, *Defiance College (Ohio)*

MICHAELE ANN BUELL, *Northwest Arkansas Community College*

DANIEL BUYNE, *South Plains College*

MICHAEL BUSBY, *Murray State University (Kentucky)*

DEAN BUTZOW, *Lincoln Land Community College (Illinois)*

MARY CAMERON, *Slippery Rock University (Pennsylvania)*

MICHAEL CAMILLE, *University of Louisiana at Monroe*

MARY CARAVELIS, *Barry University (Florida)*

DIANA CASEY, *Muskegon Community College (Michigan)*

DiANN CASTEEL, *Tusculum College (Tennessee)*

BILL CHAPPELL, *Keystone College*

MICHAEL SEAN CHENOWETH, *University of Wisconsin-Milwaukee*

STANLEY CLARK, *California State University, Bakersfield*

JOHN E. COFFMAN, *University of Houston (Texas)*

DWAYNE COLE, *Cornerstone University (Michigan)*

JERRY COLEMAN, *Mississippi Gulf Coast Community College*

JONATHAN C. COMER, *Oklahoma State University*

BARBARA CONNELLY, *Westchester Community College (New York)*

FRED CONNINGTON, *Southern Wesleyan University*

WILLIS M. CONOVER, *University of Scranton (Pennsylvania)*

OMAR CONRAD, *Maple Woods Community College (Missouri)*

ALLAN D. COOPER, *Otterbein College*

WILLIAM COUCH, *University of Alabama, Huntsville*

BARBARA CRAGG, *Aquinas College (Michigan)*

GEORGES G. CRAVINS, *University of Wisconsin-La Crosse*

ELLEN K. CROMLEY, *University of Connecticut*

JOHN A. CROSS, *University of Wisconsin-Oshkosh*

SHANNON CRUM, *University of Texas at San Antonio*

WILLIAM CURRAN, *South Suburban College (Illinois)*

KEVIN M. CURTIN, *University of Texas at San Antonio*

JOSE A. DA CRUZ, *Ozarks Technical Community College*

ARMANDO DA SILVA, *Towson State University (Maryland)*

MARK DAMICO, *Green Mountain College (Vermont)*

DAVID D. DANIELS, *Central Missouri State University*

RUDOLPH L. DANIELS, *Morningside College (Iowa)*

SATISH K. DAVGUN, *Bemidji State University (Minnesota)*

WILLIAM V. DAVIDSON, *Louisiana State University*

CHARLES DAVIS, *Mississippi Gulf Coast Community College*

JAMES DAVIS, *Illinois College*

JAMES L. DAVIS, *Western Kentucky University*

PEGGY E. DAVIS, *Pikeville College*

ANN DEAKIN, *State University of New York, College at Fredonia*

KEITH DEBBAGE, *University of North Carolina at Greensboro*

MOLLY DEBYSINGH, *California State University, Long Beach*

DENNIS K. DEDRICK, *Georgetown College (Kentucky)*

JOEL DEICHMANN, *Bentley College*

STANFORD DeMARS, *Rhode Island College*

TOM DESULIS, *Spoon River College*

THOMAS DIMICELLI, *William Paterson College (New Jersey)*

MARY DOBBS, *Highland Community College, Wamego*

SCOTT DOBLER, *Western Kentucky University*

D. F. DOEPPERS, *University of Wisconsin-Madison*

JAMES DOERNER, *University of Northern Colorado*

Ann Doolen, *Lincoln College (Illinois)*
Steven Driever, *University of Missouri-Kansas City*
William Druen, *Western Kentucky University*
Alasdair Drysdale, *University of New Hampshire*
Keith A. Ducote, *Cabrillo Community College (California)*
Elizabeth Dudley-Murphy, *University of Utah*
Walter N. Duffet, *University of Arizona*
Mike Dunning, *University of Alaska Southeast*
Christina Dunphy, *Champlain College (Vermont)*
Anthony Dzik, *Shawnee State University (Kansas)*
Dennis Edgell, *Firelands BGSU (Ohio)*
Ruth M. Ediger, *Seattle Pacific University*
James H. Edmonson, *Union University (Tennessee)*
M. H. Edney, *State University of New York-Binghamton*
Harold M. Elliott, *Weber State University (Utah)*
James Elsnes, *Western State College*
Robert W. Evans, *Fresno City College (California)*
Everson Dicken, *California State University, San Bernardino*
Michael Ferber, *Regent University*
Jeff Fesperman, *Illinois Valley Community College*
Dino Fiabane, *Community College of Philadelphia*
G. A. Finchum, *Milligan College (Tennessee)*
Ira Fogel, *Foothill College (California)*
Richard Foley, *Cumberland College*
Robert G. Foote, *Wayne State College (Nebraska)*
Ronald Foresta, *University of Tennessee*
Ellen J. Foster, *Texas State University*
G. S. Freedom, *McNeese State University (Louisiana)*
Edward T. Freels, *Carson-Newman College*
Philip Friend, *Inver Hills Community College*
James Fryman, *University of Northern Iowa*
Owen Furuseth, *University of North Carolina at Charlotte*
Richard Fusch, *Ohio Wesleyan University*
Gary Gaile, *late of University of Colorado-Boulder*
Evelyn Gallegos, *Eastern Michigan University & Schoolcraft College*
Gail Garbrandt, *Mount Union College, University of Akron*
Dale Garret, *Evangel University*
Richard Garrett, *Marymount Manhattan College*
Jerry Gerlach, *Winona State University (Minnesota)*
Mark Gismondi, *Northwest Nazarene University*
Lorne E. Glaim, *Pacific Union College (California)*
Sharleen Gonzalez, *Baker College (Michigan)*
Daniel B. Good, *Georgia Southern University*
Gary C. Goodwin, *Suffolk Community College (New York)*
S. Gopal, *Boston University*
Marvin Gordon, *Lake Forest Graduate School of Business*
Robert Gould, *Morehead State University (Kentucky)*
Mary Graham, *York College of Pennsylvania*
Gordon Grant, *Texas A&M University*
Paul Gray, *Arkansas Tech University*
Donald Green, *Baylor University (Texas)*
Gary M. Green, *University of North Alabama*
Stanley C. Green, *Laredo State University (Texas)*

Raymond Greene, *Western Illinois University*
Mark Greer, *Laramie County Community College (Wyoming)*
Walt Guttinger, *Flagler College (Florida)*
John E. Gygax, *Wilkes Community College*
Lee Ann Hagan, *College of Southern Idaho*
Ron Hagelman, *University of New Orleans*
W. Gregory Hager, *Northwestern Connecticut Community College*
Ruth F. Hale, *University of Wisconsin-River Falls*
John W. Hall, *Louisiana State University-Shreveport*
Peter L. Halvorson, *University of Connecticut*
David Hansen, *Pennsylvania State University, Harrisburg & Schuylkill*
Mervin Hanson, *Willmar Community College (Minnesota)*
Robert C. Harding, *Lynchburg College (Virginia)*
Michelle D. Harris, *Bob Jones University (South Carolina)*
Scott Harris, *CGCS Drury University*
Robert J. Hartig, *Fort Valley State College (Georgia)*
Suzanna Hartley, *Shelton State Community College*
Truman A. Hartshorn, *Georgia State University*
Carol Hazard, *Meredith College*
Harlow Z. Head, *Barton College*
Mark Healy, *William Rainey Harper College*
Doug Heffington, *Middle Tennessee State University*
James G. Heidt, *University of Wisconsin Center-Sheboygan*
Catherine Helgeland, *University of Wisconsin-Manitowoc*
Norma Hendrix, *East Arkansas Community College*
James E. Herrell, *Otero Junior College (Colorado)*
James Hertzler, *Goshen College (Indiana)*
John Hickey, *Inver Hills Community College (Minnesota)*
Thomas Higgins, *San Jacinto College (Texas)*
Eugene Hill, *Westminster College (Missouri)*
Larissa Hinz, *Eastern Illinois University*
Miriam Helen Hill, *Indiana University Southeast*
Suzy Hill, *University of South Carolina-Spartanburg*
Robert Hilt, *Pittsburg State University (Kansas)*
Sophia Hinshalwood, *Montclair State University (New Jersey)*
Priscilla Holland, *University of North Alabama*
Mark R. Hooper, *Freed-Hardeman University*
R. Hostetler, *Fresno City College (California)*
Erick Howenstine, *Northeastern Illinois University*
Lloyd E. Hudman, *Brigham Young University (Utah)*
Janis W. Humble, *University of Kentucky*
Teresa Hutchinson, *Columbus State Community College*
Juana Ibanez, *University of New Orleans*
Tony Ijomah, *Harrisburg Area Community College (Pennsylvania)*
William Imperatore, *Appalachian State University (North Carolina)*
Ed Jackiewicz, *California State University, Northridge*
Richard Jackson, *Brigham Young University (Utah)*
Mary Jacob, *Mount Holyoke College (Massachusetts)*

GREGORY JEANE, *Samford University (Alabama)*
SCOTT JEFFREY, *Catonsville Community College (Maryland)*
JERZY JEMIOLO, *Ball State University (Indiana)*
NILS I. JOHANSEN, *University of Southern Indiana*
DAVID JOHNSON, *University of Southwestern Louisiana*
INGRID JOHNSON, *Towson State University (Maryland)*
KAREN JOHNSON, *North Hennepin Community College (Minnesota)*
RICHARD JOHNSON, *Oklahoma City University*
SHARON JOHNSON, *Marymount College (New York)*
JEFFREY JONES, *University of Kentucky*
KRIS JONES, *Saddleback College (California)*
MARCUS E. JONES, *Claflin College (South Carolina)*
TAMOURA JONES, *Atlanta Technical College (Georgia)*
MOHAMMAD S. KAMIAR, *Florida Community College, Jacksonville*
MELINDA KASHUBA, *Shasta College (California)*
MATTI E. KAUPS, *University of Minnesota-Duluth*
JO ANNE W. KAY, *Brigham Young University, Idaho*
PHILIP L. KEATING, *Indiana University*
DAVID KEELING, *Western Kentucky University*
COLLEEN KEEN, *Gustavus Adolphus College (Minnesota)*
ARTIMUS KEIFFER, *Wittenberg University (Ohio)*
GORDON F. KELLS, *Mott Community College (Michigan)*
KAREN J. KELLY, *Palm Beach Community College, Boca Raton (Florida)*
RYAN KELLY, *Lexington Community College (Kentucky)*
VIRGINIA KERKHEIDE, *Cuyahoga Community College and Cleveland State University (Ohio)*
TOM KESSENGER, *Xavier University (Ohio)*
MASOUD KHEIRABADI, *Maryhurst University*
SUSANNE KIBLER-HACKER, *Unity College (Maine)*
CHANGJOO KIM, *Minnesota State University*
JAMES W. KING, *University of Utah*
JOHN C. KINWORTHY, *Concordia College (Nebraska)*
ALBERT KITCHEN, *Paine College (Georgia)*
TED KLIMASEWSKI, *Jacksonville State University (Alabama)*
ROBERT D. KLINGENSMITH, *Ohio State University-Newark*
LAWRENCE M. KNOPP, JR., *University of Minnesota-Duluth*
LYNN KOEHNEMANN, *Gulf Coast Community College*
TERRILL J. KRAMER, *University of Nevada*
JOE KRAUSE, *Community College of Indiana at Lafayette*
DEBRA KREITZER, *Western Kentucky University*
BRENDER KREKELER, *Northern Kentucky University; Miami University (Ohio)*
ARTHUR J. KRIM, *Cambridge, Massachusetts*
MICHAEL A. KUKRAL, *Rose-Hulman Institute of Technology*
CHRIS LANEY, *Berkshire Community College (Massachusetts)*
ELROY LANG, *El Camino Community College (California)*
RICHARD L. LANGILL, *Saint Martin's University (Washington)*
CHRISTOPHER LANT, *Southern Illinois University*
A. J. LARSON, *University of Illinois-Chicago*
PAUL R. LARSON, *Southern Utah University*
LARRY LEAGUE, *Dickinson State University (North Dakota)*

DAVID R. LEE, *Florida Atlantic University*
WOOK LEE, *Texas State University*
JOE LEEPER, *Humboldt State University (California)*
YECHIEL M. LEHAVY, *Atlantic Community College (New Jersey)*
SCOTT LEITH, *Central Connecticut State University*
JAMES LEONARD, *Marshall University (West Virginia)*
ELIZABETH J. LEPPMAN, *St. Cloud State University (Minnesota)*
JOHN C. LEWIS, *Northeast Louisiana University*
DAN LEWMAN JR., *Southwest Mississippi Community College*
CAEDMON S. LIBURD, *University of Alaska-Anchorage*
T. LIGIBEL, *Eastern Michigan University*
Z. L. LIPCHINSKY, *Berea College (Kentucky)*
ALLAN L. LIPPERT, *Manatee Community College (Florida)*
RICHARD LISICHENKO, *Fort Hays State University (Kansas)*
JOHN L. LITCHER, *Wake Forest University (North Carolina)*
LEE LIU, *Central Montana State University*
LI LIU, *Stephen F. Austin State University (Texas)*
WILLIAM R. LIVINGSTON, *Baker College (Michigan)*
CATHERINE M. LOCKWOOD, *Chadron State College (Nebraska)*
GEORGE E. LONGENECKER, *Vermont Technical College*
CYNTHIA LONGSTREET, *Ohio State University*
JACK LOONEY, *University of Massachusetts, Boston*
TOM LOVE, *Linfield College (Oregon)*
K. J. LOWREY, *Miami University (Ohio)*
JAMES LOWRY, *Stephen F. Austin State University (Texas)*
MAX LU, *Kansas State University*
ROBIN R. LYONS, *University of Hawai'i-Leeward Community College*
SUSAN M. MACEY, *Texas State University*
MICHAEL MADSEN, *Brigham Young University, Idaho*
RONALD MAGDEN, *Tacoma Community College (Washington)*
CHRISTIANE MAINZER, *Oxnard College (California)*
LAURA MAKEY, *California State University, San Bernardino*
MIKE MAKOWSKY, *Midland College*
HARLEY I. MANNER, *University of Guam*
ANTHONY PAUL MANNION, *Kansas State University*
GARY MANSON, *Michigan State University*
CHARLES MANYARA, *Radford University (Virginia)*
JAMES T. MARKLEY, *Lord Fairfax Community College (Virginia)*
SISTER MAY LENORE MARTIN, *Saint Mary College (Kansas)*
KENT MATHEWSON, *Louisiana State University*
PATRICK MAY, *Plymouth State College (New Hampshire)*
DICK MAYER, *Maui Community College (Hawai'i)*
SARA MAYFIELD, *San Jacinto College, Central (California)*
DEAN R. MAYHEW, *Marine Maritime Academy*
J. P. McFADDEN, *Orange Coast College (California)*
BERNARD McGONIGLE, *Community College of Philadelphia*
MOLLY McGRAW, *Southeastern Louisiana University*
PAUL D. MEARTZ, *Mayville State University (North Dakota)*
DIANNE MEREDITH, *California State University-Sacramento*
DAVID MERWIN, *Framingham State College (Massachusetts)*
GARY C. MEYER, *University of Wisconsin-Stevens Point*
JUDITH L. MEYER, *Southwest Missouri State University*

MARK MICOZZI, *East Central University (Oklahoma)*
JOHN MILBAUER, *Northeastern State University*
DALTON W. MILLER, JR., *Mississippi State University*
RAOUL MILLER, *University of Minnesota-Duluth*
ROGER MILLER, *Black Hills State University (South Dakota)*
JAMES MILLS, *State University of New York, College at Oneonta*
INES MIYARES, *Hunter College, CUNY (New York)*
BOB MONAHAN, *Western Carolina University*
KEITH MONTGOMERY, *University of Wisconsin-Madison*
DAN MORGAN, *University of South Carolina, Beaufort*
DEBBIE MORIMOTO, *Merced College (California)*
JOHN MORTRON, *Benedict College (South Carolina)*
ANNE MOSHER, *Syracuse University*
BARRY MOWELL, *Broward Community College (Florida)*
TOM MUELLER, *California University (Pennsylvania)*
DONALD MYERS, *Central Connecticut State University*
ROBERT R. MYERS, *West Georgia College*
GARY NACHTIGALL, *Fresno Pacific University (California)*
YASER M. NAJJAR, *Framingham State College (Massachusetts)*
KATHERINE NASHLEANAS, *University of Nebraska-Lincoln*
JEFFREY W. NEFF, *Western Carolina University*
DAVID NEMETH, *University of Toledo (Ohio)*
ROBERT NEWCOMER, *East Central University (Oklahoma)*
WILLIAM NIETER, *St. Johns University (New York)*
WILLIAM N. NOLL, *Highland Community College*
VALIANT C. NORMAN, *Lexington Community College (Kentucky)*
JOSEPH A. NAUMANN, *University of Missouri, St. Louis*
RAYMOND O'BRIEN, *Bucks County Community College (Pennsylvania)*
PATRICK O'SULLIVAN, *Florida State University*
DIANE O'CONNELL, *Schoolcraft College*
NANCY OBERMEYER, *Indiana State University*
JOHN ODLAND, *late of Indiana University*
DOUG OETTER, *Georgia College and State University*
ANNE O'HARA, *Marygrove College (Michigan)*
PATRICK OLSEN, *University of Idaho*
JOSEPH R. OPPONG, *University of North Texas*
LYNN ORLANDO, *Holy Family University (Pennsylvania)*
MARK A. OUMETTE, *Hardin-Simmons University (Texas)*
RICHARD OUTWATER, *California State University, Long Beach*
CISSIE OWEN, *Lamar University (Texas)*
MARY ANN OWOC, *Mercyhurst College (Pennsylvania)*
EUGENE J. PALKA, *U.S. Military Academy (New York)*
STEVE PALLADINO, *Ventura College (California)*
BIMAL K. PAUL, *Kansas State University*
SELINA PEARSON, *Northwest-Shoals Community College (Alabama)*
MAURI PELTO, *Nichols College (Massachusetts)*
JAMES PENN, *Southeastern Louisiana University*
NICK PETROPOULEAS, *Daytona State College (Florida)*
LINDA PETT-CONKLIN, *University of St. Thomas (Minnesota)*
DIANE PHILEN, *Lower Brule Community College (South Dakota)*

PAUL PHILLIPS, *Fort Hays State University (Kansas)*
MICHAEL PHOENIX, *ESRI, Redlands, California*
JERRY PITZL, *Macalester College (Minnesota)*
BRIAN PLASTER, *Texas State University*
ARMAND POLICICCHIO, *Slippery Rock University (Pennsylvania)*
ROSANN POLTRONE, *Arapahoe Community College (Colorado)*
BILLIE E. POOL, *Holmes Community College (Mississippi)*
GREGORY POPE, *Montclair State University (New Jersey)*
JEFF POPKE, *East Carolina University*
WILLIAM PRICE, *North Country Community College*
VINTON M. PRINCE, *Wilmington College (North Carolina)*
GEORGE PUHRMANN, *Drury University (Missouri)*
DONALD N. RALLIS, *Mary Washington College (Virginia)*
RHONDA REAGAN, *Blinn College (Texas)*
DANNY I. REAMS, *Southeast Community College (Nebraska)*
JIM RECK, *Golden West College (California)*
ROGER REEDE, *Southwest State University (Minnesota)*
JOHN RESSLER, *Central Washington University*
JOHN B. RICHARDS, *Southern Oregon State College*
DAVID C. RICHARDSON, *Evangel University (Missouri)*
GRAY RINGLEY, *Virginia Highlands Community College*
SUSAN ROBERTS, *University of Kentucky*
CURT ROBINSON, *California State University, Sacramento*
AMY ROCK, *John Carroll University*
WOLF RODER, *University of Cincinnati*
JAMES ROGERS, *University of Central Oklahoma*
PAUL A. ROLLINSON, *Southwest Missouri State University*
JAMES C. ROSE, *Tompkins/Cortland Community College (New York)*
THOMAS E. ROSS, *Pembroke State University (North Carolina)*
THOMAS A. RUMNEY, *State University of New York, College at Plattsburgh*
GEORGE H. RUSSELL, *University of Connecticut*
BILL RUTHERFORD, *Martin Methodist College*
RAJAGOPAL RYALI, *Auburn University at Montgomery (Alabama)*
PERRY RYAN, *Mott Community College*
JAMES SAKU, *Frostburg State University (Maryland)*
DAVID R. SALLEE, *University of North Texas*
RICHARD A. SAMBROOK, *Eastern Kentucky University*
EDUARDO SANCHEZ, *Grand Valley State University (Michigan)*
JOHN SANTOSUOSSO, *Florida Southern College*
GINGER SCHMID, *Texas State University*
BRENDA THOMPSON SCHOOLFIELD, *Bob Jones University (South Carolina)*
ADENA SCHUTZBERG, *Middlesex Community College (Massachusetts)*
ROGER M. SELYA, *University of Cincinnati*
RENEE SHAFFER, *University of South Carolina*
WENDY SHAW, *Southern Illinois University, Edwardsville*
SIDNEY R. SHERTER, *Long Island University (New York)*
HAROLD SHILK, *Wharton County Jr. College (Texas)*
NANDA SHRESTHA, *Florida A&M University*

WILLIAM R. SIDDALL, *Kansas State University*

DAVID SILVA, *Bee County College (Texas)*

STEVEN SILVERN, *Salem State College (Massachusetts)*

JOSE ANTONIO SIMENTAL, *Marshall University (West Virginia)*

MORRIS SIMON, *Stillman College (Alabama)*

ROBERT MARK SIMPSON, *University of Tennessee at Martin*

KENN E. SINCLAIR, *Holyoke Community College (Massachusetts)*

ROBERT SINCLAIR, *Wayne State University (Michigan)*

JIM SKINNER, *Southwest Missouri State University*

BRUCE SMITH, *Bowling Green State University (Ohio)*

EVERETT G. SMITH, JR., *University of Oregon*

PEGGY SMITH, *California State University, Fullerton*

RICHARD V. SMITH, *Miami University (Ohio)*

JAMES SNADEN, *Charter Oak State College (Connecticut)*

DAVID SORENSON, *Augustana College (Illinois)*

SISTER CONSUELO SPARKS, *Immaculata University (Pennsylvania)*

CAROLYN D. SPATTA, *California State University, Hayward*

M. R. SPONBERG, *Laredo Junior College (Texas)*

DONALD L. STAHL, *Towson State University (Maryland)*

DAVID STEA, *Texas State University*

ELAINE STEINBERG, *Central Florida Community College*

D. J. STEPHENSON, *Ohio University Eastern*

HERSCHEL STERN, *Mira Costa College (California)*

REED F. STEWART, *Bridgewater State College (Massachusetts)*

NOEL L. STIRRAT, *College of Lake County (Illinois)*

JOSEPH P. STOLTMAN, *Western Michigan University*

DEAN B. STONE, *Scott Community College*

WILLIAM M. STONE, *Saint Xavier University, Chicago*

GEORGE STOOPS, *Minnesota State University*

DEBRA STRAUSSFOGEL, *University of New Hampshire*

JAMIE STRICKLAND, *University of North Carolina at Charlotte*

WAYNE STRICKLAND, *Roanoke College (Virginia)*

PHILIP STURM, *Ohio Valley College, Vienna*

PHILIP SUCKLING, *Texas State University*

ALLEN SULLIVAN, *Central Washington University*

SELIMA SULTANA, *University of North Carolina at Greensboro*

RAY SUMNER, *Long Beach City College (California)*

CHRISTOPHER SUTTON, *Western Illinois University*

T. L. TARLOS, *Orange Coast College (California)*

WESLEY TERAOKA, *Leeward Community College*

MICHAEL THEDE, *North Iowa Area Community College*

DERRICK J. THOM, *Utah State University*

CURTIS THOMSON, *University of Idaho*

BEN TILLMAN, *Texas Christian University*

CLIFF TODD, *University of Nebraska, Omaha*

STANLEY TOOPS, *Miami University (Ohio)*

RICHARD J. TORZ, *St. Joseph's College (New York)*

HARRY TRENDELL, *Kennesaw State University (Georgia)*

ROGER T. TRINDELL, *Mansfield University (Pennsylvania)*

DAN TURBEVILLE, *East Oregon State College*

NORMAN TYLER, *Eastern Michigan University*

GEORGE VAN OTTEN, *Northern Arizona University*

GREGORY VEECK, *Western Michigan University*

C. S. VERMA, *Weber State College (Utah)*

KELLY ANN VICTOR, *Eastern Michigan University*

SEAN WAGNER, *Tri-State University (Indiana)*

GRAHAM T. WALKER, *Metropolitan State College of Denver*

MONTGOMERY WALKER, *Yakima Valley Community College (Washington)*

DEBORAH WALLIN, *Skagit Valley College (Washington)*

MIKE WALTERS, *Henderson Community College (Kentucky)*

LINDA WANG, *University of South Carolina, Aiken*

J. L. WATKINS, *Midwestern State University (Texas)*

DAVID WELK, *Reedley College (California)*

KIT W. WESLER, *Murray State University (Kentucky)*

PETER W. WHALEY, *Murray State University (Kentucky)*

MACEL WHEELER, *Northern Kentucky University*

P. GARY WHITE, *Western Carolina University (North Carolina)*

W. R. WHITE, *Western Oregon University*

GARY WHITTON, *Fairbanks, Alaska*

REBECCA WIECHEL, *Wilmington College (North Carolina)*

MARK WILJANEN, *Eastern Kentucky University*

GENE C. WILKEN, *Colorado State University*

FORREST WILKERSON, *Texas State University*

P. WILLIAMS, *Baldwin-Wallace College (Ohio)*

STEPHEN A. WILLIAMS, *Methodist College (North Carolina)*

DEBORAH WILSON, *Sandhills Community College (Nebraska)*

MORTON D. WINSBERG, *Florida State University*

ROGER WINSOR, *Appalachian State University (North Carolina)*

ELIZABETH WINTERNITZ-RUSSELL, *Maui Community College (Hawai'i)*

WILLIAM A. WITHINGTON, *University of Kentucky*

A. WOLF, *Appalachian State University (North Carolina)*

JOSEPH WOOD, *University of Southern Maine*

RICHARD WOOD, *Seminole Junior College (Florida)*

GEORGE I. WOODALL, *Winthrop College (North Carolina)*

STEPHEN E. WRIGHT, *James Madison University (Virginia)*

DAWN WROBEL, *Moraine Valley Community College*

LEON YACHER, *Southern Connecticut State University*

KYONG YUP CHU, *Bergen Community College (New Jersey)*

FIROOZ E. ZADEH, *Colorado Mountain College*

DONALD J. ZEIGLER, *Old Dominion University (Virginia)*

YU ZHOU, *Bowling Green State University (Ohio)*

ROBERT C. ZIEGENFUS, *Kutztown University (Pennsylvania)*

PERSONAL APPRECIATION

For assistance with the map of North American indigenous people, we are greatly indebted to Jack Weatherford, Professor of Anthropology at Macalester College (Minnesota); Henry T. Wright, Professor and Curator of Anthropology at the University of Michigan; and George E. Stuart, President of the Center for Maya Research (North Carolina). The map of Russia's federal regions could not have been compiled without the invaluable help of David B. Miller, Senior Edit

Cartographer at the National Geographic Society, and Leo Dillon of the Russia Desk of the U.S. Department of State. The map of Russian physiography was updated thanks to the suggestions of Mika Roinila of the State University of New York, College at New Paltz. Special thanks also go to Charles Pirtle, Professor of Geography at Georgetown University's School of Foreign Service for his advice on Chapter 4; to Bilal Butt (post-doctoral fellow, University of Wisconsin-Madison) for his significant help with Chapter 6; and to Charles Fahrer of Georgia College and State University for his suggestions on Chapter 6. Enormous gratitude goes to Beth Weisenborn, Virtual Course Coordinator of the Department of Geography at Michigan State University, and all the instructors of Geo 204 (World Regional Geography) for their invaluable suggestions to improve the book. And we are most grateful to Tanya de Blij, who holds a graduate degree in geography from Florida State University and whose computer and editing skills helped keep this project on track, contributing to its progress at every stage.

We also record our appreciation to those geographers who ensured the quality of this book's ancillary products. Elizabeth Muller Hames, D.O. (as well as M.A. in Geography, University of Miami) co-authored the original *Study Guide*. At the University of Miami's Department of Geography and Regional Studies, we are most grateful for the advice and support we continue to receive from faculty colleagues Richard Grant, Miguel Kanai, Mazen Labban, Shouraseni Sen Roy, and Ira Sheskin as well as GIS Lab Manager Chris Hanson. We also take this opportunity to reconfirm our appreciation of the work done on the ancillaries for earlier editions by Eugene J. Palka, Professor of Geography at the United States Military Academy, West Point, NY, whose vision and enthusiasm contributed importantly to the achievement of our objectives.

We are privileged to work with a team of professionals at John Wiley & Sons that is unsurpassed in the college textbook publishing industry. As authors we are acutely aware of these talents on a daily basis during the crucial production stage, especially the outstanding coordination and leadership skills of Senior Production Editor Janet Foxman, Illustration Editor Anna Melhorn, and Senior Photo Editor Jennifer MacMillan. Others who played a leading role in this process were Senior Designer Carole Anson, copyeditor Betty Pessagno, and Don Larson and Terry Bush of Mapping Specialists, Ltd., in Madison, Wisconsin. We much appreciated the leadership of Executive Geosciences Editor Ryan Flahive, who was the prime mover in launching and guiding this book and was superbly assisted throughout the preparations for this latest edition by Meredith Leo. Our College Marketing Manager, Margaret Barrett, advised us throughout the revision process. We also thank Veronica Armour, Lynn Pearlman, Bridget O'Lavin, Sandra Dumas, and Harry Nolan for their help and support. Beyond this immediate circle, we acknowledge the support and encouragement we have received over the years from many others at Wiley including Vice-President for Production Ann Berlin and Publisher Jay O'Callaghan.

A special note of thanks is owed to Jeanine Furino of Furino Production, one of the most gifted textbook-production professionals we have ever encountered. Not only did she oversee and coordinate every aspect of the complicated process of creating this attractive volume out of a mountain of word and graphic files, email attachments, sketches, design layouts, and myriad rounds of preliminary pages—she made us feel that we were the only authors she was working with while handling a major workload of other books.

Finally, and most of all, we express our gratitude to our spouses, Bonnie, Nancy, Jannie, and Vince, for seeing us through the challenging schedule of this latest edition of *TWT*.

H.J. de Blij
Boca Grande, Florida

Peter O. Muller
Coral Gables, Florida

Jan Nijman
Coral Gables, Florida

Antoinette M.G.A. WinklerPrins
East Lansing, Michigan

July 2, 2010

BRIEF CONTENTS

CONTENTS

FIFTH EDITION

THE WORLD TODAY

CONCEPTS AND REGIONS IN GEOGRAPHY

Searing social contrasts in Mumbai, rapidly-urbanizing India's largest city. © H. J. de Blij

IN THIS CHAPTER

Maps in our minds
A geographic approach to understanding the world today
Global climate change
The world's most dangerous places to live
Global core and periphery: power and place
Is globalization good or bad?

CONCEPTS, IDEAS, AND TERMS

INTRODUCTION

WORLD REGIONAL GEOGRAPHY:

GLOBAL PERSPECTIVES

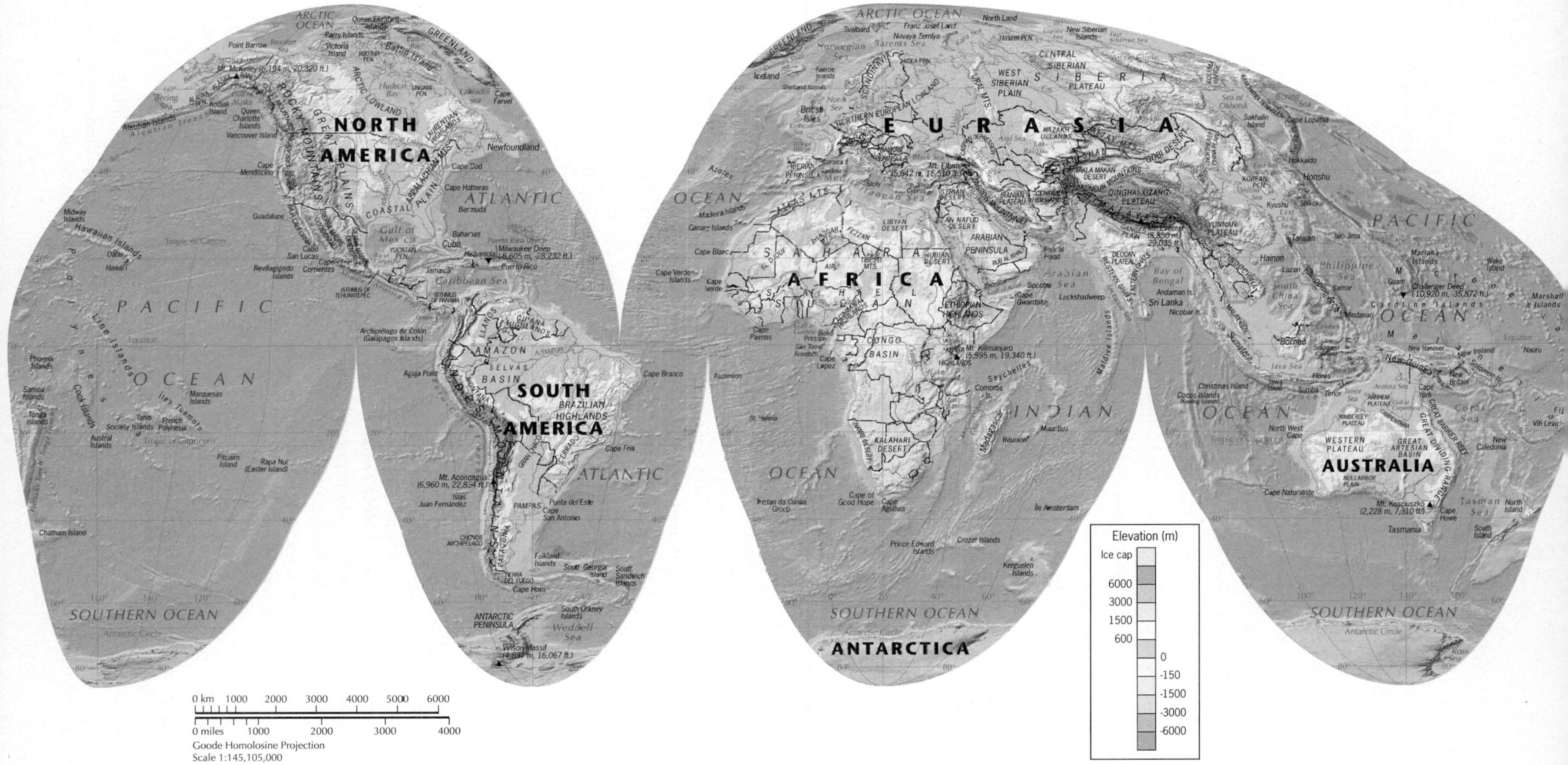

FIGURE G-1

© H. J. DE BLIJ, P. O. MULLER, AND JOHN WILEY & SONS, INC.

What are your expectations as you open this book? You have signed up for a course that will take you around the world to try to understand how it functions today. Hopefully, you will also discover how interesting and unexpectedly challenging the discipline of geography is. We hope that this course, and this book, will open new vistas and bring new perspectives, and help you navigate this increasingly complex and often daunting world.

You could not have chosen a better time to be studying geography. The world is changing on many fronts, and so is the United States. Still the most formidable of all countries, the United States remains a superpower capable of influencing nations and peoples, lives and livelihoods from pole to pole. That power confers on Americans the responsibility to learn as much as they can about those nations and livelihoods, so that the decisions of their government representatives are well-informed. But in this respect, the United States is no superpower. Geographic literacy is a measure of international comprehension and awareness, and Americans' geographic literacy ranks low among countries of consequence. For the world, that is not a good thing, because such geographic fogginess tends to afflict not only voters but also the representatives they elect, from the school board to the White House.

A WORLD OF STATES

Take a look at a map of the countries of the world today, such as the one just inside this book's front cover, and it looks like a giant jigsaw puzzle of about 200 pieces. Getting familiar with these pieces is not as daunting as it may seem at first because you are likely to have a head start. Territorially, Russia is the largest of all. Australia occupies a whole continent. Brazil covers nearly half of South America. It's at the other end of the scale, where small countries seem to vie for space, that recognition gets tougher. But size is no dependable criterion when it comes to importance. Think Switzerland and its banks. Israel and its security. Kuwait and its oil.

Officially, a sovereign country is known as a *state*, and ours is a world of states. But in the tens of thousands of years of human history, the state is a relatively recent invention. And there are signs that the state as we know it may not be here forever. States are cooperating in various ways to reduce the obstacles symbolized by their boundaries, to lower the barriers between and among them. Something different may make the state a relic of history.

That prospect seemed to brighten when, in the early 1990s, the Soviet Union collapsed even as the Europeans were expanding what they refer to as the European Union. There was much talk of a *New World Order* that would usher in an era of international amalgamation and cooperation. More than two decades later, it is all too clear that power continues to center in states and that a new world order remains a distant objective.

The Subdivided State

Meanwhile, we are all too well aware that states have subdivisions. Even the smallest states have such partitions. As all Americans (and Mexicans and Argentinians and Australians) know, some larger states call their subdivisions *States*: the State of California, the State of Chihuahua, the State of Mendoza, the State of Queensland. Note the difference: a state denotes a sovereign country, but a (capitalized) *State* signifies a subdivision. Other subdivisions have alternate names: provinces (Canada), regions (Italy), Autonomous Communities (Spain), Federal Districts (Russia), Divisions (Myanmar). And some of these subdivisions are becoming increasingly assertive, sometimes making their own decisions about their economic or social policies whether the national (state) government likes it or not. When that happens—in Scotland, in Catalonia, in California—we should pay even closer attention to the map.

A WORLD ON MAPS

Just a casual glance at the pages that follow reveals a difference between this and other textbooks: there are almost as many maps as there are pages. Geography is more closely identified with maps than any other discipline, and we urge you to give as much (or more!) attention to the maps in this book as you do to the text. It is often said that a picture is worth a thousand words, but a map can be worth a million. When we write "see Figure XX," we really mean it . . . and we hope that you will get into the habit. We humans are territorial creatures, and the boundaries that fence off our 190 or so countries reflect our divisive ways. But other, less visible borders—between religions, languages, rich and poor—partition our planet as well. When political and cultural boundaries are at odds, there is nothing like a map to summarize the circumstances. Just look at Figure 7B-5. How clear were the implications of this map to those who made the decision to send American troops into war?

Maps in Our Minds

All of us carry in our minds maps of what psychologists call our *activity space*: the apartment building or house we live in, the streets nearby, the way to school or workplace, the general layout of our hometown or city. You will know what lane to use when you turn into a shopping mall, or where to park at the movie theater. You can probably draw from memory a pretty good map of your hometown. These **1** **mental maps** allow you to navigate your activity space with efficiency, predictability, and safety. When you arrived as a first-year student on a college or university campus, a new mental map will have started forming. At first you needed an online or hard-copy map to find your way around, but soon you dispensed with that because your mental map was sufficient. And it will continue to improve as your activity space expands.

If a well-formed mental map is useful for decisions in daily life, then an adequate mental map is surely indispensable when it comes to decision making in the wider world. You can give yourself an interesting test. Choose some part of the world, beyond North America, in which you have an interest or about which you have a strong opinion—for example, Israel, Taiwan, Afghanistan, North Korea, or Venezuela. On a blank piece of paper, draw a map that reflects your impression of the regional layout there: the country, its internal divisions, major cities, neighbors, seas (if any), and so forth. That is your mental map of the place. Put it away for future reference, and try it again at the end of this course. You will have proof of your improved mental-map inventory.

The Map Revolution

The maps in this book show larger and smaller parts of the world in various contexts, some depicting political frameworks, others displaying ethnic, cultural, economic, environmental, and other features unevenly distributed across our world. But *cartography* (the making of maps) has undergone a dramatic technological revolution—a revolution that continues. Earth-orbiting satellites with special on-board scanners and television cameras transmit remotely sensed information to computers on the surface, recording the expansion of deserts, the shrinking of glaciers, the depletion of forests, the growth of cities, and myriad other geographic phenomena. Earthbound computers possess ever-expanding capabilities not only to sort this information but also to display it graphically. This allows geographers to develop *geographic information systems (GISs)*, presenting on-screen information within seconds that would have taken months to assemble just a few decades ago.

Nevertheless, satellites—even spy satellites—cannot record everything that occurs on the Earth's surface. Sometimes the transition zones between ethnic groups or cultural sectors can be discerned by satellites, for example, in changing types of houses or religious shrines, but this kind of information tends to require on-the-ground verification through field research and reporting. No satellite view of Iraq could show you the distribution of Sunni and Shia Muslim adherents. Many of the boundaries you see on the maps in this book cannot be seen from space because long stretches are not even marked on the ground. So the maps

you are about to "read" have their continued uses. They summarize complex situations and allow us to begin forming lasting mental maps of the areas they represent.

GEOGRAPHY'S PERSPECTIVE

Geography has been described as the most interdisciplinary of disciplines. That is a testimonial to geography's historic linkages to many other fields, ranging from geology to economics and from sociology to political science. And, as has been the case so often in the past, geography is in the lead on this point. Today, *interdisciplinary* studies and research are more prevalent than ever. The old barriers between disciplines are breaking down.

This should not suggest that college and university departments are no longer relevant; they are just not as exclusive as they used to be. These days, you can learn some

 FROM THE FIELD NOTES...

© H. J. de Blij

"From the observation platform atop the Seoul Tower one would be able to see into North Korea except for the range of hills in the background: the capital lies in the shadow of the DMZ (demilitarized zone), relic of one of the hot conflicts of the Cold War. The vulnerable Seoul-Incheon metropolitan area ranks among the world's largest, its population approaching 20 million. I asked my colleague, a professor of geography at Seoul National University, why the central business district, the cluster of buildings in the center of this photo, does not seem to reflect the economic power of this city (note the sector of traditional buildings in the foreground, some with blue roofs, the culture's favorite color). "Turn around and look across the [Han] River," he said. "That's the new Seoul, and there the skyline matches that of Singapore, Beijing, or Tokyo." Height restrictions, disputes over land ownership, and congestion are among the factors that slowed growth in this part of Seoul and caused many companies to build in the Samsung area. Looking in that direction, you cannot miss the large U.S. military facility, right in the heart of the urban area, on some of the most valuable real estate and right next to an upscale shopping district. In the local press, the debate over the presence of American forces dominated the letters page day after day."

useful geography in economics departments and some good economics in geography departments. But each discipline still has its own particular way of looking at the world.

In a very general way, we can visualize three key perspectives when we try to figure out how the world works. One is the historic or chronological (you have heard the expression "if you do not learn the lessons of history, you will be doomed to repeat them"). History's key question is *when*? A second perspective centers on the systems people have invented to stabilize their interactions, from the economic to the political. Here the question is *how*? The third perspective is the geographic—the *spatial*—and the key questions are *where*? and *why there*? We seek description and explanation of the patterns of human activity on this Earth. This approach is called the **2** **spatial perspective** and has defined geography from its beginnings.

Environment and Society

There is another glue that binds geography and has done so for a very long time: an interest in the relationships between human societies and the natural (physical) environment. Geography lies at the intersection of the social and natural sciences and integrates perspectives from both, being the only discipline to do so explicitly. This perspective comes into play frequently: environmental change is in the news on a daily basis in the form of worldwide climate change, but this current surge of global warming is only the latest phase of endless climatic and ecological fluctuation. Geographers are involved in understanding current environmental issues not only by considering climate change in the context of the past, but also by looking carefully at the implications of global climate change for human societies.

LOCATION AND DISTRIBUTION

Geographers, therefore, need to be conversant with the location and distribution of salient features on the Earth's surface. This includes the natural (physical) world simplified in Figure G-1 (chapter-opener map) as well as the human world, and our inquiry will view these in temporal (historical) as well as spatial perspective. We take a penetrating look at the overall geographic framework of the contemporary world, the still-changing outcome of thousands of years of human achievement and failure, movement and stagnation, stability and revolution, interaction and isolation. The spatial structure of cities, the layout of farms and fields, the networks of transportation, the configurations of rivers, the patterns of climate—all these form part of our investigation. As you will find, geography employs a comprehensive spatial vocabulary with meaning-ful terms such as area, distance, direction, clustering, proximity, accessibility, and many others we will encounter in the pages ahead. For geographers, some of these terms have more specific definitions than is generally assumed. There is a difference, for example, between *area* and *region*, and between *boundary* and *frontier*. Other terms, such as *location* and *pattern*, can have multiple meanings. The vocabulary of geography holds some surprises.

Scale and Scope

One prominent item in this vocabulary is the term **3** **scale**. Whenever a map is created, it represents all or part of the Earth's surface at a certain level of detail. Obviously, Figure G-1 has a very low level of detail; it is little more than a general impression of the distribution of land and water as well as lower and higher elevations on our planet's surface. A few prominent features such as the Himalayas and the Sahara are named, but not the Appalachians or the Gobi Desert. At the bottom of the map you can see that one inch at this scale must represent about 1650 miles of the real world, leaving the cartographer little scope to insert information.

A map such as Figure G-1 is called a *small-scale* map because the ratio between map distance and real-world distance, expressed as a fraction, is very small at 1:103,750,000. Increase that fraction, and you can represent less territory—but also enhance the amount of detail the map can represent. In Figure G-2, note how the fraction increases from the smallest (1:103,000,000) to the largest (1:1,000,000). Montreal, Canada is just a dot on Map A but an urban area on Map D.

Does this mean that world maps like Figure G-1 are less useful than larger-scale maps? It all depends on the purpose of the map. In this chapter, we often use world maps to show global distributions as we set the stage for the more detailed discussions to follow. In later chapters, the scale tends to become larger as we focus on smaller areas, even on individual countries and cities. But whenever you read a map, be aware of the scale because the scale is a guide to its utility.

The importance of the scale concept is not confined to maps. Scale plays a fundamental part in geographic research and in the ways we think about geographic problems: scale in terms of *level of analysis*. For example, if you want to investigate the geographic concentration of wealth in the United States, you can do so at a range of scales: within the neighborhood, the city, the county, the State, or at the national level. You choose the scale that is the most appropriate for your purpose, but it is not always that straightforward. Suppose you had to study patterns of ethnic segregation: what do you think would be the most relevant scale(s)?

In this book, our main purpose is to understand the geography of the world at large and how it works, and so,

EFFECT OF SCALE

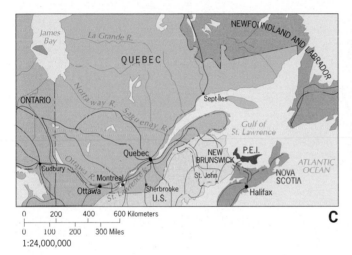

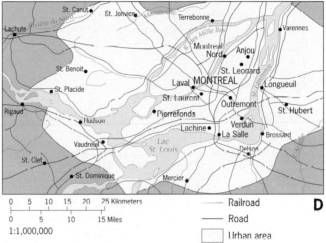

FIGURE G-2

© H. J. de Blij, P. O. Muller, and John Wiley & Sons, Inc.

inevitably, we must deal with large spatial entities. Our focus is on the world's realms and on the main regions within those realms, and in most cases we will have to forego analyses at a finer scale. For our purposes, it is the big picture that matters most.

GEOGRAPHIC REALMS

Ours may be a globalized, interconnected world, a world of international trade and travel, migration and movement, tourism and television, financial flows and Internet traffic, a world that, in some contexts, is taking on the properties of a "global village"—but that village still has neighborhoods. Their names are Europe, South America, Southeast Asia, and others familiar to us all. Like the neighborhoods of a city or town, these global neighborhoods may not have sharply defined borders, but their persistence, after tens of thousands of years of human dispersal, is beyond doubt. Geographers call such global neighborhoods **4** geographic

realms. Each of these realms possesses a particular combination of environmental, cultural, and organizational properties. These characteristic qualities are imprinted on the landscape, giving each realm its own traditional attributes and social settings. As we come to understand the human and environmental makeup of these geographic realms, we learn not only *where* they are located (as we noted, a key question in geography), but also *why they are located where they are*, how they are constituted, and what their future is likely to be in our fast-changing world. Figure G-3, therefore, forms the framework for our investigation.

REALMS AND REGIONS

Geographers, like other scholars, seek to establish order from the countless data that confront them. Biologists have established a system of classification to categorize the many millions of plants and animals into a hierarchical system of seven ranks. In descending order, we humans belong to the

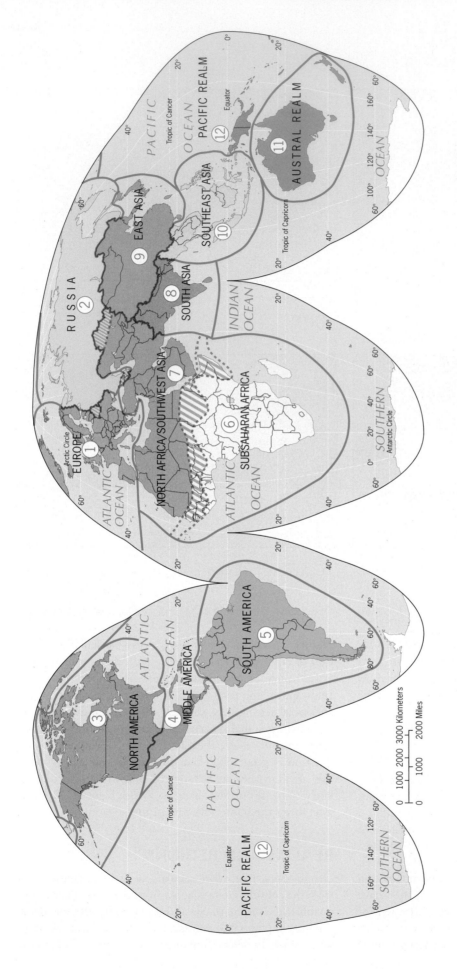

WORLD GEOGRAPHIC REALMS

① EUROPE	④ MIDDLE AMERICA	⑦ NORTH AFRICA/ SOUTHWEST ASIA
② RUSSIA	⑤ SOUTH AMERICA	⑧ SOUTH ASIA
③ NORTH AMERICA	⑥ SUBSAHARAN AFRICA	⑨ EAST ASIA

⑩ SOUTHEAST ASIA	
⑪ AUSTRAL REALM	
⑫ PACIFIC REALM	

© H. J. DE BLIJ, P. O. MULLER, AND JOHN WILEY & SONS, INC.

FIGURE G-3

animal *kingdom*, the *phylum* (division) named chordata, the *class* of mammals, the *order* of primates, the *family* of hominids, the *genus* designated *Homo*, and the *species* known as *Homo sapiens*. Geologists classify the Earth's rocks into three major (and many subsidiary) categories and then fit these categories into a complicated geologic time scale that spans hundreds of millions of years. Historians define eras, ages, and periods to conceptualize the sequence of the events they study.

Geography, too, employs systems of classification. When geographers deal with urban problems, for instance, they use a classification scheme based on the sizes and functions of the places involved. Some of the terms in this classification are part of our everyday language: megalopolis, metropolis, city, town, village, hamlet.

Regional Classification

In regional geography, the focus of this book, our challenge is different. We, too, need a hierarchical framework for the areas of the world we study, from the largest to the smallest. But our classification system is horizontal, not vertical. It is *spatial*. There are four levels:

1. *Landmasses and Oceans* (Fig. G-1). This is geography's equivalent to the biologists' overarching kingdoms (of plants and animals), although the issue arises as to whether the icy surface of the Antarctic "continent" constitutes a landmass. Since no permanent human population has become established on these glaciers, no regional geography has as yet evolved.

2. *Geographic Realms* (Fig. G-3). Based on a combination of physical and human factors, these are the most comprehensive divisions of the inhabited world but are not standard.

3. *Geographic Regions* More specific criteria divide the great geographic realms into smaller regions, as will be demonstrated at the conclusion of this chapter.

4. *Subregions, Domains, Districts* Subdivisions of regions, sometimes based on single factors, are mapped for specific purposes.

Criteria for Geographic Realms

In any classification system, criteria are the key. Not all animals are mammals; the criteria for inclusion in that biological class are more specific and restrictive. A dolphin may look and act like a fish, but both anatomically and functionally dolphins belong to the class of mammals.

- *Physical and Human* Geographic realms are based on sets of spatial criteria. They are the largest units into which the inhabited world can be divided. The criteria on which such a broad regionalization is based include both physi-

cal (that is, natural) and human (or social) yardsticks. On the one hand, South America is a geographic realm because physically it is a continent and culturally it is dominated by a set of social norms. The realm called South Asia, on the other hand, lies on a Eurasian landmass shared by several other geographic realms; high mountains, wide deserts, and dense forests combine with a distinctive social fabric to create this well-defined realm centered on India.

- *Functional* Geographic realms are the result of the interaction of human societies and natural environments, a *functional* interaction revealed by farms, mines, fishing ports, transport routes, dams, bridges, villages, and countless other features that mark the landscape. According to this criterion, Antarctica is a continent but not a geographic realm.

- *Historical* Geographic realms must represent the most comprehensive and encompassing definition of the great clusters of humankind in the world today. China lies at the heart of such a cluster, as does India. Africa constitutes a geographic realm from the southern margin of the Sahara (an Arabic word for desert) to the Cape of Good Hope and from its Atlantic to its Indian Ocean shores.

Figure G-3 displays the 12 world geographic realms based on these criteria. As we will show in more detail later, waters, deserts, and mountains as well as cultural and political shifts mark the borders of these realms. We will discuss the position of these boundaries as we examine each realm.

Geographic Realms: Margins and Mergers

Where geographic realms meet, **5** transition zones, not sharp boundaries, mark their contacts. We need only remind ourselves of the border zone between the geographic realm in which most of us live, North America, and the adjacent realm of Middle America. The line in Figure G-3 coincides with the boundary between Mexico and the United States, crosses the Gulf of Mexico, and then separates Florida from Cuba and the Bahamas. But Hispanic influences are strong in North America north of this boundary, and the U.S. economic influence is strong south of it. The line, therefore, represents an ever-changing zone of regional interaction. Again, there are many ties between South Florida and the Bahamas, but the Bahamas resemble a Caribbean more than a North American society.

In Africa, the transition zone from Subsaharan to North Africa is so wide and well defined that we have put it on the world map; elsewhere, transition zones tend to be narrower and less easily represented. In these early years of the twenty-first century, such countries as Belarus (between Europe and Russia) and Kazakhstan (between Russia and Muslim Southwest Asia) lie in inter-realm transition zones. Remember, over much (though not all) of their length, borders between realms are zones of regional change. As you will see, transition zones are often places of tension and/or conflict.

Geographic Realms: Two Categories

The world's geographic realms can be divided into two categories: (1) those dominated by one major political entity, in terms of territory or population or both (North America/United States, Middle America/Mexico, South America/Brazil, South Asia/India, East Asia/China, Southeast Asia/Indonesia as well as Russia and Australia), and (2) those that contain many countries but no dominant state (Europe, North Africa/Southwest Asia, Subsaharan Africa, and the Pacific Realm). For several decades in the twentieth century two major powers, the United States and the former Soviet Union, dominated the world and competed for global influence. Today, the United States is dominant, but its influence is dwindling. What lies ahead? Will China and/or some other power challenge U.S. supremacy? Is our map of realms a prelude to a multipolar world? We will address such questions in many chapters.

Criteria for Regions

The spatial division of the world into geographic realms establishes a broad global framework, but for our purposes a more refined level of spatial classification is needed. This brings us to an important organizing concept in geography: the **6 regional concept**. To continue the analogy with biological taxonomy, we now go from phylum to order. To establish regions within geographic realms, we need more specific criteria.

Let us use the North American realm to demonstrate the regional idea. When we refer to a part of the United States or Canada (e.g., the South, the Midwest, or the Prairie Provinces), we employ a regional concept—not scientifically but as part of everyday communication. We reveal our *perception* of local or distant space as well as our mental image of the region we are describing.

But what exactly is the Midwest? How would you draw this region on the North American map? Regions are easy to imagine and describe, but they can be difficult to outline on a map. One way to define the Midwest is to use the borders of States: certain States are part of this region, others are not. You could use agriculture as the chief criterion: the Midwest is where corn and/or soybeans occupy a certain percentage of the farmland. Look ahead to Figure 3B-5 where you will notice that a different name for this region is being used, the Heartland, because of the differing criteria (agriculture) defining it. Each method results in a different delimitation; a Midwest based on States is different from a Midwest based on farm production or on industrial location. Therein lies an important principle: regions are devices that allow us to make spatial generalizations, and they are based on artificial criteria we establish to help us construct them.

Area

Given these different dimensions of the same region, we can identify properties that all regions have in common. To begin with, all regions have **area**. This observation would seem obvious, but there is more to this idea than meets the eye. Regions may be intellectual constructs, but they are not abstractions: they exist in the real world, and they occupy space on the Earth's surface.

Boundaries

It follows that regions have **boundaries**. Occasionally, nature itself draws sharp dividing lines, for instance, along the crest of a mountain range or the margin of a forest. More often, regional boundaries are not self-evident, and we must determine them using criteria that we establish for that purpose. For example, to define a citrus-growing agricultural region, we may decide that only areas where more than 50 percent of all farmland stands under citrus trees qualify to be part of that region.

Location

All regions also have **location**. Often the name of a region contains a locational clue, as in Amazon Basin or Indochina (a region of Southeast Asia lying between India and China). Geographers refer to the **7 absolute location** of a place or region by providing the latitudinal and longitudinal extent of the region with respect to the Earth's grid coordinates. A far more practical measure is a region's **8 relative location**, that is, its location with reference to other regions. Again, the names of some regions reveal aspects of their relative locations, as in *Mainland* Southeast Asia and *Equatorial* Africa.

Homogeneity

Many regions are marked by a certain **homogeneity** or sameness. Homogeneity may lie in a region's human (cultural) properties, its physical (natural) characteristics, or both. Siberia, a vast region of northeastern Russia, is marked by a sparse human population that resides in widely scattered, small settlements of similar form, frigid climates, extensive areas of permafrost (permanently frozen subsoil), and cold-adapted vegetation. This dominant uniformity makes it one of Russia's natural and cultural regions, extending from the Ural Mountains in the west to the Pacific Ocean in the east. When regions display a measurable and often visible internal homogeneity, they are called **9 formal regions**. But not all formal regions are visibly uniform. For example, a region may be delimited by the area in which, say, 90 percent of the people or more speak a particular language. This cannot be seen in the landscape, but the region is a reality, and we can use this criterion to draw its boundaries accurately. It, too, is a formal region.

Regions as Systems

Other regions are marked *not* by their internal sameness but by their functional integration—that is, the way they work. These regions are defined as **10** **spatial systems** and are formed by the areal extent of the activities that define them. Take the case of a large city with its surrounding zone of suburbs, urban-fringe countryside, satellite towns, and farms. The city supplies goods and services to this encircling zone, and it buys farm products and other commodities from it. The city is the heart, the ***core*** of this region, and we call the surrounding zone of interaction the city's **11** **hinterland**. But the city's influence wanes on the outer periphery of that hinterland, and there lies the boundary of the functional region of which the city is the focus. A **12** **functional region**, therefore, is usually forged by a structured, urban-centered system of interaction. It has a core and a periphery. As we shall see, core-periphery contrasts in some parts of the world are becoming strong enough to endanger the stability of countries.

Interconnections

All human geographic regions are *interconnected*, being linked to other regions. We know that the borders of geographic realms sometimes take on the character of transition zones, and so do neighboring regions. Trade, migration, education, television, computer linkages, and other interactions blur regional boundaries. These are just some of the links in the fast-growing interdependence among the world's peoples, and they reduce the differences that still divide us. Understanding these differences will lessen them further.

THE PHYSICAL SETTING

This book focuses on the geographic realms and regions produced by human activity over thousands of years. But we should not forget the natural environments in which all this activity took place because we can still recognize the role of these environments in how people make their living. Certain areas of the world, for example, presented opportunities for plant and animal domestication that other areas did not. The people who happened to live in those favored areas learned to grow wheat, rice, or root crops and to domesticate oxen, goats, or llamas. We can still discern those early *patterns of opportunity* on the map in the twenty-first century. From such opportunities came adaptation and invention, and thus arose villages, towns, cities, and states. But people living in different environments found it much harder to achieve this organization. The Americas, for instance, had no large animals that could be domesticated except llamas. This meant that societies created agricultural systems that did not involve ploughing as there were no draught animals. When Europeans

 FROM THE FIELD NOTES...

© H. J. de Blij   © AP/Wide World Photos

"Flying over Iceland's volcanic topography is to see the world in the making: this is some of the youngest rock on the planet, and even at rest you can sense its impermanence. Here nature shows us what mostly goes on deep below the surface along the mid-oceanic ridges, where tectonic plates pull apart and lava pours out of fissures and vents. When that happens in plain view, as in the photo at right, the results can be catastrophic. In the 1780s, an Icelandic volcano named Laki, in a series of eruptions, killed tens of thousands and caused a global ecological crisis. In 2010, the eruption of this far smaller volcano, Eyjafjallajokull, disrupted air travel across much of the Northern Hemisphere."

introduced cows, horses, and other livestock, this change completely revolutionized the environments and cultural systems of the Western Hemisphere. The modern map carries many such imprints of the past.

Natural (Physical) Landscapes

The landmasses of Planet Earth present a jumble of **13 natural landscapes** ranging from rugged mountain chains to smooth coastal plains (Fig. G-1). Certain continents are readily linked with a dominant physical feature—for example, North America and its Rocky Mountains, South America with its Andes and Amazon River Basin, Europe with its Alps and Rhine and Danube River basins, Asia with its Himalaya Mountains and numerous river basins, and Africa with its Sahara and Congo River Basin. Physical features have long influenced human activity and movement—even today. Mountain ranges formed barriers to movement but also channeled the spread of agricultural and technical innovations. Today the Taliban, al-Qaeda, and Chechnyan fugitives still use rugged mountains to hide out. Large deserts similarly formed barriers, as did rivers, although rivers do permit accessibility and connectivity between people. As we study each of the world's geographic realms, we will find that physical landscapes continue to play significant roles in this modern world. That is one reason why the study of world regional geography is so important: it puts the human map in environmental as well as regional perspective.

Natural Hazards

Our planet may be 4.6 billion years old, but it is far from placid. As you read this chapter, earth tremors are shaking the still-thin crust on which we live, volcanoes are erupting, storms are raging. Even the very continents are moving measurably, pulling apart in some areas, colliding in others. Hundreds of thousands of human lives are lost to natural calamities in almost every decade, and such calamities have at times altered the course of history.

About a century ago a geographer named Alfred Wegener, a German scientist, used spatial analysis to explain something that is obvious even from a small-scale map like Figure G-1: the apparent, jigsaw-like fit of the landmasses, especially across the South Atlantic Ocean. He concluded that the landmasses on the map are actually pieces of a supercontinent that existed hundreds of millions of years ago (he called it **Pangaea**) that drifted away when, for some reason, that supercontinent broke up. His hypothesis of **14 continental drift** set the stage for scientists in other disciplines to search for a mechanism that might make this possible, and much of the answer to that search proved to lie in the crust beneath the ocean surface. Today we know that the continents are "rafts" of relatively light rock that rest on slabs of heavier rock called **15 tectonic plates** (Fig. G-4) whose movement is propelled by giant circulation cells in the red-hot magma below (when this molten magma reaches the surface through volcanic vents, it is called lava).

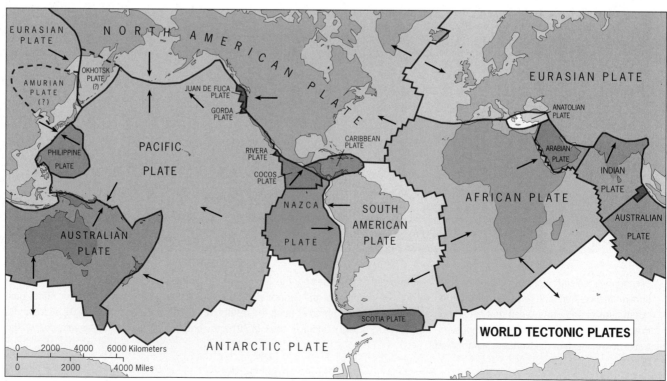

FIGURE G-4

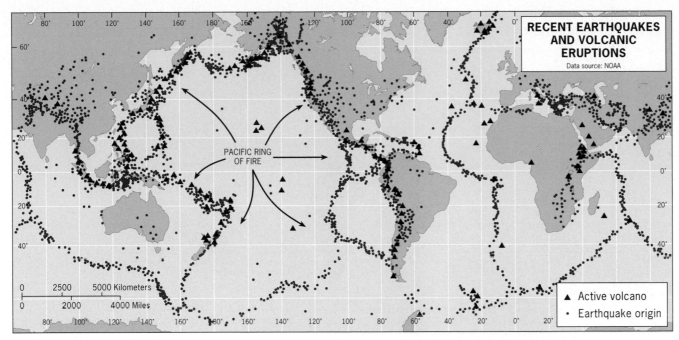

FIGURE G-5   Data courtesy U.S. National Oceanic and Atmospheric Administration (NOAA)

Inevitably, moving tectonic plates collide. When they do, earthquakes and volcanic eruptions result, and the physical landscape is thrown into spectacular relief. Compare Figures G-4 and G-5, and you can see the outlines of the tectonic plates in the distribution of these hazards to human life. The January 12, 2010 earthquake adjacent to Port-au-Prince, Haiti measured 7.0 on the Richter scale. Although a shallow quake, its epicenter was in a very densely populated region. More than 300,000 people died, a similar number were injured, and 1.3 million were made homeless and destitute. One of the Earth's oceans is almost completely encircled by active volcanoes and earthquake epicenters. Appropriately, this is called the **16 Pacific Ring of Fire**.

It is useful to compare Figure G-5 to Figure G-3 to see which of the world's geographic realms are most susceptible to the hazards inherent in crustal instability. Russia, Europe, Africa, and Australia are relatively safe; in other realms the risks are far greater in one sector than in others (western as opposed to eastern North and South America, for example). As we will find, for certain parts of the world the details on Figure G-5 present a clear and present danger. Some of the world's largest cities (e.g., Tokyo, Mexico City) lie in areas most vulnerable to sudden disaster.

Climate

The prevailing climate constitutes a key factor in the geography of realms and regions (in fact, some regions are essentially *defined* by climate). But we should always remember that climates change and that the climate predominating in a certain region today may not be the climate prevailing there several thousand years ago. Any map of climate, including the maps in this chapter, is but a still-picture of our always-changing world.

Ice Ages and Climate Change

Climatic conditions have swung back and forth for as long as the Earth has had an atmosphere. Periodically, an **17 ice age** lasting tens of millions of years chills the planet and causes massive ecological change. One such ice age occurred while Pangaea was still in one piece, between 250 and 300 million years ago. Another started about 35 million years ago, and we are still experiencing it. The current epoch of this ice age, on average the coldest yet, is called the ***Pleistocene*** and has been going on for nearly 2 million years.

In our time of global warming this may come as a surprise, but we should remember that an ice age is not a period of unbroken, bitter cold. Rather, an ice age consists of surges of cold, during which glaciers expand and living space shrinks, separated by warmer phases when the ice recedes and life spreads poleward again. The cold phases are called **18 glaciations**, and they tend to last longest, although milder spells create some temporary relief. The truly warm phases, when the ice recedes poleward and mountain glaciers melt away, are known as **19 interglacials**. We are living in one of these interglacials today. It even has a geologic name: the ***Holocene***.

Imagine this: just 18,000 years ago, great icesheets had spread all the way to the Ohio River Valley, covering most of the Midwest; this was the zenith of a glaciation that had lasted about 100,000 years, the ***Wisconsinan Glaciation*** (Fig. G-6). The Antarctic Icesheet was bigger than ever, and

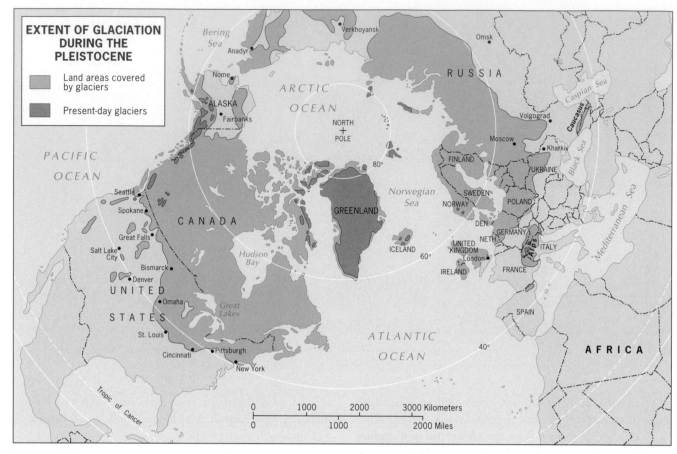

FIGURE G-6

© H. J. de Blij, P. O. Muller, and John Wiley & Sons, Inc.

even in the tropics, great mountain glaciers pushed down valleys and onto plateaus. But then Holocene warming began, the continental and mountain glaciers receded, and ecological zones that had been squeezed between the advancing ice sheets now spread north and south. In Europe particularly, where humans had arrived from Africa via Southwest Asia during one of the milder spells of the Wisconsinan Glaciation, living space expanded and human numbers grew.

Global Climate Change

Consider the transformation that our planet—and our ancestors—experienced as the ice receded. Huge slabs of ice thousands of miles across slid into the oceans from Canada and Antarctica. Rivers raged with silt-clogged meltwater and formed giant deltas. Sea level rose rapidly, submerging vast coastal plains. Animals and plants migrated into newly opened lands at higher elevations and latitudes. As climatic zones shifted, moist areas fell dry, and arid areas turned wet. In Southwest Asia, where humans had begun to cultivate crops and towns signaled the rise of early states, these changes, such as increasing aridity, often destroyed what had been achieved. In northern tropical Africa, where rivers flowed and grasses fed wildlife, a region extending from

Atlantic shores to the Red Sea fell dry in a very short time around 5000 years ago. We know it today as the Sahara. We call the process **20 desertification**.

Temperatures reached their present-day levels about 7000 years ago, but the effects of this Holocene interglacial warmth took more time to take hold, as the Sahara's later desiccation shows. Soils formed on newly exposed rock, but generally faster in warm and moist areas than in dry and still-cooler zones. Pine forests that had migrated southward during the Wisconsinan Glaciation now moved north, and equatorial rainforests expanded in all directions. Some of the world's early empires formed in areas that had been inhospitable during the Wisconsinan: the Roman Empire at the western end of Eurasia, the Han Empire at the eastern end. Europeans and Chinese traded with each other along the Silk Route, now open for business.

A great deal of this environmental history remains imprinted today (see Fig. G-8). The great river basins of East Asia, where humans exploited agricultural opportunities thousands of years ago, still anchor populous China today. The great Ganges Basin, where one of early humankind's great population explosions may have taken place, remains the core of a huge modern state, India. The Roman Empire's legacies infuse much of European culture.

The past 1000 years have witnessed some troubling environmental developments. Around 1300, the first warning signals of a return to cooler conditions caused crop failures and social dislocation, first in Europe and later in China. This episode, which worsened during the 1600s and has come to be known as the *Little Ice Age*, had major impacts on the human geography of Eurasia and on peoples elsewhere in the world as well, accompanied as it was by long-term changes in climate. A return to pre-Little Ice Age conditions commenced in the early nineteenth century, but then the exploding human population began to have a stronger impact on the atmosphere, and humanity itself became a factor in global climate change.

Today, we are living in an era of **21 global climate change**, particularly natural global warming that has been accelerated by anthropogenic (human-source) causes. Since the Industrial Revolution, we have been emitting gases that have enhanced nature's *greenhouse effect* wherein the sun's radiation becomes trapped in the Earth's atmosphere. This is leading to a series of climate changes, especially the overall warming of the globe. In 2007 the Intergovernmental Panel on Climate Change (IPCC) released an important series of documents that present indisputable scientific evidence on global climate change and the possible implications of these changes. The experts predict an increase of 2–3°C (3.6–5.4°F) overall for the globe, but with significant regional variability (e.g., more at higher latitudes, less at lower latitudes). Precipitation patterns are predicted to become more variable, particularly in regions where they are already seasonal. This change in temperature may seem small but will have significant impacts on global climate patterns, agricultural zones, and the quality of human lives. The full ramifications are not known, but scenarios are being modeled so that societies can confront the changes that are coming. Leaders of some countries are more skeptical than others, and some have already made greater adjustments than others. One of the most significant consequences of global climate change is that the Arctic icecap is melting faster than even recent models predicted, with environmental and geopolitical implications. We pick up this issue in Chapters 2A and 12.

Climatic Regions

We have just learned how variable climate can be, but in a human lifetime we see little evidence of this variability. We talk about the *weather* (the immediate state of the atmosphere) in a certain place at a given time, but as a technical term *climate* defines the aggregate, total record of weather conditions at a place, or in a region, over the entire period during which records have been kept.

Figure G-7 may appear very complicated, but this map is useful even at a glance. Devised long ago by Wladimir Köppen and later modified by Rudolf Geiger, it represents climatic regions through a combination of colors and letter symbols. In the legend, note that the *A* climates (rose, orange, and peach) are equatorial and tropical; the *B* climates (tan, yellow) are dry; the *C* climates (shades of green) are temperate, that is, moderate and neither hot nor cold; the *D* climates (purple) are cold; the *E* climates (blue) are frigid; and the *H* climates (gray) prevail in highlands like the Himalayas and the Andes.

A good way to get a sense of how this map works is to start with a climate with which you are familiar. If you live in the southeastern United States (the large green *C* area from near the Great Lakes to Florida), you would feel at home—climatically—in the green area in southeastern South America, in southeastern Australia, and in southeastern China. If you live in the *B* areas of the U.S. Southwest, either in the tan (Texas) or the yellow (Arizona) area, conditions in much of Australia, southwestern Africa, or the Middle East would be familiar to you. This can even be taken to the city level. If you enjoy the climate of San Francisco, you will rediscover it in Santiago (Chile), Cape Town (South Africa), Athens (Greece), and Perth (Australia). Details will differ, of course—higher wind incidence or maybe lower overall humidity—but in general, the similarities will be greater than such differences. If you live in Canada, you would be prepared for much of Russia and northeastern China.

Even at this relatively small scale, Figure G-7 can tell a great deal about entire regions. For instance, among the equatorial/tropical (*Af*) climates, the areas with the darkest (rose) shade get the most rain and it comes year-round, so that this is where the tropical rainforest still stands. But in the peach-colored (*Aw*) areas, the rainfall regime is subject to dry seasons, and the vegetation reflects this by its very name: *savanna*. Much more open space and more widely scattered stands of trees prevail here. And a third kind of regime is signified by the **m** after the *A*: for *monsoon*, the annual seasonal torrent on which the lives of hundreds of millions of people still depend.

Here is an important qualification to keep in mind when studying Figure G-7: this map is a still-picture of a changing scene, a single frame from an ongoing film. Global climate never stops changing, and less than a century from now those changes may compel revisions to this map. If predictions about rising sea levels turn out to be accurate, we may even have to start redrawing familiar coastlines.

You will find larger-scale maps of climate in several of the regional chapters that follow, but it is useful to refer back to the Köppen-Geiger map whenever the historical or economic geography of a region or country is under discussion. The world climatic map reflects agricultural opportunities and limitations as well as climatic regimes, and as such helps explain some enduring patterns of human distribution on our planet. We turn next to this crucial topic.

WORLD CLIMATES
After Köppen–Geiger

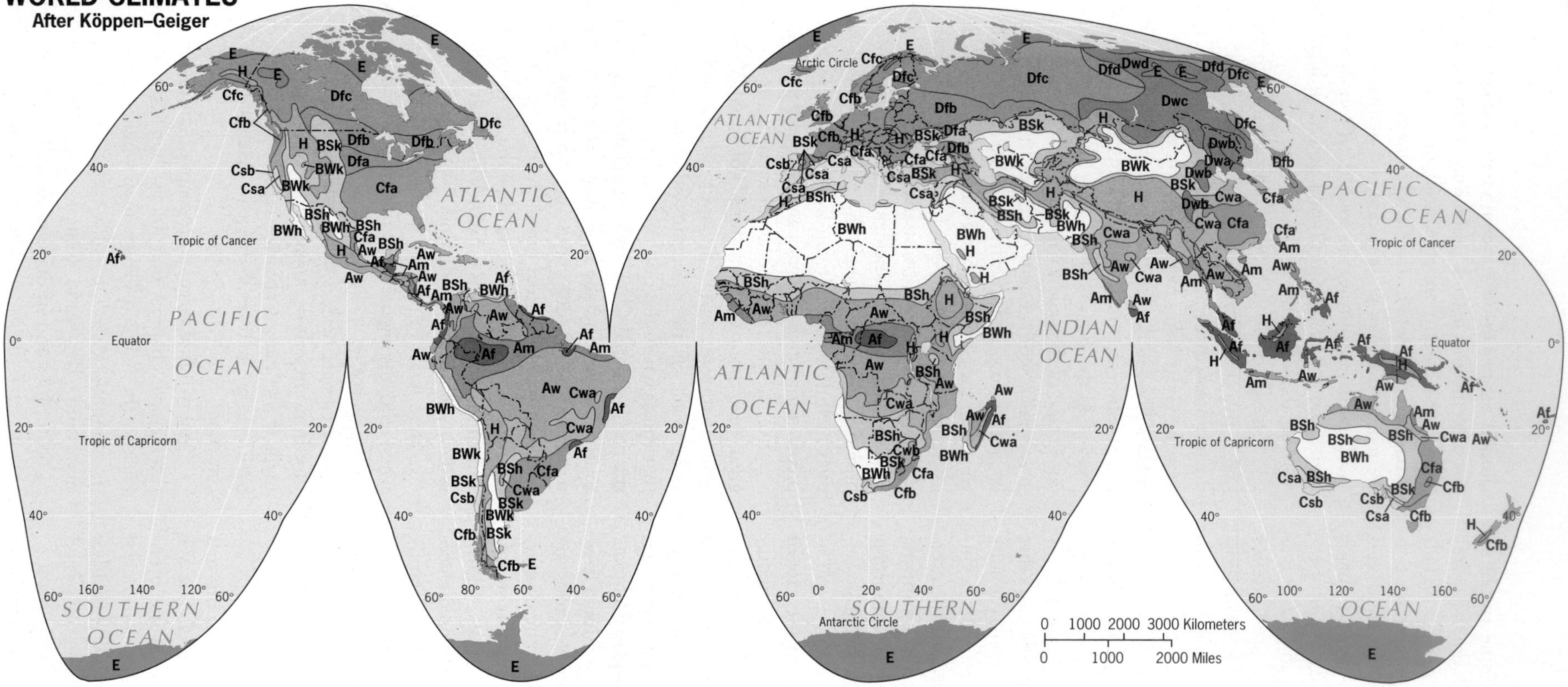

A HUMID EQUATORIAL CLIMATE

Af	No dry season
Am	Short dry season
Aw	Dry winter

B DRY CLIMATE

BS	Semiarid	}	h=hot
BW	Arid	}	k=cold

C HUMID TEMPERATE CLIMATE

Cf	No dry season
Cw	Dry winter
Cs	Dry summer

a=hot summer
b=cool summer
c=short, cool summer
d=very cold winter

D HUMID COLD CLIMATE

Df	No dry season
Dw	Dry winter

E COLD POLAR CLIMATE

E	Tundra and ice

H HIGHLAND CLIMATE

H	Unclassified highlands

FIGURE G-7

REALMS OF POPULATION

Earlier we noted that population numbers by themselves do not define geographic realms or regions. Population distributions, and the functioning society that gives them common ground, are more significant criteria. That is why we can identify one geographic realm (the Austral) with less than 30 million people and another (East Asia) with more than 1.5 billion inhabitants. Neither population numbers nor territorial size alone can delimit a geographic realm. Nevertheless, the map of world population distribution (Fig. G-8) suggests the relative location of several of the world's geographic realms, based on the strong clustering of population in certain areas. Before we examine these clusters in some detail, remember that the world's human population now totals just under 7 billion—more than six thousand nine hundred million people confined to the landmasses that constitute less than 30 percent of our planet's surface, much of which is arid desert, rugged mountain terrain, or frigid tundra. (Remember that Fig. G-8 is another still-picture of an ever-changing scene: the rapid growth of humankind continues.) After thousands of years of slow growth, world population during the nineteenth and twentieth centuries grew at an increasing rate. That rate has recently been slowing down, even imploding in some parts of the world. But consider this: it took about 17 centuries after the birth of Christ for the world to add 250 million people to its numbers; now we are adding 250 million about every three and a half years.

Major Population Clusters

One way to present an overview of the location of people on the planet is to create a map of **22** population distribution (Fig. G-8). As you see in the map's legend, every dot represents 100,000 people, and the clustering of large numbers of people in certain areas as well as the near-emptiness of others is immediately evident. There is a technical difference between population distribution and *population density*, which is another way of showing where people live. Density maps reveal the number of persons per unit area, requiring a different cartographic technique.

East Asia

Still the world's greatest population cluster, *East Asia* lies centered on China and includes the Pacific-facing Asian coastal zone from the Korean Peninsula to Vietnam. Not

 FROM THE FIELD NOTES...

© H. J. de Blij

"The Atlantic-coast city of Bergen, Norway displayed the Norse cultural landscape more comprehensively, it seemed, than any other Norwegian city, even Oslo. The high-relief site of Bergen creates great vistas, but also long shadows; windows are large to let in maximum light. Red-tiled roofs are pitched steeply to enhance runoff and inhibit snow accumulation; streets are narrow and houses clustered, conserving warmth . . . The coastal village of Mengkabong on the Borneo coast of the South China Sea represents a cultural landscape seen all along the island's shores, a stilt village of the Bajau, a fishing people. Houses and canoes are built of wood as they have been for centuries. But we could see some evidence of modernization: windows filling wall openings, water piped in from a nearby well."

www.conceptcaching.com

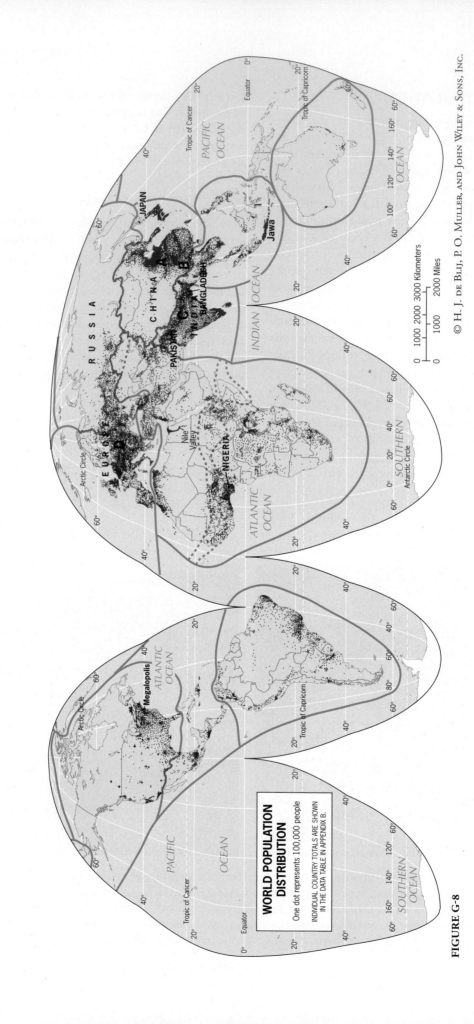

WORLD POPULATION DISTRIBUTION

One dot represents 100,000 people

INDIVIDUAL COUNTRY TOTALS ARE SHOWN IN THE DATA TABLE IN APPENDIX B.

FIGURE G-8

long ago, we would have reported this as a dominantly rural, farming population, but rapid economic growth and associated urbanization are changing this picture. In the interior river basins of the Huang (Yellow) and Chang/Yangzi (**A** and **B** on the map), and in the Sichuan Basin between these two letters, most of the people remain farmers, and, as Appendix B shows, farmers still outnumber city-dwellers in China as a whole. But the great cities of coastal and near-coastal China are attracting millions of new inhabitants, and interior cities are growing rapidly as well. By 2020, the East Asia cluster will be more urban than rural.

South Asia

The *South Asia* population cluster lies centered on India and includes its populous neighbors, Pakistan and Bangladesh. This huge agglomeration of humanity focuses on the wide plain of the Ganges River (**C** in Fig. G-8). It is nearly as large as that of East Asia and at present growth rates will overtake East Asia during 2011. A larger percentage of the people remain farmers, although pressure on the land is greater, while farming is less efficient than in East Asia.

Europe

The third-ranking population cluster, *Europe*, also lies on the Eurasian landmass but at the opposite end from China. The European cluster, including western Russia, counts over 700 million inhabitants, which puts it in a class with the two larger Eurasian concentrations—but there the similarity ends. In Europe, the key to the linear, east-west orientation of the axis of population (**D** in Fig. G-8) is not a fertile river basin but a zone of raw materials for industry. Europe is among the world's most highly urbanized and industrialized realms, its human agglomeration sustained by factories and offices rather than paddies and pastures.

The three world population concentrations just discussed (East Asia, South Asia, and Europe) account for nearly 4 billion of the world's 7 billion people. No other cluster comes close to these numbers. The next-ranking cluster, *Eastern North America,* is only about one-quarter the size of the smallest Eurasian concentrations. As in Europe, the population in this area is concentrated in major metropolitan complexes; the rural areas are now relatively sparsely settled. Geographic realms and regions, therefore, display varying levels of **23 urbanization**, the percentage of the total population living in cities and towns. Some regions are urbanizing much more rapidly than others, a phenomenon we will explain as we examine the world's realms.

REALMS OF CULTURE

Imagine yourself in a boat on the Nile River, headed upstream (south) from Khartoum, Sudan. The desert sky is blue, the heat is searing. You pass villages on the shore that look much

the same: low, square or rectangular dwellings, some recently whitewashed, others gray, with flat roofs, wooden doors, and small windows. The minaret of a modest mosque may rise above the houses, and you get a glimpse of a small central square. There is very little vegetation; here and there a hardy palm tree stands in a courtyard. People on the paths wear long white robes and headgear, also white, that looks like a baseball cap without the visor. A few goats lie in the shade. Along the river's edge lie dusty farm fields that yield to the desert in the distance. At the foot of the river's bluff lie some canoes.

All of this is part of central Sudan's rural **24 cultural landscape**, the distinctive attributes of a society imprinted on its portion of the world's physical stage. The cultural landscape concept was first articulated in the 1920s by a University of California geographer named Carl Sauer, who stated that "a cultural landscape is fashioned from a natural landscape by a culture group" and that "culture is the agent; the natural environment the medium." What this means is that people, starting with their physical environment and using their culture as their agency, fashion a landscape that is layered with forms such as buildings, gardens, and roads, and also modes of dress, aromas of food, and sounds of music.

Continue your journey southward on the Nile, and you will soon witness a remarkable transition. Quite suddenly, the square, solid-walled, flat-roofed houses of central Sudan give way to round, wattle-and-thatch, conical-roofed dwellings of the south. You may note that clouds have appeared in the sky: it rains more here, and flat roofs will not do. The desert has given way to green. Vegetation, natural as well as planted, grows between houses, flanking even the narrow paths. The villages seem less orderly, more varied. People ashore wear a variety of clothes, the women often in colorful dresses, the adult men in shirts and slacks, but shorts when they work the fields, although you see more women wielding hoes than men. You have traveled from one cultural landscape into another, from Arabized, Islamic Africa to animist or Christian Africa. You have crossed the boundary between two geographic realms.

No geographic realm, not even the Austral Realm, has just one single cultural landscape, but cultural landscapes help define realms as well as regions. The cultural landscape of the high-rise North American city with its sprawling suburbs differs from that of Brazil and South America; the organized terraced paddies of Southeast Asia are unlike anything to be found in the rural cultural landscape of neighboring Australia. Variations of cultural landscapes *within* geographic realms, such as between highly urbanized and dominantly rural (and more traditional) areas, help us define the world's regions.

The Geography of Language

Language is the essence of culture. People tend to feel passionately about their mother tongue, especially when they

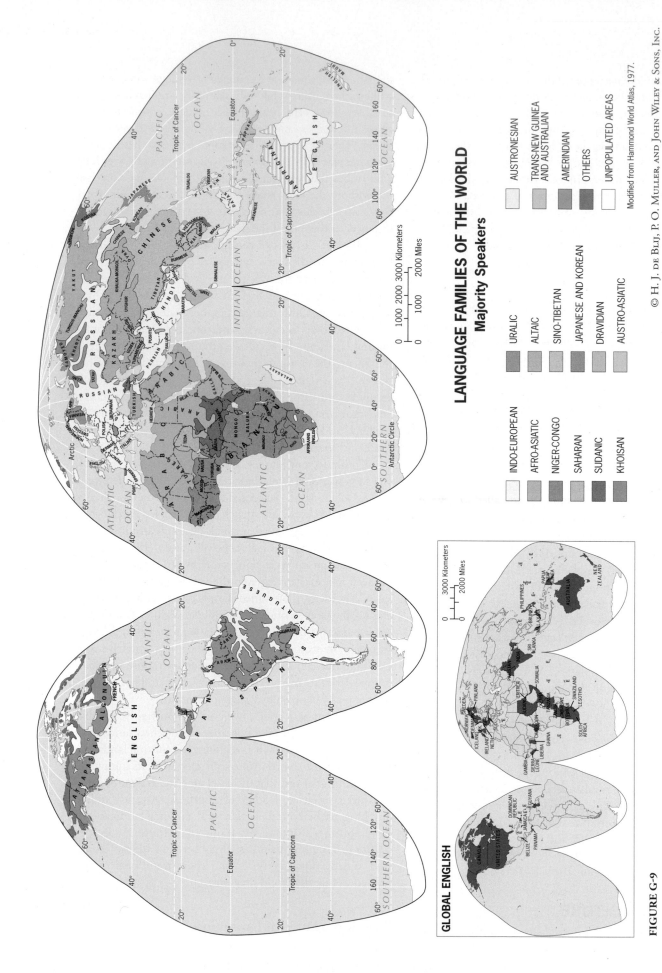

LANGUAGE FAMILIES OF THE WORLD
Majority Speakers

INDO-EUROPEAN
AFRO-ASIATIC
NIGER-CONGO
SAHARAN
SUDANIC
KHOISAN

URALIC
ALTAIC
SINO-TIBETAN
JAPANESE AND KOREAN
DRAVIDIAN
AUSTRO-ASIATIC

AUSTRONESIAN
TRANS-NEW GUINEA
AND AUSTRALIAN
AMERINDIAN
OTHERS
UNPOPULATED AREAS

Modified from Hammond World Atlas, 1977.

© H. J. de Blij, P. O. Muller, and John Wiley & Sons, Inc.

GLOBAL ENGLISH

FIGURE G-9

believe it is threatened in some way. In the United States today, movements such as "English Only" and "English First" reflect many people's fears that the primacy of English as the national language is under threat as a result of immigration. As we will see in later chapters, some governments try to suppress the languages (and thus the cultures) of minorities in mistaken attempts to enforce national unity, provoking violent reactions.

In fact, languages emerge, thrive, and die out over time, and linguists estimate that the number of lost languages is in the tens of thousands—a process that continues. One year from the day you read this, about 25 more languages will have become extinct, leaving no trace. Some major ones of the past, such as Sumerian and Etruscan, have left fragments in later languages. Others, like Sanskrit and Latin, live on in their modern successors. At present, about 7000 languages remain, half of them classified by linguists as endangered. By the end of this century, the world is likely to be left with just a few hundred languages, so billions of people will no longer be speaking their ancestral mother tongues.

Scholars have tried for many years to unravel the historic roots and branches of the "language tree," and their debates continue. Geographers trying to map the outcome of this research keep having to modify the pattern, so you should take Figure G-9 as a work in progress, not the final word. At minimum, there are some 15 so-called *language families*, groups of languages with a shared but usually distant origin. The most widely distributed language family, the Indo-European (shown in yellow on the map), includes English, French, Spanish, Russian, Persian, and Hindi. This encompasses the languages of European colonizers that were carried and implanted worldwide, English most of all. Today, English serves as the national or official language of many countries and outposts and remains the *lingua franca* of government, commerce, and higher education in many multicultural societies (see Fig. G-9 inset map). In the postcolonial era, English became the chief medium of still another wave of ascendancy now in progress: globalization.

But even English is going the way of Latin, morphing into versions you will hear (and learn to use) as you travel, forms of English that may, generations from now, be the successors that Italian and Spanish are to Latin. In Hong Kong, Chinese and English are producing a local "Chinglish" you may hear in the first taxi you enter. In Lagos, Nigeria, where most of the people are culturally and ethnically Yoruba, a language called "Yorlish" is emerging. No map can keep up with the continuing evolution of language.

REALMS, REGIONS, AND STATES

Our analysis of the world's regional geography requires data, and it is crucial to know the origin of these data. Unfortunately, we do not have a uniformly sized grid that we can superimpose over the globe: we must depend on the world's 190-plus countries to report vital information. Irregular as the boundary framework shown on the world map (see front endpaper or Fig. G-10) may be, it is all we have. Fortunately, all large and populous countries are subdivided into provinces, States, or other internal entities, and their governments provide information on each of these subdivisions when they conduct their census.

The State

The **25** **state** as a political, social, and economic entity has been developing for thousands of years, ever since agricultural surpluses made possible the growth of large and powerful towns that could command hinterlands and control peoples far beyond their walls. But the modern state is a relatively recent phenomenon. The boundary framework we see on the world political map today substantially came about only during the nineteenth century, and the independence of dozens of former colonies (which made them states as well) occurred during the twentieth century. Today, the **26** **European state model**—a clearly and legally defined territory inhabited by a population governed from a capital city by a representative government—prevails in the aftermath of the collapse of colonial and communist empires.

States and Realms

As Figures G-3 and G-10 suggest, geographic realms are mostly assemblages of states, and the borders between realms frequently coincide with the boundaries between countries—for example, between North America and Middle America along the U.S.-Mexico boundary. But a realm boundary can also cut *across* a state, as does the one between Subsaharan Africa and the Muslim-dominated realm of North Africa/Southwest Asia. Here the boundary takes on the properties of a wide transition zone, but it still divides states such as Nigeria, Chad, and Sudan. The transformation of the margins of the former Soviet Union is creating similar cross-country transitions. Newly independent states such as Belarus (between Europe and Russia) and Kazakhstan (between Russia and Muslim Southwest Asia) lie in transition zones of regional change.

Most often, however, geographic realms consist of groups of states whose boundaries also mark the limits of the realms. Look at Southeast Asia, for instance. Its northern border coincides with the political boundary that separates China (a realm practically unto itself) from Vietnam, Laos, and Myanmar (Burma). The boundary between Myanmar and Bangladesh (which is part of the South Asian realm) defines its western border. Here, the state boundary framework helps delimit geographic realms.

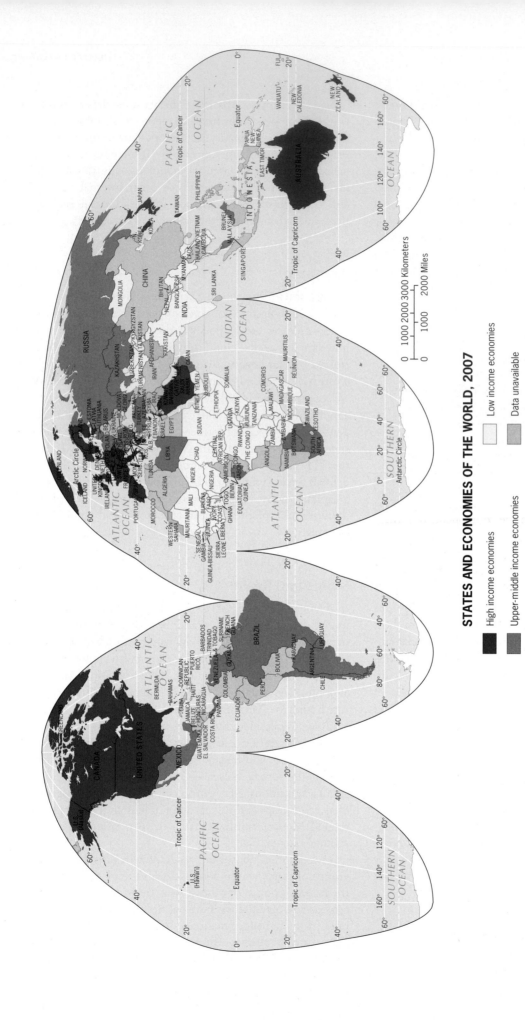

STATES AND ECONOMIES OF THE WORLD, 2007

- ■ High income economies
- ■ Upper-middle income economies
- ▨ Lower-middle income economies
- ☐ Low income economies
- ▨ Data unavailable

Data source: World Bank.

© H. J. de Blij, P. O. Muller, and John Wiley & Sons, Inc.

FIGURE G-10

States and Regions

The global boundary framework is even more useful in delimiting regions *within* geographic realms. We shall discuss regional divisions every time we introduce the regions of a geographic realm, but an example is appropriate here. In the Middle American realm, we recognize four regions. Two of these lie on the mainland: Mexico, the giant of the realm, and Central America, which consists of the seven comparatively small countries located between Mexico and the Panama-Colombia border (which is the boundary with the South American realm). Central America often is misdefined in news reports; the correct regional definition is based on the politico-geographical framework.

Political Geography

To our earlier criteria of physical geography, population distribution, and cultural geography, we now add *political geography* as a determinant of world-scale geographic regions. In doing so, we should be aware that the global boundary framework continues to change and that boundaries are created (e.g., between Serbia and Montenegro in 2006; between Serbia and Kosovo when Kosovo declared its independence in 2008) as well as eliminated (e.g., between former West and East Germany in 1990; between the components of former Czechoslovakia in 1993). But the overall system, much of it resulting from colonial and imperial expansionism, has endured, despite the predictions of some geographers that the "boundaries of imperialism" would be replaced by newly negotiated ones in the postcolonial period.

In Chapter 12, we will discuss a recent development in boundary-making: the extension of boundaries onto and into the oceans and seas. This process has been dividing up the last of the Earth's open frontiers, with uncertain consequences.

PATTERNS OF ECONOMIC DEVELOPMENT

Finally, as we prepare for our study of world regional geography, it is all too clear that realms, regions, and states do not enjoy the same level of prosperity. The field of *economic geography* focuses on spatial aspects of the ways people make their living and deals with patterns of production, distribution, and consumption of goods and services. As with all else in this world, these patterns reveal much variation. Individual states report the nature and value of their imports and exports, farm and factory output, and many other economic data to the United Nations and other international agencies. From such information, economic geographers can measure the comparative well-being of the world's countries. The concept of **27** **development** is used to gauge a state's economic, social, and institutional growth.

FROM THE FIELD NOTES...

© H. J. de Blij

"February 1, 2003. A long-held hope came true today: thanks to a Brazilian intermediary I was allowed to enter and spend a day in two of Rio de Janeiro's hillslope *favelas*, an eight-hour walk through one into the other. Here live millions of the city's poor, in areas often ruled by drug lords and their gangs, with minimal or no public services, amid squalor and stench, in discomfort and danger. And yet life in the older *favelas* has become more comfortable as shacks are replaced by more permanent structures, electricity is sometimes available, water supply, however haphazard, is improved, and an informal economy brings goods and services to the residents. I stood in the doorway of a resident's single-room dwelling for this overview of an urban landscape in transition: satellite-television disks symbolize the change going on here. The often blue cisterns catch rainwater; walls are made of rough brick and roofs of corrugated iron or asbestos sheeting. No roads or automobile access, so people walk to the nearest road at the bottom of the hill. Locals told me of their hope that they will some day have legal rights to the space they occupy. During his campaign for president of Brazil, Lula da Silva suggested that long-term inhabitants should be awarded title, and in 2003 his government approved the notion. It will be complicated: as the photo shows, people live quite literally on top of one another, and mapping the chaos will not be simple (but will be made possible with geographic information systems). This would allow the government to tax residents, but it would also allow residents to obtain loans based on the value of their *favela* properties, and bring millions of Brazilians into the formal economy. The hardships I saw on this excursion were often dreadful, but you could sense the hope for and anticipation of a better future."

www.conceptcaching.com

A Caution

The concept of development as measured by data that reflect totals and averages for entire national populations entails some risks of which we should be aware from the start. When a state's economy is growing as a whole, and even when it is "booming" by comparison to other states, this does not automatically mean that every citizen is better off and the income of every worker is rising. Averages have a way of concealing regional variability and local stagnation.

In very large states such as India and China, it is useful to assess regional, provincial, and even local economic data to discover to what extent the whole country is sharing in "development." In the case of India, we should all know that the State of Maharashtra (containing the largest city, Mumbai) is far in the lead when it comes to its share of the national economy. In China, the coastal provinces of the Pacific Rim far outstrip those of the interior. In Spain, the people of the Autonomous Community of Catalonia (focused on Barcelona) never tire of telling you that theirs is the most productive entity in the whole country. The question is: to what extent do these economic successes better the lives of all of the state's inhabitants? That information is not as easily obtained as those "national" averages.

Development in Spatial Perspective

Various schemes to group the world's states into economic-geographic categories have come and gone, and others will probably arise in the future. For our purposes, the classification scheme used by the World Bank (one of the agencies that monitor economic conditions across the globe) is the most effective. It sorts countries into four categories based on the success of their economies: (1) high-income, (2) upper-middle-income, (3) lower-middle-income, and (4) low-income. These categories, when mapped, display interesting regional clustering (Fig. G-10). Compare this map to our global framework (Fig. G-3), and you can see the role of economic geography in the layout of the world's geographic realms. Also evident are regional boundaries within realms—for instance, between Brazil and its western neighbors, between South Africa and most of the rest of Subsaharan Africa, and between west and east in Europe.

Economic geography is not the whole story, but along with factors of physical geography (such as climate), cultural geography (including resistance or receptivity to change and innovation), and political geography (history of colonialism, growth of democracy), it plays a powerful role in shaping our variable world.

Uneven Development

It has been obvious for a very long time that human success on the Earth's surface has focused on certain areas and bypassed others. The earliest cities and states of the Fertile Crescent, the empires of the Incas and the Aztecs, the dominance of ancient Rome, and many other hubs of activity tell the story of development and decay, of growth and collapse. In their heyday, such centers of authority, innovation, production, and expansion were the earliest **28 core areas**, places of dominance whose inhabitants exerted their power over their surroundings near and far.

Such core areas grew rich and, in many cases, endured for long periods because their occupants skillfully exploited these surroundings—controlling and taxing the local population, forcing workers to farm the land and mine the resources at their command. This created a **29 periphery** that sustained the core for as long as the system lasted, so that core-periphery interactions, one-sided though they were, created wealth for the former and enforced stability in the latter.

Uneven development, therefore, exists at a range of scales, from the urban to the global. Indeed, most functional regions, in essence, are *spatial networks* with nodes of various centrality and importance, and this often translates into different levels of economic development. Except for a few special cases, all countries contain core areas. These national core areas are often anchored by the capital and/or largest city in the country: for example, Paris (France), Tokyo (Japan), Buenos Aires (Argentina), or Bangkok (Thailand). Larger countries may have more than one core area, such as Australia with its East and West Coast core areas and intervening periphery.

Core-Periphery World: Diminishing Dichotomy?

Regions, individual states, even provinces and counties exhibit core-periphery economic contrasts that, in many places, are widening disparities rather than narrowing them. But advocates of *globalization* (a concept to be discussed in the next section) say that the old rules about exploiting cores and exploited peripheries are being reversed on a worldwide scale, and that a "flatter" and fairer world is in the offing. Thomas Friedman's 2005 book, *The World Is Flat*, argues that the world is becoming so mobile, so interconnected, and so integrated that historic barriers are falling, including those separating cores from peripheries. Interaction, say these observers, is now global, ever-freer trade is the new rule, and the flow of people, ideas, money, and jobs accelerates by the day. Core-periphery notions are becoming irrelevant as contrasts between them fade. "Geography is history," some of them like to say.

But is it? If the world is to become "flatter," and if regional disparities and income inequalities are to decline, it is going to take more than flowing money and freer trade (some of the richer states that advocate free trade protect their own inefficient industries, farm as well as factory, against foreign competition). It will require changes in attitude as well. Core-area residents are used to their advantages and comforts, and in general they guard against losing them. Not only do they take a dim view of cheap imports that threaten local industries; they want to control immigration, and they tend to resist cultural infusions in the form of unfamiliar languages and unconventional religions.

Looking at it this way, even the world as a whole has a core-periphery dichotomy that counters notions of a global flattening (Fig. G-11). Note that we have shifted the perspective from that used on our other world maps, and the best way to see what Figure G-11 reveals is to outline it on a globe. The *global core*, anchored by North America and flanked by Europe to the east and Japan and Australia to the west, not only constitutes an assemblage of the most affluent states and

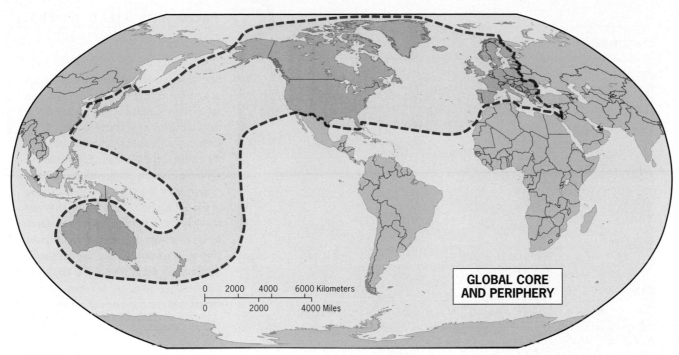

FIGURE G-11

© H. J. de Blij, P. O. Muller, and John Wiley & Sons, Inc.

the most prosperous cities, but also hosts the bulk of the financial and corporate empires that drive economic globalization. As mapped in Figure G-11, the core contains about 15 percent of the world's population, but that population earns some 75 percent of total annual income. The few core-like outliers in the periphery—for example, Singapore, Dubai (United Arab Emirates)—barely affect this disproportionate picture.

GLOBALIZATION

A Matter of Scale

The process of **30** globalization continues to change our world. It is essentially a *geographical* process in which spatial relations—economic, cultural, political—shift to ever larger scales. What that means is that what happens in one place has repercussions in places ever more remote, thereby integrating the entire world in an ever "smaller" global village. Globalization comes into our homes via television: news travels faster than ever before, and sometimes even government leaders turn to CNN to get the latest reports on international events. The process is in part driven by rapid advances in communication and transportation technologies.

Globalization is not something entirely new. The mid- and late nineteenth century, for instance, also witnessed major advances in the intensification of global interdependence. It was particularly affected by new technologies such as the steamship, the railway, and the telegraph, which subsequently were followed by the first motor vehicles and airplanes. Today's key technologies are satellite television, mobile phones, and, of course, the Internet. If the world is getting ever more interconnected, geography and our knowledge of the world's realms and regions become ever more important—because what happens elsewhere will have consequences for where *you* are!

Global Challenges, Shared Interests

Globalization plays out in various spheres, from the environmental to the cultural to the economic. Today's most pressing environmental issue, no doubt, is global warming, a threat to the world at large. It is clear that we must confront this problem together, but it is far from easy to agree on strategies: some countries are bigger polluters than others, some have more resources than others, and some are more developed than others. How to divide the burdens? At the Copenhagen Conference on Climate Change in late 2009, four countries (India, China, Brazil, and South Africa) threatened to walk out of the proceedings to protest efforts by global-core countries to impose emission reductions that these four considered too damaging to their ongoing industrial development. Why should they have to sacrifice so much while the rich countries (that are now primarily *post*industrial) never faced this kind of interference in their economic development?

Culturally, too, the world is coming ever closer together, and this is most apparent in global migration flows. Migration used to be uncommon (most people were rooted in their home environment, and that is where they lived their entire lives), and when it did occur it used to be one-way: people moved from one place to another and stayed there. In the global era, migration has picked up in terms of intensity, in part because people possess much more knowledge about opportunities elsewhere. In addition, there are

© H. J. de Blij

"Approaching the European shore of the Bosphorus was this instructive moment full of symbolism. In the far distance you can see the domed basilica of the Hagia Sophia, the Church of Holy Wisdom, built in the sixth century, when Constantinople was the center of the Byzantine Empire. On the right, the seventh-century Venetian Tower rises over the cityscape on the north side of the Golden Horn. But most of what you see here dates from modern times, when Turkey rose from the ashes of the Ottoman Empire, Constantinople was renamed Istanbul, the Asian side of the city grew exponentially, Ankara became Turkey's capital, and the trappings of globalization made this a megacity of 10.5 million whose economic gains more than compensated for its political downgrading."

www.conceptcaching.com

now many more opportunities to travel back and forth, allowing people to maintain close ties with their original home countries. ***Transnational migrants*** are highly mobile and are frequently instrumental in the spreading of cultures around the world. Examples include Algerians in Paris, Haitians in Montreal, Cubans in Miami, Mexicans in Los Angeles, Indians in Singapore, and Indonesians in Sydney.

Economic Globalization: Winners and Losers

If a geographic concept can arouse strong passions, globalization is it. To leading economists, politicians, and businesspeople, this is the best of all possible worlds—the march of international capitalism, open markets, and free trade. In theory, globalization breaks down barriers to international trade, stimulates commerce, brings jobs to remote places, and promotes social, cultural, political, and other kinds of exchanges. High-tech workers in India are employed by computer firms based in California. Japanese cars are assembled in Thailand. American shoes are made in China. Fastfood restaurant chains spread standards of service and hygiene as well as familiar (and standard) menus from Tokyo to Tel Aviv to Tijuana. If wages and standards of employment

are lower in peripheral countries than in the global core, production will shift there and the gap will shrink. Everybody wins. But that's theory.

To millions of poor farmers and powerless citizens in debt-mired African and Asian countries, it represents a system that will keep them in poverty and subservience forever. Economic geographers can prove that global economic integration allows the overall economies of poorer countries to grow faster: compare their international trade to their national income, and you will find that the ***gross national income (GNI)****** of those that engage in more international trade (and thus are more "globalized") rises, while the GNI of those with less international trade actually declines. The problem is that those poorest countries contain more than 2 billion people, and their prospects in this globalizing world are worsening, not rising. In countries such as the Philippines, Kenya, and Nicaragua, income per person has shrunk, and there globalization is seen as a culprit, not a cure.

And there is another, more complicated issue. Although many countries, even in the less-developed world where they have been able to latch on to globalization, have witnessed accelerated economic growth and rising per capita incomes, inequality within these countries has often increased just as fast. In other words, uneven development

---

*Gross national income (GNI) is the total income earned from all goods and services produced by the citizens of a country, within or outside of its borders, during a calendar year. ***Per capita GNI*** is a widely used indicator of the variation of spendable income around the globe, and is reported for each country in the world in the farthest-right column of the Data Table in Appendix B.

within countries has become more pronounced. This is quite obvious in China, the fastest-growing economy in the world over the past two decades: much of this growth takes place in its Pacific coastal zone, not in the interior of the country, and income differentials become ever wider. But the same is true in India and most other "*emerging markets*." This is why a *regional* approach is so important to understanding what is going on in the world economy.

Globalization in the economic sphere is proceeding under the auspices of the World Trade Organization (WTO), of which the United States is the leading architect. To join, countries must agree to open their economies to foreign trade and investment. As of 2010, the WTO had 153 member-states, all expecting benefits from their participation. But the leading global-core countries themselves do not always oblige when it comes to creating a "level playing field." The case of the Philippines is often cited: Filipino farmers found themselves competing against North American and European producers who receive subsidies toward the production as well as the export of their products—and losing out. Meanwhile, low-priced, subsidized U.S. corn appeared on Filipino markets. As a result, the Philippine economy lost several hundred thousand farm jobs, wages went down, and WTO membership had the effect of severely damaging its agricultural sector. Not surprisingly, the notion of globalization is not popular among rural Filipinos. Nor is rich-country protectionism confined to agriculture. When cheap foreign steel began to threaten what remained of the American steel industry in 2001, the U.S. government erected tariffs to guard domestic producers against unwanted competition.

The Future

As with all major transformations, the overall consequences of globalization are uncertain. Critics underscore that one of its outcomes is a growing gap between rich and poor, a polarization of wealth that will destabilize the world. Proponents argue that, as with the Industrial Revolution, it will take time for the benefits to spread—but that globalization's ultimate effects will be advantageous to all. Indeed, the world is functionally shrinking, and we will find evidence for this throughout the book. But the "global village" still retains its distinctive neighborhoods, and globalization did not erase their particular properties—in some cases it even sharpened the contrasts. In the chapters that follow we use the vehicle of geography to visit and investigate them.

REGIONAL FRAMEWORK AND GEOGRAPHIC PERSPECTIVE

At the beginning of this chapter, we outlined a map of the great geographic realms of the world (Fig. G-3). We then addressed the task of dividing these realms into regions, and we used criteria ranging from physical geography to economic geography. The result is Figure G-12.

FROM THE FIELD NOTES...

© H. J. de Blij

"It was an equatorial day here today, in hot and humid Singapore. Walked the three miles to the Sultan Mosque this morning and observed the activity arising from the Friday prayers, then sat in the shade of the large ficus tree on the corner of Arab Street and watched the busy pedestrian and vehicular traffic. An Indian man saw me taking notes and sat down beside me. For some time he said nothing, then asked whether I was writing a novel. 'If only I could!' I answered, and explained that I was noting locations and impressions for photos and text in a geography book. He said that he was Hindu but liked living in this area because some of what he called 'Old Singapore' still survived. 'No problems between Hindus and Muslims here,' he said. 'The police maintain strict security here around the mosques, especially after September 2001. The government tolerates all religions, but it doesn't tolerate religious conflict.' He offered to show me the Hindu shrine where he and his family joined others from Singapore's Hindu community. As we walked down Arab Street, he pointed skyward. 'That's the future,' he said. 'The symbol of globalization. Do you mention it in your book? It's in the papers every day, supposedly a good thing for Singapore. Well, maybe. But a lot of history has been lost to make space for what you see here.' I took the photo, and promised to mention his point."

www.conceptcaching.com

On this map, note that we display not only the great geographic realms but also the regions into which they subdivide. The numbers in the legend reveal the order in which the realms and regions are discussed, starting with Europe (1) and ending with the Pacific Realm (12).

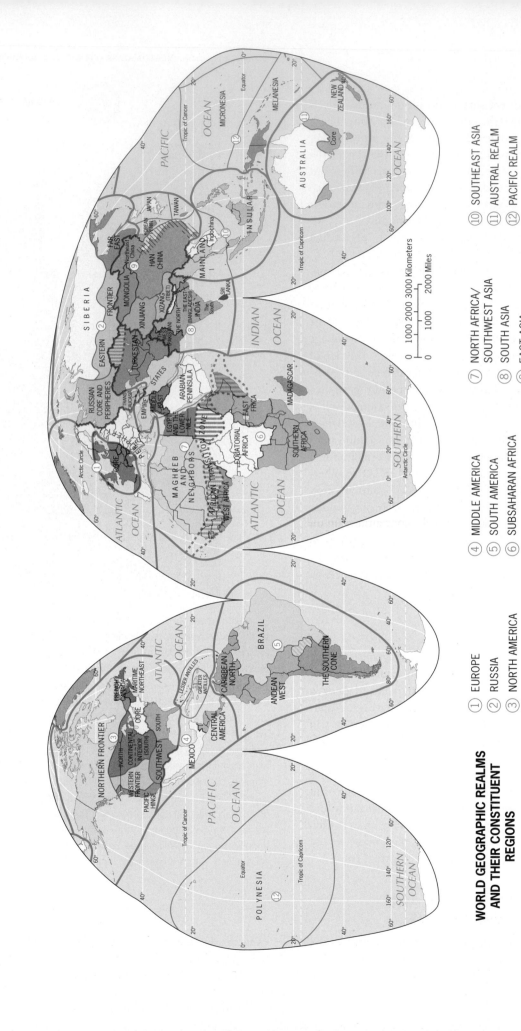

**WORLD GEOGRAPHIC REALMS
AND THEIR CONSTITUENT
REGIONS**

① EUROPE
② RUSSIA
③ NORTH AMERICA

④ MIDDLE AMERICA
⑤ SOUTH AMERICA
⑥ SUBSAHARAN AFRICA

⑦ NORTH AFRICA/
    SOUTHWEST ASIA
⑧ SOUTH ASIA
⑨ EAST ASIA

⑩ SOUTHEAST ASIA
⑪ AUSTRAL REALM
⑫ PACIFIC REALM

FIGURE G-12

© H. J. de Blij, P. O. Muller, and John Wiley & Sons, Inc..

As this introductory chapter demonstrates, our world regional survey is no mere description of places and areas. We have combined the study of realms and regions with a look at geography's ideas and concepts—the notions, generalizations, and basic theories that make the discipline what it is. We continue this method in the chapters ahead so that we will become better acquainted with the world and with geography. By now you are aware that geography is a wide-ranging, multifaceted discipline. It is often described as a social science, but that is only half the story: in fact, geography straddles the divide between the social and the physical (natural) sciences. Many of the ideas and concepts you will encounter have to do with the multiple interactions between human societies and natural environments.

Regional geography allows us to view the world in an all-encompassing way. As we have seen, regional geography borrows information from many sources to create an overall image of our divided world. Those sources are not random. They represent topical or *systematic geography*. Research in the systematic fields of geography makes our world-scale generalizations possible. As Figure G-13 shows, these systematic fields relate closely to those of other disciplines. Cultural geography, for example, is allied with anthropology; it is the spatial perspective that distinguishes cultural geography. Economic geography focuses on the spatial dimensions of economic activity; political geography concentrates on the spatial imprints of political behavior. Other systematic fields include historical, medical, behavioral, environmental, and urban geography. We will also draw on information from biogeography, marine geography, population geography, and climatology (as we did earlier in this chapter).

These systematic fields of geography are so named because their approach is global, not regional. Take the geographic study of cities, urban geography. Urbanization is a worldwide process, and urban geographers can identify certain human activities that all cities in the world exhibit in one form or another. But cities also display regional properties. The typical Japanese city is quite distinct from, say, the African city. Regional geography, therefore, borrows from the systematic field of urban geography, but it injects this regional perspective.

In the following chapters we call upon these systematic fields to give us a better understanding of the world's realms and regions. As a result, you will gain insights into the dis-

THE RELATIONSHIP BETWEEN REGIONAL AND SYSTEMATIC GEOGRAPHY

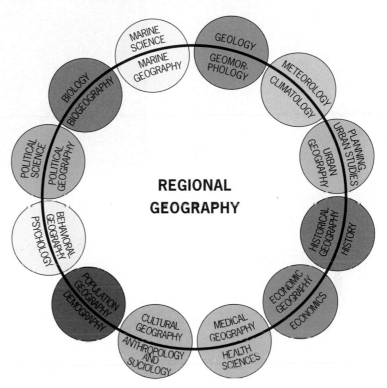

FIGURE G-13 © H. J. de Blij, P. O. Muller, and John Wiley & Sons, Inc.

cipline of geography as well as the regions we investigate. This will prove that geography is a relevant and practical discipline when it comes to comprehending, and coping with, our rapidly changing world.

POINTS TO PONDER

- The planet appears to be entering a period of increasing environmental fluctuation, posing growing threats to future economic and political stability.
- Somewhere during the middle part of 2012, the global human population will surpass 7 billion.
- At the opening of our new century's second decade, almost 1 billion people continued to survive on less than one U.S. dollar a day.
- Globalization is glorified as well as vilified, but its march is inexorable.

Amsterdam's unique urban landscape. © H.J. de Blij

IN THIS CHAPTER

Why Europe is one of the world's most complex and dynamic realms
Unification in the face of diversity
The perennial question of Europe's eastern boundary
Is Islam settling in western Europe?
The European realm as the world's home for the aged
Provinces behaving like countries rile national governments

CONCEPTS, IDEAS, AND TERMS

1A

EUROPE: DEFINING THE REALM

FIGURE 1A-1

© H. J. de Blij, P. O. Muller, and John Wiley & Sons, Inc.

t is appropriate to begin our investigation of the world's geographic realms in Europe because over the past five centuries Europe and Europeans have influenced and changed the rest of the world more than any other realm or people has done. European empires spanned the globe and transformed societies far and near. European colonialism propelled an early wave of globalization. Millions of Europeans migrated from their homelands to the Old World as well as the New, changing (and sometimes nearly obliterating) traditional communities and creating new societies from Australia to North America. Colonial power and economic incentive combined to impel the movement of millions of imperial subjects from their ancestral homes to distant lands: Africans to the Americas, Indians to Africa, Chinese to Southeast Asia, Malays to South Africa's Cape, Native Americans from east to west. In agriculture, industry, politics, and other spheres, Europe generated revolutions—and then exported those revolutions across the world, thereby consolidating the European advantage.

But throughout much of that 500-year period of European hegemony, Europe also was a cauldron of conflict. Religious, territorial, and political disputes precipitated bitter wars that even spilled over into the colonies. And during the twentieth century, Europe twice plunged the world into war. The terrible, unprecedented toll of World War I (1914–1918) was not enough to stave off World War II (1939–1945), which drew in the United States and ended with the first-ever use of nuclear weapons in Japan. In the aftermath of that war, Europe's weakened powers lost most of their colonial possessions and a new rivalry emerged: an ideological Cold War between the communist Soviet Union and the capitalist United States. This Cold War lowered an Iron Curtain across the heart of Europe, leaving most of the east under Soviet control and most of the west in the American camp. Western Europe proved resilient, overcoming the destruction of war and the loss of colonial power to regain economic strength. Meanwhile, the Soviet communist experiment failed at home and abroad, and in 1990 the last vestiges of the Iron Curtain were lifted. Since then, a massive effort has been underway to reintegrate and reunify Europe from the Atlantic coast to the Russian border, the key geographic story of this chapter.

 MAJOR GEOGRAPHIC QUALITIES

EUROPE

1. The European geographic realm lies on the western reaches of the Eurasian landmass.

2. Though territorially small, Europe is heavily populated and is fragmented into 40 states.

3. Europe's enduring world influence results mainly from advantages accrued over centuries of colonial and imperial domination.

4. European natural environments are highly varied, and Europe's resource base is rich and diverse.

5. Europe's geographic diversity, cultural as well as physical, created strong local identities, specializations, and opportunities for trade and commerce.

6. European nation-states, based in durable and powerful core areas, survived the loss of colonies and evolved into modern democratic states.

7. Europe's states are engaged in a historic effort to achieve multinational economic integration and, to a lesser degree, political coordination.

8. Europe's relatively prosperous population is highly urbanized, rapidly aging, and in demographic decline partly offset by significant immigration.

9. Local demands for greater autonomy, and cultural challenges posed by immigration, are straining the European social fabric.

10. Despite Europe's momentous unification efforts, east-west contrasts still mark the realm's regional geography.

11. Relations between Europe and neighboring Russia are increasingly problematic.

GEOGRAPHICAL FEATURES

As Figure 1A-1 shows, Europe is a realm of peninsulas and islands on the western margin of the world's largest landmass, Eurasia. It is a realm of just under 600 million people and 40 countries, but it is territorially quite small. Yet despite its modest proportions it has had—and continues to have—a major impact on world affairs. For many centuries Europe has been a hearth of achievement, innovation, invention, and domination.

Europe's Eastern Boundary

The European realm is bounded on the west, north, and south by Atlantic, Arctic, and Mediterranean waters, respectively. But where is Europe's eastern limit? Each episode in the historical geography of eastern Europe has left its particular legacy in the cultural landscape.

Twenty centuries ago the Roman Empire ruled much of it (Romania is a cartographic reminder of that period); for most of the second half of the twentieth century, the Soviet Empire controlled nearly all of it. In the intervening two millennia, Christian Orthodox church doctrines spread from the southeast, and Roman Catholicism advanced from the northwest. Turkish (Ottoman) Muslims invaded and created an empire that reached the environs of Vienna. By the time the Austro-Hungarian Empire ousted the Turks, millions of eastern Europeans had been converted to Islam. Albania and Kosovo today remain predominantly Muslim countries. Meanwhile, it is often said that western Europe's civilization had its cradle in ancient Greece, but that lies farther still to the southeast, beyond the former Yugoslavia.

Eastern Europe's tumultuous history, itself an expression of the absence of clear natural boundaries, played itself out on a physical stage of immense diversity, its landscapes ranging from open plains and wide river basins to strategic mountains and crucial corridors. Epic battles fought centuries ago remain fresh in the minds of many people living here today; pivotal past migrations are celebrated as though they happened yesterday. Nowhere in Europe is the cultural geography as complex. Illyrians, Slavs, Turks, Magyars, and other peoples converged on this region from near and far. Ethnic and cultural differences kept them in chronic conflict.

As we shall see, the geographic extent of Europe's cultural realm has always been debatable, and today it is a particularly contentious issue in terms of European Union expansion and with regard to relations with Russia. Europe's eastern boundary is a dynamic one, and it has changed with history. Some would say that there really is no clear boundary, that Europe's atmosphere, so to speak, just gets thinner toward the east. Our definition, at the present time, places Europe's eastern boundary between Russia and its numerous European neighbors to the west. This definition is based on several geographic factors including European-Russian contrasts in territorial dimensions, geopolitical developments, cultural properties, and history.

Climate and Resources

From the balmy shores of the Mediterranean Sea to the icy peaks of the Alps, and from the moist woodlands and moors of the Atlantic fringe to the semiarid prairies north of the Black Sea, Europe presents an almost infinite range of natural environments (Fig. 1A-2).

Europe's peoples have benefited from a large and varied store of raw materials. Whenever the opportunity or need arose, the realm proved to contain what was required. Early on, these requirements included cultivable soils, rich

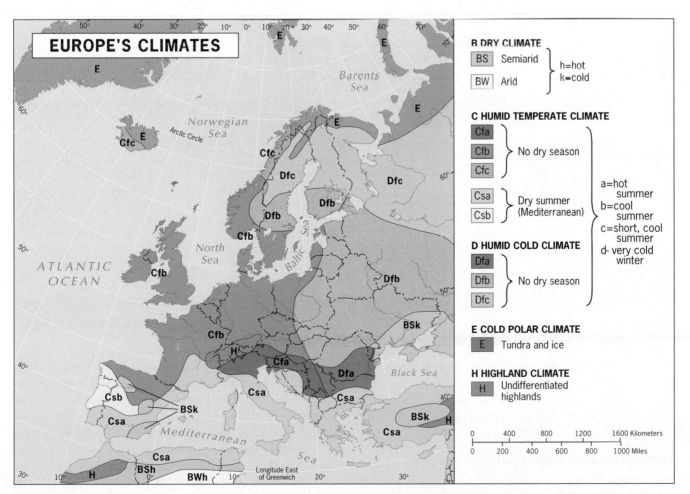

FIGURE 1A-2

© H. J. de Blij, P. O. Muller, and John Wiley & Sons, Inc.

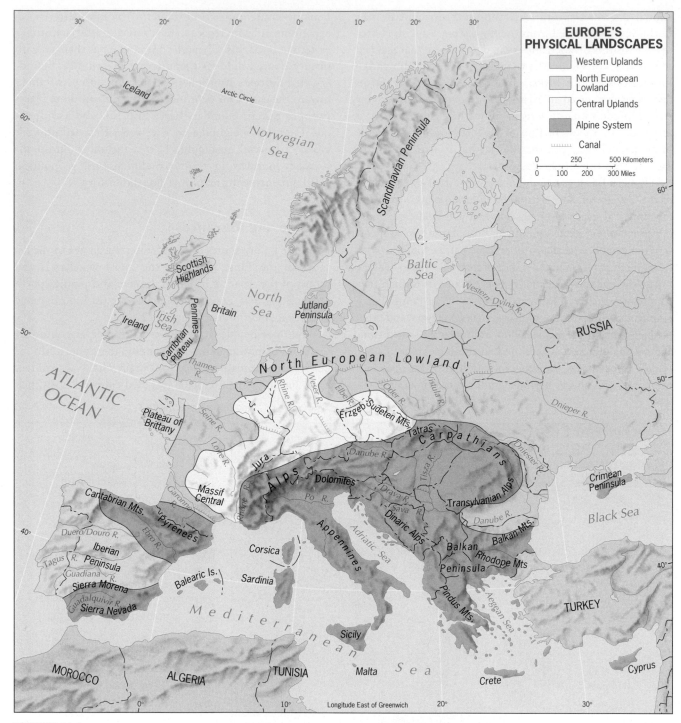

FIGURE 1A-3

© H. J. de Blij, P. O. Muller, and John Wiley & Sons, Inc.

fishing waters, and wild animals that could be domesti-cated; in addition, extensive forests provided wood for houses and boats. Later, coal and mineral ores propelled industrialization. More recently, Europe proved to contain substantial deposits of oil and natural gas.

Landforms and Opportunities

In the Introduction we noted the importance of physical geography in the definition of geographic realms. The nat-

ural landscape with its array of landforms (such as moun-tains and plateaus) is a key element in the total physical geography—or **1** **physiography**—of any part of the ter-restrial world. Other physiographic components include climate and the physical features that mark the natural landscape, such as vegetation, soils, and water bodies.

Europe's area may be small, but its landscapes are varied and complex. Regionally, we identify four broad units: the Central Uplands, the southern Alpine Mountains, the West-ern Uplands, and the North European Lowland (Fig. 1A-3).

The ***Central Uplands*** form the heart of Europe. It is a region of hills and low plateaus loaded with raw materials whose farm villages grew into towns and cities when the Industrial Revolution transformed this realm.

The ***Alpine Mountains***, a highland region named after the Alps, extend from the Pyrenees on the French-Spanish border to the Balkan Mountains near the Black Sea, and include Italy's Appennines and the Carpathians of eastern Europe.

The ***Western Uplands***, geologically older, lower, and more stable than the Alpine Mountains, extend from Scandinavia through western Britain and Ireland to the heart of the Iberian Peninsula in Spain.

The ***North European Lowland*** extends in a lengthy arc from southeastern Britain and central France across Germany and Denmark into Poland and Ukraine, from where it continues well into Russia. Also known as the Great European Plain, this has been an avenue for human migration time after time, so that complex cultural and economic mosaics developed here together with a jigsaw-like political map. As Figure 1A-3 shows, many of Europe's major rivers and connecting waterways serve this populous region, where a number of Europe's leading cities (London, Paris, Amsterdam, Copenhagen, Berlin, Warsaw) are located.

Locational Advantages

Europe is endowed with some exceptional locational advantages. Its ***relative location***, at the crossroads of the **2** land hemisphere, creates maximum efficiency for contact with much of the rest of the world (Fig. 1A-4). A "peninsula of peninsulas," Europe is nowhere far from the ocean and its avenues of seaborne trade and conquest. Hundreds of kilometers of navigable rivers, augmented by an unmatched system of canals, open the interior of Europe to its neighboring seas and to the shipping lanes of the world. The

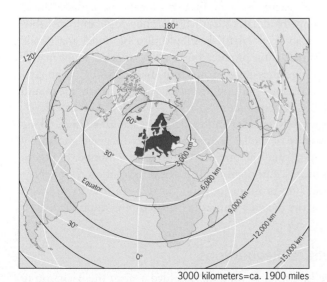

3000 kilometers=ca. 1900 miles

FIGURE 1A-4 © H. J. de Blij, P. O. Muller, and John Wiley & Sons, Inc.

Mediterranean and Baltic seas, in particular, were critical in the development of trade in early modern times, and in the emergence of Europe's early trading cities such as Venice (Italy) in the south or Lübeck (Germany) in the north.

Also consider the scale of the maps of Europe in this chapter. Europe is a realm of moderate distances and close proximities. Short distances and large cultural differences make for intense interaction, the constant circulation of goods and ideas. That has been the hallmark of Europe's geography for more than a millennium.

MODERN HISTORICAL GEOGRAPHY

The Industrial Revolution

The term **3** Industrial Revolution suggests that an agrarian Europe was suddenly swept up in wholesale industrialization that changed the realm in a few decades. In reality, seventeenth- and eighteenth-century Europe had been industrializing in many spheres, long before the chain of events known as the Industrial Revolution began. From the textiles of England and Flanders to the iron farm implements of Saxony (in present-day Germany), from Scandinavian furniture to French linens, Europe had already entered a new era of **4** local functional specialization. Nonetheless, by the end of the eighteenth century, industrial Europe took off as never before.

British Primacy

Britain was at the epicenter of this revolution. In the 1780s, the Scotsman James Watt and others devised a steam-driven engine, which was soon adopted for numerous industrial uses. At about the same time, coal (converted into carbon-rich coke) was recognized as a vastly superior substitute for charcoal in smelting iron. These momentous innovations had a rapid effect. The power loom revolutionized the weaving industry. Iron smelters, long dependent on Europe's dwindling forests for fuel, could now be concentrated near coalfields. Engines could move locomotives as well as power looms. Ocean shipping entered a new age.

Britain had an enormous advantage, for the Industrial Revolution occurred when British influence reigned worldwide and the significant innovations were achieved in Britain itself. The British controlled the flow of raw materials, they held a monopoly over products that were in global demand, and they alone possessed the skills necessary to make the machines that manufactured the products. Soon the fruits of the Industrial Revolution were being exported, and the modern industrial spatial organization of Europe began to take shape. In Britain, manufacturing regions developed near coalfields in the English Midlands, at Newcastle to the northeast, in southern Wales, and along Scotland's Clyde River around Glasgow.

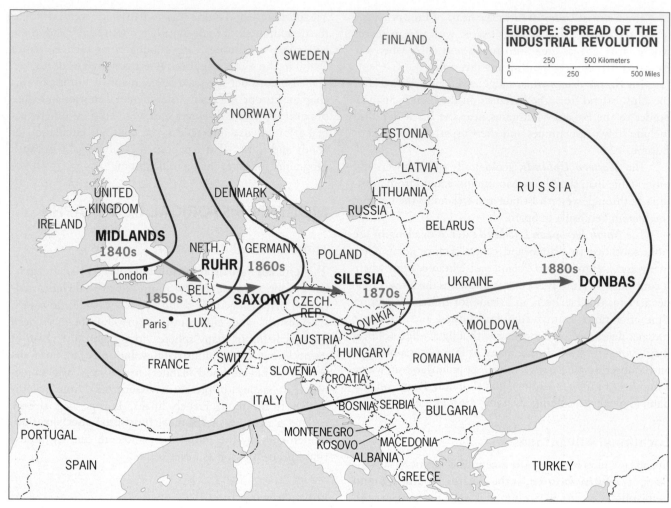

FIGURE 1A-5

© H. J. DE BLIJ, P. O. MULLER, AND JOHN WILEY & SONS, INC.

Diffusion to the Continent

The Industrial Revolution spread eastward from Britain onto the European mainland throughout the middle and late nineteenth century (Fig. 1A-5). Population skyrocketed, emigration mushroomed, and industrializing cities burst at the seams. European states already had acquired colonial empires before this revolution started; now colonialism gave Europe an unprecedented advantage in its dominance over the rest of the world.

In mainland Europe, a belt of major coalfields extends from west to east, roughly along the southern margins of the North European Lowland, due eastward from southern England across northern France and Belgium, Germany (the Ruhr), western Bohemia in the Czech Republic, Silesia in southern Poland, and the Donets Basin (Donbas) in eastern Ukraine. Iron ore is found in a broadly similar belt and together with coal provides the key raw material for the manufacturing of steel.

As in Britain, this cornerstone industry now spawned new concentrations of economic activity, growing steadily as millions migrated from the countryside to fill expanding employment opportunities. Densely populated and heavily urbanized, these emerging agglomerations became the backbone of Europe's world-scale population cluster (as shown in Fig. G-8).

Two centuries later, this east-west axis along the coalfield belt remains a major feature of Europe's population distribution map (Fig. 1A-6). It should also be noted that while industrialization produced new cities, another set of manufacturing zones arose in and near many existing urban centers. London—already Europe's leading urban focus and Britain's richest domestic market—was typical of these developments. Many local industries were established here, taking advantage of the large supply of labor, the ready availability of capital, and the proximity of so great a number of potential buyers. Although the Industrial Revolution thrust other places into prominence, London did not lose its primacy: industries in and around the British capital multiplied.

Political Revolutions

The Industrial Revolution unfolded against the backdrop of the ongoing formation of national states, a process that had been underway since the 1600s. Europe's political revolutions

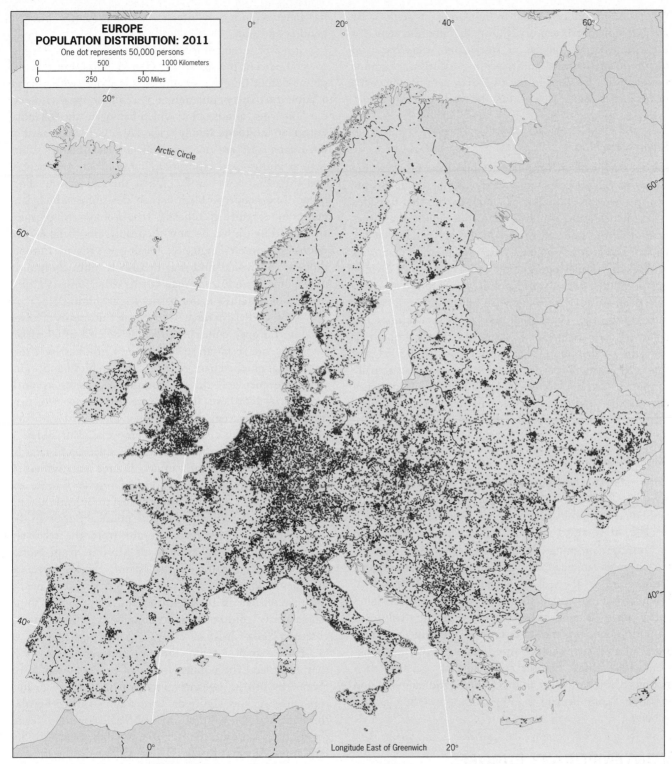

EUROPE
POPULATION DISTRIBUTION: 2011
One dot represents 50,000 persons

0   500   1000 Kilometers
0   250   500 Miles

FIGURE 1A-6

© H. J. de Blij, P. O. Muller, and John Wiley & Sons, Inc.

took many different forms and affected diverse peoples and countries, but in general it headed toward parliamentary representation and democracy. Historians often point to the Peace (Treaty) of Westphalia in 1648 as a key step in the evolution of Europe's state system, ending decades of war and recognizing territories, boundaries, and the sovereignty of countries. But this was only the beginning.

Competing Ideologies

One of the most dramatic episodes to follow was the French Revolution (1789–1795), which ended the era of absolutist states with all-powerful monarchs. More gradual political transformations occurred in the Netherlands, Britain, and the Scandinavian countries. Other parts of Europe would for much longer remain under the control of authoritarian (dictatorial) regimes headed by monarchs or despots. By the late nineteenth century, Europe became the arena for the competing ideologies of *liberalism*, *socialism*, and *nationalism* (national spirit, pride, patriotism). Extreme nationalism (*fascism*) in the twentieth century threw the realm into the most violent wars the world had ever seen.

A Fractured Map

Europe's political map has always been one of intriguing complexity. As a geographic realm, Europe occupies only about 5 percent of the Earth's land area; but that tiny area is fragmented into some 40 countries—more than one-fifth of all the states in the world today. Therefore, when you look at Europe's political map, the question that arises is how did so small a geographic realm come to be divided into so many political entities? Europe's map is a legacy of its feudal and royal periods, when powerful kings, barons, dukes, and other rulers, rich enough to fund armies and powerful enough to extract taxes and tribute from their domains, created bounded territories in which they reigned supreme. Royal marriages, alliances, and conquests actually simplified Europe's political map. In the early nineteenth century there still were 39 German states; a unified Germany as we know it today did not emerge until the 1870s.

State and Nation

Europe's political revolution produced a form of political-territorial organization known as the nation-state, a state embodied by its culturally distinctive population. But what is a **5** nation-state and what is not? The term **6** nation has multiple meanings. In one sense it refers to a people with a single language, a common history, a similar ethnic background. In the sense of *nationality* it relates to legal membership in the state, that is, citizenship. Very few states today are so homogeneous culturally that the culture is conterminous with the state. Europe's prominent nation-states of a century ago—France, Spain, the United Kingdom, Italy—have become multicultural societies, their nations defined more by an intangible "national spirit" and emotional commitment than by cultural or ethnic homogeneity.

CONTEMPORARY EUROPE: A DYNAMIC REALM

Cultural Diversity

The European realm is home to peoples of numerous cultural-linguistic stocks, including not only Latins, Germanics, and Slavs but also minorities such as Finns, Magyars (Hungarians), Basques, and Celts. This diversity of ancestries continues to be an asset as well as a liability. It has generated not only interaction and exchange, but also conflict and war.

It is worth remembering that Europe's territory is just over 60 percent the size of the United States, but that the population of Europe's 40 countries is about twice as large as America's. This population of 594 million speaks numerous languages, almost all of which belong to the **7** Indo-European language family (Figs. 1A-7, G-9). But most of those languages are not mutually understandable; some, such as Finnish and Hungarian, are not even members of this Indo-European family. When the unification effort began, one major problem was to determine which languages to recognize as "official." That problem still prevails, although English has become the realm's unofficial *lingua franca* (common language). During a visit to Europe, though, you would find that English is more commonly usable in western Europe than farther east. Europe's multilingualism remains a major barrier to integration.

Another divisive force confronting Europeans involves religion. Europe's cultural heritage is steeped in Christian traditions, but sectarian strife between Catholics and Protestants that plunged parts of the realm into bitter and widespread conflict still divides communities and, as until recently in Northern Ireland, can still arouse violence. Some political parties still carry the name "Christian," for example, Germany's Christian Democrats. But today, a new factor roils the religious landscape: the rise of Islam. In the east, this takes the form of new Islamic assertiveness in an old Muslim bastion: the (Turkish) Ottoman Empire left behind millions of converts from Bosnia to Bulgaria among whom many are demanding greater political power. In the west, this Islamic resurgence results from the relatively recent immigration of millions of Muslims from North Africa and other parts of the Islamic world. Here, as mosques overflow with the faithful, churches stand nearly empty as secularism among Europeans is on the rise. Some scholars have gone so far as to describe Europe as having entered a "post-Christian" era.

For so small a realm, Europe's cultural geography is sharply varied. The popular image of Europe tends to be formed by British pageantry, French wine country, or historic cities such as Venice or Amsterdam—but go beyond this core, and you will find isolated Slavic communities in the mountains facing the Adriatic, Muslim towns in poverty-mired Albania, Roma (Gypsy) villages in the interior of Romania, farmers using traditional methods unchanged for centuries in rural Poland. That map of 40 countries does not begin to reflect the diversity of European cultures.

Spatial Interaction

If not culture, what does unify Europe? The answer lies in this realm's outstanding opportunities for productive interaction. The European realm is best understood as an

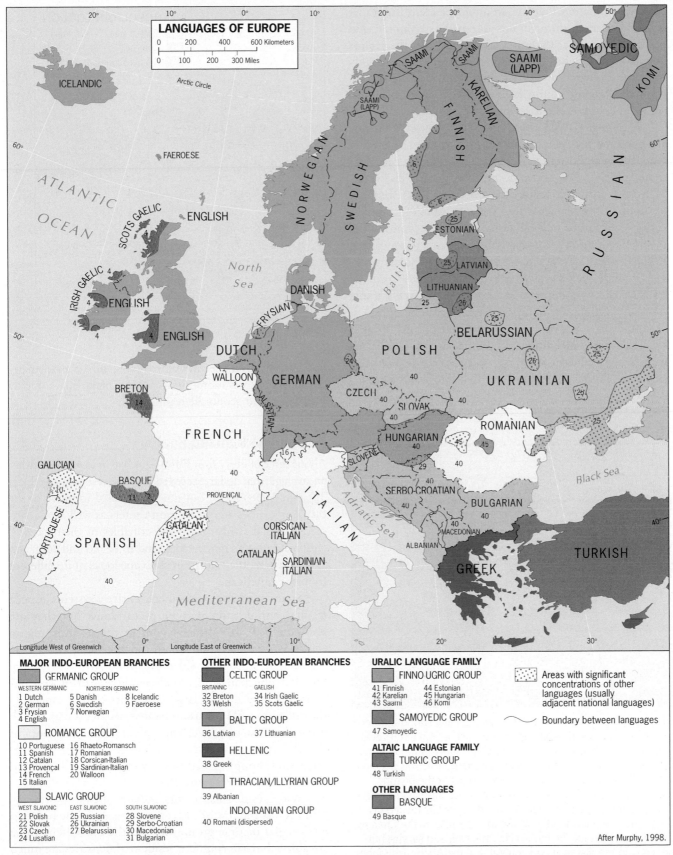

LANGUAGES OF EUROPE

MAJOR INDO-EUROPEAN BRANCHES

GERMANIC GROUP

WESTERN GERMANIC	NORTHERN GERMANIC
1 Dutch	5 Danish	8 Icelandic
2 German	6 Swedish	9 Faeroese
3 Frysian	7 Norwegian
4 English	

ROMANCE GROUP

10 Portuguese	16 Rhaeto-Romansch
11 Spanish	17 Romanian
12 Catalan	18 Corsican-Italian
13 Provençal	19 Sardinian-Italian
14 French	20 Walloon
15 Italian	

SLAVIC GROUP

WEST SLAVONIC	EAST SLAVONIC	SOUTH SLAVONIC
21 Polish	25 Russian	28 Slovene
22 Slovak	26 Ukrainian	29 Serbo-Croatian
23 Czech	27 Belarussian	30 Macedonian
24 Lusatian		31 Bulgarian

OTHER INDO-EUROPEAN BRANCHES

CELTIC GROUP

BRITANNIC	GAELISH
32 Breton	34 Irish Gaelic
33 Welsh	35 Scots Gaelic

BALTIC GROUP

36 Latvian   37 Lithuanian

HELLENIC

38 Greek

THRACIAN/ILLYRIAN GROUP

39 Albanian

INDO-IRANIAN GROUP

40 Romani (dispersed)

URALIC LANGUAGE FAMILY

FINNO UGRIC GROUP

41 Finnish	44 Estonian
42 Karelian	45 Hungarian
43 Saarni	46 Komi

SAMOYEDIC GROUP

47 Samoyedic

ALTAIC LANGUAGE FAMILY

TURKIC GROUP

48 Turkish

OTHER LANGUAGES

BASQUE

49 Basque

Areas with significant concentrations of other languages (usually adjacent national languages)

Boundary between languages

After Murphy, 1998.

FIGURE 1A-7

© H. J. de Blij, P. O. Muller, and John Wiley & Sons, Inc.

enormous functional region, an interdependent realm that is held together through intense spatial economic and political networks. Modern Europe seizes on the realm's abundant geographic opportunities to create a huge, highly intensive network of spatial interaction linking places, communities, and countries in countless ways. This interaction operates on the basis of two key principles.

First, regional **8** **complementarity** means that one area produces a surplus of a commodity required by another area. The mere existence of a particular resource or product is no guarantee of trade: it must be needed elsewhere. When two areas each require the other's products, we speak of double complementarity. Europe exhibits countless examples of this complementarity, from local communities to entire countries. Industrial Italy needs coal from western Europe; western Europe needs Italy's farm products.

Second, the ease with which a commodity can be transported by producer to consumer defines its **9** **transferability**. Distance and physical obstacles can raise the cost of a product to the point of unprofitability. But Europe is small, distances are short, and Europeans have built the world's most efficient transport system of roads, railroads, and canals linking navigable rivers. Taken together, Europe's

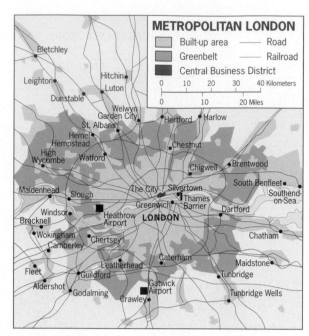

FIGURE 1A-8   © H. J. de Blij, P. O. Muller, and John Wiley & Sons, Inc.

enormously diverse economic regions and its particularly efficient transportation infrastructure make for a highly interdependent economic realm.

A Highly Urbanized Realm

About three of every four Europeans live in towns and cities, an average that is far exceeded in the west (see the Data Table in Appendix B) but not yet attained in much of the east. Large cities are production centers as well as marketplaces, and they also form the crucibles of their nations' cultures. Europe's major cities tend to be old and compact, and in general the European cityscape looks quite different from the American.

Seemingly haphazard inner-city street systems impede traffic; central cities may be picturesque, but they are also cramped. The urban layout of the London region (Fig. 1A-8) reveals much about the internal spatial structure of the European **10** **metropolis** (the central city plus its suburban ring). Such a *metropolitan area* remains focused on the large city at its center, especially the downtown **11** **central business district (CBD)**, which is the oldest part of the urban agglomeration and contains the region's largest concentration of business, government, and shopping facilities as well as its wealthiest and most prestigious residences.

Wide residential sectors radiate outward from the CBD across the rest of the central city, each one home to a particular income group. Beyond the central city lies a sizeable suburban ring, but residential densities are much higher here than in the United States because the European tradition is one of setting aside recreational spaces (in "greenbelts") and living in apartments rather than in

If you were to be asked what city might be shown here, would Paris spring to mind? Most images of France's capital show venerable landmarks such as the Eiffel Tower or the Arc de Triomphe or other landmarks of the Old City. But this is another Paris, the district called *La Défense*, where Paris shows its ultramodern face. This is where Paris escapes the height restrictions and architectural limitations of the historic center, where glass-boxed highrises reflect the vibrant global city this is, and where the landmark is the "Cube," a huge open structure admired as well as reviled (as was the Eiffel Tower in its time). Set just slightly off the axis created by the Champs Elysées beyond, the Cube is the focus for the high-tech, financial, and service hub *La Défense* has become. Look in the distance and you will see the Arc de Triomphe appropriately (as some would see it) diminished, and the Eiffel Tower to its right, just a needle rising from the vast urban tapestry that is Paris today.
© Pascal Crapet/Stone/Getty Images, Inc.

© H. J. de Blij

"After taking the train from Schiphol Airport to Rotterdam and checking into my downtown hotel, I took a train from the "Central Station" (which was in the middle of one huge construction site) to what the Dutch call "Hoek van Holland" ("Corner of Holland") on the North Sea coast, where huge ferries disembark thousands of travelers from England and elsewhere daily. A colleague had advised me to get off at Maassluis and walk the streets for a sense of the history as well as the modern changes going on here. Certainly the growing Islamic presence is transforming cultural landscapes in Europe, but not just in the cities. Even in small towns like this, the minarets of mosques rise between the spires of churches and, as is the case here, the local mosque is the centerpiece for a multipurpose Muslim cultural center that serves as a social (and political) center in ways that churches do not. A couple of generations ago, the people living in those apartments might have expected to overlook a church, but today their vista is quite different."

www.conceptcaching.com

detached single-family houses. There also is a greater reliance on public transportation, which further concentrates the suburban development pattern. That has allowed many nonresidential activities to suburbanize as well, and today ultramodern outlying business centers increasingly compete with the CBD in many parts of urban Europe (see photo at left).

A Changing Population

Negative Natural Population Growth

There was a time when Europe's population was (in the terminology of population geographers) exploding, sending millions to the New World and the colonies and still growing at home. But today Europe's indigenous population, unlike most of the rest of the world's, is actually shrinking. To keep a population from declining, the (statistically) average woman must bear 2.1 children. For Europe as a whole, that figure was 1.5 in 2009. But several countries recorded 1.3, including Germany, Romania, Portugal, and Hungary. And one eastern European country (Bosnia) even recorded 1.2—the lowest ever seen in any human population. Such negative population growth poses serious challenges for any nation. When the population pyramid becomes top-heavy, the number of workers whose taxes pay for the social services of the aged goes down, leading to reduced pensions and dwindling funds for health care.

Immigration

Meanwhile, immigration is partially offsetting Europe's population deficit. But it could be a mixed blessing, wherein demographic stability comes at the cost of social disruption. Millions of Turks, Turkish Kurds, Algerians, Moroccans, West Africans, Pakistanis, and West Indians are changing the social fabric of what once were nearly unicultural nation-states, especially France, Germany, the United Kingdom, Spain, Belgium, and the Netherlands. As noted above, one key dimension of this change is the spread of Islam in Europe (mapped in Fig. 1A-9). Muslim populations in eastern Europe (such as Albania's, Kosovo's, and Bosnia's) are indigenous communities converted during the period of Ottoman rule. The Muslim sectors of western European countries, on the other hand, represent more recent immigrations.

The majority of these immigrants are generally more religious than the Christian natives. They arrived in a Europe where native populations are stagnant or declining, where religious institutions are weakened and overshadowed by secularism, and where certain cultural norms are incompatible with Islamic traditions. Integration and assimilation of Muslim communities into the national mainstream has been slow, their education and income levels considerably lower than average.

The Growing Multicultural Challenge

The social and political implications of Europe's cultural transformation are numerous and far-reaching. Long known

EUROPE: MUSLIMS AS PERCENTAGE OF NATIONAL POPULATION, 2011

- Less than 1%
- 1–4%
- 4–7%
- 7–12%
- 12–25%
- More than 25%

0   250   500 Kilometers
0 100 200 300 Miles

FIGURE 1A-9

© H. J. de Blij, P. O. Muller, and John Wiley & Sons, Inc.

MINARETTEN VERBOT POSTER

© Goal Ag

for tolerance and openness, European societies are attempting to restrict immigration in various ways; political parties with anti-immigrant platforms are gaining ground. The new Islamic presence has been a cause of social tension in some countries, occasionally giving way to drama and political theatrics. The Swiss population approved a controversial referendum in November 2009 calling for a law that prohibits the construction of minarets. A majority of the Swiss seemed to feel that their culture was under threat from Islam and that this type of mosque architecture gave expression to an overly assertive minority—even though there were only four minarets in the entire country. Others were strongly opposed to the law, arguing that the petition organizers were playing on ungrounded fears and xenophobia, and that such a law would violate basic religious freedoms. No doubt, Switzerland's 400,000 Muslims (5 percent of the population, mainly Turks, and not known to be particularly zealous) must have felt uneasy about the vote. The poster above was distributed by organizations in favor of the law and was pasted on walls across Switzerland in the weeks prior to the referendum.

EUROPEAN UNIFICATION

Realms, regions, and countries can all be subject to dividing and unifying forces that cause them to become more or less cohesive and stable political units. Where the region or country is home to diverse populations with different political agendas, it may prove difficult to avoid divergence and territorial fragmentation. Political geographers use the term **12 centrifugal forces** to identify and measure the strength of such division, which may result from religious, racial, linguistic, political, economic, or regional factors.

Centrifugal forces are measured against **13 centripetal forces**, the binding, unifying glue of the state. General satisfaction with the system of government and administration, legal institutions, and other functions of the state (notably including its treatment of minorities) can ensure stability at the state level. In the case of the former Yugoslavia, the centrifugal forces unleashed after the end of the Cold War exceeded the weak centripetal forces in that

relatively young state, and it disintegrated with dreadful consequences.

Since the Second World War, Europe has witnessed a steady process of integration and unification at a much larger geographic scale. A growing majority of European states and their leaders recognize that closer association and regional coordination form the key to a more stable, prosperous, and secure future. A realmwide union is in the making. Despite various challenges and occasional setbacks, centripetal forces have prevailed in Europe during the past half-century.

Background

At the end of World War II, much of Europe lay shattered, its cities and towns devastated, its infrastructure wrecked, its economies ravaged. If this was one of the world's most developed economic realms at the beginning of the twentieth century, its economic prowess had been almost completely destroyed by 1945. One of the primary motives for integration and collaboration among western European countries, therefore, was rapid economic recovery.

The United States played a big role, initially, in spurring on such cooperation. In 1947, U.S. Secretary of State George C. Marshall proposed a European Recovery Program designed to help counter all this dislocation and to create stable political conditions in which democracy would survive. Over the next four years, the United States provided about $13 billion in assistance to Europe (over $100 billion in today's money). The United States was driven by economic and political motives. It had become the largest producer in the world of manufactured goods and was eager to restore European markets. Politically, the ending of World War II witnessed accelerating tensions with the Soviet Union, which had taken control over the bulk of eastern Europe. In addition, communist parties seemed poised to dominate the political life of major western European countries. The United States was also intent, through the Marshall Plan, to have a firm hand in western Europe and keep communist influences at bay. The Marshall Plan applied solely to 16 European countries, including defeated (West) Germany and Turkey.

Northwestern European countries themselves were also driven by political considerations. The two world wars had clearly shown the dangers of excessive nationalism and had laid bare the devastating problems that can arise from a lack of political cooperation and collaborative efforts (if the Allies had acted as one sooner, Hitler might have been stopped before things escalated into war in 1939). From the perspective of countries like France, the Netherlands, and Belgium, one of the key issues was to control Germany in the postwar years, and this could only be done through close political cooperation.

As the economies recovered and Germany became firmly embedded in a pan-European structure, the motives for European unification received a different emphasis. Increasingly, the process has been driven by the need to facilitate an ever larger and more efficient open market that can compete globally with the United States, Japan, and China. Politically, the goals today are more and more about stabilizing a much larger and diverse Europe that now approaches the Russian frontier. It is not just about Germany anymore but rather about a much bigger and more complicated geopolitical zone that has to accommodate a large number of (still) sovereign states and at the same time maintain good relations with the Russian giant next door.

The Unification Process

The Marshall Plan did far more than stimulate European economies. It confirmed European leaders' conclusion that their countries needed a joint economic-administrative structure. Such a structure was needed not only to coordinate the financial assistance, but also to ease the flow of resources and products across Europe's mosaic of boundaries, to lower restrictive trade tariffs, and to seek ways to improve political cooperation.

For all these needs Europe's governments had some guidelines. While in exile in Britain, the leaders of three small countries—Belgium, the Netherlands, and Luxembourg—had been discussing an association of this kind even before the end of the war. There, in 1944, they formulated and signed the Benelux Agreement, intended to achieve total economic integration. When the Marshall Plan was launched, the Benelux precedent helped speed the

creation of the Organization for European Economic Cooperation (OEEC), which was established to coordinate the investment of America's aid (see the table entitled "Supranationalism in Europe").

Soon the economic steps led to greater political cooperation as well. In 1949, the participating governments created the Council of Europe, the beginnings of what was to become a European Parliament meeting in Strasbourg, France. Europe was embarked on still another political revolution, the formation of a multinational union involving a growing number of European states. Geographers define **14 supranationalism** as the voluntary association in economic, political, or cultural spheres of three or more independent states willing to yield some measure of sovereignty for their mutual benefit.

Under the Treaty of Rome, six countries joined to become the European Economic Community (EEC) in 1958, also called the "Common Market." In 1973 the United Kingdom, Ireland, and Denmark joined, and the renamed European Community (EC) had nine members. As Figure 1A-10 shows, membership reached 15 countries in 1995, after the organization had been renamed yet one more time to become the European Union (EU). Since then, the number of member-states has climbed to 27.

The EU's administrative, economic, and even political framework has become so advanced that the organization now has a headquarters with many of the trappings of a capital city. Early on, EU planners chose Brussels (already the national capital of Belgium, a member of Benelux) as the organization's center of governance. And in order to

FROM THE FIELD NOTES...

© H. J. de Blij

"One unmistakable aspect of France: the French seem to like flying their flag. You see the tricolor everywhere, on public as well as private buildings and facilities. And here's another: I can't remember a country where the European Union flag is more commonplace, often flying beside (not below) the French flag as seen here at the regional headquarters in Rouen. This reflects the French government's enthusiasm for the "European Project" as they like to call it here: with the German government, these are the two key pillars of the EU. But surveys suggest that the general public in both countries is rather more ambivalent."

www.conceptcaching.com

SUPRANATIONALISM IN EUROPE

1944 Benelux Agreement signed.	**1987** Turkey and Morocco make first application to join EC. Morocco is rejected; Turkey is told that discussions will continue.
1947 Marshall Plan created (effective 1948–1952).	**1990** Charter of Paris signed by 34 members of the Conference on Security and Cooperation in Europe (CSCE). Former East Germany, as part of newly reunified Germany, incorporated into EC.
1948 Organization for European Economic Cooperation (OEEC) established.	
1949 Council of Europe created.	
1951 European Coal and Steel Community (ECSC) Agreement signed (effective 1952).	**1991** Maastricht meeting charts European Union (EU) course for the 1990s.
1957 Treaty of Rome signed, establishing European Economic Community (EEC) (effective 1958), also known as the Common Market and "The Six". European Atomic Energy Community (EURATOM) Treaty signed (effective 1958).	**1993** Single European Market goes into effect. Modified European Union Treaty ratified, transforming EC into EU.
	1995 Austria, Finland, and Sweden admitted into EU, creating "The Fifteen".
1959 European Free Trade Association (EFTA) Treaty signed (effective 1960).	**1999** European Monetary Union (EMU) goes into effect.
1961 United Kingdom, Ireland, Denmark, and Norway apply for EEC membership.	**2002** The euro is introduced as historic national currencies disappear in 12 countries.
1963 France vetoes United Kingdom EEC membership; Ireland, Denmark, and Norway withdraw applications.	**2003** First draft of a European Constitution is published to mixed reviews from member-states.
1965 EEC–ECSC–EURATOM Merger Treaty signed (effective 1967).	**2004** Historic expansion of EU from 15 to 25 countries with the admission of Cyprus, the Czech Republic, Estonia, Hungary, Latvia, Lithuania, Malta, Poland, Slovakia, and Slovenia.
1967 European Community (EC) inaugurated.	
1968 All customs duties removed for intra-EC trade; common external tariff established.	**2005** Proposed EU Constitution is rejected by voters in France and the Netherlands.
1973 United Kingdom, Denmark, and Ireland admitted as members of EC, creating "The Nine". Norway rejects membership in the EC by referendum.	**2007** Romania and Bulgaria are admitted, bringing total EU membership to 27 countries. Slovenia adopts the euro.
	2008 Cyprus and Malta adopt the euro.
1979 First general elections for a European Parliament held; new 410-member legislature meets in Strasbourg. European Monetary System established.	**2009** Slovakia adopts the euro.
	2010 Financial crisis strikes heavily-indebted Greece, requiring massive EU bailout and raising fears for Portugal, Spain, and Italy. The future of the EMU is clouded; the value of the euro declines after a long rise against the dollar.
1981 Greece admitted as member of EC, creating "The Ten".	
1985 Greenland, acting independently of Denmark, withdraws from EC.	**2011** Estonia becomes the seventeenth adopter of the euro.
1986 Spain and Portugal admitted as members of EC, creating "The Twelve". Single European Act ratified, targeting a functioning European Union in the 1990s.	

avoid giving Brussels too much prominence, they chose Strasbourg, in the northeast corner of France, as the seat of the European Parliament, whose elected membership represents all EU countries.

Consequences of Unification

The European Union is not just a paper organization: it has a major impact on national economies, on the role of individual states, and on the daily lives of its member-countries' citizens.

One Market

EU directives are aimed at the creation of a single market for producers and consumers, businesses and workers. Corporations should be able to produce and sell anywhere in the Union without legal impediments while workers should be able to move anywhere in the Union and find employment without legal restrictions. In order to make this happen and to keep things manageable, member-states have had to harmonize a wide range of national laws from taxation to the protection of the environment to educational standards. One major step forward came with the introduction of a single central bank (with considerable power over, for instance, interest rates) and a single currency, the *euro*. The single currency also symbolizes Europe's strengthening unity and establishes a joint counterweight to the once mighty American dollar. In 2002, twelve of the (then) 15 EU countries withdrew their currencies and began using the Euro, with only the United Kingdom (Britain), Denmark, and Sweden staying out (Fig. 1A-10). Slovenia adopted the euro in 2007, Cyprus and Malta in 2008, Slovakia in 2009, and Estonia became the seventeenth adopter in 2011.

A New Economic Geography

The establishment and continuing growth of the European Union have generated a new economic landscape that today

FIGURE 1A-10

© H. J. de Blij, P. O. Muller, and John Wiley & Sons, Inc.

not only transcends the old but is fundamentally reshaping the realm's regional geography. By investing heavily in new infrastructure and by smoothing the flows of money, labor, and products, European planners have dramatically reduced the divisive effects of their national boundaries. And by acknowledging demands for greater freedom of action by their provinces, States, departments, and other administrative units of their countries, European leaders unleashed a wave of economic energy that transformed some of these units into powerful engines of growth. Four of these growth centers are especially noteworthy, to the point that

geographers refer to them as the **15 Four Motors of Europe:** (1) France's southeastern ***Rhône-Alpes Region***, centered on the country's second-largest city, Lyon; (2) ***Lombardy*** in north-central Italy, focused on the industrial city of Milan; (3) ***Catalonia*** in northeastern Spain, anchored by the cultural and manufacturing center of Barcelona; and (4) ***Baden-Württemberg*** in southwestern Germany, headquartered by the high-tech city of Stuttgart.

Another change in regional economic geography pertains more to agricultural areas and is closely related to deliberate policies made in Brussels. Taxes tend to be high in

Europe, and those collected in the richer member-states are used to subsidize growth and development in the less prosperous ones. This is one of the burdens of membership that is not universally popular in the EU, to say the least. But it has strengthened the economies of Portugal, Spain, and other national and regional economies to the betterment of the entire organization. Some countries also object to the terms and rules of the Common Agricultural Policy (CAP), which, according to critics, supports farmers far too much and, according to others, far too little. (France in particular obstructs efforts to move the CAP closer to consensus, subsidizing its agricultural industry relentlessly while arguing that this protects its rural cultural heritage as well as its farmers.)

Diminished State Power and New Regionalism

Often, the local governments in these subregions simply bypass the governments in their national capitals, dealing not only with each other but even with foreign governments as their business networks span the globe. In this they are imitated by other provinces, all seeking to foster their local economies and, in the process, strengthen their political position relative to the state. Provinces, States, and other subnational political units on opposite sides of international boundaries can now cooperate in pursuit of shared economic goals. Such cross-border cooperation creates a new economic map that seems to ignore the older political one, creating economically powerful regions that are more or less independent of surrounding national states. Note how European unification seems to simultaneously erode the power of states from above and from below: in order to retain membership, states are more or less compelled to concede major decision-making power to Brussels while at the same time the assertion of subnational (and cross-national) regional authorities, from Catalonia to Lombardy, eats away at their territorial control at home.

Thus even as Europe's states have been working to join forces in the EU, many of those same states are confronting severe centrifugal stresses. The term **16** devolution has come into use to describe the powerful centrifugal forces whereby regions or peoples within a state, through negotiation or active rebellion, demand and gain political strength and sometimes autonomy at the expense of the center. Most states exhibit some level of internal regionalism, but the process of devolution is set into motion when a key centripetal binding force—the nationally accepted idea of what a country stands for—erodes to the point that a regional drive for autonomy, or for outright secession, is launched. As Figure 1A-11 shows, numerous European countries are affected by devolution.

States respond to devolutionary pressures in various ways, ranging from accommodation to suppression. One way to deal with these centrifugal forces is to give historic regions (such as Scotland or Catalonia) certain rights and

privileges formerly held exclusively by the national government. Another answer is for EU member-states to create new administrative divisions that will allow the state to meet regional demands, often in consultation with, or under pressure of, the *European Commission* (the name of the EU's central administration in Brussels).

Widening or Deepening?

Expansion has always been an EU objective, and the subject has always aroused passionate debate. Will the incorporation of weaker economies undermine the strength of the whole? Should the ties and cooperation between existing members be deepened and solidified before other, less prepared, countries are invited to join? Remember that member-states must adhere to strict economic policies and harmonize their political systems. This is much easier for prosperous countries with longstanding democratic traditions than for poorer nations with a volatile political past.

Despite such misgivings, negotiations to expand the EU have long been in progress and still continue. In 2004 a momentous milestone was reached: ten new members were added, creating a greater European Union with 25 member-states. Geographically, these ten came in three groups: three Baltic states (Estonia, Latvia, and Lithuania); five contiguous states in eastern Europe, extending from Poland and the Czech Republic through Slovakia, Hungary, and Slovenia; and two Mediterranean island-states, the ministate of Malta and the still-divided state of Cyprus. And in 2007, both Romania and Bulgaria were incorporated, raising the number of members to 27 and extending the EU to the shores of the Black Sea (Fig. 1A-10).

Numerous structural implications arise from this expansion, affecting all EU countries. A common agricultural policy is now even more difficult to achieve, given the poor condition of farming in most of the new members. Also, some of the former EU's poorer states, which were on the receiving end of the subsidy program that aided their development, now will have to pay up to support the much poorer new eastern members. And disputes over representation at EU's Brussels headquarters arose quickly. Even before the newest members' accession, Poland was demanding that the representative system favor medium-sized members (such as Poland and Spain) over larger ones (such as Germany and France).

The Remaining Outsiders

This momentous expansion is having major geographic consequences for all of Europe, and not just the EU. As Figure 1A-10 shows, following the admission of Romania and Bulgaria in 2007, a number of countries and territories remain outside the Union, and their prospect of joining seems a mixed bag. The first group includes the states emerging from the former Yugoslavia plus Albania. In this cluster of western

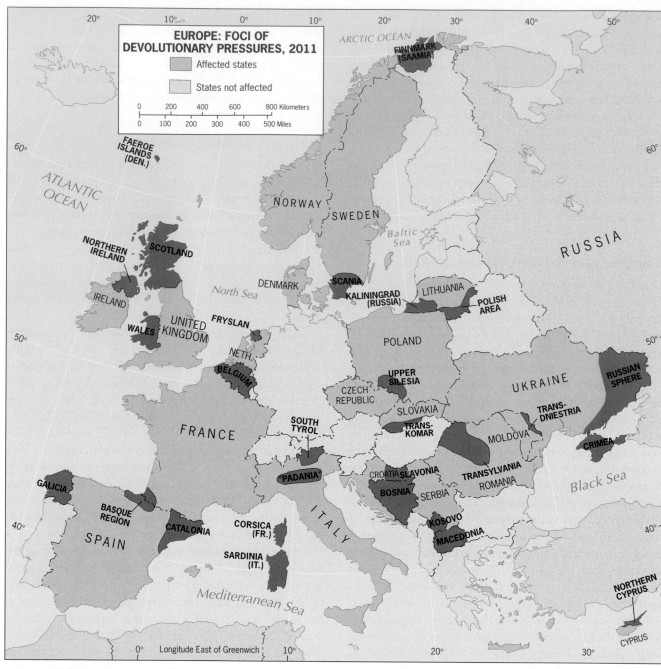

FIGURE 1A-11

© H. J. de Blij, P. O. Muller, and John Wiley & Sons, Inc.

Balkan states, most of them are troubled politically and economically; only Slovenia has achieved full membership. The remainder rank among Europe's poorest and most ethnically fractured states, but they contain nearly 25 million people whose circumstances could worsen outside "the club". Better, EU leaders reason, to move them toward membership by demanding political, social, and economic reforms, although the process will take many years. Discussions with Croatia began in 2005 and with newly independent Montenegro in 2006, but the aftermath of war and other problems are holding up Serbia's invitation. Elsewhere, ethnically and politically divided Bosnia seemed far from ready for negotiations, Alba-

nia was even less prepared, and unresolved issues continue to roil newly independent Kosovo. Note that this area of least-prepared countries also is the most Islamic corner of Europe.

The second group of outsiders is Ukraine and its neighbors. Four former Soviet republics in Europe's "far east" could some day join the EU (even Belarus, until recently the most disinterested, may now be leaning in that direction). The government of Ukraine, however, has shown interest, although its electorate is strongly (and regionally) divided between a pro-EU west and a pro-Russian east. Moldova views the EU as a potential supporter in a struggle against devolution along its eastern flank, and in 2006 a country not even on the map—Georgia, across the Black Sea from Ukraine—proclaimed its interest in EU membership for similar reasons: to find an ally against Russian intervention in its internal affairs.

Finally, we should take note of still another important candidate for EU membership: Turkey. EU leaders would like to include a mainly Muslim country in what Islamic states sometimes call the "Christian Club", but Turkey needs progress on social standards, human rights, and economic policies before its accession can be contemplated.

Once Again: In Search of Europe's Eastern Boundary

In the past, "Eastern Europe" incorporated all of Europe east of Germany, Austria, and Italy, north of Greece, and south of Finland. The Soviet communist domination of Eastern Europe (1945–1990) behind an ideological and strategic "Iron Curtain" served to reinforce the division between "West" and "East". The collapse of the Soviet Empire in the early 1990s freed the European countries that had been under Soviet rule, and those countries turned their gaze from Moscow to the west. Meanwhile, the European Union had been expanding eastward, and membership in the EU became an overriding goal for the majority of the liberated eastern states. Today, with the Soviet occupation mostly a distant memory and the European Union extending from the British Isles to the Black Sea, the boundary between Western and Eastern Europe has disappeared, at least in a political sense. Economically, there remain significant contrasts between west and east (see the Data Table in Appendix B and compare, for example, Bulgaria and France), but in this respect, too, things are evening out.

During the Cold War, Europeans (especially in the west) were preoccupied with the "Iron Curtain" and much less so with the question of an eastern boundary of Europe at large. After all, eastern Europe was drawn so tightly into the Soviet orbit politically, militarily, economically, and even culturally (the main foreign language taught in schools was Russian) that it seemed a moot point. But deep down, most of these parts remained quite European in spirit and tradition, and had historical ties that could not be easily erased. So now we are once again left to contemplate where, between Moscow and Vienna, eastern Europe's boundary lies or should be drawn. More than anything else, it is the eastward expansion of the EU that is testing this question.

Future Prospects of the EU

All these developments underscore how far-reaching and seemingly irreversible Europe's current transformation is. But Europe's ultimate supranational goals have not yet been attained. The problems of the old Eastern Europe remain evident in the Data Table in Appendix B. The sense of have and have-not, core and periphery, still pervades EU operations and negotiations. The EU consists of a large number of countries of bewildering diversity, and for all to get along infinite patience, gigantic bureaucracies, and a penchant for compromise are required. National cultures are very slow to change and must be acknowledged at all times. Just go the official European Union website (http://europa.eu): it has 23 portals each with a different language. Many Europeans, perhaps a majority, feel that the EU project is a bureaucratic edifice constructed by leaders who are out of touch with grass-roots Europe. They refer to the "democratic deficit" of the EU, which seems to get bigger with every expansion, as Brussels seems to get further away every time the Union expands. Thus Europe's quest for enduring unity continues.

But consider that the EU has now reached deeply into eastern Europe; encompasses 27 members; has a common currency and a parliament; is developing a constitution; has facilitated political stability and economic progress for half a century; and is even considering expansion beyond the realm's borders. All this constitutes a tremendous achievement in this historically volatile and fractious part of the world. The EU now has a combined population of almost 475 million constituting one of the world's richest markets; its member-states account for more than 40 percent of the world's exports.

It is remarkable that this has been accomplished in little more than five decades. Some of the EU's leaders want more than economic union: they envisage a United States of Europe, a political as well as an economic competitor for the United States. To others, such a "federalist" notion is an abomination not even to be mentioned (the British in general are especially wary of such an idea). Whatever happens, Europe is going through still another of its revolutionary transformations, and when you study its evolving map you are looking at history in the making.

POINTS TO PONDER

- Switzerland, situated at the heart of Europe but still outside the EU by choice, seems to be reconsidering its position.
- In 2010 India's automobile manufacturer Tata launched the world's lowest-priced car, the Nano, on the EU market.
- EU foreign ministers in late 2009 approved a measure that would allow Croatia to join the EU in 2012.
- A financial crisis in Greece in 2010 threatened the stability of the euro and had worldwide repercussions.

Southwest sector of central Paris, as seen from atop the Arc de Triomphe. iStockphoto

IN THIS CHAPTER

How the EU is remaking Europe's regional framework

The roles of Core and Periphery

Why Britain is different

The rebirth of East-Central Europe

The Balkan mosaic's continuing evolution

Is Europe advancing toward Russia's doorstep?

CONCEPTS, IDEAS, AND TERMS

Microstate	1
Site	2
Situation	3
Estuary	4
Conurbation	5
Landlocked location	6
Hegemony	7
World city	8
Shatter belt	9
Balkanization	10
Break-of-bulk	11
Entrepôt	12
Exclave	13
Irredentism	14

1 B

EUROPE: REGIONS OF THE REALM

≡ The Mainland Core
The Core Offshore: The British Isles
The Discontinuous South
The Discontinuous North
The Eastern Periphery

EUROPE: MODERN CORE AND HISTORIC REGIONS

- - - European Core Boundary

Western Europe

British Isles

Northern (Nordic) Europe

Mediterranean Europe

Eastern Europe

0   250   500 Kilometers

0   100   200   300 Miles

FIGURE 1B-1

Our objective in this and later regional discussions is to become familiar with the regional and national frameworks of the realms under investigation, and to discover how individual states fit into and function within the larger mosaic. Even in this era of globalization, and even in integrating Europe, the state still plays a key role in the process. Where these states are located, how they interact, and what their prospects may be in this changing world are among the key questions we will address.

EUROPE'S REGIONAL COMPLEXITY

Europe presents us with a particular challenge because of its numerous countries and territories, and because of the continuing changes affecting its geography. Readers of this book who reside in North America tend to be accustomed to a familiar and relatively stable map of two countries, the United States and Canada. In Europe, on the other hand, just since 1990 no less than 15 new national names have appeared on the political map, some of them revivals of old entities such as Estonia and Lithuania, others new and less familiar (Montenegro, Moldova). And there may be more to come: Kosovo became nominally independent in 2008, and Belgium may yet fracture the way Czechoslovakia did.

And Europe displays something else North America does not: **1** **microstates** that do not have the attributes of "complete" states but are on the map as tiny yet separate entities nonetheless, such as Monaco, San Marino, Andorra, and Liechtenstein. Add to these the Russian exclave (outlier) of Kaliningrad (on the Baltic Sea) and the British dependency of Gibraltar (at the entrance to the Mediterranean Sea), and Europe seems to display a bewildering political mosaic indeed.

Traditional Formal Regions, Modern Spatial Network

In the old days, Europe divided rather easily into Western, Northern (Nordic), Mediterranean (Southern), and Eastern regions, and countries were grouped accordingly. Some of this traditional historical geography lingers in cultural landscapes and still has relevance. But in the era of European unification, that regional scheme has become superimposed with a more dominating **core-periphery** framework, defined in terms of how central—or peripheral—countries are to the workings of the EU. Put differently, Europe can be understood as a set of formal, more or less homogeneous, cultural regions; but at the same time, the entire realm functions as a closely interdependent economic and political spatial system, a spatial network that revolves around the EU.

The original Common Market (1957) still anchors what has become a core area for the entire European realm. The British Isles form part of this Core, but the British—whose EU membership was delayed by a French veto and who never adopted the euro—still are not the full-fledged members the Germans, the French, and the Irish are. It is therefore appropriate to consider the United Kingdom as a geographically distinct component of the European Core.

Take a close look at the map in Fig. 1B-1 and note that today the core area of the EU does not coincide with national borders—the core is primarily defined on the basis of regional economic performance, not political geography. For example, Northern Italy (but not the South) or Southern Sweden (but not the North) are part of this core. You see how complicated and challenging the work of geographers can be when trying to understand the world's spatial organization. And the European realm illustrates that challenge like no other. But this chapter will bring clarity to what at first glance may appear to be a chaotic regional landscape.

In the discussion that follows, we will focus first on countries such as Germany and France that lie entirely within the Core region. Next, we discuss countries with significant regions inside as well as outside the Core such as Sweden, Spain, and Italy. Finally, our attention focuses on Europe's Periphery.

≡ THE MAINLAND CORE

Not counting the microstate of Liechtenstein, eight states form the Mainland Core of Europe: dominant Germany and France, the three Benelux countries, the two landlocked mountain states of Switzerland and Austria, and the Czech Republic (Fig. 1B-2). This is the European region sometimes still referred to as Western Europe, and its most populous country (Germany) also has Europe's largest economy.

Germany

Twice during the twentieth century Germany plunged Europe and the world into war, until, in 1945, the defeated and devastated German state was divided into two parts, West and East (see the delimitation in red on Fig. 1B-3), the latter of which had to surrender some important industrial areas to Poland.

West and East Germany then set forth on widely different economic and political trajectories. Soviet rule in East Germany was established on the Russian-communist model and, given the extreme hardships the USSR had suffered at German hands during the war, harshly punitive. The American-

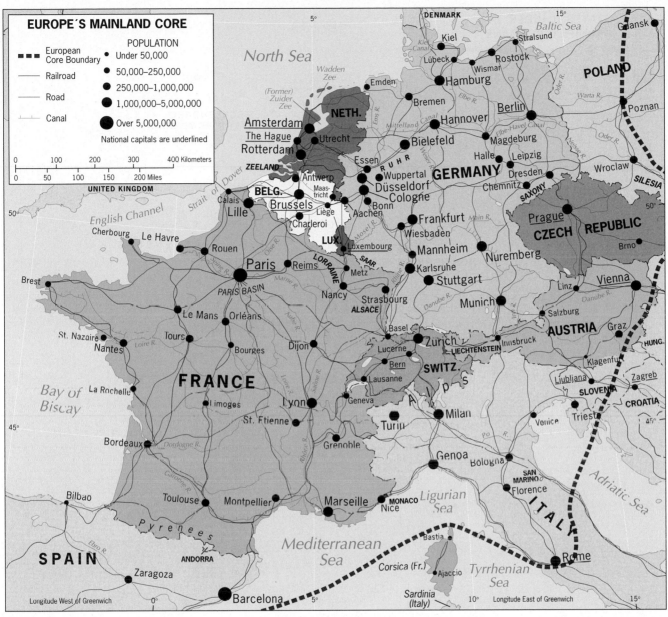

EUROPE'S MAINLAND CORE

POPULATION
- Under 50,000
- 50,000–250,000
- 250,000–1,000,000
- 1,000,000–5,000,000
- Over 5,000,000

National capitals are underlined

- - - - European Core Boundary
——— Railroad
——— Road
——├— Canal

0   100   200   300   400 Kilometers
0   50   100   150   200 Miles

FIGURE 1B-2

© H. J. de Blij, P. O. Muller, and John Wiley & Sons, Inc.

led authority in West Germany was less strict and aimed more at rehabilitation. When the Marshall Plan was instituted, West Germany was included, and its economy recovered rapidly. Meanwhile, West Germany was reorganized politically into a modern federal state on democratic foundations.

West Germany's economy thrived. Between 1949 and 1964 its gross national income (GNI) tripled while industrial output rose 60 percent. Simultaneously, West Germany's political leaders participated enthusiastically in the negotiations that led to the six-

member Common Market in 1957. Geography worked in West Germany's favor: it had common borders with all but one of the initial member-states. Its transport infrastructure, rapidly rebuilt, was second to none in the realm. More than compensating for its loss of Saxony and Silesia were the expanding Ruhr (in the hinterland of the Dutch port of Rotterdam) and the newly emerging industrial complexes centered on Hamburg in the north, Frankfurt (the leading financial hub as well) in the center, and Stuttgart in the south. West Germany exported huge quanti-

ties of iron, steel, motor vehicles, machinery, textiles, and farm products. To this day, Germany has the largest export economy by value in the world.

In 1990, West Germany had a population of about 62 million and East Germany 17 million. Communist misrule in the East had yielded outdated factories, crumbling infrastructures, polluted environments, drab cities, inefficient farming, and inadequate legal and other institutions. Reunification was more a rescue than a merger, and the cost to West Germany was enormous. When

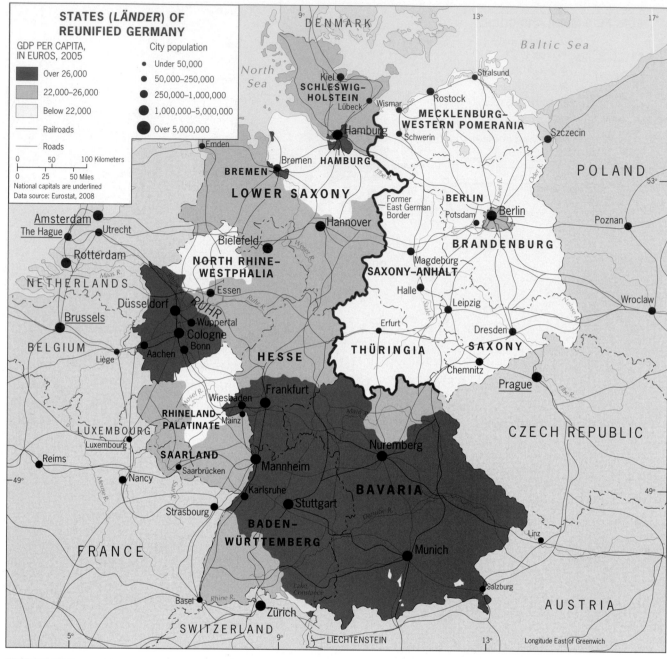

FIGURE 1B-3

© H. J. de Blij, P. O. Muller, and John Wiley & Sons, Inc.

the West German government imposed sales-tax increases and an income-tax surcharge on its citizens, many Westerners doubted the wisdom of reunification. It was projected that it would take decades to reconstruct Virginia-sized East Germany: ten years later, exports from the former East still contributed only about seven percent of the national total.

Regional disparity would afflict Germany for a very long time to come.

Upon reunification, East Germany was reorganized into six new States to fit the federal system of West Germany. Today's reunified Federal Republic of Germany consists of 16 States (or *Länder*). Figure 1B-3 makes a key point: regional disparity in terms of gross domestic product (GDP) per

person* remains a serious problem between the former East and West. Note that five of former East Germany's six States (Berlin being the sole exception) are in the lowest income category, while most of the ten former West German States rank in the two higher income categories. In the first decade of this century, Germany's economy was stagnant, raising unemployment and slowing former East Germany's recovery. Nonetheless, the gap continues to narrow, and with 81.9 million inhabitants including

Gross domestic product (GDP) is the total value of all goods and services produced in a country (or subnational entity) by that political unit's economy during a given calendar year. GDP per capita is that total divided by the resident population.

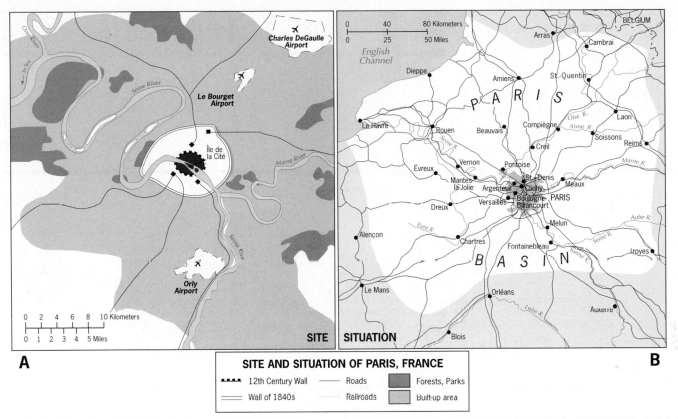

A

SITE

SITUATION

B

SITE AND SITUATION OF PARIS, FRANCE

■■■■ 12th Century Wall   ── Roads   ▨ Forests, Parks

══ Wall of 1840s   ── Railroads   ░ Built-up area

FIGURE 1B-4

© H. J. de Blij, P. O. Muller, and John Wiley & Sons, Inc.

over 7 million foreigners and more than 4 million ethnic Germans born outside the country, Germany is again exerting its dominance over a mainland Europe in which it has no peer.

France

German dominance in the European Union is a constant concern in the other leading Mainland Core country. The French and the Germans have been rivals in Europe for centuries. France (population: 62.6 million) is an old state, by most measures the oldest in this region. Germany is a young country, created in 1871 after a loose association of German-speaking states had fought a successful war against . . . the French.

Territorially, France is larger than Germany, and the map suggests that France has a superior relative location, with coastlines on the Mediterranean, the Atlantic Ocean, and, at Calais,

even a window on the North Sea. But France does not have any good natural harbors, and oceangoing ships cannot navigate its rivers and other waterways far inland.

The map of the Mainland Core region (Fig. 1B-2) reveals a significant contrast between France and Germany in terms of population distribution. France has one dominant city, Paris, and no other city comes close either in size or centrality: Paris has 9.9 million residents, whereas its closest rival, Lyon, has only 1.4 million. Germany has no city to match Paris, but it does have a range of medium-sized cities and it is more highly urbanized overall (89 percent) than France (77 percent).

Why should Paris, without major raw materials nearby, have grown so large? Whenever geographers investigate the evolution of a city, they focus on two important locational qualities: its **2** **site** (the physical attributes of the place it occupies) and its **3** **situa-**

tion (its location relative to surrounding areas). The site of the original settlement at Paris lay on an island in the Seine River, a defensible place where the river was often crossed. This island, the *Île de la Cité*, was a Roman outpost 2000 years ago; for centuries its security ensured continuity. Eventually the island became overcrowded, and the city expanded along the banks of the river (Fig. 1B-4).

Soon the settlement's advantageous situation stimulated its growth and prosperity. Its fertile agricultural hinterland thrived, and, as an enlarging market, Paris's focality increased steadily. The Seine River is joined near Paris by several navigable tributaries (the Oise, Marne, and Yonne). When canals extended these waterways even farther, Paris was linked to the Loire Valley, the Rhône-Saône Basin, Lorraine (an industrial area in the northeast), and the northern border with Belgium. When Napoleon reorganized France and built a radial system of roads—followed later by railroads—that focused on Paris from all parts of the country, the city's *primacy** was assured (Fig. 1B-4).

---

*When a country's leading city is disproportionately large and exceptionally expressive of its national culture, it is called a *primate city*. Paris is the quintessential example, not only personifying France but also serving as its capital.

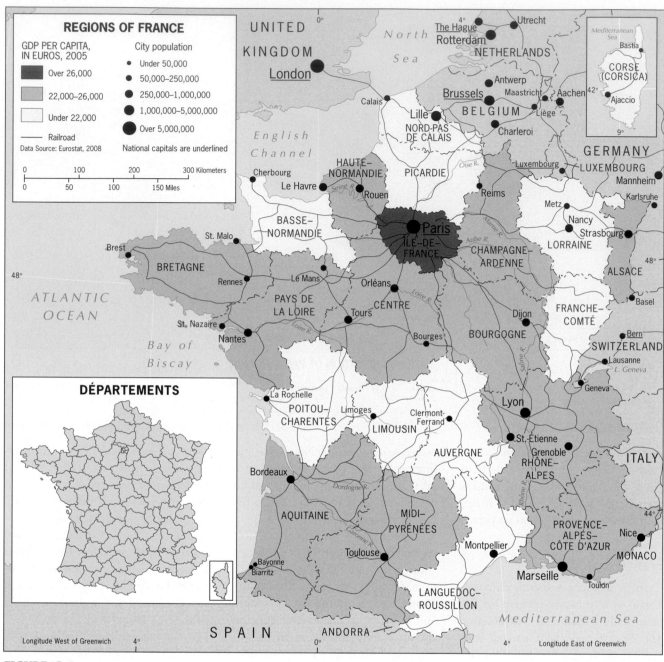

FIGURE 1B-5

© H. J. de Blij, P. O. Muller, and John Wiley & Sons, Inc.

In addition, France developed early on a strong tradition of a highly centralized state, where the central government in Paris maintained tight control over the provinces near and far. With Napoleon's rule in the early 1800s, the country was cut up into a great number of small *départements* (eventually 96), each of which had representation in Paris, but with power concentrated in the capital. It remained so for nearly two centuries, with the entire country tightly integrated and without notable regional opposition (see Fig. 1B-5 inset map). Only the island *département* of Corsica produced a rebel movement, whose violent opposition to French rule continued for decades and even touched the mainland. In 2003 the voters in Corsica rejected an offer of special status for their island, including limited autonomy. They want more, and trouble lies ahead.

Today, France is decentralizing. A new layer of governance, consisting of 26 larger provinces named *régions*, has been inserted at the level between Paris and the *départements* to accommodate devolutionary forces and to seize the opportunities of regional and local growth throughout the country (Fig. 1B-5). These regions (one of which is Corsica and four of which are overseas) have substantial autonomy in such spheres as taxation, borrowing, and development spending. The cities that anchor them benefit because they are the seats of governing regional councils that can attract investment, not only within France but also from abroad.

France has one of the world's most productive and diversified economies, based in one of humanity's richest cultures and vigorously promoted and protected (notably its heavily subsidized agricultural sector). Northern French agriculture remained Europe's most productive and varied, exploiting the country's wide range of soils and climates and enjoying state subsidies and protections.

Lyon, France's second city and headquarters of the region named Rhône-Alpes, has become a focus for growth industries and multinational firms. This region is evolving into a self-standing economic powerhouse that is becoming a driving force in the European economy. Indeed, it is one of the Four Motors of Europe with its own international business connections to countries as far away as China and Chile.

Today France's economic geography is marked by new high-tech industries. It is a leading producer of high-speed trains, aircraft, fiber-optic communications systems, and space-related technologies. It also is the world leader in nuclear power, which currently supplies more than 80 percent of its electricity and thereby reduces its dependence on foreign oil imports.

Although France and Germany agree on many aspects of EU integration, they tend to differ on important issues. Old, historically centralized France is less eager than young, federal Germany to push political integration in supranational Europe.

Benelux

Three small countries are crowded into the northwest corner of the Mainland Core—the Netherlands, Belgium, and Luxembourg—and are collectively referred to by their first syllables (**Be-Ne-Lux**). Their total population, 27.7 million, reminds us that this is one of the most densely peopled corners of our planet. Not coincidentally, the Low Countries, as they are also sometimes called, are situated at the **4** estuary of the great Rhine and Scheldt rivers and have access to the seas that has, throughout history, been the envy of the Germans and the French. They are a highly productive trio: both the Netherlands and Belgium rank among the top 20 economies of the world, and tiny Luxembourg has the world's highest per-capita gross national income. The seafaring Dutch had a thriving agricultural economy and a rich colonial empire; the Belgians forged ahead during the Industrial Revolution; and Luxembourg came into its own after the Second World War as a financial center for the evolving European Union and, indeed, the world.

The Netherlands, one of Europe's oldest democracies and a constitutional monarchy today, has for centuries been expanding its living space—not by warring with its neighbors but by wresting land from the sea. Its greatest project so far, the draining and reclaiming of almost the entire Zuider Zee (Southern Sea), began in 1932 and continues. In the southwestern province of Zeeland, islands are

 FROM THE FIELD NOTES...

© Jan Nijman

"The southwestern part of the Netherlands is marked by major waterways, low-lying islands, and brackish ecosystems teeming with wildlife. Because the area is largely below sea level it is protected by the so-called Delta Works, an engineering marvel created to keep the land from flooding during North Sea storms, and to simultaneously maintain the natural habitat. This is the Haringvliet Sluizen, one of a series of adjustable storm surge barriers. Having these types of structures across the inlets provides safety and avoids the need to build protective barriers along all the shorelines further inland. The last major flood, which killed almost 2000, occurred more than half a century ago, before the Delta Works were built."

www.conceptcaching.com

being connected by dikes and the water is being pumped out, adding still more **polders** (as the Dutch call inhabited land claimed from the sea lying behind dikes and below sea level) to the national territory. Among the technologies in which the Dutch lead the world, not surprisingly, is the engineering of coastal systems to control the sea and protect against storms.

The regional geography of this highly urbanized country (16.5 million) is noted for the **Randstad**, a roughly triangular urban core area anchored by Amsterdam, the constitutional capital, Rotterdam, Europe's largest port, and The Hague, the seat of government. This **5** **conurbation**, as geographers call large urban areas when two or more cities merge spatially, now forms a ring-shaped complex that surrounds a still-rural center (in Dutch, *rand* means edge or margin; *stad* means city).

The economic geography of the Netherlands, like that of the other states in this subregion, is heavily dominated by services, finance, and trade; manufacturing contributes about 15 percent of the value of the GDP annually, and farming, once a mainstay, less than 3 percent. Amsterdam's airport, Schiphol, is regularly recognized as Europe's best; Rotterdam, one of the world's busiest ports, is rated as the most efficient as well.

Belgium also has a thriving economy and a major port in the city of Antwerp. But Belgium is a much younger state than the Netherlands, becoming independent within its present borders as recently as 1830. With 10.7 million people, Belgium's regional geography is dominated by a cultural fault line that cuts diagonally across the country, separating a Flemish-speaking majority (58 percent) centered on Flanders in the northwest from a French-speaking minority in southeastern Wallonia (31 percent). Brussels, the mainly French-speaking capital, lies like a cultural island in the Flemish-speaking sector; but the city also is one of Belgium's greatest assets

because it serves as the headquarters, and in many ways as the functional capital, of the European Union (though it is important to note that the EU does not have an official capital city). Still, devolution is a looming problem for Belgium, with political parties espousing Flemish separatism roiling the social landscape.

Luxembourg, one of Europe's many ministates, lies between Germany, Belgium, and France, with a Grand Duke as head of state (its official name is the Grand Duchy of Luxembourg), a territory of only 2600 square kilometers (1000 sq mi), and a population of just half a million. Luxembourg has translated sovereignty, relative location, and stability into a haven for financial, service, and information-technology industries. In 2007 there were more than 160 banks in this tiny country and nearly 14,000 holding companies (corporations that hold controlling stock in other businesses in Europe and worldwide). With its unmatched per-capita income (U.S. $64,400 in 2009), Luxembourg is in some ways the greatest beneficiary of the advent of the European Union. And in no country in Europe is support for the EU stronger than it is here.

The Alpine States

Switzerland, Austria, and the **microstate** of Liechtenstein on their border share an absence of coasts and the mountainous terrain of the Alps—and little else (Fig. 1B-2). Austria speaks one language; the Swiss speak German in their north, French in the west, Italian in the southeast, and even a bit of Rhaeto-Romansch in the remote central highlands (Fig. 1A-7). Austria has a large primate city; multicultural Switzerland does not. Austria has a substantial range of domestic raw materials; Switzerland does not. Austria is twice the size of Switzerland and has a larger population, but far more trade crosses the Swiss Alps between western and Mediterranean Europe than crosses Austria.

Switzerland, not Austria, is in most ways the leading state in the Alpine subregion of the Mainland Core. Mountainous terrain and **6** **landlocked location** can constitute crucial barriers to economic development, tending to inhibit the dissemination of ideas and innovations, obstruct circulation, constrain farming, and divide cultures. That is why Switzerland is such an important lesson in the complexities of human geography. Through the skillful maximization of their limited opportunities (including the transfer needs of their neighbors), the Swiss have transformed their seemingly restrictive environment into a prosperous state. They deftly utilized the waters cascading from their mountains to generate hydroelectric power to develop highly specialized industries. Swiss farmers perfected ways to optimize the productivity of mountain pastures and valley soils. Swiss leaders converted their country's isolation into stability, security, and neutrality, making it a world banking giant, a global magnet for money. Zürich, in the German-speaking sector, is the financial center; Geneva, in the French-speaking sector, is one of the world's most internationalized cities. The Swiss feel that they do not need to join the EU—and they have not done so to this day.

Austria, which joined the EU in 1995, is a remnant of the Austro-Hungarian Empire and has a historical geography that is far more reminiscent of unstable eastern Europe than that of Switzerland. Even Austria's physical geography seems to demand that the country look eastward: it is at its widest, lowest, and most productive in the east, where the Danube links it to Hungary, its old ally in the anti-Muslim wars of the past.

Vienna, by far the Alpine subregion's largest city, also lies on the country's eastern perimeter. One of the world's most expressive primate cities with magnificent architecture and monumental art, Vienna today is the Mainland Core's easternmost city, but Vienna's relative location changed dramatically with EU enlargement in 2004 and 2007. Peripheral to the EU and a vanguard of the Core until 2004, Vienna found itself in a far more centralized position when the EU border shifted eastward. But many Austrians had their doubts about their neighbors to the east becoming EU members of potentially equal standing, and in Austria public support for the European Union has fallen to the lowest level of any member-state.

The Czech Republic

The Czech Republic, product of the 1993 Czech-Slovak "velvet divorce," centers on the historic province of Bohemia, the mountain-encircled national core area that focuses on the capital, Prague. This is a classic primate city, its cultural landscape faithful to Czech traditions; but it is also an important industrial center. The surrounding mountains contain many valleys with small towns that specialize, Swiss-style, in fabricating high-quality goods. In the old Eastern Europe, even during the communist period, the Czechs always were leaders in technology and engineering; their products could be found on markets in foreign countries near and far.

Bohemia always was cosmopolitan and Western in its exposure, outlook, development, and linkages; Prague lies in the basin of the Elbe River, its traditional outlet through northern Germany to the North Sea. Today, the Czech Republic (10.4 million) is reclaiming its position at the center of European action.

≡ THE CORE OFFSHORE: THE BRITISH ISLES

As Figure 1B-1 shows, two countries form the maritime portion of the European core area: the United Kingdom and Ireland. These countries lie on two major islands, surrounded by a constellation of tiny ones. The larger island, a mere 34 kilometers (21 mi)

off the mainland at the closest point, is **Britain**; its smaller neighbor to the west is **Ireland** (Fig. 1B-6).

The names attached to these islands and the countries they encompass are the source of some confusion. They still are called the British Isles, even though British dominance over most of Ireland ended in 1921. The state that occupies Britain and the northeastern corner of Ireland is officially called the United Kingdom—UK by abbreviation. But this country often is referred to simply as Britain, and its people are known as the British. The state of Ireland officially is known as the Republic of Ireland (Eire in Irish Gaelic), but it does not include the entire island of Ireland.

During the long British occupation of Ireland, which is overwhelmingly Catholic, many Protestants from northern Britain settled in northeastern Ireland. In 1921, when British domination ended, the Irish were set free—except in that corner in the north, where London kept control to protect the area's Protestant settlers. That is why the country to this day is officially known as the United Kingdom of Great Britain and Northern Ireland.

Northern Ireland was home not only to Protestants from Britain, but also to a substantial population of Irish Catholics who found themselves on the wrong side of the border when Ireland was liberated. Ever since, conflict has intermittently engulfed Northern Ireland and spilled over into Britain and even, in the form of terrorism, into nearby mainland Europe.

Although all of Britain lies in the United Kingdom, political divisions exist here as well. England is the largest of these units, the center of power from which the rest of the region was originally brought under unified control. The English conquered Wales in the Middle Ages, and Scotland's link to England, cemented when a Scottish king ascended the English throne in 1603, was ratified by the Act of Union of 1707. Thus England, Wales,

Scotland, and Northern Ireland became the United Kingdom.

Having united the Welsh, Scots, and Irish, the British set out to forge what would become the world's largest colonial empire. An era of mercantilism (competitive accumulation of wealth among countries) and domestic manufacturing (the latter based on water power from streams flowing off the Pennines, Britain's highland backbone) foreshadowed the momentous Industrial Revolution that transformed Britain—and much of the world.

But British **/ hegemony** (political dominance) came to an end in the early twentieth century, challenged by the rise of other powers such as Russia, Germany, Japan, and the United States. World War II marked the end of the age of colonial empires and the beginning of U.S. global dominance—now with Britain as a "junior partner". For Britain, change after 1945 was profound: not only was it forced to let go of most of its overseas territories (though many remained part of the largely symbolic British Commonwealth), but it also had to deal with the rapid resurgence of the European mainland. Always ambivalent about the EC and EU, and with its first membership application vetoed by the French in 1963, Britain (admitted in 1973) has worked to restrain moves toward tighter integration. When most member-states adopted the new euro in favor of their national currencies, the British kept their pound sterling and delayed their participation in the EMU. As for a federalized Europe, to Britain this is out of the question. In this as in other respects, Britain's historic, insular standoffishness continues.

The United Kingdom

The UK, with an area about the size of Oregon and a population of 61.8 million, is by European standards quite a large country. Based on a combination of physiographic, historical, cultural, economic, and political criteria,

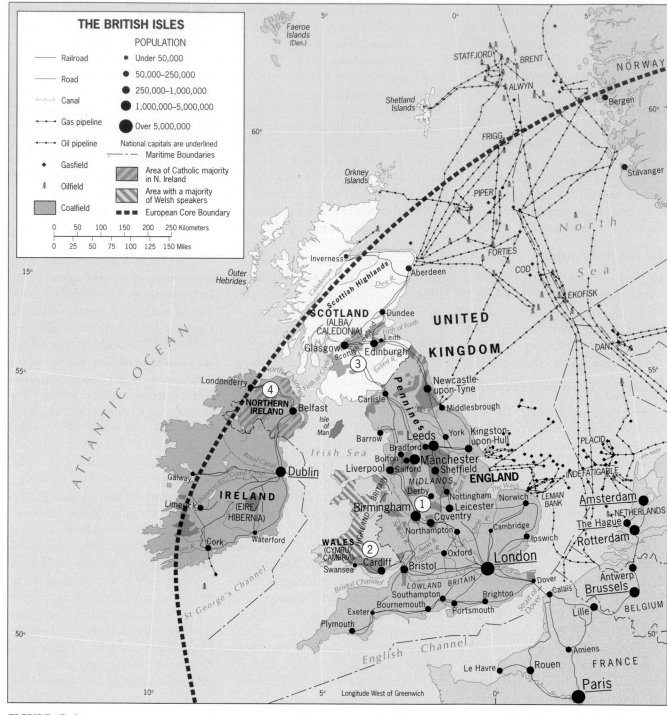

THE BRITISH ISLES

POPULATION

- Railroad
- Road
- Canal
- Gas pipeline
- Oil pipeline
- ◆ Gasfield
- Oilfield
- Coalfield

- • Under 50,000
- • 50,000–250,000
- ● 250,000–1,000,000
- ● 1,000,000–5,000,000
- ● Over 5,000,000

National capitals are underlined
—·—· Maritime Boundaries
Area of Catholic majority in N. Ireland
Area with a majority of Welsh speakers
▪▪▪▪ European Core Boundary

| 0 | 50 | 100 | 150 | 200 | 250 Kilometers |
| 0 | 25 | 50 | 75 | 100 | 125 | 150 Miles |

FIGURE 1B-6

© H. J. de Blij, P. O. Muller, and John Wiley & Sons, Inc.

the United Kingdom can be divided into four subregions (numbered in Fig. 1B-6):

1. **England.** So dominant is this subregion of the United Kingdom that the entire country is sometimes referred to by this name. Small wonder: England is anchored by the huge London metropolitan area, which by itself contains more than one-seventh of the UK's total population. Indeed, along with New York and Tokyo, London is regarded as one of the three leading **8 world cities** of the current era of globalization, with financial, high-technology, communications, engineering, and related industries reflecting the momentum of its long-term growth and agglomeration. To the north and west, England was the hearth of the Industrial Revolution, its cities synonymous with manufactures of matchless variety.

2. **Wales.** This nearly rectangular, rugged territory was a refuge for ancient Celtic peoples, and in its western counties more than half the inhab-

itants still speak Welsh. Because of the high-quality coal reserves in its southern tier, Wales too was engulfed by the Industrial Revolution, and Cardiff, the capital, was once the world's leading coal exporter. But the fortunes of Wales also declined, and many Welsh emigrated. Among the 3 million who remained, however, the flame of Welsh nationalism survived, and in 1997 the voters approved the establishment of a Welsh Assembly to administer public services in Wales, a first devolutionary step.

3. *Scotland.* Nearly twice as large as the Netherlands and with a population about the size of Denmark's, Scotland is a major component of the United Kingdom. Most of Scotland's more than 5 million people live in the Scottish Lowlands anchored by Edinburgh, the capital, in the east and Glasgow in the west. Attracted there by the labor demands of the Industrial Revolution (coal and iron reserves lay in the area), the Scots developed a world-class shipbuilding industry. Decline and obsolescence were followed by high-tech development, notably in the hinterland of Glasgow, and Scottish participation in the exploitation of oil and gas reserves under the North Sea, which transformed the eastern ports of Aberdeen and Leith (Edinburgh). But many Scots feel that they are disadvantaged within the UK and should play a major role in the EU. Therefore, when the British government put the option of a Scottish Parliament before the voters in 1997, 74 percent approved. Many Scots still hope that sovereignty lies in their future, and in local elections in 2007 the independence-minded Scottish National Party won more seats in the Parliament than any of the other parties.

4. *Northern Ireland.* A declining majority of the overall population of 1.8 million, now about 53 percent, are Protestants and trace their

"The Troubles" between Protestants and Catholics, pro- and anti-British factions that have torn Northern Ireland apart for decades at a cost of more than 3000 lives, are etched in the cultural landscape. The so-called "Peace Wall" across West Belfast, shown here separating Catholic and Protestant neighborhoods, is a tragic monument to the failure of accommodation and compromise, a physical manifestation of the emotional divide that still runs deep—despite the political compromise engineered by former Prime Minister Tony Blair and his negotiators. In May 2007, a Northern Ireland Assembly began functioning again, after hardline Catholic and Protestant political parties were persuaded to join in the effort. But what an analyst said to the author back in 2005 still is the view across the cityscape you see here: "Good fences make good neighbors . . . the Peace Wall will have to stay up for a few more decades yet."
© Barry Chin/The Boston Globe/Landov LLC

ancestry to Scotland or England; a growing minority, currently around 46 percent, are Roman Catholics and share their religion with the Irish Republic on the other side of the border. Although Figure 1B-6 suggests that there are majority areas of Protestants and Catholics in Northern Ireland, no clear separation exists; mostly they live in clusters throughout the territory, including walled-off neighborhoods in the major cities of Belfast and Londonderry. Partition is no solution to a conflict that has raged for more than four decades at a cost of thousands of lives; Catholics accuse London as well as the local Protestant-dominated administration of discrimination, whereas Protestants accuse Catholics of seeking union with the Republic of Ireland. The

Northern Ireland Assembly, to which powers are supposed to be devolved from London, had to be suspended from 2002 to 2007 before the parties could be brought together at the table again. The prospects for greater autonomy are clouded.

Republic of Ireland

What Northern Ireland has been missing through its conflicts is evident to the south, in the Irish Republic itself. Here participation in the EU, adoption of the euro, business-friendly tax policies, comparatively low wages, an English-speaking workforce, and an advantageous relative location combined to produce, around the turn of this century, the highest rate of economic growth in the entire European Union. This booming, service-based economy, accompanied by burgeoning cities and towns, fast-rising real estate prices, mushrooming industrial parks, and bustling traffic, transformed a country long known for emigration into a magnet for industrial workers, producing new social challenges for a closely knit, long-isolated society. Among these immigrants were thousands of people of Irish descent returning from foreign places to take jobs at home, workers from elsewhere in the European Union (including large numbers of Poles), and job-seekers from African and Caribbean countries.

The Republic of Ireland fought itself free from British colonial rule just three generations ago. Its cool, moist climate had earlier led to the adoption of the American potato as the staple crop, but excessive rain and a blight in the late 1840s, coupled with colonial mismanagement, caused famine and cost over a million lives. Another 2 million Irish left for North America and other shores.

Hard-won independence in 1921 did not bring real economic prosperity until the last few decades of the twentieth century, when Ireland became known as the **Celtic Tiger** (likening it to the miraculous economic development of the so-called *Asian Tigers*). By 2009, however, Ireland's first economic boom had faded: the real estate market declined, service industries found more favorable conditions in eastern Europe, and unemployment rose. It is clear that the Irish economy has become strongly linked to, and highly dependent on, global developments. It remains to be seen whether its current economic problems are merely part of a cyclical downturn in the global economy or require more structural solutions.

THE DISCONTINUOUS SOUTH

As Figures 1B-1 and 1B-7 show, the northern sectors of two major southern states form parts of the European Core: northern Italy and northern Spain. Italy and Spain are two of the four countries that constitute southernmost Europe, and in both important urbanized and industrialized subregions have become integral parts of Europe's core area.

Portugal, on the western flank of the Iberian Peninsula, remains outside the European core, far less urbanized, much more agrarian, and not strongly integrated into it. The fourth national entity in this southern domain is the island ministate of Malta, south of Sicily, an historically important crossroads with a population of about 400,000 and a booming tourist industry.

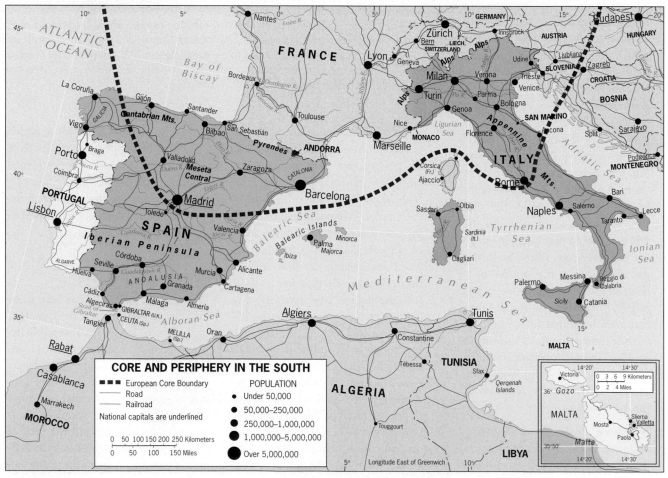

FIGURE 1B-7

© H. J. de Blij, P. O. Muller, and John Wiley & Sons, Inc.

Italy

Centrally located in Mediterranean Europe, most populous of the realm's southern states, best connected to the European Core, and economically most advanced is Italy (population: 59.8 million), a charter member of Europe's Common Market.

Administratively, Italy is organized into 20 internal regions, many with historic roots dating back centuries [FIG. 1B-8]. Several of these regions have become powerful economic entities centered on major cities, such as Lombardy (Milan) and Piedmont (Turin); others are historic hearths of Italian culture, including Tuscany (Florence) and Veneto (Venice). These regions in the northern half of Italy stand in strong social, economic, and political contrast to such southern regions as Calabria (the "toe" of the Italian "boot") and Italy's two major Mediterranean islands, Sicily and Sardinia. Not surprisingly, Italy is often described as two countries—a progressive north and a stagnant south (known as the ***Mezzogiorno***). The urbanized, industrialized north is part of Europe's Core; the low-income south typifies the Periphery.

North and south are bound by the ancient headquarters, Rome, which lies astride the narrow transition zone between Italy's contrasting halves. This clear manifestation of Europe's Core-Periphery contrast is referred to in Italy as the ***Ancona Line***, named after the city on the Adriatic coast where it reaches the other side of the peninsula (Fig. 1B-8, blue line). Whereas Rome remains Italy's capital and cultural focus, the functional core area of Italy has shifted northward into Lombardy in the basin of the Po River. Here lies southern Europe's leading manufacturing complex. The Milan–Turin–Genoa triangle exports appliances, instruments, automobiles, ships, and many specialized products. Meanwhile, the Po Basin, lying on the margins of southern Europe's dominant Mediterranean climatic regime (with

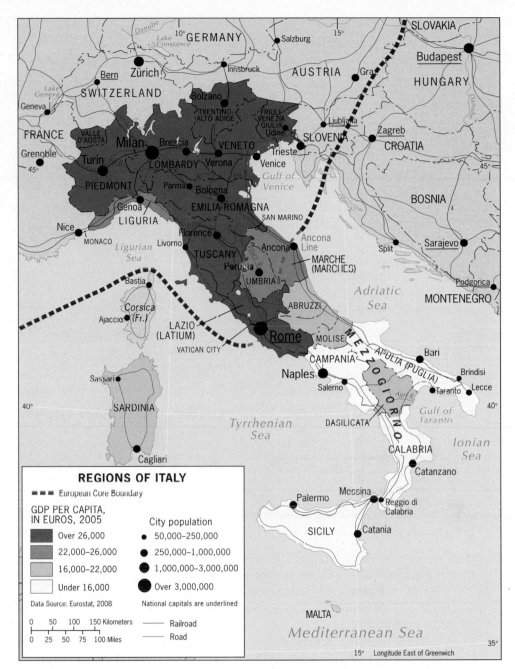

REGIONS OF ITALY

- - - European Core Boundary

GDP PER CAPITA, IN EUROS, 2005

■ Over 26,000	
▨ 22,000–26,000	
▨ 16,000–22,000	
□ Under 16,000	

City population
- • 50,000–250,000
- ● 250,000–1,000,000
- ⬤ 1,000,000–3,000,000
- ⬤ Over 3,000,000

Data Source: Eurostat, 2008

National capitals are underlined

0 50 100 150 Kilometers
0 25 50 75 100 Miles

—— Railroad
—— Road

FIGURE 1B-8   © H. J. DE BLIJ, P. O. MULLER, AND JOHN WILEY & SONS, INC.

its hot, dry summers), enjoys a more even pattern of rainfall distribution throughout the year, making it a productive agricultural zone as well.

Metropolitan Milan embodies the new, modern Italy. Not only is Milan (at 4 million) Italy's largest city and leading manufacturing center—making Lombardy one of Europe's Four Motors—but it also is the country's financial and service-industry headquarters. Today the Milan area, a cor-

nerstone of the European Core, has just 7 percent of Italy's population but accounts for fully one-third of the entire country's national income.

Spain, Portugal, and Malta

At the western end of southern Europe lies the Iberian Peninsula, separated from France and western Europe by the rugged Pyrenees and from North Africa by the narrow Strait of Gibraltar. Spain

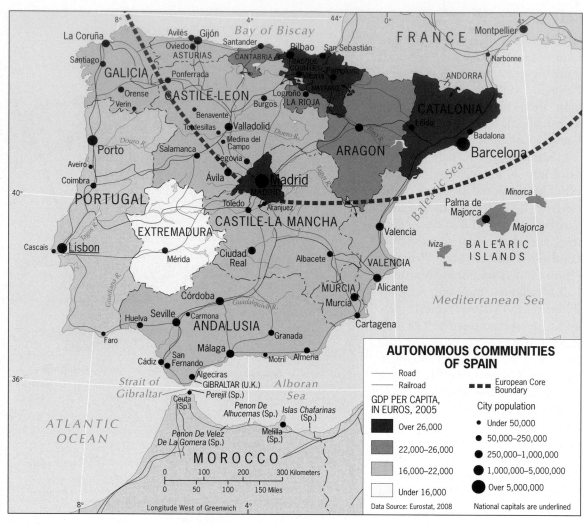

FIGURE 1B-9

© H. J. de Blij, P. O. Muller, and John Wiley & Sons, Inc.

(population: 46.7 million) occupies most of this compact Mediterranean landmass, and peripheral Portugal lies in its southwestern corner.

Both countries benefited enormously from their admission to the European Union in 1986. **Spain** followed the leads of Germany and France and decentralized its administrative structure in response to devolutionary pressures. These pressures were especially strong in Catalonia, the Basque Country, and Galicia, and in response the Madrid government created so-called *Autonomous Communities* (ACs) for all 17 of its regions (Fig. 1B-9). Every AC has its own parliament and administration that control planning, public works, cultural affairs, education, environmental policy, and even, to some extent,

international commerce. Each AC can negotiate its own degree of autonomy with the central government in the capital. Some Spanish observers feel that devolution has gone too far and that a federal system (such as Germany's) would have been preferable, but now there is no turning back.

Even so, the AC framework has not totally defused the most violent of the secessionist-minded movements, propelled by a small minority of Basques. On the other hand, relations between Madrid and Catalonia have improved. Centered on the prosperous, productive coastal city of Barcelona, the AC named Catalonia is Spain's leading industrial area and has become one of Europe's Four Motors. Catalonia is endowed with its own distinctive language and culture that

find vivid expression in Barcelona's urban landscape.

As Catalans like to remind visitors to their corner of Spain, most of the country's industrial raw materials are found in the northwest, but most of its major industrial development has occurred in the northeast, where innovations and skills drive a high-technology regional economy. In recent years, Catalonia—with 6 percent of Spain's territory and 16 percent of its population—has annually produced 25 percent of all Spanish exports and nearly 40 percent of its industrial exports. Such economic strength translates into political power, and in Spain the issue of Catalonian separatism is never far from the surface.

As Figure 1B-9 shows, Spain's capital and largest city, Madrid, lies near the geographic center of the state. It also lies within an economic-geographic transition zone. In terms of people's annual income, Spain's north is far more affluent than its south, a direct result of the country's distribution of resources, climate (the south suffers from drought and poorer soils), and overall development opportunities. The most prosperous ACs are Barcelona-centered Catalonia and Madrid; between them, the contiguous group of ACs extending from the Basque Country to Aragon rank next. In the ACs of northern Spain, economies tend to do better in the east, where tourism and winegrowing (especially in La Rioja, a famous name in Spanish wine) are among the mainstays, than in the west, where industrial obsolescence, dwindling raw-material sources, the decline of the fishing industry, and emigration plague local economies. Spain's southernmost ACs have long been the least developed economically and are more emblematic of Europe's Periphery.

Much the same can be said of **Portugal** (10.6 million), which occupies the southwestern corner of the Iberian Peninsula. Because EU agricultural policies entail transfers from rich to poor regions in the Union and because the EU funds major infrastructural projects in hitherto isolated areas, Portugal has benefited enormously from its admission to the EU—but it remains a far cry from Europe's leading economies.

Unlike Spain, which has major population clusters on its interior plateau as well as its coastal lowlands, the Portuguese are concentrated along and near the Atlantic coast. Lisbon and the second city, Porto, are coastal cities. The best farmlands lie in the moister western and northern zones of the country; but the farms are small and inefficient, and even though Portugal remains dominantly rural, it must import as much as half of its foodstuffs.

Southernmost Europe also contains the ministate of **Malta**, located in the central Mediterranean Sea just south of Sicily. Malta is a small archipelago of three inhabited and two uninhabited islands with a population of just over 400,000 (Fig. 1B-7, inset map). An ancient crossroads and culturally rich with Arab, Phoenician, Italian, and British infusions, Malta became a British dependency and served British shipping and its military. It suffered terribly during World War II bombings, but despite limited natural resources recovered strongly during the postwar era. Today Malta has a booming tourist industry and a relatively high standard of living, and was one of the ten new member-states to join the European Union as part of the historic expansion of 2004.

Greece and Cyprus

If the label *discontinuous geography* applies anywhere, it is the area between Austria and the Aegean Sea, covering the Balkans (former Yugoslavia) and Greece. This is an area encompassing various (and conflicting) religions and language groups; prosperous and very poor nations; lands with volatile and violent histories; EU members and nonmembers (some without any prospect of gaining entry anytime soon); and, furthest removed from the Core, is Greece, the birthplace of European civilization and a longtime EU member and Western ally. We turn first to Greece and the Greek-connected island-state of Cyprus, and then to the Balkans.

Seemingly dangling from the southern end of eastern Europe, **Greece** was the wellspring of one of the ancient world's greatest civilizations, its scientists and philosophers still cited to this day, its famous tragedies still staged in the amphitheatres built more than two thousand years ago. Its familiar peninsulas and islands are bounded by Turkey to the east and by Bulgaria, Macedonia, and Albania to the north (Fig. 1B-10). Note that some of Greece's islands in the Aegean Sea lie on Turkey's very doorstep; in addition, Greeks represent the majority of the population of Cyprus in the northeast corner of the Mediterranean Sea.

FROM THE FIELD NOTES...

© H. J. de Blij

"The photograph says it all. From thousands of vantage points in Athens, the view of the Acropolis is marred by globalization's intrusion."

www.conceptcaching.com

A member of the EU since 1981, Greece has had a turbulent modern history, with alternating communist and fascist dictatorships yielding to more democratic government in the 1980s and economic upheavals marking the EU period. Economic stagnation in the 1980s was followed by such progress that Greece came to be called the "locomotive of the Balkans," an EU beacon in its remote corner of the Union. Metropolitan Athens burgeoned, benefiting from EU assistance, the hosting of the 2004 Olympics, and membership in the EMU. Arriving by air, you find yourself at one of the world's most modern airports followed by a ride on world-class superhighways or subways. The metropolitan area of Athens and its busy port, Piraeus, contain about one-third of the country's entire population of 11.2 million.

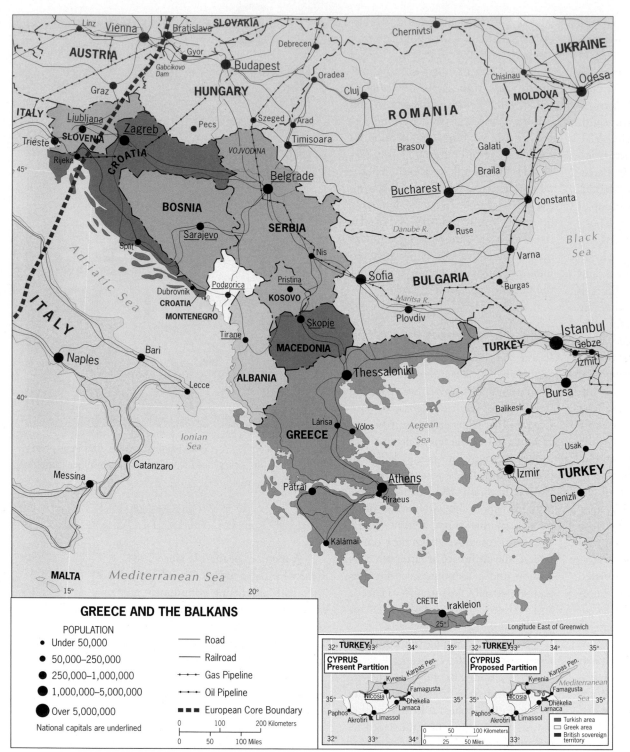

GREECE AND THE BALKANS

POPULATION
- • Under 50,000
- ● 50,000–250,000
- ● 250,000–1,000,000
- ⬤ 1,000,000–5,000,000
- ⬤ Over 5,000,000

National capitals are underlined

——— Road
——— Railroad
—•—•— Gas Pipeline
—•—•— Oil Pipeline
▬ ▬ ▬ European Core Boundary

0   100   200 Kilometers
0   50   100 Miles

CYPRUS Present Partition

CYPRUS Proposed Partition

■ Turkish area
□ Greek area
■ British sovereign territory

0   50   100 Kilometers
0   25   50 Miles

FIGURE 1B-10

© H. J. de Blij, P. O. Muller, and John Wiley & Sons, Inc.

close, Greece itself was the subject of a tragedy as its inefficient, hamstrung government was mired in debt, deadly riots in the streets greeted efforts at painful reform, and a proud, historic nation found itself having to conform to the terms imposed by the EU powers providing the necessary loans.

Also in this southeastern corner of the European Periphery lies the island country of **Cyprus** (an EU member since 2004), whose political geography merits special attention because of the complications it created, and continues to create, for the EU. It is useful to elaborate on this issue because it is such a good illustration of the political intricacies and challenges of the unification process. As the map shows, Cyprus lies closer to Turkey than it does to any part of Europe, but it is peopled dominantly by Greeks. In 1571, the Turks conquered Cyprus, then ruled by Venice, and controlled it until 1878 when the British took over. Most of the island's Turks arrived during the Ottoman period; the Greeks have been there longest.

When the British were ready to give Cyprus independence after World War II, the 80-percent-Greek majority mostly preferred union with Greece. Ethnic conflict followed, but in 1960 the British granted Cyprus independence under a constitution that prescribed majority rule but guaranteed minority rights.

This fragile order broke down in 1974, and civil war engulfed the island. Turkey sent in troops and massive dislocation followed, resulting in the

But EU membership had other, less favorable consequences. When the EU expanded to include poorer states such as Bulgaria and Romania, Greece lost much of its EU subsidy as Brussels had to divert its equalization funds to needier members. This meant that Greece had to tighten its financial belt, but most citizens had become used to early retirement, loose tax enforcement, lifetime government employment, and protection against competition (for example, the notion of private universities elicits angry street protests). And so, as the first decade of the new century came to a partition of Cyprus into northern Turkish and southern Greek sectors (Fig. 1B-10, inset map). In 1983, the 40 percent of Cyprus under Turkish control, with about 100,000 inhabitants (plus some 30,000 Turkish soldiers), declared itself the independent **Turkish Republic of Northern Cyprus**. Only Turkey recognizes this ministate (which today contains a population of about 200,000); the international community recognizes the government on the Greek side as legitimate. This legitimacy, however, is questionable in light of the Greek side's rejection of the United Nations plan to allow both the Greek and the Turkish side of the island to join the EU in 2004. The Turkish-Cypriot voters accepted it, yet they were left out when the Greek side of the island was admitted to the EU in 2004. Resentment was high on the Turkish side and in Turkey itself as well—just as discussions on Turkey's own admission to the EU were getting under way. It was—and remains—a reminder that Cyprus's "Green Line" separating the Greek and Turkish communities constitutes not just a regional border but a boundary between geographic realms.

The Balkans

We now turn to a part of Europe that has undergone wrenching changes since 1990, a process that started with the violent dismantling of communist Yugoslavia during the last decade of the twentieth century and continues, fortunately less violently, today. Here the European Union has made only slight progress, and just one of the countries that emerged from Yugoslavia's disintegration, Slovenia, has joined the EU. Discussions with Croatia, Slovenia's

neighbor, have repeatedly been put on hold. No other state in this area is a serious candidate for admission at present.

This part of Europe has had a particularly volatile history, and it is a good example of what geographers call a **9 shatter belt**, a zone of persistent splintering and fracturing. Geographic terminology uses several expressions to describe the breakup of established order, and these tend to have their roots in this part of the world. One of those expressions is **10 balkanization**. The southern half of eastern Europe is referred to as the Balkans or Balkan Peninsula (named after a mountain range in Bulgaria). Balkanization denotes the recurrent division and fragmentation of a region.

The key state on the new map is **Serbia**, the name of what is left of a much larger domain once ruled by the Serbs—who were dominant in the

former Yugoslavia. Centered on the historic capital of Belgrade on the Danube River, Serbia (7.3 million people) is the largest and potentially the most important new country in the area. But the Serbs are having to accommodate some major changes. First, more than 1 million of them live in neighboring Bosnia, where they have an uneasy relationship with the local Muslims and Croats (Fig. 1B-11). Second, the coastal province named **Montenegro** broke away in 2006, when voters there opted to form an independent state. Third, its Muslim-majority province of **Kosovo** declared independence in 2008, its sovereignty immediately recognized by the United States and a majority of (but not all) European governments. And fourth, Serbia still incorporates a Hungarian minority of some 400,000 in its northern province of **Vojvodina** on the northern side of the Danube at

FIGURE 1B-11

© H. J. DE BLIJ, P. O. MULLER, AND JOHN WILEY & SONS, INC.

a time when Hungary has already joined the EU. Clearly, the Balkans represent a virtual mosaic of ethnic identities that severely complicates any kind of territorial arrangement (see Fig. 1B-11).

If EU expansion occurs here in the south, it may involve **Croatia**, the crescent-shaped country with prongs along the Hungarian border as well as the Adriatic coast. But here, too, there are problems. About 90 percent of Croatia's 4.4 million citizens are Croats, but the country's Serb minority has faced discrimination and has declined from 12 percent of the population to under 5. The EU took a dim view of human rights issues in Croatia, but these improved when EU membership beckoned. Meanwhile, about 800,000 Croats live outside Croatia, in Bosnia, where their relationships with Muslims and Serbs are not always cooperative.

When the former Yugoslavia collapsed, **Bosnia** was the cauldron of calamity. No ethnic group was overwhelmingly dominant here, and this multicultural, effectively landlocked triangle of territory, lying between the Serbian stronghold to the east and the Croatian republic to the west and north, fell victim to disastrous conflict among Serbs, Croats, and ***Bosniaks*** (now the official name for Bosnia's Muslims, who constitute about 50 percent of the population of 3.8 million). As many as 250,000 people perished in concentration camps associated with ***ethnic cleansing*** practices; in 1995, a U.S. diplomatic effort resulted in a truce that partitioned the country as shown in Figure 1B-11. This is one rough corner of Europe.

The southernmost "republic" of former Yugoslavia was **Macedonia**, which emerged from the collapse as a state with a mere 2 million inhabitants, of whom two-thirds are Macedonian Slavs. As the map shows, Macedonia adjoins Muslim Albania and Kosovo, and its northwestern corner is home to the 30 percent of the Macedonians who are nominally Muslims. The remainder of this culturally diverse population are Turks, Serbs, and Roma (see footnote p. 72). Macedonia is one of Europe's poorest countries, landlocked and powerless. Even its very name caused it problems: Macedonia's Greek neighbors argued that this name was Greek property and so would not recognize it. Next, Macedonia faced an autonomy movement among its Albanian citizens, requiring allocation of scarce resources in the effort to hold the fledgling state together. Macedonians cling to the hope that eventual EU admission will bring it subsidies and better economic times.

The ministate of **Montenegro** has a mere 620,000 inhabitants, one-third of them Serbs, a small capital (Podgorica), some scenic mountains, and a short but spectacular Adriatic coastline—and very little else to justify its position as one of Europe's 40 countries. Tourism, a black market, some Russian investment, and a scattering of farms sum up the assets of this country.

In 1999, amid the chaos of disintegrating Yugoslavia, NATO took charge in **Kosovo**, then still a formal Serbian province, after a brief but damaging military campaign. With an overwhelmingly Albanian-Muslim population of more than 2 million and a small Serb minority in its northern corner, landlocked Kosovo has few of the attributes of a full-fledged state, but balkanization continues in this part of Europe. NATO eventually turned over Kosovo's administration to the United Nations, and in 2008 the capital of Pristina witnessed independence celebrations as the UN administration yielded to a newly elected national government.

The only other dominantly Muslim country in Europe is **Albania**, where some 70 percent of the population of 3.2 million adhere to Islam. Albania also shares with one other country—Moldova—the status of being Europe's poorest state. Albania has one of Europe's highest rates of natural population growth, and many Albanians try to emigrate to the EU by crossing the Adriatic to Italy. Most Albanians subsist on livestock herding and farming, but the poverty-stricken Gegs in the north lag behind the somewhat better-off Tosks in the south, with the capital of Tirane lying close to this cultural divide. For all of Europe's globalization, Albania would represent the symptoms of the global periphery anywhere in the world.

THE DISCONTINUOUS NORTH

The north's remoteness, isolation, and environmental severity, especially at higher latitudes, have had positive binding effects for part of this domain and seem to have fostered similar cultures. The Scandinavian Peninsula lay removed from most of the wars of mainland Europe and developed in relatively tranquil circumstances (although Norway was overrun by Nazi Germany during World War II). The three major languages—Swedish, Norwegian, and Danish—are mutually intelligible, and in terms of religion there is overwhelming adherence to the same Lutheran church in the three Scandinavian countries as well as Iceland, Finland, Estonia, and Latvia. Lithuania and Latvia have distinct languages, and in Lithuania the Roman Catholic religion prevails—but all the Baltic states have a shared history on the shores of the Baltic Sea and in proximity to Russia. In Scandinavia, democratic and representative governments emerged early, and individual rights and social welfare have long been carefully protected. Women participate more fully in government and politics here than in any other part of the world.

But consider the implications of Figures 1B-1 and 1B-12, and it is obvious why we refer to the *Discontinuous North*: the southern, coastal, and urban areas of the region form part of the Core, but the rest does not. In the aggregate, economic indicators for

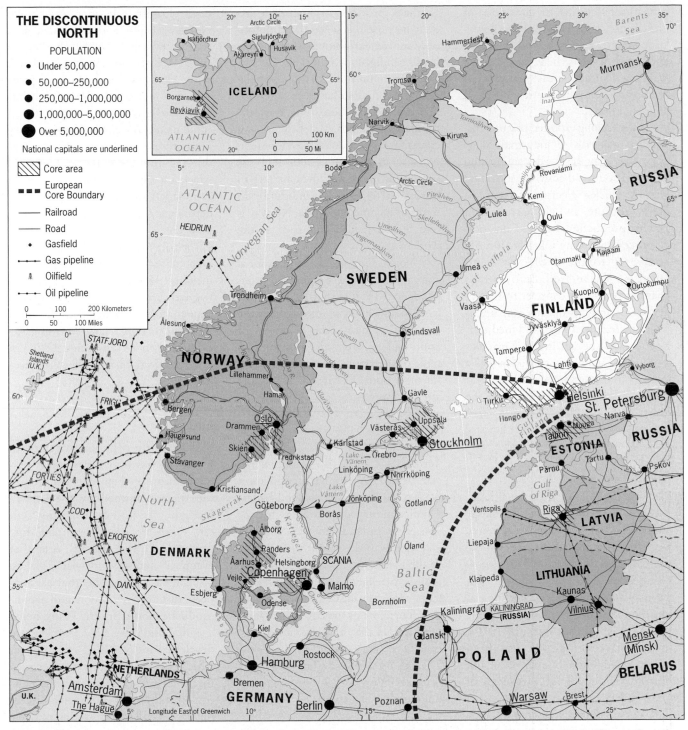

FIGURE 1B-12

© H. J. de Blij, P. O. Muller, and John Wiley & Sons, Inc.

most of the region are very strong, but almost all development is concentrated within that smaller Core zone. Territorially, most of northern Norway, Sweden, and Finland form part of the European Periphery.

The countries of this northern domain of Europe have a combined population of only 31 million, about half of Italy's. Nothing here compares to Italy's north or Spain's Catalonia, but the core areas of the three Scandinavian countries (Sweden, Norway, and Denmark) make up in prosperity and external linkages what they lack in dimensions.

Sweden

Sweden is the largest Nordic country in terms of both population (9.2 million) and territory. Most Swedes live south of 60° North latitude (which passes through Uppsala just north of the capital, Stockholm), in what is climatically the most moderate part of the country (Fig. 1B-12). Here lie the primate city, core area, and the main industrial districts; here, too, are the main agricultural areas that benefit from the lower relief, better soils, and milder climate.

Sweden long exported raw or semi-finished materials to industrial countries, but today the Swedes are making finished products themselves, including motor vehicles, electronics, stainless steel, furniture, and glassware. Much of this production is based on local resources, including a major iron ore reserve at Kiruna in the far north (there is a steel mill at Luleå). Swedish manufacturing, in contrast to that of several western European countries, is based in dozens of small and medium-sized towns specializing in particular products. Energy-poor Sweden was a pioneer in the development of nuclear power, but a national debate over the risks involved has reversed that course.

Norway

Norway does not need a nuclear power industry to supply its energy needs. It has found its economic opportunities on, in, and beneath the sea. Norway's fishing industry, now augmented by highly efficient fish farms, long has been a cornerstone of the economy, and its merchant marine spans the world. But since the 1970s, Norway's economic life has been transformed by the bounty of oil and natural gas discovered in its sector of the North Sea.

With its limited patches of cultivable soil, high relief, extensive forests, frigid north, and spectacularly fjorded coastline, Norway has nothing to compare to Sweden's agricultural or industrial development. Its cities, from the capital Oslo and the North Sea port of Bergen to the historic national focus of Trondheim as well as Arctic Hammerfest, lie on the coast and have difficult overland connections. The isolated

northern province of Finnmark has even become the scene of an autonomy movement among the reindeer-herding indigenous Saami (Fig. 1A-11). The distribution of Norway's population of 4.8 million has been described as a necklace, its beads linked by the thinnest of strands. But this has not constrained national development. Norway in the late 2000s had the second-lowest unemployment rate in Europe (after tiny Luxembourg). In terms of income per capita, Norway is one of the richest countries in the world.

Norwegians have a strong national consciousness and a spirit of independence. In the mid-1990s, when Sweden and Finland voted to join the European Union, the Norwegians again said no. They did not want to trade their economic independence for the regulations of a larger, even possibly safer, Europe.

Denmark

Territorially small by Scandinavian standards, Denmark has a population of 5.5 million, the second-largest Nordic country after Sweden. It consists of the Jutland Peninsula and several islands to the east at the gateway to the Baltic Sea; it is on one of these islands, Sjaelland, that the capital of Copenhagen is located. Copenhagen, the "Singapore of the Baltic," has long been a port that collects, stores, and transships large quantities of goods. This **11 break-of-bulk** function exists because many oceangoing vessels cannot enter the shallow Baltic Sea, making the city an **12 entrepôt** where transfer facilities and activities prevail. The completion of the Øresund bridge-tunnel link to southern Sweden in 2000 enhanced Copenhagen's situation (Fig. 1B-12).

Denmark remains a kingdom, and in centuries past Danish influence spread far beyond its present confines. Remnants of that period now challenge Denmark's governance. Greenland came

under Danish rule after union with Norway (1380) and continued as a Danish possession when that union ended (1814). In 1953, Greenland's status changed from colony to province, and in 1979 the 60,000 inhabitants were given home rule with an Inuit name: ***Kalaallit Nunaat***. They promptly exercised their rights by withdrawing from the European Union, of which they had become a part when Denmark joined. Another restive dependency is the Faroe Islands, located between Scotland and Iceland. These 17 small islands and their 45,000 inhabitants were awarded self-government in 1948, complete with their own flag and currency, but even this was not enough to defuse demands for total independence. A referendum in 2001 confirmed that even Denmark is not immune from Europe's devolutionary forces.

Finland

Finland, territorially almost as large as Germany, has only 5.3 million residents, most of them concentrated in the triangle formed by the capital, Helsinki, the textile-producing center, Tampere, and the shipbuilding center, Turku (Fig. 1B-12). A land of evergreen forests and glacial lakes, Finland has an economy that has long been sustained by wood and wood-product exports. But the Finns, being a skillful and productive people, have developed a diversified economy in which the manufacture of precision machinery and telecommunications equipment (prominently including cellphones) as well as the growing of staple crops are key.

As in Norway and Sweden, environmental challenges and relative location have created Nordic cultural landscapes in Finland, but the Finns are not a Scandinavian people; their linguistic and historic links are instead with the Estonians to the south across the Gulf of Finland. Ethnic groups speaking Finno-Ugric languages are widely dispersed across what is today western Russia.

Estonia

The most northern of the three "Baltic states," Estonia has longstanding ethnic and linguistic ties to Finland. But during the period of Soviet control from 1940 to 1991, Estonia's demographic structure changed drastically: today about 25 percent of its 1.3 million inhabitants are Russians, most of whom came there as colonizers.

After a difficult period of adjustment, Estonia today is forging ahead of its Baltic neighbors (see Appendix B) and catching up with its Nordic counterparts. Busy traffic links Tallinn, the capital, with Helsinki, and a new free-trade zone at Muuga Harbor facilitates commerce with Russia. But more important for Estonia's future was its entry into the European Union in 2004. Estonia has gained attention in recent years due to its economic advances and especially its creative entrepreneurship in the high-tech and software industries. Skype, for example, was originally an Estonian company. Having joined the euro zone in 2011, Estonia is on track to soon join the ranks of the European Core.

Latvia

As Figure 1B-12 reminds us, the boundary of the European Core that traverses Europe's North includes southern Norway and Sweden but excludes Estonia and the other two Baltic states. The latter—Latvia and Lithuania—have experienced economic improvement in recent years, but are not quite yet on a par with Estonia, let alone far-ahead Scandinavia. Latvia, the middle Baltic state centered on the port and capital city of Riga, was tightly integrated into the Soviet system during Moscow's long domination. Still today, only 59 percent of the population of 2.3 million is Latvian, and about 28 percent is Russian. After independence eth-

nic tensions arose between these Baltic and Slav sectors, but the prospect of EU membership required the end of discriminatory practices. Latvians concentrated instead on the economy, which had been left in dreadful shape, with the result that it qualified for 2004 admission. Consider this: 20 years ago, virtually all of Latvia's trade was with the Soviet Union. Today its principal trading partners are Germany, the United Kingdom, and Sweden. Russia figures in only one category: Latvia's import of oil and gas.

Lithuania

This southernmost Baltic state of 3.4 million has a residual Russian minority of only about 7 percent, but relations with its giant neighbor are worse than Latvia's—this despite Lithuania's greater dependence on Russia as a trading partner. One reason for this bad relationship has to do with neighboring Kaliningrad,* Russia's **13** exclave (small territorial outlier separated from the main body of a state) on the Baltic Sea (Fig. 1B 12). When Kaliningrad became a Russian territory after World War II, Lithuania was left with only about 80 kilometers (50 mi) of Baltic coastline and a small port that was not even connected by rail to the interior capital of Vilnius. Lithuania in 2005 called for the demilitarization of Kaliningrad as a matter of national security. Despite these problems, Lithuania's economy during 2003 had the highest growth rate in Europe, spurred by foreign investment and by profits from its oil refinery at Mazeikiai, facilitating EU admission in 2004.

Iceland

Iceland, the volcanic, glacier-studded island in the frigid waters of the North Atlantic just south of the Arctic Circle, is the sixth Nordic country. Inhabited by people with Scandinavian ancestries (population: 340,000), Iceland and its small neighboring archipelago, the Westermann Islands, are of special scientific interest because they lie on the Mid-Atlantic Ridge; here the Eurasian and North American tectonic plates of the Earth's crust are diverging (see Fig. G 4), new land can be seen forming, and spectacular volcanic eruptions are periodically on display (as occurred memorably in 2010).

Iceland's population is almost totally urban, and the capital, Reykjavik, contains about half the country's inhabitants. The nation's economic geography is traditionally oriented to the surrounding waters, whose seafood harvests have given Iceland a high standard of living. In the 1990s, Iceland embarked on economic liberalization policies and its financial industries (banks) grew rapidly. For a while, Iceland was referred to as the *Nordic Tiger*—but, not unlike Ireland, which went through a similar experience, the economy came crashing down with the global financial crisis in 2008. The Icelandic government, closely involved with some of the troubled banks, had to be bailed out by the International Monetary Fund.

THE EASTERN PERIPHERY

Europe has changed fundamentally in the two decades since the end of the Cold War. The old East Europe, sharply defined along the Iron Curtain,

---

*__Kaliningrad__ is located on the Baltic Sea between Poland and Lithuania. This Russian exclave, acquired following the end of World War II, is potentially important as an outpost in Europe of Russian influence at a time when the EU as well as NATO (the North Atlantic Treaty Organization, led by the United States) are advancing toward Russia's borders in what was once the Soviet Union's front yard.

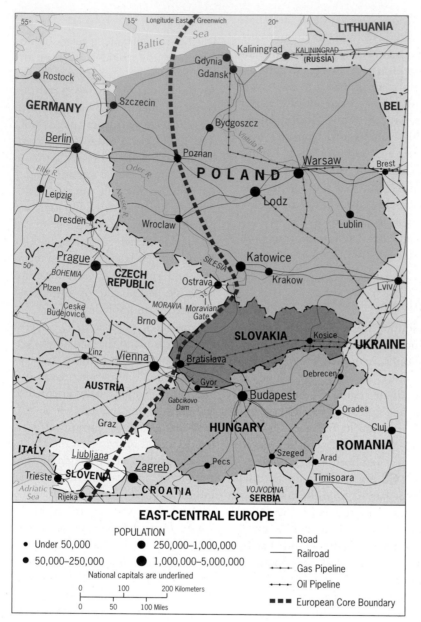

FIGURE 1B-13   © H. J. de Blij, P. O. Muller, and John Wiley & Sons, Inc.

no longer exists, and in some ways we have seen the rebirth of the even older cultural region known as Central Europe or, in the prevailing German language, *Mitteleuropa*. Many of these changes are consolidated with the eastward expansion of the EU. As we saw, the Czech Republic has already become integrated into the Core region, and Slovenia is likely to be the first post-Yugoslavia state to merit similar inclusion. A relatively large number of Eastern countries have already joined the Union while others are knocking on the door.

Things are changing fast in the eastern Periphery.

East-Central Europe

As Figures 1B-1 and 1B-13 show, the four states in this group all share borders with EU-Core country Germany or Austria. Of these four, the most important state is **Poland**, which

was also the largest and most populous country among the ten that joined the EU in 2004. With 38.1 million people and lying between two historic enemies, Poland has borders that have shifted time and again, but its current status may be more durable than in the past. As the map shows, the historic and once-central capital of Warsaw now lies closer to Russia than to Germany, but the country looks west, not east. During the Soviet-communist period, Silesia was its industrial heartland, and Katowice, Wroclaw, and Krakow grew into major industrial cities amid some of the world's worst environmental degradation. The Soviets invested far less in agriculture, collectivizing farms without modernizing technologies and leaving post-Soviet farming in abysmal condition. All this made governing Poland difficult, but the prospect of EU membership (with the promise of EU subsidies) motivated the government to get its house in order. After entering the EU, Poland saw hundreds of thousands of workers leave for jobs in the European Core, but many have returned and the economy is growing.

Government also was the problem in neighboring, landlocked **Slovakia**, where during the communist period the people were far more pro-Soviet than the Czechs next door. Many observers of the 2004 EU expansion wondered whether misgoverned Slovakia should be admitted because the capital, Bratislava, had become synonymous with corruption and inefficiency. Slovakia's Hungarian minority, comprising about 10 percent of the population of 5.4 million and concentrated in the south along the Danube River, was at odds with the Slovak regime. Moreover, additional concerns were raised over reports of mistreatment of the smaller Roma (Gypsy) minority.* But

---

*The Roma constitute Europe's stateless nation, some 8 million people often referred to as "gypsies" who retain a sort of nomadic life-style in a world that rarely welcomes them. Believed to have originated in India, they are found mainly in Bulgaria, Romania, Hungary, Slovakia, and the Czech Republic.

again the promise of EU membership led to some reforms, and after 2004 the economy perked up. Following another downturn in 2007, a more enlightened administration took over, and in 2008 Slovakia surprised many by meeting the terms of admission to the European Monetary Union (EMU), adopting the euro in 2009 before the Poles or the Hungarians could.

Many economic geographers anticipated a bright future for also-landlocked **Hungary** following the collapse of the Soviet Empire and the opening of the door to Europe. The Hungarians (Magyars) moved into the middle Danube River Basin more than a thousand years ago from an Asian source; they have neither Slavic nor Germanic roots. They converted their fertile lowland into a thriving nation-state and created an imperial power that held sway over an area far larger than present-day Hungary. Ethnic Magyar remnants of this Greater Hungary can still be found in parts of Romania, Serbia, and Slovakia (Fig. 1B-14), and the government in the twin-cities capital astride the Danube River, Budapest, has a history of irredentism toward these external minorities.

The concept of **14 irredentism**—a government's support for ethnic or cultural cohorts in neighboring or distant countries—derives from a nineteenth-century campaign by Italy to incorporate the territory inhabited by an Italian-speaking minority of Austria, calling it *Italia Irredenta* or "Unredeemed Italy." One advantage of joining the EU, obviously, was that much of the reason for Hungarian irredentism disappeared. When and if Serbia joins the EU, it may disappear altogether.

With a population of 9.9 million, a distinctive culture, and a considerable and varied resource base, Hungary should have good prospects, and its economic potential was a strong factor in its 2004 admission

Europe's largest minority, the stateless Roma, also are the realm's poorest. Slovakia is one of several European countries with substantial Roma populations, and its government has been criticized by the EU for its treatment of Roma citizens. This depressing photograph of a Roma settlement in Hermanovce shows a cluster of makeshift dwellings virtually encircled by a moat bridged only by a walkway. The village of Hermanovce may itself not be very prosperous, but that moat represents a social chasm between comparative comfort and inescapable deprivation. © Tomasz Tomaszewski/ngs/Getty Images, Inc.

into the EU. But mismanagement, political corruption, and growing indebtedness have set Hungary back. Behind the elegance of Budapest, a primate city nearly ten times the size of Hungary's next-largest urban center, lies an economy still mired in its rural past.

The success story among the four countries in this grouping remains modest, with most of it accounted for by progressive **Slovenia**, which lies wedged against Austria and Italy in the hilly terrain near the head of the Adriatic Sea. With two million people, a nearly homogeneous ethnic complexion, and a productive economy, Slovenia was the first "republic" of the seven that emerged from the collapse of Yugoslavia to be invited to join the EU. Shortly thereafter, Slovenia became part of the euro zone.

The Southeast: Romania and Bulgaria

Just how **Romania** managed to persuade EU leaders to endorse its 2007 accession remains a question for many Europeans. As the Data Table in Appendix B indicates, Romania has some of Europe's worst social indicators: its economy is weak, its incomes are low, its political system has not been sufficiently upgraded from communist times (a number of "apparatchiks" have acquired state assets under the guise of "privatization" and are controlling the political process), and political infighting and corruption are endemic.

But Romania is an important country, located in the lower basin of the Danube River and occupying much of the heart of eastern Europe. With 21.4 million inhabitants and a pivotal situation on the Black Sea, Romania is a bridge between central Europe

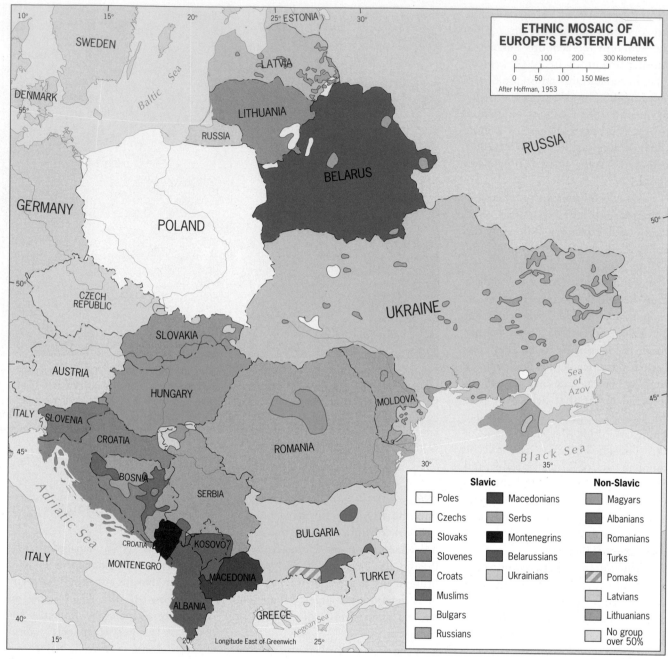

FIGURE 1B-14

© H. J. de Blij, P. O. Muller, and John Wiley & Sons, Inc.

and the realm's southeastern corner, where EU member Greece and would-be member Turkey face each other across land and water (Fig. 1B-15).

Romania's drab and decaying capital of Bucharest (once known as the Paris of the Balkans) and its surrounding core area lie in the interior, linked by rail to the Black Sea port of Constanta. The country's once-productive oilfields have now been fully depleted, about a third of the labor force works in agriculture, poverty is widespread in the countryside as well as the towns, and unemployment is high. Many talented Romanians continue to leave the country in search of opportunities elsewhere.

Across the Danube lies Romania's southern neighbor, **Bulgaria**. The rugged Balkan Mountains form Bulgaria's physiographic backbone, separating the Danube and Maritsa basins. As the map shows, Bulgaria has five neighbors, several of which are in political turmoil.

The Bulgarian state appeared in 1878, when the Russian czar's armies drove the Turks out of this area. The Slavic Bulgars, who form 85 percent of the population of 7.5 million, were (unlike the Romanians) loyal allies of Moscow during the Soviet period. But they did not treat their Turkish minority, about 10 percent of the population, very kindly, closing mosques, prohibiting use of the Turkish language, and forcing Turkish families to adopt Slavic names. Conditions for

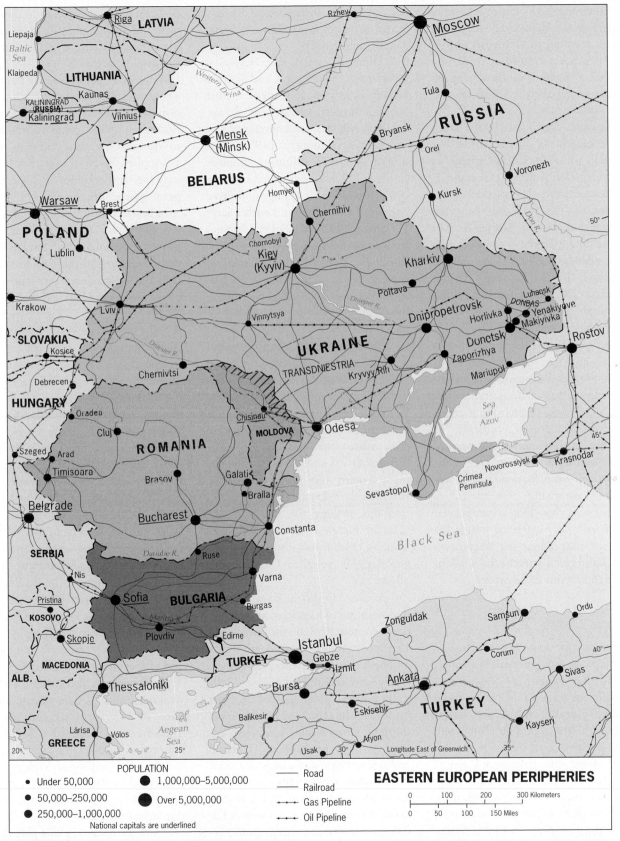

© H. J. de Blij, P. O. Muller, and John Wiley & Sons, Inc.

the remaining Turks improved somewhat after the demise of the Soviet Union. They improved further as Bulgaria became a candidate and then in 2007 a member of the European Union, which required judicial and other social reforms.

Bulgaria has a Black Sea coast and an outlet, the port of Varna, but the main advantage it derives from its coastal location is the tourist trade its beaches generate. The capital, Sofia, lies not on the coast but near the opposite border with Serbia, and in its core-area hinterland some foreign investment is changing the economic landscape—but slowly (Fig. 1B-15). Bulgaria's GNI is on a par with Romania's, although the telltale agricultural sector is much smaller here.

Bulgarians, too, are emigrating in droves—and this worries the countries of western Europe, where an uncontrolled influx of immigrants with newly won access to their job markets would create serious problems. In 2006 the United Kingdom announced that it would place severe restrictions on workers from Romania and Bulgaria—a surprising reversal for a country that has long championed openness in the EU job market. It was a signal that admitting these two states had been a stretch, the wisdom of which remains to be proven.

Europe's Inland Edge

In Europe's far east, and on Russia's doorstep, are three countries of which one, the key state of **Ukraine**, is territorially the largest in all of Europe. Demographically, with a population of 45.6 million, Ukraine also firmly ranks in the second tier of European states, along with Poland and Spain. As Figure 1B-15 shows, Ukraine's relative location is crucial. Not only does it link the core of Russia to the periphery of the European Union: it forms the northern shore of the Black Sea from Russia to Romania, including the strategically important Crimea Peninsula. And most importantly,

oil and gas pipelines connect Russian fields to European markets across Ukrainian territory.

Ukraine's capital, Kiev (Kyyiv), is a major historic, cultural, and political focus. Briefly independent before the communist takeover in Russia, Ukraine regained its sovereignty as a much-changed country in 1991. Once a land of farmers tilling its famously fertile soils, Ukraine emerged from the Soviet period with a huge industrial complex in its east—and with a large (17 percent) Russian minority. Ukraine's boundaries also changed during the Soviet era. In 1954, a Soviet dictator capriciously transferred the entire Crimea Peninsula, including its Russian inhabitants, to Ukraine as a reward for its productivity.

The Dnieper River forms a useful geographic reference to comprehend Ukraine's spatial division (Fig. 1B-15). To its west lies agrarian, rural, mainly Roman Catholic Ukraine; in its great southern bend and eastward lies industrial, urban, Russified (and Russian Orthodox) Ukraine. Soviet planners made eastern Ukraine a communist Ruhr based on abundant local coal and iron ore, making the Donets Basin (*Donbas* for short) a key industrial complex. Meanwhile, the Russian Soviet Republic supplied Ukraine with oil and gas.

As the map shows, Ukraine, with the exception of its eastern and urban-concentrated Russian minority (still 17 percent today), has an ethnically homogeneous population by eastern European standards. Ukraine is a critically important country for Europe's future as it has access to international shipping lanes, a large resource base, massive farm production, educated and skilled labor, and a large domestic market. But it suffers from numerous problems ranging from political mismanagement and corruption to a faltering economy and rising crime.

A presidential election in 2004 first drew international attention to Ukraine's unique political geography. In effect, Ukraine's electorate is divided

between a pro-Western (and pro-EU) west and a pro-Russian east, resulting in a situation reminiscent of Czechoslovakia's before its "velvet divorce." In Ukraine's case, however, the transit of energy supplies creates a complication. Ukraine's own dependence on Russian supplies makes it vulnerable to Moscow's decisions on prices as well as supplies. Some of Ukraine's leaders urge speedy integration into the European sphere; others prefer a middle road between Europe and Russia.

Moldova, Ukraine's small and impoverished neighbor (which is by many measures Europe's poorest country), was a Romanian province seized by the Soviets in 1940 and made into a landlocked "Soviet Socialist Republic." A half-century later, along with other such "republics," Moldova gained independence when the Soviet Union disintegrated. Romanians remain in the majority among its 4.1 million people, but most of the Russians and Ukrainians (each about 13 percent) have moved across the Dniester River to an industrialized strip of land between that river and the Ukrainian border, proclaiming there a "Republic of Transdniestria" (see Fig. 1A-11). But such separatist efforts constitute only one of Moldova's several problems. Its economy, dominated by farming, is in decline; an estimated 40 percent of the population works outside its borders because unemployment in Moldova is typically as high as 30 percent; smuggling and illegal arms trafficking are rife. These many misfortunes translate into weakness, and Russia's malign influence in the form of support for Transdniestria's separatists keeps the country in turmoil. In fact, in 2007 the elected president of Moldova was persuaded by Russian pressure to recognize the regime of breakaway Transdniestria as legitimate.

The third state in this part of the European Periphery in many ways is the most peripheral of all: **Belarus**. Landlocked, autocratic, and misgoverned, sustained in part by the transit of energy supplies from Russia to Europe's Core countries, Belarus has few functional links to Europe and little prospect of progress. More than 80 percent of Belarus's 9.6 million inhabitants are Belarussians ("White" Russians), a West Slavic people; only some 11 percent are (East Slavic) Russians. A small Polish minority occasionally complains of mistreatment and discrimination. Devastated during World War II, Belarus became one of Moscow's most loyal satellites, and the Soviets built Minsk (Mensk), the country's capital, into a large industrial center. But in the post-Soviet era, Belarus has lagged badly, and its government functioned in ways reminiscent of the communist period.

Unlike Ukrainians, Belarussians so far express little interest in joining the EU, and Belarus cannot even be viewed as a functioning part of the European Periphery. On the contrary, its longtime authoritarian leader has made overtures toward Moscow, seeking to join the Russian Federation in some formal way. The Russian leadership, however, has not responded with enthusiasm.

Finally, there is **Turkey**, a mighty country of 77 million that straddles Europe and Asia. Turkey is the successor of the great Ottoman Empire that ruled large parts of Asia, North Africa, and southeastern Europe, but it also experienced notable secularization and Westernization beginning in the 1920s under the charismatic leadership of Kemal Atatürk. It is a Muslim country with considerable influence in the Arab and Islamic worlds, but also a key military ally of the West and a longtime member of NATO. Bordering no less than eight countries, controlling the straits between the Black Sea and the Mediterranean, and facing three seas, Turkey holds a key geostrategic position (Fig. 1B-15). It has been officially associated with the EU for decades and has sought formal entry since 2005. But is Turkey European enough? Or is it perhaps too big (second in population size only to Germany) and potentially destabilizing? Or is the Turkish case less urgent than others because it is already firmly embedded in Western security structures (NATO) led by the United States? At any rate, Turkish membership is on hold. The main reasons, it is said, are the poor treatment of Kurds in eastern Turkey and the apparent inability of the Turkish government to conform to EU economic standards—issues that are not likely to be resolved in the foreseeable future.

With nearly 600 million inhabitants in 40 countries, including some of the world's highest-income economies, a politically stable and economically integrated Europe would be a superpower in any new world order. However, Europe's political geography is anything but stable as devolutionary forces and cultural stresses continue to trouble the realm even as EU expansion proceeds. Europeans have not yet found a way to give voice to collective viewpoints in the world, or to generate collective action in times of crisis. Importantly, even after the expansion of 2007, the EU incorporates only two-thirds of the realm's national economies, and additional enlargement will become increasingly difficult for economic, political, and cultural reasons. Europe has always been a realm of revolutionary change, and it remains so today.

POINTS TO PONDER

- In 2009, the Swiss voted to ban the further construction of minarets on mosques. In 2010, the French were considering a ban on women wearing the *burqa* in public places.
- Anti-immigrant violence rocked southern Italy in 2010, where its agriculture-dependent regions have failed to cope with globalization and competition.
- In 2010, the Autonomous Community of Catalonia outlawed bullfighting, a pointed rebuke to Spain of which it is a part.
- Nicosia (Cyprus) is now "Europe's Last Divided Capital"—and seems ever more likely to stay that way.

St. Petersburg's spectacular canalside Church
of the Bleeding Savior. © H. J. de Blij

IN THIS CHAPTER

Challenges and opportunities of difficult high-latitude environments
Atmospheric warming and changing Arctic prospects
Complexities of managing a multicultural ethnic mosaic
Opening a new era of oil and natural gas production
How Russian nationalism shapes relations with the Near Abroad
Urban continuity and change in post-Soviet Russia

CONCEPTS, IDEAS, AND TERMS

Core area	1
Climatology	2
Continentality	3
Permafrost	4
Tundra	5
Taiga	6
Weather	7
Near Abroad	8

2A

RUSSIA: DEFINING THE REALM

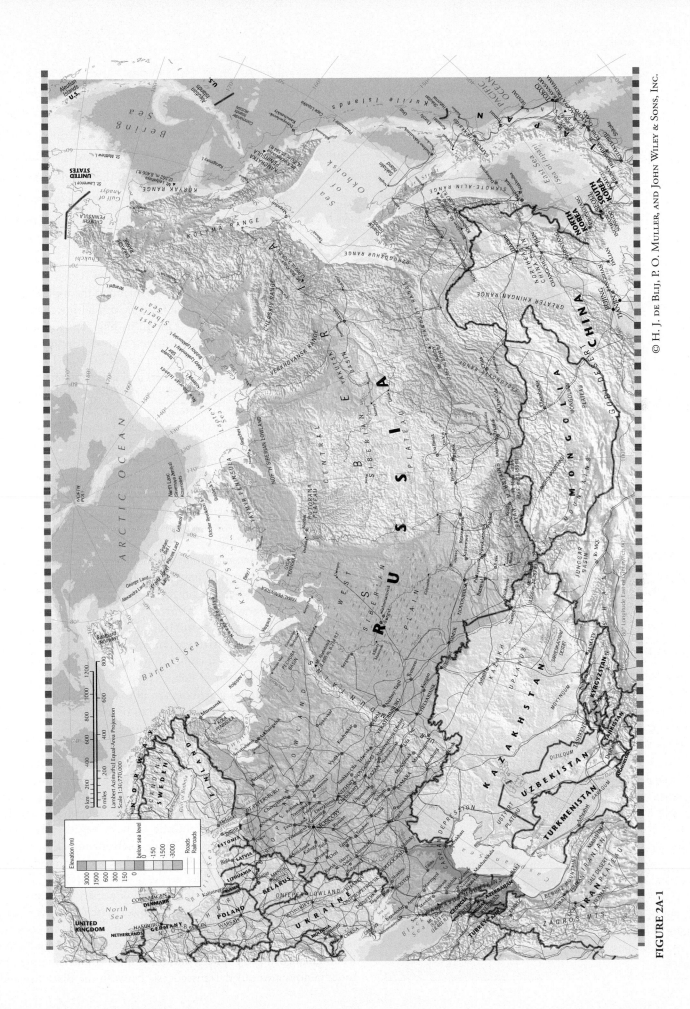

FIGURE 2A-1

Every geographic realm has a dominant, distinguishing feature: Europe's jigsaw-puzzle map of more than 40 countries, South America's familiar triangle, Subsaharan Africa's straddling of the equator, Southeast Asia's thousands of islands and peninsulas. But the Russian geographic realm has what is perhaps the most distinctive hallmark of all—its gigantic territorial size. Almost (but not quite) all of this realm is constituted by the state that dominates it, which is why we call this the Russian realm. By itself, Russia is nearly twice as large as Canada, the world's second-largest country. Russia is three times as large as neighboring Europe. Stretching east-west from the Pacific Ocean to the Baltic Sea, Russia has *eleven* time zones. When Russians have breakfast in Vladivostok, they're getting ready for bed in Moscow.

 MAJOR GEOGRAPHIC QUALITIES

RUSSIA

1. Russia is the largest territorial state in the world. Its area is nearly twice as large as that of the next-ranking country (Canada).

2. Russia is the northernmost large and populous country in the world; much of it is cold and/or dry. Extensive rugged mountain zones separate Russia from warmer subtropical air, and the country lies open to Arctic air masses.

3. Russia was one of the world's major colonial powers. Under the czars, the Russians forged the world's largest contiguous empire; the Soviet rulers who succeeded the czars took over and expanded that empire.

4. For so large a territory, Russia's shrinking population of just over 140 million is comparatively small. The population remains heavily concentrated in the westernmost one-fifth of the country.

5. Development in Russia is concentrated west of the Ural Mountains; here lie the major cities, leading industrial regions, densest transport networks, and most productive farming areas. National integration and economic development east of the Urals extend mainly along a narrow corridor that stretches from the southern Urals region to the southern Far East around Vladivostok.

6. Russia is a multicultural state with a complex domestic political geography. Twenty-one internal Republics, originally based on ethnic clusters, function as politico-geographical entities.

7. Its large territorial size notwithstanding, Russia suffers from land encirclement within Eurasia; it has few good and suitably located ports.

8. Regions long part of the Russian and Soviet empires are realigning themselves in the postcommunist era. Eastern Europe and the heavily Muslim Southwest Asia realm are encroaching on Russia's imperial borders.

9. The failure of the Soviet communist system left Russia in economic disarray. Many of the long-term components described in Chapters 2A and 2B (food-producing areas, railroad links, pipeline connections) broke down in the transition to the post-Soviet order.

10. Russia long has been a source of raw materials but not a manufacturer of export products, except weaponry. Few Russian automobiles, televisions, cameras, or other consumer goods reach world markets.

As Figure 2A-1 reveals and a globe shows even better, Russia's entire northern coast, from the border with Finland in the west to the Bering Strait in the east, faces the Arctic Ocean, with all the climatic consequences you would expect. And for all its huge dimensions, most of the Russian realm lies at high latitudes, shut out by mountains and deserts from the warmer and moister parts of Eurasia. That has been a geographic problem for Russian rulers and governments for as long as Russia has existed as a state. Russia's ports on the Pacific Ocean in the east lie about as far from where most Russians live as they could be. And in the west, where Russia's core area is situated, every maritime exit is hamstrung in some way. The Arctic ports operate on seasonal access. The Baltic Sea ports lie far from the open ocean. The Black Sea ports (some leased from Ukraine) require navigation through narrow straits across Turkish territory just to reach the Mediterranean Sea.

Not surprisingly, a sense of encirclement has much to do with the way many Russians, leaders and ordinary citizens alike, came to view their place in the world. In centuries past, the czars' armies pushed southward in the mountainous corridor between the Black and the Caspian seas, reaching the borders of Turkey and Iran. During the twentieth century, when the Soviet Union put much of Central Asia under Moscow's sway, Russian forces tried to subdue Afghanistan—less than 500 kilometers (300 mi) from the shores of the Indian Ocean. And today, when the Russian government seems to defy the "international community" when it comes to proposed pressure and sanctions on neighbors such as Iran or Belarus, a look at the map can explain a great deal. When those neighbors are all that stands between you and the sea, your geopolitical options tend to be limited.

The Russian realm is vast, but its human numbers are modest. As the Data Table in Appendix B reports, this entire realm's population of less than 160 million is about equal to those of such individual countries as Nigeria and Bangladesh—and what is more, this population has been shrinking rapidly. For a combination of reasons we will examine later, the collapse of the Soviet Union created political, economic, and social conditions that continue to have negative demographic effects today. This issue has come to the forefront among Russia's internal concerns at a time when Russia's leaders want to restore Russia's power and prestige in the wider world. To accomplish this, Russia will need to strengthen its social institutions, diversify its economy, and restore the confidence lost when the Soviet system disintegrated. The Russian geographic realm as defined by our framework consists of four political entities: Russia, essentially the defining component because of its dominance in every sphere, and three small countries located in the mountainous area called Transcaucasia between the Black Sea and the

Caspian Sea: Georgia, Armenia, and Azerbaijan. Small as these three countries are, their combined population exceeds 10 percent of the realm's total, and they are located in a historically turbulent zone of conflict between Russian and non-Russian peoples. They may be overshadowed by Russia, but they are not unimportant.

Take a look at the Russian realm in Figure G-3, and you will note that many of Russia's neighbors, even those countries that formed part of the former Soviet Union, are now absorbed into other realms. The Baltic states, for example, have become an integral part of Europe and the European Union. The Central Asian countries once under Moscow's sway, such as Kazakhstan and Uzbekistan, now form a discrete region in a realm in which Islam is resurgent. But the three Transcaucasian states are not collectively European (although Georgia has European aspirations), or Islamic (only in Azerbaijan does Islam have a strong foothold), or Russian (despite the protection Moscow gave Armenia during Soviet times). What these three states have in common is Russia's still-powerful influence in one form or another. Georgia endures Russian political, economic, and even military intervention; Azerbaijan seeks ways to avoid having to depend on Russian transit for its energy exports; and Armenia continues to view Russia as its most dependable ally. For these reasons, elaborated as we proceed below, we map these states as part of the Russian geographic realm.

Even though our first task is to look in some detail at the giant physical stage on which the Russian geographic realm is built, we should note that the human geography of this realm is not neatly defined by sharp boundaries. Look again at Figure G-3, and you will see that the margins of the Russian realm in two prominent places (and elsewhere not shown at that map's small scale) are marked by ***transition zones***. There, geographic features of this realm spill over into neighboring realms, sometimes creating social and political problems for the adjacent countries affected. But first, let us examine the physical landscapes, climate, and ecologies of the Russian realm.

PHYSICAL GEOGRAPHY OF THE RUSSIAN REALM

Physiographic Regions

The first feature you notice when looking at the physiographic map of the Russian realm is a prominent north-south trending mountain range that extends from the Arctic Ocean southward to Kazakhstan (Figs. 2A-1 and 2A-2), dividing Russia into two parts: the Russian Plain to the west and Siberia to the east. This range, the Ural Mountains, is sometimes designated as the "real" eastern boundary of Europe, but as we noted in Chapters 1A and 1B, Russia is not Europe. Cultural life to the east of the Urals is pretty much the same as it is to the west, and there is no geographic justification for putting the city of Samara (see Fig. 2A-1) in Europe but Chelyabinsk, on the other side of the Urals, outside of it. Neither is European; both are Russian.

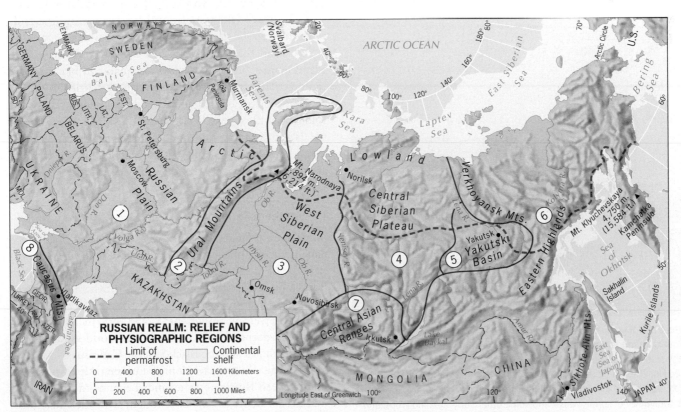

FIGURE 2A-2

© H. J. de Blij, P. O. Muller, and John Wiley & Sons, Inc.

The Russian Plain

The Russian Plain ①, west of the Ural Mountains, is the eastward continuation of the North European Lowland, and here lies Russia's **1** core area. Travel northward from the centrally located capital of Moscow, and the countryside soon changes to coniferous (needleleaf) forests like those of Canada. Centuries ago, those forests afforded protection to the founders of the Russian state when horse-riding invaders from the east swept into this region. The Volga River, Russia's Mississippi, flows southward in a wide arc across the plain from the forested north—but unlike the Mississippi, the Volga drains into a landlocked lake, the Caspian "Sea." Travel southward from Moscow today, and the land is draped in grain fields and pastures. Eastward, the Ural Mountains ② form a range not tall enough to create a major barrier to transportation, but it is prominent because it separates two vast expanses of low *relief* (elevation range).

Siberia

The West Siberian Plain ③, on the other side of the Urals, is often described as the world's largest unbroken lowland, and here the rivers flow north, not south. Over the last 1600 kilometers (1000 mi) of its course to the Arctic Ocean, the Ob River falls less than 90 meters (300 ft) in huge meanders flanked by forests until the trees give way in the northern cold to mosses and lichens. Follow an imaginary trail eastward, and in just about the middle of Russia the West Siberian Plain gives way to the higher relief of the Central Siberian

The village of Tara on the Irtysh River in the Omsk Region in the Siberian Federal District displays little or no evidence of modernization. Czarist-era wooden houses and wobbly sheds are seen here on a sunny summer day; but although this is one of Russia's southernmost administrative units (see Fig. 2B-3), it lies exposed to the harsh Siberian winter, when the river freezes over and you can walk to the inhabited meander knoll seen in the distance. The short growing season allows cultivation of spring wheat and some other early-ripening crops, and small farm settlements like this dot the flat landscape. © Jon Arnold Images Ltd/Alamy

Plateau ④, one of the most sparsely populated areas in the habitable world. Where the West Siberian Plain meets the Central Siberian Plateau, the Yenisey River, also flowing north into the Arctic Ocean, marks the change of landscape. Continuing to go eastward, we encounter the Yakutsk Basin ⑤, drained by the Lena River and forming the last vestige of moderate ***topography*** (surface configuration). Now we approach the realm's mountainous eastern perimeter (see again Fig. 2A-1), where the few roads and railroads must wend their way through tunnels and hairpin valley turns. We map this jumble of ranges collectively as the Eastern Highlands ⑥, but this is one of the most spectacular, diverse, and still-remote regions on Earth.

Kamchatka and Sakhalin

In its farthest eastern reaches, the Russian realm makes contact with the Pacific Ring of Fire (refer to Fig. G-5). There are no active volcanoes in Siberia and earthquakes are rare, but the Kamchatka Peninsula has plenty of both. This is one of the most volatile segments of the Pacific Ring of Fire, with nearly 70 volcanoes, many of them active or dormant, forming the spine of the peninsula. Nothing here resembles Siberia (the climate is moderated by the Pacific waters offshore; the vegetation is mixed, not coniferous). It is one of the most difficult places on Earth to live, and the major town here, Petropavlovsk, has the reputation of being the largest place anywhere without any surface link (other than ocean) to the outside world. There is no road off the peninsula. Petropavlovsk's population has been declining rapidly since the collapse of the old (communist) order.

Also on the Pacific fringe of the Russian realm is an island named Sakhalin, where earthquakes rather than volcanic activity are the main risk. Long a battleground between Russians and Japanese, Sakhalin finally fell to the Russians at the end of World War II. In the postwar period the island and its maritime environs proved to contain major reserves of oil and natural gas, making it a major asset in Russia's energy-based export economy.

The Southern Perimeter

A final look at Figures 2A-1 and 2A-2 reminds us that mountainous topography encircles much more than just the eastern edge of the Russian realm. Some of the highest relief in Eurasia prevails in the southern interior, where the Eastern Highlands meet the Central Asian Ranges ⑦; here lies Lake Baykal in a tectonic trough that is over 1500 meters (5000 ft) deep—the deepest lake of its kind in the world. And between the Caspian Sea and the Black Sea, the Caucasus Mountains ⑧, geologically an extension of Europe's Alps, rise like a wall between Russia and the lands beyond—a multi-tiered wall over which Russians have fought with their neighbors for centuries. When you approach the forbidding Caucasus ranges from the low relief of the Russian Plain to the north, they seem impenetrable, and you realize why so few routes cross them, even today. It is also clear why Russia's armies had such difficulty pushing across the Caucasus into the areas beyond. Even today, these mountains shelter insurgents defying Russia's authority in such territories as Chechnya and Ingushetiya. And the territory beyond is no easy target: to the south of the Caucasus, high relief dominates all the way into neighboring Turkey and Iran.

Harsh Environments

The historical geography of Russia is the story of Slavic expansion from its populous western heartland across interior Eurasia to the east, and into the mountains and deserts of the south. This eastward march was hampered not only by vast distances but also by harsh natural conditions. As the northernmost populous country on Earth, Russia has virtually no natural barriers against the onslaught of Arctic air. Moscow lies farther north than Edmonton, Canada, and St. Petersburg lies at latitude 60° North—the latitude of the southern tip of Greenland. Winters are long, dark, and bitterly cold in most of Russia; summers are short and growing seasons limited. Many a Siberian frontier outpost was doomed by cold, snow, and hunger.

It is therefore useful to view Russia's past, present, and future in the context of its **2** climatology. This field of geography investigates not only the distribution of climatic conditions over the Earth's surface but also the processes that generate this spatial arrangement. The Earth's atmosphere traps heat received as radiation from the Sun, but this greenhouse effect varies over planetary space—and over time. As we noted in the introductory chapter, much of what is today Russia was in the grip of a glaciation until the onset of the warmer Holocene. But even today, with a natural warming cycle in progress augmented by human activity, Russia still suffers from severe cold and associated drought.

Currently, precipitation totals, even in western Russia, range from modest to minimal because the warm, moist air carried across Europe from the North Atlantic Ocean loses much of its warmth and moisture by the time it reaches Russia. Figures G-7, 1A-2, and 2A-3 reveal the consequences. Russia's climatic **3** continentality (inland climatic environment remote from moderating and moistening maritime influences) is expressed by its prevailing *Dfb* and *Dfc* conditions. Compare the Russian map to that of North America (Fig. G-7), and you note that, except for a small corner off the Black Sea, Russia's climatic conditions resemble those of the Upper Midwest of the United States and interior Canada. Along its entire northern edge, Russia has a zone of *E* climates, the most frigid on the planet. In these Arctic latitudes originate the polar air masses that dominate its environments.

The Russian realm's harsh northern climates affect people, animals, plants—and even the soil. In Figure 2A-2 you

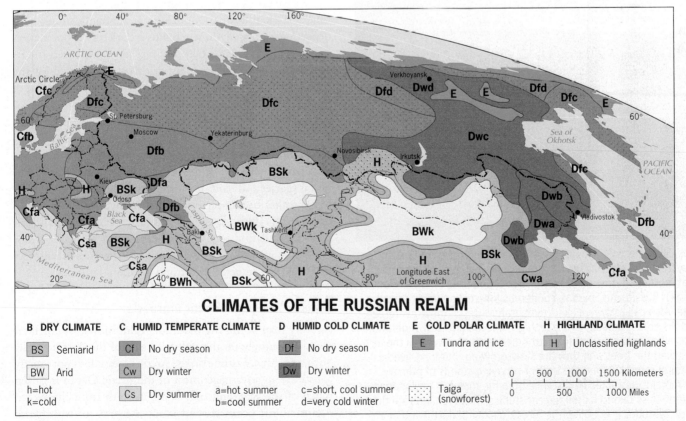

CLIMATES OF THE RUSSIAN REALM

B DRY CLIMATE	**C HUMID TEMPERATE CLIMATE**	**D HUMID COLD CLIMATE**	**E COLD POLAR CLIMATE**	**H HIGHLAND CLIMATE**
BS Semiarid	Cf No dry season	Df No dry season	E Tundra and ice	H Unclassified highlands
BW Arid	Cw Dry winter	Dw Dry winter		
h=hot k=cold	Cs Dry summer			

a=hot summer b=cool summer

c=short, cool summer d=very cold winter

Taiga (snowforest)

0   500   1000   1500 Kilometers

0   500   1000 Miles

FIGURE 2A-3

can find a direct consequence of what you see in Figure 2A-3: a dashed blue line that starts at the shore of the Barents Sea, crosses Siberia eastward, and reaches the coast of the Pacific Ocean north of the neck of the Kamchatka Peninsula. North of this line, water in the ground is permanently frozen, creating an even more formidable obstacle to settlement and infrastructure than the severe weather alone. It is referred to as **4** **permafrost**, and it affects other high-latitude environments as well (Alaska, for example). Where the permafrost ends, seasonal temperature changes cause alternate thawing and freezing, with destructive impacts on buildings, roads, railroad tracks, and pipelines.

Examine Figure 2A-3 carefully, and you will see two unfamiliar terms: *tundra* and *taiga*. **5** **Tundra**, as the map suggests, refers to both climate and vegetation. The blue area mapped as *E* marks the coldest and ice-affected environmental zone in Russia and elsewhere in the high Arctic: this is frigid, treeless, windswept, low-elevation terrain where bare ground and rock prevail and mosses, lichens, bits of low grass, and a few hardy shrubs are all that grows. **6** **Taiga**, the stippled area on the map, a Russian word meaning "snowforest" (also called boreal forest), extends over vast reaches of Eurasia as well as northern North America and is dominated by conif-

erous (as in pine *cones*) trees. As the map shows, the taiga extends southward into relatively moderate environs, and there it becomes a mixed forest of coniferous and deciduous trees. Although taiga prevails from northern Scandinavia to the Russian Far East, this is the vegetative landscape most often associated with Siberia: endless expanses of rolling countryside draped in dense stands of evergreen pine trees.

Climates and Peoples

Climate and weather (there is a distinction: *climate* refers to a long-term average, whereas **7** **weather** describes a set of existing atmospheric conditions at a given place and time) have always challenged the peoples of this realm. The high winds that drive the bitter Arctic cold southward deep into the landmass, the blizzards of Siberia, the temperature extremes, rainfall variability, and short and undependable growing seasons even in the more moderate western parts of this realm have always made farming difficult. That was true during the time of the ruling czars, when the threat of famine never receded. It was also a problem during the Soviet-communist period, when Moscow was the capital of an empire far larger than this realm and the non-Russian sectors of that empire (such as Ukraine and parts of Central Asia) produced food staples Russia needed. Despite a massive effort to restructure agriculture throughout that empire by means of collectivization and irrigation projects, the Russians often had to import grain.

As we noted in the Introduction, and as we will observe time and again throughout this book, humanity's long-term dependence on agriculture remains etched on the population map, even as our planet becomes ever more urbanized and its economy more globalized. By studying the climates of the Russian realm, we can begin to understand what the map of population distribution shows (Fig. 2A-4). The overwhelming majority of the realm's nearly 160 million people remain concentrated in the west and southwest, where environmental conditions were least difficult at a time when farming was the mainstay of most of the people. To the east, the population is sparser and tends to cluster along the southern margin of the realm, becoming even more thinly distributed east of Lake Baykal. If you consider that nearly three-quarters of this realm's population today lives in cities and towns, it is not difficult to imagine just how empty vast stretches of countryside must be—especially in frigid northern latitudes.

Climate Change and Arctic Prospects

As every map of this high-latitude realm shows, its northern coast lies entirely on the poleward side of the Arctic Circle. Such a lengthy coastline on the Arctic Ocean does not exactly constitute an advantage: most of the Arctic Ocean is frozen much of the year, and only some warmth from the North Atlantic Drift ocean current keeps the ports of Murmansk and Arkhangelsk open a bit longer. Ports like this (and even St. Petersburg on the Gulf of Finland, a branch of the Baltic

As Figure 2A-3 shows, Siberia is cold and, in Russia's far northeast, dry as well. In the north lies the *tundra*, treeless and windswept, and beyond is the ice of the Arctic. But where somewhat more moderate conditions prevail (moderate being a relative concept in northwestern Russia and Siberia), coniferous forests known as *taiga* cover the countryside. Also called *boreal* (cold-temperate) forests, these evergreen, needleleaf pines and firs create a dense and vast high-latitude girdle of vegetation across northern Eurasia as well as northern North America. This view from the air shows how tightly packed the trees are; they are slow-growing, but long-lived. Most of the world's taiga forest, among the largest surviving stands of primary forest on the planet, remains protected by distance from the threat of exploitation—but lumbering is nevertheless taking its toll. Recent studies, however, indicate that climate change in high latitudes is enabling the forest to expand northward at a faster rate than it diminishes due to lumbering activities elsewhere, a rare case of good news relating to global warming. © Arcticphoto/Alamy

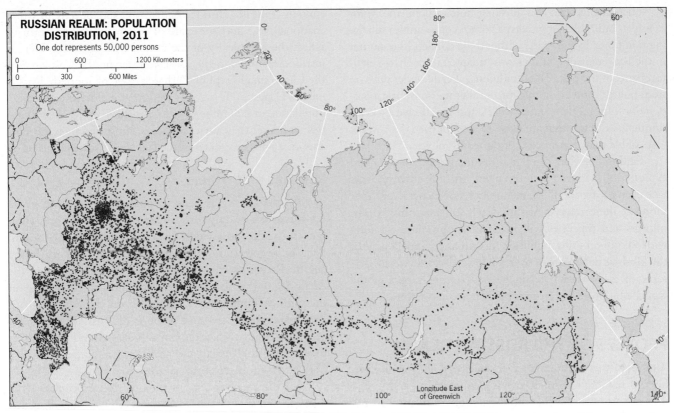

RUSSIAN REALM: POPULATION DISTRIBUTION, 2011
One dot represents 50,000 persons

FIGURE 2A-4

Sea) would never have developed the way they did if Russia had better access to the world oceans, but as we noted, this has been one of the country's historic impediments.

Now it looks as though nature will give Russia a helping hand. If, as most climatologists anticipate, global warming causes long-term melting of large parts of the Arctic Ocean's ice, this body of water will come to play a new and different role in this realm's future. Milder atmospheric conditions may shrink the area of permafrost mapped on Figure 2A-2; moister air masses may improve agriculture on the Russian Plain; warmer water may keep Arctic ports open year-round; and a so-called Russian Maritime Passage may even open up between the Bering Strait and the North Sea, shortening some international sea routes by thousands of kilometers and opening a new era of trade.

That, at least, is how many Russians see global warming—as a potential advantage nature has long denied them. And indeed, the Arctic provides much evidence for such warming: the rapid melting of the Greenland Ice Cap, the shrinking of the average size of the permanent Arctic Ocean ice, and the reduced incidence of icebergs. Russian economic planners look at the shallow waters offshore along the Arctic Ocean coastline (see the white- and light-blue areas in Fig. 2A-1) and hope that oil and gas reserves beneath those waters will come within reach of exploitation.

These developments may even have the effect of expanding the Russian geographic realm far into the Arctic. As we note when we discuss the polar areas in more detail in our final chapter, states with coastlines in the Arctic are likely to

demand certain exclusive rights not only over waters offshore but also the resources on and beneath the ocean floor. It is therefore in their interest to extend these rights as far from shore as possible, even hundreds of (nautical) miles outward. In 2007, the Russian government even placed a metal Russian flag at the North Pole on the seafloor, under the permanent ice of the Arctic Ocean, symbolizing its intentions. The map of the northern perimeter of the Russian realm is changing.

Ecologies at Risk

If, as computer models predict, global warming continues and accelerates even (and perhaps especially) in these polar latitudes, ecologically sensitive environments will be severely affected. Animals as well as humans have long adapted to prevailing climatic conditions, and such adaptations are severely disrupted by environmental change. The intricate web of relationships among species and their environments on the one hand, and among species themselves on the other, can be quickly damaged by temperature change. The polar bear is just one prominent example: it depends on ample floating sea ice to hunt and raise cubs. When sea ice diminishes and longer stretches of open water force polar bears to swim greater distances during the cub-rearing season, fewer cubs make it to adulthood. If ice-free Arctic summers are indeed in the offing, the polar bear could become extinct in your lifetime. Rapid ecosystem change will also endanger seal, bird, fish, and other Arctic populations.

Such changes are also affecting human populations, for example, Inuit (formerly called Eskimo) communities still living traditional lives in parts of the Arctic domain (though not in the Russian realm) and likewise adapted to the harsh environments of the past. Their traditions, already under pressure from political and economic forces resulting from their incorporation into modern states, will be further affected by environmental changes that are likely to alter, or even destroy, the ways of life they developed over thousands of years.

And if a new era of oil and natural gas exploration and exploitation is indeed about to open, offshore environments face even greater peril. As technologies of recovery put ever more of these reserves within reach, oil platforms, drills, pumps, and pipelines will make their appearance—along with risks of oil spills, pollution, disturbance, and damage of the kind we have seen all over the lower-latitude world. The force of globalization is finally penetrating a part of the world long protected from it by distance and nature.

PEOPLES OF THE RUSSIAN REALM

Although Russia's dominance of this geographic realm justifies our naming it as such, this is a culturally and ethnically diverse part of the world whose traditions and customs spill over into neighboring realms even as neighbors have come to live here. Russians still form the majority, but large parts of this realm are home to non-Russian peoples—and not just along the borders. As Figure 2A-5 shows, the Russian realm

contains Finnish, Turkic, Armenian, and dozens of other "nationalities," and the scale of this map cannot begin to reflect the complexity of the overall ethnic mosaic. Here is just one example: in the southwest, facing the Caspian Sea, there is a small "republic" that is part of modern Russia, named Dagestan. It is about half the size of Maine and has a population of 2.8 million—with nearly 30 ethnic "nationalities" speaking their own languages, most of which are variants of languages spoken in the Caucasus, Turkey, and Iran.

Russians and Others

As the map shows, the Slavic peoples known collectively as the Russians not only form the majority of the population but also are the most widely dispersed. Although the Russian Plain is the core area of the Russian state and was its historic hearth, Russian settlement extends from the shores of the Arctic Sea to the Black Sea coast and from St. Petersburg on the Gulf of Finland to Vladivostok on the Sea of Japan. Discontinuous nuclei and ribbons of Russian settlement are scattered across Siberia, but as we noted in Figure 2A-4, the population, Russian and non-Russian, tends to concentrate in the southern sector of the realm.

But the Russians are not the only Slavic peoples. More than a thousand years ago, when many ethnic groups struggled to establish themselves in Europe, the early Slavs, whose original home may have been on the North European Plain in the area north of the Carpathians, achieved stability and security in a homeland that expanded steadily, not only eastward into

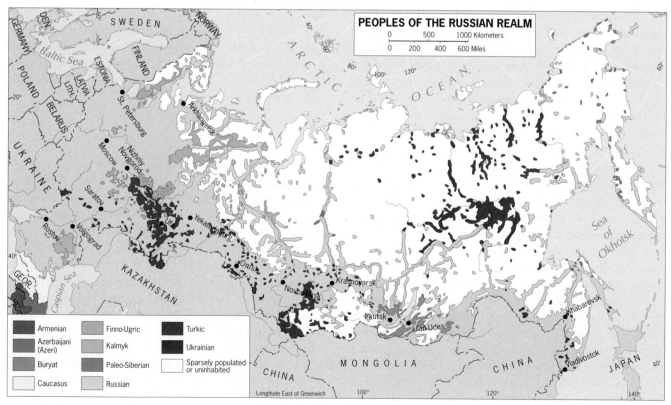

FIGURE 2A-5

© H. J. de Blij, P. O. Muller, and John Wiley & Sons, Inc.

what is today Russia, but also westward and southwestward into the valley of the Danube and beyond. As we noted in Chapters 1A and 1B, the Serbs, Croats, Slovaks, and Czechs are Slavic peoples, as are the Poles, Ukrainians, and Belarussians—among others. Today, you can discern evidence of this distant past in the Slavic languages these descendants speak, which still have much in common despite centuries of divergence.

Eventually, present-day Ukraine became the stage for another Slavic success story, the founding of small, well-defended states (each of which was referred to as a *Rus*) from which, in a geographic saga on which we will focus later, the vast Russian state was to evolve. But by that time, many other peoples had migrated into the Russian domain, and from a distant territory, Mongolia, came invaders who tried to stamp the Russes out. Figure 2A-5 is a partial record of those invasions and attacks (of which the Nazi assault during World War II was only the latest).

A Complex Cultural Mosaic

But even as the early Russian state consolidated, non-Russian peoples incubated in (and beyond) the Caucasus Mountains in the corridor between the Black and Caspian seas. Georgians, Armenians, Azeris, and many others created a jigsaw puzzle of nationalities in this region (the aforementioned Dagestan is only one small part of it). Then, after about AD 1400, Islam propagated by the Ottoman Empire pushed northward here, and by 1500 Islam's vanguards were challenging Slavic peoples along the north shore of the Black Sea. Remnants of such advances and invasions still remain on the cultural map (Fig. 2A-5): look at the crescent of Turkic peoples (shaded red) extending from near the city of Nizhniy

Novgorod to the border of Kazakhstan, and you see part of the evidence. Today, Russia has a far higher percentage of Muslims in its population than western European countries do; although data are not precise, the consensus is that about 12 percent of Russia's inhabitants are Muslims.

Long before Islam appeared on the scene, Slavic peoples from one end of their domain to the other had accepted the teachings of the Eastern Orthodox Church, and in the Russian realm the Russian church became Eastern Orthodoxy's dominant institution. It stayed that way until the triumph of communist revolutionaries who, in 1917, put an end to rule by the Russian czars and began the disestablishment of the church. For seven decades, atheism was official policy—until the communist system disintegrated. Since the early 1990s, the Russian Orthodox Church has made a vigorous comeback, attended by unrestrained nationalist and ethnic propaganda. The postcommunist-era Patriarch Alexy II in his sermons often asserted that Orthodox beliefs and "the Slavic soul" are one and the same, so that the former is the spiritual and cultural foundation of the latter. In an increasingly heterogeneous society with Islamic, Buddhist, and other non-Christian constituencies, such doctrinaire assertions have worrisome implications for the future.

CITIES NEAR AND FAR

When the czars ruled Russia, people living in countryside villages and homesteads far outnumbered those residing in cities and towns. It is not that the czars wanted to keep it that way, but their policies—social, economic, and otherwise—obstructed change and impeded opportunities that draw

The landscape of Tula, a city of 500,000 about 150 kilometers (100 mi) south of Moscow, is typical of urban centers in western Russia. Tula's townscape today is a mixture of pre-Soviet historic buildings, drab Soviet-era tenements, and scattered post-Soviet high rises and single-family houses. The urban cultivators in the foreground are a common sight in a country with limited agricultural land and short growing seasons. Fresh produce is particularly coveted and is widely used for bartering among friends and neighbors in exchange for other hard-to-get commodities. Informal farming on fertile plots of urban land has increased since 1991 as the redistribution of population has channeled rural migrants into cities, where their slow absorption into urban society often requires them to fall back on their food production and preservation skills to supplement their diets and make meager incomes stretch further. © MauroGalligani/Contrasto/ReduxPictures

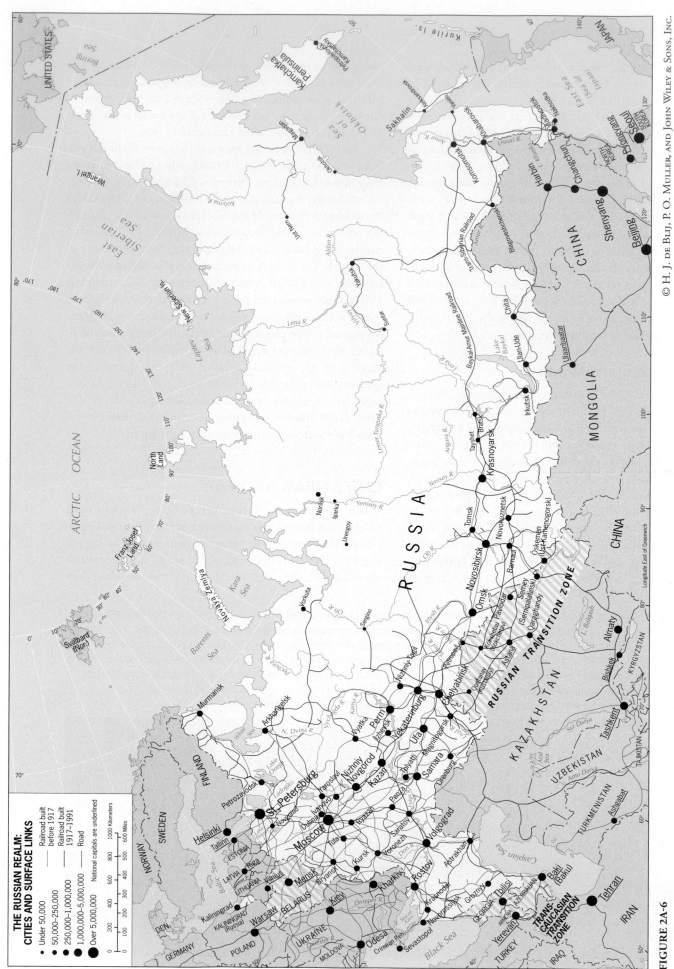

THE RUSSIAN REALM: CITIES AND SURFACE LINKS

- Under 50,000
- 50,000–250,000
- 250,000–1,000,000
- 1,000,000–5,000,000
- Over 5,000,000

Railroad built before 1917
Railroad built 1917–1991
Road
National capitals are underlined

0   100 200 300 400 500 600 Miles
0   200 400 600 800 1000 Kilometers

FIGURE 2A-6

© H. J. de Blij, P. O. Muller, and John Wiley & Sons, Inc.

people to urban areas. Even under Peter the Great, who admired urbanizing Western Europe and wanted to make St. Petersburg a glittering Russian window on the wider world, Russia's urbanization lagged far behind Europe's. Compare Russia's level of urbanization today (73 percent) to that of western Europe (84 percent), and this contrast endures. Figure 1A-5 reminds us that the Industrial Revolution came late to this realm, and that its impact was slowed—until the revolutionary communist organizers who succeeded the czars in the 1920s undertook a massive effort to catch up.

The three countries that form the Transcaucasus region are even less urbanized than Russia: in Azerbaijan and Georgia barely over 50 percent of the people live in urban areas, and even in Armenia urbanization lags (see the Data Table in Appendix B to compare these figures to, say, East Asia or South America). Here the twentieth-century period of Russian influence did little to dislodge rural traditions or stimulate change.

Nevertheless, the cities, when mapped by size (Fig. 2A-6), reveal much about this realm's regional geography. Dominant Moscow and northerly but coastal St. Petersburg anchor the Russian core area. Historic urban centers like Novgorod, Kazan, and Yekaterinburg mark crucial episodes in Russia's historical geography. The cities on the Volga River spearheaded the country's post-czarist transformation. As the ribbon of cities and towns thins out eastward,

there are the names we associate with the communist Soviet Union's industrial might and heroic resistance against Nazi penetration: Omsk, Krasnoyarsk, and chill-inducing Novosibirsk. In the Russian Far East, Vladivostok symbolizes Soviet naval power and more recent Russian indifference. On the Kamchatka Peninsula, people by the tens of thousands are leaving Petropavlovsk, once an outpost favored by the Soviets, now abandoned by Moscow.

In Transcaucasia, the three leading cities, also the capitals of their respective republics, all signify a singular and dominant issue: oil and its export routes in Baki (Baku), Azerbaijan; conflict with Russia in Tbilisi, Georgia; landlocked weakness in Yerevan, Armenia. Just as we could not understand the regional geography of Europe without regarding London as a world city, Rome as a historic center, and Brussels as a symbol of economic union, we should take note of the major cities of the Russian realm and their roles in the human saga.

A REALM IN TRANSITION

Realms and regions, as we noted in our Introduction, change over time. Three decades ago, then-modern Russia lay at the center of an empire, ruled from Moscow, that incorporated 14 Soviet Socialist Republics (SSRs) in addition to Russia itself (Fig. 2A-7). This contiguous, essentially colonial empire,

FIGURE 2A-7 © H. J. de Blij, P. O. Muller, and John Wiley & Sons, Inc.

known as the Soviet Union, was administered and economically guided by Russian expatriates and local collaborators who created an integrated and centrally planned economy. As in the case of other colonial powers, Russians by the millions moved from their homeland to such places as Estonia, Latvia, Ukraine, and Kazakhstan to further Soviet objectives and to forge a new ideological and material era. They built cities and towns, dams and irrigation systems, they laid out collective farms and transport networks, and they forged what was a Soviet geographic realm built on ideology and socialist design. The Soviet geographic realm seemed stable and secure, its influences radiating far beyond its borders into Eastern Europe and Asia.

When the Soviet Union collapsed in 1991, its members quite suddenly became sovereign states, but this did not mean that Russian influences—or indeed the Russian presence—ended simultaneously. Although many Russian nationals returned from the former SSRs, millions of others stayed on. They lived in areas where Russians were in the majority (for example, in eastern Ukraine and in northern Kazakhstan) and where post-Soviet life was not all that different from what it was during Moscow's rule. But elsewhere, Russians who stayed in the former SSRs found themselves mistreated in various ways, and often they appealed to Moscow for help. Thus the concept of a Russian "Near Abroad"

found its way into Russian discourse. Beyond Russia's borders but of concern to Russia's leaders and people, the **8 Near Abroad** became geographic shorthand for a Russian sphere of influence, where Moscow stood ready to help its kin in case of trouble. On the map, the most vivid example of this is the so-called Russian Transition Zone in northern Kazakhstan (Fig. 2A-6), which was tightly integrated into the adjacent Russian sphere during Soviet times. (A larger-scale map would show similar, but smaller, Russian clusters in Estonia, Latvia, Moldova, and elsewhere.)

In addition, the Soviet period produced minority issues similar to those arising in former Western colonial empires. During Moscow's rule, some local minorities cooperated with the communist leaders while others resisted. Some paid dearly for their opposition, for instance the Chechens, who were summarily exiled from their homeland in and near the Caucasus, at a huge cost in lives, to Central Asia. Others benefited from cooperating, for example the South Ossetians, who would have preferred continued Soviet rule to independence under Georgian government. After the disintegration of the Soviet Union, these minorities formed another concern for Moscow in the Near Abroad. In the case of South Ossetia, in 2008 the Russians actually went to war on behalf of this minority, their first armed intervention in the Near Abroad.

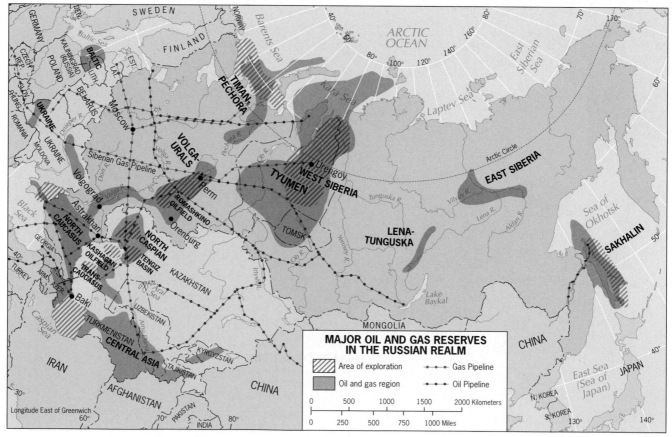

FIGURE 2A-8

© H. J. de Blij, P. O. Muller, and John Wiley & Sons, Inc.

Thus the boundaries of the Russian geographic realm are in some places transitional and in others quite sharply defined. As we will see in Chapter 2B, there is potential for further adjustment elsewhere on the map.

RESOURCES OF THE REALM

Given the huge territorial dimensions of the Russian realm, it is not surprising that its natural resources are vast and varied. The Russians today and excited about the energy reserves that are to be found and exploited beneath Arctic waters, but the Russian realm already is a major producer and exporter of oil and natural gas from a growing number of fields dispersed across the realm from the North Caucasus in the west to the island of Sakhalin in the east and from western Siberia in the north to the Caspian Basin in the south (Fig. 2A-8). And oil and natural gas are not the only underground riches of this realm. Deposits of so many minerals lie within this realm that virtually all of the raw materials required by modern industry are present. Nor are oil and natural gas the only energy reserves. Major coalfields are found east as well as west of the Urals, and in Siberia as well as in more southerly latitudes (the string of mines producing high-quality coal along the Trans-Siberian Railroad corridor were vital in Russia's industrial development and its successful was against Nazi Germany). Large deposits of iron ore, too, lie widely scattered across this enormous realm, from the so-called Kursk Magnetic Anomaly near the border with Ukraine to the Kola Peninsula in the Arctic north. And when it comes to other metals, this realm has it all—or almost all—from gold to lead and from platinum to zinc. Moreover, one of the world's great aggregations of *nonferrous* (non-iron) metals lies in and around the Ural Mountains, where Russia's metallurgical industries emerged.

And there is likely to be much more. Vast reaches of this far-flung realm have not yet been fully explored; in particular, the potential of the region numbered ⑥ in Figure 2A-2 is significant in global terms. As we shall see in Chapter 2B, the comparatively nearby Japanese, always in search of raw materials, have long tried to persuade the Russians to allow them to search for and develop new reserves, but political obstacles have deferred these initiatives. Perhaps the time will yet come.

Given such material assets, you would imagine that this realm has always been in the forefront of industrial production and economic diversification. But as we will also see, the availability of raw materials is only one part of such development. You are likely to buy Chinese and Japanese products on an almost daily basis—but you are likely to own few if any items of Russian manufacture. That is part of the story we relate in Chapter 2B.

POINTS TO PONDER

- High-speed train travel, badly needed in this enormous realm, has been slow to develop. In 2010 the first such route opened between the capital, Moscow, and the leading port and second city, St. Petersburg.

- Global warming produces winners as well as losers, and the Russians regard themselves as potential winners as Siberia's harsh environments moderate and the Arctic Ocean's ice recedes, opening up sea lanes and submerged energy reserves.

- Although the population of this realm has been declining overall, its three southernmost (Transcaucasian) countries continue to show an annual increase.

- The National Geographic Society's famous *Atlas of the World* says that the western portion of this realm is geographically a part of Europe, but the eastern segment is not.

Moscow's heart, with the Kremlin (right) fronting on adjacent Red Square. © Travelwide/Alamy

IN THIS CHAPTER

Russia's colonies before, during, and after the reign of the Soviet Empire
The post-Soviet experiment with federal governance
How Russia copes with distance and isolation
Where Russia meets China, Japan, and the Korean Peninsula
Moscow and the challenge of terrorism
Trouble in Transcaucasia

CONCEPTS, IDEAS, AND TERMS

Forward capital	1
Colonialism	2
Imperialism	3
Russification	4
Federation	5
Collectivization	6
Command economy	7
Unitary state system	8
Distance decay	9
Population implosion	10
Core area	11
Centrality	12
Double complementarity	13

2 B

RUSSIA: REGIONS OF THE REALM

The Russian Core
The Eastern Frontier
Siberia
The Russian Far East
Transcaucasia

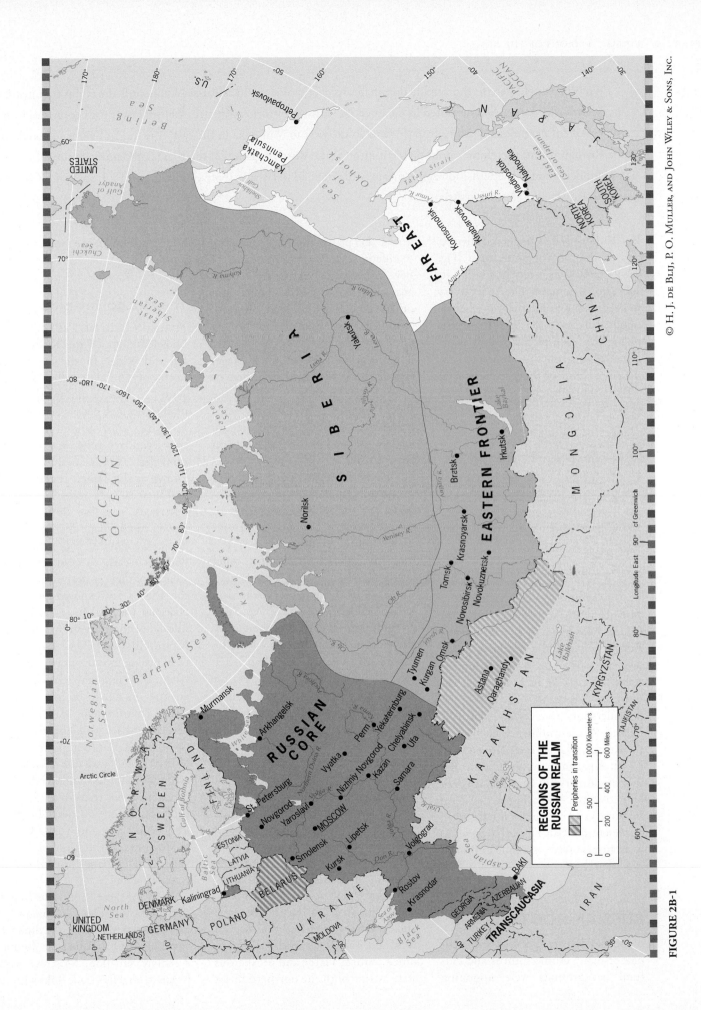

REGIONS OF THE RUSSIAN REALM

Peripheries in transition

1000 Kilometers
600 Miles

500   400
200
0   0

© H. J. de Blij, P. O. Muller, and John Wiley & Sons, Inc.

FIGURE 2B-1

No other single state so thoroughly dominates its geographic realm as Russia commands the realm that justifiably carries its name. And, as noted in Chapter 2A, Russia's historic influences still spill over into countries forming parts of neighboring realms from Estonia in the west to Mongolia in the east. The most obvious regionalization of the Russian realm, therefore, would divide it into (1) the Russian giant, (2) the three comparatively tiny Transcaucasian republics, and (3) the transition zones most vividly defined in Belarus (between Russia and Europe) and Kazakhstan (between Russia and Southwest Asia).

But Russia itself is huge and diverse, and as the physiographic map (Fig. 2A-2) showed us, even Siberia has its subregional components. So let us try to keep it simple and delimit what we might call first-order regions, recognizing that each of this realm's enormous regions contains numerous subdivisions. Four of the five regions we discuss lie entirely within Russia (Fig. 2B-1):

1. The Russian Core
2. The Eastern Frontier
3. Siberia
4. The Russian Far East

and only one lies outside the Russian state:

5. Transcaucasia

Environment—in the form of climate, relief, vegetation, and other factors—has much to do with the way Russia evolved as the dominant power over this vast realm. To better understand the map, let us briefly trace the historical geography of Russia, its defenders and rulers who eventually pushed Russian influence far beyond the Russian Plain—even into western North America. It comes as a surprise to many Americans that Russians not only colonized Alaska, but even reached California and built a fort not far north of present-day San Francisco. So let us focus first on place and time in Russia's evolution.

RUSSIAN ROOTS

A thousand years ago, Eurasian peoples of many ethnic sources and cultural backgrounds were migrating across the plains south of the *taiga* (coniferous snowforest) in search of new and secure homelands; Scythians, Sarmatians, Goths, Huns, and others came, fought, settled, survived, absorbed neighbors, or were driven off. Eventually the Slavs, comparative latecomers to this turbulent stage, established settlements in the area of present-day Ukraine, north of the Black Sea and in the southwestern corner of the physiographic region we have defined as the Russian Plain (Fig. 2A-2).

Here the Slavic peoples found fertile soils, a relatively moderate climate, and a physical landscape with advantages (the Dnieper River and its tributaries, stands of forest) and opportunities (relatively low relief and easy contact among settlements). The Slavs used the name **Rus** to designate such settlements, and the largest and most successful of these early Russes was the one located where the capital of modern Ukraine, Kiev (Kyyiv), lies today. This is one reason why many "Russians" today cannot imagine an independent, Europe-oriented Ukraine: this, after all, is their historic heartland. And from this southern base, the Slavs expanded their domain into the Russian Plain, establishing their northern headquarters at Novgorod on Lake Ilmen (see Fig. 2A-1). This northern Rus was well positioned to benefit from the trade between the Hanseatic ports on the Baltic Sea and the trading centers on the Black and Mediterranean seas. During the eleventh and twelfth centuries, the Kievan Rus and the Novgorod Rus combined to form a large and prosperous state astride both the northern forest and southern *steppe* (semiarid grassland).

The Mongol Invasion

Prosperity attracts attention, and knowledge about the Russes spread far and wide. In the distant east, north of China, another successful state had been building: the empire of the Mongol peoples under Genghis Khan. Also under the sway of this legendary ruler was a group of nomadic, Turkic-speaking peoples known as the Tatars. Together, Mongol-Tatar armies rode westward on horseback into the domain of the Russes to challenge the power of the Slavs. Geography had much to do with their early success: on the open steppes of the southern Russian Plain, the Russes lay exposed to the fast-charging Mongol forces, and by the middle of the thirteenth century, the Kievan Rus had fallen. Slavic refugees were fleeing into the northern forests, where they reorganized to face their enemies in newly built Russes. The forest environment was their ally: Mongol tactics were effective for the open plain, but not in the woods. What ensued was not a victory for either side, but a standoff: the Tatars threatened and tried to lay siege, but they could not win outright. So the leaders of the forest-based Russes paid tribute to the Mongol-Tatar invaders, in exchange for which they were left alone.

One of these Russes was Moscow, deep in the forest on the Moscow River and destined to become the capital of a vast empire. On a defensible site and in a remote situation, Moscow's leaders began to establish trade links with even

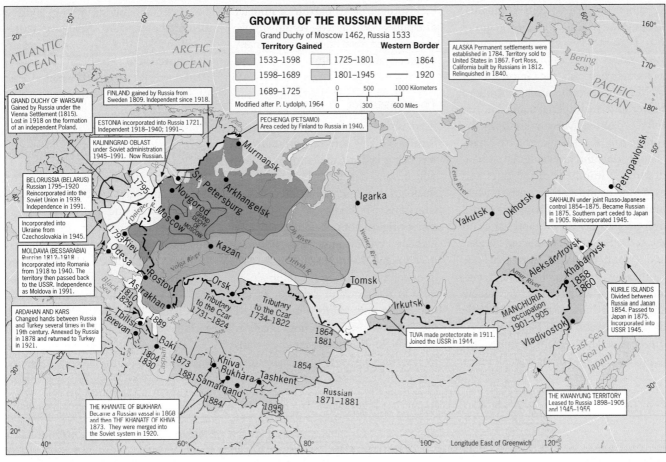

FIGURE 2B-2

© H. J. de Blij, P. O. Muller, and John Wiley & Sons, Inc.

safer Novgorod, and the city grew and thrived. When the Mongols, worried about Moscow's growing power and influence, attacked and were successfully repulsed, Moscow emerged as the leading Rus among Russes, its destiny assured. The Mongol-Tatar campaign had essentially failed.

Before we go on, we should take note of a crucial geographic development. Whereas the Mongols and Tatars fought together, their failure to conquer the Slavic-Russian heartland had both cultural and geographic consequences that have endured into the present. Although most of the Mongols withdrew following their failed fourteenth-century assault on Moscow, many Tatars remained on what, at the time, was the periphery around the Slavic/Russian core in such places as the Volga River Basin, the Crimean Peninsula, and other

smaller pockets. There they were converted in the wave of Islam that was spreading into the Black Sea area and beyond, creating a new kind of division between Christian Slavs and their old-time adversaries.

Grand Duchy of Muscovy: Dukes and Czars

During the fourteenth century, as we noted, the "Grand Duchy of Muscovy," ruled by leaders called princes or dukes, rose to preeminence among the Russes which, following the defeat of the Tatars, subdued other Russes whose rulers saw themselves as candidates for Russian leadership; this extended Moscow's trade links from the Baltic to Black Sea shores. Moscow's religious links with the leadership of the Eastern Orthodox Church in Constantinople also helped secure

its role despite some serious internal strife. But in the mid-fifteenth century there began a period lasting more than three centuries during which powerful, often despotic Russian rulers etched their imprints on the map of Russia, of Eurasia, and, by extension, the world (Fig. 2b-2).

No account of Russia's historical geography should ignore the role of one of Russia's less famous but enormously influential rulers. Ivan III, also known as *Ivan the Great* (ruled 1462–1505), did what the emerging Russian state needed: he consolidated Moscow's dominance not just over the Russes of the old forested heartland but also over Slavic peoples far and wide, including those that had drifted away during and following the Mongol invasion (for example, the Kievan Russes) as well as those never effectively under Moscow's sway, such as those in Belarus and Lithuania. During the sixteenth-century reign of Ivan IV, better known as the infamous *Ivan the Terrible* and often referred to as Russia's first czar, the Grand Duchy of Muscovy became a major military power and an imperial state. He not only expanded Moscow's empire by conquering the Islamic (Tatar) Khanate of Kazan and annexing Astrakhan, destroying hundreds of mosques and executing countless thousands of Muslims, but he also tried to gain control over present-day Estonia and Latvia on the Baltic Sea, thereby drawing Sweden and Poland into war. After that, the Crimean-based Muslim Tatars retaliated by pushing all the way to Moscow and burning most of the city. The rule of Ivan IV was a time of almost continuous warfare, but he acquired his reputation because of the reign of terror he instituted in his pursuit of military discipline, centralized administrative control, and retaliation against many members of the nobility.

The Cossacks

Even as these events were unfolding in Russia's evolving core area, groundwork was being laid for the empire's expansion beyond the Urals by a relatively small group of seminomadic peoples who came to be known as Cossacks and whose original home was in present-day Ukraine. Opportunists and pioneers, they sought the riches of the eastern frontier, chiefly fur-bearing animals, as early as the sixteenth century. By the mid-seventeenth century they reached the Pacific Ocean, defeating Tatars in their path and consolidating their gains by constructing *ostrogs* (strategic fortified waystations) along key river courses. Before the eastward expansion terminated in 1812 (Fig. 2B-2), the Russians had moved across the Bering Strait into Alaska and far down the western coast of North America into what is now northern California.

Czar Peter the Great

When Peter the Great became czar (he ruled from 1682 to 1725), Moscow already lay at the center of a great empire—great, at least, in terms of the territories it controlled. The Islamic threat had been ended with the defeat of the Tatars. The influence of the Russian Orthodox Church was represented by its distinctive religious architecture and powerful bishops.

Peter consolidated Russia's gains and endeavored to make a modern,

 FROM THE FIELD NOTES...

© H. J. de Blij

"Not only the city of St. Petersburg itself, but also its surrounding suburbs display the architectural and artistic splendor of czarist Russia. The czars built opulent palaces in these outlying districts (then some distance from the built-up center), among which the Catherine Palace, begun in 1717 and completed in 1723 followed by several expansions, was especially majestic. During my first visit in 1994, the palace, parts of which had been deliberately destroyed by the Germans during World War II, was still being restored; large black and white photographs in the hallways showed what the Nazis had done and chronicled the progress of the repairs during communist and post-communist years. A return visit in 2000 revealed the wealth of sculptural decoration on the magnificent exterior (left) and the interior detail of a set of rooms called the 'golden suite', of which the ballroom (right) exemplifies eighteenth-century Russian Baroque at its height."

www.conceptcaching.com

European-style state out of his loosely knit country. He built St. Petersburg as a **1 forward capital** on the doorstep of Swedish-held Finland, fortified it with major military installations, and made it Russia's leading port.

Czar Peter the Great, an extraordinary leader, was in many ways the founder of modern Russia. In his desire to remake Russia—to pull it from the forests of the interior to the waters of the west, to open it to outside influences, and to relocate its population—he left no stone unturned. Prominent merchant families were forced to move from other cities to St. Petersburg. Ships and wagons entering the city had to bring building stones as an entry toll. The czar himself, aware that to become a major power Russia had to be strong at sea as well as on land, went to the Netherlands to work as a laborer in the famed Dutch shipyards to learn the most efficient method for building ships. Meanwhile, the czar's forces continued to conquer people and territory: Estonia was incorporated in 1721, widening Russia's window to the west, and in Siberia major expansion soon occurred south of the city of Tomsk (Fig. 2B-2).

Czarina Catherine the Great

Under Czarina Catherine the Great, who ruled from 1760 to 1796, Russia's empire in the Black Sea area grew at the expense of the Ottoman Turks. The Crimea Peninsula, the port city of Odesa (Odessa), and the entire northern coastal zone of the Black Sea fell under Russian control. Also during this period, the Russians made a fateful move: they penetrated the area between the Black and Caspian seas, the mountainous Caucasus with its dozens of ethnic and cultural groups, many of which were Islamized. The cities of Tbilisi (now in Georgia), Baki (Baku) in Azerbaijan, and Yerevan (Armenia) were captured. Even-

tually, the Russian push toward an Indian Ocean outlet was halted by the British, who held sway in Persia (modern Iran), and also by the Turks.

Meanwhile, Russian colonists had entered Alaska, founding their first North American settlement at Kodiak Island in 1784. As they moved southward, numerous forts were built to protect their tenuous holdings against indigenous peoples. Eventually, they reached nearly as far south as San Francisco Bay where, in the fateful year of 1812, they erected Fort Ross.

Catherine the Great had made Russia a colonial power, but the Russians eventually gave up on their North American outposts. The sea-otter pelts that had attracted the early pioneers were running out, European and white American hunters were cutting into the profits, and indigenous American resistance was growing. When U.S. Secretary of State William Seward offered to purchase Russia's Alaskan holdings in 1867, the Russian government immediately agreed—for $7.2 million. Thus Alaska and its Russian-held panhandle became U.S. territory and, ultimately in 1959, the forty-ninth State.

A Russian Empire

Although Russia had withdrawn from North America, Russian expansionism during the nineteenth century continued in Eurasia. While extending their empire southward, the Russians also took on the Poles, old enemies to the west, and succeeded in taking most of what is today the Polish state, including the capital of Warsaw. To the northwest, Russia took over Finland from the Swedes in 1809.

Penetration of Muslim Asia

During most of the nineteenth century, however, the Russians were preoccupied with Central Asia—the

region between the Caspian Sea and western China—where Tashkent and Samarqand (Samarkand) came under St. Petersburg's control (Fig. 2B-2). The Russians here were still bothered by raids of nomadic horsemen, and they sought to establish their authority over the Central Asian steppe country as far as the edges of the high mountains that lay to the south. Thus Russia gained many Muslim subjects, because this was Islamic Asia they were penetrating. Under czarist rule, though, these people retained some autonomy.

Confronting China and Japan

Much farther to the east, a combination of Japanese expansionism and a decline of Chinese influence led Russia to annex from China a number of provinces beyond the Amur River. Soon thereafter, in 1860, the Russians founded the port of Vladivostok on the Pacific.

Now began the events that were to lead to the first involuntary halt in the Russian drive for territory. As Figure 2B-1 shows, the most direct route from western Russia to the port of Vladivostok lay across northeastern China, the territory then still called Manchuria. The Russians had begun construction of the Trans-Siberian Railroad in 1892, and they wanted China to permit the track to cross Manchuria. But the Chinese resisted. Then, taking advantage of the Boxer Rebellion in China in 1900 (see Chapter 9B), Russian forces occupied Manchuria so that railway construction might proceed.

This move, however, threatened Japanese interests in this area, and the Japanese confronted the Russians in the Russo-Japanese War of 1904–1905. Not only was Russia defeated and forced out of Manchuria: Japan even took possession of the southern part of Sakhalin Island, which they named Karafuto and retained until 1945.

THE COLONIAL LEGACY

Thus Russia, like Britain, France, and other European powers, expanded through **2** colonialism. Yet whereas the other European powers expanded overseas, Russian influence traveled overland into Central Asia, Siberia, China, and the Pacific coastlands of the Far East. What emerged was not the greatest empire but the largest territorially contiguous empire in the world. At the time of the Japanese war, the Russian czar controlled more than 22 million square kilometers (8.5 million sq mi), just a tiny fraction less than the area of the Soviet Union after the 1917 Revolution. Accordingly, the communist empire, to a large extent, was the legacy of St. Petersburg and European Russia, not the product of Moscow and the socialist revolution.

The czars embarked on their imperial conquests in part because of Russia's relative location: Russia always lacked warm-water ports. Had the Revolution not intervened, their southward push might have reached the Persian Gulf or even the Mediterranean Sea. Czar Peter the Great envisaged a Russia open to trading with the entire world; he developed St. Petersburg on the Baltic Sea into Russia's leading port. But in truth, Russia's historical geography is one of remoteness from the mainstreams of change and progress, as well as one of self-imposed isolation.

An Imperial, Multinational State

The map showing the expansion of the Russian state and its acquisition of an empire (Fig. 2B-2) also reveals how the modern layout of Russia's regions evolved. Initially, most of the action was on the Russian Plain, which was to become the state's core area. The czars' armies advanced southward across the Caucasus, gaining Christian allies and challenging Muslim enemies to the very borders of Iran and Turkey.

Eastward expansion beyond the Urals naturally followed a southerly path, creating the settlements and transport routes that define today's Eastern Frontier region (Fig. 2B-1). Still farther eastward, the Russians reached the Pacific Ocean, pushing into Alaska and then southward along the western coast of North America. By the time the imperial armies were at war with the Japanese in the Russian Far East, Moscow had already sold Alaska to the United States and the Russian Empire ended where the Pacific Ocean began.

Centuries of Russian expansionism did not confine itself to empty land or unclaimed frontiers. The Russian state became an imperial power that annexed and incorporated many nationalities and cultures. This was done by employing force of arms, by overthrowing uncooperative rulers, by annexing territory, and by stoking the fires of ethnic conflict. By the time the ruthless Russian regime had begun to confront revolution among its own people, czarist Russia was a hearth of **3** imperialism, and its empire controlled peoples representing more than 100 nationalities. The winners in the ensuing revolutionary struggle—the communists who forged the Soviet Union—did not liberate these subjugated peoples. Rather, they changed the empire's framework, binding the peoples colonized by the czars into a new system that would in theory give them autonomy and identity. In practice, it doomed those peoples to bondage and, in some cases, extinction.

When the Soviet system failed and the Soviet Socialist Republics became independent states in 1991, Russia was left without the empire that had taken centuries to build and consolidate—and that contained crucial agricultural and mineral resources. No longer did Moscow control the farms of Ukraine and the oil and natural gas reserves of Central Asia. But look again at Figure 2B-2 and you will see that, even without its European and Central Asian colonies,

Russia remains an empire. Russia lost the "republics" on its periphery, but Moscow still rules over a domain that extends from the borders of Finland to North Korea.

Inside that domain Russians are in the overwhelming majority, but many subjugated nationalities, from Tatars to Yakuts, still inhabit ancestral homelands. Accommodating these many indigenous peoples is one of the challenges facing the Russian Federation today.

In the 1990s, Russia began to reorganize in the aftermath of the collapse of the Soviet Union. This reorganization cannot be understood without reference to the seven decades of Soviet communist rule that went before. We turn next to this crucial topic.

THE SOVIET LEGACY

The era of communism may have ended in the Soviet Empire, but its effects on Russia's political and economic geography will long remain. Seventy years of centralized planning and implementation cannot be erased overnight; regional reorganization toward a market economy cannot be accomplished in a day.

While the world of capitalism celebrates the failure of the communist system in the former Soviet realm, it should remember why communism found such fertile ground in the Russia of the 1910s and 1920s. In those days Russia was infamous for the wretched serfdom of its peasants, the cruel exploitation of its workers, the excesses of its nobility, and the ostentatious palaces and riches of the czars. Ripples from the European Industrial Revolution introduced a new age of misery for those laboring in factories. There were workers' strikes and ugly retributions, but when the czars finally tried to better the lot of the poor, it was too little too late. There was no democracy, and the people had no way to express or channel their grievances. Europe's democratic revolution had passed Russia by, and its economic revolution touched the czars' domain only slightly. Most Russians, as well as tens of millions of non-Russians under the czars' control, faced exploitation, corruption, starvation, and harsh subjugation. When the people began to rebel in 1905, there was no hint of what lay in store; even after the full-scale Revolution of 1917, Russia's political future hung in the balance.

The Political Framework

Russia's great expansion had brought many nationalities under czarist control; now the revolutionary government sought to organize this heterogeneous ethnic mosaic into a smoothly functioning state. The czars had conquered, but they had done little to bring Russian culture to the peoples they ruled. The Georgians, Armenians, Tatars, and residents of the Muslim states of Central Asia were among dozens of individual cultural, linguistic, and religious groups that had not been "Russified." In 1917, however, the Russians themselves constituted only about one-half of the population of the entire empire. Thus it was impossible to establish a Russian state instantly over this vast political region because these diverse national groups had to be accommodated.

The question of the nationalities became a major issue in the young Soviet state after 1917. Lenin, who brought the philosophy of Karl Marx to Russia, talked from the beginning about the "right of self-determination for the nationalities." The first response by many of Russia's subject peoples was to proclaim independent republics, as they did in Ukraine, Georgia, Armenia, Azerbaijan, and even in Central Asia. But Lenin had no intention of permitting the Soviet state to break up. In 1923, when his blueprint for the new Soviet Union went into effect, the last of these briefly independent units was fully absorbed into the sphere of the Moscow regime. Ukraine, for example, had declared itself independent in 1917 and managed to sustain this initiative until 1919. But in that year the Bolsheviks set up a provisional government in the Ukrainian capital of Kiev, thereby ensuring the incorporation of the country into Lenin's Soviet framework.

The Communist System

The political framework for the Soviet Union was based on the ethnic identities of its many incorporated peoples. Given the size and cultural complexity of the empire, it was impossible to allocate territory of equal political standing to all the nationalities; the communists controlled the destinies of well over 100 peoples, both large nations and small isolated groups. It was decided to divide the vast realm into *Soviet Socialist Republics* (*SSRs*), each of which was delimited to correspond broadly to one of the major nationalities. At the time, Russians constituted about half of the developing Soviet Union's population, and, as Figure 2A-5 shows, they also were (and still are) the most widely dispersed ethnic group in the realm. The Russian Republic, therefore, was by far the largest designated SSR, comprising just under 77 percent of total Soviet territory.

Within the SSRs, smaller minorities were assigned political units of lesser rank. These were called Autonomous Soviet Socialist Republics (ASSRs), which in effect were republics within republics; other areas were designated Autonomous Regions or other nationality-based units. It was a complicated, cumbersome, often poorly designed framework, but in 1924 it was launched officially under the banner of the **Union of Soviet Socialist Republics** (**USSR**).

A Soviet Empire

Eventually, the Soviet Union came to consist of 15 SSRs (shown in Fig. 2A-7), including not only the original republics of 1924 but also such later acquisitions as Moldova (formerly Moldavia in Romania), and what are now the Baltic states of Estonia, Latvia, and Lithuania. The internal political layout often was changed, sometimes at the whim of the communist empire's dictators. But no communist apartheid-like system of segregation could accommodate the shifting multinational mosaic of the Soviet realm. The republics quarreled among themselves over boundaries and territory.

Demographic shifts, forced migrations, war, and economic factors soon made much of the layout of the 1920s obsolete. Moreover, the communist planners made it Soviet policy to relocate entire peoples from their homelands in order to better fit the grand design, and to reward or punish—sometimes capriciously. The overall effect, however, was to move minority peoples eastward and to replace them with Russians. This **4** **Russification** of the Soviet Empire produced substantial ethnic Russian minorities in all the non-Russian republics.

Phantom Federation

The Soviet planners called their system a **5** **federation**. We focus in more detail on this geographic concept in Chapter 11, but we note here that *federalism* involves the sharing of power between a country's central government and its political subdivisions (provinces, States, or, in the Soviet case, "Socialist Republics"). Study the map of the former Soviet Union (Fig. 2A-7) and an interesting geographic corollary emerges: every one of the 15 SSRs had a boundary with a non-Soviet neighbor. Not one was spatially locked within the others. This seemed to give geographic substance to the notion that any Republic was free to leave the USSR if it so desired. Reality, of course, was different, and Moscow's control over the SSRs made the Soviet Union a federation in theory only.

The centerpiece of the tightly controlled Soviet "federation" was the Russian Republic. With half the vast state's population, the capital city, the realm's core area, and over three-quarters of the Soviet Union's territory, Russia was the empire's nucleus. In other republics, "Soviet" often was simply equated with "Russian"—it was the reality with which the lesser republics lived. Russians came to the other republics to teach (Russian was taught in the colonial schools), to organize (and frequently dominate) the local Communist Party, and to implement Moscow's economic decisions. This was colonialism, but somehow the communist disguise—how could socialists, as the communists called themselves, be colonialists?—and the contiguous spatial nature of the empire made it appear to the rest of the world as something else. Indeed, on the world stage the Soviet Union became a champion of oppressed peoples, a force in the decolonization process. It was an astonishing contradiction that would, in time, be fully exposed.

The Soviet Economic Framework

The geopolitical changes that resulted from the establishment of the Soviet Union were accompanied by a gigantic economic experiment: the conversion of the empire from a czarist autocracy with a capitalist veneer to communism. From the early 1920s onward, the country's economy would be centrally planned—the communist leadership in Moscow would make all decisions regarding economic planning and development. Soviet planners had two principal objectives: (1) to accelerate industrialization and (2) to **6 collectivize** agriculture. For the first time ever on such a scale, and for the first time in accordance with Marxist-Leninist principles, an entire country was organized to work toward national goals prescribed by a central government.

Land and the State

The Soviet planners believed that agriculture could be made more productive by organizing it into huge state-run enterprises. The holdings of large landowners were expropriated, private farms were taken away from the farmers, and the land was consolidated into collective farms. Initially, all such land was meant to be part of a *sovkhoz*, literally a grain-and-meat factory in which agricultural efficiency, through maximum mechanization and minimum labor requirements, would be at its peak. But many farmers opposed the Soviets and tried to sabotage the program in various ways, hoping to retain their land.

The farmers and peasants who obstructed the communists' grand design suffered a dreadful fate. In the 1930s, for instance, Stalin confiscated Ukraine's agricultural output and then ordered a stretch of the border between the Russian and Ukrainian republics sealed—thereby leading to a famine that killed millions of farmers and their families. In the Soviet Union under communist totalitarianism, the ends justified the means, and untold hardship came to millions who had already suffered under the czars. In his book *Lenin's Tomb*, David Remnick estimates that between 30 and 60 million people lost their lives from imposed starvation, constant political purges, Siberian exile, and forced relocation. It was an incalculable human tragedy, but the secretive character of Soviet officialdom made it possible to hide it from the world.

The Soviet planners hoped that collectivized and mechanized farming would free hundreds of thousands of workers to labor in factories. Industrialization was the prime objective of the regime, and here the results were superior. Productivity rose rapidly, and when World War II engulfed the empire in 1941, the Soviet manufacturing sector was able to produce the equipment and weapons needed to repel the German invaders.

A Command Economy

Yet even in this context, the Soviet grand design entailed liabilities for the future.

The USSR practiced a **7 command economy**, in which state planners assigned the production of particular manufactures to particular places, often disregarding the rules of economic geography. For example, the manufacture of railroad cars might be assigned (as indeed it was) to a factory in Latvia. No other factory anywhere else would be permitted to produce this equipment—even if supplies of raw materials would make it cheaper to build them near, say, Volgograd 2000 kilometers (1250 mi) away. Yet, despite an expanded and improved transport network (see Fig. 2A-6), such practices made manufacturing in the USSR extremely expensive, and the absence of competition made managers complacent and workers far less productive than they could be.

Of course, the Soviet planners never imagined that their experiment would fail and that a market-driven economy would replace their command economy. When that happened, the transition was predictably difficult; indeed, it is far from over and continues to severely stress the now more democratic state.

RUSSIA'S CHANGING POLITICAL GEOGRAPHY

When the USSR dissolved in 1991, Russia's former empire devolved into 14 independent countries, and Russia itself was a changed nation. Russians now made up about 83 percent of the population of just under 150 million, a far higher proportion than in the days of the Soviet Union. But numerous minority peoples remained under Moscow's new flag, and millions of Russians found themselves under new governments in the former Republics.

Soviet planners had created a very complicated administrative structure for their "Russian Soviet Federative Socialist Republic," and Russia's postcommunist leaders had to use this framework to make their country function (Fig. 2B-3). In 1992, most of Russia's internal "republics," autonomous

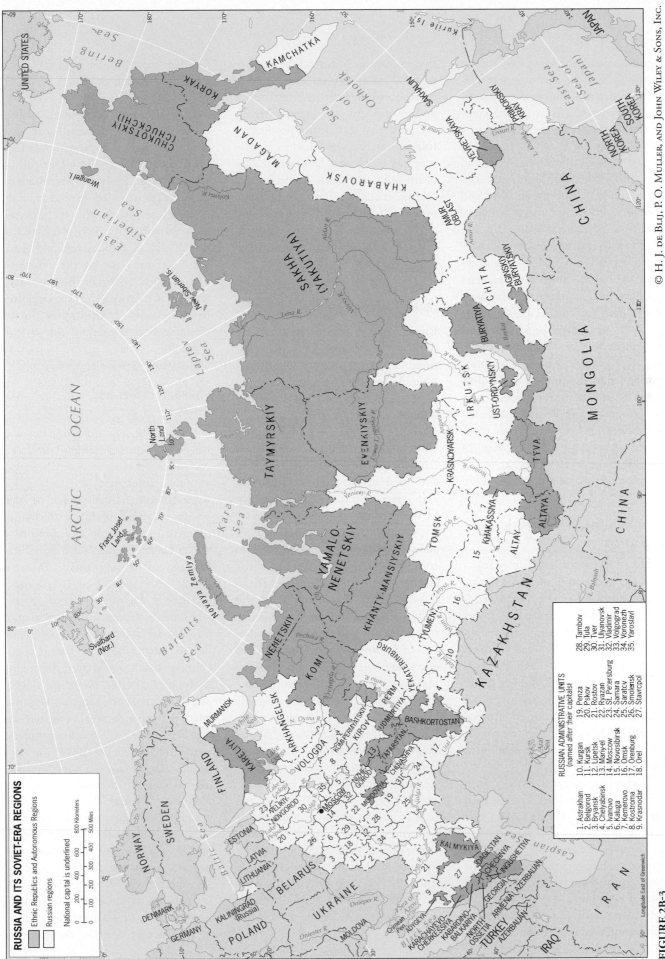

RUSSIA AND ITS SOVIET-ERA REGIONS

Ethnic Republics and Autonomous Regions

Russian regions

National capital is underlined

0   100 200 300 400 500 Miles
0 200 400 600 800 Kilometers

RUSSIAN ADMINISTRATIVE UNITS
(named after their capitals)

1. Astrakhan	10. Kurgan	19. Penza	28. Tambov
2. Belgorod	11. Kursk	20. Pskov	29. Tula
3. Bryansk	12. Lipetsk	21. Rostov	30. Tver
4. Chelyabinsk	13. Mariy-el	22. Ryazan	31. Ulyanovsk
5. Ivanovo	14. Moscow	23. St. Petersburg	32. Vladimir
6. Kaluga	15. Novosibirsk	24. Samara	33. Volgograd
7. Kemerovo	16. Omsk	25. Saratov	34. Voronezh
8. Kostroma	17. Orenburg	26. Smolensk	35. Yaroslavl
9. Krasnodar	18. Orel	27. Stavropol	

Longitude East of Greenwich

FIGURE 2B-3

© H. J. de Blij, P. O. Muller, and John Wiley & Sons, Inc.

regions, oblasts, and krays (all of them components of the administrative hierarchy) signed a document known as the Russian Federation Treaty, which committed them to cooperate in the new federal system. At first a few units refused to sign, including Tatarstan, scene of Ivan the Terrible's brutal conquest more than four centuries ago, and a republic in the Caucasus periphery, Chechnya-Ingushetiya, where Muslim rebels waged a campaign for independence (Fig. 2B-3). Chechnya-Ingushetiya next split into two separate republics, whose names at the time were spelled Chechenya and Ingushetia. Eventually, only Chechnya refused to sign the Russian Federation Treaty, and subsequent Russian military intervention led to a prolonged and violent conflict, with disastrous consequences for Chechnya's people and infrastructure (the capital, Groznyy, was completely destroyed). The conflict in Chechnya continues today and is a disaster for Russia's government as well.

The Federal Framework of Russia

Figure 2B-3 reflects the complex Soviet administrative system of Russia during communist times. There were 89 entities: two Autonomous Federal Cities (Moscow and Leningrad), 21 Republics, 11 Autonomous Regions (Okrugs), 49 Provinces (Oblasts), and six Territories (Krays). The 21 Republics, established and recognized to accommodate substantial ethnic minorities, included a cluster in the far south, another to the east of Moscow, and still another in the Mongolia border zone.

When Russia's post-1991 government took over, it faced complicated administrative problems. Not only were some entities reluctant to sign the Russian Federation Treaty, but despite the ranking implied by the list above, some regions were more equal than others—that is, some regions were used to privilege under the old system, and their local leaders expected

that to continue. Meanwhile, a multinational, multicultural state that had been accustomed to authoritarian rule and government control over virtually everything—from factory production to everyday life—now had to be governed in a new way. Democratization of the political system, transition to a market economy, the sale of state-owned industries (privatization), and other far-reaching changes had to come quickly or the country risked chaos.

Unitary and Federal Options

Russia's leaders recognized that their options were limited. They could continue to hold as much power as possible at the center, making decisions in Moscow that would apply to all the Republics, Regions, and other subdivisions of the state. Such a **8 unitary state system**, with its centralized government and administration, marked authoritarian kingdoms of the past and serves totalitarian dictatorships of the present. Or they could share power with the Republics and Regions, allowing elected regional leaders to come to Moscow to represent the interests of their people. This is the federal system Russia chose as the only way to accommodate the country's economic and cultural diversity.

In a *federal system*, the national government usually is responsible for matters such as defense, foreign policy, and foreign trade. The Regions (or provinces, States, or other subdivisions) retain authority over affairs ranging from education to transportation. A federal system does not create unity out of diversity, but it does allow diverse components of the state to coexist, their common interests represented by the national government and their regional interests by their local administrations. Some countries owe their survival as coherent states to their federal frameworks. India and Australia are cases in point.

But to maintain a generally acceptable balance of power between the center and Regions (or States) is difficult. Disputes concerning "States' rights" even continue to roil the American political scene more than two centuries after the Constitution was adopted. Early on, the Russian government decided to end, for all practical purposes, the hierarchical regional system the Soviets had established. The Republics retained their special status, but all the others—okrugs, krays, and so forth—were designated as Regions. This was intended to address the favoritism of Soviet times and to streamline the system generally.

Problems of Size and Distance

The new Russian government also faced an old Soviet problem: the sheer size of the country, its vast distances, and the remoteness of many of its Regions. Geographers refer to the principle of **9 distance decay** in order to explain how increasing distances between places tend to reduce interactions among them. Because Russia is the world's largest country, distance is a significant factor in the relationships between the capital and outlying areas. Furthermore, Moscow lies in the far west of the gigantic country, half a world away from the shores of the Pacific. Not surprisingly, one of the most obstreperous Regions has been remote Primorskiy, the Region of Vladivostok. Still another problem is reflected in Figure 2B-3: the stupendous size variation among the Republics (and to a lesser extent, the Regions). Whereas the (territorially) smallest Republics are concentrated in the Russian core area, the largest lie far to the east, where Sakha is nearly a thousand times as large as Ingushetiya. On the other hand, the populations of the enormous eastern Republics are tiny compared to those of the smaller ones in the Russian

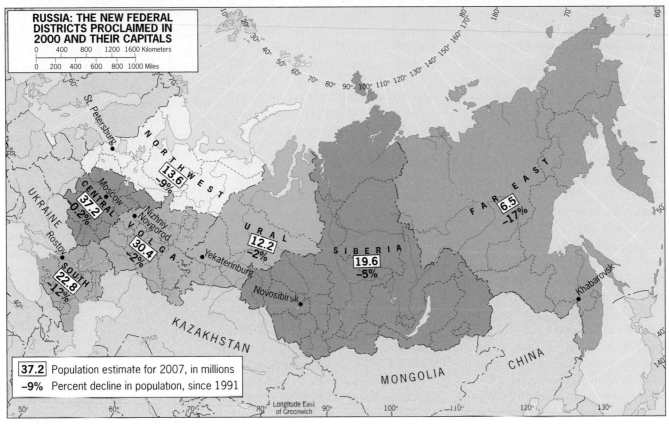

FIGURE 2B-4

© H. J. de Blij, P. O. Muller, and John Wiley & Sons, Inc.

Core. Such diversity spells administrative difficulty.

But perhaps the most serious current problem for Russia is the intensifying social disharmony between Moscow and the subnational political entities. Almost wherever one goes in Russia today, Moscow is disliked and often berated by angry locals. The capital is seen as the privileged playground for those who have benefited most from the post-Soviet transition—bureaucrats and hangers-on whose economic policies have driven down standards of living, whose greed and corruption have hurt the economy, whose actions in Chechnya have been a disaster, who have allowed foreigners (prominently the United States) to erode Russian power and prestige, who fail to pay the wages of workers toiling in industries still owned by the state, and who do not represent the Russian people. Complaints about the capital are not uncommon in free countries, but in Russia the mistrust between capital and subordinate areas has become so serious that it constitutes a key challenge for the government.

Federal Administrative Districts

In 2000, the new Putin administration moved to diminish the influence of the Regions by creating a new geographic framework that combined the 89 Regions, Republics, and other entities into seven new administrative units—not to enhance their influence in Moscow, but to increase Moscow's authority over them. As Figure 2B-4 shows, each of these new Federal Districts has a capital, elevating such cities as Rostov and Novosibirsk to a status secondary only to Moscow. In a related move, the Russian president proposed, and the parliament approved, that Regional governors would be appointed rather than elected—thereby concentrating still more power in Moscow.

Russia's federal system now appears to be moving in a unitary direction.

RUSSIA'S DEMOGRAPHIC DILEMMA

All this pales, however, against what some geographers refer to as Russia's demographic disaster: its rapid decline in population coupled with the deteriorating health of its citizens. When the Soviet Union disintegrated in 1991, Russia's population totaled about 149 million. Today, two decades later, this number is down to less than 141 million—despite the in-migration of several million ethnic Russians from the former Soviet "republics" beyond Russia's borders. Since the end of communist rule, Russia has witnessed over 10 million more deaths than births. Severe social dislocation is the cause.

Undoubtedly, the transition away from Soviet rule is one reason: uncertainty tends to cause families to have fewer children, and abortion is widespread in Russia. But the birthrate has stabilized at around 12 per thousand; it is Russia's death rate that has skyrocketed, now exceeding more than 15 per thousand. This produces an annual loss of population of over 0.3 percent, or nearly half a million per year.

Russian males are the ones most affected. Male life expectancy dropped from 71 in 1991 to 61 in 2009 (female life expectancy has also declined, but markedly less, to 74). Males are more likely to be afflicted by alcoholism and related diseases, by AIDS (which is significantly underreported in Russia, according to international agencies), by heavy smoking, and by suicide, accidents, and murder. On average, a Russian male is nine times more likely to die a violent or an accidental death than his European counterpart. Fewer than half of today's Russian male teenagers will survive to age 60.

If this **10 population implosion** continues, Russia will have only about 110 million citizens by 2050, possibly even fewer, raising doubts about the future of the state itself. Look again at that map of the Districts (Fig. 2B-4) and consider this: since 1990, the Far East District has lost 17 percent of its population, Siberia 5 percent, the Northwest 9 percent, and the South 12 percent. Only the Central District, the one around Moscow, has a loss on a par with European countries (0.2 percent).

What is the answer? Russia's leaders hope that improvements in the social circumstances of the average Russian will reduce the rate of population loss. Campaigns against alcoholism and careless lifestyles are under way. Immigration is another option, and should Russia relax its rules, hundreds of thousands of Koreans and Chinese would move into the Russian Far East and counter the outflow of Russians there. But Moscow is not eager to see its Pacific Rim transformed into an extension of East Asia.

Tellingly, during an address to the nation in January 2010, President Medvedev claimed that Russia "had not lost any people between July 2008 and July 2009." Although the long-term evidence indicates otherwise, it is clear that demographic concerns are now at the forefront among Russian priorities.

It is now time to turn our attention to the realm's regional geography. As was noted at the beginning of this chapter, so vast is Russia, so varied its physiography, and so diverse its cultural landscape that regionalization requires a small-scale perspective and a high level of generalization. We will proceed to outline the five-region framework that is mapped in Figure 2B-1: (1) the Russian Core, everything lying west of the Urals; (2) the Eastern Frontier, eastward extension of the Core into the heart of Eurasia; (3) Siberia, the frigid territory sprawling across the country's huge northeast quadrant; (4) the Far East, Russia's long window on the Pacific Ocean; and (5) Transcaucasia, whose three small republics lie just outside Russia in the corridor between the Black and Caspian seas. As we shall see, each of these first-order regions contains major subregions.

THE RUSSIAN CORE

The heartland of a state is its **11 core area.** Here much of the population is concentrated, and here lie its biggest cities, leading industries, densest transportation networks, most intensively cultivated lands, and other key components of the country. Core areas of long standing strongly reflect the imprints of culture and history. The Russian Core, broadly defined, extends from the western border of the Russian realm to the Ural Mountains in the east (Fig. 2B-1). This is the Russia of Moscow and St. Petersburg, of the Volga River and its industrial cities.

Central Industrial Region

At the heart of the Russian Core lies the Central Industrial Region (Fig. 2B-5). The precise definition of this subregion varies, for all regional definitions are subject to debate. Some geographers prefer to call this the Moscow Region, thereby emphasizing that for over 400 kilometers (250 mi) in all directions from the capital, everything is oriented toward this historic focus of the state. As Figures 2B-5 and, even better, 2A-6 show, Moscow has maintained its decisive **12 centrality:** roads and railroads converge in all directions from Ukraine in the south; from Mensk (Belarus) and the rest of eastern Europe in the west; from St. Petersburg and the Baltic coast in the northwest; from Nizhniy Novgorod and the Urals in the east; from the cities and waterways of the Volga Basin in the southeast (a canal links Moscow to the Volga, Russia's single most important navigable river); and even to the subarctic northern periphery that faces the Barents Sea, where the strategic naval port of Murmansk and the lumber-exporting outpost of Arkhangelsk lie.

Two Great Cities

Moscow (population: 10.5 million) is the megacity hub of an area comprising some 50 million inhabitants (more than one-third of the country's entire population), most of them concentrated in cities that, during the Soviet period, specialized in assigned industries such as Nizhniy Novgorod (the "Soviet Detroit" where automobiles were made) and Ivanovo (known for its textiles). But today, Russia is less a diversified manufacturing country than a commodity producer and exporter (mainly oil and natural gas). The tall steel-and-glass towers you see rising above the central business districts, such as the one now being built in the heart of St. Petersburg's historic district, for the most part belong

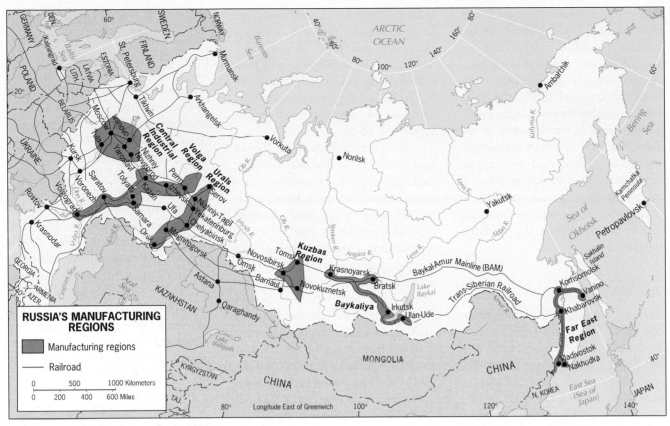

FIGURE 2B-5

to state-controlled energy companies such as Gazprom. And the cars being built are likely to have foreign names such as Fiat or Volkswagen.

St. Petersburg (the former Leningrad) remains Russia's second city, with a population of 4.6 million. Under the czars, St. Petersburg was the focus of Russian political and cultural life, and Moscow was a distant second city. Today, however, St. Petersburg possesses none of Moscow's locational advantages, at least not with respect to the domestic market. It lies well outside the Central Industrial Region near the northwestern corner of the country, 650 kilometers (400 mi) from the capital. Neither is it better off than Moscow in terms of resources: fuels, metals, and foodstuffs must all be brought in, mostly from far away. The former Soviet emphasis on self-sufficiency even reduced St. Petersburg's asset of being on the Baltic seacoast because some raw materials could have been

imported much more cheaply across the Baltic from foreign sources than from domestic sites in distant Central Asia (only bauxite deposits lie nearby, at Tikhvin).

Povolzhye: The Volga Region

Another major subregion within the Russian Core is the ***Povolzhye***, the Russian name for an area that extends along the middle and lower valley of the Volga River as well as its tributary, the Kama, above the city of Kazan. It would be appropriate to call this the Volga Region, for this greatest of Russia's rivers is its lifeline and most of the Povolzhye's cities lie on its banks (Fig. 2B-5). In the 1950s, a canal was completed to link the lower Volga with the lower Don River (and thereby the Black Sea).

The Volga corridor was an important historic route in old Russia, but for a long time neighboring regions

overshadowed it. The Moscow area and Ukraine were far ahead in industry and agriculture. The Industrial Revolution that came late in the nineteenth century to the Central Industrial Region did not have much effect in the Povolzhye. Its major function remained the transit of foodstuffs and raw materials to and from other regions.

The elongated Povolzhye (the *vol* refers to the Volga) forms the eastern flank of the Central Industrial Region, its cities and farms sustained by Russia's greatest river. For a long time this subregion was bypassed in the Soviet industrialization drive, but during World War II it suddenly took on much greater importance because its cities lay relatively remote from the German armies invading from the west. Accelerated industrial development put the Volga Region's cities on the map as never before. Then, in the 1950s, the Volga-Don Canal was opened, creating new linkages and opportunities. Next, the real dimensions of the oil- and gasfields in and near the Povolzhye became clear (see Fig. 2A-8), and for some time these reserves were the largest in the entire Soviet Union. And finally, the Volga Region's northwestward linkages, via the Moscow and Mariinsk canals to the Baltic Sea, were improved and supplemented by rail and road connections. Again, the industrial composition of the Volga's riverside cities continues to change, but if the Central Industrial Region is Russia's heart, the Povolzhye is its key artery.

The Urals Region

The Ural Mountains form the eastern edge of the Russian Core. They are not particularly high; in the north they consist of a single range, but southward they broaden into a hilly zone. Nowhere are they an obstacle to east-west surface transportation. An enormous storehouse of metallic mineral resources located in and near the Urals has made this area a natural place for industrial development. Today, the Urals Region, well connected to the nearby Volga and Central Industrial Regions, extends from Serov in the north to Orsk in the south (Fig. 2B-5).

As Figure 2B-5 reveals, the Soviet legacy survives on Russia's economic landscape, despite the country's conversion from a domestic manufacturing system to what some observers now call a "petro-ruble state." Not only the Central Industrial Region but also the great Volga corridor, the Urals manufacturing complex, the industrial nodes of the Eastern Frontier, and even the outliers of the Far East still mark the manufacturing map, even if many factories have fallen silent and people are leaving. Urban dwellers tend to find new ways to make their living, and such cities as Volgograd, Chelyabinsk, Novosibirsk, Irkutsk, or Vladivostok will never be empty.

THE EASTERN FRONTIER

From the eastern flanks of the Ural Mountains to the headwaters of the Amur River, and from the latitude of Tyumen to the northern zone of neighboring Kazakhstan, lies Russia's vast Eastern Frontier Region, product of a gigantic experiment in the eastward extension of the Russian Core (Fig. 2B-1). As the maps of cities and transportation (Fig. 2A-6) as well as population distribution (Fig. 2A-4) suggest, this region is more densely peopled and more fully developed in the west than in the east. To the east of the Yenisey River, linear settlement prevails, marked by ribbons and clusters along the east-west railroads. Two subregions dominate the human geography: the Kuznetsk Basin in the west and the Lake Baykal area in the east.

The Kuznetsk Basin (Kuzbas)

Some 1500 kilometers (950 mi) east of the Urals lies another of Russia's primary regions of heavy manufacturing resulting from the communist period's national planning: the Kuznetsk Basin, or *Kuzbas* (Fig. 2B-5). In the 1930s, it was opened up as a supplier of raw materials (especially coal) to the Urals, but that function became less important as local industrialization accelerated. The original plan was to move coal from the Kuzbas west to the Urals and allow the returning trains to carry iron ore east to the coalfields—a classic case of **13 double complementarity**. However, good-quality iron ore deposits were subsequently discovered near the Kuznetsk Basin itself. As the new resource-based Kuzbas industries expanded, so did its urban centers. The leading city, located just outside the region, is Novosibirsk, which stands at the intersection of the Trans-Siberian Railroad and the Ob River as the symbol of Russian enterprise in the vast eastern interior. To the northeast lies Tomsk, one of the oldest

The container port at Krasnoyarsk, where the south-to-north-flowing Yenisey River meets the east-west Trans-Siberian Railroad near the center of the Eastern Frontier region (see Fig. 2B-1). Supplies brought by train from the west are shipped northward to settlements in the Siberian interior; raw materials from Siberia arrive by boat and barge for despatch to Russia's factories and markets. Large storage facilities stand across the river. © Wolfgang Kaehler/Alamy

Russian towns east of the Urals, founded in the seventeenth century and now caught up in the modern development of the Kuzbas. Southeast of Novosibirsk lies Novokuznetsk, a city that produces steel for this subregion's machine and metal-working plants as well as aluminum products from Urals bauxite.

The Lake Baykal Area (*Baykaliya*)

East of the Kuzbas, development becomes more insular, and distance becomes a stronger adversary. North of the central segment of the Mongolian border zone and eastward around Lake Baykal, larger and smaller settlements cluster along the two railroads to the Pacific coast (Fig. 2B-5). West of the lake, these rail corridors lie in the headwater zone of the Yenisey River and its tributaries. A number of dams and hydroelectric projects serve the valley of the Angara River, particularly the city of Bratsk. Mining, lumbering, and some farming sustain life here, but isolation dominates it. The city of Irkutsk, near the southern end of Lake Baykal, is the principal service center for the vast Siberian region to the north and for a lengthy east-west stretch of southeastern Russia.

Beyond Lake Baykal, the Eastern Frontier really lives up to its name: this is southern Russia's most rugged, remote, and forbidding country. Settlements are rare, with many being mere camps. The Buryat Republic (see Fig. 2B-3) is part of this zone; the territory bordering it to the east was taken from China by the czars and may become an issue in the future. Where the Russian-Chinese boundary turns southward, along the Amur River, the region called the Eastern Frontier ends and Russia's Far East begins.

SIBERIA

Before we assess the potential of Russia's Pacific Rim, we should remember that the ribbons of settlement just discussed hug the southern perimeter of this gigantic country, avoiding the vast Siberian region to the north (Fig. 2B-1). Siberia extends from the Ural Mountains to the Kamchatka Peninsula—a vast, bleak, frigid, forbidding land. Larger than the conterminous United States but inhabited by only an estimated 15 million people, Siberia quintessentially symbolizes the Russian environmental plight: vast distances, cold temperatures worsened by strong Arctic winds, difficult terrain, poor soils, and limited options for survival.

But Siberia also possesses resources. From the days of the first Russian explorers and Cossack adventurers, Siberia's riches have beckoned. Gold, diamonds, and other precious minerals were found. Later, metallic ores including iron and bauxite were discovered. Still more recently, the Siberian interior proved to contain massive quantities of oil and natural gas (Fig. 2A-8) that now contribute significantly to Russia's energy supply and exports.

As the physiographic map (Fig. 2A-2) shows, major rivers—the Ob, Yenisey, and Lena—flow gently northward across Siberia and the Arctic Lowland into the Arctic Ocean. Hydroelectric power development in the basins of these rivers has generated electricity used to extract and refine local ores, and run the lumber mills that have been set up to exploit the vast Siberian forests.

The Future

The human geography of Siberia is fragmented, and most of the region is virtually uninhabited (Fig. 2A-5). Ribbons of Russian settlement have developed; the Yenisey River, for instance, can be traced on this map of Soviet peoples (a series of small settlements north of Krasnoyarsk), and the upper Lena Valley is similarly fringed by ethnic Russian settlement. Yet hundreds of kilometers of empty territory separate these thin ribbons and other islands of habitation.

Siberia, Russia's freezer, is stocked with goods that may become mainstays of future national development. Already, precious metals and mineral fuels are bolstering the Russian economy. In time, we may expect Siberian resources to play a growing role in the economic development of the Eastern Frontier and the Russian Far East as well. One step in that process was already taken during Soviet times: the completion of the BAM (Baykal-Amur Mainline) Railroad in the 1980s. This route, lying north of and parallel to the old Trans-Siberian Railroad, extends 3500 kilometers (2200 mi) eastward from near the important center of Krasnoyarsk directly to the Far East city of Komsomolsk (Fig. 2B-5). In the post-Soviet era, the BAM Railroad has been beset by equipment breakdowns and workers' strikes. Nonetheless, it is a key element of the infrastructure that will serve the economic growth of easternmost Russia in the twenty-first century.

THE RUSSIAN FAR EAST

When the government in Moscow in 2000 superimposed its framework of Federal Districts over Russia's matrix of Soviet-era regions (Fig. 2B-4), it gave the name *Far East* to the largest one (in terms of territory) of all. In one stroke, the Far East incorporated four residual ethnic republics and six Russian regions (see Fig. 2B-3). But many Russian citizens have a different view of what constitutes the geographic region they call the Far East. Justifiably, they see most of the northern zone of the official Far East as a continuation of Siberia. To them, the real Far East is constituted by the area beyond the Eastern Frontier to the Pacific coast, the island of Sakhalin, the Kamchatka Peninsula, and a narrow stretch of land along the northern shore of the Sea of Okhotsk (Figs. 2B-1 and 2B-6).

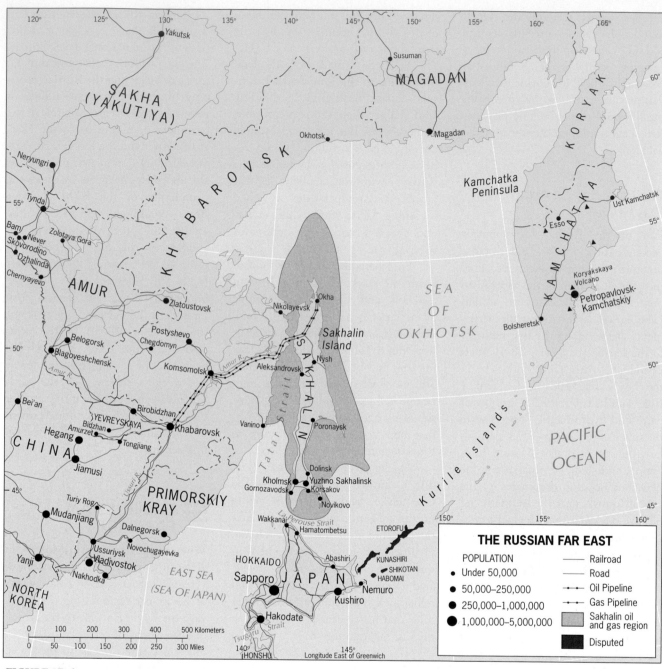

FIGURE 2B-6

© H. J. de Blij, P. O. Muller, and John Wiley & Sons, Inc.

In Soviet times, citizens who agreed to move to this distant region to work in factories or for the administration were rewarded with special privileges to compensate for their hardship. Make no mistake: the Pacific Ocean's proximity tends to temper natural environments here, but life in the Russian Far East is tough. And in the postcommunist period, Moscow's favors were withdrawn, so that hundreds of thousands of inhabitants in

the new democracy voted with their feet and departed. Take the city of Petropavlovsk on the Kamchatka Peninsula. In 1991, just before the Soviet Union collapsed, it had a population of 248,000. By 2001, it was down to 168,000. Today, it contains barely more than 120,000 residents.

Almost everywhere in the geographic Far East, people now feel abandoned by Moscow. The once-great naval base of Vladivostok (the city's

name means "We Own the East") lies dormant. Factories have closed. And when the locals tried something new—clandestinely importing used Japanese cars—the government sent its federal police to close down the business. Meanwhile, the nearby container port at Nakhodka suffers repeated breakdowns. Cross-border trade with China is minimal. Trade with Japan is inconsequential. The two railroads that were to stimulate cross-country traffic and

© H. J. de Blij

"The Russian Far East is losing people at the fastest rate among all Russian Federal Districts, and my field trip to Petropavlovsk on the Kamchatka Peninsula only reinforced what I had learned in Vladivostok some years ago. During Soviet times, Russians willing to move to the Far East were given special privileges (an older man told me that 'being as far away from Moscow as possible was reward enough'), but today locals here feel abandoned, even repressed. In 1991, Petropavlovsk had about 250,000 residents; by 2001 the population was down to under 170,000, and today it is estimated to be no more than 125,000. It is not just the bleak environment—the ubiquitous ash and soot from dozens of active volcanoes on the peninsula, covering recent snowfalls and creating a pervasive black mud—it is also loss of purpose people feel here. There was a time when the Soviet windows on the Pacific were valuable and worthy; now the people scramble to make a living and when they find a way, for example by selling used cars bought cheaply in Japan and brought to the city by boat, Moscow sends law enforcement to stop the illegal trade. Petropavlovsk has no surface links to the rest of the world, and a springtime (April 2009) view of the city at the foot of the volcano Koryakskaya (left) and a street scene near the World War II museum (right) suggest why the Far East is depopulating."

www.conceptcaching.com

trade cannot operate efficiently without huge subsidies—available during Soviet times but not in the twenty-first century's market economy.

The Far East does have significant reserves of what underpins the Russian economy—oil and natural gas in and around the island of Sakhalin. And, as Figure 2B-6 reminds us, the southern sector of the Russian Far East lies between other reserves and potential markets: China next door and Japan across a narrow sea. Already, the Japanese and the Chinese are both bidding for Russia's energy supplies, and undoubtedly there will be construction of pipelines and other energy-related infrastructure. But what the region needs is a diversified economy of the kind that is reviving China's Northeast across the Amur and Ussuri rivers.

Such development, however, is currently not in sight.

THE SOUTHERN PERIPHERY

Having followed Russia's eastward expansion and organization, let us take another look at the realm's regional map (Fig. 2B-1) to focus on the most complicated region of them all: the area where Russia's leaders, communists as well as czars, pushed the expanding state's influence across and beyond the Caucasus Mountains. On this map, it looks as though the Russian Core extends to the international boundary (which coincides, more or less, with the crestlines of this massive range). But it is not that simple. When you travel southward from Volgograd or Rostov,

the landscape—physical as well as cultural—does not change; the Russian Plain's gently rolling countryside with its villages and farms is very much like that south of Moscow. Taking the train south from Astrakhan in the Volga River Delta, you soon see the Caucasus looming in the distance, but you may not realize that you have entered a different world until you get off at Groznyy, the capital of Chechnya, one of Russia's ethnic "republics" (Fig. 2b-7). Just observing the mix of people makes you aware that you have left the heartland and reached Russia's periphery.

Figure 2B-7 is important enough to spend some serious time studying it. Here the communist leaders who succeeded the czars faced multiple challenges. First, inside the Russian state itself, there were many minority

SOUTHERN RUSSIA:
INTERNAL AND EXTERNAL PERIPHERIES

POPULATION

- • Under 50,000
- • 50,000–250,000
- • 250,000–1,000,000
- • 1,000,000–5,000,000
- • Over 5,000,000

——	Railroad
——	Road
++++	Canal
•—•—•	Oil pipeline
○—○—○	Proposed oil pipeline

National capitals are underlined

FIGURE 2B-7

© H. J. de Blij, P. O. Muller, and John Wiley & Sons, Inc.

peoples (such as the Chechens of Chechnya) who had fought the czars' armies and who now faced new rulers. Second, there were peoples who had sided with the czars and who wanted to cooperate with the Soviet regime, such as the Ossetians. To recognize the aspirations of these peoples, some (but not all) of whom were Muslim, the communist administration designated ethnic "republics" for them. On our map, these are colored pink to distinguish them from the rest of Russia, which is shown in tan.

Russia's Internal Periphery

From Buddhist-infused Kalmykia on the Caspian side westward to mainly Orthodox Christian Adygeya near the Black Sea, the Internal Periphery of southern Russia is defined by eight ethnic republics, each with its own identity (Fig. 2B-7). In the east, Kalmykia is Buddhist-infused, and, as noted earlier, Dagestan is a hodgepodge of small culturally discrete communities. Physiographically as well as culturally divided Chechnya, as the inset map suggests, lies in a transition zone between plain and mountains, the mountains serving as refuge for those opposed to Moscow's rule. When the post-Soviet Russian government asked all Regions and Republics within Russia to sign its proposed Russian Federation Treaty, Chechnya's leaders refused, believing that they now had the opportunity to finally end Russian control. Its western neighbor, Ingushetiya also got caught up in anti-Russian activism, and the ensuing conflict took a terrible toll in North Ossetia where, in 2004, more than 350 students, teachers, and parents died in a terrorist attack on a school. Still farther west, things remained calmer, and no problems at all arose in generally pro-Moscow Adygeya.

The tier of ethnic republics on Russia's southern border display all the properties of a disadvantaged periphery: not only do they share a history of subjugation by the powerful core to the north, but they also lag behind the rest of Russia by almost any measure of social progress. In terms of health, education, income, opportunity, and other such indices, the peoples of the Internal Periphery, whether they are pro-Russian or not, have a long way to go to reach Russian living standards. And there is much simmering resentment of Russian rule, past and present, in such republics as Chechnya and Ingushetiya.

In the case of Chechnya, such anger has its most recent roots in the Second World War, when Soviet leaders accused the Chechens of sympathizing, and even collaborating, with the Nazi invaders. Josef Stalin, then the Soviet dictator, ordered the entire Chechen population loaded on trains and exiled to the desert of Kazakhstan across the Caspian Sea. Tens of thousands died along the way or after arriving there, and even though Stalin's successor Nikita Khrushchev in 1957 allowed the survivors to return to their homeland, the Chechens never forgave their tormentors. Then in 1991, when the Soviet Union disbanded, they saw their opportunity, refusing to sign the Russian Federation Treaty that would make their republic an integral part of the new Russian state and mounting a campaign for independence that led to a bitter and costly war that destabilized a number of neighboring republics as well. Chechen terrorists carried this campaign to the very heart of Moscow, killing hundreds of civilians there and changing Russia's political landscape. When presidential candidate Vladimir Putin promised victory, the Russians elected him overwhelmingly in 2000. Massive military intervention did overpower the Chechens, but the periphery remains restive and unstable.

TRANSCAUCASIA: RUSSIA'S EXTERNAL PERIPHERY

Russian (and later Soviet) manipulation of the ethnic map did not end at the boundary we see in Figures 2B-7 and 2A-5, but it did take a different turn. Beyond the Russian border, the Soviet Empire extended its power over three adjoining Transcaucasian entities where the Russians had for centuries exerted influence and waged war. These three units were **Georgia** (mapped in yellow in Fig. 2B-7), facing the Black Sea; **Azerbaijan** (green), bordering on the Caspian Sea; and **Armenia** (lavender), landlocked in the middle. During Soviet times, Georgia was a loyal partner; Josef Stalin was a Georgian by birth. Christian Armenia appreciated membership in the Soviet Union because of the security it provided next door to Islamic Turkey. And Islamic Azerbaijan, whose ethnic majority is a people called the *Azeris*—who have close kinship with Iranian Azeris directly across their southern boundary with Iran—was a valued member of the USSR because of its rich oil reserves.

As if this were not enough, note in Figure 2B-7 that Azerbaijan has a large exclave on the Iranian border (Naxcivan), separated from the main part of the country by Armenian territory—and that the Armenians hold another exclave called Nagorno-Karabakh inside what would seem to be Azerbaijan's territory (which is home to 150,000 Christian Armenian citizens who were placed under Islamic Azerbaijan's jurisdiction by Soviet mandate).

Soviet rule kept the lid on potential conflict in this historically turbulent part of Transcaucasia, and after the collapse of the Soviet Union the Russians sought to retain their influence and maintain stability in this sector of their Near Abroad. The Russians wanted to keep Azerbaijan's oil flowing to and through Russia (see

© H.J. de Blij

"We had flown into Yerevan, Armenia in by far the most rickety airplane I had ever seen, a rattling Aeroflot twin-prop with broken seats, cracked windows, and, it seemed, sputtering engines. Two days later we were on our way by what our Moscow University colleague called the Georgian Military Highway to Tbilisi and across the Caucasus. What we were able to see of Armenia suggested an impoverished economy but public approval of Soviet Union membership: Moscow provided security against neighboring Turkey while meddling little, people said, in local-level politics. On our way north you could tell why the road had its name. Strategic, dangerous, and seemingly lined with wrecks, it afforded great views (here across the northern margin of retreating Lake Sevan, old beach lines in the foreground, and the town of Sevan in the distance) while reminding us—see the military caravan glimpsed at the bottom of the photo—why this was called a *military* road. (First Soviet field trip, July 1964)."

www.conceptcaching.com

the pipeline that leads from Baki [Baku] via Chechnya's Groznyy to the Black Sea terminal at Novorossiysk in Fig. 2B-7); they also wanted to keep Georgia in the Slavic fold, and they sought ways to mitigate the lingering conflict between Azerbaijan and Armenia over the enclave of Nagorno-Karabakh.

But things have not gone well for Moscow. Spurred by Western investments, Azerbaijan in 2006 began exporting oil via a new pipeline across Georgia and Turkey to the Mediterranean coast, providing it with an alternative to Russian transit. Armenia and Azerbaijan persisted in their sometimes-violent quarrel over Nagorno-Karabakh. And things really

fell apart in Georgia, where the government pursued pro-European policies, allegedly mistreated its pro-Russian Ossetian minority, and where Russia openly supported devolutionary initiatives in the northwestern corner of the country, a small ethnic entity known as Abkhazia (Fig. 2B-7).

Soon after the USSR's disintegration in 1991, the argument between Russia and Georgia escalated into serious discord. The Russians closed the Georgian border, shutting Georgian farmers out of their crucial market. Then the Russians started issuing Russian passports to Abkhazian citizens of Georgia. Next, the Russians signaled

even stronger support for their Ossetian allies, and in a still-disputed sequence of events during 2008 Russian forces entered South Ossetia and engaged the Georgian military in a brief but costly war. After a cease-fire was brokered by international action, Moscow took the extraordinary step of recognizing the two pro-Russian autonomous regions of Georgia, Abkhazia and South Ossetia, as independent states.

This action sent shudders through Europe and much of the world. If Russia could intervene militarily in this part of its Near Abroad, where else might Russian forces appear to enforce Russia's will? Georgia, as the map shows, has a lengthy border with

Russia, but so does Ukraine, where Russia could claim to have even stronger interests including leased military bases. Unlike Azerbaijan, mostly agricultural Georgia has no energy card to play—although it agreed to allow the new pipeline from Azerbaijan to Turkey to traverse its territory, thereby angering the Russians even more. And the Georgian government has courted the European Union and even NATO, seeking the security Georgians have lacked for centuries. The Azeris of Azerbaijan, whose dominant religion is Shi'ite Islam like that of Iran, are caught in the global web (and political geography) of energy demand and supply, but most ordinary people have benefited little from their country's oil wealth. And the landlocked 3 million Armenians live in a world of unsettled scores: with the Turks over their decimation by Ottoman Muslims, with the Azeris over their exclave of Nagorno-Karabakh, with the international community over the recognition of their plight. Transcaucasia may be the smallest region of the Russian realm, but it is also the second-most populous and undoubtedly the most volatile.

POINTS TO PONDER

- Terrorism originating in the Caucasus area continues to threaten Russia. In early 2010, suicide bombers killed 39 passengers on two subway trains in Moscow; a few months earlier, a similar number were killed when terrorists bombed the country's main railroad linking the capital and St. Petersburg.

- In early 2010 Russia signed an agreement to establish a military base on the coast of the separatist Georgian province of Abkhazia, proclaiming this to be a step on Abkhazia's road to independence and international recognition.

- The economy of Russia, the dominant state of this realm, depends largely on oil and gas exports. The problem: how to avoid having to depend on unreliable neighbors for transit to reach the international energy market.

- Russia is by far the world's largest state territorially. But in terms of population, it barely makes the top ten.

San Francisco from the north: Fisherman's Wharf (foreground), Telegraph Hill crowned by Coit Tower (middle), and the downtown skyline beyond. © H. J. de Blij

IN THIS CHAPTER

Spectacular scenery, natural wealth
Ties and differences between the United States and Canada
The North American creed
Bilingualism in Canada
The evolving geography of the information economy
Immigration and the future of multiculturalism

CONCEPTS, IDEAS, AND TERMS

3A
NORTH AMERICA: DEFINING THE REALM

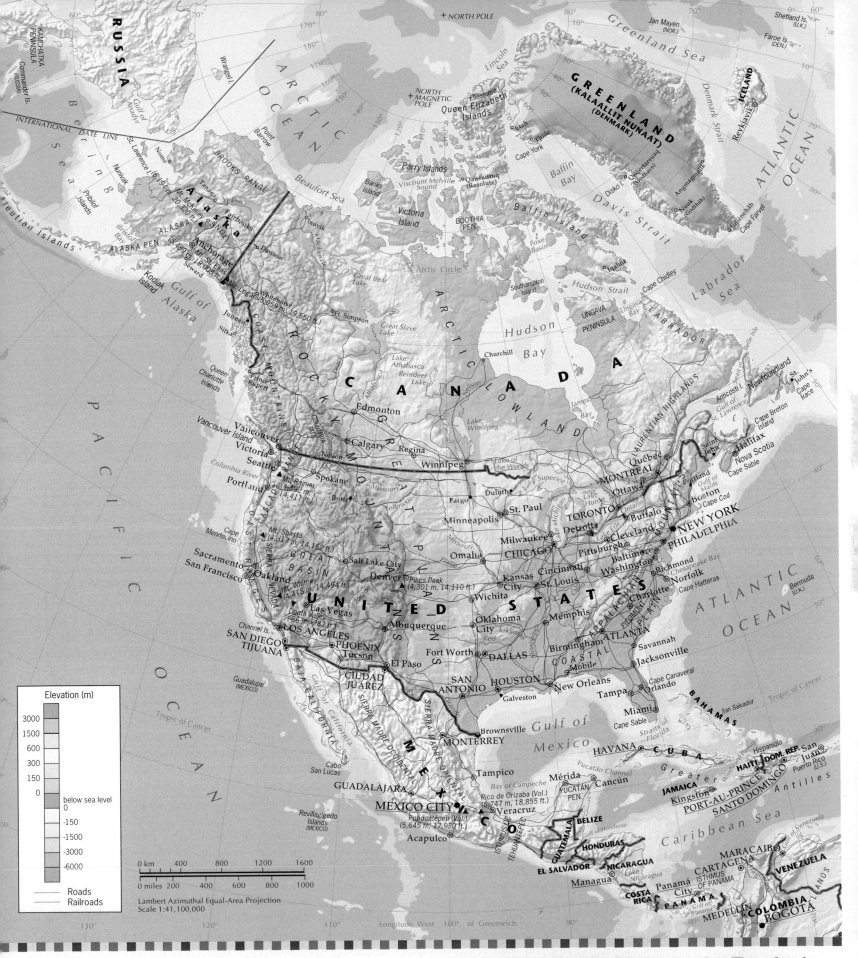

FIGURE 3A-1

© H. J. de Blij, P. O. Muller, and John Wiley & Sons, Inc.

We turn next to the Western Hemisphere, the two great interconnected continents that separate the Atlantic and Pacific oceans and extend, very nearly, from pole to pole, flanked by numerous islands large and small, indented by gulfs and bays of historic and economic import, and endowed with an enormous range of natural resources. Two continents—North and South—form the Americas, but three geographic realms blanket them (North, Middle, and South). In the context of physical (natural) geography, North America from Canada's Ellesmere Island in the far north to Panama in the south is a continent. In terms of modern human geography, the northern continent is divided into North American and Middle American realms along a transition zone marked by a political as well as physical boundary between the United States and Mexico (Fig. 3A-1). From the Gulf of Mexico to El Paso/Ciudad Juarez, the Rio Grande forms this border. From El Paso westward to the Pacific Ocean, straight-line boundaries, reinforced by fences and walls, separate the North from the Middle. Here, global core and global periphery meet, sometimes contentiously. We begin in North America.

▲ MAJOR GEOGRAPHIC QUALITIES

NORTH AMERICA

1. North America encompasses two of the world's biggest states territorially (Canada is the second-largest in size; the United States is third).

2. Both Canada and the United States are federal states, but their systems differ. Canada's is adapted from the British parliamentary system and is divided into ten provinces and three territories. The United States separates its executive and legislative branches of government, and it consists of 50 States, the Commonwealth of Puerto Rico, and a number of island territories under U.S. jurisdiction in the Caribbean Sea and the Pacific Ocean.

3. Both Canada and the United States are plural societies. Canada's pluralism is most strongly expressed in regional bilingualism. In the United States, divisions occur largely along ethnic, racial, and income lines.

4. A substantial number of Quebec's French-speaking citizens support a movement that seeks independence for the province. The movement's high-water mark was reached in the 1995 referendum in which (minority) non-French-speakers were the difference in the narrow defeat of separation. Prospects for a breakup of the Canadian state have diminished since 2000 but have not disappeared.

5. North America's population, not large by international standards, is one of the world's most highly urbanized and mobile. Largely propelled by a continuing wave of immigration, the realm's population total is expected to grow by more than 40 percent over the next half-century.

6. By world standards, this is a rich realm where high incomes and high rates of consumption prevail. North America possesses a highly diversified resource base, but nonrenewable fuel and mineral deposits are consumed prodigiously.

7. North America is home to one of the world's great manufacturing complexes. The realm's industrialization generated its unparalleled urban growth, but today both of its countries are experiencing the emergence of a new postindustrial society and economy.

8. The two countries heavily depend on each other for supplies of critical raw materials (e.g., Canada is a leading source of U.S. energy imports) and have long been each other's chief trading partners. Today, the North American Free Trade Agreement (NAFTA), which also includes Mexico, is linking all three economies ever more tightly as the remaining barriers to international trade and investment flow are dismantled.

9. North America is home to the world's most mobile people and also attracts more international migrants than any other geographic realm. Continued immigration and high transnational mobility make for an exceptionally multicultural realm.

North America is constituted by two of the world's most highly advanced countries by virtually every measure of social and economic development. Blessed by an almost unlimited range of natural resources and bonded by trade as well as culture, Canada and the United States are locked in a mutually productive embrace that is reflected by the economic statistics: in an average year of the recent past, more than 80 percent of Canadian exports went to the United States and about two-thirds of Canada's imports came from its southern neighbor. For the United States, Canada is not quite as dominant, but it is still its leading export—and import—market.

The realm is also defined on the basis of a range of broad cultural traits, from urban landscapes and a penchant for mobility to religious beliefs, language, and political persuasions. Both countries rank among the world's most highly urbanized: nothing symbolizes the North American city as strongly as the skyscrapered panoramas of New York, Chicago, and Toronto—or the vast, beltway-connected suburban expanses of Los Angeles, Washington, and Houston. North Americans are also the most mobile people in the world. Each year about one out of every eight individuals changes his or her residence, a proportion that has recently declined but still leads the world.

Despite notable multilingualism, in most areas English is both countries' *lingua franca*. And notwithstanding all the religions followed by minorities, the great majority of both Canadians and Americans are Christians. Both states are stable democracies, and both have federal systems of government.

Canadians and Americans also share a passion for the same sports. Baseball and a modified version of American football are popular sports in Canada; in 1992, the Toronto Blue Jays became the first team outside the United States to win the World Series. Canada's national sport, ice hockey, now has more NHL teams in America than in Canada, though curling, a slower-paced Canadian pastime, has not exactly captured the American imagination.

Driving from Michigan through southern Ontario to upstate New York does not involve a sharp contrast in cultural landscapes. Americans visiting Canadian cities (and vice versa) find themselves in mostly familiar settings. Canadian cities have fewer impoverished neighborhoods, no ethnic ghettos (although low-income ethnic districts do exist), lower crime rates, and, in general, better public transportation; but rush hour in Toronto very much resembles rush hour in Chicago. The boundary between the two states is porous, even after the events of 9/11. Americans living near the border shop for more affordable medicines in Canada; Canadians who can afford it seek medical treatment in the United States.

None of this is to suggest that Canadians and Americans don't have their differences—they do. Most of these differences may seem subtle to people from outside the realm, but they are important nonetheless, particularly so to Canadians who at times feel dominated by Americans whose political or cultural values they don't always share. You can even argue over names: if this entire hemisphere is called the Americas, aren't Canadians, Mexicans, and Brazilians, and others Americans too?

POPULATION CLUSTERS

Although Canada and the United States share many historical, cultural, and economic qualities, they also differ in significant ways, diversifying this realm. The United States, somewhat smaller territorially than Canada, occupies the heart of the North American continent and, as a result, encompasses a greater environmental range. The U.S. population is dispersed across most of the country, forming major concentrations along both the (north-south trending) Atlantic and Pacific coasts (Fig. 3A-2). The overwhelming majority of Canadians, on the other hand, live in an interrupted east-west corridor that extends across southern Canada, mostly within 300 kilometers (200 mi) of the U.S. border. The United States, again unlike Canada, is a fragmented country in that the peninsula of Alaska is part of it (offshore Hawai'i, however, belongs in the Pacific Realm).

Figure 3A-2 reveals that the great majority of both the U.S. and Canadian population resides to the east of a line drawn down the middle of the realm, reflecting the historic core-area development in the east and the later and still-continuing shift to the west, and, in the United States, to the south. Certainly this map shows the urban concentration of North America's population: you can easily identify cities such as Toronto, Chicago, Denver, Dallas-Fort Worth, San Francisco, and Vancouver. Just under 80 percent of the realm's population is concentrated in cities and towns, a higher proportion even than Europe's.

As the Data Table in Appendix B indicates, the population of the United States, which recently passed the

The central business districts (CBDs) of North America's large cities have distinctive skylines, often featuring architectural icons like the Empire State Building. Indeed, you should have little trouble identifying this CBD: the black profile of Hancock Tower on the shoreline of Lake Michigan just to the right of center is one of the signature structures of downtown Chicago, America's third-largest city. In fact, the only skyscraper more famous than Hancock Tower is Willis Tower, the Western Hemisphere's tallest building, from whose 110-story-high Skydeck this photo was taken. Willis Tower??? Absolutely—although it is still far more widely known by its former name of Sears Tower (the original 1973 tenant that relocated its corporate headquarters to suburban Hoffman Estates well over a decade ago). Today, according to its website, Willis Tower is reinventing itself as one of the world's greenest buildings, in the vanguard of a movement that marks Chicago as the U.S. leader in working toward a more sustainable urban future. © Vito Palmisano/Photographer's Choice/Getty Images, Inc.

300-million mark, is growing at a rate 50 percent higher than Canada's, and the 400-million mark may be reached as soon as 2040. This is an unusually high rate of growth for a high-income economy, resulting from a combination of natural increase (about 60 percent) and substantial immigration.

Although Canada's overall growth rate is significantly lower than that of the United States, immigration contributes proportionally even more to this increase than in the United States. With just under 34 million citizens, Canada, like the United States, has been relatively open to legal immigration, and as a result both societies exhibit a high degree of **1** cultural pluralism (a diversity of ancestral and traditional backgrounds). Indeed, Canada recognizes two official languages, English and French (the United States does not even designate English as such); and by virtue of its membership in the British Commonwealth, East and South Asians form a larger sector of Canada's population than Asians and Pacific Islanders do in the United States. On the other hand, cultural pluralism in the United States reflects large Hispanic (15 percent) and African American (13 percent) minorities and a wide range of other ethnic backgrounds.

Robust urbanization, substantial immigration, and cultural pluralism are defining properties of the human geography of the North American realm. We turn now to the physical stage on which the human drama is unfolding.

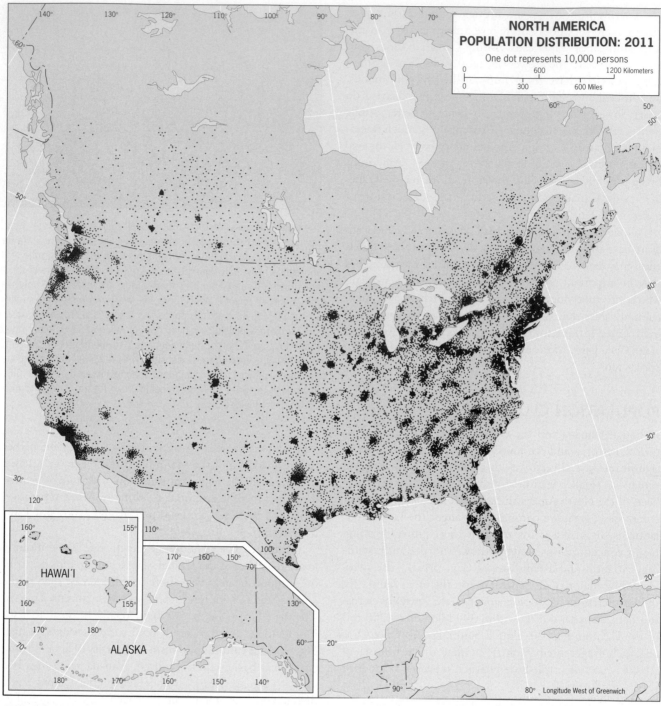

FIGURE 3A-2

© H. J. DE BLIJ, P. O. MULLER, AND JOHN WILEY & SONS, INC.

NORTH AMERICA'S PHYSICAL GEOGRAPHY

Physiographic Regions

One of the distinguishing properties of the North American landmass, all of which lies on the North American tectonic plate, is its remarkable variegation into regional landscapes. So well defined are many of these landscape regions that we use their names in everyday language—for example, when we say that we flew over the Rocky Mountains or drove across the Great Plains or hiked in the Appalachians. These landscapes are called **2** **physiographic regions**, and in the entire hemisphere their diversity is greatest to the north of the Rio Grande (Fig. 3A-3).

In the Far West, the Pacific Mountains extend all the way from Southern California through coastal Canada to Alaska. In the western interior, the Rocky Mountains form a continental backbone from central Alaska to New Mexico. Around the Great Lakes, the low-relief landscapes of the

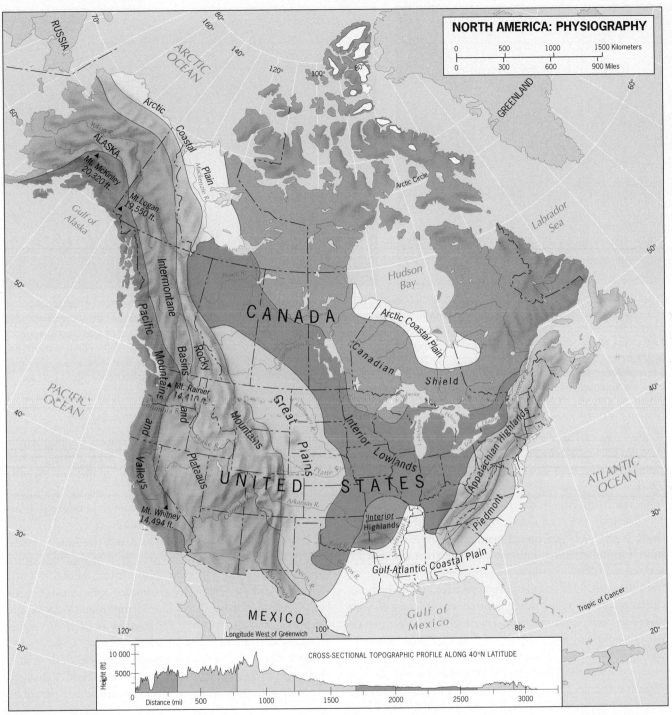

NORTH AMERICA: PHYSIOGRAPHY

CROSS-SECTIONAL TOPOGRAPHIC PROFILE ALONG 40°N LATITUDE

FIGURE 3A-3

© H. J. de Blij, P. O. Muller, and John Wiley & Sons, Inc.

Interior Lowlands and the Great Plains to the west are shared by Canada and the United States, and the international boundary even divides the Great Lakes. In the east, the Appalachian Mountains (as the cross-sectional inset shows, no match for the Rockies) form a corridor of ridges, valleys, and plateaus that represent a familiar topography from Alabama and Georgia to Nova Scotia and Newfoundland. If there is a major physiographic province that belongs to one of the realm's two countries, it is the Canadian Shield, scoured bare by the glaciers that deposited the

pulverized rocks as fertile soil in the U.S. Midwest, that is, in the Interior Lowlands.

Climate

This diversity of landscape is matched by a variety of climates. Take another look at Figure G-7, and you can see the lineaments of landscape mirrored in the contrasts of climate. North America may not have it all—there are no areas of true tropical environment in the North American realm except at the

With the exception of the Pacific coastal strip, the western half of the continental United States is virtually blanketed by dry (**B**) climates. This is the result of high mountain ranges, dominated by the Cascades and Sierra Nevada, that parallel the coast from the Canadian border southward to just northeast of Los Angeles. As the text indicates, this mountain wall functions as a barrier to wring out the maritime moisture in the eastward-flowing air, turning the leeward environment arid as it creates a far-reaching rain shadow. The photo on the left shows the eastern foothills and slopes of California's Sierra (capped by Mount Whitney, the lower-48's highest peak), desert-like and barren just a few miles east of the inland terminus of the coastal-strip moisture zone that has spent itself producing the snow in the higher elevations at the rear. The photo on the right, taken in South America's Peruvian Andes, reminds us that this transition from humid to arid climates can occasionally be abrupt: the razor-sharp boundary that follows the ridgeline unmistakably divides the lush tropical forest of the terrain in the rear from the foreground slopes totally devoid of vegetation. (Left) © Bruce Heinemann/Getty Images, Inc.; (Right) © Shouraseni Sen Roy

southern tip of Florida—but it has a lot of variation. North America has moist coastal zones and arid interiors, well-watered plains, and even bone-dry deserts. On the world map, *Cf* and *Df* climates are especially good for commercial farming; note how large North America's share of these environments is.

The map leaves no doubt: the farther north you go, the colder it gets, and even though coastal areas derive moderation from warmer offshore waters, which is why there is a Vancouver and a Halifax, the rigors of continentality set in not far from the coast. Hot summers, frigid winters, and limited precipitation make high-latitude continental interiors difficult places to make it on the land. Figure G-7 has much to do with the southerly concentration of Canadian population and with the lower densities in the interior throughout the realm.

In the west, especially in the United States, you can see what the Pacific Mountains do to areas inland. Moisture-laden air arrives from the ocean, the mountain wall forces the air upward, cools it, condenses the moisture in it, and produces rain—the rain for which Seattle and Portland and other cities of the Pacific Northwest are (in)famous. By the time the air crosses the mountains and descends on the landward side, most of the moisture has been drawn from it, and the forests of the ocean side give way to scrub and brush (see photo above). This **3** **rain shadow effect** extends all the way across the Great Plains; North America does not turn moist again until the Gulf of Mexico sends humid tropical air into the eastern interior via the Mississippi-Missouri River Basin.

In a very general way, therefore, and not including the coastal areas on the Pacific, nature divides North America into an arid west and a humid east. Again the population map reveals more than a hint of this: draw a line approximately from Lake Winnipeg to the mouth of the Rio Grande, and look at the contrast between the (comparatively humid) east and drier west when it comes to population density. Water is a large part of this story.

Between the Rocky Mountains and the Appalachians, North America lies open to air masses from the frigid north and tropical south. In winter, southward-plunging polar fronts send frosty, bone-dry air masses deep into the heart of the realm, turning places like Memphis and Atlanta into iceboxes; in summer, hot and humid tropical air surges northward from the Gulf of Mexico, giving Chicago and Toronto a taste of the tropics. Such air masses clash in low-pressure systems along weather fronts loaded with lightning, thunder, and, frequently, dangerously destructive tornadoes.

Great Lakes and Great Rivers

Two great drainage systems lie between the Rockies and the Appalachians: (1) the five Great Lakes that drain into the St. Lawrence River and the Atlantic Ocean, and (2) the mighty Mississippi-Missouri system that carries water from a vast interior watershed to the Gulf of Mexico, where the Mississippi forms one of the world's major *deltas*. Both natural systems have been modified by human engineering. In the case of the St. Lawrence Seaway, a series of locks and canals has created a direct shipping route from the Midwest to the Atlantic. The Mississippi and Missouri rivers have

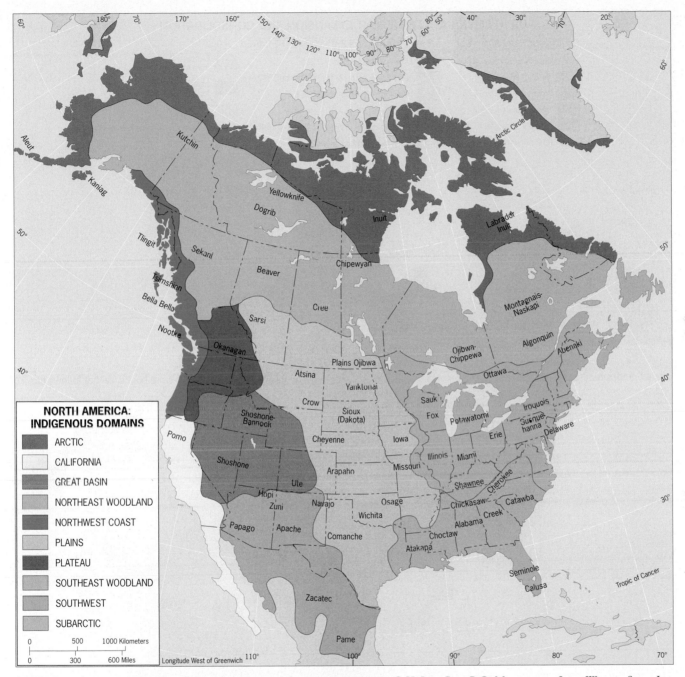

FIGURE 3A-4

© H. J. de Blij, P. O. Muller, and John Wiley & Sons, Inc.

been fortified by artificial levees that, while failing to contain the worst of flooding, have enabled farmers to cultivate the most fertile of American soils.

EUROPEAN SETTLEMENT AND EXPANSION

When the first Europeans set foot on North American soil, the continent was inhabited by millions of people whose ancestors had reached the Americas from Asia via Alaska, and possibly also across the Pacific, more than 14,000 years before (and perhaps as long as 30,000 years ago). In search of Asia, the Europeans misnamed them Indians. These Native Americans or *First Nations**—as they are now called in the United States and Canada, respectively—were organized into hundreds of nations with a rich mosaic of languages and a great diversity of cultures (Fig. 3A-4).

The eastern nations were the first to bear the brunt of the European invasion. By the end of the eighteenth

---

*The indigenous peoples of Canada, numerically dominated by First Nations, also include Métis (of mixed native and European descent) and the Inuit (formerly Eskimos) of the far north.

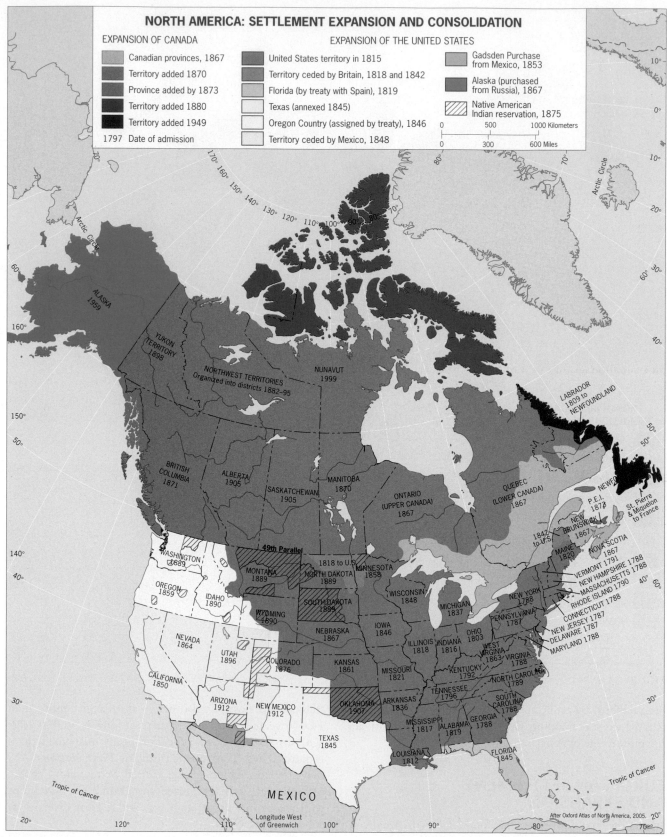

NORTH AMERICA: SETTLEMENT EXPANSION AND CONSOLIDATION

EXPANSION OF CANADA

- Canadian provinces, 1867
- Territory added 1870
- Province added by 1873
- Territory added 1880
- Territory added 1949
- 1797 Date of admission

EXPANSION OF THE UNITED STATES

- United States territory in 1815
- Territory ceded by Britain, 1818 and 1842
- Florida (by treaty with Spain), 1819
- Texas (annexed 1845)
- Oregon Country (assigned by treaty), 1846
- Territory ceded by Mexico, 1848
- Gadsden Purchase from Mexico, 1853
- Alaska (purchased from Russia), 1867
- Native American Indian reservation, 1875

0   500   1000 Kilometers
0   300   600 Miles

After Oxford Atlas of North America, 2005.

FIGURE 3A-5

© H. J. DE BLIJ, P. O. MULLER, AND JOHN WILEY & SONS, INC.

century, ruthless and land-hungry settlers had driven most of the Native American peoples living along the Atlantic and Gulf coasts from their homes and lands, initiating a westward push that was to devastate indigenous society. North America's native nations are now left with about 4 percent of U.S. territory in the form of mostly impoverished reservations.

The current population geography of North America is rooted in the colonial era of the seventeenth and eighteenth centuries that was dominated by Britain and France. The French sought mainly to organize a lucrative fur-trading network, while the British established settlements along the coast of what is today the eastern seaboard of the United States (the oldest, Jamestown, was founded in Virginia more than four centuries ago).

The British colonies soon became differentiated in their local economies, a diversity that was to endure and later shape much of North America's cultural geography. The northern colony of New England (Massachusetts Bay and environs) specialized in commerce; the southern Chesapeake Bay colony (Virginia and Maryland) emphasized the plantation farming of tobacco; the Middle Atlantic area lying in between (southeastern New York, New Jersey, eastern Pennsylvania) formed the base for a number of smaller, independent-farmer colonies.

These neighboring colonies thrived and yearned to expand, but the British government responded by closing the inland frontier and tightening economic controls. This policy provoked general opposition in the colonies, followed by the revolutionary challenge that was to lead to independence and the formation of the United States of America. The western frontier of the fledgling nation now swung open, and the area north of the Ohio River was occupied soon after it was discovered that soils and climate in these Interior Lowlands (Fig. 3A-3) were more favorable to farming than those of the Atlantic Coastal Plain and the Appalachian foothills of the Piedmont to its west. Deforestation cleared huge tracts for farming and livestock, and new cultural landscapes spread apace. The acquisition of more distant western frontiers followed in short order (Fig. 3A-5).

CULTURAL FOUNDATIONS

The modern North American creed has been characterized as exhibiting an adventurous drive, a liking for things new, an ability to move, a sense of individualism, an aggressive pursuit of goals and ambition, a need for societal acceptance, and a firm sense of destiny. These qualities are not unique to modern North American cultures, of course, but in combination they create a particular and pervasive mind-set that is reflected in many ways in this geographic realm. Generally speaking, these tendencies tend to translate into an intense pursuit of educational and other goals and high aspirations for upward socioeconomic mobility.

Facilitating these aspirations is *language*. None of these goals would be within reach for so many were it not for the use of English throughout most of the realm. In Europe, by contrast, language inhibits the mobility so routinely practiced by Americans: workers moving from Poland to Ireland find themselves at a disadvantage in competition with immigrants from Ghana or Sri Lanka. In North America, a worker from Arkansas would not even consider language to be an issue when applying for a job in Calgary.

The near-universality of English in North America is diminishing, as we shall see, but English remains the dominant medium of interaction. Interestingly, English in the United States is undergoing a change that is affecting it worldwide: in areas where it is the second language it is blending with the local tongue, producing hybrids just as is happening in Nigeria ("Yorlish"), Singapore ("Singlish"), and the Philippines ("Taglish"). In this respect, English may become the Latin of the present day: Latin, too, produced blends that eventually consolidated into Italian, French, Spanish, and the other Romance languages.

Also reflecting the cultural values cited earlier is the role of *religion*, which sets American (more so than Canadian) culture apart from much of the rest of the Western world. The overwhelming majority of Americans express a belief in God, and a large majority regularly attend church, in contrast to much of Europe. Religious observance is a virtual litmus test for political leaders; no other developed country prints "In God We Trust" on its currency. Sects and denominations are important in the life of Christian-dominated North America.

Figure 3A-6 shows the mosaic of Christian faiths that blankets the realm. A map at this scale can only suggest the broadest outlines of a much more intricate pattern: Protestant denominations are estimated to number in the tens of thousands in North America, and no map could show them all. Southern (and other) Baptists form the majority across the U.S. Southeast from Texas to Virginia, Lutherans in the Upper Midwest and northern Great Plains, Methodists in a belt across the Lower Midwest, and Mormons in the interior West centered on Utah. Roman Catholicism prevails in most of Canada as well as the U.S. Northeast and Southwest, where ethnic Irish and Italian adherents form majorities in the Northeast and Hispanics in the Southwest. Behind these patterns lie histories of proselytism, migration, conflict, and competition, but tolerance of diverse religious (even nonreligious) views and practices is a hallmark of this realm.

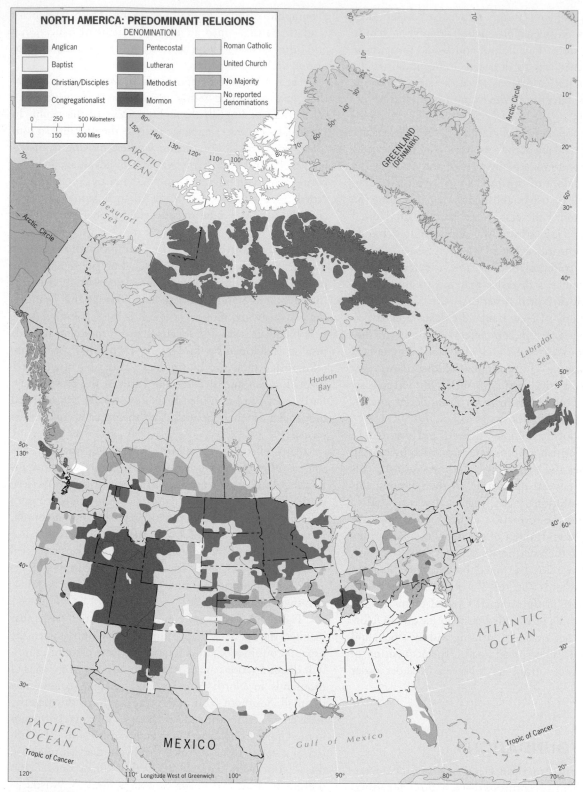

FIGURE 3A-6

© H. J. de Blij, P. O. Muller, and John Wiley & Sons, Inc.

THE FEDERAL MAP OF NORTH AMERICA

The two states that constitute North America may have arrived at their administrative frameworks with different motives and at different rates of speed, but the result is unmistakably similar: their internal political geographies are dominated by straight-line boundaries of administrative convenience (Fig. 3A-7). In comparatively few places, physical features such as rivers or the crests of mountain ranges mark internal boundaries, but by far the greater

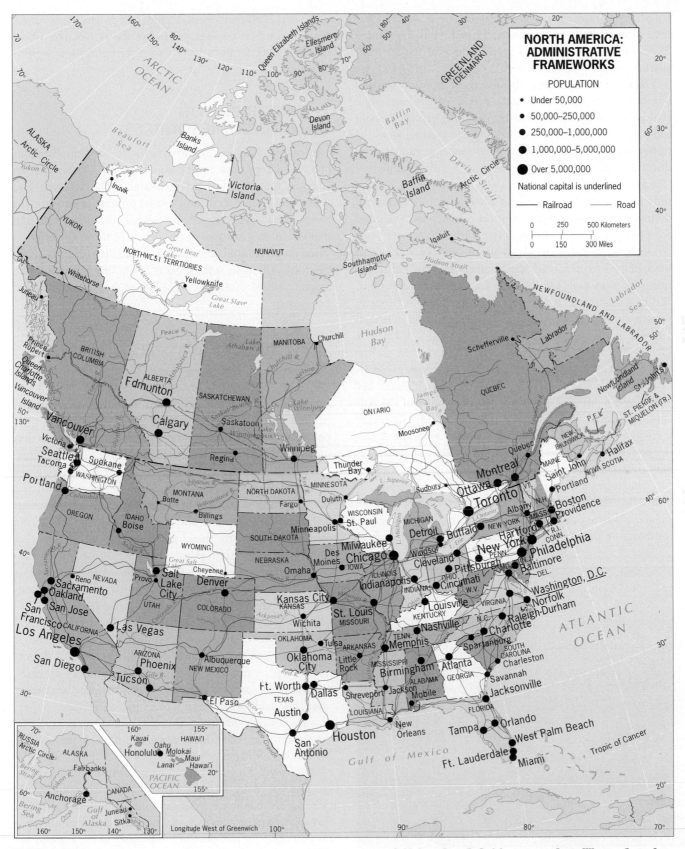

NORTH AMERICA: ADMINISTRATIVE FRAMEWORKS

POPULATION

- Under 50,000
- 50,000–250,000
- 250,000–1,000,000
- 1,000,000–5,000,000
- Over 5,000,000

National capital is underlined

Railroad   Road

0   250   500 Kilometers
0   150   300 Miles

FIGURE 3A-7   © H. J. de Blij, P. O. Muller, and John Wiley & Sons, Inc.

length of boundaries is ruler-straight. Even most of the international boundary between Canada and the United States west of the Great Lakes coincides with a parallel of latitude, 49°N.

The reasons are all too obvious: the framework was laid out before, in some areas long before, significant white settlement occurred. There was something arbitrary about these delineations, but by delimiting the internal administrative units early and clearly, governments precluded later disputes over territory and resources. In any case, neither Canada nor the United States was to become a unitary state: both countries are *federal states*. The Canadians call their primary subdivisions provinces, whereas in the United States they are, as the country's name implies, States.

It is no matter of trivial pursuit to take a careful look at some of the provinces and States that compartmentalize this realm, because some are of greater consequence than others. We have already noted the special significance of mainly French-cultured Quebec, still in a special position in federal Canada. In many ways the key Canadian province is Ontario, containing the country's largest and most globalized city (Toronto), and with the capital (Ottawa) on its eastern river border with Quebec. In the United States, key States in the political and economic geography of the federation include Texas bordering Mexico, California facing the Pacific, New York with its urban window on the Atlantic, and Florida pointing toward the Caribbean (Miami is often referred to as the gateway to Latin America).

The rectangular layout of North America seems to symbolize the realm's modernity and its stability, while the federal structure itself (with considerable autonomy at the State or provincial level) reflects the culture of free-

Oil-drilling always poses ecological risks, and the massive blowout of BP's notorious Deepwater Horizon well in the Gulf of Mexico just south of the Mississippi Delta on April 20, 2010 quickly became one of the worst environmental disasters in American history. This appalling scene is the shoreline of Louisiana's Grand Isle, the barrier island on the west side of the delta, taking the brunt of the first onslaught of oil about a month into the spill. © Julie Dermansky/© Corbis

dom and independence. If federalism works in these high-income economies, shouldn't the rest of the world emulate it? Federalism has its assets but also some significant liabilities.

THE DISTRIBUTION OF NATURAL RESOURCES

One of these liabilities has to do with the variable allotment of natural resources among the States and provinces. In Canada, Alberta is favored with large energy reserves (and more may be in the offing), but its provincial government is not always eager to see the federal administration take that wealth to assist poorer provinces. In the United States, when oil prices are high, Texas and Oklahoma boom while California's budget suffers. So it is important to view the map of natural resources (Fig. 3A-8) with that of States and provinces.

Water certainly is a natural resource, and North America as a realm is comparatively well supplied with it despite concerns over long-term prospects in the American Southwest and the Great Plains, where some States depend on sources in other States. Another concern focuses on lowering water tables in some of North America's most crucial aquifers (underground water reserves) in which a combination of overuse and decreasing replenishment suggests that supply problems may arise.

North America is endowed with abundant reserves of *minerals* that are mainly found in three zones: the Canadian Shield north of the Great Lakes, the Appalachian Mountains in the east, and the mountain ranges of the west. The Canadian Shield contains substantial iron ore, nickel, copper, gold, uranium, and diamonds. The Appalachians yield iron ore, lead, and zinc. And the western mountain zone has significant deposits of copper, lead, zinc, molybdenum, uranium, silver, and gold.

In terms of **4** fossil fuel energy resources (oil, natural gas, coal), North America is also quite well endowed, though the realm's voracious demand, the highest in the world, cannot be met by domestic supplies and necessitates an enormous need for imports. Figure 3A-8 displays the distribution of oil, oil-yielding tar sands, natural gas, and coal.

The leading *oil*-production areas lie (1) along and offshore from the Gulf Coast, where the floor of the Gulf of Mexico is yielding a growing share of the output; (2) in the Midcontinent District, from western Texas to eastern Kansas; and (3) along Alaska's North Slope facing the Arctic Ocean. An important development is taking place in Canada's northeastern Alberta, where oil is being drawn from deposits of *tar sands* in the vicinity of the boomtown of Fort McMurray. The process is expensive and can reward investors only when the price of oil is comparatively high, but the reserves of oil estimated to be contained in the tar sands may exceed those of Saudi Arabia. In 2008, when the price of oil skyrocketed, activity in this area soared. When

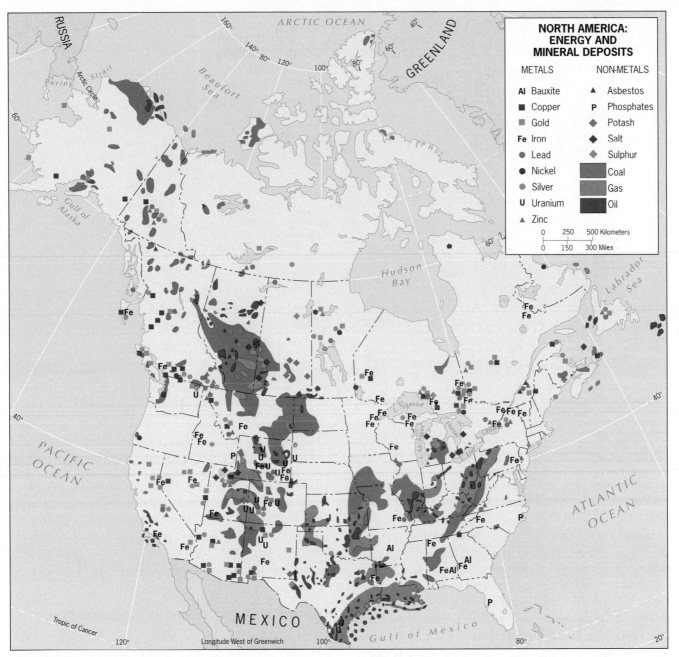

FIGURE 3A-8   © H. J. de Blij, P. O. Muller, and John Wiley & Sons, Inc.

the price collapsed later that year, production slowed. It is a pattern that will probably continue for decades to come.

The distribution of ***natural gas*** reserves resembles that of oilfields because petroleum and natural gas tend to be found in similar geological formations (that is, the floors of ancient, shallow seas). What the map cannot reveal is the volume of production, in which this realm leads the world (Russia and Iran lead in terms of proven reserves). Natural gas has risen rapidly as the fuel of choice for electricity generation in North America, raising fears that the supply will not keep up with growing demand.

The ***coal*** reserves of North America, perhaps the largest on the planet, are found in Appalachia, under the Great Plains of the United States as well as Canada, and in the southern Midwest among other places. These reserves guarantee an adequate supply for centuries to come, though coal has become a less desirable fuel due its polluting emissions that contribute to global warming.

CITIES, INDUSTRIES, AND THE INFORMATION ECONOMY

Industrial Cities

When the Industrial Revolution crossed the Atlantic and touched down in America in the 1870s, it took hold so successfully and advanced so robustly that only 50 years later North America was surpassing Europe as the world's mightiest

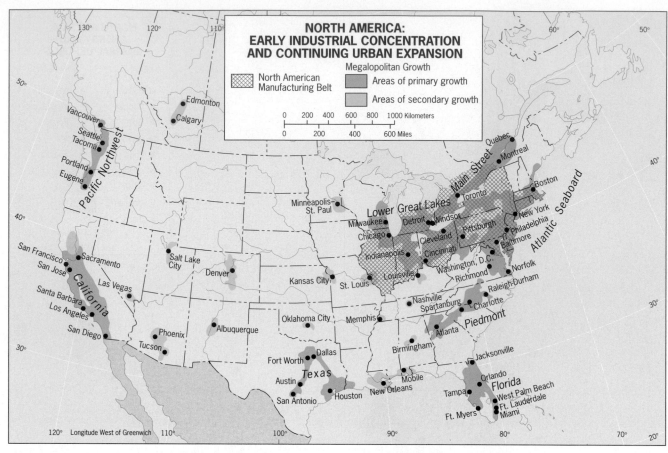

FIGURE 3A-9

© H. J. DE BLIJ, P. O. MULLER, AND JOHN WILEY & SONS, INC.

industrial complex. Industrialization moved in tandem with urbanization, as manufacturing plants in need of large labor supplies were built in or near cities. This, in turn, propelled rural-urban migration (including the migration of African Americans from the South to northern cities during the first half of the twentieth century) and the growth of individual cities. At a larger scale, a *system* of new cities emerged that specialized in the collection and processing of raw materials and the distribution of manufactured products. This **5** **urban system** was interconnected through a quickly expanding network of railroads (themselves a product of the Industrial Revolution).

As new technologies and innovations emerged, and specializations such as Detroit's automobile industry strengthened, an **6** **American Manufacturing Belt** (that extended into southern Ontario) evolved into the foundation of a North American Core (Fig. 3A-9). This core area, on the way to becoming the world's most productive and important, contained the majority of the realm's industrial activity and leading cities, including New York, Chicago, Toronto, and Pittsburgh (the "steel city").

In the course of intertwined urbanization and industrialization, the spatial economy of North America underwent profound changes. The **primary** sector, involving the extraction of raw materials from nature (agriculture, mining, fishing), was rapidly mechanized, and the workforce in this sector shrank considerably. Employment in the **secondary** sector, using the input of raw materials and involving manufacturing, grew rapidly. The **tertiary** sector, entailing all kinds of services to support production and consumption (banking, retail, transport) expanded as well. Both the secondary and tertiary sectors were concentrated in cities.

Deindustrialization and Suburbanization

Cities continued to evolve, along with the nature of the economy and with the development of new transportation and communication technologies. The mass introduction of automobiles and the accompanying construction of a large-scale highway system, especially after World War II, had two major effects on cities. First, it resulted in a much higher mobility and interconnectedness between cities. Second, it propelled a process of suburbanization: the change from compact city to widely dispersed metropolis, the evolution of residential suburbia into a complete **7** **outer city** with its own businesses and industries, schools, hospitals, and other amenities. As the newly urbanized suburbs increasingly captured major economic activities, many central cities saw their comparative status diminish.

Importantly, this development coincided with the ***deindustrialization*** that hit American cities in the 1960s: the loss of manufacturing (and jobs) due to automation and relocation of production to countries with lower wages. While the suburban outer cities became the destination of intrametropolitan migrants and of economic activity and employment in the service sector, many once-thriving and dominant CBDs were now all but reduced to serving the less affluent populations that increasingly dominated the central city's close-in neighborhoods. The fate of Detroit is often cited as exemplifying this sequence of events, but to a certain degree all major U.S. cities suffered.

The Information Economy and City Regions

Since the early 1980s, many North American cities (not all) have bounced back from the destructive effects of deindustrialization and unemployment. The *information economy*—embodied in the *quaternary* sector—had arrived. Many northern cities had lost population and had been "hollowed out" for nearly two decades, but the tide turned again. Employment in the tertiary sector started to grow rapidly, especially in high-technology, producer services (such as consulting, advertising, and accounting), finance, research and development,

and the like. Indeed, with the advent of computerization, most of these information-rich services evolved into quaternary activities. The geography of the information economy is certainly very different from that of the old manufacturing industries and is still not fully understood. Some of it concentrates in established CBDs like Midtown Manhattan, some of it locates around important hubs of the Internet infrastructure to benefit from maximum bandwidth (for example, Dallas), and yet other activities cluster in areas with a high density of highly skilled workers such as Silicon Valley or North Carolina's Research Triangle.

Northern California's *Silicon Valley*—the world's leading center for computer research and development and the headquarters of the U.S. microprocessor industry—illustrates the locational dynamics of the new (high-tech) economy. Proximity to Stanford, a world-class research university, not far from cosmopolitan San Francisco, with the presence of a large pool of highly educated and skilled workers, a strong business culture, a lot of available capital, good housing, and a scenic area with good weather, combined to make Silicon Valley a prototype for similar developments elsewhere, and not just in North America. Under different

 FROM THE FIELD NOTES...

© Andy Ryan Photography

"Monitoring the urbanization of U.S. suburbs for the past four decades has brought us to Tyson's Corner, Virginia on many a field trip and data-gathering foray. It is now hard to recall from this 2008 view that less than 50 years ago this place was merely a near-rural crossroads. But as nearby Washington, D.C. steadily decentralized, *Tyson's* capitalized on its unparalleled regional accessibility (its Capital Beltway location at the intersection with the radial Dulles Airport Toll Road) to attract a seemingly endless parade of high-level retail facilities, office complexes, and a plethora of supporting commercial services." [Today, this suburban downtown ranks among the largest business districts in all of North America. But it also exemplifies urban sprawl, and in 2009 its developers, tenants, and county government formed a coalition to transform Tyson's into a true city characterized by smarter, greener development. The elements of the new plan include much higher residential densities, mass transit in the form of four stations on a new Metro line connecting to Washington and Dulles Airport, and major new facilities to accommodate pedestrians and bicycles. Details can be found in a prominent article in the June 22, 2009 issue of *Time*.]

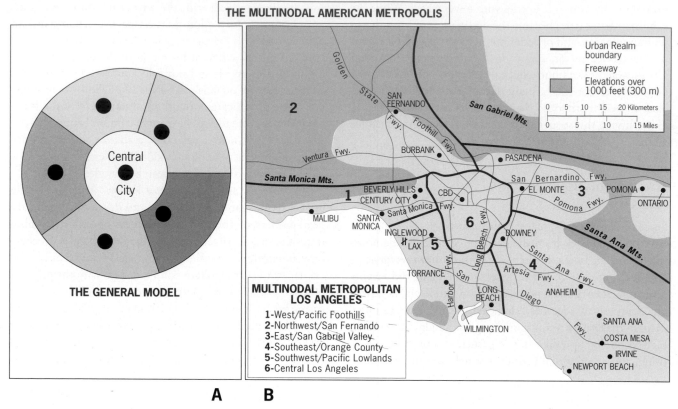

FIGURE 3A-10

© H. J. de Blij, P. O. Muller, and John Wiley & Sons, Inc.

names (*technopolis* is used in Brazil, France, and Japan; *science park* in China, Taiwan, and South Korea) such ultramodern, campus-like complexes symbolize the postindustrial era just as the smoke-belching factory did the industrial age.

Polycentric Cities

Fly into a contemporary U.S. or Canadian city, and you will see the high-rises of suburban downtowns encircling the old central-city CBD, some of them boasting their own impressive skylines. In Canada, where the process has had more to do with lower land values outside the city centers than with ethnic separation, the spatial effect has become similar, although most Canadian cities still remain more compact than the far-flung U.S. metropolis.

Thus the overall structure of the modern North American metropolis is polycentric and resembles a pepperoni pizza (Fig. 3A-10) in its typical form. The traditional CBD still tends to be situated at the center, much of its former cross-traffic diverted by beltways; but the outer city's CBD-scale nodes are ultramodern and thriving. Efforts to attract businesses and more affluent residents back to the old CBD sometimes involve the construction of multiple-use high-rises that displace low-income residents, resulting in conflicts and lawsuits. The **8 gentrification** of nearby, crumbling neighborhoods raises real estate values as well as taxes and tends to drive lower-income locals from their homes. Restoring the old CBDs may be attainable in some

cities, but the formation of large interconnected urban regions with multiple centers will continue.

IMMIGRANTS AND THE MAKING OF A MULTICULTURAL REALM

The Virtues of Mobility and Immigration

Given the mobility of North America's population, the immigrant streams that continue to alter it, the economic changes affecting it, and the forces acting upon it, Figure 3A-2 should be viewed as the latest still in a motion picture, one that has been unreeling for four centuries. Slowly at first, then with accelerating speed after 1800, North Americans pushed their settlement frontier westward to the Pacific. Even today, such shifts continue. Not only does the "center of gravity" of population continue to move westward, but within the United States it is also moving southward—the latter a drift that gained momentum in the 1970s as the advent of universal air conditioning made the so-called ***Sunbelt*** States of the U.S. southern tier ever more attractive to internal migrants.

The current population map is the still-changing product of numerous forces. For centuries, North America has attracted a pulsating influx of immigrants who, in the faster-growing United States, were rapidly assimilated into the societal mainstream. Throughout the realm, people have sorted themselves out to maximize their proximity to evolving economic opportunities, and they have shown little resistance to relocating as these opportunities successively favored new locales.

During the past century such transforming forces have generated a number of major migrations, of which the still-continuing shift to the west and south is only the latest. Five others are: (1) the persistent growth of metropolitan areas, first triggered by the late-nineteenth-century Industrial Revolution's impact in North America; (2) the large-scale movement of African Americans from the rural South to the urban North during the height of the industrial era; (3) the shift of tens of millions of urban residents from central cities to suburbs and subsequently to exurbs even farther away from the urban core; (4) the return migration of millions of African Americans from the deindustrializing North back to the growing opportunities in the South (in metropolises such as Atlanta); and (5) the strong and steady influx of immigrants from outside North America including, in recent times, Cubans, Mexicans, and other Latinos; South Asians from India and Pakistan; and East and Southeast Asians from Japan, Hong Kong, and Vietnam.

Several of the migration streams affecting North America rank among the largest in the history of the world, creating the pair of plural societies that characterize this realm. North America functions, in some ways, as a global magnet for human resources, and its continuous prosperity hinges at least in part on wave after wave of ambitious immigrants, be they engineers from China, physicians from India, nurses from Jamaica, or agricultural workers from Mexico.

The Challenge of Multiculturalism

But growing pluralism, especially at a time of globalization and **transnationalism**, comes with challenges. The sheer dimensions of immigration into the United States, from virtually all corners of the world, creates an increasingly complex ethnic and cultural mosaic that tests **9 melting pot** assertions at every geographic scale. Today there are more people of African descent in the United States than live in Kenya. The number of Hispanic residents in America is nearing half the population of Mexico. Miami is the second-biggest Cuban city after Havana, and Montreal the largest French-speaking city in the world after Paris. In short, there are sufficient immigrant numbers in the United States for them to create durable societies within the national society. The challenge is to ensure that the great majority of the immigrants become full participants in that larger society.

The issue is compounded, and politicized, by the fact that never in its history has the realm received so large an undocumented (illegal) immigration flow in addition to the legitimate stream. The ongoing influx from Middle to North America, principally from or via Mexico, is one of the largest population shifts in human history. A considerable part of it is illegal, engendering a sometimes acrimonious debate over border security and law enforcement. In 2010, an estimated 12 million illegal immigrants from Mexico were in the country at a time when the Hispanic minority had already become the country's largest, transforming cities and neighborhoods.

Such are the numbers and so diverse are the cultural traditions that the melting pot that America once was, is changing into something different. The United States is becoming a *mosaic culture*, an increasingly heterogeneous complex of separate, more or less uniform "tiles" that spend less time interacting and "melting." This does not apply only to new immigrant groups but also to existing communities, underscored by the proliferation of "gated communities" in metropolitan areas across the realm. There is a serious downside to this, as Bill Bishop writes in his book *The Big Sort: Why the Clustering of Like Minded America Is Tearing Us Apart*. As is the case in the world at large, balkanization fueled by people wanting to interact only with those closely resembling themselves leads to misunderstanding of others, miscalculations when it comes to decisions and policies, and widening fissures between communities. This could threaten the very survival of the democratic values that have underpinned the evolution of American society.

In Canada, the melting-pot notion was put to the test during the 1990s when East Asian immigrants from Hong Kong arrived during and following the 1997 takeover of the British colony by Beijing's communist government. Affluent Chinese families arriving in Vancouver proceeded to purchase and renovate (or replace) traditional homes in long-stable Vancouver neighborhoods, arousing the ire of some locals. But in general, Canadian views of immigration differ from those in the United States. Canada (before the economic downturn starting in 2008) has long faced critical labor shortages, especially in its western provinces including energy-booming Alberta and Asian-trade-burgeoning British Columbia. In need not only of professionals but also truck drivers and dock workers, Canada tries to balance its legal immigration process to keep it compatible with the country's employment as well as demographic needs. Across the realm, multicultural tendencies now vary on a regional basis depending on the subtle interplay of economics, immigration, local politics, and existing cultural patterns.

POINTS TO PONDER

- The U.S. population passed the 300-million mark in 2006. Get ready for 400 million in 2043.

- The United States does not have an official language. Should it?

- Whereas more than 70 percent of Canada's population lives within 200 kilometers (125 mi) of the U.S. border, only about 12 percent of the Mexican population lives that close to the U.S. border.

- The North Atlantic Ocean for generations was the most important external trade route for North America, by volume and value. Now it is the Pacific Ocean.

Small fishing ports still dot the New England coast: Chatham, Massachusetts, at the elbow of Cape Cod.
© H. J. de Blij

IN THIS CHAPTER

Four Canadas or more?
Where are America's minorities?
Economic fortunes of nine regions
The destructive force of hurricanes
China's impact on North America's west coast
The Alaskan frontier and U.S. geopolitics

CONCEPTS, IDEAS, AND TERMS

Devolution	1
Cross-border linkages	2
Technopole	3
Pacific Rim	4

3B

NORTH AMERICA: REGIONS OF THE REALM

≡ North American Core
Maritime Northeast
French Canada
South
Southwest
Pacific Hinge
Western Frontier
Continental Interior
Northern Frontier

NORTH AMERICAN REGIONS: CORE AND PERIPHERIES

ARCTIC OCEAN

PACIFIC OCEAN

ATLANTIC OCEAN

Anchorage

Hudson Bay

NORTHERN FRONTIER

FRENCH CANADA

Fort McMurray

Edmonton

Vancouver

Seattle

Calgary

Winnipeg

Portland

(North)

Montreal

MARITIME NORTHEAST

Halifax

Boston

Toronto

New York

Milwaukee

Detroit

Philadelphia

Washington, D.C.

WESTERN FRONTIER

CONTINENTAL

Minneapolis

Chicago

CORE

PACIFIC HINGE

INTERIOR

Omaha

St. Louis

Norfolk

San Francisco

Salt Lake City

Denver

(South)

Kansas City

Charlotte

Las Vegas

Memphis

Atlanta

Los Angeles

SOUTHWEST

Dallas

SOUTH

Jacksonville

San Diego

Phoenix

Houston

Tampa

San Antonio

New Orleans

Miami

Tropic of Cancer

Gulf of Mexico

Longitude West of Greenwich

FIGURE 3B-1

© H. J. de Blij, P. O. Muller, and John Wiley & Sons, Inc.

Geology, climate, culture, and history have combined to create an interesting set of regions in North America (Fig. 3B-1). Mountain ranges, coastal zones, deserts, low-lying plains, and a range of climate types all demand different kinds of human adaptation. Successive waves of immigrants brought with them a variety of cultures; sometimes the differences were subtle, but sometimes they were quite stark. The emergence of two large countries, the United States and Canada, provided two major political and institutional territorial structures within which regional differences have been accommodated. As Figure 3B-1 shows, a majority of North America's regions traverse both countries irrespective of political boundaries. But some other regional developments are best understood within the national context of each country. As much as Canada and the United States have in common, there is ample reason to focus the first part of this chapter on each individually. In the discussion that follows, therefore, we begin by highlighting the regional geographies of two concordant neighbors—in the knowledge that Canadians tend to know more about the United States than Americans know about Canada. After that, we will examine the nine evolving regions of the realm as a whole.

REGIONALISM IN CANADA

Canada's Spatial Structure

Territorially the second-largest state in the world after Russia, Canada is administratively divided into only 13 subnational entities—10 provinces and 3 territories (the United States has 50). The provinces—where 99.7 percent of all Canadians live—range in dimensions from tiny, Delaware-sized Prince Edward Island to Quebec, nearly twice the size of Texas (Fig. 3B-2).

As in the United States, the smallest provinces lie in the northeast. Canadians call these four the Atlantic Provinces (or simply Atlantic Canada): Prince Edward Island, Nova Scotia, New Brunswick, and the mainland-and-island province named Newfoundland and Labrador. Gigantic, mainly Francophone (French-speaking) Quebec and populous, heavily urbanized Ontario, both flanking Hudson Bay, form the heart of the country. Most of western Canada (which is what Canadians call everything west of Ontario) is organized into the three Prairie Provinces: Manitoba, Saskatchewan, and Alberta. In the far west, beyond the Canadian Rockies and extending to the Pacific Ocean, lies the tenth province, British Columbia.

Of the three territories in Canada's Arctic North, Yukon is the smallest—but nothing is small here in Arctic Canada. With the Northwest Territo-ries and recently created Nunavut, the three Territories cover almost 40 percent of Canada's total area. The population, however, is very small: only about 100,000 people inhabit this vast, frigid frontier zone.

The case of Nunavut merits special mention. Created in 1999, this newest Territory is the outcome of a major aboriginal land claim agreement between the *Inuit* people (formerly called Eskimos) and the federal government, and encompasses all of Canada's eastern Arctic as far north as Ellesmere Island (Fig. 3B-2), an area far larger than any other province or territory. With about one-fifth of Canada's total area, Nunavut—which means "our land"—has 31,000 residents, of whom 80 percent are Inuit.

Canada's population is clustered in a discontinuous ribbon, 300 kilometers (ca. 200 mi) wide, nestled against the U.S. border and along ocean shores, whereas most of the country's energy reserves lie far to the north of this ribbon. As such, the map of Canada's human geography bears out the generalization and emphasizes the importance of environment. People choose to live in the more agreeable areas to the south, but territorial control of abundant northern resources is very important. The population pattern creates cross-border affinities with major American cities in several places (Toronto-Buffalo, Windsor-Detroit, Vancouver-Seattle), although none of this happens in the three Prairie Provinces of the Canadian West, which for the most part adjoin North Dakota and Montana.

Cultural Contrasts

Canada's capital, Ottawa, is symbolically located astride the Ottawa River, which marks the boundary between English-speaking Ontario and French-speaking Quebec. Though comparatively small, Canada's population is markedly divided by culture and tradition, and this division has a pronounced regional expression. Just under 60 percent of Canada's citizens speak English as their mother tongue, 23 percent speak French, and the remaining 17 percent other languages. Only about 18 percent of the population is thoroughly bilingual in English and French. The spatial concentration of more than 85 percent of the country's Francophones in Quebec accentuates this division (Fig. 3B-2). Quebec's population is 80 percent French-Canadian, and Quebec remains the historic, traditional, and emotional focus of French culture in Canada.

Over the past half-century a strong nationalist movement has emerged in Quebec, and at times it has demanded outright separation from the rest of Canada. This ethnolinguistic division continues to be the litmus test for Canada's federal system, and the issue continues to pose a latent threat to the country's future national unity. The historical roots of the matter are related

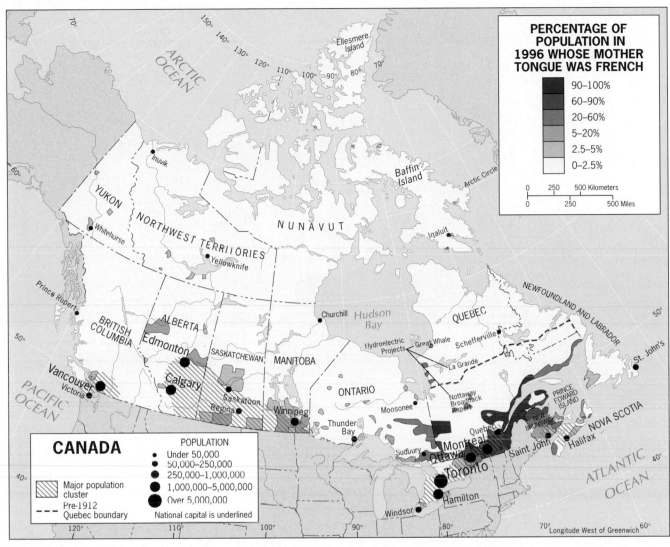

FIGURE 3B-2

© H. J. de Blij, P. O. Muller, and John Wiley & Sons, Inc.

to French-British competition since the seventeenth century in which Britain gained the upper hand. To this day, the British monarch remains the official head of state (as Canada remains part of the British Commonwealth), represented by the governor general. It is one of the issues that riles Quebec nationalists.

There does seem to be an overall decline in support for Quebec's secession. This is in part due to a number of laws that reassure the primacy of Québécois culture in the province (a 2006 parliamentary resolution effectively recognizes the Québécois as a *distinct nation*). In addition, there appears to be a growing realization among the French-speaking people that continued integration in Canada combined with a high degree of autonomy

is the best possible option. But that struggle for maximum autonomy continues, along with the accompanying political and ethnic tensions.

The Ascendancy of Indigenous Peoples

On the wider Canadian scene, the culture-based Quebec struggle has stirred ethnic consciousness among the country's 1.4 million native (First Nations, Métis, and Inuit) peoples. They, too, have received a sympathetic hearing in Ottawa, and in recent years breakthroughs have been achieved with the creation of Nunavut and local treaties for limited self-government in northern British Columbia.

A foremost concern among these peoples is that their aboriginal rights

be protected by the federal government against the provinces of which they are a part. This is especially true for the First Nations of Quebec, the Cree, whose historic domain covers the northern half of the province. As Figure 3B-2 shows, the territory of the Cree is no unproductive wilderness: it contains the James Bay Hydroelectric project, a massive scheme of dikes and dams that has transformed much of northwestern Quebec and generates hydroelectric power for a huge market within and outside the province. In 2002, after the federal courts supported its attempts to block construction of the project, the Cree negotiated a groundbreaking treaty with the Quebec government whereby this First Nation dropped its opposition in return for a portion of the income earned from electricity

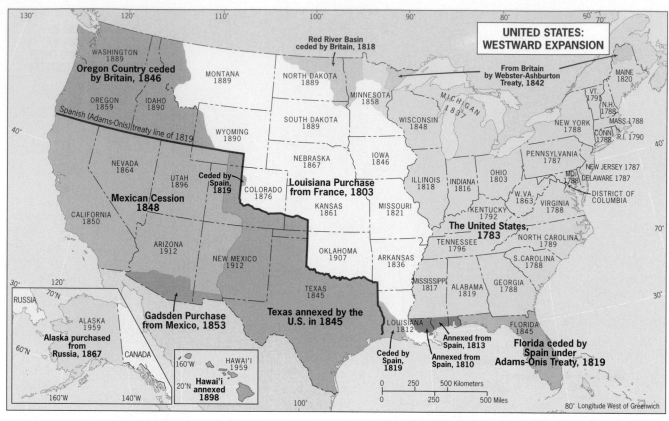

FIGURE 3B-3

© H. J. de Blij, P. O. Muller, and John Wiley & Sons, Inc.

sales. The Cree also secured the right to control their own economic and community development.

Centrifugal Forces

The events of the past four decades have had a profound impact on Canada's national political geography, and not only in Quebec. Leaders of the western provinces objected to federal concessions to Quebec, insisting that the equal treatment of all ten provinces is a basic principle that precludes the designation of special status for any single one. Recent federal elections and opinion surveys have clearly revealed the emerging fault lines that surround Quebec and set the western provinces and British Columbia off from the rest of the country. Thus, with national unity always a challenge in this comparatively young federation, Canada confronts forces of **1** devolution that threaten to weaken the state.

In addition, the continuing economic integration of Canada and the United States (galvanized by the North American Free Trade Agreement of 1994) tends to pull the southern parts of Canada's most populous provinces into an ever closer relationship with its United States neighbor. Local **2** cross-border linkages are already well developed and are likely to intensify in the future: the Atlantic Provinces with neighboring New England; Quebec with New York State; Ontario with Michigan and adjacent Midwestern States; the Prairie Provinces with the Upper Midwest; and British Columbia with the (U.S.) Pacific Northwest. Such functional reorientations constitute yet another set of potentially powerful devolutionary challenges confronting the Ottawa government.

ETHNICITY, RACE, AND REGIONALISM IN THE UNITED STATES

Nothing in the United States compares with the devolutionary challenges Canada has faced throughout its evolution as a modern state. The United States has about four times as many subnational units as Canada, but none since the Civil War has mounted a serious campaign for *secession* of the kind Quebec has, and in only one (Hawai'i) have indigenous people demanded recognition of a part of the State as sovereign territory.

In Canada, indigenous peoples were able to hold on to sparsely populated regions in the north. In the United States, on the other hand, the relentless push westward by European settlers and the U.S. government (see Fig. 3B-3) in effect dispossessed—and decimated—the original inhabitants, in the end confining them to some 300 reservations located mainly in the western half of the country (Fig. 3B-4, upper left map). There are only about 4.8 million Native Americans left, with the Navajo and Cherokee as the largest surviving nations. Whereas in Canada the First Nations hold significant territorial power, in the United States they are much less in a position to challenge the federal government.

In the United States, much more so than in Canada, the social fabric

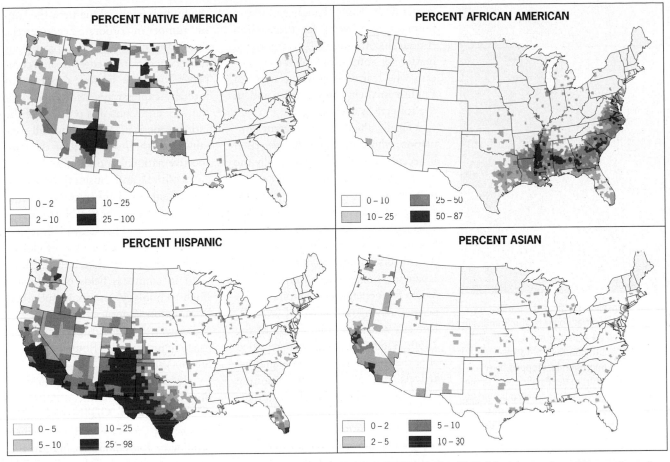

FIGURE 3B-4   © H. J. de Blij, P. O. Muller, and John Wiley & Sons, Inc.

and ethnic relations are determined by past and present waves of immigration. In certain cases, immigration has helped forge a particular regional geography within the United States. This was certainly the case with hundreds of thousands of Africans who arrived in the United States between the seventeenth and nineteenth centuries in bondage, against their will, and put to work as slaves on plantations in the American South.

Slavery was abolished in 1865 but segregation continued in various legal forms until the civil rights movement in the 1960s. In violation of the Constitution, a wide range of State and local laws, also known as Jim Crow laws, were enacted to systematize the repression of African Americans. As a result of this history, there is still today a geography of race in the United States, with a clear regional concentration of African Americans in the Southeast, from eastern Texas

to Maryland (Fig. 3B-4, upper right map). We noted before that considerable numbers of African Americans migrated to find work in the industrial cities of the north during the first half of the twentieth century, so there too we find African American population clusters, but they are not clearly visible on the national map.

Race continues to be a sensitive issue in the United States, more so than in Canada, because of this terrible history and because until recently many African Americans were effectively denied equal opportunities. The election of the first African American president, Barack Obama, appears to be serving as a further catalyst of progressive change.

Another highly visible regional expression of ethnic migration concerns the presence of Hispanics in the southwestern parts of the United States (Fig. 3B-4, lower left map). This migration, too, should be under-

stood in historical and geographical contexts. In these parts of the country, many Hispanic place names reflect Mexico's ouster from a vast region acquired in 1848 by the United States at the end of the Mexican War through the Treaty of Guadalupe Hidalgo. This treaty established the Mexican-American border at the Rio Grande and assigned to the United States no less than 1.3 million square kilometers (525,000 sq mi) of Mexican territory, including present-day California, Arizona, New Mexico, Nevada, southern Colorado, Texas (which had already declared independence from Mexico), and Utah (Fig. 3B-3). In Mexico, this disastrous event (which precipitated that country's civil war) is much more vivid in the national consciousness than it is in the United States. Some Mexican nationalists even proclaim that much of the current migration across the U.S.-Mexican border, illegal though

People have walled and fenced themselves off since the earliest days, as the Great Wall of China and Hadrian's Wall (built by the Romans in northern Britain) remind us. But no wall—whether the Berlin Wall during the Cold War or the Israeli "Security Barrier" of today—has ever halted all migration. Neither will migration be stopped by a fortified fence along the Mexican border from the Pacific coast to the Gulf of Mexico. However, long stretches of the U.S.–Mexico boundary already look like this, and more construction will take place. This photo shows a ground-level view of the improvised wall that slashes westward, between Tijuana, Mexico and the southernmost suburbs of San Diego, California, as it makes its way toward the Pacific shore on the horizon. The poverty-stricken landscape of this part of Tijuana extends almost to the rusty barrier itself, while development on the U.S. side has kept its distance from this well-patrolled section of the border. © Diane Cook & Ken Jenshel.

it may be, is a matter of resettlement of unjustly lost lands.

Since 1980, the number of Hispanics in the United States has more than tripled and now exceeds 45 million, outnumbering the African American minority and changing the ethnic-cultural map of America. Although Hispanics are now dispersing into many parts of the country—including rural areas in the Southeast and Midwest as well as urban communities from Massachusetts to Washington State—they remain strongly clustered in the Southwest, California, and Florida (see Fig. 3B-4, lower left map). In fact, along virtually the entire multi-State tier north of the Mexican border, America has become a cultural transition zone (think of the expression Tex-Mex), where immigration, legal as well as illegal, has become a contentious issue. Language, not a point of controversy for the African American community, is an emotional matter for some Anglophone Americans who fear erosion of the status of English in their communities and in the country as a whole.

Finally, the Asian immigrant population, whose arrival rate has also risen in recent decades, is geographically the most agglomerated of the main ethnic minorities (Fig. 3B-4, lower right map). Chinese remain the most numerous among the 15-plus million Americans who identify themselves as Asian, but this diverse minority also includes Japanese who have been in the United States for generations, Filipinos who began arriving after World War II, Koreans who have migrated since the 1950s, Vietnamese who were resettled after the Indochina War, more gradual arrivals of Pacific Islanders including Hawaiians, and considerable numbers of South Asian Indians in the past few decades. Whatever their origins, the great majority have remained in California, where many have achieved notable economic success and upward mobility.

Such is the regional geography of ethnicity within the national context of the United States, one that is largely driven by histories of migration and processes of voluntary or involuntary inclusion and exclusion. If we broaden the scope and consider the North American realm as a whole, combining both formal and functional regions, we can make a number of more comprehensive distinctions on the basis of physical geography, culture, and history.

NORTH AMERICA'S REGIONS

NORTH AMERICA'S HISTORIC CORE (1)

The historic Core region of the North American realm integrates what have long been the most prominent parts of the United States and Canada (Fig. 3B-1). They contain the largest cities and federal capitals of both countries, the leading financial markets, the largest number of corporate headquarters, dominant media centers, prestigious universities, cutting-edge research complexes, and the busiest airports and intercity expressways. Moreover, both the U.S. and Canadian portions of the North American Core region still contain more than one-third of their respective national populations.

Political and business decisions, investments, and other commitments made here affect not just North America, but the world at large. This region is one of the world's centers of globalization, with innovations radiating outward in countless fields of endeavor. High-technology centers in suburban Washington (Internet Alley lining the expressway to Dulles Airport) and Boston (circumferential Route 128) rival Silicon Valley.

Nonetheless, the dominance of this historic Core has been declining. As was true of the American Manufacturing Belt, it once had unmatched supremacy. But deindustrialization and the subsequent rise of the information economy have eroded that dominance as competitors to the south and west continue to siphon away some of its key functions.

THE MARITIME NORTHEAST (2)

When you travel north from Boston into New Hampshire, Vermont, and Maine, you move across one of North America's historic culture hearths whose identity has remained strong for four centuries. Even though Massachusetts, Connecticut, and Rhode Island—traditional New England States—share these qualities, they have become part of the Core, so that on a functional as well as formal basis, the Maritime Northeast extends from the northern border of Massachusetts to Newfoundland, thus incorporating all four Canadian Atlantic Provinces (Fig. 3B-1).

Difficult environments, a maritime orientation, limited resources, and a persistent rural character have combined to slow economic development here. Primary industries—fishing, logging, some farming—are mainstays, although recreation and tourism have boosted the regional economy in recent decades. Upper New England experiences the spillover effect of the Boston area's prosperity to some degree, but not enough to ensure steady and sustained economic growth. Atlantic Canada has also endured economic hard times in the recent past. For example, overfishing has depleted fish stocks. Alternate opportunities focus on the region's spectacular scenery and the tourism it attracts. Economic prospects have also been boosted in Canada's poorest province, Newfoundland and Labrador, by the discovery of significant offshore oil reserves.

FRENCH CANADA (3)

Francophone Canada constitutes the inhabited, southern portion of Quebec from near Montreal to the mouth of the St. Lawrence River (Fig. 3B-1). It includes the French-speaking Acadians who reside in neighboring New Brunswick (Fig. 3B-2). The French cultural imprint on the region's cities and towns is matched in the rural areas by the narrow, rectangular *long lots* perpendicular to the river, also of French origin.

Montreal is the cultural capital of French Canada, which with 3.7 million people is the largest French-speaking city in the world after Paris. It is without a doubt one of the realm's most cosmopolitan cities, its strong European flavor pervading its stately CBD, lively neighborhoods, and bustling street life. To escape the long, harsh winters here, much of downtown Montreal has become an underground city with miles of handsome passageways lined with shops, restaurants, and cinemas, all of it served by one of the world's most modern and attractive subway systems.

The economy of French Canada was historically less industrial and more rural than the adjacent Core region (and Toronto far outranks Montreal here). Montreal's economic role has

An unmistakably French cultural landscape in North America: the Rue Saint-Louis in Quebec City, the capital of Quebec Province. Not an English sign in sight in this center of French/Québécois culture. The green spires in the background belong to Le Chateau Frontenac, a grand hotel built on the site of the old Fort St. Louis (the Count of Frontenac was a prominent governor of what was then known as New France). French-speaking, Roman Catholic Quebec lies at the heart of French Canada. © SUPERSTOCK

gathered pace in recent times, with information technology, telecommunications, biopharmaceutical industries, tourism, and recreation all growing in importance.

But Quebec's economic prospects tend to be shadowed by the region's political uncertainties, and bursts of nationalism (such as laws prohibiting shops from advertising in English) have at times eroded business confidence. When the provincial government sought to address the declining birth rate resulting from accelerating urbanization by encouraging immigration, tolerance for multiculturalism in Quebec proved to be so low that a special commission began to hold hearings in 2007 to investigate the matter. Parochialism has cost Quebec dearly and sets this region off from the rest of the Canada that virtually encircles it.

The Acadians in the neighboring province of New Brunswick, Canada's largest cluster of French-speakers outside Quebec, take a different view. Not only do they reject the notion of independence for themselves, but they also actively promote all efforts

to keep Quebec within the Canadian federation. Accommodation with the Anglophone majority and acceptance of multiculturalism in New Brunswick set an example for the rest of French Canada.

THE SOUTH (4)

For more than a century following the Civil War, the South (Fig. 3B-1) remained in economic stagnation and cultural isolation from the rest of the country. In the 1970s, however, things changed so fast that a New South arose almost overnight. The Sunbelt migration drove people and enterprises into the long-stagnant cities. Core-region companies looking for headquarters or subsidiary offices found Atlanta, Charlotte, Miami-Fort Lauderdale, Tampa, and other urban areas economical and attractive, turning them into boomtowns virtually overnight. Racial segregation had been dismantled in the wake of the civil rights movement. A new social order was matched by new facilities ranging from airports (Atlanta's quickly became one of the world's busiest) to theme parks open to all.

Soon the nation was watching Atlanta-based CNN on television and following space flights that originated at Cape Canaveral.

The geography of development in the South, however, is very uneven. While many cities and some agricultural areas have benefited, others have not; the South contains some of the realm's poorest rural areas, and the gap between rich and poor here is wider than in any other region in the realm. Virginia's Washington suburbs, North Carolina's Research Triangle, Tennessee's Oak Ridge complex, and Atlanta's corporate campuses are locales that typify the New South; but Appalachia and rural Mississippi still represent the Old South, where depressed farming areas and stagnant small industries restrict both incomes and change.

Climatically, the South is warmer than the Core region to the north and more humid than the Southwest. It is not for nothing that this is where the tobacco, cotton, and sugar plantations thrived. But the summer heat brings a notable threat to the Gulf-Atlantic Coastal Plain of the South: hurricanes capable of inflicting catastrophic devastation on low-lying areas. Hurricanes are fueled by the rapidly rising air over the warm waters of the Gulf of Mexico or the Caribbean that reach peak temperatures in late summer. South Florida and southern Louisiana have been particularly hard hit in recent decades.

≡ THE SOUTHWEST (5)

The Southwest has taken on its own regional identity in recent years, but today it is firmly established in the North American consciousness for several reasons. One is physiographic: here, steppes and deserts dominate. Another is cultural: Anglo-American, Hispanic, and Native American cultures coexist in uneasy accommodation.

As Figure 3B-1 shows, by our definition the Southwest begins in eastern Texas and extends all the way to the eastern part of Southern California. Texas leads this region in most respects. Its economy, once dependent on oil

and natural gas, has been restructured and diversified so that today the Dallas-Forth Worth–Houston–San Antonio triangle has become one of the world's most productive postindustrial complexes, with several **3** technopoles (state-of-the-art, high-technology centers) including Austin at the heart of it. This is also a hub of international trade and the northern anchor of a NAFTA-generated transnational growth corridor that extends into Mexico as far as Monterrey.

New Mexico, in the middle of the Southwest region, is economically least developed and ranks low on many U.S. social indices, but its environmental and cultural attractions benefit Albuquerque and Santa Fe, with Los Alamos a famous name in research and development. In the west, Arizona's technologically transformed desert now harbors two large coalescing metropolises, Phoenix and Tucson. Growth continues, but climate change adds to the uncertainties of water supplies already stretched to the limit.

≡ THE PACIFIC HINGE (6)

The Southwest meets the Far West near San Diego, California, and from there the region we call the Pacific Hinge extends all the way up to Vancouver, its northern anchor in southwestern British Columbia (Fig. 3B-1). We include almost all of California in this region but only (for environmental as well as economic reasons) the western portions of Oregon and Washington State. It is a hugely important part of North America and increasingly a counterweight to the historic core in the northeast.

In many parts of the world, this region would constitute a top-ranked core area, with such leading cities as Los Angeles, San Francisco, Portland, and Seattle plus America's most populous State, whose economy in recent times has ranked among the world's ten largest—by country. The Pacific Hinge also includes one of the realm's most productive agricultural areas in

California's Central Valley, magnificent scenery, agreeable climates, and a culturally diverse population drawn by its spectacular economic growth and pleasant living conditions.

Today this region is in a still-emerging phase of development because of its deep involvement in the economic growth of countries on opposite shores of the Pacific Ocean. Earlier, Japan's success had a salutary impact here, but what has since happened in China, South Korea, Taiwan, Vietnam, and other distant **4** Pacific Rim economies has created unprecedented opportunities.

The term *Pacific Rim* has come into use to describe the discontinuous regions surrounding the great Pacific Ocean that have experienced miraculous economic growth and progress in the past few decades: coastal China and different parts of East and Southeast Asia but also Australia, South America's Chile, and the western shores of Canada and the United States. That is why in this chapter we speak of the *Pacific Hinge* of North America, now the key interface between the North American realm and the Pacific Rim.

Los Angeles is the region's leading trade, manufacturing, and financial center, and with 18.1 inhabitants the second largest metropolis of the entire realm. In the global arena, Los Angeles is the eastern Pacific Rim's biggest city and the origin of the largest number of transoceanic flights and sailings to Asia.

Farther north, the San Francisco Bay Area has seen rather more orderly expansion, with the Silicon Valley technopole the key component of its high-technology success. In what Americans refer to as the Pacific Northwest, the Seattle-Tacoma area originally benefited from cheap hydroelectricity generated by Columbia River dam projects, attracting aluminum producers and aircraft manufacturers. Boeing remains one of the world's two leading aircraft manufacturers, the East Asian market having boosted its sales. Meanwhile, the technopole centered on suburban Redmond (Microsoft's headquarters) heralds this area's transformation into a prototype metropolis of the postindustrial economy.

Vancouver, on the northern flank of the Pacific Hinge, has the locational advantage of being closer, on air and sea routes, to East Asia than any other large North American city. Ethnic Chinese constitute more than 20 percent of the metropolitan population, and the total number of Asians now approaches 40 percent.

THE WESTERN FRONTIER (7)

Where the forests of the Pacific Hinge yield to the scrub of the rain shadow on the inland slopes of the mountain wall that parallels the west coast, and the relief turns into intermontane ("between the mountains") basins and plateaus, lies a region stretching from the Sierra Nevada and Cascades to the Rockies, encompassing parts of southern Alberta and British Columbia, eastern Washington and Oregon, all of Nevada, Utah, and Idaho plus western Montana, Wyoming, and Colorado. In Figure 3B-1, you can see Edmonton, Calgary, and Denver situated at the margin of this aptly named region, and Salt Lake City, anchoring the Software Valley technopole and symbolizing its new high-tech era, at the heart of it.

Remoteness, dryness, and sparse population typified this region for a very long time. It became the redoubt of the Mormon faith and a place known for boom-and-bust mining, logging, and, where possible, livestock raising. However, in recent decades advances in communications and transportation technologies, sunny climates, open spaces, lower costs of living, and growing job opportunities have constituted effective *pull* factors here. Every time an earthquake struck in California, eastward migration got a boost. Over the past decade, Californians by the millions have emigrated from the overcrowded, overpriced Pacific coast to the high desert.

Development in this region centers on its widely dispersed urban

For years the hottest growth area in the Western Frontier region was its southernmost corner—the Las Vegas Valley. Nevada was one of America's fastest-growing States, attracting new residents by the hundreds of thousands and visitors to its burgeoning recreation industry by the tens of millions annually. Between 1990 and 2000, this Las Vegas suburb of Henderson ranked as the fastest-growing municipality in the United States. Sprawling development on a massive scale converted this patch of high desert into a low-density urban checkerboard whose expansion seemed without end. But by 2009 the growth cycle had ceased as thousands of homeowners were forced into foreclosure, unemployment was rising, and national and global economic crises were clouding the future of Boomtown USA. © Ethan Miller/Getty Images, Inc.

areas, which are making this the realm's fastest-growing region (albeit from a low base)—and where thousands of high-tech manufacturing and specialized service jobs drove a two-decade-long influx, slowed but not reversed by the economic crisis that began in 2008. This growth has not come without problems as locals see land values rise (photo above), conservative ideas challenged, and lifestyles endangered.

The fastest-growing large metropolis in North America, Las Vegas, lies on the border between the Western Frontier and Southwest regions. Far more than a gambling and an entertainment center that welcomes some 40 million visitors annually, Las Vegas attracts immigrants because of its history of job creation, relatively cheap land and low-cost housing, moderate taxes, and sunny if seasonally hot weather. No doubt, proximity to Southern California has had much to do with its explosive growth, but Las Vegas is also known far and wide

as the ultimate frontier city—where (almost) anything goes.

THE CONTINENTAL INTERIOR (8)

The Western Frontier region meets the Continental Interior region along a line marked by cities that influence both regions, from Denver in the south to Edmonton in the north. This vast North American heartland extends from interior Canada to the borders of the South and Southwest (Fig. 3B-1). This region contains many noteworthy cities, including Kansas City, Omaha, Minneapolis, and Winnipeg, but agriculture is the dominant story here. As the map of farm resource regions (Fig. 3B-5) shows, this is North America's breadbasket. This is the land of the Corn Belt (which extends into the western half of the Core region), of the soybean (the fastest-expanding crop in the past century), of spring wheat in the Dakotas and Canada's Prairie Provinces and winter wheat in Kansas, of

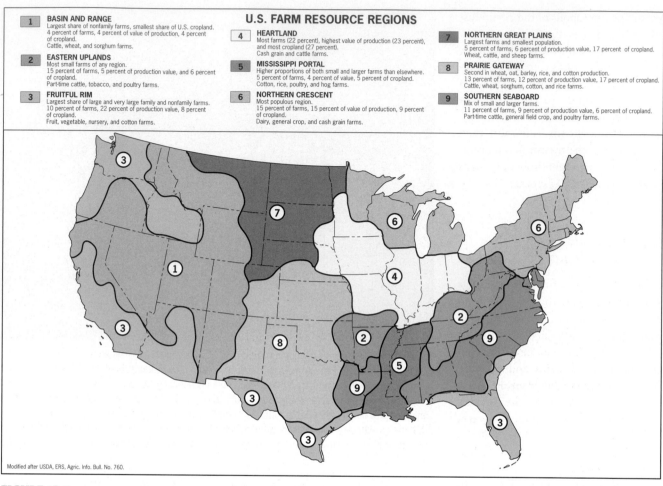

U.S. FARM RESOURCE REGIONS

1 BASIN AND RANGE
Largest share of nonfamily farms, smallest share of U.S. cropland.
4 percent of farms, 4 percent of value of production, 4 percent of cropland.
Cattle, wheat, and sorghum farms.

2 EASTERN UPLANDS
Most small farms of any region.
15 percent of farms, 5 percent of production value, and 6 percent of cropland.
Part-time cattle, tobacco, and poultry farms.

3 FRUITFUL RIM
Largest share of large and very large family and nonfamily farms.
10 percent of farms, 22 percent of production value, 8 percent of cropland.
Fruit, vegetable, nursery, and cotton farms.

4 HEARTLAND
Most farms (22 percent), highest value of production (23 percent), and most cropland (27 percent).
Cash grain and cattle farms.

5 MISSISSIPPI PORTAL
Higher proportions of both small and larger farms than elsewhere.
5 percent of farms, 4 percent of value, 5 percent of cropland.
Cotton, rice, poultry, and hog farms.

6 NORTHERN CRESCENT
Most populous region.
15 percent of farms, 15 percent of value of production, 9 percent of cropland.
Dairy, general crop, and cash grain farms.

7 NORTHERN GREAT PLAINS
Largest farms and smallest population.
5 percent of farms, 6 percent of production value, 17 percent of cropland.
Wheat, cattle, and sheep farms.

8 PRAIRIE GATEWAY
Second in wheat, oat, barley, rice, and cotton production.
13 percent of farms, 12 percent of production value, 17 percent of cropland.
Cattle, wheat, sorghum, cotton, and rice farms.

9 SOUTHERN SEABOARD
Mix of small and larger farms.
11 percent of farms, 9 percent of production value, 6 percent of cropland.
Part-time cattle, general field crop, and poultry farms.

Modified after USDA, ERS, Agric. Info. Bull. No. 760.

FIGURE 3B-5

© H. J. de Blij, P. O. Muller, and John Wiley & Sons, Inc.

beef and pork. Indeed, the cities and towns of this region share histories of food processing, packing, and marketing, of flour milling, and of soybean, sunflower, and canola oil production. It is also the scene of the struggle for survival of the family farm. The unrelenting, aggressive incursion of large-scale corporate farming threatens a longstanding way of life.

This may be farmland, but it is not immune from what might seem to be the nonagricultural realities of the contemporary world, not even the energy issue. Corn used to be grown for people and pigs; now much of the crop is used to make ethanol, the gasoline substitute (or additive). Farmers hope that rising corn prices will revitalize the Corn Belt. Others worry that rising prices will put a strain on consumers who eat it, mostly in the form of meat produced from corn-fed hogs and beef cattle. In any case, the

amount of corn needed to make a real dent in our gasoline consumption is many times larger than the entire annual corn crop. In 2007, when around half of all corn grown in the Midwest was turned into biofuels, it replaced only about 4 percent of gasoline consumption while raising concerns about possible food shortages.

Take a look at population statistics, and it is clear that many States in the Continental Interior are losing people or gaining far fewer than the national average. The Great Plains, the western zone of this region, is especially hard hit: younger people and those better off are leaving, and older and poorer residents are staying behind. Villages and small towns are dying, and the notion of abandoning certain areas and allowing them to return to their natural state is being seriously discussed.

But it is not the same everywhere. Alberta, in the northwestern corner of

this region, is experiencing one of the most spectacular economic booms in the realm's history. Recent estimates suggest that Alberta's oil reserves, including the tar sands around Fort McMurray (photo at right), are second only to those of Saudi Arabia, and possibly even larger. Alberta has become Canada's fastest-growing province, Calgary its fastest-expanding large city. The boom makes Calgary a magnet for business and highly skilled workers; the city's skyline is in spectacular transformation, comparable to what is happening on the Arabian Peninsula. The question is: how long will the good times last?

THE NORTHERN FRONTIER (9)

Figure 3B-1 leaves no doubt as to the dimensions of this final North American region: it is by far the largest of the realm, covering nearly 90 percent of Canada and all of the biggest U.S. State, Alaska. Not only does this region include the northern parts of seven of Canada's provinces, but it also comprises the Yukon and Northwest Territories as well as recently established Nunavut.

The sparsely peopled Northern Frontier remains a land of isolated settlement based on the exploitation of newly discovered resources. The Canadian Shield, underlying most of the eastern half of the region, is a rich storehouse of mineral resources, including metallic ores such as nickel, uranium, copper, gold, silver, lead, and zinc. The Yukon and Northwest Territories have proven especially bountiful, with gold- and diamond-mining (Canada now ranks among the world's top five producers); at the opposite, eastern edge of the Shield, near Voisey's Bay on the central coast of Labrador, the largest body of high-grade nickel ore ever discovered is being opened to production.

Productive locations in the Northern Frontier comprise a far-flung network of mines, oil and gas installations, pulp mills, and hydropower stations that have spawned hundreds of small settlements as well as thousands of kilometers of interconnecting transportation and communications lines. Inevitably, these activities have infringed on the lands of indigenous peoples without the preparation of treaties or agreements, which has led to recent negotiations between the government and leaders of First Nations over resource development in the Northern Frontier. But essentially this vast region remains a frontier in the truest sense of the term.

Alaska's regional geography differs from the rest of the Northern Frontier in that the State contains several urban settlements and an incipient core area in the coastal zone around Anchorage. Alaska's population, which now exceeds 700,000, accounts for more than one-third of the region's total. Here also lies the only city of any size,

Anchorage (290,000). The State's internal communications (air and surface) also are rather more developed.

Another difference lies in Alaska's North Slope oil exploitation, one of the hemisphere's key energy sources since the 1300-kilometer (800-mi) Trans-Alaska Pipeline began operating in 1977. The North Slope is the area north of the Brooks Range and along the Arctic Ocean. Dwindling supplies at the main reserve at Prudhoe Bay are now compelling producers to turn to huge additional reserves in Alaska's north, but opposition from preservationist groups has slowed their plans. The issue remains in contention—as is another project to construct a natural gas pipeline to parallel the oil pipeline to send gas, a byproduct of oil drilling, to the growing market in the Lower 48.

Climate change is likely to affect the Northern Frontier in as-yet unforeseeable ways. As the map shows, this region extends all the way north to Ellesmere Island and thus includes waters that, in the event of longer-term sustained global warming and associated recession of Arctic ice, will open, possibly year-round, to create natural waterways that would dramatically alter shipping routes and intercontinental distances (a topic discussed in Chapter 12). The question of ownership of these waters—and the submerged seafloor beneath them, coming within reach of exploitation—will be a matter of significant international contention in the decades ahead.

As the price of a barrel of oil (and a gallon of gasoline) rises, it becomes profitable to derive oil from sources other than liquid reserves. The Canadian province of Alberta contains vast deposits of "oil sands" in which the petroleum is mixed with sand, requiring a relatively expensive and complicated process to extract it. The quantity of oil locked in these Athabasca Tar Sands is estimated to constitute one of the world's largest reserves, and this huge open-pit mining project is under way around the town of Fort McMurray to recover it. Those who assume that Canada will sell this oil to the United States should be aware that the Chinese are offering to fund construction of a pipeline across the mountains to Canada's Pacific coast; the rising cost of oil has political as well as economic ramifications. © Jim Wark

POINTS TO PONDER

- Spaceport America, the first commercial facility designed specifically for space-tourism flights, is being built northeast of Las Cruces, New Mexico.

- Nearly 110,000 foreign students attend American universities and colleges annually, and more than 40 percent of them come from just three countries: China, India, and South Korea.

- California's Central Valley, the State's agricultural heartland, is in severe economic decline because of water shortages and the continued influx of impoverished and often unemployed immigrants.

- Counting all sources—onshore and offshore, known oil and gas reserves, and particularly Alberta's tar-sands deposits—Canada ranks first among the world's states in actual and potential energy resources.

The Dutch colonial impress, vividly etched on the townscape of Willemstad, Curaçao. © H. J. de Blij

IN THIS CHAPTER

Why "Latin America" is a misnomer
Building a wall between two realms
Mexico's dominance of the realm
The natural vulnerability of Haiti and other Caribbean islands
Small is beautiful? The predicament of tiny island-nations
Supranational efforts to overcome economic fragmentation

CONCEPTS, IDEAS, AND TERMS

4A

MIDDLE AMERICA: DEFINING THE REALM

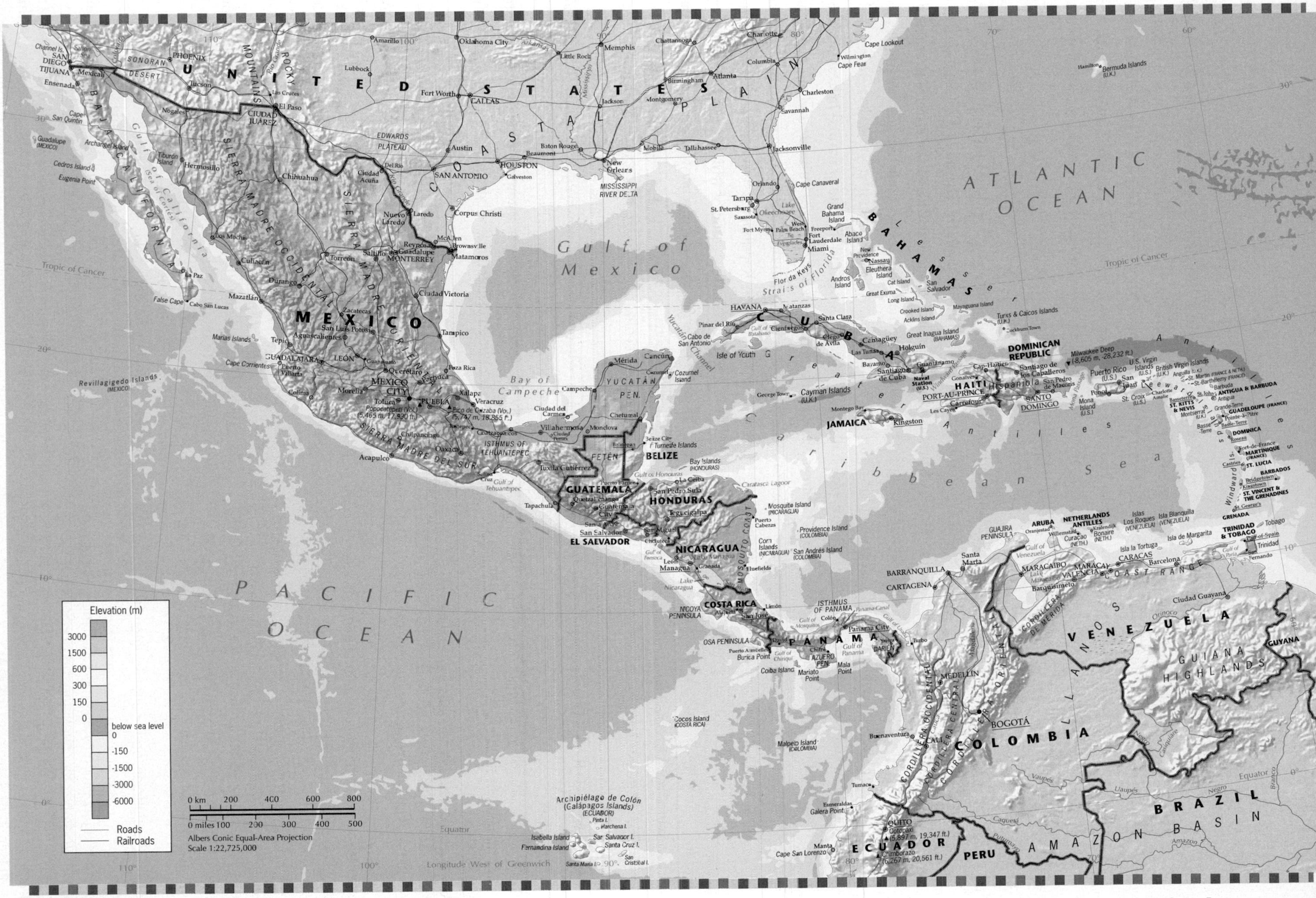

FIGURE 4A-1

© H. J. de Blij, P. O. Muller, and John Wiley & Sons, Inc.

L ook at a world map, and it is obvious that the Americas comprise two landmasses: North America extending from Alaska to Panama and South America from Colombia to Argentina. But here we are reminded that continents and geographic realms do not usually coincide. In Chapters 3A and 3B we discussed a North American realm whose southern boundary is the U.S.-Mexican border and the Gulf of Mexico. Between North America and South America lies a small but important geographic realm known as Middle America. Consisting of a mainland corridor and numerous Caribbean islands, Middle America is a highly fragmented realm.

From Figure 4A-1 it is clear that Middle America is much wider longitudinally than it is long latitudinally. The distance from Baja California to Barbados is about 6000 kilometers (3800 mi), but from the latitude of Tijuana to Panama City is only half that distance. In terms of total area, Middle America is the second-smallest of the world's geographic realms. As the map shows, the dominant state of this realm is Mexico, larger than all its other countries and territories combined.

Middle America may be a small geographic realm by global standards, but comparatively it is densely peopled. Its population in 2011 was a bit less than 200 million, over half of it in Mexico alone, and the rate of natural increase (at 1.6 percent) remains above the world average. Figure 4A-2 shows Mexico's populous interior core area quite clearly, but note that population in Guatemala and Nicaragua tends to cluster toward the Pacific rather than the Caribbean coast. Among the islands, this map reveals how crowded Hispaniola (containing Haiti and the Dominican Republic) is, but at this scale we cannot clearly see the pattern on the smaller islands of the eastern Caribbean and the Bahamas, some of which are also densely populated.

What Middle America lacks in size it makes up in physiographic and cultural diversity. This is a realm of soaring volcanoes and spectacular shorelines, of tropical forests and barren deserts, of windswept plateaus and scenic islands. It holds the architectural and technological legacies of ancient indigenous civilizations. Today it is a mosaic of immigrant cultures from Africa, Europe, and elsewhere, richly reflected in music and the visual arts. Material poverty, however, is endemic: island Haiti is the poorest country in the Americas; Nicaragua, on the mainland, is almost as badly off. As we will discover, a combination of factors has produced a distinctive but challenged realm between North and South America.

 MAJOR GEOGRAPHIC QUALITIES

MIDDLE AMERICA

1. Middle America is a relatively small realm consisting of the mainland countries from Mexico to Panama and all the islands of the Caribbean Basin.

2. Middle America's mainland constitutes a crucial barrier between Atlantic and Pacific waters. In physiographic terms, this is a land bridge connecting the continental landmasses of North and South America.

3. Middle America is a realm of intense cultural and political fragmentation. The presence of many small, insular, and remote countries poses major challenges to economic development.

4. Middle America's cultural geography is complex. Various African and European influences dominate the Caribbean, whereas Spanish and Amerindian traditions survive on the mainland.

5. The realm contains the Americas' least-developed territories. New economic opportunities may help alleviate Middle America's endemic poverty.

6. In terms of area, population, and economic potential, Mexico leads the realm.

7. Mexico is reforming its economy and has experienced major industrial growth. Its hopes for continuing this development are tied to overcoming its remaining economic problems and to expanding trade with the United States and Canada under the North American Free Trade Agreement (NAFTA).

GEOGRAPHICAL FEATURES

Sometimes you will see Middle and South America referred to, in combination, as "Latin" America, alluding to their prevailing Spanish-Portuguese heritage. This is an inappropriate regional name, just as Anglo-America, a term once commonly used for North America, was also improper. Such culturally based terminologies reflect historic power and dominance, and they tend to make outsiders out of those people they do not represent.

In North America, the term *Anglo* (as a geographic appellation) was offensive to many Native Americans, African Americans, Hispanics, Quebecers, and others. In Middle (and South) America, millions of people of indigenous American, African, Asian, and European ancestries do not fit under the "Latin" rubric. You will not find the cultural landscape very "Latin" in the Bahamas, Barbados, Jamaica, Belize, or large parts of Guatemala and Mexico. So let us adopt the geographic neutrality of North, Middle, and South.

But is Middle America sufficiently different from either North or South America to merit distinction as a realm? Certainly, in this age of globalization and migration, border areas are becoming transition zones, as is happening along the U.S.-Mexican boundary. Still, consider this:

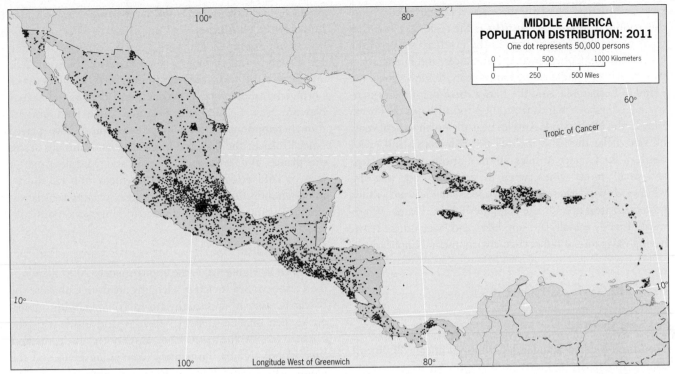

FIGURE 4A-2

© H. J. de Blij, P. O. Muller, and John Wiley & Sons, Inc.

North America encompasses only a pair of states, and the entire continent of South America only 12 (plus France's dependency on the northeast coast). But far smaller Middle America, as we define it, incorporates more than three-dozen political entities, including several dependencies (or quasi-dependencies) of the Netherlands, the United Kingdom, and France, but not counting constituent territories of the United States. Therefore, unlike South America, Middle America is a multilingual patchwork of independent states, territories in political transition, and residual colonial dependencies, with strong continuing ties to the United States and non-Iberian Europe. Middle America is defined in large measure by its vivid cultural-geographic pluralism.

The Realm's Northern Boundary

The 3200-kilometer (2000-mi) land border between North America and Middle America is the longest in the world separating a rich realm from a poor one. The U.S.-Mexican boundary crosses the continent from the Pacific to the Gulf, but Mexican cultural influences penetrate deeply into the southwestern States and American impacts reach far into Mexico. To Mexicans the border is a reminder of territory lost to the United States in historic conflicts; to Americans it is a symbol of economic contrasts and illegal immigration. Along the Mexican side of this border the effects of NAFTA (the North American Free Trade Agreement between Canada, the United States, and Mexico) are transforming Mexico's economic geography.

Every day, more than 20,000 trucks and nearly one million people cross the border in either direction, overwhelming evidence of close cross-border ties. But it is also a troubled boundary. Illegal crossings from Mexico into the United States are countless. In the last decade, U.S. authorities apprehended an average of about 1 million would-be (illegal) migrants per year, many of them repeat offenders. Some were so desperate to make it across that they risked their lives: the number of deaths in the border zone since 2000 has varied from 200 to 500 annually. It is a painful reminder of the enormous gap in living standards and life opportunities in the two realms.

The U.S. government, responding in part to homeland security considerations (potential terrorists are thought to enter the country from Mexico as well) and to drug (mainly cocaine) smuggling, has initiated the construction of a fortified fence along the entire length of the border. The very idea seems at odds with ideals of closer economic and political cooperation, and many security experts question its efficacy. It is unclear what the fence will do to the emergence, in recent decades, of a dynamic border region.

The Regions

As Figure 4A-1 suggests, Middle America can be divided into four distinct regions. Dominant *Mexico* occupies the largest part of the collective territory, with much of its northern border defined by the Rio Bravo (Rio Grande) from El Paso to the Gulf of Mexico. To the southeast, Mexico yields to the region called *Central America*, consisting of the seven republics of Guatemala, Belize, Honduras, El Salvador, Nicaragua, Costa Rica, and Panama. Sometimes the entire mainland portion of the realm is incorrectly referred to as

Central America, but only these seven countries constitute this region. The *Greater Antilles* is the regional name that refers to the four large islands in the northern sector of the Caribbean Sea: Cuba, Jamaica, Hispaniola, and Puerto Rico; two countries, Haiti and the Dominican Republic, share the island of Hispaniola. And the *Lesser Antilles* form an extensive arc of smaller islands from the Virgin Islands off Puerto Rico to the Netherlands Antilles near the northwestern coast of Venezuela; they also include the Bahamas island chain north of the Greater Antilles. In the Caribbean islands large and small, moist winds sweep in from the east, watering windward (wind-facing) coasts while leaving leeward (wind-protected) areas dry. On the map, the "Leeward" and "Windward" Islands actually are not "dry" and "wet," these terms having navigational rather than environmental implications.

PHYSICAL GEOGRAPHY

A Land Bridge

The funnel-shaped mainland, a 4800-kilometer (3000-mi) connection between North and South America, is wide enough in the north to contain two major mountain chains and a vast interior plateau, but narrows to a slim 65-kilometer (40-mi) ribbon of land in Panama. Here this strip of land—or *isthmus*—bends eastward so that Panama's orientation is east-west. Thus mainland Middle America is what

This shocking street scene in central Port-au-Prince two weeks after the January 2010 earthquake is typical of the unimaginable devastation that afflicted the Haitian capital. Despite the massive human casualties, the greatly heightened daily struggle for survival, the wrecked economy, the thousands of uninhabitable buildings and debris-clogged streets, and the near-total reliance on assistance from foreign governments and relief agencies—life somehow found a way to go on as these resourceful individuals demonstrate. There will be no end to this human drama anytime soon as the hemisphere's poorest country confronts the colossal struggle to rebuild, and it is certain to remain one of the Caribbean's most important geographical stories far into the future. © Craig Ruttle/Alamy

physical geographers call a **1** land bridge, an isthmian link between continents.

If you examine a globe, you can see other present and former land bridges: Egypt's Sinai Peninsula between Asia and Africa, the (now-broken) Bering land bridge between northeasternmost Asia and Alaska, and the shallow waters between New Guinea and Australia. Such land bridges, though temporary features in geologic time, have played crucial roles in the dispersal of animals and humans across the planet. But even though mainland Middle America forms a land bridge, its internal fragmentation has always inhibited movement. Mountain ranges, swampy coastlands, and dense rainforests make contact and interaction difficult.

Island Chains

As shown in Figure 4A-1, the approximately 7000 islands of the Caribbean Sea stretch in a lengthy arc from Cuba and the Bahamas eastward and southward to Trinidad, with numerous outliers outside (such as Barbados) and inside (e.g., the Cayman Islands) the main chain. As we noted above, the four large islands—Cuba, Hispaniola (containing Haiti and the Dominican Republic), Puerto Rico, and Jamaica—are called the Greater Antilles, and all the remaining smaller islands constitute the Lesser Antilles. The entire Antillean **2** archipelago (island chain) consists of the crests and tops of mountain chains that rise from the floor of the Caribbean, the result of collisions between the Caribbean Plate and its neighbors (Fig. G-4). Some of these crests are relatively stable, but elsewhere they contain active volcanoes, and almost everywhere in this realm earthquakes are an ever-present danger—in the islands as well as on the mainland (Fig. G-5).

Dangerous Landscapes

As Figure 4A-1 shows, fragmented Middle America is a realm of high relief, studded with active volcanoes. Figure G-4 reminds us why: the Caribbean, North American, South American, and Cocos tectonic plates converge here, creating dangerous landscapes subject to earthquakes and landslides. Add to this the realm's seasonal exposure to Atlantic hurricanes, and it amounts to some of the highest-risk real estate on Earth.

No part of Middle America is safe from environmental onslaught, but certain areas are riskier than others—and when disaster strikes, the poorest inhabitants, often living in the least desirable and most hazardous locales, suffer disproportionately. On January 12, 2010, Haiti was hit by a massive earthquake; as the inset map in Figure 4B-10 shows, that country is laced with fault lines associated with the nearby boundary that separates the North American and Caribbean plates. The epicenter of the 7.0 quake was just 10 miles outside Port-au-Prince and virtually destroyed this teeming, impoverished capital city of 2.2 million. At least 300,000 people died, and within a week more than a

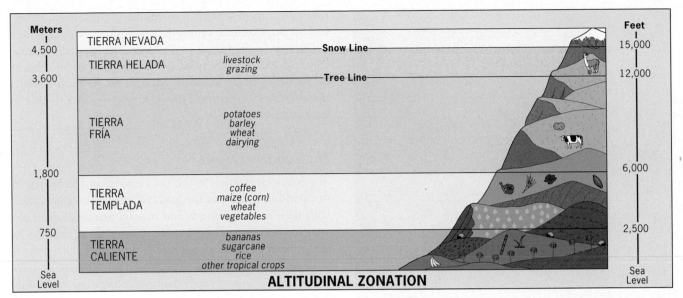

FIGURE 4A-3

million had fled to the countryside. It was the worst natural disaster in Haiti's history, but certainly not the first. In 2008 it was struck by three hurricanes and a tropical storm, repeatedly flooding cities and towns, destroying homes, and dislocating hundreds of thousands of people.

Altitudinal Zonation of Environments

Continental Middle America and the western margin of South America are areas of high relief and strong environmental contrasts. Even though settlers have always favored temperate intermontane basins and valleys, people also cluster in hot tropical lowlands as well as high plateaus just below the snow line in South America's Andes Mountains. In each of these zones, distinct local climates, soils, vegetation, crops, domestic animals, and modes of life prevail. Such **3** altitudinal zones (diagrammed in Fig. 4A-3) are known by specific names as if they were regions with distinguishing properties—as, in reality, they are.

The lowest of these vertical zones, from sea level to 750 meters (2500 ft), is known as the **4** *tierra caliente*, the "hot land" of the coastal plains and low-lying interior basins where tropical agriculture predominates. Above this zone lie the tropical highlands containing Middle and South America's largest population clusters, the **5** *tierra templada* of temperate land reaching up to about 1800 meters (6000 ft). Temperatures here are cooler; prominent among the commercial crops is coffee, while corn (maize) and wheat are the staple grains. Still higher, from about 1800 to 3600 meters (6000 to nearly 12,000 ft), is the **6** *tierra fría*, the cold country of the higher Andes where hardy crops such as potatoes and barley are mainstays. Above the tree line, which marks the upper limit of the *tierra fría*, lies the **7** *tierra helada*; this fourth altitudinal zone, extending from about

3600 to 4500 meters (12,000 to 15,000 ft), is so cold and barren that it can support only the grazing of sheep and other hardy livestock. The highest zone of all is the **8** *tierra nevada*, a zone of permanent snow and ice associated with the loftiest Andean peaks. As we will see, the varied human geography of Middle and western South America closely reflects these diverse environments.

Tropical Deforestation

Before the Europeans arrived, two-thirds of continental Middle America (at lower altitudes) was covered by tropical rainforests. It is estimated that at present only about 10 percent remains. Over the past decade alone, about 1.2 million hectares (3 million acres) of Central American and Mexican woodland disappeared annually. El Salvador has already lost nearly all of its forests, and most of the six other republics will soon approach that stage.

The causes of **9** tropical deforestation are related to the persistent economic and demographic problems of disadvantaged countries. In Central America, the leading cause has been the need to clear rural lands for cattle pasture as many countries, especially Costa Rica, became meat producers and exporters. Because tropical soils are so nutrient-poor, newly deforested areas can function as pastures for only a few years at most. These fields are then abandoned for other freshly cut lands and quickly become a ravaged landscape (see the photo on the next page). Without the protection of trees, local soil erosion and flooding immediately become problems, affecting still-productive areas nearby. A second cause of deforestation is the rapid logging of tropical woodlands as the timber industry increasingly turns from the exhausted forests of the midlatitudes to harvest the rich tree resources of the equatorial

Despite scattered attempts to reverse this landscape scourge, deforestation continues to plague Central America. Its worst effects are often seen on the steeper slopes of interior highlands, as here in western Panama. In the wake of recent deforestation, the land near the top of this hill has already begun to erode, because in the absence of binding tree roots the copious tropical rains are making short work of the unprotected topsoil. © Humberto Olarte Cupas/Alamy

zones, responding to accelerating global demands for housing, paper, and furniture. The third major contributing factor is related to the region's population explosion: as more and more peasants are required to extract a subsistence from inferior lands, they have no choice but to cut down the remaining forest for both firewood and additional crop-raising space, and their intrusion prevents the trees from regenerating.

CULTURAL GEOGRAPHY

Mesoamerican Legacy

Mainland Middle America was the scene of the emergence of a major ancient civilization. Here lay one of the world's true **10** **culture hearths**, a source area from which new ideas radiated and whose population could expand and make significant material and intellectual progress. Agricultural specialization, urbanization, and transport networks developed, and writing, science, art, and other spheres of achievement saw major advances. Anthropologists refer to the Middle American culture hearth as **Mesoamerica**, which extended southeast from the vicinity of present-day Mexico City to central Nicaragua. Its development is especially remarkable because it occurred in highly different geographic environments, each presenting obstacles that had to be overcome in order to unify and integrate large areas. First, in the low-lying

tropical plains of what is now northern Guatemala, Belize, and Mexico's Yucatán Peninsula, and perhaps simultaneously in Guatemala's highlands to the south, the Maya civilization arose more than 3000 years ago. Later, far to the northwest on the high plateau in central Mexico, the Aztecs founded a major civilization centered on the largest city ever to exist in pre-Columbian times.

The Lowland Maya

The Maya civilization is the only major culture hearth in the world that arose in the lowland tropics. Its great cities, with their stone pyramids and massive temples, still yield archeological information today. Maya culture reached its zenith from the third to the tenth centuries AD.

The Maya civilization, anchored by a series of city-states, unified an area larger than any of the modern Middle American countries except Mexico. Its population probably totaled between 2 and 3 million; certain Maya languages are still used in the area to this day. The Maya city-states were marked by dynastic rule that functioned alongside a powerful religious hierarchy, and the great cities that now lie in ruins were primarily ceremonial centers. We also know that this culture produced skilled artists and scientists, and the Maya achieved a great deal in agriculture and trade as well. They grew cotton, created a rudimentary textile industry, and exported cotton cloth by seagoing canoes to other parts of Middle America in return for valuable raw materials. They domesticated the turkey and grew cacao, developed writing systems, and studied astronomy.

The Highland Aztecs

In what is today the intermontane highland zone of Mexico, significant cultural developments were also taking place. Here, just north of present-day Mexico City, lay Teotihuacán, the first true urban center in the Western Hemisphere, which prospered for nearly seven centuries after its founding around the beginning of the Christian era.

The Aztec state, the pinnacle of organization and power in pre-Columbian Middle America, is thought to have originated in the early fourteenth century with the founding of a settlement on an island in a lake that lay in the *Valley of Mexico* (the area surrounding what is now Mexico City). This urban complex, a functioning city as well as a ceremonial center, named Tenochtitlán, was soon to become the greatest city in the Americas and the capital of a large powerful state.

The Aztec state soon began to conquer territories to the east and south. The Aztecs' expansion of their empire was driven by their desire to subjugate peoples and towns in order to extract taxes and tribute. As Aztec influence spread throughout Middle America, the state grew ever richer, its population mushroomed, and its cities thrived and expanded.

The Aztecs produced a wide range of impressive accomplishments, although they were better borrowers and refiners than they were innovators. They developed irrigation systems, and they built elaborate walls to terrace slopes where soil erosion threatened. Indeed, the greatest contributions of Mesoamerica's Amerindians surely came from the agricultural sphere. Corn (maize), the sweet potato, various kinds of beans, the tomato, squash, cacao beans (the raw material of chocolate), and tobacco are just a few of the crops that grew in Mesoamerica when the Europeans first made contact.

Spanish Conquest

Spain's defeat of the Aztecs in the early sixteenth century opened the door to Spanish penetration and supremacy. The Spaniards were ruthless colonizers but not more so than other European powers that subjugated other cultures. True, the Spaniards first enslaved the Amerindians and were determined to destroy the strength of indigenous society. But biology accomplished what ruthlessness could not have achieved in so short a time: diseases introduced by the Spaniards and slaves imported from Africa killed millions of Amerindians.

Middle America's cultural landscape was drastically modified. Unlike the Amerindians, who had used stone as their main building material, the Spaniards employed great quantities of wood and used charcoal for heating, cooking, and smelting metal. The onslaught on the forests was immediate, and rings of deforestation swiftly expanded around the colonizers' towns. The Spaniards also introduced large numbers of cattle and sheep, and people and livestock now had to compete for available food (requiring the opening of vast areas of marginal land that further disrupted the region's food-production balance). Moreover, the Spaniards introduced their own crops (notably wheat) and farming equipment, and soon large fields of wheat began to encroach upon the small plots of corn that the Amerindians cultivated.

The Spaniards' most far-reaching cultural changes derived from their traditions as town-dwellers. Amerindians were moved off their land into nucleated villages and towns that the Spaniards established and laid out. In these settlements, the Spaniards could exercise the kind of rule and administration to which they were accustomed (Fig. 4A-4). The internal focus of each Spanish town was the central *plaza* or market square, around which both the local church and government buildings were located. The surrounding street pattern was deliberately laid out in gridiron form, so that any insurrections could be contained by having a small military force seal off the affected blocks and then root out the troublemakers. Each town was located near what was thought to be good agricultural land (which was often not so good), so that the Amerindians could go

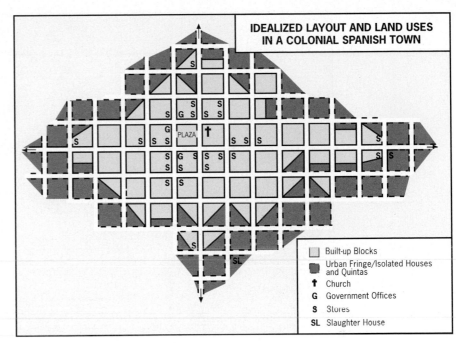

IDEALIZED LAYOUT AND LAND USES IN A COLONIAL SPANISH TOWN

☐ Built-up Blocks
◨ Urban Fringe/Isolated Houses and Quintas
✝ Church
G Government Offices
S Stores
SL Slaughter House

FIGURE 4A-4   © H. J. DE BLIJ, P. O. MULLER, AND JOHN WILEY & SONS, INC.

out each day and work in the fields. Packed tightly into these towns and villages, they came face to face with Spanish culture. Here they (forcibly) learned the Europeans' Roman Catholic religion and Spanish language, and they paid their taxes and tribute to a new master. Many of Middle America's leading cities still bear this Spanish imprint.

Collision of Cultures

But Middle America is not Spain. The cultural fabric of Middle America reflects the collision of Amerindian, Spanish, and other European influences. Indeed, in more remote areas in southeastern Mexico and inner Guatemala, the nucleated indigenous village survived and to this day native languages prevail over Spanish.

In Middle America outside Mexico, only Panama, with its twin attractions of interoceanic transit and gold deposits, became an early focus of Spanish activity. From there, following the Pacific side of the isthmus, Spanish influence radiated northwestward through Central America and into Mexico. The major arena of international competition in Middle America, however, lay not on the Pacific side but on the islands and coasts of the Caribbean Sea. Here the British gained a foothold on the mainland, controlling a narrow coastal strip that extended southeast from Yucatán to what is now Costa Rica. As the colonial-era map (Fig. 4A-5) shows, in the Caribbean the Spaniards faced not only the British but also the French and Dutch, all interested in the lucrative sugar trade, all searching for instant wealth, and all seeking to expand their empires.

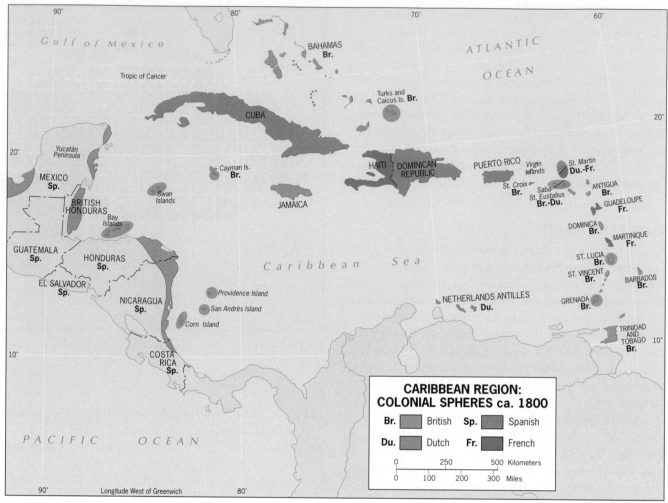

FIGURE 4A-5

© H. J. de Blij, P. O. Muller, and John Wiley & Sons, Inc.

Much later, after centuries of European colonial rivalry in the Caribbean Basin, the United States entered the picture and made its influence felt in the coastal areas of the mainland, not through colonial conquest but through the introduction of widespread, large-scale, banana plantation agriculture. The effects of these plantations were as far-reaching as the impact of colonialism on the Caribbean islands. Because the diseases the Europeans had introduced were most rampant in these hot, humid lowlands (as well as the Caribbean islands to the east), the Amerindian population that survived was too small to provide a sufficient workforce. This labor shortage was remedied through the trans-Atlantic slave trade from Africa that transformed the Caribbean Basin's demography.

The cultural variety of the Caribbean Basin is especially striking, and it is hardly an arena of exclusive Hispanic cultural heritage. For example, Cuba's southern neighbor, Jamaica (population 2.8 million, mostly of African ancestry), has a legacy of British involvement, while to the east in Haiti (9.4 million, overwhelmingly of African ancestry) the strongest imprints have been African and French. The Lesser Antilles also exhibit great cultural diversity. There are the

(once Danish) U.S. Virgin Islands; French Guadeloupe and Martinique; a group of British-influenced islands, including Barbados, St. Lucia, and Trinidad and Tobago; and the Dutch St. Maarten (shared with the French) and the A-B-C islands of the Netherlands Antilles—Aruba, Bonaire, and Curaçao—off the northwestern Venezuelan coast.

POLITICAL AND ECONOMIC FRAGMENTATION

Independence

Independence movements stirred Middle America at an early stage. On the mainland, revolts against Spanish authority (beginning in 1810) achieved independence for Mexico by 1821 and for the Central American republics by the end of the 1820s, resulting in the creation of eight different countries. The United States, concerned over European designs in the realm, proclaimed the Monroe Doctrine in 1823 to deter any European power from reasserting its authority in the newly independent republics or from further expanding its existing domains.

© Jan Nijman

"Driving around the small Caribbean island of Bonaire and coming face to face with this surreal landscape, I first thought I had experienced a mirage. These glistening white, perfectly cone-shaped hills are actually salt piles. This small, Dutch-controlled island off the coast of Venezuela possesses the perfect geographic conditions for salt production: it has a series of salt water inlets, it is very hot and dry, and lies in the zone of persistent tradewinds. Remember that salt, nowadays taken for granted in every household, has long been one of the world's most precious spices, especially when used for the preservation of meat and fish. The Dutch began large-scale production of salt in the 1620s; today this local industry is in the hands of the Antilles International Salt Company and is still an important source of Bonaire's foreign revenues."

www.conceptcaching.com

By the end of the nineteenth century, the United States itself had become a major force in Middle America. The Spanish-American War of 1898 made Cuba independent and put Puerto Rico under the U.S. flag; soon afterward, the Americans were in Panama constructing the Panama Canal. Meanwhile, with U.S. corporations driving a boom based on huge banana plantations, the Central American republics had become colonies of the United States in all but name.

Independence came to the Caribbean Basin in fits and starts. Afro-Caribbean Jamaica as well as Trinidad and Tobago, where the British had brought a large South Asian population, attained full sovereignty from the United Kingdom in 1962; other British islands (among them Barbados, St. Vincent, and Dominica) became independent later on. France, however, retains Martinique and Guadeloupe as overseas *départements* of the French Republic, and the Dutch islands are at various stages of autonomy. No less than 33 states are found on the map of the Caribbean Basin today.

Regional Contrasts

There are some striking contrasts, socially and economically, between the Middle American highlands on the one hand and the Caribbean coasts and islands on the other (Fig. 4A-6). These were conceptualized by cultural geographer John Augelli into the **Mainland-Rimland framework**. The Euro-Amerindian *Mainland*, from Mexico down to Panama, is dominated by European (Spanish) and Amerindian influences and also includes **11 mestizo** sectors where the two ancestries mixed. As Figure 4A-6 shows, the Mainland is subdivided into several areas based on the strength of the Amerindian legacy. The Mainland's Caribbean coastal strip and all of the Basin's islands to the east

constitute the *Rimland*, which for the most part has a very different cultural heritage, based on a fusion of European and African influences.

Supplementing these contrasts are regional differences in outlook and orientation. The Caribbean Rimland was an area of sugar and banana plantations, of high accessibility, of seaward exposure, and of maximum cultural contact and mixture. The Middle American Mainland, being farther removed from these contacts, was an area of greater isolation. The Rimland was the domain of the great **12 plantation**, and its commercial economy was therefore susceptible to fluctuating world markets and tied to overseas investment capital. The Mainland was dominated by the **13 hacienda**, which was far more self-sufficient and less dependent on external markets.

This contrast between plantation and hacienda land tenure in itself constitutes strong evidence for the Rimland-Mainland dichotomy. The hacienda was a Spanish institution, whereas the modern plantation was the concept of northwestern Europeans. In the hacienda, Spanish landowners possessed a domain whose productivity they might never push to its limits: the very possession of such a vast estate brought with it social prestige and a comfortable lifestyle. Native workers lived on the land—which may once have been their land—and had plots where they could grow their own subsistence crops. All this is written as though it is mostly in the past, but the legacy of the hacienda system, with its inefficient use of land and labor, still exists throughout mainland Middle America.

The plantation, in contrast, is all about efficiency and profit. Foreign ownership and investment is the norm, as is production for export. Most plantations grow only a single crop, be it sugar, bananas, or coffee. Much of the labor is

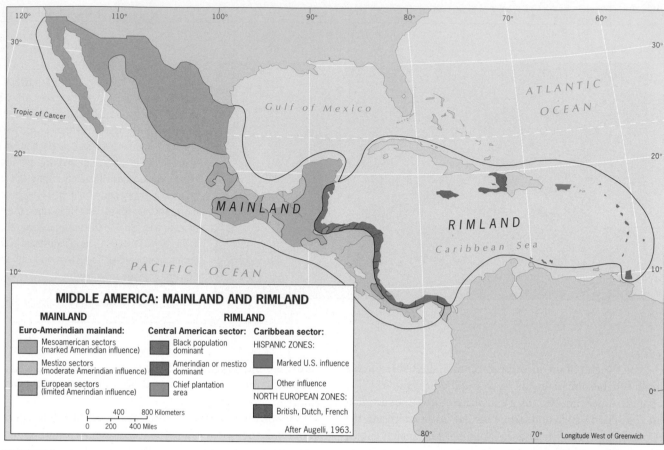

MIDDLE AMERICA: MAINLAND AND RIMLAND

MAINLAND   **RIMLAND**

Euro-Amerindian mainland: **Central American sector:**   **Caribbean sector:**

Mesoamerican sectors   Black population   HISPANIC ZONES:
(marked Amerindian influence)   dominant

Mestizo sectors   Amerindian or mestizo   Marked U.S. influence
(moderate Amerindian influence) dominant

European sectors   Chief plantation   Other influence
(limited Amerindian influence) area
                                                             NORTH EUROPEAN ZONES:

0   400   800 Kilometers   British, Dutch, French
0   200   400 Miles
                                                             After Augelli, 1963.

FIGURE 4A-6

© H. J. DE BLIJ, P. O. MULLER, AND JOHN WILEY & SONS, INC.

seasonal—needed in large numbers mainly during the harvest period—and such labor has been imported because of the scarcity of Amerindian workers. With its "factory-in-the-field" operations (see photo below), the plantation is more efficient in its use of land and labor than the hacienda. Profit and wealth, rather than social prestige, are the dominant motives for the plantation's establishment and operation. The Rimland is also subdivided, with the most obvious division the one between the mainland-coastal plantation zone and the islands to the east. Notice, too, that the islands themselves are classified according to their cultural heritage (Figs. 4A-5 and 4A-6).

Contrasting land uses in the Middle American Rimland and Mainland give rise to some very different rural cultural landscapes. Huge stretches of the realm's best land continue to be controlled by (often absentee) landowners whose haciendas yield export or luxury crops, or foreign corporations that raise fruits for transport and sale on their home markets. The banana plantation shown here (left) lies near the Caribbean coast of Costa Rica. The vast fields of banana plants stand in strong contrast to the lone peasant who ekes out a bare subsistence from small cultivable plots of land, often in high-relief countryside where grazing some goats or other livestock is the only way to use most of the land. (Left): © John Coletti/Getty Images, Inc. (Right): © Jean-Gerard Sidaner/Photo Researchers.

If in the past the plantations brought their owners considerable wealth, little of that ever reached their workers, and inequality seemed a structural part of life. In recent decades, the near monopoly of some of the Caribbean crops (particularly sugar) came to an end with competition from elsewhere (for example, the United States). Revenues declined and unemployment increased. Millions of workers were pushed into a life of subsistence, and many took refuge in the cities, where conditions and opportunities are hardly better.

Small Is Beautiful?

As noted, the Middle American realm is exceptional in its modest size, in terms of both territory and population. But the number of countries is considerable, and they tend to be quite small. Caribbean islands often invoke images of beautiful scenery, white sandy beaches, palm trees, tropical cocktails, and shiny blue waters. But the economic realities are often harsh. The limited land areas of the Lesser Antilles each average well below half a million people, and that, combined with insularity and remoteness, poses formidable challenges that are common to these **14** **small-island developing economies**. First, natural resources are often limited; this means a heavy reliance on imports, adding the expense of transport costs. Second, government is relatively expensive on a per-capita basis: even the smallest population will require services such as schools, hospitals, waste disposal, military capabilities, and general administration. Third, specialized services must often be brought in from elsewhere. And fourth, local production cannot really benefit from economies of scale; this means that local producers can be put out of business by cheaper imports, which in turn causes unemployment.

Given the Caribbean Basin's limited economic options, does the tourist industry offer better opportunities? Opinions on this question are divided. The resort areas, scenic treasures, and historic locales of Caribbean America attract well over 20 million visitors annually, with about half of them traveling on Florida-based cruise ships. Certainly, Caribbean tourism is a prospective money-maker for many islands. In Jamaica alone, this industry accounts for about one-sixth of the gross domestic product and employs more than one-third of the labor force.

But Caribbean tourism also has serious drawbacks. The invasion of poor communities by affluent tourists contributes to rising local resentment, which is further fueled by the glaring contrasts of shiny new hotels towering over substandard housing and luxury liners gliding past poverty-stricken villages. At the same time, tourism can have the effect of debasing local culture, which often is adapted to suit the visitors' tastes at hotel-staged "culture" shows. In addition, the cruise industry tends to monopolize revenues (accommodations, meals, drinks, entertainment) with relatively few dollars flowing into the local economy. Finally, while tourism does generate income in the Caribbean, the intervention of island governments and multinational corporations removes opportunities from local entrepreneurs in favor of large operators and major resorts.

The Push for Regional Integration

Another challenge for the Middle American realm is to foster greater economic integration. Many of the countries on the mainland as well as in the Caribbean have few ties within the realm and are highly dependent on big outside countries, the United States in particular. For the large majority of these countries, the United States is the primary trading partner. Consider this: of all trade involving Middle American countries, less than 10 percent takes place within the realm. And less than 1 percent takes place between the Caribbean Basin and the Middle American mainland.

Over the years, efforts have been made to advance economic integration and to turn this realm into more of a *functional region*. The mainland saw the creation of the Central American Common Market in 1960, but it became moribund within a decade because of the 1969 war between El Salvador and Honduras and further intraregional conflict in the 1970s and 1980s (some with U.S. involvement). The organization was revived in the 1990s but since 2005 has been overshadowed by **CAFTA**, the Central American Free Trade Agreement with the United States. CAFTA seems a mixed blessing: it may increase access to U.S. markets and lead to cheaper imports, but it may also galvanize the singular domination of the United States at the cost of greater intraregional integration.

In the Basin, CARICOM (Caribbean Community) was established in 1989 and now consists of 15 full members, including Guyana and Suriname in South America. CARICOM follows the example, in some respects, of the European Union and in 2009 even introduced a common passport. But economic change has been slow. The geography of the Middle American realm poses many significant challenges, and these are not easily overcome even with the best political intentions.

POINTS TO PONDER

- The 3200-kilometer (2000-mi) land border between North America and Middle America is the longest in the world separating a rich realm from a poor one.
- The United States is the single most important economic partner of just about every country and territory in this realm—is that a positive or a negative for Middle America?
- "Poor Mexico . . . so far from God and so close to the United States"—former Mexican President Porfirio Diaz
- Some of the small Caribbean island-nations count less than a few hundred thousand people—is that enough to sustain a viable economy?

One of Central America's signature volcanoes looms over the city of Antigua, Guatemala.
© Antoinette M.G. A. WinklerPrins

IN THIS CHAPTER

How NAFTA changed the economic geography of Mexico
Is Mexico a narco-state?
Indigenous peoples demand recognition and rights
Projected Panama Canal expansion fuels boom in Panama City
The 2010 killer earthquake in Haiti
The debate over Puerto Rico's status continues

CONCEPTS, IDEAS, AND TERMS

4B

MIDDLE AMERICA: REGIONS OF THE REALM

Mexico
Central America
Caribbean Basin
Greater Antilles
Lesser Antilles

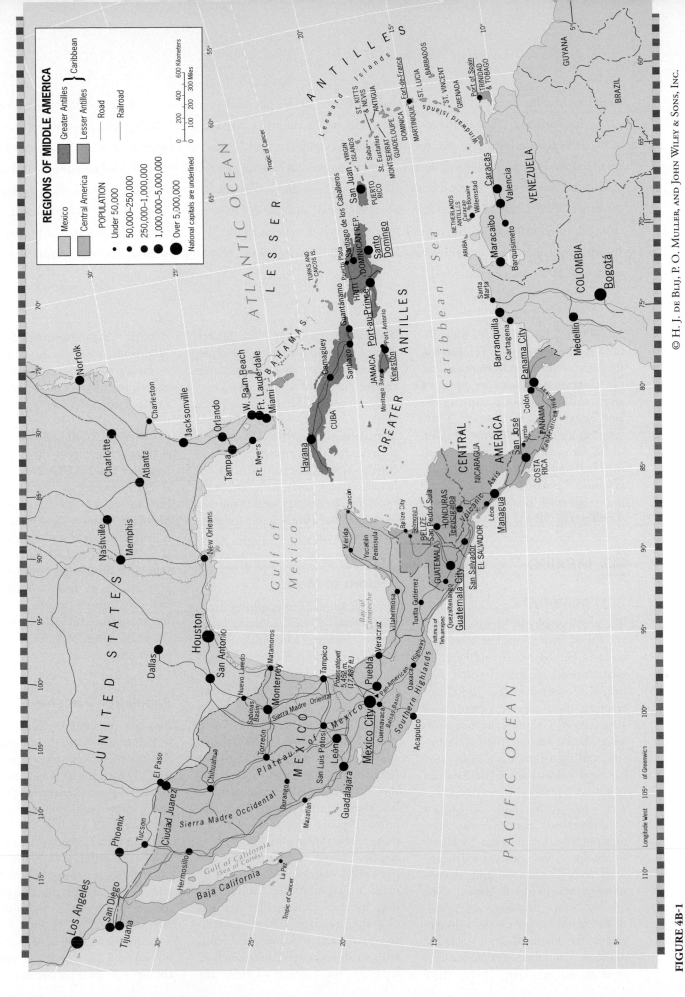

FIGURE 4B-1

The Middle American realm stands out in terms of its overall small population, limited land cover, territorial fragmentation, and the insulated position of many its countries. Mexico is the obvious exception when it comes to size, accounting for 57 percent of the realm's population and 72 percent of its area; all other countries vary from small to tiny. Fragmentation and isolation tend to go hand in hand in this realm. All territories are either islands or have long coastlines (most of them on both sides). Only one country, Guatemala, shares borders with as many as four other countries; Mexico and Honduras have three neighbors, and the five remaining Central American republics have only two. All other territories in the realm have either one or none at all as they are completely surrounded by water. Not surprisingly, regional integration, a condition for political stability and economic progress, has been one of the foremost challenges in this part of the world.

Middle America consists of four distinct geographic regions: (1) *Mexico*, the giant of the realm in every respect; (2) *Central America*, the string of seven small republics occupying the land bridge to South America; (3) the four islands of the *Greater Antilles*, Cuba, Jamaica, Hispaniola (containing Haiti and the Dominican Republic), and Puerto Rico; and (4) the numerous small islands of the *Lesser Antilles* (Fig. 4B-1).

≡ MEXICO

Physiography

The physiography of Mexico is reminiscent of that of the conterminous western United States, although its environments are more tropical. Figure 4B-2 shows several prominent features: the elongated peninsula of Baja (Lower) California in the northwest, the far eastern Yucatán Peninsula, and the Isthmus of Tehuantepec in the southeast where the Mexican landmass tapers to its narrowest extent. Here in the southeast, Mexico most resembles Central America physiographically; a mountain backbone forms the isthmus, curves southeastward into Guatemala, and extends northwestward toward Mexico City. Shortly before reaching the capital, this mountain range divides into two chains, the Sierra Madre Occidental in the west and the Sierra Madre Oriental in the east (Fig. 4B-2). These

diverging ranges frame the funnel-shaped Mexican heartland, the center of which consists of the rugged, extensive Plateau of Mexico (the Valley of Mexico lies near its southeastern end). As Figure G-7 reveals, Mexico's climates are characterized by dryness, particularly in the broad, mountain-flanked north. The majority of the better-watered areas lie in the southern half of the country where a number of major population concentrations have developed.

Regions of Mexico

Physiographic, demographic, economic, historical, and cultural criteria combine to reveal a regionally diverse Mexico extending from the lengthy ridge of Baja California to the tropical lowlands of the Yucatán Peninsula, and from the economic frenzy of the NAFTA North to the Amerindian traditionalism of the Chiapas southeast (Fig. 4B-3). In the Core Area, anchored by Mexico City, and the West, centered on Guadalajara, lies the transition zone from the more Hispanic-mestizo north to the more Amerindian-infused mestizo south. East of the Core Area lies the Gulf Coast, once dominated by major irrigation projects and sprawling livestock-raising schemes but now the mainland center of Mexico's petroleum industry. The semi-arid, scrub-vegetated Balsas Lowland separates the Core Area from the

rugged, Pacific-fronting Southern Highlands, where Acapulco's luxurious resorts stand in stark contrast to the Amerindian villages and communal-farm settlements of the interior.

The dry, vast north stands in the sharpest contrast to these southern regions. The NAFTA North region is still formative and discontinuous, but, as we will see, it is changing northern Mexico significantly. This is true even in Yucatán, where the city of Mérida and environs are substantially affected by NAFTA development. To go from comparatively well-off northern Yucatán to poverty-stricken southern Chiapas is to see the whole range of Mexico's regional geography.

Population Patterns

Mexico's population expanded rapidly during the last three decades of the twentieth century, doubling in just 28 years; but demographers have recently noted a sharp drop in fertility, and they are predicting that Mexico's population (currently at 111 million) will stop growing altogether by about 2050. This will have enormous implications for the country's economy, and it will reduce the cross-border migration that is of so much concern at present.

The distribution of Mexico's population and its association with the country's 31 internal States is shown

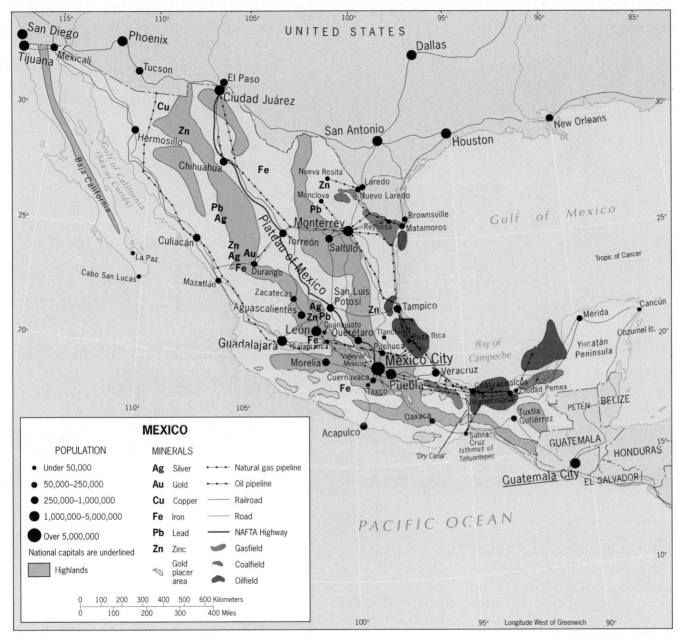

FIGURE 4B-2

© H. J. de Blij, P. O. Muller, and John Wiley & Sons, Inc.

in Figure 4B-4 (and even more precisely in Fig. 4A-2). The largest concentration, containing more than half the Mexican people, extends across the densely populated "waist" of the country from Veracruz State on the eastern Gulf Coast to Jalisco State on the Pacific. The center of this corridor is dominated by the most populous State, Mexico (**3** on the map), at whose heart lies the Federal District of Mexico City (**9**). In the dry and rugged terrain to the north of this

central corridor lie Mexico's least-populated States. Southern Mexico also exhibits a sparsely peopled periphery in the hot and humid lowlands of the Yucatán Peninsula, but here most of the highlands of the continental spine contain sizeable populations.

Another major feature of Mexico's population map is urbanization, driven by the pull of the cities (with their perceived opportunities for upward mobility) in tandem with the

push of the economically stagnant countryside. Today, 76 percent of the Mexican people reside in towns and cities, a surprisingly high proportion for a developing country. Undoubtedly, these numbers are affected by the explosive recent growth of Mexico City, which now totals almost 29 million (making it the largest urban concentration on Earth) and is home to 26 percent of the national population. Urbanization rates at the other end of Mexico, however, are at their lowest in

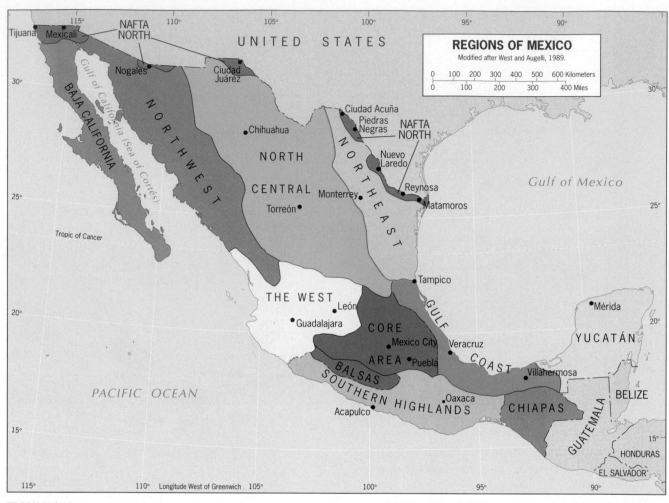

FIGURE 4B-3

© H. J. de Blij, P. O. Muller, and John Wiley & Sons, Inc.

those remote uplands where Amerindian society has been least touched by modernization (Fig. 4B-4).

A Mix of Cultures

Nationally, the Amerindian imprint on Mexican culture remains strong. Today, 60 percent of all Mexicans are mestizos, 22 percent are predominantly Amerindian, and about 8 percent are full-blooded Amerindians; most of the remaining 10 percent are Europeans. The Spanish influence in Mexico has been profound, but it has been met with an equally powerful thrust of Amerindian culture. It has therefore not been a case of one-way European-dominated **1** acculturation but rather **2** transculturation—the two-way exchange of culture traits between societies in close contact. In the southeastern periphery (Fig. 4B-5), several hundred thousand Mexicans still speak only an Amerindian language, and millions more still utilize these languages daily even though they also speak Mexican Spanish. The latter has been strongly shaped by Amerindian influences, as have Mexican modes of dress, foods and cuisine, artistic and architectural styles, and folkways. This fusion of heritages, which makes Mexico unique, is the product of an upheaval that began to reshape the country a century ago.

Agriculture: Fragmented Modernization

Modern Mexico was forged in a revolution that began in 1910 and set into motion events that are still unfolding today. At its heart, this revolution was about the redistribution of land, an issue that had not been resolved after Mexico freed itself from Spanish colonial control in the early nineteenth century. As late as 1900, only about 8000 haciendas (large, traditional, family-owned farms) blanketed virtually all of Mexico's better farmland. About 95 percent of all rural families owned no land whatsoever and toiled as *peones* (landless, constantly indebted serfs) on the haciendas. The triumphant revolution produced a new constitution in 1917 that launched a program of expropriation and parceling out of the haciendas to rural communities.

Since 1917, more than half the cultivated land in Mexico has been redistributed, mostly to peasant communities consisting of 20 families or more. On such farmlands, known as

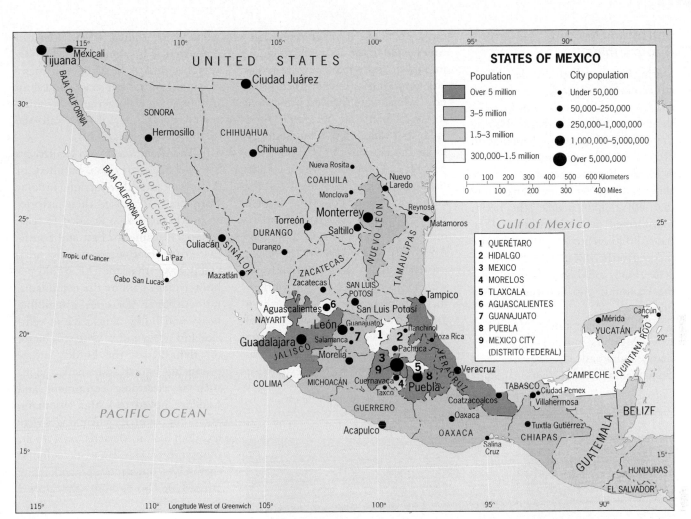

FIGURE 4B-4 © H. J. de Blij, P. O. Muller, and John Wiley & Sons, Inc.

3 ejidos, the government holds title to the land, but the rights to use it are parceled out to village communities and then to individuals for cultivation. This system of land management is an Amerindian legacy, and not surprisingly most *ejidos* lie in central and southern Mexico where Amerindian social and agricultural traditions are strongest. About half of Mexico's land continues to be held in such "social landholdings." But the reforms did not lead to increased production. Some of the land is excessively fragmented, causing both low farm yields and widespread rural poverty. In the 1990s the Mexican government attempted to privatize *ejidos*, hoping to promote consolidation and increase productivity, but that effort did not work. To date, less than 10 percent of the *ejidos* have been privatized.

Although traditional subsistence agriculture and the output of the inefficient *ejidos* have not changed a great deal in the poorer areas of rural Mexico, larger-scale commercial agriculture has diversified during the past three decades and made major gains with respect to both domestic and export markets. The country's arid northern tier has led the way as major irrigation projects have been built on streams flowing down from the interior highlands. Along the booming northwest coast of the mainland, which lies within a day's drive of Southern California, mechanized large-scale cotton production now supplies an increasingly profitable export trade. Here, too, wheat and winter vegetables are grown, with fruit and vegetable cultivation attracting myriad foreign investors. Still, many small farmers are having

a tough time of it due to competition of cheap (and subsidized!) corn from the United States and lack of access to the U.S. market.

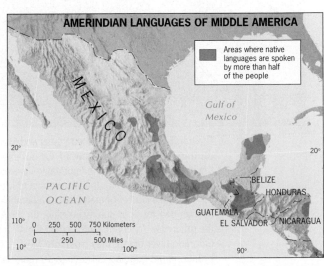

FIGURE 4B-5 © H. J. de Blij, P. O. Muller, and John Wiley & Sons, Inc.

Shifting Economic Geographies

During the last three decades, Mexico's economic geography has changed, and in some respects progressed—though not without setbacks. During the early 1990s, the implementation of NAFTA led to an economic boom as Mexico became part of a free-trade zone and market comprising more than 400 million people. This boom transformed urban landscapes along the 3170-kilometer (1970-mi) border that separates Mexico and the United States, but it could not, of course, close the economic gap between the two sides.

With the inception of NAFTA in 1994, Mexico's industrial geography changed dramatically, together with the liberalization of the national econ-omy. Under the new agreement, factories based in Mexico could assemble imported, duty-free raw materials and components into finished prod-ucts, which were then exported back to the U.S. market. Logically, these factories, called **4** **maquiladoras**, located as close to the U.S. border as possible (see photo below). As a result, manufacturing employment in the cities and towns along that border, from Tijuana in the west to Matam-oros in the east, expanded rapidly. After only seven years of NAFTA's existence, there were 4000 factories with more than 1.2 million workers in the border zone and in northern Yucatán (where Mérida was part of the process), accounting for nearly one-third of Mexico's industrial jobs and 45 percent of its total exports. It should be noted that Mexican employees work long hours for low wages with few benefits, and many live in the most basic shacks and slum dwellings encircling the burgeoning towns of the region we call the NAFTA North (Fig. 4B-3).

And there was no job security. In the past five years or so, hundreds of American and other foreign corpora-tions that had moved their factories to northern Mexico decided to relo-cate once again—to East and South-east Asia where wages were consid-erably lower than the U.S. $2.00 per hour average paid by the maquilado-ras. Although factories assembling heavy and bulky items such as vehi-cles and refrigerators were still better off right across the border, others producing lighter and smaller goods such as electronic equipment and cameras moved to China, Vietnam, and other countries where wages were less than half of Mexico's. As a result, thousands of Mexican workers found themselves unemployed—and many crossed the border into the United States.

The only way to counter this trend, it seems, is by adding higher-wage jobs in more advanced sectors, such as electronics, that are less likely to be lost to other parts of the devel-oping world. Education in high-tech and management fields is part of it. In the Northeast, the city of Mon-terrey in the relatively high-income State of Nuevo Léon has become a model of success: it has a substantial international business community and modern industrial facilities that have attracted major multinational companies. Here lies hope for Mexi-co's future.

States of Contrast

Countries with strong regional dis-parities face serious challenges that can be difficult to overcome and may worsen over time. Mexico's southern-most States—Chiapas, Oaxaca, and Guerrero, all bordering the Pacific

The border between the United States and Mexico occasionally provides some stunning contrasts. Here, looking westward, we observe the opposing economic geographies that mark the border landscape of the Imperial Valley east of the urban area formed by Mexicali, Mexico and Calexico, California. The larger town of Mexicali (on the left) sprawls eastward along the border, with a cluster of maquiladora assembly plants, visible in the left foreground, surrounded by high-density, poor-quality housing. On the American side, lush irrigated croplands blanket the otherwise dry countryside, fed by the All-American Canal that taps the waters of the Colorado River before they cross into Mexico. At this point, the canal swings northward to bypass border-hugging Calexico (right rear) before rejoining the international boundary beyond the town's western edge. © Alex McLean/Landslides

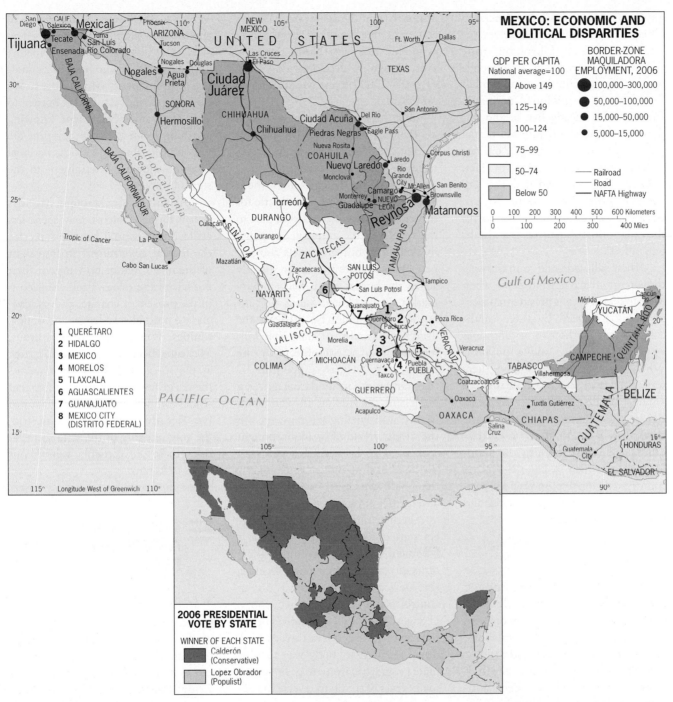

FIGURE 4B-6

© H. J. de Blij, P. O. Muller, and John Wiley & Sons, Inc.

Ocean—are by far the poorest. States bordering the United States in the north, including Nuevo Léon, Chihuahua, and Baja California, have the highest incomes. Using rural poverty as a measure, only about 10 percent of people in the countryside in the north are in the poorest category, but nearly 50 percent in the south.

Mexico's north-south divide is especially evident from the economic data in Figure 4B-6. In very general terms, the annual per-capita income in the northern States exceeds U.S. $10,000; in the southern States it falls below $5000. Economic growth in the northern States has averaged more than 4 percent in recent years; in the southern States it is less than 2 percent. Mexico's infrastructure, already inadequate, serves the south far less well than the north. Whatever the index—

literacy, electricity use, water availability—the south lags.

This is a serious problem for Mexico because the south is not only culturally the most Amerindian part of the country, but also the least well educated, the most isolated and remote, the least productive agriculturally, and lagging badly in terms of infrastructure investment and overall development. Since 1994, a radical group of Maya peasant farmers in Chiapas State calling themselves the Zapatista National Liberation Army (ZNLA) have engaged in guerrilla warfare, demanding better treatment for Mexico's 33 million Amerindian citizens (who mostly live in the impoverished south). Despite substantial public support, their struggle has not yielded significant results. In fact, NAFTA has only widened the gap between north and south. To upwardly mobile Mexicans in the north, the cause of the Zapatistas is a distant one indeed.

These still-widening contrasts were thrown into sharp relief in 2006, when Mexico's most recent presidential election was contested by three candidates, of whom the two leading ones were a conservative and a populist. When the ballots were counted, there was a near-tie, a result so close that the populist at first refused to concede. But on the map, it was clear that the north was won by the conservative candidate and the south by the populist (see inset map in Fig. 4B-6). The country's regional economic disparities were producing serious social and political consequences.

Mexico's Future

The future of the country depends on the government's efforts to close the gap between rich and poor in order to reduce the vast regional inequalities. Put differently, it must spread the positive effects of NAFTA from north to south. This can include federal anti-poverty programs, investment in education, and infrastructural projects.

Mexico's future is inextricably bound up with the United States, in good times and bad. The good refers to the two countries' economic interaction: the United States is Mexico's most important trade partner. For all their differences over such issues as cross-border trucking, migration, and corporate practices in the maquiladora cities, the two governments have found ways to cooperate. There is no

doubt that, in the aggregate, Mexico's economy has benefited from closer economic ties with North America, even though the economic downturn of 2008–2009 hit the Mexican economy hard.

But recent years have witnessed another major obstacle. Over the past decade, the drug cartels that were centered in Colombia established new bases in U.S.-border cities in northern Mexico and began a vicious battle for supremacy. They responded in part to the success of the antidrug campaign in Colombia, but the cartels also saw new opportunities in Mexico directly across the border from their primary market in the United States. So serious was the situation by 2010 that the drug barons and their allies were killing law-enforcement officers by the hundreds as well as competitors by the thousands, intimidating government and threatening all who stood in their way. Fueling this rampage was the steady stream of weapons, bought with ease in the United States with the flood of drug money generated by that trade, which crossed the border and further empowered the cartels.

In the United States some observers started referring to Mexico as a **5** **failed state** because, in the words of its director of national intelligence, the Mexican government was not in control of parts of its own territory. U.S. government representatives began referring to Mexico as a threat to national security. It surely was an ultimate geographic irony that, following many years of trying to reverse the devolutionary tide in its poverty-afflicted south, Mexico's government now faced what some called the state's "Colombianization" in the comparatively prosperous north.

In the meantime, the drug cartels have worked to diversify their activities: for instance, during the two-year period ending in January 2010, the "Zetas" (a cartel founded by former military commandos) siphoned more than $1 billion worth of oil from Mexico's pipelines in the State of

Veracruz. The Zetas do not tap the oil themselves—they are said to effectively own long stretches of pipeline and tax anybody who has the technology to siphon it off. Most of the stolen oil is then sold across the border to U.S. companies (who deny knowing anything about it). These activities are not only bleeding the Mexican national treasury, they are a testament to the state's impotence.

Against this background, Mexico's longer-term problems may seem less acute, but it would be a mistake for the federal government to lose sight of them, no matter how urgent their pursuit of the drug cartels. Mexico is in the process of change, and its leaders face multiple challenges and opportunities. When times are prosperous, Mexicans dream of opportunities such as the so-called **6** **dry canal** across the narrowest part of the country that would compete with the Panama Canal (see Fig. 4B-2). When times are bad, the very integrity of the state seems at risk. These are certainly tumultuous times in Mexico.

THE CENTRAL AMERICAN REPUBLICS

A Land Bridge

Crowded onto the narrow segment of the Middle American land bridge between Mexico and the South American continent are the seven countries of Central America (Fig. 4B-7). Territorially, they are all quite small; their population sizes range from Guatemala's nearly 15 million down to Belize's 300,000. The land bridge here consists of a highland belt flanked by coastal lowlands on both the Caribbean and Pacific sides (Fig. 4B-8). These highlands are studded with volcanoes, and local areas of fertile volcanic soils are scattered throughout them.

The land bridge has a fascinating geologic and evolutionary history, and one famous study refers to it as the

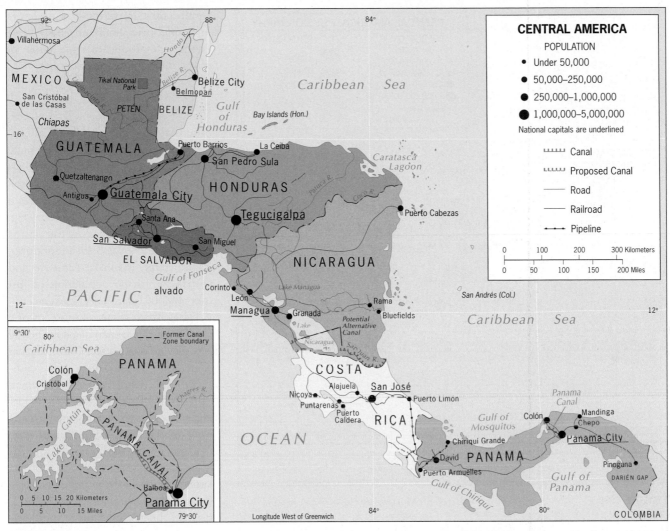

FIGURE 4B-7

© H. J. DE BLIJ, P. O. MULLER, AND JOHN WILEY & SONS, INC.

Monkey Bridge. For some 50 million years, North and South America were separated; the land bridge was formed "only" 3 million years ago, becoming a highway of sorts for evolutionary exchange. The region contains only 1 percent of the Earth's land but 7 percent of all the world's natural species. The southern part (Costa Rica and Panama) is known as a global **7 biodiversity hotspot**, even though deforestation has been a major problem. Human inhabitants have always been concentrated in the upland zone, where tropical temperatures are moderated by elevation and rainfall is sufficient to support a variety of crops.

Central America is not a large region, but because of its physiography it contains many isolated, comparatively inaccessible locales. The population tends to concentrate in the much cooler uplands—**8 tierra templada**—and population densities are generally greater toward the Pacific than toward the Caribbean side (see Fig. 4A-2). The most significant exception is El Salvador, whose political boundaries confine its inhabitants mostly to its tropical coastal lowlands—**9 tierra caliente**—tempered here by the somewhat cooler Pacific offshore.

Central America, we noted earlier, actually begins within Mexico, in Chiapas and in Yucatán, and the region's republics face many of the same problems as less-developed parts of Mexico. Population pressure is one of these problems. A population explosion began in the mid-twentieth century, increasing the region's human inhabitants from 9 million to almost 45 million by 2010.

Today, the region also continues to struggle to emerge from a period of turmoil that lasted through the 1980s into the mid-1990s. Devastating inequities, repressive governments, external interference, and the frequent unleashing of armed forces have destabilized Central America for much of its modern history.

Guatemala

The westernmost of Central America's republics, Guatemala has more land neighbors than any other. Straight-line boundaries across the tropical forest mark much of the border with Mexico, creating the box-like region of Petén between Chiapas State on

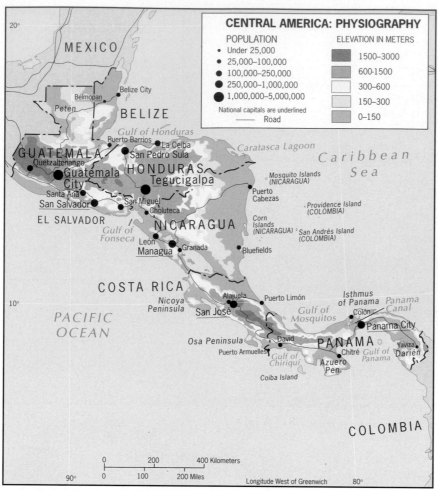

FIGURE 4B-8   © H. J. de Blij, P. O. Muller, and John Wiley & Sons, Inc.

the west and Belize on the east; also to the east lie Honduras and El Salvador (Fig. 4B-7). This heart of the ancient Maya Empire, which remains strongly infused by Amerindian culture and traditions, has just a small window on the Caribbean but a longer Pacific coastline. Guatemala was still part of Mexico when the Mexicans threw off the Spanish yoke, and though independent from Spain after 1821, it did not become a separate republic until 1838. Mestizos, not the Amerindian majority, secured the country's independence.

Most populous of the seven republics with 14.5 million inhabitants (mestizos are in the majority with 55 percent, Amerindians 43 percent), Guatemala has seen much conflict. Repressive regimes brokered deals with U.S. and other foreign economic interests that stimulated development, but at a high social cost. Over the past half-century, military regimes have dominated political life. The deepening split between the wretchedly poor Amerindians and the better-off mestizos, who here call themselves *ladinos*, generated a civil war that started in 1960 and claimed more than 200,000 lives as well as 50,000 "disappearances" before it ended in 1996. An overwhelming number of the victims were of Maya descent; the mestizos continue to control the government, army, and land-tenure system.

Guatemala's tragedy is that its economic geography has considerable potential but has long been shackled by the unceasing internal conflicts that have kept the income of 52 percent of the population below the poverty line. The country's mineral wealth includes nickel in the highlands and oil in the lower-lying north. Agricul-

turally, soils are fertile and moisture is ample over highland areas large enough to produce a wide range of crops, including excellent coffee.

Belize

Strictly speaking, Belize is not a Central American republic in the same tradition as the other six. Until 1981, this country, a wedge of land between northern Guatemala, Mexico's Yucatán Peninsula, and the Caribbean Sea (Fig. 4B-7), was a dependency of the United Kingdom known as British Honduras. Slightly larger than Massachusetts and with a minuscule population of just over 300,000 (many of African descent), Belize is much more reminiscent of a Caribbean island than of a continental Middle American state. Today, all that is changing as the demographic complexion of Belize is being reshaped. Thousands of residents of African descent have recently emigrated (many went to the United States) and were replaced by tens of thousands of Spanish-speaking immigrants. Most of the latter are refugees from strife in nearby Guatemala, El Salvador, and Honduras, and their proportion of the Belizean population has risen from 33 to just under 50 percent since 1980. Within the next few years the newcomers will be in the majority, Spanish will become the *lingua franca*, and Belize's cultural geography will exhibit a trend toward Hispanicization.

The Belizean transformation extends to the economic sphere as well. No longer just an exporter of sugar and bananas, Belize is producing new commercial crops, and its seafood-processing and clothing industries have become major revenue earners. Also important is tourism, which annually lures more than 150,000 vacationers to the country's Mayan ruins, resorts, and newly legalized casinos; a growing specialty is ecotourism, based on the natural attractions of the country's near-pristine environment. Belize is also known as a center for **10 offshore banking**—a financial haven for

© Jan Nijman

"Invited by the Honduran government to survey the country's ecotourism potential, I toured the beautiful countryside with a National Geographic Society delegation in 2006. There is no doubt that Honduras is exceptionally endowed by nature and that there is considerable potential for a vibrant tourist industry. But this is a poor country and the government has great difficulty making the required investments—in infrastructure, for example. We flew in these small propeller planes to get from the capital, Tegucigalpa, to the Mayan ruins of Copán. There was no airport nearby and overland travel is arduous and time-consuming. Ecotourism by its very nature is small-scale and the revenues tend to be relatively modest, so it is a major challenge to raise the substantial funds needed for costly infrastructural improvements."

www.conceptcaching.com

foreign companies and individuals who want to avoid paying taxes in their home countries.

Honduras

Honduras is a country still on hold as it struggles to rebuild its battered infrastructure and economy more than a decade after taking a direct hit from Category-5 Hurricane Mitch in 1998. With 7.6 million inhabitants, about 90 percent mestizo, bedeviled Honduras still has years to go even to restore what was already the third-poorest economy in the Americas (after Haiti and Nicaragua). Agriculture, livestock, forestry, and limited mining formed the mainstays of the pre-1998 economy, with the familiar Central American products—bananas, coffee, shellfish, and apparel—earning most of the external income.

Honduras, in direct contrast to Guatemala, has a lengthy Caribbean coastline and a small window on the Pacific (Fig. 4B-7). The country also occupies a critical place in the political geography of Central America,

flanked as it is by Nicaragua, El Salvador, and Guatemala—all continuing to grapple with the aftermath of years of internal conflict and, more recently, natural disaster. Comparable in natural beauty and biodiversity to Costa Rica, Honduras is seeking to exploit its potential for ecotourism but is hampered by a poor infrastructure and lack of funds for new facilities.

El Salvador

This is Central America's smallest country territorially, smaller even than Belize, but with a population 25 times as large (7.5 million) it is the most densely peopled. With Belize, it is one of only two continental republics that lack coastlines on both the Caribbean and Pacific sides (Fig. 4B-7). El Salvador adjoins the Pacific in a narrow coastal plain backed by a chain of volcanic mountains, behind which lies the country's heartland. Unlike neighboring Guatemala, El Salvador has a quite homogeneous population (90 percent mestizo and just 1 percent Amerindian). Yet ethnic homogeneity

has not translated into social or economic equality or even opportunity. Whereas other Central American countries were called banana republics, El Salvador was a coffee republic, and the coffee was produced on the huge landholdings of a few landowners and on the backs of a subjugated peasant labor force. The military supported this system and repeatedly suppressed violent and desperate peasant uprisings.

From 1980 to 1992, El Salvador was torn by a devastating civil war that was worsened by outside arms supplies from the United States (supporting the government) and Nicaragua (aiding the Marxist rebel forces). But ever since the negotiated end to that war, efforts have been under way to prevent a recurrence because El Salvador is having difficulty overcoming its legacy of searing inequality. The civil war did have one positive result: affluent citizens who left the country and did well in the United States and elsewhere send substantial funds back home, which now provide the largest single source of foreign revenues. This has helped stimulate such industries as apparel and footwear manufacturing as well as food processing. But a major stumbling block to revitalization of the agricultural sector has again been land reform, and El Salvador's future still hangs in the balance.

Nicaragua

This country is best approached by reexamining the map (Fig. 4B-7), which underscores Nicaragua's pivotal position in the heart of Central America. The Pacific coast follows a southeasterly direction here, but the Caribbean coast is oriented north-south so that Nicaragua forms a triangle of land with its lakeside capital, Managua, located in a valley on the mountainous, earthquake-prone, Pacific side (the country's core area has always been located here). The Caribbean side, where the uplands yield to a coastal plain of tropical rainforest, savanna, and swampland, has for centuries been home to Amerindian peoples such as the Miskito, who have been remote from the focus of national life.

Until the 1990s, Nicaragua had a checkered history of political instability, conflict, and economic backwardness. The strife ended in 1990 when the first democratic government in decades was voted into office. But economic progress has been scant. For the past two decades Nicaragua's economy has ranked as continental Middle America's poorest. Hurricane Mitch struck here too, devastating the country's farms and driving tens of thousands into the impoverished towns just at a time when the agricultural sector was recovering and land reform promised a better life for some 200,000 peasant families.

The country's options are limited: agriculture dominates the economy and manufacturing is weak. There is a growing reliance on remittances from Nicaraguans who have emigrated (now 15 percent of the economy) as well as on foreign aid. For years there has been talk of a *dry canal* transit role

for this country, but other potential overland routes are more promising. Meanwhile, the accelerated growth of Nicaragua's population continues (2010 total: 5.9 million), dooming hopes for a rise in standards of living.

Costa Rica

If there is one country that underscores Middle America's endless variety and diversity it is Costa Rica—because it differs significantly from its neighbors and from the norms of Central America as well. Bordered by two volatile countries (Nicaragua to the north and Panama to the east), Costa Rica is a nation with an old democratic tradition and, in this cauldron, no standing army for the past six decades! Although the country's Hispanic imprint is similar to that found elsewhere on the mainland, its early independence, its good fortune to lie remote from regional strife, and its leisurely pace of settlement allowed Costa Rica the luxury of concentrating on its economic development. Perhaps most important, internal political stability has prevailed over much of the past 175 years.

Like its neighbors, Costa Rica is divided into environmental zones that parallel its coastlines. The most densely settled is the central highland zone, lying in the cooler *tierra templada*, whose heartland is the *Valle Central* (Central Valley), a fertile basin that contains Costa Rica's main coffee-growing area and the leading population cluster focused on the capital, San José (Fig. 4B-7)—the most cosmopolitan urban center between Mexico City and the primate cities of northern South America.

The long-term development of Costa Rica's economy has given it the region's highest standard of living, literacy rate, and life expectancy (though even here, one-quarter of the population is trapped in poverty). Agriculture continues to dominate (with bananas, coffee, tropical fruits, and seafood the leading exports), and

tourism has expanded steadily. Costa Rica is widely known for its superb scenery and for its efforts to protect what is left of its diverse tropical flora and fauna. (Nonetheless, deforestation is a major threat.)

Panama

Panama owes its existence to the idea of a canal connecting the Atlantic and Pacific oceans to avoid the lengthy circumnavigation of South America. The Panama Canal (see the inset map in Fig. 4B-7) was opened in 1914, a symbol of U.S. power and influence in Middle America. The Canal Zone was held by the United States under a treaty that granted it "all the rights, powers, and authority" in the area "as if it were the sovereign of the territory." Such language might suggest that the United States held rights over the Canal Zone in perpetuity, but the treaty nowhere stated specifically that Panama permanently yielded its own sovereignty in that transit corridor. In the 1970s, as the canal was transferring more than 14,000 ships per year (that number today is only slightly lower, but the cargo tonnage is up significantly) and generating hundreds of millions of dollars in tolls, Panama sought to terminate U.S. control in the Canal Zone. Delicate negotiations began. In 1977, an agreement was reached on a staged withdrawal by the United States from the territory, first from the Canal Zone and then from the Panama Canal itself (a process completed on December 31, 1999).

Panama today reflects some of the usual geographic features of the Central American republics. Its population of 3.5 million is about 70 percent mestizo and also contains substantial Amerindian, white, and black minorities. Spanish is the official language, but English is also widely used. Ribbonlike and oriented east-west, Panama's topography is mountainous and hilly. Eastern Panama, especially Darien Province adjoin-

FROM THE FIELD NOTES...

© H. J. de Blij

"The Panama Canal remains an engineering marvel 90 years after it opened in August 1914. The parallel lock chambers each are 1000 feet long and 110 feet wide, permitting vessels as large as the Queen Elizabeth II to cross the isthmus. Ships are raised by a series of locks to Gatún Lake, 85 feet above sea level. We watched as tugs helped guide the QEII into the Gatún Locks, a series of three locks leading to Gatún Lake, on the Atlantic side. A container ship behind the QEII is sailing up the dredged channel leading from the Limón Bay entrance. The lock gates are 65 feet wide and 7 feet thick, and range in height from 47 to 82 feet. The motors that move them are recessed in the walls of the lock chambers. Once inside the locks, the ships are pulled by powerful locomotives called mules that ride on rails that ascend and descend the system. It was still early morning, and a major fire, probably a forest fire, was burning near the city of Colón, where land clearing was in progress. This was the beginning of one of the most fascinating days ever."

www.conceptcaching.com

ing Colombia, is densely forested, and here is the only remaining gap in the intercontinental Pan American Highway (Fig. 4B-7). Most of the rural population lives in the uplands west of the canal; there, Panama produces bananas, shrimps and other seafood, sugarcane, coffee, and rice. Much of the urban population is concentrated in the vicinity of the waterway, anchored by the cities at each end of the canal.

Near the northern end of the Panama Canal lies the city of Colón, site of the Colón Free Zone, a huge trading *entrepôt* designed to transfer and distribute goods bound for South America. It is augmented by the Manzanillo International Terminal, an ultramodern port facility capable of transshipping more than 1000 containers a day. By 2002, China had become the third-largest user of the Canal (after the United States and Japan) and accounted for more than 20 percent of the cargo entering the Colón Free Zone. Near the southern end lies Panama City, called the Miami of the Caribbean because of its waterfront location

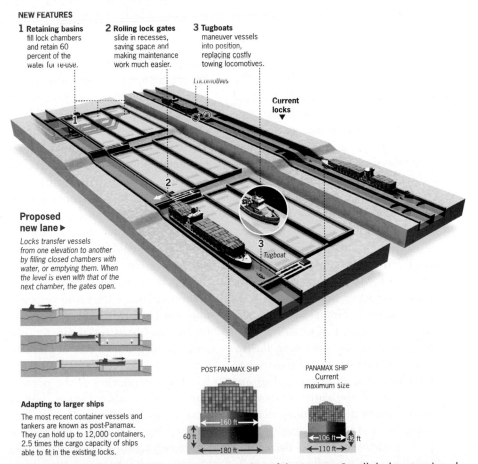

NEW FEATURES

1 Retaining basins fill lock chambers and retain 60 percent of the water for re-use.

2 Rolling lock gates slide in recesses, saving space and making maintenance work much easier.

3 Tugboats maneuver vessels into position, replacing costly towing locomotives.

Locomotives

Current locks ▼

Proposed new lane ▶

Locks transfer vessels from one elevation to another by filling closed chambers with water, or emptying them. When the level is even with that of the next chamber, the gates open.

Tugboat

Adapting to larger ships

The most recent container vessels and tankers are known as post-Panamax. They can hold up to 12,000 containers, 2.5 times the cargo capacity of ships able to fit in the existing locks.

POST-PANAMAX SHIP
160 ft
60 ft
180 ft

PANAMAX SHIP
Current maximum size
106 ft — 42 ft
110 ft

Diagram showing the planned expansion and upgrading of the Panama Canal's lock system in order to accommodate ships that currently exceed the waterway's ("Panamax") capacity. The target date for completing the project is August 15, 2014, the one-hundredth anniversary of the first interocean transit of the Canal by the passenger-cargo vessel, *S.S. Ancon.* © NG Image Collection and skyscrapered skyline. The capital is the financial center that handles the revenues generated by the Canal, but its high-rise profile also reflects the proximity of Colombia's illicit drug industry and associated money-laundering and corruption. The city's modern image presents a stark contrast to the (rural) poverty that afflicts 40 percent of the total population.

THE CARIBBEAN BASIN

Fragmentation and Insularity

As Figure 4B-1 reveals, the Caribbean Basin, Middle America's island region, consists of a broad arc of numerous islands extending from the western tip of Cuba to the southern coast of Trinidad. The four larger islands or **Greater Antilles** (Cuba, Hispaniola, Jamaica, and Puerto Rico) are clustered in the western segment of this arc. The smaller islands or **Lesser Antilles** extend to the east in a crescent-shaped zone from the Virgin Islands to Trinidad and Tobago. (Breaking this tectonic-plate-related regularity are the Bahamas and the Turks and Caicos, north of the Greater Antilles, and numerous other islands too small to appear on a map at the scale of Fig. 4B-1.)

On these islands, whose combined land area constitutes only 9 percent of Middle America, lie 33 states and many other political entities. (Europe's colonial flags have not totally disappeared from this region, and the U.S. flag flies over Puerto Rico.) The populations of these states and territories, however, constitute 21 percent of the entire geographic realm, making this the most densely peopled part of the Americas.

The island-nations of the Caribbean Basin are generally (very) small, and their territories are separated and often at considerable distances from other islands. These geographic conditions create a set of circumstances that has proved to be challenging. Economic opportunities are few, most things are relatively expensive, and due to limited interaction with the outside world the island societies tend to be static.

Ethnicity and Class

For one thing, **11 social stratification** in most Caribbean islands is rigid and social mobility is limited. Class structures tend to be closely associated with ethnicity and as such the region still carries imprints of colonial times. The historical geography of Cuba, the Dominican Republic, and Puerto Rico is suffused with Hispanic culture; Haiti and Jamaica carry stronger African legacies. But the reality of this ethnic diversity is that European lineages still hold the advantage. Hispanics tend to be in the best positions in the Greater Antilles; people who have mixed European-African ancestries, and who are described as **12 mulatto**, rank next. The largest part of this social pyramid is also the least advantaged: the Afro-Caribbean majority. In virtually all Caribbean societies, the minorities hold disproportionate power and exert overriding influence. In Haiti, the mulatto minority accounts for less than 6 percent of the population but has long held most of the power. In the adjacent Dominican Republic, the pyramid of power puts Hispanics (16 percent) at the top, the mixed sector (73 percent) in the middle, and the Afro-Caribbean minority (11 percent) at the bottom. In the Caribbean social mosaic, historic advantage has a way of perpetuating itself.

The composition of the population of the islands is further complicated by the presence of Asians from both China and India. During the nineteenth century, the emancipation of slaves and subsequent local labor shortages brought some far-reaching solutions. Some 100,000 Chinese emigrated to Cuba as indentured laborers; and Jamaica, Guadeloupe, and especially Trinidad saw nearly 250,000 South Asians arrive for similar purposes. To the African-modified forms of English and French heard in the Caribbean Basin, therefore, can be added several Asian languages. The ethnic and cultural variety of the plural societies of Caribbean Middle America is indeed endless.

THE GREATER ANTILLES

The four islands of the Greater Antilles (whose populations constitute 90 percent of the Caribbean Basin's total) contain five political entities: Cuba, Haiti, the Dominican Republic, Jamaica, and Puerto Rico (Fig. 4B-1). Haiti and the Dominican Republic share the island of Hispaniola.

Cuba

The largest Caribbean island-state in terms of both territory (111,000 square kilometers/43,000 sq mi) and population (11.3 million), Cuba lies only 145 kilometers (90 mi) from the southern tip of Florida (Fig. 4B-9). Havana, the now-dilapidated capital, lies almost directly across from the Florida Keys on the northwest coast of the elongated island. Cuba was a Spanish possession up until the late 1890s when, with American help in the Spanish-American War, it achieved independence. Fifty years later, a U.S.-backed dictator was in control, and by the 1950s Havana had become an American playground. The island was ripe for revolution, and in 1959 Fidel Castro's insurgents gained control, thereby converting Cuba into a communist dictatorship and a Soviet client. Nearly a million Cubans fled the island for the United States, and Miami overnight became the second-largest "Cuban" city after Havana.

In the end, Castro's rule survived the 1991 collapse of the Soviet Empire

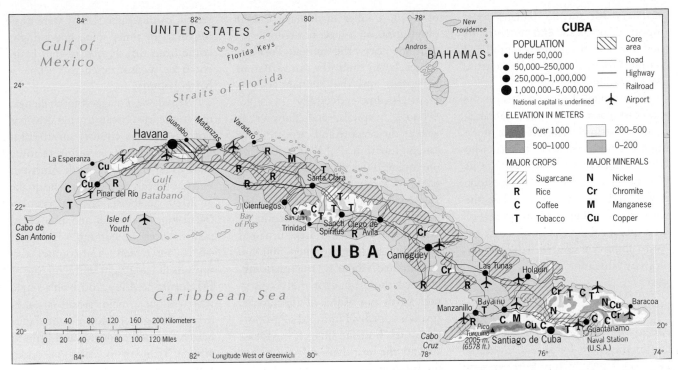

FIGURE 4B-9

despite the loss of subsidies and sugar markets on which it had long relied. As the map shows, sugar was Cuba's economic mainstay for many years; the plantations, once the property of rich landowners, extend all across the island. But sugarcane is losing its position as the leading Cuban foreign exchange earner. Mills have been closing down, and canefields are being cleared for other crops as well as for pastures. Cuba has other economic opportunities, however, especially in its highlands. There are three mountainous areas, of which the southeastern chain, the Sierra Maestra, is the highest and most extensive. These highlands create considerable environmental diversity as reflected by extensive timber-producing tropical forests and varied soils on which crops ranging from tobacco to coffee to rice to subtropical and tropical fruits are grown. Rice and beans are the staples, but Cuba is unable to meet its dietary needs and so must import food. The savannas of the center and west support livestock. Even though Cuba has only limited mineral reserves, its

nickel deposits are extensive and have been mined for a century.

Early in the twenty-first century, Cuba found a crucial new supporter in Venezuela's leader, Hugo Chávez. Cuba has no domestic petroleum reserves, but Venezuela is oil-rich and, since 2003, has been providing all of the fuel that Cuba needs. In return, Castro sent 30,000 health workers and other professionals to Venezuela, where they aid the poor.

Poverty, crumbling infrastructure, crowded slums, and unemployment mark the Cuban cultural landscape, but Cuba's regime still has support among the general population. In the United States, the view is that Cuba could be the shining star of the Caribbean, its people free, its tourist economy booming, its products flowing to North American markets. But much will have to change for that vision to become reality.

Jamaica

Across the deep Cayman Trench from southern Cuba lies Jamaica, and a

cultural gulf separates these two countries as well. A former British dependency, Jamaica has an almost entirely Afro-Caribbean population. As a member of the British Commonwealth, it still recognizes the British monarch as the chief of state, represented by a governor-general. The effective head of government in this democratic country, however, is the prime minister. English remains the official language here, and British customs still linger.

Smaller than Connecticut and with 2.8 million people, Jamaica has experienced a steadily declining GNI over the past few decades despite its relatively slow population growth. Tourism has become the largest source of income, but the markets for bauxite (aluminum ore), of which Jamaica is a major exporter, have dwindled. And like other Caribbean countries, Jamaica has trouble making money from its sugar exports. Jamaican farmers also produce crops ranging from bananas to tobacco, but the country faces the disadvantages on world markets common to those in the global periphery. Meanwhile, Jamaica must import all of its oil and much of its food because the densely populated coastal flatlands suffer from overuse and shrinking harvests.

The capital, Kingston, lies on the southeast coast and reflects Jamaica's economic struggle. Almost none of the hundreds of thousands of tourists who visit the country's beaches, explore its Cockpit Country of (karst) limestone towers and caverns, or populate the cruise ships calling at Montego Bay or other points along the north coast get even a glimpse of what life is like for the ordinary Jamaican.

Haiti

Already the poorest state in the Western Hemisphere for several decades, Haiti was struck by four hurricanes in 2008. The storms and accompanying floods killed 800 people and dislocated some 800,000. Then, on January 12 of 2010, a killer earthquake caused devastation on a scale not seen in recent memory. The capital, Port-au-Prince, lay in ruins, the infrastructure collapsed, and the Haitian government became effectively invisible. Schools, hospitals, and railways as well as the seaport and airport all no longer functioned. Millions became unemployed overnight. But all of that paled in comparison to the immediate human suffering. In the months following the earthquake, the estimated death toll reached and then surpassed 300,000; at least another 300,000 were injured. About 1.5 million homeless fled the capital for the countryside, partly out of fear of aftershocks. It was a tragedy of unheard of proportions, its horrific images broadcast around the world on television and the Internet. Large-scale international emergency aid arrived, but it became clear that, in the long run, much of the affected area would have to be rebuilt from scratch. Many were hoping that, in time, some good might come from this disaster in that Haiti could be reconstructed to become a better country.

It will be a staggering challenge, not in the least because geography has not handed the country any breaks. Haiti lies in a very dangerous tectonic fault zone where the North American and Caribbean plates meet. The quake of 2010, geologists say, had been in the making for decades, and more could easily follow in coming years (Fig. 4B-10, inset map). The country also finds itself lying right in the middle of **13** Hurricane Alley, and even a single year without a major storm is something to be grateful for. Finally, Haiti has few natural resources and little to offer in terms of international trade. Few countries in the world have a more checkered history than Haiti, where political instability, repression, and material deprivation have been

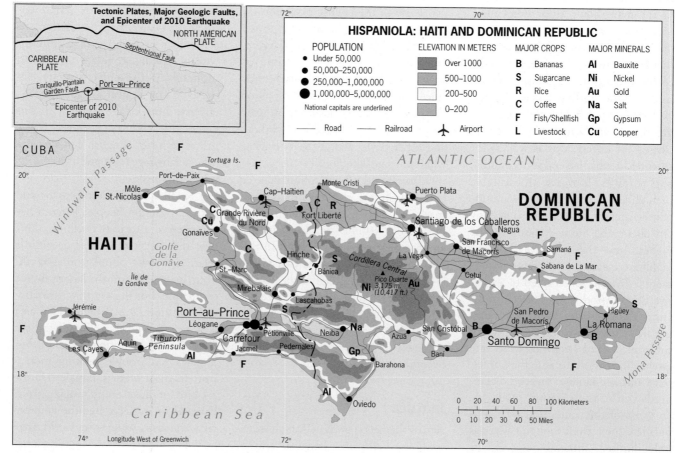

FIGURE 4B-10

© H. J. de Blij, P. O. Muller, and John Wiley & Sons, Inc.

FROM THE FIELD NOTES...

"Fly along the political boundary between Haiti and the Dominican Republic, and you see long stretches of the border marked by a stark contrast in vegetation: denudation prevails to the west in Haiti while the forest survives on the Dominican (eastern) side. Overpopulation, lack of governmental control, and mismanagement on the Haitian side combine to create one of the region's starkest spatial contrasts."

www.conceptcaching.com

constants for more than a century. In 2009, Haiti's GNI per capita had fallen to barely one-fifth of Jamaica's, a level lower than that of many poor African countries. Compared to other countries in Middle and South America, its infant mortality rate was nearly three times as high. Only 50 percent of elementary school-age girls actually received an education. Most of the country's limited public expenditures were made possible through foreign aid, and private consumption relied heavily on remittances from the Haitian diaspora in places like Miami, Paris, New York, and Montreal. In mid-2010, the world kept its fingers crossed for Haiti.

Dominican Republic

The Dominican Republic has a larger share of the island of Hispaniola than Haiti (Fig. 4B-10) in terms of both territory and population. Fly along the north-south border between the two countries, and you see a crucial difference: to the west, Haiti's hills and plains are treeless and gullied, its soils eroded, and its streams silt-laden; to the east, forests drape the countryside and streams run clear (see photo this page).

The mountainous Dominican Republic (2010 population: 10.3 million) has a wide range of natural environments and a far stronger resource base than Haiti. Nickel, gold, and silver have long been exported along with sugar, tobacco, coffee, and cocoa, but tourism (the great opportunity lost to Haiti) is the leading industry. A long period of dictatorial rule punctuated by revolutions and U.S. military intervention ended in 1978 with the first peaceful transfer of power following a democratic election.

Political stability brought the Dominican Republic handsome rewards, and during the late 1990s the economy, based on manufacturing, high-tech industries, and remittances from Do-minicans abroad as well as tourism, grew at an average of 7 percent per year. But in the early 2000s the economy imploded, not only because of the downturn in the world economy but also because of bank fraud and corruption in government. Suddenly the Dominican peso collapsed, inflation skyrocketed, jobs were terminated, and blackouts prevailed. As the people protested, lives were lost and the self-enriched elite blamed foreign financial institutions that were unwill-

ing to lend the government additional money. Once again the hopes of ordinary citizens were dashed by greed and corruption among those in power.

Puerto Rico

The largest U.S. domain in Middle America, this easternmost and smallest island of the Greater Antilles group covers 3500 square miles and has a population of 4 million. Puerto Rico is larger than Delaware and more populous than Oregon. Most of the population is concentrated in the urbanized northeastern sector of this rectangular island (Fig. 4B-11).

Puerto Rico fell to the United States more than a century ago during the Spanish-American War of 1898. Since the Puerto Ricans had been struggling for some time to free themselves from Spanish hegemony, this transfer of power was, in their view, only a change from one colonial master to another. As a result, the first half-century of U.S. administration was difficult, and it was not until 1948 that Puerto Ricans were permitted to elect their own governor.

When the island's voters approved the creation of a Commonwealth in a

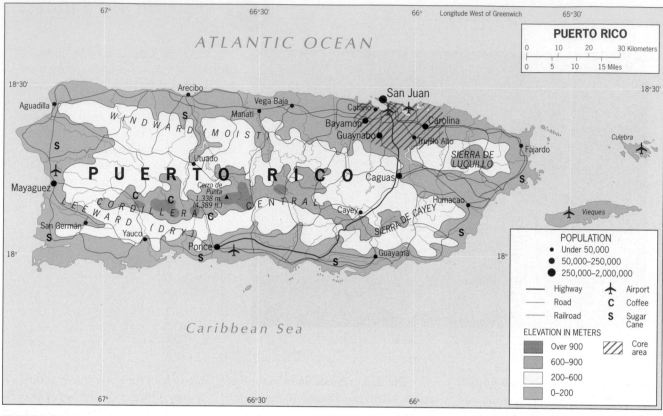

FIGURE 4B-11

© H. J. de Blij, P. O. Muller, and John Wiley & Sons, Inc.

1952 referendum, Washington, D.C. and San Juan, the two seats of government, entered into a complicated arrangement. Puerto Ricans are U.S. citizens but pay no federal taxes on their local incomes. The Puerto Rican Federal Relations Act governs the island under the terms of its own constitution and awards it considerable autonomy. Puerto Rico also receives a sizeable annual subsidy from Washington, totaling more than U.S. $4 billion in recent years.

Despite these apparent advantages in the poverty-mired Caribbean, Puerto Rico has not thrived under U.S. administration. Long dependent on a single-crop economy (sugar), the island based its industrialization during the 1950s and 1960s on its comparatively cheap labor, tax breaks for corporations, political stability, and special access to the U.S. market. Consequently, pharmaceuticals, electronic equipment, and apparel top today's list of exports, not sugar or bananas. But this industrialization failed to stem a tide of emigration that carried more than 1 million Puerto Ricans to the New York City area alone. The same wages that favored corporations kept many Puerto Ricans poor or unemployed. By some estimates, unemployment on the island stands at about 45 percent today. Most receive federal support. Another 30 percent work in the public sector, that is, in government. The private sector is severely underdeveloped.

Although Puerto Ricans are vocal in demanding political change, it is not surprising that successive referendums during the 1990s resulted in retention of the status quo: continuation of Commonwealth status rather than either Statehood or independence.

THE LESSER ANTILLES

As Figure 4B-1 shows, the Greater Antilles are flanked by two clusters of islands: the extensive Bahamas-Turks/ Caicos archipelago to the north and the Lesser Antilles to the east and south. The Bahamas, the former British colony that is now the closest Caribbean neighbor to the United States, alone consists of nearly 3000 coral islands, most of them rocky, barren, and uninhabited, but about 700 carrying vegetation, of which some 30 are inhabited. Centrally positioned New Providence Island houses most of the country's 320,000 inhabitants and contains the capital, Nassau, a leading tourist attraction.

The Lesser Antilles are grouped geographically into the Leeward Islands and the Windward Islands, a (climatologically incorrect) reference to the prevailing airflows in this tropical area. The Leeward Islands extend from the U.S. Virgin Islands to the French dependencies of Guadeloupe and Martinique, and the Windward Islands from St. Lucia to the Netherlands Antilles off the northwestern Venezuelan coast.

It would be impractical to detail the individual geographic characteristics of

The capital city of Trinidad and Tobago may have an historic colonial name (Port of Spain), but after nearly three centuries of Spanish rule, the British took control here. English became the *lingua franca*, democratic government followed independence in 1962, and natural gas reserves propelled a thriving economy. As can be seen here, the harbor of Port of Spain is bursting at the seams as a car carrier delivers automobiles from Japan, and containers are stacked high on the docks. © Boutin/Sipa Press.

each of the Lesser Antilles, but we should note that most share the insularity and environmental risks of this region—earthquakes, volcanic eruptions, and hurricanes; that they confront, to varying degrees, similar socioeconomic problems in the form of limited domestic resources, overpopulation, soil deterioration, land fragmentation, and market limitations; and that tourism has become the leading industry for many.

Under these circumstances one looks for certain hopeful signs, and such a sign comes from **Trinidad and Tobago**, the two-island republic at the southern end of the Lesser Antilles (Fig. 4B-1). This country (population: 1.3 million) has embarked on a natural-gas-driven industrialization boom that could turn it into an economic tiger. Trinidad has long been an oil producer, but lower world prices and dwindling supplies in the 1990s forced a reexamination of its natural gas deposits to help counter the downturn. That quickly resulted

in the discovery of major new supplies, and this cheap and abundant fuel has sparked a local gas-production boom as well as an influx of energy, chemical, and steel companies from Europe, Canada, and even India. (Meanwhile, Trinidad has become the largest supplier of liquefied natural gas for the United States.) Many of the new industrial facilities have agglomerated at the ultramodern Point Lisas Industrial Estate outside

the capital, Port of Spain, and they have propelled Trinidad to become the world's leading exporter of ammonia and methanol. Natural gas is also an efficient fuel for the manufacturing of metals, and steelmakers as well as aluminum refiners have been attracted to locate here. With Trinidad lying only a few kilometers from the Venezuelan coast of South America, it is also a sea-lane crossroads that is well connected to the vast, near-coastal supplies of iron ore and bauxite that are mined in nearby countries, particularly the Brazilian Amazon.

But Trinidad is an exception. It is a reflection of the predicament of **14 small-island developing economies** that, after the initial wave of decolonization around 1960, several territories decided that they were probably better off staying with the European country that had ruled them. Thus, for example, Guadeloupe remained with France, the Cayman Islands are still British, and the Netherlands Antilles forewent independence to continue their affiliation with the Dutch. Whereas economic logic seems to dictate continued affiliation with the former colonial powers, local politics and pride demand a minimal degree of autonomy and avoidance of paternalistic European interference. The combined result is often a complicated legal framework for the political status of the islands.

POINTS TO PONDER

- Northern Mexico highlights the volatility of globalization: the inception of NAFTA led to foreign investment and the proliferation of maquiladoras, but ten years later many companies were leaving for East Asian destinations that offered even lower wages.
- Costa Rica has known political stability for more than 175 years, and it has been a continuous democracy for over half a century. What could other Central American countries learn from this example?
- Residents of Panama like to refer to the fast-developing Panama Canal corridor as the "Singapore of Latin America."
- Parts of Haiti will have to be rebuilt almost from the ground up: will its economic and geographic conditions ever allow self-sustained development?

The old city in historic Salvador, Brazil, former capital and port of entry for tens of thousands of Africans.
© H.J. de Blij

IN THIS CHAPTER

An explorers' realm
200 years of independence
The growing power of indigenous peoples: "Latin" America no more
South America's green heart under stress
The Chinese are coming . . .
Efforts toward economic integration

CONCEPTS, IDEAS, AND TERMS

5A
SOUTH AMERICA: DEFINING THE REALM

FIGURE 5A-1

© H. J. de Blij, P. O. Muller, and John Wiley & Sons, Inc.

Of all the continents, South America has the most familiar shape—a giant triangle connected by mainland Middle America's tenuous land bridge to its neighbor in the north. South America also lies not only south but mostly east of its northern counterpart. Lima, the capital of Peru—one of the continent's westernmost cities—lies farther east than Miami, Florida. Thus South America juts out much more prominently into the Atlantic Ocean toward southern Europe and Africa than does North America. But lying so far eastward means that South America's western flank faces a much wider Pacific Ocean, with the distance from Peru to Australia nearly twice that from California to Japan.

As if to reaffirm South America's northward and eastward orientation, the western margins of the continent are rimmed by one of the world's longest and highest mountain ranges, the Andes, a gigantic wall that extends unbroken from Tierra del Fuego near the continent's southern tip in Chile to northeastern Venezuela in the far north (Fig. 5A-1). The other major physiographic feature of South America dominates its central north—the Amazon Basin; this vast humid-tropical amphitheater is drained by the mighty Amazon River, which is fed by several major tributaries. Much of the remainder of the continent can be classified as plateau, with the most important components being the Brazilian Highlands that cover most of Brazil southeast of the Amazon Basin, the Guiana Highlands located north of the lower Amazon Basin, and the cold Patagonian Plateau that blankets the southern third of Argentina. Figure 5A-1 also reveals two other noteworthy river basins beyond Amazonia: the Paraná-Paraguay Basin of south-central South America, and the Orinoco Basin in the far north that drains interior Colombia and Venezuela.

PHYSIOGRAPHY

Explorers' Continent

It was here in northern South America that the great German explorer and scientist Alexander von Humboldt, one of the founders of the modern discipline of geography, embarked on his legendary expeditions in the early nineteenth century. After landing on the coast of Venezuela and trekking across the continent's northern interior, the 30-year-old von Humboldt was struck by the area's biodiversity, its majestic natural beauty, and the adaptive abilities of the human populations. He discovered and named many species of flora and fauna, traversed tropical grasslands and jungles, met with indigenous peoples, crossed dangerous rivers, and reached the summit of the highest mountain that had been climbed by Europeans to that time (Ecuador's Chimborazo). He compiled large numbers of maps and was one of the first scientists to note how the coastlines of eastern South America and western Africa fitted together like pieces of a jigsaw puzzle, speculating about the continents' geologic movements.

Von Humboldt was important in the rise of the discipline because of his many discoveries as well as his views on the **1** unity of place: that in a particular locale or region intricate connections exist between climate, geology, biology, and human cultures. As such, he laid the foundation for modern geography as an *integrative discipline* with a spatial perspective. It took him about four decades to produce his magnum opus that articulated this holistic perspective, the highly ambitious and appropriately titled *Cosmos*, published in five volumes in 1845.

Almost three centuries earlier, the opposite end of South America was the scene of a crucial stage in the first circumnavigation of the globe and expedition led by Ferdinand Magellan. Once it became clear that Columbus had stumbled upon America, not India, Spanish and Portuguese explorers continued their efforts to discover a western passage from the Atlantic to the Pacific. This took them to the far south along what is now the Argentinean coast. Magellan and his crew spent

MAJOR GEOGRAPHIC QUALITIES

SOUTH AMERICA

1. South America's physiography is dominated by the Andes Mountains in the west and the Amazon Basin in the central north. Much of the remainder is plateau country.

2. Almost half of the realm's area and just over half of its total population are concentrated in one country—Brazil.

3. South America's population remains concentrated along the continent's periphery. Most of the interior is sparsely peopled, but sections of it are now undergoing significant development.

4. Interconnections among the states of the realm are improving rapidly. Economic integration has become a major force, particularly in southern South America.

5. Regional economic contrasts and disparities, both in the realm as a whole and within individual countries, remain strong.

6. Cultural pluralism exists in almost all of the realm's countries and is often expressed regionally.

7. Rapid urban growth continues to mark much of the South American realm, and urbanization overall is today on a par with the levels of the United States and western Europe.

© Jan Nijman

"Traveling on Peru's Cuzco *altiplano* in 2009, it was easy to get a sense of the advantages of this highland environment to the Incas of the past. The soils are fertile, temperatures are moderate, and the surrounding snowcapped ranges of the Andes feed the Urubamba River and other streams. The Urubamba Range in the background, partially hidden behind the clouds, rises up to nearly 6000 meters (19,000 ft) and the city of Cuzco itself sits at 3350 meters. For centuries, this area has produced abundant corn (seen in the foreground), potatoes, and other *tierra fría* crops. But the retreating glaciers in the higher elevations are now a cause of concern. Global warming may jeopardize what has been nature's gift to this *altiplano* for many centuries: a reliable water supply."

www.conceptcaching.com

some five months in Patagonia, named by Magellan for "big-foot people" (official, but never-verified, reports of the voyage told of the crew's mysterious encounters with 8-foot giants). With the onset of spring in 1520, Magellan's five ships set sail for the treacherous waters of what he named the *Estrecho de Todos los Santos*—now called the Strait of Magellan (Fig. 5A-1). Of the five ships in Magellan's fleet, only three survived the daring, 600-kilometer (375-mile)-long passage; the other two crashed on the rocks in the icy waters of the Southern Ocean.

Myriad Climates and Habitats

If the Russian realm is the widest in east-west extent, the South American realm is the longest measured from north to south. Within South America, no other country is more emblematic of this elongated geography than Chile, averaging only about 150 kilometers (90 mi) in width but 4000 kilometers (2500 mi) in length. As a consequence of this latitudinal span, fully one-tenth of our planet's circumference, the realm contains an enormous range of climates and vegetation. Combine this with substantial variation in relief from west to east, and it is clear why South America has such an impressive range of natural habitats.

Take another look at the map showing the global distribution of climates (Fig. G-7) and note the variation in climatic types, particularly in the realm's northwest and southern half. Travel northeast from Lima, Peru for about 600 kilometers (375 mi) and you encounter no less than four different climate zones: arid, highland, and two varieties of humid tropical. A transect of similar length from Santiago, Chile eastward across the continent to Buenos Aires, Argentina will take you through five climate zones: interior

highland, arid, and semiarid environments bracketed by a different humid temperate climate along each coast. Vegetation in South America varies accordingly, from lush tropical rainforests to rocky and barren snow-covered mountain tops to grasslands, fertile as well as parched. This natural diversity also makes for considerable cultural differences, as we shall soon see.

STATES ANCIENT AND MODERN

Thousands of years before the first European invaders appeared on the shores of South America, peoples now referred to as **Amerindians** had migrated into the continent via North and Middle America and founded societies in coastal valleys, in river basins, on plateaus, and in mountainous locales. These societies achieved different and remarkable adaptations to their diverse natural environments, and by about 1000 years ago, a number of regional cultures thrived in the elongated valleys between mountain ranges of the Andes from present-day Colombia southward to Bolivia and Chile. These high-altitude valleys, called **2** *altiplanos*, provided fertile soils, reliable water supplies, building materials, and natural protection to their inhabitants.

The Inca State

One of these *altiplanos*, at Cuzco in what is now Peru, became the core area of South America's greatest indigenous empire, that of the **Inca**. The Inca were expert builders whose stone structures (among which Machu Picchu near Cuzco is the most famous), roads, and bridges helped unify their vast empire; they also proved themselves to be efficient administrators, successful farmers and herders,

FROM THE FIELD NOTES...

© Philip L. Keating

"From this high vantage point I got a good perspective of a valley near Pisac in the Peruvian Andes, not far from Cuzco. This was part of the Incan domain when the Spaniards arrived to overthrow the empire, but the terraces you can see actually predate the Inca period. Human occupation in these rugged mountains is very old, and undoubtedly the physiography here changed over time. Today these slopes are arid and barren, and only a few hardy trees survive; the stream in the valley bottom is all the water in sight. But when the terrace builders transformed these slopes, the climate may have been more moist, the countryside greener."

www.conceptcaching.com

and skilled manufacturers; scholars studied the heavens, and its physicians even experimented with brain surgery. Great military strategists, the Inca integrated the peoples they vanquished into a stable and well-functioning state, an amazing accomplishment given the high-relief terrain they had to contend with.

As a minority ruling elite in their far-flung empire, the Inca were at the pinnacle in their rigidly class-structured, highly centralized society. So centralized and authoritarian was their state that a takeover at the top was enough to gain immediate power over all of it—as a small army of Spanish invaders discovered in the 1530s. The European invasion brought a quick end to thousands of years of Amerindian cultural development and changed the map forever.

The Iberian Invaders

The modern map of South America started to take shape when the Iberian colonists began to understand the location and economies of the Amerindian societies. The Inca, like Mexico's Maya and Aztec peoples, had accumulated gold and silver at their headquarters, possessed productive farmlands, and constituted a ready labor force. Not long after the defeat of the Aztecs in 1521, Francisco Pizarro sailed southward along the continent's northwestern coast, learned of the existence of the Inca Empire, and withdrew to Spain to organize its overthrow. He returned to the Peruvian coast in 1531 with 183 men and two dozen horses, and the events that followed are well known. In 1533, his party rode victorious into Cuzco.

At first, the Spaniards kept the Incan imperial structure intact by permitting the crowning of an emperor who was under their control. But soon the breakdown of the old order began. The new order that gradually emerged in western South America placed the indigenous peoples in serfdom to the Spaniards. Great haciendas were formed by **3** **land alienation** (the takeover of indigenously held land by foreigners), taxes were instituted, and a forced-labor system was introduced to maximize the profits of exploitation.

Lima, the west-coast headquarters of the Spanish conquerors, soon became one of the richest cities in the world, its wealth based on the exploitation of vast Andean silver deposits. The city also served as the capital of the viceroyalty of Peru, as the Spanish authorities quickly integrated the new possession into their colonial empire (Fig. 5A-2). Subsequently, when Colombia and Venezuela came under Spanish control and, later, when Spanish settlement expanded into what is now Argentina and Uruguay, two additional viceroyalties were added to the map: New Granada and La Plata.

Meanwhile, another vanguard of the Iberian invasion was penetrating the east-central part of the continent, the coastlands of present-day Brazil. This area had become a Portuguese sphere of influence because Spain and Portugal had signed a treaty in 1494 to recognize a north-south line 370 leagues west of the Cape Verde Islands as the boundary between their New World spheres of influence. This border ran approximately along the meridian of 50°W longitude, thereby cutting off a sizeable triangle of eastern South America for Portugal's exploitation (Fig. 5A-2). But a brief look at the political boundaries of South America (Fig. 5A-1) shows that this treaty did not limit Portuguese colonial territory to the east of the 50th meridian. Instead, Brazil's boundaries were bent far inland to include almost the entire Amazon Basin, and the country came to be only slightly smaller in territorial size than all the other South American countries combined. This westward thrust was the result of Portuguese and Brazilian penetration, particularly by the **Paulistas**, the settlers of São Paulo who needed Amerindian slave labor to run their plantations.

Independence and Isolation

Despite their adjacent location on the same continent, their common language and cultural heritage, and their shared national problems, the countries that arose out of South America's Spanish viceroyalties (together with Brazil) until quite recently existed in a considerable degree of isolation from one another. Distance and physiographic barriers reinforced this separation, and the realm's major population agglomerations still adjoin the coast, mainly the eastern and northern coasts (Fig. 5A-3). The viceroyalties existed primarily to extract riches and fill Spanish coffers. In Iberia there was little interest in developing the American lands for their own sake. Only after those who had made Spanish and Portuguese America their new home and who had a stake there rebelled against Iberian authority did things begin to change, and then very slowly. Thus South America was saddled with the values, economic outlook, and social attitudes of eighteenth-century Iberia—not the best tradition from which to begin the task of forging modern nation-states.

Certain isolating factors had their effect even during the wars for independence. Spanish military strength was always concentrated at Lima, and those territories that lay farthest from their center of power—Argentina as well as Chile—were the first to gain independence from Spain (in 1816 and 1818, respectively). In the north Simón Bolívar led the burgeoning independence movement, and in 1824 two decisive military defeats there spelled the end of Spanish power in South America.

This joint struggle, however, did not produce unity because no fewer than nine countries emerged from the three former viceroyalties. It is not difficult to understand why this fragmentation took place. With the Andes intervening between Argentina and Chile and the Atacama Desert between Chile and Peru, overland distances seemed even greater than they really were, and these obstacles to contact proved quite effective. Hence, from their outset the new countries of South America began to grow apart amid friction and even wars. Only within the past two decades have the countries of this realm finally begun to recognize the mutual advantages of increasing cooperation and to make lasting efforts to steer their relationships in this direction.

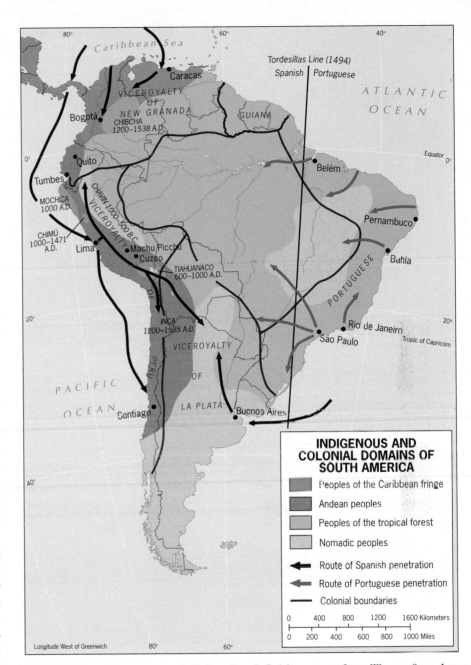

FIGURE 5A-2   © H. J. DE BLIJ, P. O. MULLER, AND JOHN WILEY & SONS, INC.

THE CULTURAL MOSAIC

When we speak of the interaction of South American countries, it is important to keep in mind just who does the interacting. The fragmentation of colonial South America into ten individual republics, and the subsequent postures of each of these states, was the work of a small minority that constituted the landholding, upper-class elite. Thus in every country a vast majority—be they Amerindians in Peru or people of African descent in Brazil—could only watch as their European masters struggled with one another for supremacy.

SOUTH AMERICA: POPULATION DISTRIBUTION, 2011

One dot represents 50,000 persons

| 0 | 600 | 1200 Kilometers |
| 0 | 300 | 600 Miles |

FIGURE 5A-3

© H. J. de Blij, P. O. Muller, and John Wiley & Sons, Inc.

The Population Map—Then and Now

If we were able to reconstruct a map of South America's population before the arrival of the Europeans (a "pre-Columbian" map, as it would be called), it would look quite different from the current one in Figure 5A-3. Indigenous Amerindian societies inhabited not only the Andes and adjacent lowlands but also riverbanks in the Amazon Basin, where settlements numbering in the thousands subsisted on fishing and farming. They did not shy away from harsh environments such as those of the island of Tierra del Fuego in the far south, where the fires they kept going against the bitter cold led the Europeans to name the place "land of fire."

Today the map looks quite different. Many of the indigenous societies succumbed to the European invaders, not just through warfare but also because of the diseases the Hispanic conquerors brought with them. Geographers estimate that 90 percent of native Amazonians died within a few years of contact, and the peoples of Tierra del Fuego also are no longer there to build their fires. From one end of South America to the other, the European arrival spelled disaster.

Spanish and Portuguese colonists penetrated the interior of South America, but the great majority of the settlers stayed on or near the coast, a pattern still visible today. Almost all of the realm's major cities have coastal or near-coastal locations, and the current population distribution map gives you the impression of a continent yet to be penetrated and inhabited. But look carefully at Figure 5A-3, and you will see a swath of population located well inland from the coastal settlements, most clearly in Peru but also extending northward into Ecuador and southward into Bolivia. That is the legacy of the Inca Empire and its incorporated peoples, surviving in their mountainous redoubt and still numbering in the millions.

Amerindian Reawakening

Today, South America's long-downtrodden Amerindians are staging a social, political, and economic reawakening. In several South American countries where their numbers are large enough to translate into political strength, they have begun to realize their potential. Amerindians in Peru constitute 45 percent of the national population, and in Bolivia they are in the majority at 55 percent.

Newly empowered Amerindian political leaders are emerging and bringing the plight of the realm's indigenous peoples to both local and international attention. Amerindians were conquered, decimated by foreign diseases, robbed of their best lands, subjected to involuntary labor, denied the right to grow their traditional crops, socially discriminated against, and swindled out of their fair share of the revenues from resources in their traditional domains. They may still be the poorest of the realm's poor, but they are increasingly asserting themselves. For some South American states, the consequences of this movement will be far-reaching.

The Amerindian reawakening is in part related to changing religious practices in South America. Officially, more than 90 percent of the population is Roman Catholic, and traditionally South Americans tend to be viewed as devout followers of the Vatican. But recent surveys show that many do not attend church regularly and that more than half of all Catholics describe themselves as believers of the doctrine, rather than adherents of the Church. Since the late nineteenth century, the Catholic Church has often been criticized for its conservative position on social issues and for siding with the establishment. In the 1950s, a

© H. J. de Blij

"Near the waterfront in Belém lie the now-deteriorating, once-elegant streets of the colonial city built at the mouth of the Amazon. Narrow, cobblestoned, flanked by tiled frontages and arched entrances, this area evinces the time of Dutch and Portuguese hegemony here. Mapping the functions and services here, we recorded the enormous diversity of activities ranging from carpentry shops to storefront restaurants and from bakeries to clothing stores. Dilapidated sidewalks were crowded with shoppers, workers, and people looking for jobs (some newly arrived, attracted by perceived employment opportunities in this growing city of 2.3 million). The diversity of population in this and other tropical South American cities reflects the varied background of the region's peoples and the wide hinterland from which these urban magnets have drawn their inhabitants."

www.conceptcaching.com

powerful movement known as ***liberation theology*** emerged in South America, and subsequently gained followers around the world. The movement was a blend of Christian religion and socialist thinking that read the teachings of Christ as a quest to liberate the impoverished masses from oppression. Although the Church has lately been trying to make amends, it could not avoid losing popular support, especially among Amerindians. In recent years, Protestant-evangelical faiths have also gained adherents, mainly at the expense of Catholicism.

African Descendants

As Figure 5A-2 shows, the Spaniards initially got very much the better of the territorial partitioning of South America—not just in land quality but also in the size of the aboriginal labor force. When the Portuguese began to develop their territory, they turned to the same lucrative activity that their Spanish rivals had pursued in the Caribbean—the plantation cultivation of sugar for the European market. And they, too, found their labor force in the same source region, as millions of Africans (nearly half of all who came to the Americas) were brought in bondage to the tropical Brazilian coast north of Rio de Janeiro (see Fig. 6A-7). Not surprisingly, Brazil now has South America's largest black population, which is still heavily concentrated in the country's poverty-stricken

northeastern States. With Brazilians of direct or mixed African ancestry today accounting for nearly half of the population of 201 million, the Africans decidedly constitute a major immigration of foreign peoples into South America.

Ethnic Landscapes

The cultural landscape of South America, similar to that of Middle America, is a layered one. Amerindians cultivated and crafted diverse landscapes throughout the continent, some producing greater impacts than others. When the Europeans arrived, the cultural transformation that resulted from depopulation severely impacted the environment. Native peoples became minorities in their own lands, and Europeans introduced crops, animals, and ideas about land ownership and land use that irreversibly changed South America. They also brought in Africans from various parts of Subsaharan Africa. Europeans from non-Iberian Europe also started immigrating to South America, especially during the first half of the twentieth century. Japanese settlers arrived in Brazil and Peru during the same era. All of these elements have contributed to the present-day ethnic complexion of this realm. Figure 5A-4 shows the distinct concentrations of Amerindian and African cultural dominance, as well as areas where these groups are largely absent and people of European ancestry dominate.

Of course, ethnic origins are not always so straight-forward, and patterns can change as a result of internal migrations and ethnic mixing. In recent years, research on individual DNA and genetics has taught us a lot about the regional and group origins of people. On the map, Argentina is indicated as having predominantly European ancestry. More specifically, recent research shows that, for the average Argentine, nearly 80 percent of his or her genetic structure is European, 18 percent Amerindian, and 2 percent African. If nobody had any mixed ancestors, these percentages would translate perfectly into European, Amerindian, and African shares of the population—but of course that is not the case. Many people do not have perfect knowledge about their ethnic ancestry, and while some may be strictly of one single origin, many others will have some degree of mixed ancestry. In the aggregate, however, there is no doubt that the population of cone-shaped southern South America is predominantly of European origin.

South America, then, is a realm of **4 plural societies**, where Amerindians of different cultures, Europeans from Iberia and elsewhere, Africans mainly from western tropical Africa, and even some Asians from India, Japan, and Indonesia cluster in adjacent areas but generally do not mix. The result is a cultural mosaic of almost endless variety.

It is probably true that most visitors to Brazil's Foz do Iguaçu, across the Paraná River from Ciudad del Este in Paraguay, expect to see one of the world's most spectacular waterfalls—but not this evidence of the substantial Muslim presence among the population of this border town. Several impressive mosques, set in carefully manicured grounds and reflecting considerable communal wealth, diversify the cultural landscape in this corner of the contentious Triple Frontier (a regional term discussed in Chapter 5B). © Norberto Duarte/Getty Images, Inc.

FIGURE 5A-4   © H. J. de Blij, P. O. Muller, and John Wiley & Sons, Inc.

ECONOMIC GEOGRAPHY

Agricultural Land Use and Deforestation

The internal divisions of this cultural kaleidoscope are further reflected in the realm's economic landscape. In South America, larger-scale **5 commercial** or market (for-profit) and smaller-scale **6 subsistence** (primarily for household use) **agriculture** exist side by side to a greater degree than anywhere else in the world (Fig. 5A-5). The geography of plantations and other commercial agricultural systems was initially tied to the distribution of landholders of European background, while subsistence farming (such as highland mixed subsistence-market, agroforestry, and shifting cultivation) is historically associated with the spatial patterns of indigenous peoples as well as populations of African and Asian descent (see map on next page).

spills over into adjacent areas of Paraguay, Uruguay, and Argentina (Fig. 5A-5).

Deforestation is a particularly acute problem in northern Brazil. In the past, deforestation was attributed mainly to small-scale landholders, colonists who had made their way into the Amazon rainforest to eke out a new living. The usual pattern of settlement often went like this. As main and branch highways were cut through the forest, settlers followed (often in response to government-sponsored land occupation schemes), moving out laterally to clear land for farming. Subsistence crops were planted, but within a year or two weed infestation and declining soil fertility made these plots unproductive. As soil fertility continued to decline, the settlers planted grasses and then sold their land to cattle ranchers. The peasant farmers then moved on to newly opened nearby areas, cleared more land for planting, and the cycle repeated itself. Unfortunately, this activity not only entrenches low-grade land uses across widespread areas, but also entails the burning and clearing of vast stands of tropical woodland (in fact, since the 1980s, an area of rainforest about the size of Ohio has been lost *annually* in northern Brazil). To make matters even worse, the deforestation crisis is now intensifying as *agro-industrial* operations engaged in large scale production for export markets increasingly penetrate Amazonia and transform its land cover.

As South American economies modernize, agriculture remains highly important. More than one-fifth of this realm's workforce is still employed in the primary sector that includes farming, cattle-raising, and fishing—a much greater share than in North America. The South American contribution to global trade in grains, soybeans, coffee, orange juice, sugar, and many other crops is significant and growing.

Industrial Development

Manufacturing is growing rapidly in South America, ranging from chemicals to electronics and from textiles to biofuels. But the geography of industrial production varies markedly. Brazil, Chile, and Argentina have traditionally been ahead, while countries like Peru, Ecuador, and Bolivia have long struggled to modernize their economies and improve standards of living. Within countries, there is considerable **7** uneven development as well. Most manufacturing is concentrated in and around major urban centers, leaving vast empty spaces in the interior of the realm.

Brazil, in particular, is increasingly drawing attention for its momentous growth and the rapid sophistication of a number of economic sectors. In the past, typical Brazilian exports were foods and footwear. Today, the leading exports include oil (thanks to the fortuitous discovery of large offshore reserves in the past few years), steel, and state-of-the-art Embraer aircraft. For almost a decade now, Brazil has

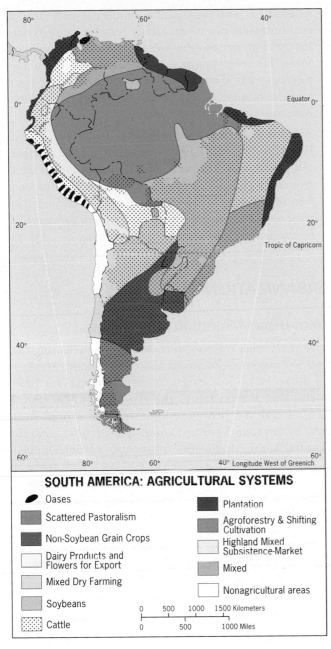

FIGURE 5A-5 © H. J. de Blij, P. O. Muller, and John Wiley & Sons, Inc.

At a broader level of generalization, agricultural systems and land use in South America vary in close relationship to physiography, as one would expect. Forestry and agroforestry prevail in the Amazon Basin; ranching dominates in the grasslands of the Southern Cone; and different forms of agriculture, with or without irrigation, are found in a wide zone from northeastern Brazil to northern Argentina, as well as scattered among pockets of moderate elevation in the Andean highlands. Land uses throughout South America are changing rapidly today, mainly due to ongoing deforestation and the introduction or expansion of new crops. The fastest-growing crop is soybeans, the cultivation of which now dominates much of southeastern Brazil and been included as one of the four biggest emerging markets in the world, along with Russia, India, and China—the so-called **BRICs.**

Despite recent, impressive economic growth across much of the realm, it is clear that the longstanding isolation of its individual countries remains a serious obstacle. Nonetheless, most governments agree that tighter integration, political as well as in terms of transportation infrastructure, would greatly enhance economic opportunities.

Economic Integration

The separatism that has so long characterized international relations in this realm is giving way as South American countries discover the benefits of forging new partnerships with one another. With mutually advantageous trade the catalyst, a new continentwide spirit of cooperation is blossoming at every level. Periodic flare-ups of boundary disputes now rarely escalate into open conflict. Cross-border rail, road, and pipeline projects, stalled for years, are multiplying steadily. In southern South America, five formerly contentious nations are developing the *hidróvia* (water highway), a system of river locks that is opening most of the Paraná-Paraguay Basin to barge transport; similar proposals have been advanced to connect this basin to the Amazon River system. And importantly, investments today flow more freely than ever from one country to another, particularly in the agricultural sector.

Recognizing that free trade may well solve many of the realm's economic-geographic problems, governments are now pursuing several avenues of economic supranationalism. In 2011, South America's republics were affiliating with the following major trading blocs:

- *Mercosur/l* (*Mercosur* in Spanish; *Mercosul* in Portuguese): Launched in 1995 by countries of the Southern Cone and Brazil, this Common Market established a free-trade zone and customs union linking Brazil, Argentina, Uruguay, and Paraguay. Bolivia, Chile, Colombia, Ecuador, Peru, and Venezuela are associate members (Venezuela's full accession into the organization has been blocked for years by Brazil and now by Paraguay). This organization is becoming the dominant free-trade organization for South America.

- *Andean Community:* Formed as the Andean Pact in 1969 but restarted in 1995 as a customs union with common tariffs for imports, this bloc is made up of Colombia, Peru, Ecuador, and Bolivia. Venezuela was a member until it withdrew in 2006.

- *Union of South American Nations (UNASUR):* Founded in 2008 in Brasília, Brazil, the 12 independent countries of South America signed a treaty to create a union envisioned as similar to the European Union (see Chapter 1A) with the goal of a continental parliament, a coordinated defense effort, a single passport for all its citizens, and greater cooperation on infrastructure development. However, significant disagreement between the member-states about details still puts these efforts years into the future. UNASUR was preceded by the South American Community of Nations.

- *Free Trade Area of the Americas (FTAA):* The United States and other NAFTA proponents have tried to move this hemispheric free-trade idea forward, but it has been resisted by peasants and workers in South America, and formally opposed by Mercosur/l. As long as the terms of trade remain set by the North, the Southern partners will be reluctant to participate in this initiative.

URBANIZATION

Rural-Urban Migration

As in most other realms, South Americans are leaving the land and migrating to towns and cities. South America's **8** urbanization process got an early start and has been intensifying since 1950 as the growth of urban settlements averaged 5 percent a year. Meanwhile, rural areas grew by less than 2 percent annually over those same six decades. As a result, the realm's urban population climbed to its current level of 81 percent, ranking it with those of western Europe and the United States. These numbers underscore not only the dimensions but also the durability of the **9** rural-to-urban migration from the countryside to the cities.

In South America, as in Middle America, Africa, and Asia, people are attracted to the cities and driven from the poverty of the rural areas. Both *pull* and *push* factors are at work. Rural land reform has been slow in coming, and for this and other reasons every year tens of thousands of farmers simply give up and leave, seeing little or no possibility for economic advancement. The urban centers lure them because they are perceived to provide opportunity—the chance to earn a regular wage. Visions of education for their children, better medical care, upward social mobility, and the excitement of life in a big city draw hordes to places such as São Paulo and Lima.

But the actual move can be traumatic. Cities in developing countries are surrounded and often invaded by squalid slums, and this is where the urban immigrant most often finds a first—and frequently permanent—abode in a makeshift shack without even the most basic amenities and sanitary facilities. Unemployment remains persistently high, often exceeding 25 percent of the available labor force. But still the people come, hopeful for a better life, the overcrowding in the shantytowns worsens, and the threat of regional-scale disease (and other disasters) rises.

© H. J. de Blij

"Two unusual perspectives of Rio de Janeiro form a reminder that here the wealthy live near the water in luxury high-rises, such as these overlooking Ipanema Beach, while the poor have million-dollar views from their hillslope *favelas*, such as Rocinho."

www.conceptcaching.com

Regional Patterns

The generalized spatial pattern of South America's urban transformation is displayed in Figure 5A-6, which shows a *cartogram* of the continent's population.* Here we see not only the realm's countries in population space relative to each other, but also the proportionate sizes of individual large cities within their total national populations.

Regionally, the Southern Cone is the most highly urbanized. Today in Argentina, Chile, and Uruguay, almost all of the population resides in cities. Ranking next in urbanization is Brazil. The next highest group of countries borders the Caribbean in the north. Not surprisingly, the Andean countries constitute the realm's least urbanized zone. Figure 5A-6 tells us a great deal about the relative positions of major metropolises in their countries. Three of them—Brazil's São Paulo and Rio de Janeiro, and Argentina's Buenos Aires—rank among the world's **10** megacities (cities whose populations exceed 10 million). But even in the Amazon Basin the population is now more than 70 percent urban.

The "Latin" American City Model

The urban experience in the South and Middle American realms varies because of diverse historical, cultural, and economic influences. Nevertheless, there are many common threads that have prompted geographers to search for useful generalizations. One is the model of the intraurban spatial structure of the **11** "Latin" American city proposed by Ernst Griffin and Larry Ford (Fig. 5A-7).

As noted in Chapter 1A, the idea behind a *model* is to create an idealized representation of reality, displaying as many key real-world elements as possible. In the case of South America's cities, the basic spatial framework of city structure, which blends traditional elements of South and Middle American culture with modernization forces now reshaping the urban scene, is a composite of radial sectors and concentric zones. Anchoring the model is the *CBD*, which is the primary business, employment, and entertainment focus of the surrounding metropolis. The CBD contains many modern high-rise buildings but also mirrors its colonial beginnings. As shown in Figure 4A-4, by colonial law Spanish colonizers laid out their cities around a central square, or plaza, dominated by a church and government buildings. Santiago's *Plaza de Armas*, Bogotá's *Plaza Bolívar*, and Buenos Aires's *Plaza de Mayo* are classic examples. The plaza was the hub of the city, which later outgrew its old center as new commercial districts formed nearby; but to this day the plaza remains an important link with the past.

---

*A cartogram is a specially transformed map in which countries and cities are represented in proportion to their populations. Those containing large numbers are blown up in population space, while those containing lesser numbers are shrunk in size accordingly.

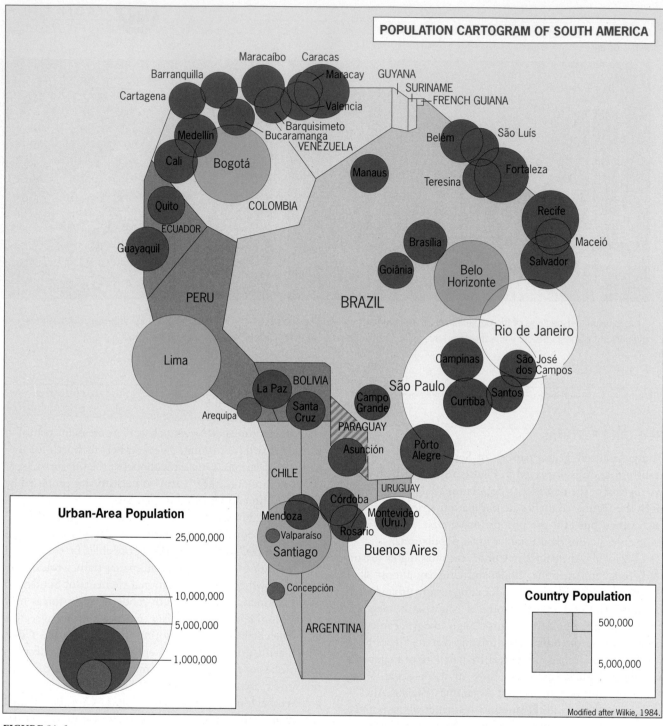

FIGURE 5A-6

© H. J. de Blij, P. O. Muller, and John Wiley & Sons, Inc.

Radiating outward from the urban core along the city's most prestigious axis is the commercial *spine*, which is adjoined by the *elite residential sector* (shown in green in Fig. 5A-7). This widening corridor is essentially an extension of the CBD, featuring offices, retail facilities, and housing for the upper and upper-middle classes.

The three remaining concentric zones are home to the less fortunate residents of the city, with income level and housing quality decreasing as distance from the CBD increases. The *zone of maturity* in the inner city contains housing for the middle class, who invest sufficiently to keep their aging dwellings from deteriorating. The adjacent *zone of in situ accretion* is one of much more modest housing interspersed with unkempt areas, representing a transition from inner-ring affluence to outer-ring poverty and taking on slum characteristics.

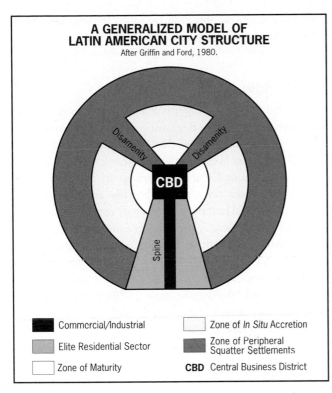

A GENERALIZED MODEL OF LATIN AMERICAN CITY STRUCTURE
After Griffin and Ford, 1980.

- Commercial/Industrial
- Elite Residential Sector
- Zone of Maturity
- Zone of *In Situ* Accretion
- Zone of Peripheral Squatter Settlements
- **CBD** Central Business District

FIGURE 5A-7 © H. J. de Blij, P. O. Muller, and John Wiley & Sons, Inc.

The outermost *zone of peripheral squatter settlements* is home to the millions of comparatively poor and unskilled workers who have recently migrated to the city. Here many newcomers earn their first cash income by becoming part of the **12** **informal sector**, in which workers are undocumented and money transactions are beyond the control of government. The settlements consist mostly of self-help housing, vast shantytowns known as **13** *barrios* in Spanish-speaking South America and *favelas* in Brazil. Some of their entrepreneurial inhabitants succeed more than others, transforming parts of these shantytowns into beehives of activity that can propel resourceful workers toward a middle-class existence.

A final structural element of many South American cities forms an inward, narrowing sectoral extension of the zone of peripheral squatter settlements and is known as the *zone of disamenity*. It consists of undesirable land along highways, rail corridors, riverbanks, and other low-lying areas; people are so poor that they are forced to live in the open. Thus this realm's cities present enormous contrasts between poverty and wealth, squalor and comfort—harsh contrasts all too frequently seen in the cityscape.

FUTURE PROSPECTS

The Need for Stability

In 2010, a number of South American countries were celebrating two centuries of independence. What mattered most, for nationalistic purposes, was the date when independence was declared, even if it took additional years for Spain to recognize that new political status. The wars with Spain lasted from 1808 (not coincidentally, soon after Spain had been weakened by Napoleon's invasion

 FROM THE FIELD NOTES...

© H. J. de Blij

"At the heart of Buenos Aires lies the Plaza de San Martín, flanked by impressive buildings but, unusual for such squares in Iberian America, carpeted with extensive lawns shaded by century-old trees. Getting a perspective of the plaza was difficult until I realized that you could get to the top of the 'English Tower' across the avenue you see in the foreground. From there, one can observe the prominent location occupied by the monument to the approximately 700 Argentinian military casualties of the Falklands War of 1982 (center), where an eternal flame behind a brass map of the islands symbolizes Argentina's undiminished determination to wrest these islands from British control."

www.conceptcaching.com

of Iberia) to 1838. Brazil's independence from Portugal was achieved during the same period, in 1822.

But 200 years of freedom have not seen the evolution of the kind of stable, mature political climates and institutions that one might expect. South American countries have been in frequent political turmoil. Dictatorial regimes ruled from one end of the realm to the other; unstable governments fell with damaging frequency. Widespread poverty, harsh regional disparities, poor internal surface connections, limited international contact, and economic stagnation prevailed.

Today, South America is a realm in dramatic transition. Entering the second decade of the twenty-first century, democracy seems to be taking hold almost everywhere. Long-isolated countries are becoming more interconnected through new transport routes and trade agreements. New settlement frontiers are being opened. Energy resources, some long exploited and others newly discovered, are boosting national economies as world prices have risen. Foreign states and corporations have appeared on the scene to buy commodities and invest in infrastructure. The pace of globalization has increased from Bogotá to Buenos Aires.

But these exciting developments must be seen against a backdrop of persistent problems. The realm's giant, Brazil, is embarked on a program of land reform, a campaign against poverty, and an effort to maintain financial rigor that have all run up against endemic corruption in government. The economy of Argentina in mid-2010 was recovering after an implosion that shook the country to its core. Whereas Chile has been the realm's success story, emerging from the horrors of the Pinochet era as a stable and vibrant democracy with a thriving economy, neighboring Bolivia remains in the grip of a social revolution arising in part from the realization of its energy riches. And on the north coast lies Venezuela, its oil reserves among the largest in the world (and by far the largest in the realm) and its political life dominated by a one-time coup leader whose closest ideological ally is Cuba's communist ruler and whose major adversary is the U.S. government.

Even if economic growth continues, several South American governments will have to address, more effectively than they have done in the past, the enormous inequality among their social classes. By some measures, the disparity between rich and poor is still wider in this realm than in any other, and wealth is disproportionately concentrated in the hands of a small minority (the richest 20 percent of the people control 70 percent of all wealth, while the poorest 20 percent own 2 percent). South American leaders must walk a fine line between fueling growth while improving the lives of the masses, and some (such as Lula, in Brazil) have seemed to master that balancing act better than others (Chávez, in Venezuela). Moreover, in certain parts of the realm the internal divisions are complicated by the resurgence of Amerindian identity among

native peoples, and this, too, requires delicate and even-handed maneuvering by national governments.

The Shadow of the United States

The United States has long played a key role in this realm, for better or for worse. Beginning with the Monroe Doctrine in 1823 (which stated that European colonial powers had no rights to South America), the United States has consistently claimed a special interest in this part of the hemisphere. Over time, and especially during the Cold War in the second half of the twentieth century, the United States became politically involved in a number of countries. Preoccupation with efforts to keep Soviet influence out of the Western Hemisphere sometimes led the American government to choose the wrong side in local political struggles (as in Chile during the early 1970s); such things are not easily forgotten by the people of these countries. Indeed, anti-Americanism never seems far from the surface in this realm, and is mainly based on past U.S. support for dictatorships or perceived imperialistic behavior.

To be sure, relations with the United States have never been smooth or always asymmetrical. For instance, South America today represents only about 4 percent of all U.S. foreign trade, and it is far less important to the United States than, say, Canada or western Europe. But the United States remains the biggest trading partner for South America as a whole, accounting for almost one-fifth of the exports and imports of all South American countries.

It was hardly a coincidence that **14** *dependencia* **theory** originated here in South America in the 1960s. It was a new way of thinking about economic development and underdevelopment that explained the persistent poverty of some countries (or of the masses inside those countries) in terms of the unequal relations with other (rich) countries. In this view, South American nations may have become officially independent of Spain in the early nineteenth century, but effectively they wound up becoming subservient to the United States (and other rich industrialized countries). In the present political and economic climate, however, most would consider *dependencia* theory as dated or at least unable to provide a way forward.

Nevertheless, the asymmetry remains, and U.S. foreign policy, as seen from South America, is not always credible. Interestingly, U.S. special interests in the realm are now under increasing pressure from another major power that would have been furthest from President Monroe's mind back in 1823: China.

China Calling

By 2010, China had displaced the United States as the biggest trading partner of Brazil and Chile, and had become the second-largest of both Colombia and Peru. It is a remarkable development that has yet to fully register in the United States: China is on the way to becoming the most important great power in this realm.

Nearly invisible only ten years ago, China is now making its presence felt across South America in a variety of ways that include establishing new embassies and consulates, buying up companies, partnering joint ventures, financing infrastructure projects and development assistance, and sending and inviting high-level trade delegations. Their motives are clear: the Chinese need raw materials such as cement, oil, copper, and minerals to fuel the enormous economic growth of their country. They also are seeking to expand the markets for Chinese exports, and the growing middle classes of Brazil and other countries make enticing targets.

China has none of the historical "baggage" that handicaps America's efforts to maintain a decisive influence in the realm. It has never had any political designs on South America, and its objectives are seen as strictly economic. From the point of view of Brazil, at least, it also represents a far more equal relationship: as *BRICs*, both rank among the biggest emerging economies in the world.

Whether it is the United States or China, South Americans will need to carefully consider how and where to focus their economic attention. Within the realm itself, there is still much to be done in terms of political and economic integration. Perhaps those Chinese investments in the infrastructural projects mentioned above will help to build the bridges so badly needed to bring this realm together.

POINTS TO PONDER

- The Amazonian rainforest is mainly located inside Brazil, but sometimes is referred to as "the lungs of the Earth." Should its protection be a global responsibility?

- Is it a coincidence that both *dependencia* theory and liberation theology originated in this realm?

- The map of South America's religions is changing. Brazil is still the world's largest Roman Catholic country, but the number of adherents is declining and now accounts for less than 75 percent of the national population (down from 90 percent three decades ago).

- Should China's growing involvement in South America be a cause of concern for the United States?

Large-scale corporate farming of soybeans in Interior Brazil's *cerrado*: one of the world's greatest agricultural frontiers—flat, fertile, well-watered, and ideally suited for mechanized production.
© PauloFridman/SambaPhoto/Getty Images, Inc.

IN THIS CHAPTER

The cocaine curse
Oil and politics in Venezuela
The transformative powers of oil
Amerindian resurgence
Chile: Star of the realm
Brazil: Great power in the making

CONCEPTS, IDEAS, AND TERMS

Insurgent state	1
Failed state	2
Altiplano	3
Triple Frontier	4
Elongation	5
Buffer state	6
BRIC	7
El Niño	8
Forward capital	9
Cerrado	10
Growth-pole concept	11

5B

SOUTH AMERICA: REGIONS OF THE REALM

The Caribbean North
The Andean West
The Southern Cone
Brazil

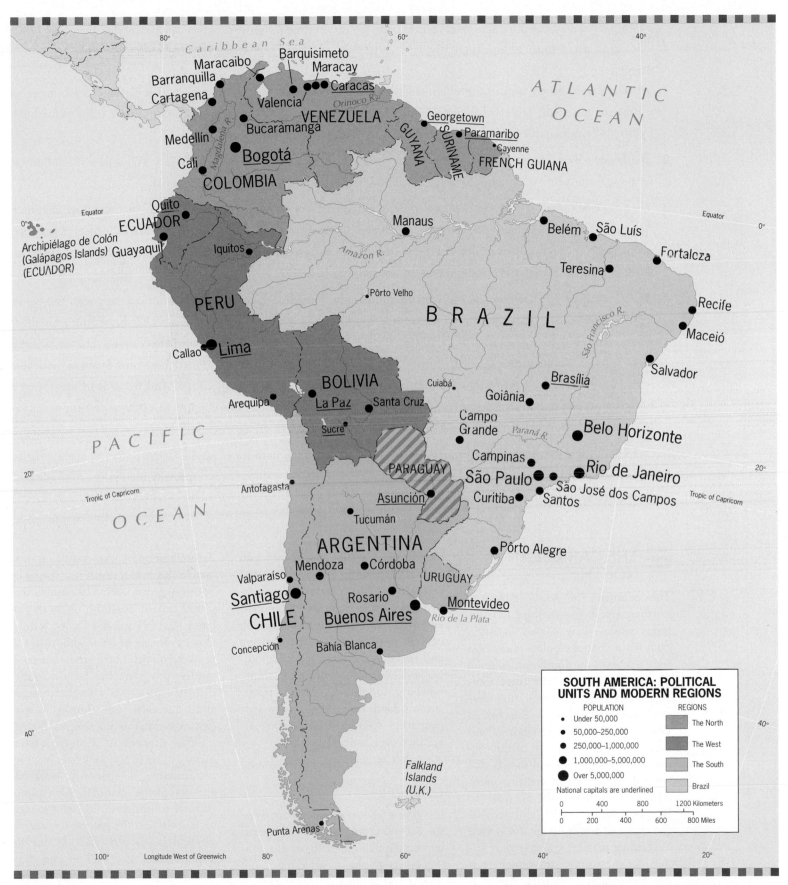

Caribbean Sea

Barranquilla
Maracaibo
Barquisimeto
Maracay
Cartagena
Valencia
Caracas
Orinoco R.

VENEZUELA

Bucaramanga

Medellín

Bogotá

Cali

COLOMBIA

GUYANA
Georgetown
SURINAME
Paramaribo
Cayenne
FRENCH GUIANA

ATLANTIC
OCEAN

Equator

Quito

ECUADOR

Archipiélago de Colón
(Galápagos Islands)
(ECUADOR)

Guayaquil

Iquitos

Manaus

Amazon R.

Belém
São Luís

Fortaleza

Teresina

PERU

Pôrto Velho

B R A Z I L

São Francisco R.

Recife

Maceió

Callao
Lima

Salvador

Arequipa

BOLIVIA
La Paz

Sucre

Santa Cruz
Cuiabá

Goiânia

Brasília

Belo Horizonte

PACIFIC

Antofagasta

Campo
Grande
Paraná R.

Campinas

PARAGUAY

São Paulo

Rio de Janeiro

Asunción

Curitiba
Santos
São José dos Campos

OCEAN

Tucumán

Tropic of Capricorn

Tropic of Capricorn

ARGENTINA

Pôrto Alegre

Mendoza
Córdoba

URUGUAY

Valparaíso

Rosario

Montevideo

Santiago

Buenos Aires

Rio de la Plata

CHILE

Concepción
Bahía Blanca

Falkland
Islands
(U.K.)

**SOUTH AMERICA: POLITICAL
UNITS AND MODERN REGIONS**

POPULATION | REGIONS
Under 50,000 | The North
50,000–250,000 | The West
250,000–1,000,000 | The South
1,000,000–5,000,000 | Brazil
Over 5,000,000

National capitals are underlined

0   400   800   1200 Kilometers
0 200 400 600 800 Miles

Punta Arenas

Longitude West of Greenwich

n the basis of physiographic, cultural, and political criteria, South America can be divided into four rather clearly defined regions (Fig. 5B-1):

1. **The Caribbean North**, located almost entirely north of the equator, consists of five entities that display a combination of Caribbean and South American features: Colombia, Venezuela, and those that represent three historic colonial footholds by Britain (Guyana), the Netherlands (Suriname), and France (French Guiana).

2. **The Andean West** is formed by four republics that share a strong Amerindian cultural heritage as well as powerful influences resulting from their Andean physiography: Ecuador, Peru, Bolivia, and, transitionally, Paraguay.

3. **The Southern Cone**, for the most part located south of the Tropic of Capricorn (latitude 23½°S), includes three countries: Argentina, Chile, and Uruguay (all with strong European imprints and little remaining Amerindian influence) plus aspects of Paraguay. The far northern part of this region that lies east of the Andes is occupied by the *Gran Chaco*, a sparsely populated, hot, semiarid, grassy lowland centered on western Paraguay.

4. **Brazil** occupies an enormous part of interior and eastern South America. Its own interior overlaps with the continent's "green heart"—sometimes called the world's "lungs"—the largest tropical rainforest on Earth. As South America's giant, accounting for just about half the realm's territory as well as population, it is swiftly developing into the Western Hemisphere's second superpower. In Brazil the dominant Iberian influence is Portuguese, not Spanish, and here Africans, not Amerindians, form a significant component of demography and culture. And Brazil is in some ways a bridge between the Americas and Africa, which is why we discuss it last in this chapter, just before we turn to Subsaharan Africa.

Each of these four regions shares some important commonalities in terms of physical environment, ethnic origins, cultural milieu, and international outlook. But there also are important differences, especially involving economic performance, democratic functioning, and social stability.

THE CARIBBEAN NORTH

The countries of South America's northern tier have something in common besides their coastal location: each has a coastal tropical-plantation zone based on the Caribbean colonial model. Especially in the three Guianas, early European plantation development encompassed the forced immigration of African laborers and eventually the absorption of this element into the population matrix. Far fewer Africans were brought to South America's northern shores than to Brazil's Atlantic coasts, and tens of thousands of South Asians also arrived as contract laborers and stayed as settlers, so the overall situation here is not comparable to Brazil's. And it is also distinctly different from that of the rest of South America.

Today, Guyana, Suriname, and French Guiana still display the coastal orientation and plantation dependency with which the colonial period endowed them, although the logging of their tropical forests is penetrating and ravaging the interior. In Colombia and Venezuela, however, farming, ranching, and mining drew the population inland, overtaking the coastal-plantation economy and creating diversified economies (Fig. 5B-2).

Colombia

Imagine a country more than twice the size of France, not at all burdened by overpopulation, with an environmental geography so varied that it can produce crops ranging from the temperate to the tropical, and possessing world-class oil reserves as well as other natural resources. This country is situated in the crucial northwest corner of South America, with 3200 kilometers (2000 mi) of coastline on both Atlantic (Caribbean) and Pacific waters, closer than any of its neighbors to the markets of the north and sharing a border with giant Brazil to the southeast. Its nation uses a single language and adheres to one dominant religion. Wouldn't such a country thrive?

The answer is no. Colombia has a history of strife and violence, its politics unstable, its economy damaged, and its future clouded. Colombia's cultural uniformity did not produce social cohesion. Its spectacular, scenic physical geography also divides its population of 45.7 million into clusters not sufficiently interconnected to foster integration; even today, this huge country has less than 800 kilometers

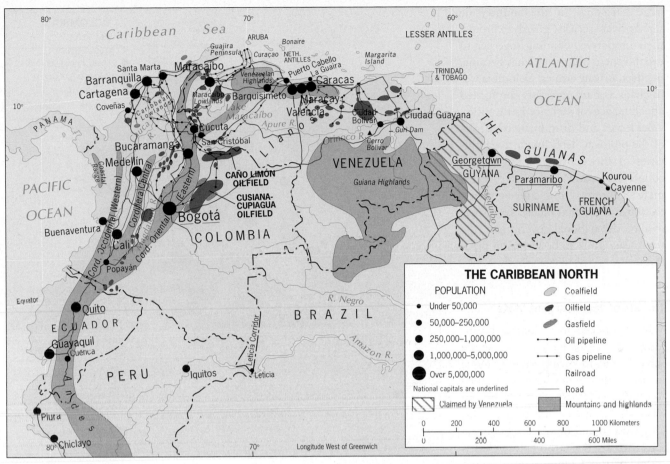

FIGURE 5B-2

© H. J. de Blij, P. O. Muller, and John Wiley & Sons, Inc.

(500 mi) of four-lane highways. Its proximity to U.S. markets is a curse as well as a blessing: at the root of Colombia's latest surge of internal conflict lies its role as one of the world's largest producers of illicit drugs.

People and Resources

As Figure 5B-2 reveals, Colombia's physiography is mountainous in its Andean west and north and comparatively flat in its interior. Colombia's scattered population (see Fig. 5A-3) tends to cluster in the west and north, where the resources and the agricultural opportunities (including the coffee for which Colombia is famous) lie.

Colombia's population clusters are not well interconnected. Several lie on the Caribbean coast, centered on Barranquilla, Cartagena, and Santa Marta, old colonial entry points. Others are anchored by major cities such as

Medellín and Cali. What is especially interesting about Figure 5B-2 in this context is how little development has taken place on the country's lengthy Pacific coast, where the port city of Buenaventura, across the mountains from Cali, is the only place of any size. The map suggests that Colombia's Pacific Rim era, already in full force in Chile, Peru, and Ecuador, has yet to arrive.

The map also shows that Colombia and neighboring Venezuela share the oil and gas reserves in and around the "Lake" Maracaibo area ("Lake" because this is really a gulf with a narrow opening to the sea), but Venezuela has the bigger portion. Recent discoveries, however, have boosted Colombia's production along the base of the easternmost Andean cordillera, and our map shows a growing system of pipelines from the interior to the coast. Meanwhile, the vast and remote

interior proved fertile ground for that other big money-maker in Colombia: drugs.

Cocaine's Curse

The rise of the narcotics industry, fueled by outside (especially U.S. but also Brazilian and European) demand, and coupled with its legacy of violence, has crippled the state for decades and threatened its very survival. Drug cartels based in the cities controlled vast networks of producers and exporters; they infiltrated the political system, corrupted the army and police, and waged wars with each other that cost tens of thousands of lives and destroyed Colombia's social order. The drug cartels also organized their own armed forces to combat attempts by the Colombian government to control the illegal narcotics economy. Meanwhile, the owners of large haciendas in the countryside hired private security guards to protect their properties, banding together to expand these units into what became, in effect, private armies. Colombia was in chaos as narcoterrorists committed appalling acts of violence in the cities, rebel forces and drug-financed armies of the political "left" fought paramilitaries of the political "right" in the countryside, and Colombia's legitimate economy, from coffee-growing to tourism, suffered fatally. To further weaken the national government, the rebels even took to bombing oil pipelines.

Threats of the Insurgent State

What happened in Colombia beginning in the 1970s and escalating in the three decades following was not unique in the world—states have succumbed to chaos in the past—but this was an especially clear-cut case of a process long studied and modeled by political scientists. By the turn of this century, certain parts of Colombia were beyond the control of its government and armed forces. There, insurgents of various stripes created their own domains, successfully resisting interference and demanding to be left alone to pursue their goals, illegitimate though they might be. Leaders of some of these insurgent domains even sent emissaries to Bogotá to negotiate their terms for "independence."

Viewing this process institutionally, political scientists recognized three evolutionary stages. During the first, **contention**, a rebellion erupts, sustains itself, and becomes based in some part of a country. During the second stage, **equilibrium**, the rebels gain effective control over a territorially defined sector of the state, establish a capital, impose rules of conduct, and have sufficient strength to bring the national government to the negotiating table. If they succeed, they may achieve secession from the state or even bring the government down and take control of all of it. But equilib-

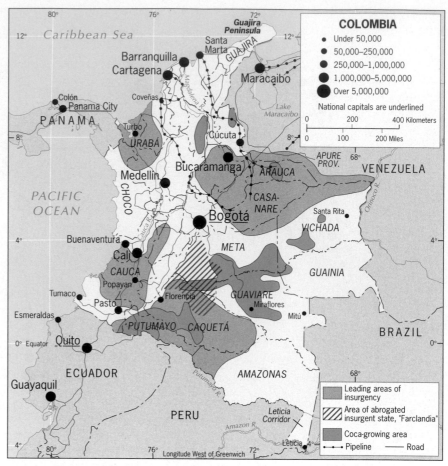

FIGURE 5B-3   © H. J. de Blij, P. O. Muller, and John Wiley & Sons, Inc.

rium may also be followed by a third stage, **counteroffensive**, in which the state, perhaps with the support of outside forces, resumes the conflict and ultimately defeats the insurgents. We will encounter instances of all three stages in this model in later chapters.

The geographer Robert McColl looked at the sequence in spatial context and formulated the concept of an **1 insurgent state** to represent the second stage, that of equilibrium. He argued that this equilibrium implied the formation of more than just a rebel base: this is the moment of truth when the rebel domain takes on the elements of a nascent state complete with boundaries, a core area as well as a capital, a local government, and schools and other social services that substitute for those the formal state may have provided previously.

Colombia in the 1990s and early 2000s contained several entities that

were taking on the properties of insurgent states, and one of these, the red-striped zone shown in Figure 5B-3, even acquired a name—"Farclandia"—after the initials of the Revolutionary Armed Forces of Colombia (FARC), one of the most brutal and by far the most powerful among Colombia's insurgent groups. At the end of 1999, FARC had the Colombian government on the ropes, forcing it to demilitarize its area south of Bogotá (about the size of Switzerland) and announcing plans for a second insurgent state centered on the remote southeastern town of Mitú. It appeared to be just a matter of time before the national government in Bogotá would lose control, equilibrium would turn into disintegration, and Colombia would devolve into a **2 failed state** (a country whose institutions have collapsed and in which anarchy prevails).

But in the past decade, Colombia mounted a counteroffensive. In 2002, newly elected President Alvaro Uribe began a twin campaign to defeat the rebels militarily and to persuade them through legal means to give up their arms. He secured significant assistance from the United States and was able to increase the pressure on armed rebels as well as coca growers, invaded "Farclandia," and scored some noteworthy successes in killing and arresting leaders as well as freeing hostages held by the rebels. Colombia's nascent insurgent states weakened, and disarray marked the once-formidable rebel armies.

A New Economic Future?

The government matched its domestic counteroffensive with an international campaign to help revive the economy, promoting market-oriented, business-friendly policies. Under the political circumstances, it is remarkable that economic growth before the global downturn of 2008 was as high as 8 percent annually, led by revenues from energy, metals, and agricultural products; the comparative lull in violence boosted coffee exports, cut-flower exports (Colombia ranks second in the world in this business), and tourism.

But Colombia remains a nation in crisis, not all of it of its own making. About 70 percent of the cocaine consumed in the United States comes from Colombia; most of the remainder goes to Brazil or Europe. American support for Uribe's antidrug campaign (*Plan Colombia*), logistical as well as financial, raises political issues in Bogotá, where extraditions of Colombian drug lords to the United States are not universally popular. Conversely, the U.S. Congress refused to approve a free-trade agreement with Colombia because of the Uribe government's alleged tilt toward right-wing paramilitaries in conjunction with the amnesty program. Colombians argue that the cocaine keeps flowing because Americans continue to buy it,

and that Americans should look in the mirror before they accuse others of bad behavior. In 2011, Colombia's newly-elected government faced a clouded future.

Venezuela

A long and tortuous boundary separates Colombia from Venezuela, its neighbor to the east. Much of what is important in Venezuela is concentrated in the northern and western parts of the country, where the Venezuelan Highlands form the eastern spur of the north end of the Andes system. Most of Venezuela's 29.1 million people are concentrated in these uplands, which include the capital of Caracas, its early rival Valencia, and the commercial/industrial centers of Barquisimeto and Maracay.

The Venezuelan Highlands are flanked by the Maracaibo Lowlands and "Lake" Maracaibo to the northwest and by an immense plainland of savanna country, known as the *Llanos*, in the Orinoco Basin to the south and east (Fig. 5B-2). The Maracaibo Lowlands, at one time a disease-infested, sparsely peopled coastland, today constitute one of the world's leading oil-producing areas; much of the oil is drawn from reserves that lie beneath the shallow waters of the lake itself. The country's third-largest city, Maracaibo, is the focus of the petroleum industry that transformed the Venezuelan economy in the 1970s; however, as we shall see, since then oil has been more of a curse than a blessing.

The *Llanos* on the southern side of the Venezuelan Highlands and the Guiana Highlands in the country's southeast, like much of Brazil's interior, are in a nascent stage of development. The 300- to 650-kilometer (180- to 400-mi)-long *Llanos* slope gently from the base of the northernmost Andean spur to the Orinoco's floodplain. Their mixture of savanna grasses and scrub woodland supports cattle grazing on higher ground, but

widespread wet-season flooding of the more fertile lower-lying areas has thus far inhibited the plainland's commercial farming potential (much of the development of the *Llanos* to date has been limited to the exploitation of its substantial oil reserves). Crop-raising conditions are more favorable in the **tierra templada** areas at more moderate altitudes in the Guiana Highlands. Economic integration of this more remote interior zone with the heartland of Venezuela has been spearheaded by the discovery of rich iron ores on the northern side of the Guiana Highlands southwest of Ciudad Guayana. Local railroads now connect with the Orinoco, and from there ores are shipped directly to foreign markets.

Oil and Politics

Despite these opportunities, Venezuela since 1998 has been in upheaval as longstanding economic and social problems finally intensified to the point where the electorate decided to push the country into a new era of radical political change. For nearly two decades since the euphoric 1970s, oil had not bettered the lives of most Venezuelans. A primary reason was that the government unwisely acquired the habit of living off oil profits, forcing the country to suffer the consequences of the prolonged global oil depression that began in the early 1980s. Venezuela found itself heavily burdened by a huge foreign debt it had incurred through borrowing against future oil revenues that were not materializing fast enough. By the mid-1990s, the government was required to sharply devalue the currency, and a political crisis ensued that resulted in a severe recession and widespread social unrest. With more and more Venezuelans enraged at the way their oil-rich country was approaching bankruptcy without making progress toward the more equitable distribution of the national wealth, voters resoundingly turned in an extreme direction in the 1998 presidential election. Expressing their disgust with Venezuela's ruling elite of both political parties, they elected Hugo Chávez, a former colonel who in 1992 had led a failed military coup. Reaffirming their decision in 2000, Venezuelans gave nearly 60 percent of their votes to Chávez, apparently providing him with a mandate to act as strongman on behalf of the urban poor and the increasingly penurious middle class.

Venezuela's Socialist Turn

Chávez has indeed pursued this course since entering office in 1999, sweeping aside Congress and the Supreme Court, supervising the rewriting of the Venezuelan constitution in his own image, and proclaiming himself the leader of a "peaceful leftist revolution" that will transform the country. Although he professes that social equality ranks highest on his agenda, Chávez has stirred up racial divisions by actively promoting mestizos (68 percent of the population) over people of European background (21 percent).

In the international arena, Chávez sparks controversy at every turn: angering the government of neighboring Colombia by expressing his "neutrality" in its confrontation with cocaine-producing insurgent forces; unsettling another neighbor, Guyana, by aggressively reviving a century-old territorial claim to the western zone of that country (the striped area in Fig. 5B-2); embracing the Cuban regime; proclaiming his solidarity with Iran's president; and, in January 2010, openly declaring himself a Marxist. More generally, Chávez has obstructed the globalization plans of the United States throughout South America, using the windfall income from high-priced oil to reward allies in this effort with substantial subsidies. For example, considerable financial support has been provided to Bolivia's President Evo Morales, the realm's first

national leader of Amerindian ancestry, who faced powerful opposition from the country's affluent Hispanic minority.

Chávez has positioned himself as the champion of Venezuela's (and South America's) poor, trumpeting his "Bolivarian Revolution" as a local alternative to what he describes as the insidious encroachment of U.S. imperialism. He has taken Venezuela to center stage in the ideological contest in this realm, casting doubt on the virtues of democracy and free enterprise, castigating elites for not doing enough to reduce the gap between rich and poor, and urging the downtrodden to assert themselves.

In 2012, Chavez's second presidential term will come to an end, but in 2009 he won a referendum that abolished term limits. It was a huge victory, paving the way for his continued reign in years to come. Vilified by much of the Western world and by many of Venezuela's entrepreneurial citizens (a good number of whom have fled the country), he sounds a message that resonates with the South American masses, in itself testimony to the continuing inequality that marks this realm.

The "Three Guianas"

Three small entities form the eastern flank of the realm's northern region: Guyana, Suriname, and French Guiana. They are good reminders of why the name "Latin" America is inappropriate: the first is a legacy of British colonialism and employs English as its official language; the second is a remnant of Dutch influence where Dutch is still official among its polyglot of tongues; and the third is still a dependency—of France. None has a population exceeding 1 million, and all three exhibit social indices and cultural landscapes more representative of the Caribbean islands than South America. Here British, Dutch, and French colonial powers acquired possessions and established plantations,

brought in African and Asian workers, and created economies similar to those that mark a number of the Caribbean islands.

Guyana, with 820,000 people, still has more inhabitants than Suriname and French Guiana combined. When Guyana became independent in 1966, its British rulers left behind an ethnically and culturally divided population in which people with South Asian (Indian) ancestry make up about 50 percent and those with African heritage (including African-European ancestry) 36 percent. This makes for contentious politics, given the religious mix, which is approximately 50 percent Christian and 45 percent Hindu and Muslim. Guyana remains dominantly rural, and plantation crops continue to figure prominently among exports (gold from the interior is the most valuable single product). Oil may soon become a factor in the economy, though, because a recently discovered reserve that lies offshore from Suriname extends westward beneath Guyana's waters. Still, Guyana is among the realm's poorest and least urbanized countries, and it is strongly affected by its neighbor's narcotics industry. Its thinly populated interior, beyond the reach of antidrug campaigns, has become a staging area for drug distribution to Brazil, North America, and even Europe. A recent official report suggests that drug money amounts to as much as one-fifth of Guyana's total economy.

Venezuela's territorial claim against Guyana may be in abeyance, but the United Nations in 2007 settled a dispute with Suriname over a potentially oil-rich maritime zone in Guyana's favor (Fig. 5B-2). Some geologists believe that this offshore basin may hold more oil than Europe's North Sea, which would—if properly managed—transform Guyana's economy. Exploratory drilling began in 2009.

Suriname actually progressed more rapidly than Guyana did after it was granted independence in 1975, but persistent political instability soon ensued. The Dutch colonists brought South Asians, Indonesians, Africans, and even some Chinese to their colony, making for a fractious nation. More than 100,000 residents—about one-quarter of the entire population—emigrated to the Netherlands, and were it not for support from its former colonial ruler, Suriname would have collapsed. Nonetheless, its rice farms give Suriname self-sufficiency and even allow for some exporting, and plantation crops continue to figure among the exports. Suriname's leading income producer, however, is its bauxite (aluminum ore) mined in a zone across the middle of the country. And the recent oil finds (Fig. 5B-2) could provide important revenues in the years ahead.

Suriname, most of whose population of 480,000 resides along the northern coastal strip, is one of South America's poorest countries, and its prospects remain bleak. Along with Guyana, Suriname is involved in an environmental controversy centering on its luxuriant tropical forests. Timber companies from Asia's Pacific Rim offer lucrative rewards for the right to cut down the magnificent hardwood trees, but conservationists are trying to slow the destruction by buying up concessions before they can be opened to logging.

Suriname's cultural geography is enlivened by its many languages. Dutch may be the official tongue, but other than by officials and in schools it is not heard much. A mixture of Dutch and English, *Sranan Tongo*, serves as a kind of common tongue, but you can also hear Amerindian, Hindi, Chinese, Indonesian, and even some French Creole in the streets. Suriname may have a Dutch past, but today it is a member of CARICOM, the English-speaking, supranational Caribbean economic organization.

French Guiana, the easternmost outpost of the Caribbean North, is a dependency—mainland South America's only one. This territory is an anomaly in other ways as well. Consider this: it is almost as large as South Korea (population: nearly 50 million) with a population of just over 200,000. Its status is an *Overseas Département* of France, and its official language is French. Nearly half the population resides in the immediate vicinity of the coastal capital, Cayenne.

In 2011 there still was no prospect of independence for this severely underdeveloped relic of the former French Empire. Gold remains the most valuable export, and the small fishing industry sends some exports to France. But what really matters here in French Guiana is the European Space Agency's launch complex at Kourou on the coast, which accounts for more than half of the territory's entire economic activity. From plantation farming to spaceport . . . this must be the ultimate story of globalization.

≡ THE ANDEAN WEST

The second regional grouping of South American states—the Andean West (Fig. 5B-4)—is dominated physiographically by the great Andean mountain chain and historically by Amerindian peoples. This region encompasses Peru, Ecuador, Bolivia, and transitional Paraguay, the last with one foot in the West and the other in the South (Fig. 5B-1). Bolivia and Paraguay also constitute South America's only two landlocked countries. For several decades now, the three main countries in this region, Peru, Ecuador, and Bolivia, have joined in the *Andean Community*, a trading bloc that also includes Colombia.

Spanish conquerors overpowered the Amerindian nations, but they did not reduce them to small minorities as happened to so many indigenous peoples in other parts of the world. Today, roughly 45 percent of the people in Peru, the region's most populous country, are Amerindian; in Bolivia, Amerindians are in the majority at 55 percent. About 25 percent of Ecuador's population identifies itself as Amerindian, and in Paraguay the ethnic mix, not regionally clustered as in the other three countries, is overwhelmingly weighted toward Amerindian ancestry.

As the Data Table in Appendix B reports, this is South America's poorest region economically, with lower incomes, higher numbers of subsistence farmers, and fewer opportunities for job-seekers. For a very long time, the urbane lives of the land-owning elite have been worlds away from the hard-scrabble existence of the landless peonage (the word **peon** is an old Spanish term for an indebted day laborer).

But today, this region, like the realm as a whole, is stirring, and oil and natural gas are part of the story. In Bolivia, the first elected president of Amerindian ancestry is trying to gain control over an energy industry not used to his aggressive tactics. In Ecuador's 2006 elections, a populist gained the presidency on a promise to divert more of the country's oil revenues toward domestic needs. In Peru, where an energy era seems to be opening as new reserves are discovered, the government is under pressure to protect Amazonian peoples and environments, limit foreign involvement, and put domestic needs and rights first.

Peru

Peru straddles the Andean spine for more than 1600 kilometers (1000 mi) and is the largest of the region's four republics in both territory and population (28.7 million). Physiographically and culturally, Peru divides into three subregions: (1) the desert coast, the European-mestizo region; (2) the Andean highlands or *Sierra*, the Amerindian region; and (3) the eastern slopes, the sparsely populated Amerindian-mestizo interior (Fig. 5B-4).

Three Subregions

Lima and its port, Callao, lie at the center of the desert *coastal strip*, and

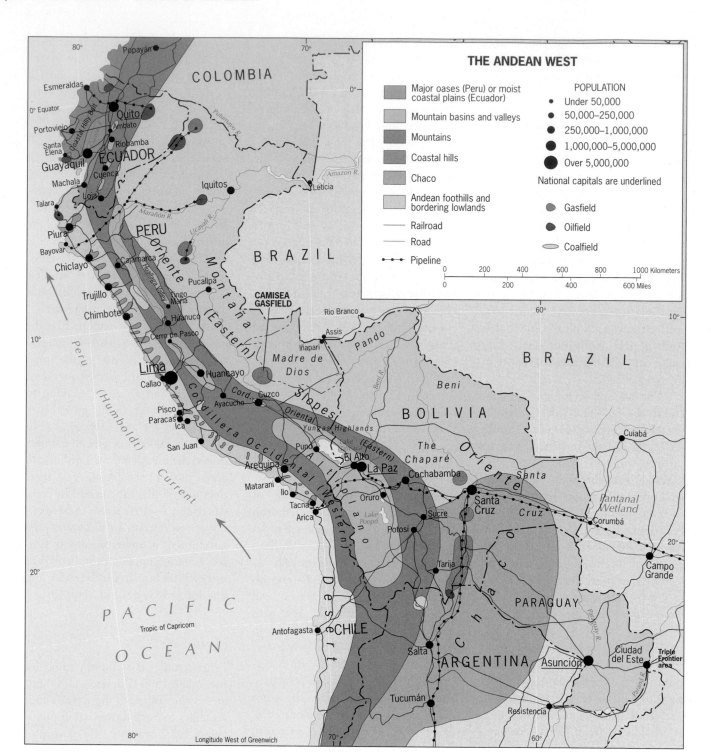

FIGURE 5B-4

© H. J. de Blij, P. O. Muller, and John Wiley & Sons, Inc.

it is symptomatic of the cultural division prevailing in Peru that for nearly 500 years the capital city has been positioned on the periphery, not in a central location in a basin of the Andes. From an economic point of view, however, the Spaniards' choice of a headquarters on the Pacific coast

proved to be sound, for the ***coastal*** subregion has become commercially the most productive part of the country. A thriving fishing industry contributes significantly to the export trade; so do the products of irrigated agriculture in some 40 oases distributed all along the arid coast, which

include fruits such as citrus, olives, and avocados, and vegetables such as asparagus (a big money-maker) and lettuce.

The ***Sierra*** (Andean) subregion occupies just about one-third of the country and is the ancestral home of the largest component in the total population, the speakers of Quechua, the *lingua franca* that emerged during the Inca Empire. Their physical survival during the harsh colonial regime of the Spanish was made possible by their adaptation to the high-altitude environments they inhabited, but their social fabric was ripped apart by communalization, forced cropping, religious persecution, and outward migration to towns and haciendas where many became serfs.

Although Amerindian people constitute almost half of the population of Peru, their political influence remains slight. Nor is the Andean subregion a major factor in Peru's commercial economy—except, of course, for its huge mineral storehouse, which yields copper, zinc, and lead from mining centers, the largest of which is Cerro de Pasco. In the high valleys and intermontane basins, the Amerindian population is clustered either in isolated villages, around which people practice a precarious subsistence agriculture, or in the more favorably located and fertile areas where they are tenants, peons on white- or mestizo-owned haciendas. It was a very different story at the time of the Incas, when the indigenous social structure was yet unharmed and the region provided bountiful harvests. Today, there is a painful contrast between the prosperous coast, booming Lima, and the thriving north on the one hand, and this poverty-stricken southern Andean zone on the other. That division could be a threat to Peru's long-term stability.

Of Peru's three subregions, the **Oriente**, or East—the inland slopes of the Andean ranges and the Amazon-drained, rainforest-covered *montaña*—is the most isolated. The focus of the eastern subregion, in fact, is Iquitos, a city that looks east rather than west and can be reached by ocean-going vessels sailing 3700 kilometers (2300 mi) up the Amazon River across northern Brazil. Iquitos grew rapidly during the Amazon wild-rubber boom of just a century ago and then declined; now it is finally growing again and reflects Peruvian plans to open up the eastern interior.

A New Energy Era

Today, the *Oriente* subregion, and perhaps Peru as a whole, is on the threshold of a new era. Petroleum had been discovered west of Iquitos as long ago as the 1970s, when oil began to flow through a pipeline across the Andes to the Pacific port of Bayóvar. But major new discoveries of oil and gas reserves since the turn of the millennium are promising (some say the verb should be *threatening*) to change the country, with associated impacts on the Amerindian peoples still living an isolated existence in this remote, forested East. Already, pipelines carry natural gas from the Camisea reserve (north of Cuzco) to a conversion plant on the Paracas Peninsula south of Lima, from where an ocean-floor pipeline sends it to an offshore loading platform for tankers taking it to the U.S. market (Fig. 5B-4). Other reserves will come "on line" later, but already environmentalists and supporters of indigenous rights are raising issues in the interior even as political activists are arguing against the terms of trade that Peru has accepted with the oil companies. They argue that the proceeds will further benefit the already-favored coastal, northern, and urban residents of Peru, and leave the disadvantaged residents of the interior even further behind. In 2009, deadly clashes between Amerindians and industry-supporting government forces killed dozens.

Peru has never had an Amerindian president. The country's current leader, Alan Garcia, is, like the others before him, supported by the traditional Hispanic (and mestizo) establishment. New presidential elections in 2011 will tell if Peru is ready for the transition already underway in neighboring countries. Amerindians form a near-majority, they are restive, and they now have models of empowerment not previously seen in this region. The question is what course Peru will take.

Ecuador

On the map, Ecuador, smallest of the three Andean West republics, appears to be just a northern corner of Peru. But that would be a misrepresentation because Ecuador possesses a complete range of regional variations (Fig. 5B-4). It has a coastal belt; an Andean zone that may be narrow (less than 250 kilometers [150 mi]) but by no means of lower elevation than elsewhere; and an *Oriente*—an eastern subregion that is as sparsely settled and as economically marginalized as that of Peru. As in Peru, nearly half of Ecuador's population (which totals 14.4 million) is concentrated in the Andean intermontane basins and valleys, and the most productive region is the coastal strip. Here, however, the similarities end.

Ecuador's Pacific coastal zone consists of a belt of hills interrupted by lowlands, of which the most important lies in the south between the hills and the Andes, drained by the Guayas River. Guayaquil—the country's largest city, main port, and leading commercial center—forms the focus of this subregion. Unlike Peru's coastal strip, Ecuador's is not a desert: it consists of fertile tropical plains not afflicted by excessive rainfall. Seafood (especially shrimp) is a leading product, and these lowlands support a thriving commercial agricultural economy built around bananas, cacao, cattle-raising, and coffee on the hillsides. Moreover, Ecuador's western subregion is also far less Europeanized than Peru's because its white component of the national population is only about 7 percent.

A greater proportion of whites is engaged in administration and hacienda ownership in the central Andean zone, where most of the Ecuadorians who are Amerindian also reside—and, not surprisingly, where land-tenure reform is an explosive issue. The

© H. J. de Blij

"I can't remember being hotter anyplace on Earth, not in Kinshasa, not in Singapore . . . not only are you near the equator here in steamy Guayaquil, but the city lies in a swampy, riverine lowland too far from the Pacific to benefit from any cooling breezes and too far from the Andes foothills to enjoy the benefits of suburban elevation. But Guayaquil is not the disease-ridden backwater it used to be. Its port is modern, its city-center waterfront on the Guayas River has been renovated, its international airport is the hub of a commercial center, and it has grown into a metropolis of 2.7 million. Ecuador's oil revenues have made much of this possible, but from the White Hill (a hill that serves as a ceme-tery, with the most elaborate vaults near its base and the poorest at the feet of the giant statue of Jesus that tops it) you can see that globaliza-tion has not quite arrived here. High-rise development remains limited, international banks, hotels, and other businesses remain comparatively few, and the "middle zone" encircling the city shows little evidence of prosperity (left). Beating the heat and glare is an everyday priority: whole streets have been covered by makeshift tarpaulins and more permanent awnings to protect shoppers (right). Talk to the locals, though, and you find that there is another daily concern: the people 'up there in the mountains who rule this country always put us in second place.' Take the 45-minute flight from Guayaquil to cool and comfortable Quito, the capital, and you're in another world, and you quickly forget Guayaquil's problems. That's just what the locals here say the politicians do."

www.conceptcaching.com

differing interests of the Guayaquil-dominated coastal lowland and the Andean-highland subregion focused on the capital (Quito) have long fostered a deep regional cleavage between the two. This schism has intensified in re-cent years, and autonomy and other devolutionary remedies are now being openly discussed in the coastlands.

In the rainforests of the *Oriente* subregion, oil production is expanding as a result of the discovery of additional reserves. Some analysts are predicting that interior Ecuador as well as Peru will prove to contain reserves compa-rable to those of Venezuela and Colom-bia, and that an "oil era" will soon transform their economies. As it is, oil already tops Ecuador's export list, and

the industry's infrastructure is being modernized. This is essential because much ecological damage has already been done: the trans-Andean pipeline to the port of Esmeraldas, constructed in 1972, was the source of numerous oil spills, and toxic waste was dumped along its route in a series of dreadful environmental disasters. A second, more modern pipeline went into oper-ation in 2003, but like Peru, Ecuador faced growing opposition from envi-ronmentalists and activists who, in 2005, shut it down for a week. With revenues from oil and gas exports exceeding 50 percent of all exports, Ecuador's leaders are hearing an increas-ingly familiar refrain from its people: demand more from the companies and

give these funds to those who need it. But it is not as simple as that: oil and gas exploration and exploitation require huge investments, and foreign compa-nies can afford to spend what the gov-ernment's own state company, Petro Ecuador, cannot. That leads to difficult choices because the state needs income to cover its obligations. Clearly, energy riches are a double-edged sword.

Bolivia

Nowhere are the problems so typical of the Andean West more acute than in landlocked, volatile Bolivia. Before reading further, take a careful look at Bolivia's regional geography in both Figures 5B-4 and 5B-5. Bolivia is

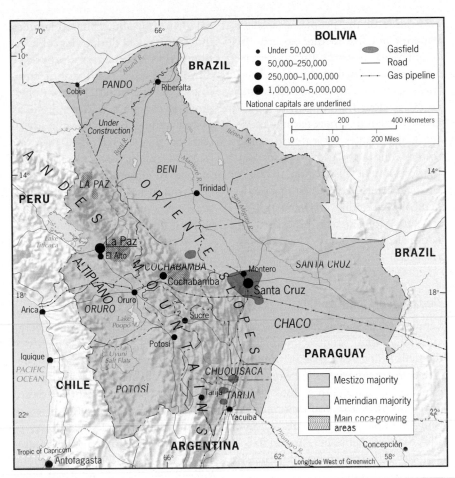

FIGURE 5B-5

© H. J. de Blij, P. O. Muller, and John Wiley & Sons, Inc.

Bolivia, freshwater Lake Titicaca lies at 3700 meters (12,500 ft) above sea level and helps make the adjacent *Altiplano* (see map at left) livable by ameliorating the coldness in its vicinity, where the snow line lies just above the plateau surface. On the surrounding cultivable land, potatoes and grains have been raised for centuries dating back to pre-Inca times, and the Titicaca Basin still supports a major cluster of Amerindian (Aymara) subsistence farmers. This portion of the *Altiplano* is the heart of modern Bolivia and also contains the capital city, La Paz.

The European/Amerindian Divide

The Bolivian state is the product of the Hispanic impact, and the country's Amerindians (who still make up at least 55 percent of the national population of 10.4 million) no more escaped the loss of their land than did their Peruvian or Ecuadorian counterparts. What made the richest Europeans in Bolivia wealthy, however, was not land but minerals. The town of Potosí in the eastern cordillera became a legend for the immense deposits of silver in its vicinity; tin, zinc, copper, and several ferroalloys were also discovered there. Amerindian workers were forced to work in the mines under the most appalling conditions.

bounded by remote parts of Brazil and Argentina, mountainous Andean highlands and intermontane **3** *altiplanos* (high-elevation basins and valleys) of Peru, and coveted coastal zones of northern Chile. As the maps show, the Andes in this area broaden into a vast mountain complex some 700 kilometers (450 mi) wide. On the boundary between Peru and

Bolivia's political geography is as divided as its physical geography, the power base of the indigenous population lying in the *altiplanos* and mountains, and that of the mestizo minority centered in the lowlands (see the map above). These photos, taken a day apart in Santa Cruz, where mestizo strength is concentrated, show the crucial difference of opinion: a majority of mestizos want autonomy for their eastern provinces (left); virtually all indigenous people protest the prospect of such autonomy, seen as a prelude to secession and independence, as fatally fracturing their country. (Left) © AFP/Getty Images, Inc. (Right) © Rodrigo Buendia/AFP/Getty Images, Inc.

These buckets of salt brine sitting in the moonscape of southwestern Bolivia's Uyuni Salt Flats represent one of the country's most promising new development opportunities. Once the liquid evaporates, the residue consists mainly of lithium. This lightest of all metals is used to manufacture the lithium-ion batteries that power the portable electronic devices that dominate our daily lives, such as cellphones, smart phones, and laptops. Larger versions of these batteries are also used to power hybrid motor vehicles, and Bolivia contains as much as half of the world's lithium reserves—enough to support close to 5 billion electric cars. The facility under construction here in this remote part of the Altiplano is a pilot lithium-extraction plant that will begin operation in 2011. The Bolivian government is the major player in this effort and is determined to keep lithium production a domestic enterprise—despite considerable pressure by foreign companies and investors to become involved. Another fascinating angle is that much of land of the Salt Flats is the property of Amerindian groups, who are seeking a share in the income to be earned from mineral exploitation. Their claims are consistent with the country's new (2010) constitution, which may allow indigenous communities to have control over resources in their territory—and potentially shape the future course of development.
© Martin Bernetti/AFP/GettyImages, Inc.

Today natural gas and oil, exported to Argentina and Brazil, are leading sources of foreign revenues, and zinc has replaced tin as the leading metal export. But Bolivia's economic prospects will always be impeded by the loss of its outlet to the Pacific Ocean during its war with Chile in the 1880s, despite its transit rights and dedicated port facilities at Antofagasta.

More critical than its economic limitations or its landlocked situation is Bolivia's social predicament. The government's history of mistreatment of indigenous people and harsh exploitation of its labor force hangs heavily over a society where about two-thirds of the people, almost all Amerindian, live in dire poverty. In recent years, however, this underrepresented aboriginal majority has been making an impact on national affairs. In 2003, violent opposition to a government plan to export natural gas to the United States via a new pipeline to the Chilean coast led to chaos and the government's resignation. In 2005, Bolivian voters elected their first president of Amerindian (Aymara) ancestry, Evo Morales, who proceeded to nationalize the country's natural gas resources.

Departments and Capitals

Landlocked, physiographically bisected, culturally split, and economically divided, Bolivia is a state in trouble. The country's prospects are worsened by its political geography: look again at Figure 5B-5 and you can see that Bolivia's nine provinces are regionally divided between Amerindian-majority *Departments* (as subnational units are called here) in the west and those with mestizo majorities in the east. The capital city, La Paz, lies on the Amerindian-majority Andean Altiplano, but many mestizo Bolivians do not recognize it as such: historically, the functions of government have been divided between La Paz (the administrative headquarters) and Sucre (which lost most government branches in 1899 during a civil war but retained the Supreme Court, calling itself the "constitutional capital"). And some mestizo Bolivians even suggest that the eastern city of Santa Cruz should be considered as a candidate for the "compromise capital" of their country.

The Santa Cruz *Department*, like the others in the *Oriente*, stands in sharp contrast to those of the Andes in the west. Here the hacienda system (see Chapter 4A) persists almost unchanged from colonial times, its profitable agriculture supporting a wealthy aristocracy. Nearly 90 percent of Bolivia's agriculturally productive land is still owned by about 50,000 families, much of it taken during the colonization period and the rest expropriated during subsequent military dictatorships. Now this eastern zone of Bolivia proves to contain rich energy resources as well, adding to its economic advantage. So there is talk (and there are frequent public demonstrations) in support of autonomy, even secession here, but neither the haciendas nor the energy industry could function without the Amerindian labor force.

When Evo Morales was elected and began trying to gain control of the energy industry while making moves to alter the division of national wealth, the struggle between east and west took a new turn. When energy prices were high, there was plenty of money to spend on the social programs President Morales had promised to expand. But when, toward the end of the 2000s, the price of natural gas declined, the specter of state failure rose again. The combination of rising expectations among Amerindian citizens and devolutionary hopes among the minority mestizos continues to cast an ominous cloud over Bolivia's future.

Paraguay

Paraguay, Bolivia's landlocked neighbor to the southeast, is one of those transitional countries in-between regions, exhibiting properties of each (see Fig. 5B-1). Certainly, Paraguay (population: 6.5 million) is not an "Andean" country: it has no highlands of consequence. Its well-watered eastern plains give way to the dry scrub of the *Gran Chaco* in the west. Nor does it have clear, spatially entrenched ethnic divisions between Amerindians, mestizos, and others as do Bolivia and Peru. But Amerindian ancestry dominates the ethnic complexion of Paraguay, and Amerindian Guaraní is so widely spoken in the country that this is one of the world's most thoroughly bilingual societies. And continuing protests by landless peasants mirror those taking place elsewhere in the Andean West. Looking south, there is little in Paraguay's economic geography to compare to Argentina, Uruguay, or Chile, as Appendix B confirms. In a sense, Paraguay is a bridge between the Andean West and the Southern Cone, but not yet a heavily traveled one.

And Paraguay is transitional in yet another way: as many as 300,000 Brazilians have crossed the border to settle in eastern Paraguay, where they have created a thriving commercial agricultural economy that produces soybeans, livestock, and other farm products exported to or through Brazil. Brazil, of course, is the giant in the Mercosur/l free-trade zone, but Paraguay is in a geographically difficult position, often complaining that Brazil does not live up to its regional-trade obligations and creates unacceptable difficulties for Paraguayan exporters. Meanwhile, politicians raise fears that increasing Brazilian immigration is creating a foreign enclave within Paraguay, where people speak Portuguese (including in the local schools), the Brazilian rather than the Paraguayan flag flies over public buildings, and a Brazilian cultural landscape is evolving. Like Bolivia, Paraguay pays heavily for its landlocked weakness.

Paraguay's low GNI resembles that of countries of the Andean West, not the more advantaged Southern Cone region. This is also one of South America's least urbanized states, and poverty dominates the countryside as well as the slums encircling the capital, Asunción, and other towns (Fig. 5B-4). In 2007, nearly two-thirds of the population lived at or below the official poverty level. Even though records are inadequate, research suggests that 1 percent of the population owns about 75 percent of the land. That may be a record for inequality in the South American realm.

In 2008, Paraguay ended its history of ruthless dictatorial rule with the election of a radical, pro-Guaraní priest, Fernando Lugo, who in his days as a missionary supported hacienda invasions by landless peasants and promised land reform and other remedies for the poor. It was a momentous departure from the past. Lugo's election was a "first" in many ways: the first president in half a century who did not come from the establishment Colorado party; the first who was a cleric (in fact, a Roman Catholic bishop); the first who had been involved in, and inspired by, *liberation theology* (a blend of Christian religion and socialist ideology that interprets the teachings of Christ as a quest to liberate the poor from oppression); and the first who openly declared the relief of poverty of the masses as his overriding objective. No wonder, then, that Lugo's election brought great new hope to much of Paraguay's population.

Money for these remedies will be hard to come by, and it is probably too soon to judge his success. But in 2009 President Lugo scored a major success when he negotiated better terms with Brazil for the sale of Paraguay's share of electricity generated by Itaipu Dam located on the Paraná River between the two countries. When this huge dam's turbines began producing power in 1984, each country (then ruled by dictators) got 50 percent of the electricity, but severely underdeveloped Paraguay needed just a tiny fraction of that. So Paraguay's ruler agreed to sell to Brazil the remainder of his country's share at prices far below market value. Democratically elected Lugo persuaded Brazil's (also democratic) government to begin paying market rates for the 40 percent of Paraguay's Itaipu-generated electricity it buys each year, thereby adding substantially to the Paraguayan national income. Furthermore, Brazil will assist Paraguay in building a modern transmission line from Itaipu to Asunción, to be ready in 2012 and mark the start of a crucial expansion of Paraguay's electrical-power grid.

Another, and quintessentially geographical, problem stemming from Paraguay's long-term weakness and misrule lies in the southeast, where the borders of Brazil, Argentina, and Paraguay converge in a chaotic scene of smuggling, money-laundering, political intrigue, and even terrorist activity, centered on the town of Ciudad del Este. Locals call this the **4** Triple Frontier, and warning flags went up when reports of Hizbullah (Iranian-backed terrorist) activity were confirmed by the discovery of maps of the area in a Taliban safe house in Afghanistan. Paraguay's state system decidedly needs strengthening, and not just for domestic reasons.

THE SOUTHERN CONE

This region of South America, consisting of Argentina, Chile, and Uruguay, acquired its appellation from its tapered, ice-cream-cone shape (Fig. 5B-6). As noted earlier, Paraguay has strong links to this region and in some ways forms part of it, although social and economic contrasts between Paraguay and those in the Southern Cone remain sharp.

Since 1995, the countries of the Southern Cone have been drawing closer together in an economic union named **Mercosur** (*Mercado Común del Sur*), the hemisphere's second-largest trading bloc after NAFTA. Despite setbacks and disputes, Mercosur has expanded and today encompasses Argentina, Uruguay, Paraguay, and Brazil—and (pending ratification) Venezuela; Chile, Bolivia, Peru, Ecuador, Colombia, and Venezuela participate as associate members. In Portuguese-speaking Brazil, the organization is known as **Mercosul** (*Mercado Comum do Sul*).

Argentina

The largest Southern Cone country by far is Argentina, whose territorial size ranks second only to Brazil in this geographic realm; its population

FIGURE 5B-6

© H. J. de Blij, P. O. Muller, and John Wiley & Sons, Inc.

of 40.6 million ranks third after Brazil and Colombia. Argentina exhibits a great deal of physical-environmental variety within its boundaries, and the vast majority of the Argentines are concentrated in the physiographic subregion known as the **Pampa** (a word meaning "plain"). Figure 5A-3 underscores the degree of clustering of Argentina's inhabitants on the land and in the cities of the Pampa. It also shows the relative emptiness of the other six subregions (mapped in Fig. 5B-6): the scrub-forest **Chaco** in the northwest; the mountainous **Andes** in the west, along whose crestline lies the boundary with Chile; the arid plateaus of **Patagonia** south of the Rio Colorado; and the undulating transitional terrain of intermediate **Cuyo, Entre Rios** (also known as "Mesopotamia" because it lies between the Paraná and Uruguay rivers), and the **North**.

A Culture Urban and Urbane

Argentina once was one of the richest countries in the world. Its historic affluence is still reflected in its architecturally splendid cities whose plazas and avenues are flanked by ornate public buildings and private mansions. This is true not only of the capital, Buenos Aires, at the head of the Rio de la Plata estuary—it also applies to interior cities such as Mendoza and Córdoba. The cultural imprint is dominantly Spanish, but the cultural landscape was diversified by a massive influx of Italians and smaller but influential numbers of British, French, German, and Lebanese immigrants.

Argentina has long been one of the realm's most urbanized countries: 91 percent of its population is concentrated in cities and towns, a higher percentage even than western Europe or the United States. Almost one-third of all Argentinians live in metropolitan Buenos Aires, by far the leading industrial complex where processing Pampa products dominates. Córdoba has become the second-ranking industrial center and was chosen by foreign automobile manufacturers as the car-assembly center for the expanding Mercosur/l market. But what concentrates the urban populations is the processing of products from the vast, sparsely peopled interior: Tucumán (sugar), Mendoza (wines), Santa Fe (forest products), and Salta (livestock). Argentina's product range is enormous. There is even an oil reserve near Comodoro Rivadávia on the coast of Patagonia.

Economic Volatility

Despite all these riches, Argentina's economic history is one of boom and bust. With just over 40 million inhabitants, a vast territory with diverse natural resources, adequate infrastructure, and good international linkages, Argentina should still be one of the world's wealthiest countries, as it once was. But political infighting and economic mis-

management have combined to ruin a vibrant and varied economy.

Part of the problem, it appears, lies in the country's lopsided political geography. Argentina is a federal state consisting of the Buenos Aires Federal District and 23 provinces (Fig. 5B-7). As would be expected from what we have just learned, the urbanized provinces are populous, while the mainly rural ones have smaller populations— but the gap between Buenos Aires Province (whose capital is La Plata), with nearly 15 million, and Tierra del Fuego (capital: Ushuaia), with barely over 100,000, is wide indeed. Several other provinces contain under 500,000 inhabitants, so that the larger ones in addition to dominant Buenos Aires are also disproportionately influential in domestic politics, especially Córdoba and Santa Fe, both with capitals of the same name.

The never-ending problem for Argentina has been corrupt politics and associated mismanagement, at least since the mid-twentieth century. The darkest episode was the "Dirty War" of 1976–1983, in which a repressive military junta made more than 10,000 (and perhaps as many as 30,000) Argentinians disappear without a trace. In 1982, this ruthless military clique launched an invasion of the British-held Falkland Islands (which the Argentines call the Malvinas), resulting in a major defeat for Argentina (Fig. 5B-6). By the time civilian government replaced the discredited junta, inflation and national debt were soaring. Economic revival during the 1990s was followed by another severe downturn that exposed the flaws in Argentina's fiscal system, including scandalously inefficient tax collection and unconditional federal handouts to the politically powerful provinces.

The administration in power at the opening of this decade, led by the country's first elected female president, Cristina Fernandez de Kirchner (whose husband preceded her as

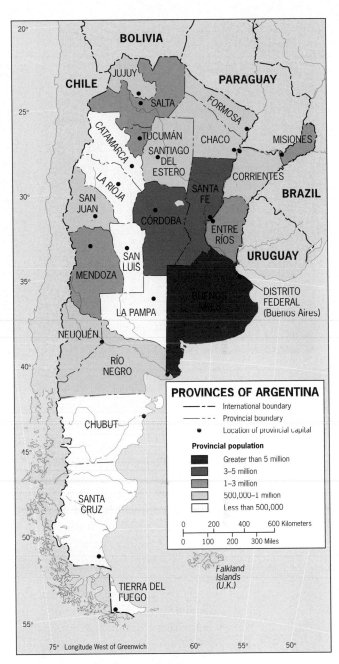

FIGURE 5B-7 © H. J. de Blij, P. O. Muller, and John Wiley & Sons, Inc.

president), was primarily focused on returning Argentina to fiscal responsibility (taming inflation) and economic stability. It also had been seeking to heal the festering wounds of the Dirty War through prosecution of the guilty and some form of reconciliation. But the president's popularity ratings in 2010 were quite low, and the majority of Argentines did not show signs of renewed hope. Although Argentina has the highest

© H. J. de Blij

"In a small, admission-charge private museum I found a treasure of historic information about this place in Chile—its environmental challenges ranging from droughts to earthquakes, its growth as a center for irrigated agriculture augmented by mining in the hinterland, its religious importance as the site of many churches and convents. But what matters today is tourism and retirement. Not only is La Serena (just inland from Coquimbo) on the international visitor circuit, but it lies at the center of a growing cluster of second-home and retirement complexes, benefiting from its equable climate and attractive scenery, and from the Southern Cone's comparative economic prosperity. So La Serena has become an amusement park, its stores converted to boutique shops, its historic homes on visitor tours, its plaza throbbing with the music of competing boomboxes, its streets clogged by buses. An older resident sitting on a park bench had his doubts. "All this started when they built this Pan-American Highway right past us," he said. "Now you can get here by sea, by road, even by air. Is this your first visit? You came too late." No doubt about it: what I saw here wasn't what I read in those faded pages in the museum."

www.conceptcaching.com

per-capita GNI in the realm, rising inflation and growing labor troubles cost her much public support. Moreover, foreign investors showed reluctance to assume risks in a country whose government has habitually ignored contract terms. And added to that were Argentina's continued uncertainties with regard to its energy supplies.

In the spring of 2010, the old Falklands conflict erupted once again when Britain began preparations for oil-drilling operations in the potentially rich seabed surrounding these distant South Atlantic islands. Buenos Aires registered strong protests and repeated its claims that the Malvinas should be returned to Argentina.

Chile

For 4000 kilometers (2500 mi) between the crestline of the Andes and the coastline of the Pacific lies the narrow strip of land that is the Republic of Chile (Fig. 5B-8). On average just 150 kilometers (90 mi) wide (and only rarely over 250 kilometers or 150 mi in width), Chile is

the world's quintessential example of what **5** elongation means to the functioning of a state. Accentuated by its north-south orientation, this severe territorial attenuation not only results in Chile extending across numerous environmental zones; it has also contributed to the country's external political, internal administrative, and general economic challenges. Nevertheless, throughout most of their modern history, the Chileans have made the best of this potentially centrifugal force: from the beginning, the sea has constituted an avenue of longitudinal communication; the Andes Mountains continue to form a barrier to encroachment from the east; and when confrontations loomed at the far ends of the country, Chile proved to be quite capable of coping with its northern rivals, Bolivia and Peru, as well as Argentina in the extreme south.

Three Subregions

As Figures 5B-6 and 5B-8 indicate, Chile is a three-subregion country. About 90 percent of its 17.1 million

people are concentrated in what is called Middle Chile, where Santiago, the capital and largest city, and Valparaíso, the chief port, are located. North of Middle Chile lies the Atacama Desert, wider, drier, and colder than the coastal desert of Peru. To the south of Middle Chile, the coast is punctuated by a plethora of fjords and islands, the topography is mountainous, and the climate—wet and cool near the Pacific—soon turns drier and colder against the Andean interior. South of the latitude of Chiloé Island, there are few permanent overland routes and hardly any settlements.

Not surprisingly, the land in Middle Chile is the most fertile and valuable, with hardly any agriculture in either North or South (Fig. 5A-5). The three subregions are also apparent on the realm's cultural map, with Europeans dominating the Middle, mestizos prevailing in the North, and Amerindians being the majority in the South (Fig. 5A-4).

Still, prior to the 1990s, the arid Atacama region in the north accounted for more than half of Chile's foreign revenues. It holds the world's largest exploitable deposits of nitrates, which was the country's economic mainstay before the discovery of methods of synthetic nitrate production a century ago. Subsequently, copper became the dominant export (Chile again possesses the world's largest reserves, which in 2008 accounted for more than half of all export revenues). It is mined in several places, but the main concentration lies on the eastern margin of the Atacama near Chuquicamata, not far from the port of Antofagasta. For most of the past decade, commodity prices were high and Chile's export revenues soared; but prices plunged sharply as global recession deepened during 2009, which underscored the risks involved—and the continuing need for economic diversification.

Political and Economic Success

Following the withdrawal of its brutal military dictatorship in 1990, Chile embarked on a highly successful program of free-market economic reform that brought stable growth, lowered inflation as well as unemployment, reduced poverty, and attracted massive foreign investment. The last is of particular significance because these new international connections enabled the export-led Chilean economy to diversify and develop in some badly needed new directions. Copper remains the single leading export, but many other mining ventures have been launched. In the agricultural sphere, fruit and vegetable production for export has soared because Chile's harvests coincide with the winter farming lull in the affluent countries of the Northern Hemisphere. Industrial expansion is taking place as well, though at a more leisurely pace, and new manufactures include a modest array of goods that range from basic chemicals to computer software.

Chile's increasingly globalized economy has propelled the country into a prominent role on the international economic scene. The United States, long Chile's leading trade partner, now is in second place compared to the Asian Pacific Rim, where China takes the bulk of Chile's exports followed by Japan and South Korea (Argentina remains Chile's leading source of imports, mostly energy, which is a potential problem given Argentina's own rising needs). Chile's regional commerce also is growing, and Chile has affiliated with Mercosur/l as an associate member.

The January 2010 elections in Chile saw Sebastián Piñera assume the presidency. Although a member of the Conservative Party whose election marked the end of two decades of center-left government (former President Michelle Bachelet had served the maximum two terms), as his first act he paid tribute to the successful policies that have made Chile South America's richest economy. Even though the global recession that began in 2008 did not leave the Chilean economy unaffected, it looks like Chile's new government will not modify what has been a winning formula for two decades: a pragmatic approach that eschews ideological extremism, that prioritizes transparency, and that balances economic growth with poverty reduction.

Indeed, the new regime's energies were required to totally focus on the aftermath of another major event of early 2010—one of most powerful earthquakes in the realm's modern history. This catastrophic temblor and accompanying tsunamis struck the coastal zone of southern Middle Chile on February 27, killing hundreds, causing massive destruction, and disrupting the lives of more than 2 million Chileans. Although recovery efforts were well under way by mid-2010, this disaster has clearly slowed progress toward Chile's goal of becoming the first South American country to join the ranks of the world's most developed economies before the end of this decade.

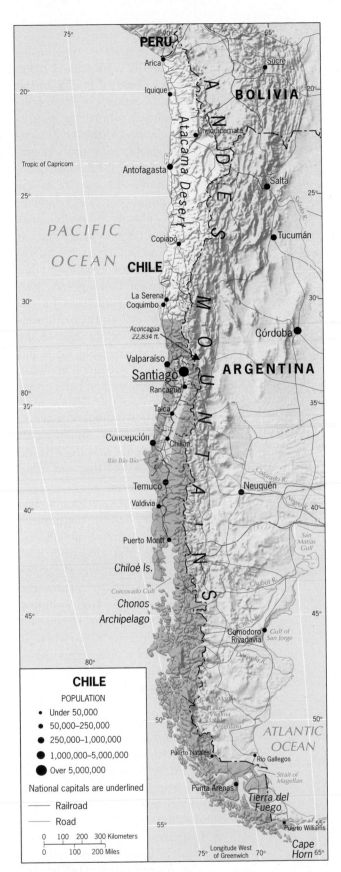

FIGURE 5B-8   © H. J. de Blij, P. O. Muller, and John Wiley & Sons, Inc.

Uruguay

Uruguay, in contrast to Argentina or Chile, is compact, small, and rather densely populated. This **6 buffer state** of old became a fairly prosperous agricultural country, in effect a smaller-scale Pampa (though possessing less favorable soils and topography). Montevideo, the coastal capital, contains almost 40 percent of the country's population of 3.3 million; from here, railroads and roads radiate outward into the productive agricultural interior (Fig. 5B-6). In the immediate vicinity of Montevideo lies Uruguay's major farming area, producing vegetables and fruits for the metropolis as well as wheat and fodder crops. Most of the rest of the country is used for grazing cattle and sheep, with beef products, wool and textile manufactures, and hides dominating the export trade. Tourism is another major economic activity as Argentines, Brazilians, and other visitors increasingly flock to the Atlantic beaches at Punta del Este and other thriving resort towns.

About the size of Florida but with less than one-fifth the population, there is plenty of agricultural potential here. But Uruguay's government seeks to diversify the economy, and one of its plans—the construction of two large cellulose (paper) factories on the Uruguayan side of the Uruguay River where it forms the border with Argentina—has caused a quarrel between the two Mercosur/l neighbors that has exposed some serious rifts. Montevideo is the administrative headquarters of Mercosur/l, but, like Paraguay, Uruguay has long felt neglected and even obstructed by its much larger neighbors. When Argentinians across the river began demonstrating, barricaded a bridge, and stopped taking vacations on Uruguay's beaches, they severely impacted the Uruguayan economy.

The Argentinians, who in any case dislike major foreign investments (the factories were to be built by Finnish and Spanish companies), argued that the paper mills would accelerate deforestation, cause river pollution, create acid rain, and damage the farming, fishing, and tourist industries. When the presidents of the two countries met to negotiate a solution and Uruguay's leader agreed to halt construction for a "study period," Uruguayan public opinion showed strong opposition to the compromise.

This issue reveals just how quickly nationalist feelings can overwhelm the need for international cooperation. It also shows that Mercosur/l partners are far from united on economic matters, and indicates that the collaboration implied by Mercosur/l membership is still a distant reality.

≡ BRAZIL: GIANT OF SOUTH AMERICA

The next time you board an airplane, don't be surprised if the aircraft was built in Brazil. When you go to the supermarket to buy provisions, take a look at their sources. Chances are some of them will come from Brazil. When you listen to your car radio, some of the best music you hear may have originated in Brazil. The emergence of Brazil as the regional superpower of South America and an economic superpower in the world at large is going to be one of the main stories of the first quarter of the twenty-first century. Along with Russia, India, and China, Brazil is designated as one of the four biggest emerging markets in the world, the **7 BRICs** (an acronym based on the first letter of this quartet of countries).

Why is Brazil so upward-bound today? In large measure it is because, after a long period of dictatorial rule by a minority elite that used the military to stay in power, Brazil embraced democratic government in 1989 and has not looked back since. The era of military coups and repeated crushing of civil liberties is over. In some ways, Brazil's political and economic turn-around runs parallel to that of Chile, but because of its sheer size, Brazil matters more to the rest of the realm and, indeed, the world.

The last ten years, in particular, have witnessed spectacular progress in Brazil, and much of this must be attributed to President Luiz Inácio Lula da Silva, popularly known as Lula, who was elected into office in 2002 and completed his second term in 2010. The leader of the Workers' Party, Lula was expected by many to pursue a leftist agenda that would harm business interests (in a show of displeasure at the election result in 2002, the United States did not send a single high-ranking representative to Lula's inauguration—an example of the United States' occasional poor political judgment in this realm). In fact, Lula so skillfully managed Brazil's numerous countervailing forces—thereby unleashing the country's regional and global potential—that many Brazilians dreaded his leaving office in 2010.

By any measure, Brazil is South America's giant. It is so large that it has common boundaries with all the realm's other countries except Ecuador and Chile (Fig. 5B-1). Its tropical and subtropical environments range from the equatorial rainforest of the Amazon Basin to the humid temperate climate of the far south. Territorially as well as in terms of population, Brazil ranks fifth in the world, and on both counts it represents just about half of the entire realm. The Brazilian economy, with its ultramodern industrial base, is now the ninth-largest on Earth—and climbing.

Population and Culture

Brazil's population grew rapidly during the world's twentieth-century population explosion. But over the past three decades, the rate of natural increase has slowed from nearly 3.0 percent to 1.4 percent, and the average number of children born to a Brazilian woman has been more than halved from 4.4 in

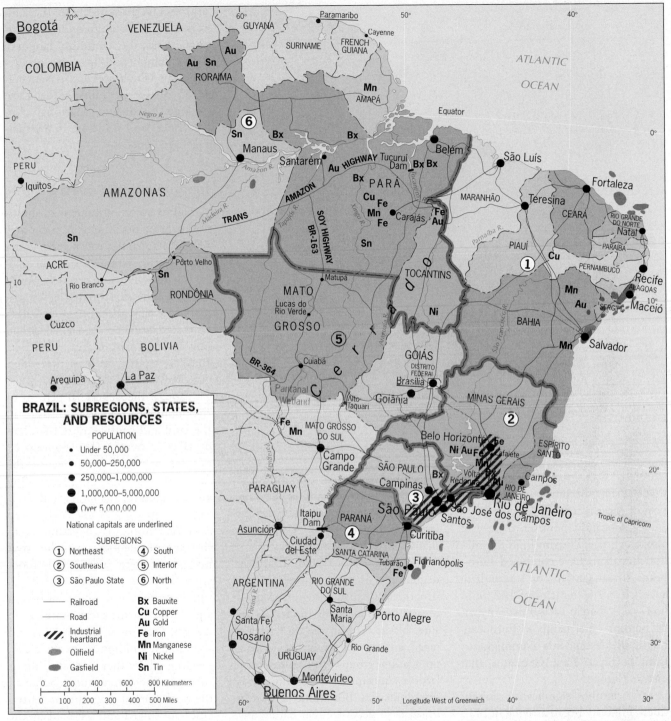

FIGURE 5B-9   © H. J. de Blij, P. O. Muller, and John Wiley & Sons, Inc.

1980 to 2.0 in 2009. It is a trend in line with Brazil's overall modernization in recent decades.

The population of just over 200 million is as diverse as that of the United States. In a pattern quite familiar in the Americas, the indigenous inhabitants of the country were decimated following the European inva-sion (fewer than 200,000 Amerindians now survive deep in the Amazonian interior). Africans came in great num-bers, too, and today there are about 13 million Afro-Brazilians in Brazil.

Brazil's culture is infused with African themes, a quality that has marked it from the very beginning. Three centuries ago, the Afro-Brazil-ian sculptor and architect affection-ately called Aleijadinho was Brazil's most famous artist. The world-famous composer Heitor Villa-Lobos used nu-merous Afro-Brazilian folk themes in his music. So many Africans were brought in bondage to the city and hinterland of Salvador in Bahia State (Fig. 5B-9) from what is today Benin

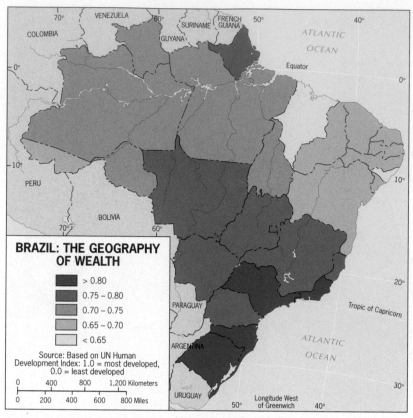

FIGURE 5B-10   © H. J. de Blij, P. O. Muller, and John Wiley & Sons, Inc.

(formerly Dahomey) in West Africa that Bahia has become a veritable outpost of African culture.

Significantly, however, there was also much racial mixing, and 86 million Brazilians have combined European, African, and minor Amerindian ancestries. The remaining 102 million—now barely in the majority at 51 percent—are mainly of European origin, the descendants of immigrants from Portugal, Italy, Germany, and eastern Europe.

Yet another significant, though small, minority began arriving in Brazil in 1908: the Japanese, who today are concentrated in São Paulo State. The more than 1 million Japanese-Brazilians form the largest ethnic Japanese community outside Japan, and in their multicultural environment they have risen to the top ranks of Brazilian society as business leaders, urban professionals, commercial farmers—and even as politicians in

the city of São Paulo. Committed to their Brazilian homeland as they are, the Japanese community also retains its contacts with Japan, resulting in many a trade connection.

Brazilian society, to a greater degree than is true elsewhere in the Americas, has made progress in dealing with its racial divisions. To be sure, blacks are still the least advantaged among the country's leading population groups, and community leaders continue to complain about discrimination. But ethnic mixing in Brazil is so pervasive that hardly any group is unaffected, and official census statistics concerning "blacks" and "Europeans" are meaningless.

What the Brazilians do have is a true national culture, expressed in an historic adherence to the Roman Catholic faith (its adherents now constitute about 73 percent of the population, a number that is steadily eroding under Protestant-evangelical

and secular pressures); in the universal use of a modified form of Portuguese as the common language ("Brazilian"); and in a set of lifestyles in which soccer, "beach culture," distinctive music and dance, and an intensifying national consciousness and pride are fundamental ingredients.

Inequality and Poverty

For all its accomplishments in multiculturalism, Brazil remains a country of stark, appalling social inequalities (Fig. 5B-10). Although such inequality is hard to measure precisely, South America is frequently cited as the geographic realm exhibiting the world's sharpest division between affluence and poverty. And within South America, Brazil is reputed to have the widest gap of all.

Today, the richest 10 percent of the population own two-thirds of all the land and control more than half of Brazil's wealth. The poorest 20 percent of the people live in the most squalid conditions prevailing anywhere on the planet, even including the megacities of Africa and Asia. According to UN estimates, in this age of adequate available (but not everywhere affordable) food, about half the population of Brazil suffers some form of malnutrition and, in the northeastern States, even hunger. Packs of young orphans and abandoned children roam the cities, sleeping where they can, stealing or robbing when they must. Some of the world's most magnificent central cities are ringed and sectored by the most wretched *favelas* (shantytowns) where poverty, misery, and crime converge.

The new left-of-center government in 2002 focused on two key areas: land reform and poverty reduction. At the turn of this century, 1 percent of Brazilians owned 50 percent of all productive land. But the government was able to settle only about 60,000 landless families per year, far below what was needed. By 2006, the government had found the resources to increase the rate of settlement, but the voters who elected President Lula da Silva had expected so much more.

In the poverty arena, the administration began a *Fome Zero* (Zero Hunger) program, but in the process created a huge, corruption-riddled bureaucracy. Officially, Brazil counts some 40 million people who live on half the minimum wage of about U.S. $80 per month or less, and the question was how to aid these poorest of the poor with government handouts that had to be modest because of budgetary constraints. By 2006, this muddled plan had been replaced by a much more efficient subsidy program that had significant results in the poorest States. This *Bolsa Familia* (Family Fund) plan gives families small amounts of cash to keep their children in school and to ensure their vaccinations against diseases that especially afflict the poor. In just a few years this program has become so successful that it now serves as a model for antipoverty campaigns in many other parts of the world.

Development Prospects

Brazil is richly endowed with mineral resources, including enormous iron and aluminum ore reserves, extensive tin and manganese deposits, and sizeable oil- and gasfields (Fig. 5B-9). Other significant energy developments involve massive new hydroelectric facilities and the successful substitution of sugarcane-based alcohol (gasohol) for gasoline, allowing well over half of Brazil's cars to use this biofuel instead of more expensive imported petroleum. Besides these natural endowments, Brazilian soils sustain a bountiful agricultural output that makes the country a global leader in the pro-

duction and export of coffee, soybeans, and orange juice concentrate. Commercial agriculture, in fact, is now the fastest growing economic sector, driven by mechanization and the opening of a major new farming frontier in the fertile grasslands of southwestern Brazil.

Industrialization has propelled Brazil's rise as a global economic power. Much of the momentum for this continuing development was unleashed in the early 1990s after the government opened the country's long-protected industries to international competition and foreign investment. These new policies are proving to be highly effective because productivity has risen by over a third since 1990, as Brazilian manufacturers attain world-class quality. During the mid-1990s, the revenues from industrial exports surpassed those from agriculture. Commerce with Argentina made that member of Mercosur/l one of Brazil's leading trade partners. On the global stage, Brazil became a formidable presence in other ways. The country's enormous and easily accessible iron ore deposits, the relatively low wages of its workers, and the mechanized efficiency of its steelmakers enable Brazil to produce that commodity at half the cost of steel made in the United States. This causes American steel producers to demand government protection through higher tariffs, which goes against purported U.S. principles of free trade.

A Promising Oil Future

Compared to the other BRICs, Brazil stands out in terms of its enormous and diverse natural resources. It has been paying for government programs with revenues mainly derived from commodities—iron ore, soybeans, coffee, orange juice, beef, sugar—that during the first decade of this century commanded high prices

on international markets. Economic geographers note that dependence on a singular commodity is always risky, as a country's well-being is highly reliant on world market prices for that product (which are, in turn, the combined result of supply and demand across the world). As Brazil diversified its economy in the past decade and increased its production in the manufacturing and high-tech sectors, it greatly improved its sustainability and competitiveness. In the global recession that began in 2008, Brazil was the last country in and the first one out: by early 2010, the country was on a roll, way ahead of, for instance, the United States or Europe.

On top of this sterling performance, Brazil can now add oil, lots of it, to its portfolio. In the past, a leading concern for Brazil was energy and its costs. Recessions were deepened by high oil and gas prices, and other than some small domestic reserves and those in neighboring Bolivia there were few nearby sources from which to acquire what was needed. But in 2009 the state-owned (but publicly traded) oil company, Petrobras, confirmed the discovery of an "enormous" oil reserve, quite possibly among the world's three largest, providing for an estimated 2 billion barrels. It was proof, said President Lula with a broad smile on his face, that "God is Brazilian."

Brazil in the 2010s is already self-sufficient in oil, and within the next few years it will become a major exporter, adding significantly to its revenues. The newfound reserves lie off the coast in a series of oilfields, fortuitously located not far from Rio de Janeiro and São Paulo (see Fig. 5B-9), and are being explored further as you read this. Although these oil deposits are huge and reserve estimates are constantly revised upward as new discoveries are made, they are buried far beneath the ocean floor. Thus exploitation will be costly and

São Paulo may not have an imposing skyline to match New York or a skyscraper to challenge the Sears (now Willis) Tower of Chicago, but what this Brazilian megacity does have is mass. São Paulo is more than just another city—it is a vast conurbation of numerous cities and towns containing more than 26 million inhabitants who constitute the third-largest human agglomeration on Earth. This view shows part of the razor-sharp edge of the CBD, a concrete jungle with little architectural distinction but vibrant urban cultures. In the foreground, juxtaposed against the affluence of downtown, the teeming inner city reflects the opposite end of the social spectrum—a chaotic jumble of *favelas* that mainly house the newest urban migrants, especially from Brazil's hard-pressed Northeast. São Paulo, demographers say, gives us a glimpse of the urban world of the future. © AP/Wide World Photos

require substantial foreign investment, but is likely to take Brazil's technology sector to an even more advanced level.

Brazil's Subregions

Brazil is a federal republic consisting of 26 States and the federal district of the capital, Brasília (Fig. 5B-9). As in the United States, the smallest States lie in the northeast and the larger ones farther west; their populations range from about 400,000 in the northernmost, peripheral Amazon State of Roraima to more than 40 million in burgeoning São Paulo State. Although Brazil is about as large as the 48 contiguous United States, it does not exhibit a clear physiographic regionalism. Even the Amazon Basin, which covers almost 60 percent of the country, is not entirely a plain: between the tributaries of the great river lie low but extensive tablelands. Given this physiographic ambiguity, the six Brazilian subregions discussed next have no absolute or even generally accepted boundaries. In Figure 5B-9, those boundaries have been drawn to coincide with the borders of States, making identifications easier.

The **Northeast** ① was Brazil's source area, its culture hearth. The plantation economy took root here at an early date, attracting Portuguese planters, who soon imported the country's largest group of African slaves to work in the sugar fields. But the ample rainfall occurring along the coast soon gives way to lower and more variable patterns in the interior, which is home to about half of the region's 50-plus million people. This drier inland backcountry—called the *sertão*—is not only seriously overpopulated but also contains some of the worst poverty to be found anywhere in the Americas. The Northeast produces less than one-sixth of Brazil's gross domestic product, but its inhabitants constitute more than one-fourth of the national population. Given this staggering imbalance, it is not surprising that this subregion contains half of the country's poor, a literacy rate 20 percent below Brazil's mean, and an infant mortality rate twice the national average.

Much of the Northeast's misery is rooted in its unequal system of land tenure. Farms must be at least 100 hectares (250 acres) to be profitable in the hard-scrabble *sertão*, a size that only large landowners can afford. Moreover, the Northeast is plagued by a monumental environmental problem: the recurrence of devastating droughts at least partly attributable to **8 El Niño** (periodic events of sea-surface warming off the continent's northwestern coast that skew weather patterns).

The Northeast today is Brazil's great contradiction. In cities such as Recife and Salvador, hordes of peasants driven from the land constantly arrive to expand the surrounding shantytowns. As yet, few of the generalizations about emerging Brazil apply here, but there are some bright spots. A petrochemical complex has been built near Salvador, creating thousands of jobs and luring foreign investment. Irrigation projects have nurtured a number of productive new commercial agricultural ventures. Tourism is booming along the entire Northeast coast, whose thriving beachside resorts attract tens of thousands of vacationing Europeans. Recife has spawned a budding software industry and a major medical complex. And Fortaleza is the center of new clothing and shoe industries that have already put the city on the global economic map.

The **Southeast** ② has been modern Brazil's *core area*, with its major cities and leading population clusters. Gold first drew many thousands of settlers, and other mineral finds also contributed to the influx—with Rio de Janeiro itself serving as the terminus of the "Gold Trail" and as the long-term capital of Brazil until 1960. Rio de Janeiro became the cultural capital as well, the country's most international center, *entrepôt*, and tourist hub. The third quarter of the twentieth century brought another mineral age to the Southeast, based on the iron ores around Lafaiete carried to the steelmaking complex at Volta Redonda (Fig. 5B-9).

The surrounding State of Minas Gerais (the name means "General Mines") formed the base from which industrial diversification in the Southeast has steadily expanded. The burgeoning metallurgical center of Belo Horizonte paved the way and is now the endpoint of a rapidly developing, ultramodern manufacturing corridor that stretches 500 kilometers (300 mi) southwest to metropolitan São Paulo (Fig. 5B-9).

São Paulo State ③ is the leading industrial producer and primary focus of ongoing Brazilian development. This economic-geographic powerhouse accounts for nearly half of the country's gross domestic product, with an economy that today matches Argentina's in overall size. Not surprisingly, this subregion is growing phenomenally (it already contains more than 20 percent of Brazil's population) as a magnet for migrants, especially from the Northeast.

The wealth of São Paulo State was built on its coffee plantations (known as *fazendas*), and Brazil is still the world's leading producer. But coffee today has been eclipsed by other farm commodities. One of them is orange juice concentrate (here, too, Brazil leads the world). São Paulo State now produces more than double the annual output of Florida, thanks to a climate all but devoid of winter freezes, to ultramodern processing plants, and to a fleet of specially equipped tankers that ship the concentrate to foreign markets. Another leading pursuit is soybeans, in which Brazil ranks second among the world's producers.

Matching this agricultural prowess is the State's industrial strength. The revenues derived from the coffee plantations provided the necessary investment capital, ores from Minas Gerais supplied the vital raw materials, São Paulo City's outport of Santos facilitated access to the ocean, and immigration flows from Europe, Japan, and other parts of Brazil contributed the increasingly skilled labor

force. As the capacity of the domestic market grew, the advantages of central location and large-scale agglomeration secured São Paulo's primacy. This also resulted in metropolitan São Paulo becoming the country's—and South America's—leading industrial complex and megacity (population: 26.2 million).

The South ④ consists of three States, whose combined population exceeds 27 million: Paraná, Santa Catarina, and Rio Grande do Sul (Fig. 5B-9). Southernmost Brazil's excellent agricultural potential has long attracted sizeable numbers of European immigrants. Here the newcomers introduced their advanced farming methods to several areas. Portuguese rice farmers clustered in the valleys of Rio Grande do Sul, where tobacco production has now propelled Brazil to become the world's number-one exporter. The Germans, specialists in raising grain and cattle, occupied the somewhat higher areas to the north and in Santa Catarina. The Italians selected the highest slopes, where they established thriving vineyards. All of these fertile lands proved highly productive, and with growing markets in the large urban areas to the north, this tri-state subregion became Brazil's most affluent corner.

With the South firmly rooted in the European/commercial agricultural sphere (Figs. 5A-4, 5A-5), European-style standards of living match the diverse Old World heritage that is reflected in the towns and countryside (where German and Italian are spoken alongside Portuguese). This has led to hostility against non-European Brazilians, and many communities actively discourage poor job-seeking migrants from the North by offering to pay return bus fares or even blocking their household-goods-laden vehicles. Moreover, extremist groups have proliferated to openly espouse the secession of the South from the rest of Brazil.

Economic development in the South is not limited to the agricul-

tural sector. Coal from Santa Catarina and Rio Grande do Sul is shipped north to the steel plants of Minas Gerais. Local manufacturing is growing as well, especially in Pôrto Alegre and Tubarão. During the 1990s, a major center of the computer software industry was established in Florianópolis, the island-city as well as State capital just off Santa Catarina's coast. Known as *Tecnópolis*, this budding technopole continues to grow by capitalizing on its seaside amenities, skilled labor force, superior air-travel and global communications linkages, and government and private-sector incentives to support new companies.

The Interior ⑤ subregion—constituted by the States of Goiás, Mato Grosso, and Mato Grosso do Sul—is also known as the **Central-West**. This is the region that Brazil's developers have long sought to make a part of the country's productive heartland, and in 1960 the new capital of Brasília was deliberately situated on its margins (Fig. 5B-9).

By locating the new capital city in the wilderness 650 kilometers (400 mi) inland from its predecessor, Rio de Janeiro, the nation's leaders dramatically signaled the opening of Brazil's development thrust toward the west. Brasília is noteworthy in another regard because it represents what political geographers call a **9 forward capital**. A state will sometimes relocate its capital to a sensitive area, perhaps near a peripheral zone under dispute with an unfriendly neighbor, in part to confirm its determination to sustain its position in that contested zone. Brasília does not lie near a contested area, but Brazil's interior was an internal frontier to be conquered by a growing nation. Spearheading that drive, the new capital occupied a decidedly forward position.

Despite the subsequent growth of Brasília to 3.8 million inhabitants today (which includes a sizeable ring of peripheral squatter settlements),

This is what the Amazon's equatorial rainforest looks like from an orbiting satellite after the human onslaught in preparation for settlement. The colors on this Landsat image emphasize the destruction of the trees, with the dark green of the natural forest contrasted against the pale green and pinks of the leveled forest. The linear branching pattern of deforestation here in Rondônia State's Highway BR-364 corridor is explained in the text. But farming here is not likely to succeed for very long, and much of the cleared land is likely to be abandoned. Then the onslaught will resume to clear additional land—as an entire ecosystem comes ever closer to failing forever. © NRSC LTS/Photo Researchers, Inc.

it was not until the 1990s that the Interior began its economic integration with the rest of Brazil. The catalyst was the exploitation of the vast **10** *cerrado*—the fertile savannas that blanket the Central-West and make it one of the world's most promising agricultural frontiers (at least two-thirds of its arable land still awaits development). As with the U.S. Great Plains, the flat terrain of the *cerrado* is one of its main advantages because it facilitates the large-scale mechanization of farming with a minimal labor force. Another advantage is rainfall, which is more prevalent there than in the Great Plains or Argentine Pampa.

The leading crop is soybeans, whose output per hectare here exceeds even that of the U.S. Corn Belt. Other grains and cotton are also expanding across the farmscape of the *cerrado*, but the current pace of regional development is inhibited by a serious accessibility problem. Thus the Interior's products must travel along poor roads and intermittent railroads to reach the markets and ports of the Atlantic seaboard. Today several projects are finally underway to alleviate these bottlenecks, including the privately financed *Ferronorte* railway that links Santos to the southeastern corner of Mato Grosso State, and the much-improved, so-called **Soy Highway** north to the Amazon River port city of Santarém.

The North ⑥ is Brazil's territorially largest and most rapidly developing subregion, which consists of the seven States of the Amazon Basin (Fig. 5B-9). This was the scene of the great rubber boom a century ago, when the wild rubber trees in the *selvas* (tropical rainforests) produced huge profits and the central Amazon city of Manaus enjoyed a brief period of wealth and splendor. But the rubber boom ended in 1910, and for most of the seven decades that followed, Amazonia was a stagnant hinterland lying remote from the centers of Brazilian settlement. However, all that changed quite dramatically during the 1980s as new development began to stir throughout this awakening subregion, which currently is the scene of the world's largest migration into virgin territory. More than 200,000 new settlers arrive each year. The North of Brazil is well known for its high **deforestation** rates (an issue discussed in Chapter 4A). Removing the forest results directly from logging operations, but more of it is a matter of urbanization, land occupation, use by settlers, and the arrival of large-scale agribusiness.

Development projects abound in the Amazonian North. One of the most durable is the **Grande Carajás Project** in southeastern Pará State, a huge multifaceted scheme centered on one of the world's largest known deposits of iron ore in the hills around Carajás (Fig. 5B-9). In addition to a vast mining complex, other new construction here includes the Tucuruí Dam on the nearby Tocantins River and an 850-kilometer (535-mi) railroad to the Atlantic port of São Luís. This ambitious development project further emphasizes the exploitation of additional minerals, cattle raising, crop farming, and forestry. What is taking place here is a manifestation of the **11** **growth-pole concept**. A growth pole is a location where a set of activities, given a start, will expand and generate widening ripples of development in the surrounding area. In this case, the stimulated hinterland could one day cover one-sixth of all Amazonia.

Understandably, tens of thousands of settlers have descended on this part of the Amazon Basin. Those seeking business opportunities have been in the vanguard, but they have been followed by masses of lower-income laborers and peasant farmers in search of jobs and land ownership. The initial stage of this colossal enterprise has boosted the fortunes of many towns, particularly Manaus northwest of Carajás. Here, a thriving industrial complex (specializing in the production of electronic goods) has emerged in the free-trade zone adjoining the city thanks to the outstanding air-freight operations at Manaus's ultramodern airport. But many problems have also arisen as the tide of pioneers rolled across central Amazonia. One of the most tragic involved the Yanomami people, whose homeland in Roraima State was overrun by thousands of claim-stakers (in search of newly discovered gold), who triggered violent confrontations that ravaged the fragile aboriginal way of life.

Another leading development scheme, known as the **Polonoroeste Plan**, is located about 1600 kilometers (1000 mi) to the southwest of Grande Carajás in the 2400-kilometer (1500-mile)-long Highway BR-364 corridor that parallels the Bolivian border and connects the western Brazilian towns of Cuiabá, Pôrto Velho, and Rio Branco (Fig. 5B-9). Even though the government had planned for the penetration of western Amazonia to proceed via the east-west Trans-Amazon Highway, the migrants of the 1980s and 1990s preferred to follow BR-364 and settle within the Basin's southwestern rim zone, mostly in Rondônia State. Agriculture has been the dominant activity here, but in the quest for land, bitter conflicts continue to break out between peasants and landholders as the Brazilian government pursues the volatile issue of land reform.

Brazil is the cornerstone of South America, the dominant economic force in Mercosur/l, the only dimensional counterweight to the United States in the Western Hemisphere, a maturing democracy, and an emerging global giant. The future of the entire South American realm depends on Brazil's stability and social as well as economic progress.

POINTS TO PONDER

- The United States consumes the bulk of cocaine produced in Colombia. Is the United States to blame for the dislocation caused by cocaine in that country?
- Political discord and economic contrasts are threatening the landlocked state of Bolivia.
- Watch for Argentina's renewed claims to "their" Malvinas as the British drill for oil near the Falkland Islands.
- If you are of voting age in Brazil, you must vote. It's compulsory. What are the (dis)advantages?

South Africa's Cape Town, whose site is dominated by Table Mountain, was founded by the Dutch in 1652 as a waystation for empire building in the East Indies. © H. J. de Blij

IN THIS CHAPTER

Heart of the world, cradle of humanity
Africa's unique physical geography
Environmental dimensions of health and disease
Evolution of the rich and varied African cultural mosaic
Why Africa remains in the grip of poverty
Consequences of rapid urban growth in the world's least urbanized realm

CONCEPTS, IDEAS, AND TERMS

6A

SUBSAHARAN AFRICA: DEFINING THE REALM

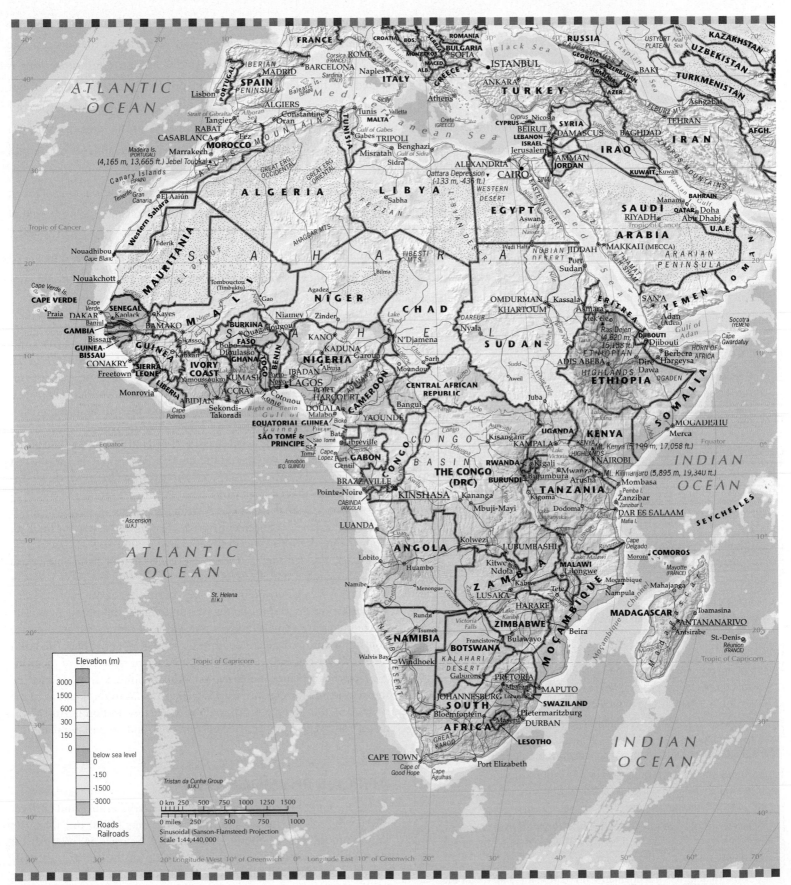

FIGURE 6A-1

© H. J. de Blij, P. O. Muller, and John Wiley & Sons, Inc.

The African continent occupies a special place in the physical as well as the human world. You need a globe to confirm the former: the Earth has a land hemisphere and a water hemisphere, and Africa lies at the center of the land hemisphere, surrounded in all directions by other landmasses. As for Africa's human significance, this is where **1** **human evolution** began. In Africa we formed our first communities, spoke our first words, and created our first art. From Africa our ancestor hominids spread outward into Eurasia more than 2 million years ago. From Africa our own species emigrated, beginning perhaps 130,000 years ago, northward into present-day Europe and eastward via southern Asia into Australia and, much later, farther afield into the Americas. Disperse our forebears did, but we should remember that, at the source, we are all Africans.

For millions of years, therefore, Africa served as the cradle for humanity's emergence. For tens of thousands of years, Africa was the source of human cultures. For thousands of years, Africa led the world in countless spheres ranging from tool manufacture to plant domestication to forging trade relationships. But in Chapters 6A and 6B we will encounter an Africa that has been struck by a series of disasters ranging from environmental deterioration to human dislocation on a scale unmatched anywhere in the world. When we assess Africa's misfortunes, though, we should remember that they have lasted hundreds, not thousands or millions of years. Africa's catastrophic interlude will end, and Africa's time and turn will come again.

 ## MAJOR GEOGRAPHIC QUALITIES

SUBSAHARAN AFRICA

1. Physiographically, Africa is a plateau continent without a major mountain range, with a set of Great Lakes, several major river basins, variable rainfall, generally low-fertility soils, and mainly savanna and steppe vegetation.

2. Dozens of nations, hundreds of ethnic groups, and many smaller entities make up Subsaharan Africa's culturally rich and varied population.

3. Most of Subsaharan Africa's peoples depend on farming for their livelihood.

4. Health and nutritional conditions in Subsaharan Africa need improvement as the incidence of disease remains high and diets are often unbalanced. The AIDS pandemic continues to be a major health crisis in this realm.

5. Africa's boundary framework is a colonial legacy; many boundaries were drawn without adequate knowledge of or regard for the human and physical geography they divided.

6. The realm is rich in raw materials vital to industrialized countries, but many economies continue to rely on primary activities—the extraction of resources—and not the better income-generating activities of assembly and manufacturing.

7. Patterns of raw-material exploitation and export routes set up in the colonial period still prevail in most of Subsaharan Africa. Interregional and international connections are poor.

8. During the Cold War, great-power competition magnified conflicts in several Subsaharan African countries, with results that will be felt for generations.

9. Severe dislocation affects many Subsaharan African countries, from Liberia to Rwanda to Zimbabwe. This realm has the largest refugee population in the world today.

10. Government mismanagement and poor leadership afflict the economies of many Subsaharan African countries.

The focus in this chapter will be on Africa south of the Sahara, for which the unsatisfactory but convenient name **Subsaharan Africa** has come into use to signify not physically "under" the great desert but directionally south of it. The African continent contains two geographic realms: the African, extending from the southern margins of the Sahara to the Cape of Good Hope, and the western flank of the realm dominated by the Muslim faith and Islamic culture whose heartland lies in the Middle East and the Arabian Peninsula. The great desert forms a formidable barrier between the two, but the powerful influences of Islam crossed it centuries before the first Europeans set foot in West Africa. By that time, the African kingdoms in what is known today as the Sahel had been converted, creating an Islamic foothold all along the northern periphery of the African realm (see Fig. G-3). As we note later, this cultural and ideological penetration had momentous consequences for Subsaharan Africa.

The African continent may be partitioned into two human-geographic realms, but the landmass is indivisible. Before we investigate the human geography of Subsaharan Africa, therefore, we should take note of the entire continent's unique physical geography (Figs. 6A-1 and 6A-2). We have already noted Africa's situation at the center of the planet's land hemisphere; moreover, no other landmass is positioned so squarely astride the equator, reaching almost as far to the south as to the north. This location has much to do with the distribution of Africa's climates, soils, vegetation, agricultural potential, and human population.

AFRICA'S PHYSIOGRAPHY

Africa accounts for about one-fifth of the Earth's entire land surface. The north coast of Tunisia lies 7700 kilometers (4800 mi) from the southernmost coast of South Africa. Coastal Senegal, in West Africa, lies 7200 kilometers (4500 mi) from the tip of the *Horn* in easternmost Somalia. These distances have major environmental implications. Much of Africa is far from maritime sources of moisture. In addition, as Figure G-7 shows, large parts of the landmass lie in latitudes where

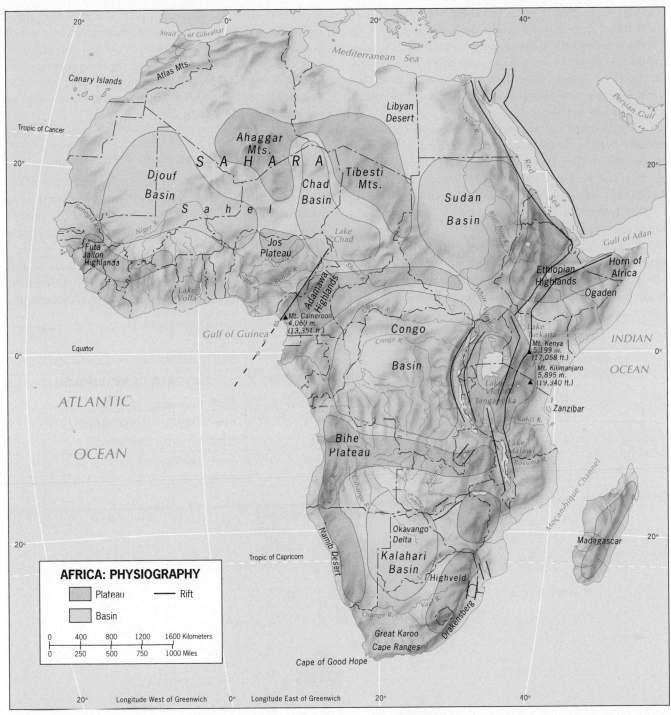

FIGURE 6A-2

© H. J. de Blij, P. O. Muller, and John Wiley & Sons, Inc.

global atmospheric circulation systems produce arid conditions. The Sahara in the north and the Kalahari in the south form part of this globe-girdling desert zone. Water supply is one of Africa's great problems.

Rifts and Rivers

Africa's topography reveals several properties that are not replicated on other landmasses. Alone among the conti-

nents, Africa does not have a mountain backbone; neither the northern Atlas nor the southern Cape Ranges are in the same league as the Andes or Himalayas. Where Africa does have high mountains, as in Ethiopia and South Africa, these are deeply eroded plateaus or, as in East Africa, high snowcapped volcanoes. Furthermore, Africa is one of only two continents containing a cluster of Great Lakes, and the only one whose lakes result from powerful tectonic forces in the Earth's crust. These lakes (with the exception of Lake

Victoria) lie in deep trenches called **2** **rift valleys**, which form when huge parallel cracks or faults appear in the Earth's crust and the strips of crust between them sink, or are pushed down, to form great, steep-sided, linear valleys. In Figure 6A-2 these rift valleys, which stretch more than 9600 kilometers (6000 mi) from the Red Sea to Swaziland, are marked by red lines.

Africa's rivers, too, are unusual: their upper courses frequently bear landward, seemingly unrelated to the coast toward which they eventually flow. Several rivers, such as the Nile and the Niger, have inland as well as coastal deltas. Major waterfalls, notably Victoria Falls on the Zambezi, or lengthy systems of cataracts, separate the upper from the lower courses.

Finally, Africa may be described as the plateau continent. Except for some comparatively limited coastal plains, almost the entire continent lies above 300 meters (1000 ft) in elevation, and fully half of it lies over 800 meters (2500 ft)

high. As Figure 6A-2 shows, the plateau surface has sagged under the weight of accumulating sediments into a half dozen major basins (three of them in the Sahara). The margins of Africa's plateau are marked by escarpments, often steep and step-like. Most notable among these is the Great Escarpment of South Africa, marking the eastern edge of the Drakensberg Mountains.

Continental Drift and Plate Tectonics

Africa's remarkable and unusual physiography was a key piece of evidence that geographer Alfred Wegener used to construct his hypothesis of **3** **continental drift**. The present continents, Wegener reasoned, lay assembled as one giant landmass called *Pangaea* not very long ago in geologic time (220 million years ago). The southern part of this supercontinent was *Gondwana*, of which Africa formed the core (Fig. 6A-3). When, about 200 million years ago, tectonic

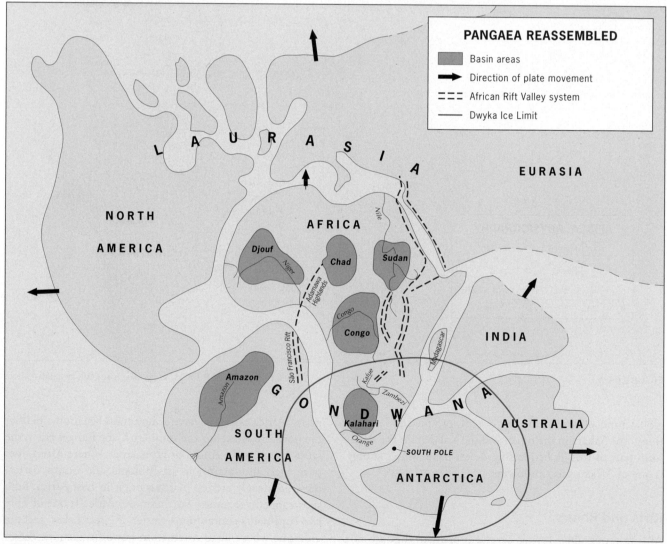

FIGURE 6A-3

© H. J. de Blij, P. O. Muller, and John Wiley & Sons, Inc.

forces began to split Pangaea apart, Africa (and the other landmasses) acquired their present configurations. That process, now known as **plate tectonics**, continues and is marked by earthquakes and volcanic eruptions. By the time it started, however, Africa's land surface had begun to acquire some of the features that mark it today—and make it unique. The rift valleys, for example, demarcate the zones where plate movement continues—hence the linear shape of the Red Sea, where the Arabian Plate is separating from the African Plate (Fig. G-4). And yes, the rift valleys of East Africa probably mark the further fragmentation of the African Plate (some geophysicists have already referred to a Somali Plate, which will separate, Madagascar-like, from the rest of Africa).

Africa's ring of escarpments, its rifts, its river systems, its interior basins, and its lack of significant mountains, all relate to the continent's central location in Pangaea, all pieces of the puzzle that led to the plate-tectonics solution.

NATURAL ENVIRONMENTS

Only the southernmost tip of Subsaharan Africa lies outside the tropics. Although African elevations are comparatively high, they are not high enough to ward off the heat that comes with tropical location except in especially favored locales such as the Kenya Highlands and parts of Ethiopia. And, as we have noted, Africa's bulky shape means that much of the continent lies far from maritime moisture sources. Variable weather and frequent droughts are among Africa's environmental challenges.

It is useful at this point to refer back to Figure G-7. As that map shows, Africa's climatic regions are distributed almost symmetrically about the equator, though more so in the center of the landmass than in the east, where elevation changes the picture. The hot, rainy climate of the Congo Basin merges gradually, both northward and southward, into climates with distinct winter-dry seasons. Winter, however, is marked more by drought than by cold. In parts of the area mapped **Aw** (savanna), the annual seasonal cycle produces two rainy seasons, often referred to locally as the long rains and the short rains, separated by two winter dry periods. As you go farther north and south, away from the moist Congo Basin, the dry season(s) grow longer and the rainfall diminishes and becomes less and less dependable.

End of an Era

Africa's shrinking rainforests and vast savannas form the world's last refuges for wildlife ranging from primates to wildebeests. Gorillas and chimpanzees survive in dwindling numbers in threatened forest habitats, while millions of herbivores range in great herds across the savanna plains where people compete with them for space. European colonizers, who introduced hunting as a "sport" (a practice that was not part of African cultural traditions) and who brought their capacities for mass destruction to animals as well as people in Africa, helped clear vast areas of wildlife and push species to near-extinction. Later, they laid out game reserves and other types of conservation areas, but these were not sufficiently large or well enough connected to allow herd animals to follow their seasonal and annual migration routes. The same climatic variability that affects farmers also affects wildlife, and when the rangelands wither, the animals seek better pastures. When the fences of a game reserve wall them off, they cannot survive. When there are no fences, the wildlife invades neighboring farmlands and destroys crops, and the farmers retaliate. After thousands of years of equilibrium, the competition between humans and animals in Africa has taken a new turn. It is the end of an era.

Wildlife Management and Tourism

African governments recognize the importance of nature conservation and benefit from the revenues generated by tens of thousands of tourists who annually come to view Africa's incomparable combination of natural landscapes and varied wildlife. But managing the sometimes conflicting needs and demands of sedentary local people, migratory animals, and tourist facilities (such as game lodges and artificial water holes to attract animals) is often difficult. The simple notion of fencing animals in and people out obviously does not always work; not only do many wild animals migrate seasonally, but so must the livestock of pastoralists. Maintain a game reserve on ancestral land, and you had better ensure that the local inhabitants have a financial stake in its preservation. In Kenya and Tanzania you can see local herders drive their livestock inside game reserves, across the same pastures that sustain wildlife. And governments have to combat poaching to protect the value of their wildlife heritage.

It might appear obvious that tourism is a benign industry whose revenues help sustain local ecologies, but there is another side to this. Tourist interest tends to focus on seeing the "big four" during short visits, including elephants. But attracting elephant herds through artificial water supply results in habitat change that affects (and can drive away) other wildlife. From ongoing research work by geographers and other scientists, it is clear that ecosystem modification by tree-trampling elephant herds is converting African woodland into shrub or grassland, driving away grazing animals that require particular plant species and allowing less choosy browsers to take over. Thus the concentration of one species, itself in need of protection and much in tourist demand, has a negative effect on wildlife diversity and richness over wide areas.

When ongoing climate change creates severe and prolonged droughts, as parts of Subsaharan Africa have experienced over the past decade, the problems of wildlife

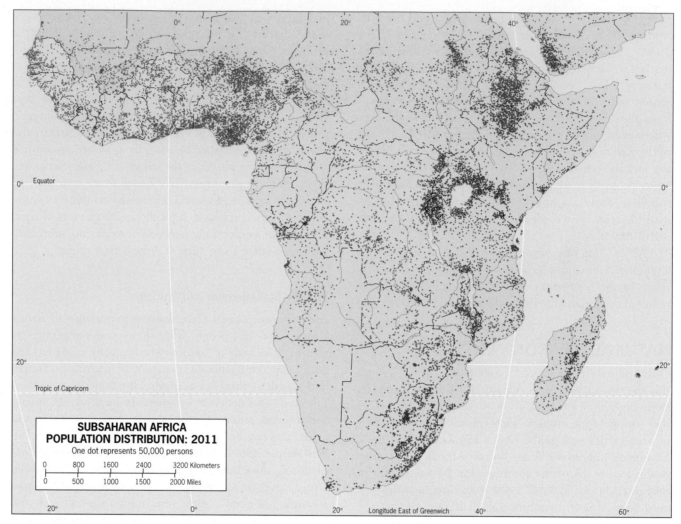

FIGURE 6A-4

© H. J. de Blij, P. O. Muller, and John Wiley & Sons, Inc.

management become even more difficult. The ability of national governments to protect this natural resource is severely constrained.

People, Farmlands, and Environments

And yet it would seem that there could be space for wildlife as well as humans in Subsaharan Africa. Figure 6A-4 is not a picture of a densely peopled realm: while there are major clusters of population in West Africa (where Nigeria is the realm's most populous country), East Africa (encircling Lake Victoria), and the Horn, where Ethiopia's highlands sustain a large concentration, most of the rest of the realm seems relatively sparsely populated. Our Data Table in Appendix B bears this impression out: all the countries of Subsaharan Africa *combined* have a population only about half of China's alone.

But much of this population continues to depend on farming for their livelihood, and we have already noted that African environments are difficult for millions of farmers. Not everywhere, of course: Africa has its areas of good soil,

ample water, and high farm productivity. But on the map, these areas are not extensive. The major population concentration in the Great Lakes region reflects the volcanic soils of Mount Kilimanjaro and those of the Western Rift Valley zone. The Ethiopian Highlands were once known as the "breadbasket of northeast Africa" and could be capable of much greater productivity than is presently the case. Higher-latitude, cooler and moister areas of South Africa are exceptionally productive with a large range of farm produce. And parts of West Africa sustain intensive farming. Taken together, though, these productive areas do not add up to the huge, fertile alluvial basins of China or India, or even the (North African) Nile Valley and Delta. Most African farmers face daunting challenges.

These challenges include: (1) climatic variability, (2) the economic policies of national governments, and (3) the difficulties African farmers have in reaching world markets. Look again at Figure G-7, and you can see the relatively short distances involved as the moist tropics of the equatorial Congo Basin give way to the deserts of the north and south. Moreover, as the annual rainfall total declines, its variability increases, so that farmers in the drier zones cannot count on sufficient moisture in any given year.

The economic policies of national governments frequently disadvantage farmers as the prices of their products are kept artificially low to please urban (and politically powerful) consumers. We return to this topic later, but the regional effect on African agriculture has been to constrain productivity.

And in this era of globalization, African farmers often get a raw deal on world markets. For all the talk of free trade in the wealthy global core, many governments favor their local farmers over those of Africa (and other areas of the global periphery), from rice subsidies in Japan to "market support" in France. Economists estimate that such unfree trade costs African farmers as much as U.S. $200 billion annually.

AFRICANS AND THEIR LAND

While rainfall is a critical physical criterion for farming, a number of political and economic factors are also influential. Among these factors are land tenure; type of agricultural system (e.g., subsistence or commercial); type of farming system (rotational, shifting cultivation, intercropping, etc.); crop prices; government policies promoting the planting of one crop in preference to another; indigenous agricultural knowledge of particular crops; and the degree of technology adoption and mechanization.

The issue of land tenure is crucial in Africa because most Africans are still farmers. **4 Land tenure** refers to the way people own, occupy, and use land. African traditions of land tenure are different from those of Europe or the Americas today. In most of Subsaharan Africa, communities, not individuals, customarily hold land. Occupants of the land have temporary, custodial rights to it and cannot sell it. Land may be held by large (extended) families, by a village community, or even by a traditional chief who holds the land in trust for the people. His subjects may house themselves on it and farm it, but in return they must follow his rules.

Stolen Lands

At the onset of **5 colonialism**, colonial administrators intended to control the most fertile areas of the occupied colonies through eviction of indigenous peoples. In some cases this was done by physical force (mostly through military conquest), and in other areas by coercion. Prior to colonialism, many traditional African livelihood systems adhered to sustainable land management practices such as leaving the land fallow or rotational grazing. Viewing this land as unused, colonial planners initiated a process of **6 land alienation** not unlike what happened in Middle and South America. Many of the most fertile and productive areas were placed under the control of colonial settlers and governments. Over time, these lands were bought and sold and became private legal property. At the end of colonialism, many newly independent African countries started a program whereby land would revert to traditional forms of ownership and management. However, the legacy of colonialism has been difficult to overcome, and African governments have adopted several neocolonial policies on land management that continue to marginalize farmers.

Rapid population growth, as Africa experienced early in the postcolonial period, makes access to land even more complicated. Traditional systems of land use, which involve subsistence farming in various forms ranging from shifting cultivation to pastoralism, work best when the population is fairly stable and tenure is communal. Land must be left fallow to recover from cultivation, and pastures must be kept free of livestock so that the grasses can revive. The population explosion that Africa experienced during the mid-twentieth century set in motion a cycle of land overuse. When soils cannot rest and pastures are overgrazed, the land becomes degraded and yields decline.

Persistent Subsistence

Although there is commercial farming in parts of Africa, most African farmers remain subsistence farmers who grow grain crops (maize [corn], millet, sorghum) in drier areas and root crops (yams, manioc [cassava], sweet potatoes) where moisture is ample. Others herd livestock, mostly cattle and goats, as they counter environmental and climatic variability. Farmers and pastoralists alike have been unable to secure access to regular markets and stable prices for their products because of government policies that often promote one particular type of export-oriented crop (for example, peanuts in Gambia, tea and coffee in Kenya, or cacao—cocoa—in the Ivory Coast) over other crops that do not fetch high prices on the world market.

Many African farmers have had to adapt to these problems; nonetheless, farm yields in Africa have been modest for many years. Governments, in response to World Bank policies, paid greater attention to high-profile industrial projects and as a result have neglected agriculture. The decline of African agriculture has been disproportionally hard on Africa's women who, according to current estimates, produce 75 percent of local food in Subsaharan Africa. Development policies often pay little attention to this situation and the related household dynamics, thereby dooming many well-intended projects.

The **7 Green Revolution**—the development of more productive, drought-tolerant, pest-resistant, higher-yielding types of grain (discussed at greater length in Chapter 8B)—has had less impact in Africa than elsewhere. Where people depend mainly on rice and wheat, the Green Revolution pushed back the prospect of increased hunger. But little research was done on Africa's dominant crops, mostly tubers, so the advances of the Green Revolution were minimal for

The cell phone revolution has made a dramatic impact on farmers in many parts of the developing world, especially in Subsaharan Africa where distances can be far, land lines absent, and market information scarce. Now farmers and market women, such as the Kenyan woman pictured here, can be in touch and can better gauge when it is best to market crops. © AP/Wide World Photos

Africa. The Green Revolution is also not an unqualified remedy: the poorest farmers, who need help the most, can least afford the more expensive, higher-yielding seeds or the pesticides that are often required.

In spite of the obstacles that most African farmers face, a second revolution is occurring in much of Subsaharan Africa and elsewhere in the developing world. The **8 cell phone revolution** is allowing many farmers to send text (or SMS— short-message-system) messages to obtain critical information such as weather alerts or forecasts, and the prices for agricultural products such as the most up-to-date price estimate for a kilogram of maize or a sack of potatoes at the nearest market. In the past, African farmers were constrained by poor access to reliable market information. The cell phone revolution allows them to negotiate higher prices for their products.

HIV/AIDS (to be discussed in the next section) presents yet another serious challenge to farming and natural resource management in Subsaharan Africa. This disease has caused the rapid debilitation and death of so many farmers of working age that both productivity and the knowledge systems that contributed to agriculture and natural resource management are seriously threatened.

ENVIRONMENT AND HEALTH

The study of human health in spatial context is the field of **9 medical geography**, and medical geographers employ modern methods of analysis (including geographic information systems) to track disease outbreaks, identify their

sources, detect their carriers, and prevent their repetition. Alliances between medical personnel and medical researchers, and geographers have already yielded significant results. Doctors know how a disease ravages the body; geographers know how climatic conditions such as wind direction or variations in river flow can affect the distribution and effectiveness of disease carriers. This collaboration helps protect vulnerable populations.

Tropical Africa is the source area of many serious illnesses and has thereby become the focus of much research in medical geography. Researchers look at the carriers (*vectors*) of infectious diseases, the environmental conditions that give rise to them, and also the cultural and social geography of disease dispersion and transmission. Comparing medical, environmental, and social/cultural maps can lead to crucial evidence that helps combat the scourge.

In Africa today, hundreds of millions of people carry one or more maladies, often without knowing exactly what ails them. A disease that infects many people (the *hosts*) in a kind of equilibrium, without causing rapid and widespread deaths, is said to be **10 endemic** to that population. People affected may not die suddenly or dramatically, but their quality of life and productive capacity are hindered as their overall health is weakened and can deteriorate quickly when an acute illness strikes. In tropical Africa, hepatitis, venereal diseases, and hookworm are among many public-health threats in this category.

Epidemics and Pandemics

When a disease outbreak has local or regional dimensions, it is called **11 epidemic**. It may claim thousands, even tens of thousands, of lives, but it remains confined to a certain area, perhaps one defined by the range of its vector. In tropical Africa, trypanosomiasis, the disease known as sleeping sickness and vectored by the tsetse fly, has regional proportions. The great herds of savanna wildlife form the *reservoir* of this disease, and the tsetse fly transmits it to livestock and people. It is endemic to wildlife, but it also kills cattle, so Africa's herders try to keep their animals in tsetse-free zones. African sleeping sickness appears to have originated in a West African source area during the fifteenth century, and it spread throughout much of tropical Africa. Its epidemic range was limited by that of the tsetse fly: where there are no tsetse flies, there is no sleeping sickness. More than anything else, the tsetse fly has kept significant parts of Subsaharan Africa's savannas free of livestock and open to wildlife.

When a disease spreads worldwide, it is described as a **12 pandemic**. Africa's and the world's most deadly vectored disease is malaria, transmitted by a mosquito and killer of as many as 1 million children each year. Malaria is an ancient affliction. Hippocrates, the Greek physician of the fifth century BC, mentions it in his writings. Apes, monkeys, and several other species also suffer from it. Fever attacks, anemia, and enlargement of the spleen are its symptoms. Malaria has dispersed around the world and prevails not only in tropical but also in temperate areas. Eradication campaigns against the mosquito vector have had some success, but always the carrier has come back with renewed vigor. At present, as many as 250 million people are affected by malaria globally. Current efforts at fighting the disease through an increased use of mosquito nets appears to be having good results, yet malaria causes at least 1 million deaths each year. The short life expectancies for tropical African countries reported in Appendix B in part reflect infant and child mortality from malarial infection.

Africa's Latest Scourge

Malaria remains Africa's most deadly disease to this day, but during the past three decades another disease, known as AIDS, has dominated the medical news from this part of the world. Although AIDS first erupted in Subsaharan Africa and quickly assumed epidemic proportions there, this terrible malady is now a true pandemic. No geographic realm has been spared.

AIDS stands for Acquired Immune Deficiency Syndrome, the body's failure to protect itself against a virus. That virus, for want of a better name at the time researchers were trying to identify it, is called the Human Immunodeficiency Virus—HIV. Thus the disease is properly called HIV/AIDS. Since this disease was first recorded in the early 1980s, more than 30 million people worldwide have died of it, between 75 and 80 percent of them Africans.

By the early 1990s, HIV/AIDS had spread most virulently in Equatorial and East Africa, and medical geographers referred to an "AIDS Belt" from The Congo* to Kenya. A decade later, however, the worst-afflicted countries lay in Southern Africa, where more than 25 percent of the population aged 15 to 49 were infected with HIV. More than 60 percent of all those infected are women, reflecting cultural and social circumstances. Overall, no part of tropical Africa was spared. Life expectancies plummeted. Children by the millions were orphaned. Companies lost workers and could not replace them. National economies shrank. Associated costs—benefits, treatment, medicines—skyrocketed.

Why is Africa suffering so disproportionately? First, HIV/AIDS originated in tropical African forest margins and quickly spread through all segments of society. Second, the social stigma associated with HIV/AIDS, which is sexually transmitted, makes acknowledging and treating it especially problematic. Third, life-prolonging medications are expensive and especially difficult to provide in remote rural areas. Fourth, governmental leadership during the AIDS crisis has varied from highly aggressive and effective (as in Uganda, where political and medical officials cooperated in a massive campaign to distribute free condoms and advocate their use) to catastrophically negligent, as in South Africa where government ministers misled the public and unnecessarily delayed the medical intervention that was needed.

In 2011 the situation still was serious, but some progress was being made. A new South African government is addressing the AIDS crisis vigorously. Public health campaigns elsewhere are having good effect. Lower-cost, generic anti-HIV medicines are becoming more widely available through international help. But the national data in Appendix B form a reminder that Africa will long suffer from this, only the latest in its long series of death-dealing diseases.

AFRICA'S HISTORICAL GEOGRAPHY

Africa is the cradle of humanity. Archeological research has chronicled 7 million years of transition from Australopithecenes to Hominids to *Homo sapiens*. It is therefore ironic that we know comparatively little about Subsaharan Africa from 5000 to 500 years ago—that is, before the onset of European colonialism. This is partly due to the colonial period itself, during which African history was neglected, numerous African traditions and artifacts were destroyed, and many misconceptions about African cultures and institutions became entrenched. It is also a result of the absence of a written history over most of Africa south of the Sahara until the sixteenth century—and over a large part of it until much later than that.

African Genesis

Africa on the eve of the colonial period was a continent in transition. For several centuries, the habitat in and near one of the continent's most culturally and economically productive areas—West Africa—had been changing. For 2000 years, probably more, Africa had been innovating as well as adopting ideas from outside. In West Africa, cities were developing on an impressive scale; in central and Southern Africa, peoples were moving, readjusting, sometimes struggling with each other for territorial supremacy. The Romans had penetrated to southern Sudan, North African peoples were trading with West Africans, and Arab *dhows* (wooden boats with triangular sails) were sailing the waters along the eastern coasts, bringing Asian goods in exchange for gold, copper, and a comparatively small number of slaves. These same dhows today are increasingly used in tourism.

It is known that African cultures had been established in all the environmental settings shown in Figure G-7 for

---

*Two countries in Africa have the same short-form name, *Congo*. In this book, we use *The Congo* for the larger Democratic Republic of the Congo, and *Congo* for the smaller Republic of Congo.

© H. J. de Blij

"Got up before dawn this date in the hope of being the first visitor to the Great Zimbabwe ruins, a place I have wanted to see ever since I learned about it in a historical geography class. Climbed the hill and watched the sun rise over the great elliptical 'temple' below, then explored the maze of structures of the so-called fortification on the hilltop above. Next, for an incredible 90 minutes I was alone in the interior of the great oval walls of the temple. What history this site has seen—it was settled for perhaps six centuries before the first stones were hewn and laid here around AD 750, and from the 11th to the 15th centuries this may have been a ceremonial or religious center of a vast empire whose citizens smelted gold and copper and traded them via Sofala and other ports on the Indian Ocean for goods from South Asia and even China. No cement was used: these walls are built with stones cut to fit closely together, and in the valley they rise over 30 feet (nearly 10 m) high. And what secrets Great Zimbabwe still conceals—where are the plans, the tools, the quarries? The remnants of an elaborate water-supply system suggest a practical function for this center, but a decorated conical tower and adjacent platform imply a religious role. Whatever the answers, history hangs heavily, almost tangibly, over this African site."

www.conceptcaching.com

thousands of years and thus long before Islamic or European contact. One of these, the Nok culture, endured for over eight centuries on the Benue Plateau (north of the Niger-Benue confluence in present-day Nigeria) from about 500 BC to the third century AD. The Nok people made stone as well as iron tools, and they left behind a treasure of art in the form of clay figurines representing humans and animals. But we have no evidence that they traded with distant peoples. The opportunities created by environments and technologies still lay ahead.

Early Trade

West Africa, over a north-south span of a few hundred kilometers, displayed an enormous contrast in environments, economic opportunities, modes of life, and products. The peoples of the tropical forest produced and needed goods that were different from the products and requirements of the peoples of the dry, distant north. For example, salt is a prized commodity in the forest, where humidity precludes

its formation, but it is plentiful in the desert and semiarid steppe. This enabled the desert peoples to sell salt to the forest peoples in exchange for ivory, spices, and dried foods. Thus there evolved a degree of *regional complementarity* between the peoples of the forest and those of the drylands. And the savanna peoples—those located in between—found themselves in a position to channel and handle the trade (which is always economically profitable).

The markets in which these goods were exchanged prospered and grew, and urban centers arose in the savanna belt of West Africa. One of these old cities, now an epitome of isolation, was once a thriving center of commerce and learning and one of the leading urban places in the world—Timbuktu (now located in Mali). In fact, its university is one of the oldest in the world, with a library that holds valuable documents that are being preserved. Others, predecessors as well as successors of Timbuktu, have declined, some of them into oblivion. Still other savanna cities, such as Kano in the northern part of Nigeria, are still important today.

Early States

Strong and durable states arose inland in West Africa. The oldest state we know anything about is ancient Ghana, located to the northwest of the modern country of Ghana. It covered parts of present-day Mali, Mauritania, and adjacent territory. Ghana lay astride the upper Niger River and included gold-rich streams flowing off the Futa Jallon highlands, where the Niger River has its origin. Between the ninth and twelfth centuries AD, and perhaps longer, old Ghana managed to weld various groups of people into a stable state. The country had a large capital city complete with markets, suburbs for foreign merchants, religious shrines, and, some distance from the city center, a fortified royal retreat. Taxes were collected from the citizens, and tribute was extracted from subjugated peoples on Ghana's periphery; tolls were levied on goods entering Ghana, and an army maintained control. Muslims from the northern drylands invaded Ghana in 1067, when it may already have been in decline. Even so, the Ghanaians managed to protect their capital for 14 years. However, the invaders had ruined the farmlands and destroyed the trade links with the north. Ancient Ghana could not survive. It finally broke into smaller units.

Eastward Shift

In the centuries that followed, the focus of politico-territorial organization in the West African *culture hearth* (source area of culture) shifted almost continuously to the east—first to ancient Ghana's successor state of Mali, which was centered on Timbuktu and the middle Niger River Valley, and then to the state of Songhai, whose focus was Gao, a city on the Niger that still exists. This eastward movement may have been the result of the growing influence and power of Islam. Traditional religions prevailed in ancient Ghana, but Mali and its successor states sent huge, gold-laden pilgrimages to Mecca along the savanna corridor south of the Sahara, passing through present-day Khartoum and Cairo. Of the tens of thousands who participated in these pilgrimages, some remained behind. Today, many Sudanese trace their ancestry to the West African savanna kingdoms.

Beyond the West

West Africa's savanna region undoubtedly experienced momentous cultural, technological, and economic developments, but other parts of Africa also progressed. Early states emerged in present-day Sudan, Eritrea, and Ethiopia. Influenced by key innovations from the Egyptian culture hearth, these kingdoms were stable and durable: the oldest, Kush, lasted 23 centuries (Fig. 6A-5). The Kushites built elaborate irrigation systems, forged iron tools, and built impressive structures as the ruins of their long-term capital and industrial center, Meroe, reveal. Nubia, to the southeast of Kush, was Christianized until the Muslim wave overtook it in the eighth century. And Axum was the richest market in northeastern Africa, a powerful kingdom that controlled Red Sea trade and endured for six centuries. Axum, too, was a Christian state that confronted Islam, but Axum's rulers deflected the Muslim advance and gave rise to the Christian dynasty that eventually shaped modern Ethiopia.

The process of **13** state formation spread throughout Africa and was still in progress when the first European contacts occurred in the late fifteenth century. Large and effectively organized states developed on the equatorial west coast (notably Kongo) and on the southern plateau from the southern part of The Congo to Zimbabwe. East Africa had several city-states, including Mogadishu, Kilwa, Mombasa, and Sofala.

Bantu Migration

A crucial event affected virtually all of Equatorial, West, and Southern Africa: the great Bantu migration from present-day Nigeria and Cameroon southward and eastward across the continent. This migration appears to have occurred in waves starting as long as 5000 years ago, populating the Great Lakes area and penetrating South Africa, where it resulted in the formation of the powerful Zulu Empire in the nineteenth century (Fig. 6A-5).

All this reminds us that, before European colonization, Africa was a realm of rich and varied cultures, diverse lifestyles, technological progress, and external trade. It was, however, also a highly fragmented realm, its cultural mosaic (Fig. 6A-6) spelling weakness when European intervention came to change the social and political map forever.

The Colonial Transformation

European involvement in Subsaharan Africa began in the fifteenth century. It would interrupt the path of indigenous African development and irreversibly alter the entire cultural, economic, political, and social makeup of the continent. It started quietly in the late fifteenth century, with Portuguese ships groping their way along the west coast and rounding the Cape of Good Hope. Their goal was to find a sea route to the spices and riches of the Orient. Soon other European countries were sending their vessels to African waters, and a string of coastal stations and forts sprang up. In West Africa, the nearest part of the continent to European spheres in Middle and South America, the initial impact was strongest. At their coastal control points, the Europeans traded with African intermediaries for the slaves who were destined to work New World plantations, for the gold that had been flowing northward across the desert, and for ivory and spices.

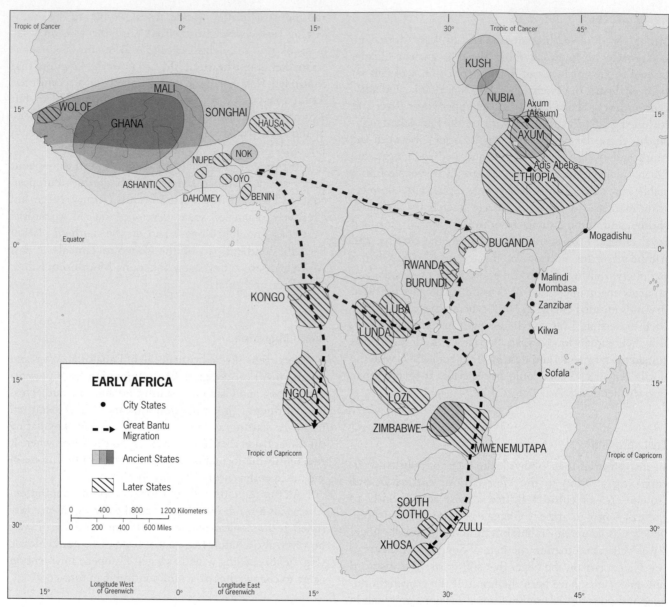

FIGURE 6A-5

© H. J. de Blij, P. O. Muller, and John Wiley & Sons, Inc.

Coastward Reorientation

Suddenly, the centers of activity lay not with the inland cities of the savanna but in the foreign stations on the Atlantic coast. As the interior declined, the coastal peoples thrived. Small forest states gained unprecedented wealth, transferring and selling slaves captured in the interior to the European traders on the coast. Dahomey (now called Benin) and Benin (now part of neighboring Nigeria) were states built on the slave trade. When slavery was eventually abolished in Europe, those who had inherited the power and riches it had brought vigorously opposed abolition on both continents.

Horrors of the Slave Trade

As discussed elsewhere in the book, millions of Africans were forced to migrate from their homelands to the Americas,

especially Brazil, the Caribbean Basin, and the United States. The slave trade was one of those African disasters alluded to earlier, and it was facilitated in part by what we may call the peril of proximity. The northeastern tip of Brazil, by far the largest single destination for the millions of Africans forced from their homes in bondage, lies about as far from the nearest West African coast as South Carolina lies from Venezuela. This is a short maritime intercontinental journey indeed (it is more than twice as far from West Africa to South Carolina). That proximity facilitated the forced migration of millions of West Africans to Brazil, which in turn contributed to the emergence of an African cultural diaspora in Brazil that is without equal in the New World.

Although slavery was not new to West Africa, the *kind* of slave raiding and trading the Europeans introduced certainly was. In the interior of Africa and within city-states,

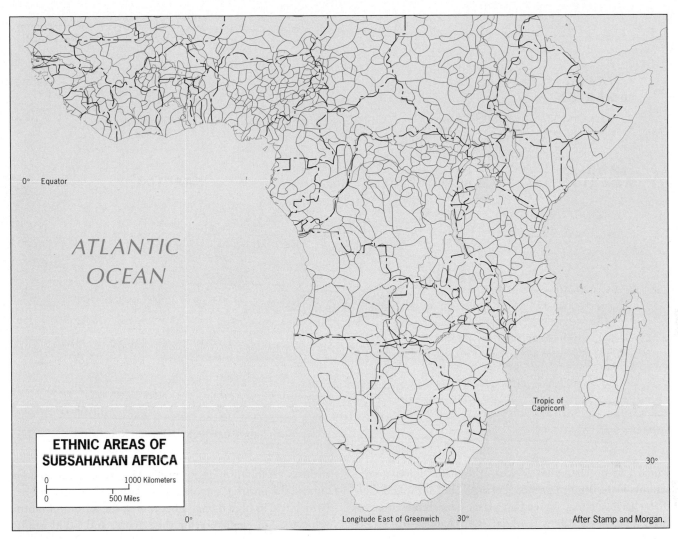

FIGURE 6A-6

© H. J. de Blij, P. O. Muller, and John Wiley & Sons, Inc.

kings, chiefs, and prominent families traditionally took a few slaves, but the status of those slaves was unlike anything that lay in store for those who were shipped across the Atlantic. In fact, large-scale slave trading had been introduced in East Africa long before the Europeans brought it to West Africa. African intermediaries from the coast raided the interior for able-bodied men and women and marched them in chains to the Arab markets on the coast (the island of Zanzibar was one such slave trading market). There, packed in specially built dhows, they were carried off to Arabia, Persia, and India. When the European slave trade took hold in West Africa, however, its volume was far greater. Europeans, Arabs, and collaborating Africans ravaged the continent, forcing perhaps as many as 30 million persons away from their homelands in captivity (Fig. 6A-7). Families were destroyed, as were whole villages and cultures; those who survived their exile suffered unfathomable misery.

The European presence on the West African coast completely reoriented its trade routes, for it initiated the decline of the interior savanna states and strengthened the coastal

forest states. Moreover, the Europeans' insatiable demand for slaves ravaged the population of the interior. But it did not lead to any major European thrust toward the interior or produce colonies overnight. The African intermediaries were well organized and strong, and they held off their European competitors, not just for decades but for centuries. Although the Europeans first appeared in the fifteenth century, they did not carve up West Africa until nearly 400 years later, and they did not conquer many other areas until after the beginning of the twentieth century.

Colonization

In the second half of the nineteenth century, whether or not they had control, the European powers finally laid claim to virtually all of Africa. Colonial competition was intense, and spheres of influence began to overlap. It was time for negotiation among the powerful, and in 1884 a conference was convened in Berlin to divide up Africa. Fourteen states participated (including the United States,

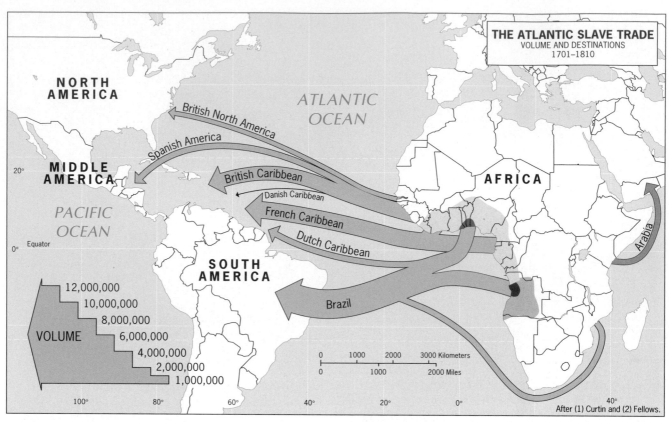

FIGURE 6A-7

© H. J. de Blij, P. O. Muller, and John Wiley & Sons, Inc.

which had no claims on Africa). The major colonial contestants were Britain, France, Portugal, Belgium, and Germany itself. On maps spread on a large table, representatives from these powers drew boundaries, exchanged real estate, and forged a new map that would become a liability in Africa decades later. As Figure 6A-8 indicates, when the three-month conference was in progress most of Africa remained under traditional African rule. Not until after 1900 did the colonial powers manage to control all the areas they had marked off on their new maps.

It is important to examine Figure 6A-8 carefully because the colonial powers governed their new dependencies in very different ways, and their contrasting legacies remain in evidence to this day in the countries their colonies spawned. Some colonial powers were democracies (Great Britain and France); others were dictatorships (Portugal and Spain). The British established a system of indirect rule over much of their domain, leaving indigenous power structures in place and making local rulers representatives of the British Crown. This was unthinkable in the Portuguese colonies, where harsh, direct control was the rule. The French sought to create culturally assimilated elites that would represent French ideals in the colonies. Unlike the British, French, or Portuguese, King Leopold II of Belgium sought to claim Congo Free State as his own personal empire. After financing the expeditions that staked Belgium's claim in Berlin, he embarked on a campaign of

ruthless exploitation and murder. His enforcers mobilized almost the entire Congolese population to gather rubber (which was in high demand because of the advent of motor vehicles), kill elephants for their ivory, and build public works to improve export routes. For failing to meet production quotas, entire communities were executed and tortured. Killing and maiming became routine in a colony in which horror was the only common denominator. After the impact of the slave trade, King Leopold's reign of terror was Africa's most severe demographic disaster. By the time it ended, after growing condemnation from all over the world, as many as 10 million Congolese had been murdered. In 1908, the Belgian government took over and began to mirror Belgium's own internal divisions: corporations, government administrators, and the Roman Catholic Church each pursued their sometimes competing interests. But no one thought to change the name of the colonial capital: it was Leopoldville until the Belgian Congo achieved independence in 1960.

Colonialism transformed Africa, but in its post-Berlin form it lasted less than a century. In Ghana, for example, the Ashanti (Asante) Kingdom still was fighting the British in the early years of the twentieth century; by 1957, Ghana was independent again. These days, much of Subsaharan Africa is marking half a century of independence, and, in retrospect, the colonial period should be seen as an interlude rather than a paramount chapter in modern African

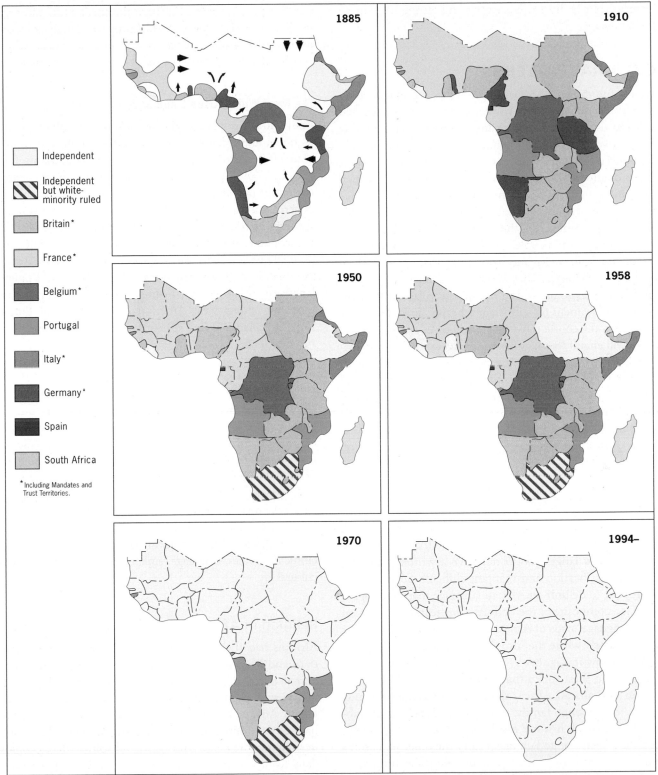

Independent

Independent but white-minority ruled

Britain*

France*

Belgium*

Portugal

Italy*

Germany*

Spain

South Africa

* Including Mandates and Trust Territories.

1885

1910

1950

1958

1970

1994–

FIGURE 6A-8
© H. J. de Blij, P. O. Muller, and John Wiley & Sons, Inc.

COLONIZATION AND DECOLONIZATION SINCE 1885

history—but it did leave a strong and lasting imprint on the landscape.

CULTURAL PATTERNS

We tend to think of Africa in terms of its prominent countries and famous cities, its development problems and political dilemmas, but Africans themselves have another perspective. The colonial period created states and capitals, introduced foreign languages to serve as the *linguae francae*, and brought railways and roads. The colonizers stimulated labor movements to the mines they opened, and they disrupted other migrations that had been part of African life for many centuries. But they did not change the ways of life of most of the people. Fully 65 percent of the realm's population still live in, and work near, Africa's hundreds of thousands of villages. They speak one of more than a thousand languages in use in the realm. The villagers' concerns are local; they focus on subsistence, health, and safety. They worry that the conflicts over regional power or political ideology will engulf them, as has happened to millions in Liberia, Sierra Leone, Ethiopia, Rwanda, The Congo, and Angola since the 1970s. Africa's (numerically) largest groups of people are major nations, such as the Yoruba of Nigeria and the Zulu of South Africa. Africa's smallest groups of people number just a few thousand. As a geographic realm, Subsaharan Africa has the most complex cultural mosaic on Earth.

African Languages

Africa's linguistic geography is a key component of that cultural intricacy. Most of Subsaharan Africa's more than one thousand languages do not have a written tradition, making classification and mapping difficult. Scholars have attempted to delimit an African language map, and Figure 6A-9 is a composite of their efforts. One feature is common to all language maps of Africa: the geographic realm begins approximately where the Afro-Asiatic language family (mapped in yellow in Fig. 6A-9) ends, although the correlation is sharper in West Africa than to the east.

In Subsaharan Africa, the dominant language family is the Niger-Congo family, of which the Kordofanian subfamily is a small, historic northeastern outlier (Fig. 6A-9), and the Niger-Congo languages carry the other subfamily's name. This subfamily extends across the realm from West to East and Southern Africa. The Bantu language forms the largest branch of this subfamily, but Niger-Congo languages in West Africa, such as Yoruba and Akan, also have millions of speakers. Another important language family is the Nilo-Saharan family, extending from Maasai in Kenya northwest to Teda in Chad. No other language families are of similar extent or importance: the Khoisan family, of ancient origins, now survives among the dwindling Khoi

and San peoples of the Kalahari; the small white minority in South Africa speak Indo-European languages; and Malay-Polynesian languages prevail in Madagascar, which was peopled from Southeast Asia before Africans reached it.

The Most Widely Used Languages

About 40 African languages are spoken by 1 million people or more, and a half-dozen by about 10 million or more: Hausa (50 million), Yoruba (23 million), Ibo, Swahili, Lingala, and Zulu. Although English and French have become important *linguae francae* in multilingual countries such as Nigeria and Ivory Coast (where officials even insist on spelling the name of their country—Côte d'Ivoire—in the Francophone way), African languages also serve this purpose. Hausa is a common language across the West African savanna; Swahili is widely used in East Africa. And pidgin languages, mixtures of African and European tongues, are spreading along West Africa's coast. Millions of Pidgin English (called *Wes Kos*) speakers use this medium in Nigeria and Ghana.

Language and Culture

14 **Multilingualism** can be a powerful centrifugal force in society, and African governments have tried with varying success to establish national alongside local languages. Nigeria, for example, made English its official language because none of its 250 languages, not even Hausa, had sufficient internal interregional use. But using a European, colonial language as an official medium invites criticism, and Nigeria remains divided on the issue. On the other hand, making a dominant local language official would invite negative reactions from ethnic minorities. Language remains a potent force in Africa's cultural life.

Religion in Africa

Africans had their own belief systems long before Christians and Muslims arrived to convert them. And for all of Subsaharan Africa's cultural diversity, Africans had a consistent view of their place in nature. Spiritual forces, according to African tradition, are manifest everywhere in the natural environment, not in a supreme deity that exists in some remote place. Thus gods and spirits affect people's daily lives, witnessing every move, rewarding the virtuous and punishing (through injury or crop failure, for example) those who misbehave. Ancestral spirits can inflict misfortune on the living. They are everywhere: in the forest, rivers, and mountains.

As with land tenure, the religious views of Africans clashed fundamentally with those of the colonists. Monotheistic ***Christianity*** first touched Africa in the northeast when Nubia and Axum were converted, and Ethiopia has

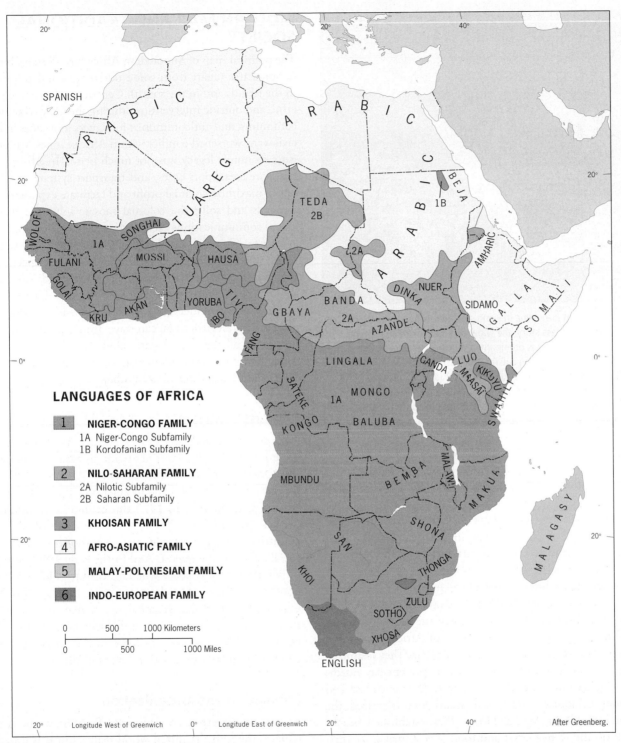

LANGUAGES OF AFRICA

1 **NIGER-CONGO FAMILY**
1A Niger-Congo Subfamily
1B Kordofanian Subfamily

2 **NILO-SAHARAN FAMILY**
2A Nilotic Subfamily
2B Saharan Subfamily

3 **KHOISAN FAMILY**

4 **AFRO-ASIATIC FAMILY**

5 **MALAY-POLYNESIAN FAMILY**

6 **INDO-EUROPEAN FAMILY**

0   500   1000 Kilometers
0   500   1000 Miles

After Greenberg.

FIGURE 6A-9   © H. J. de Blij, P. O. Muller, and John Wiley & Sons, Inc.

been a Coptic Christian stronghold since the fourth century AD. But the Christian churches' real invasion did not commence until the onset of colonialism after the turn of the sixteenth century. Christianity's various denominations made inroads in different areas: Roman Catholicism in much of Equatorial Africa, mainly at the behest of the Belgians; the Anglican Church in British colonies; and Presbyterians and others elsewhere. Evangelical churches are now gaining adherents rapidly. Some of these churches today are more conservative than those found in Europe or North America, appealing to conservative congregations in the United States. A split in the Episcopal (Anglican) Church in the United States regarding the treatment of homosexuals has some congregations aligning themselves under the Bishop of Uganda. But almost everywhere, Christianity's penetration led to a blending of traditional and Christian

The faithful kneel during Friday prayers at a mosque in Kano, northern Nigeria. The survival of Nigeria as a unified state is an African success story; the Nigerians have overcome strong centrifugal forces in a multi-ethnic country that is dominantly Muslim in the north, Christian in the south. In the 1990s, some Muslim clerics began calling for an Islamic Republic in Nigeria, and after the death of the dictator Abacha and the election of a non-Muslim president, the Islamic drive intensified. A number of Nigeria's northern States adopted Sharia (strict Islamic) law, which led to destructive riots between the majority Muslims and minority Christians who felt threatened by this turn of events. Can Nigeria avoid the fate of Sudan (see Chapter 7B)? © Marc and Evelyn Bernheim/Woodfin Camp & Associates.

beliefs, so that much of Subsaharan Africa is nominally, though not exclusively, Christian (see Fig. 7A-3). Go to a church in Gabon or Uganda or Zambia, and you may hear drums instead of church bells, sing African music rather than hymns, and see African carvings alongside the usual statuary.

Islam had a different arrival and impact. By the time of the colonial invasion, Islam had advanced out of Arabia, across the desert, and part-way down the coasts of Africa. Muslim clerics converted the rulers of African states and commanded them to convert their subjects. They Islamized the savanna states and penetrated into present-day northern Nigeria, Ghana, and Ivory Coast. They encircled and isolated Ethiopia's Coptic Christians and Islamized the Somali people in Africa's Horn. They established beachheads on the Kenya coast and took over Zanzibar. Arabizing Islam and European Christianity competed for African minds, and Islam proved to be a far more pervasive force. From Senegal to Somalia, the population is overwhelmingly Muslim (see Fig. 6B-9), and Islam's rules dominate everyday life. The Sunni *mullahs* would never allow the kind of marriage between traditional and Christian beliefs seen in much of formerly colonial Africa. This fundamental contradiction between Islamic dogma and Christian accommodation creates a potential for conflict in countries where both religions have adherents.

MODERN MAP AND TRADITIONAL SOCIETY

The political map of Subsaharan Africa has 45 states but no nation-states (apart from some microstates and ministates in the islands and in the south). Centrifugal forces are powerful, and outside interventions during the Cold War, when communist and anticommunist foreigners took sides in local civil wars, worsened conflict within African states. Colonialism's economic legacy was not much better. In Africa, capitals, core areas, port cities, and transport systems were laid out to maximize colonial profit and facilitate exploitation of minerals and soils; the colonial mosaic inhibited interregional communications except where cooperation enhanced efficiency. Colonial Zambia and Zimbabwe, for example (then called Northern and Southern Rhodesia, respectively), were landlocked and needed outlets, so railroads were built to Portuguese-owned ports. But such routes did little to create intra-African linkages. The modern map reveals the results: in West Africa you can travel from the coast into the interior of all the coastal states along railways or adequate roads. But no high-standard roadway was ever built to link these coastal neighbors to each other.

Supranationalism

To overcome such disadvantages, African states must cooperate internationally, continentwide as well as regionally. The Organization of African Unity (OAU) was established for this purpose in 1963 and in 2001 was superseded by the African Union (AU). In 1975 the Economic Community of West African States (ECOWAS) was established by 15 countries to promote trade, transportation, industry, and social affairs in the region. And in the early 1990s another important step was taken when 12 countries joined in the Southern African Development Community (SADC), organized to facilitate regional commerce, inter-country transport networks, and political interaction. The AU has been an active peacekeeping force in the ongoing Darfur conflict in Sudan (discussed in Chapter 7B).

Population and Urbanization

As can be discerned in Appendix B, Subsaharan Africa remains the least urbanized world realm, but it is urbanizing at a fast pace. By the time you read this, the percentage of urban dwellers will have passed 35. This means that more than 270 million people now live in towns and cities, of which many were founded and developed by the colonial powers. But the infrastructure of the cities has not kept pace with the number of arrivals. African cities became centers of embryonic national core areas, and of course they served as government headquarters. This *formal sector* of the city used to be the dominant one, with government control and regulations affecting civil service, business,

A recent view over the Jankara Market area on Lagos Island, in the heart of this realm's only megacity, Lagos, Nigeria (10.6 million). In the distance, high-rises stand comparatively isolated (no clustered CBD here); note the minarets of an impressive mosque rising above the cityscape. In the foreground, the street scene is comparatively well-ordered and spacious, in contrast to the city's chaotic ring of shantytowns. Lagos Island, the water-buffered aggregation of office and residential buildings seen here, is relatively modern—but remarkably small for a metropolis with a population as large as that of Belgium. © AFP/ Getty Images, Inc.

industry, and workers. Today, however, African cities look different. From a distance, the skyline still resembles that of a modern center. But in the streets, on the sidewalks right below the shopwindows, there are hawkers, basket weavers, jewelry sellers, garment makers, wood carvers—a second economy, most of it beyond government control. This *informal sector* now dominates many African cities. It is peopled by the rural immigrants, who also work as servants, apprentices, construction workers, and in countless other menial jobs.

Millions of urban immigrants, however, cannot find work, at least not for months or even years at a time. They live in squalid circumstances, in desperate poverty, and governments cannot assist them. As a result, the squatter rings around (and also within) many of Africa's cities are unsafe—uncomfortable, unhealthy slums without adequate shelter, water supply, or basic sanitation. Garbage-strewn (no solid-waste removal here), muddy, and insect-infested during the rainy season, and stifling and smelly during the dry period, they are incubators of disease. Yet few of their residents return to their villages. Every new day brings hope.

In our regional discussion we refer to some of Subsaharan Africa's cities, all of which, to varying degrees, are stressed by the rate of population influx. Despite the plight of the urban poor and the poverty of Africa's rural areas, some of Africa's capitals remain the strongholds of privileged elites who, dominant in governments, fail to address the needs of other ethnic groups. Discriminatory policies and artificially low food prices disadvantage farmers and create even greater urban-rural disparities than the colonial period saw. But today the prospect of democracy brings hope that Africa's rural majorities will be heard and heeded in the capitals.

POINTS TO PONDER

- Widespread drought across Subsaharan Africa is hurting industries from farming to tourism.
- Subsaharan Africa remains the most severely AIDS-afflicted realm in the world.
- If the political process is allowed to run its course, a new state may soon emerge in southern Sudan in the realm's White Nile Basin.
- China's impact on Subsaharan Africa is growing rapidly through activities ranging from trade and investment to education.

Village amid farms along the
fertile, terraced wall of Kenya's
Eastern Rift Valley. © H. J. de Blij

6B

SUBSAHARAN AFRICA: REGIONS OF THE REALM

Southern Africa
East Africa
Equatorial Africa
West Africa
African Transition Zone

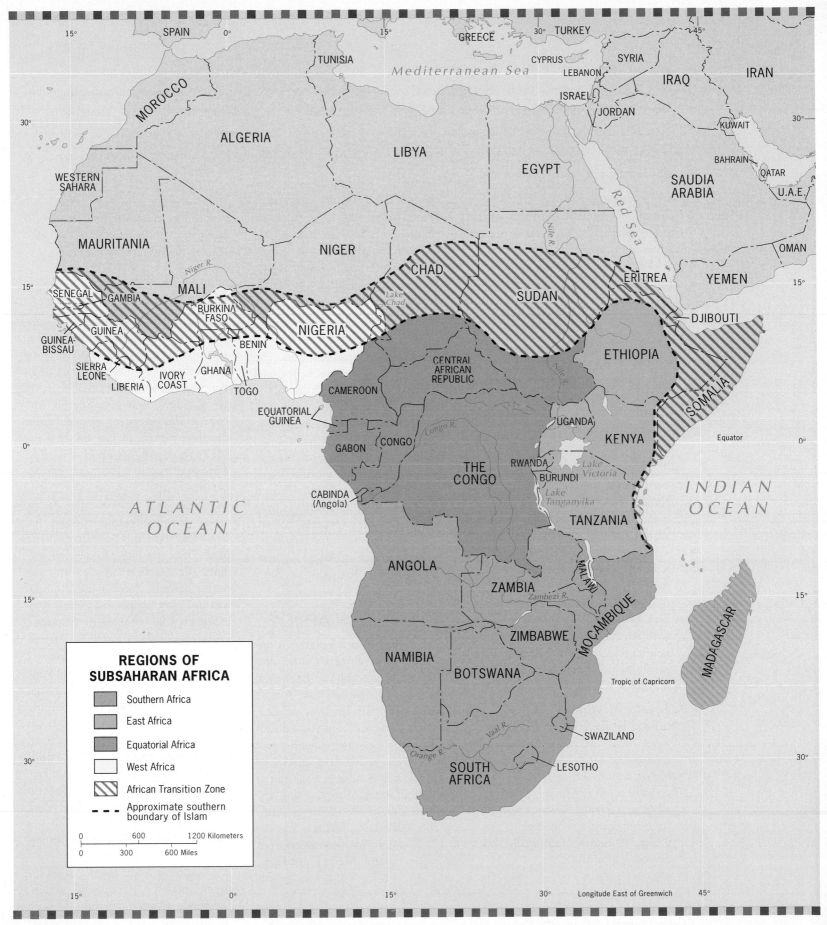

REGIONS OF SUBSAHARAN AFRICA

- Southern Africa
- East Africa
- Equatorial Africa
- West Africa
- African Transition Zone
- Approximate southern boundary of Islam

0   600   1200 Kilometers

0   300   600 Miles

© H. J. de Blij, P. O. Muller, and John Wiley & Sons, Inc.

On the face of it, Africa seems to be so massive, compact, and unbroken that any attempt to justify a contemporary regional breakdown is doomed to fail. No deeply penetrating bays or seas create peninsular fragments as they do in Europe. No major islands (other than Madagascar) provide the broad regional contrasts we see in Middle America. Nor does Africa really taper southward to the peninsular proportions of South America. And Africa is not cut by an Andean or a Himalayan mountain barrier. Given Africa's colonial fragmentation and cultural mosaic, is regionalization possible? Indeed it is.

Maps of environmental distributions, ethnic patterns, cultural landscapes, historic culture hearths, and other spatial data yield a four-region structure complicated by a fifth, overlapping zone as shown in Figure 6B-1. Beginning in the south, we identify the following regions:

1. **Southern Africa**, extending from the southern tip of the continent to the northern borders of Angola, Zambia, Malawi, and Moçambique. Ten countries constitute this region, which extends beyond the tropics and whose giant is South Africa.

2. **East Africa**, where natural (equatorial) environments are moderated by elevation and where plateaus, lakes, and mountains, some carrying permanent snow, define the countryside. Six countries, including the highland part of Ethiopia, comprise this region. The island state of Madagascar, with Southeast Asian influences, is neither Southern nor East African, but is included here because of its location.

3. **Equatorial Africa**, much of it defined by the basin of the Congo River, where elevations are lower than in East Africa, temperatures are higher and moisture more ample, and where most of Africa's surviving rainforests remain. Among the eight countries that form this region, The Congo* dominates territorially and demographically.

4. **West Africa**, which includes the countries of the western coast and those on the margins of the Sahara in the interior, a populous region anchored in the southeast by Africa's demographic giant, Nigeria. Fifteen countries form this crucial African region.

5. **The African Transition Zone**, the complicating factor on the regional map of Africa. In Figure 6B-1, note that this striped zone of increasing Islamic influence completely dominates some countries (e.g., Somalia in the east and Senegal in the west) while cutting across others, thereby creating Islamized northern areas and non-Islamic southern zones (Nigeria, Chad, Sudan), and even encompassing only a strip of others (e.g., lowland Ethiopia, Kenya, Tanzania).

SOUTHERN AFRICA

Southern Africa, as a geographic region, consists of all the countries and territories lying south of Equatorial Africa's The Congo and East Africa's Tanzania (Fig. 6B-2). Thus defined, the region extends from Angola and Moçambique (on the Atlantic and Indian Ocean coasts, respectively) to South Africa and includes a half-dozen landlocked states. Also marking the northern limit of the region are Zambia and Malawi. Zambia is nearly cut in half by a long land extension from The Congo, and Malawi penetrates deeply into Moçambique. The colonial boundary framework, here as elsewhere, produced many liabilities.

Africa's Richest Region

Southern Africa constitutes a geographic region in both physiographic and human terms. Its northern zone marks the southern limit of the Congo Basin in a broad upland that stretches across Angola and into Zambia (the tan corridor extending eastward from the Bihe Plateau in Fig. 6A-2). Lake Malawi is the southernmost of the East African rift-valley lakes; Southern Africa has none of East Africa's volcanic and earthquake activity. Most of the region is plateau country, and the Great Escarpment is much in evidence here. There are two pivotal river systems: the Zambezi (which forms the border between Zambia and Zimbabwe) and the Orange-Vaal (South African rivers that combine to demarcate southern Namibia from South Africa).

Southern Africa is the continent's richest region materially. A great zone of mineral deposits extends through the heart of the region from Zambia's Copperbelt through Zimbabwe's Great Dyke and South Africa's Bushveld Basin and Witwatersrand to the goldfields and diamond mines of the Orange Free State and northern Cape Province in the heart of South Africa. Ever since these minerals began to be exploited in colonial times, many migrant laborers have come to work in the mines.

Southern Africa's agricultural diversity matches its mineral wealth. Vineyards drape the slopes of South Africa's Cape Ranges; tea plantations hug the eastern escarpment slopes of

---

*Two countries in Africa have the same short-form name, *Congo*. In this book, we use **The Congo** for the larger Democratic Republic of the Congo, and **Congo** for the smaller Republic of Congo.

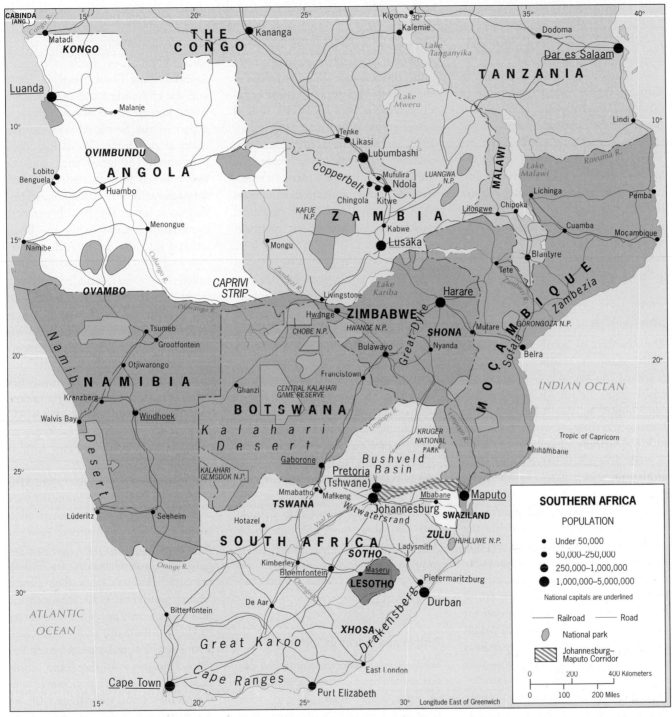

FIGURE 6B-2

© H. J. de Blij, P. O. Muller, and John Wiley & Sons, Inc.

Zimbabwe. Before civil war destroyed its economy, Angola was one of the world's leading coffee producers. South Africa's relatively high latitudes and its range of altitudes create environments for apple orchards, citrus groves, banana plantations, pineapple farms, and many other crops.

Despite this considerable wealth and potential, not all of the ten countries of Southern Africa have prospered. As Figure G-10 shows, four of them remain mired in the low-income category (Malawi, Moçambique, Zambia, and Zimbabwe); although South Africa, Angola, Lesotho, Swaziland, Namibia,

and Botswana fall into the two middle-income ranks, the (desert-dominated) last two rank among the realm's three most sparsely populated countries. A period of rapid population growth, followed by the devastating onslaught of AIDS, civil conflict, political instability, incompetent government, widespread corruption, unfair competition on foreign markets, and environmental problems, have combined to constrain regional development.

Still, with the tragic exception of Zimbabwe (the realm's basket case in 2011), Southern Africa as a region is better off than any other in Subsaharan Africa: as Figure G-10 shows, six countries here have risen above the lowest-income rank among world states. Some international cooperation including a regional association and a customs union are emerging. And South Africa, the realm's most important country by many measures, provides hope for a better future.

South Africa

The Republic of South Africa (RSA) is the giant of Southern Africa, its economy by far the largest of this entire realm. South Africa is an African country at the center of world attention, a bright ray of hope not only for Africa but for all humankind.

For 40 years, between about 1950 and 1990, multicultural South Africa was in the grip of the world's most notorious racist policy, **1** **apartheid**, the word meaning "apartness" but the practice involving strict racial segregation and severe discrimination. Out of the concept of apartheid grew a notion, promulgated by the white (European) minority then in control of the state, called **2** **separate development**. This would apply apartheid to the entire country, carving it up into racially-based entities whose inhabitants would be citizens of those ethnic domains, but not of South Africa as a whole. Predictably, such racist social engineering aroused strong opposition within South Africa and beyond, leading to general condemnation and international sanctions.

South Africa seemed headed for a violent revolution, but disaster was averted by what was, at the time, an almost inconceivable turn of events. A leader of the white-minority government that for decades had ruthlessly pursued its apartheid policies, and a revered leader of the multicultural majority who had for 28 years languished in an island prison not far from Cape Town, struck an accord that, in effect, created a new South Africa virtually overnight. Nelson Mandela walked out of prison a free man in February 1990, and in the RSA's first democratic election of 1994 Mandela became the country's president on a platform representing the African National Congress (ANC), the multiracial anti-apartheid movement that had fought apartheid for many years. The ANC's old foes, the architects of apartheid, were seated as part of the loyal opposition in a parliament that was now indeed a rainbow assembly.

South Africa stretches from the warm subtropics in the north to Antarctic-chilled waters in the south. With a land area in excess of 1.2 million square kilometers (470,000 sq mi) and a heterogeneous population of 49.1 million, South Africa contains the bulk of the region's minerals, most of its good farmlands, its largest cities, best ports, most productive factories, and most developed transport networks. Mineral exports from Zambia and Zimbabwe move through South African ports. Workers from as far away as Malawi and as close as Lesotho work in South Africa's mines, factories, and fields.

People and Places in South Africa

South Africa's historical geography differs somewhat from much of the rest of Subsaharan Africa. Its lands were fought over by various African nations before the Europeans arrived and the colonial "scramble for Africa" took place. Peoples migrated southward—first the Khoisan-speakers and then the Bantu peoples—into the South African cul-de-sac. The Zulu and Xhosa nations fought over lands at about the time of the first European arrivals. Europeans arrived via the oceans and claimed the southernmost Cape. It is one of the most strategic places on Earth, the gateway from the Atlantic to the Indian Ocean, a source of provisions on the route to Asia's riches. The Dutch East India Company founded Cape Town as early as 1652, and the Hollanders and their descendants, known as the **Boers**, have been a part of the South African cultural makeup ever since. The British took over about 150 years later, and the two colonial powers vied for power throughout South Africa's early history as a state. The British came to dominate the Cape, while the Boers trekked into the South African interior and, on the high plateau they called the *highveld*, founded their own republics. By 1910, the Boers and the British had negotiated a power-sharing arrangement, although the Boers eventually achieved hegemony between 1948 and 1994. Having long since shed their European links, they came to call themselves **Afrikaners**, their word for Africans.

The Ethnic Mosaic

In addition to the various native African nations and the Europeans that settled in South Africa are peoples from Asia. The Hollanders brought Southeast Asians to the Cape to serve as domestics and laborers. The British imported laborers from their South Asian colonies to work in sugarcane plantations, adding cultural diversity to this multiethnic state. Moreover, a substantial population of mixed ancestry, clustered at the Cape, comprises today's so-called *Coloured* sector of the country's citizenry. In the process, South Africa became Africa's most pluralistic and heterogeneous society. Even so, Africans now outnumber non-native Africans by about 4 to 1.

Although heterogeneity marks the spatial demography of South Africa, regionalism pervades the human mosaic. The Zulu nation still is largely concentrated in today's Kwazulu-Natal Province (Fig. 6B-3). The Xhosa still cluster in the Eastern Cape, from the city of East London to the

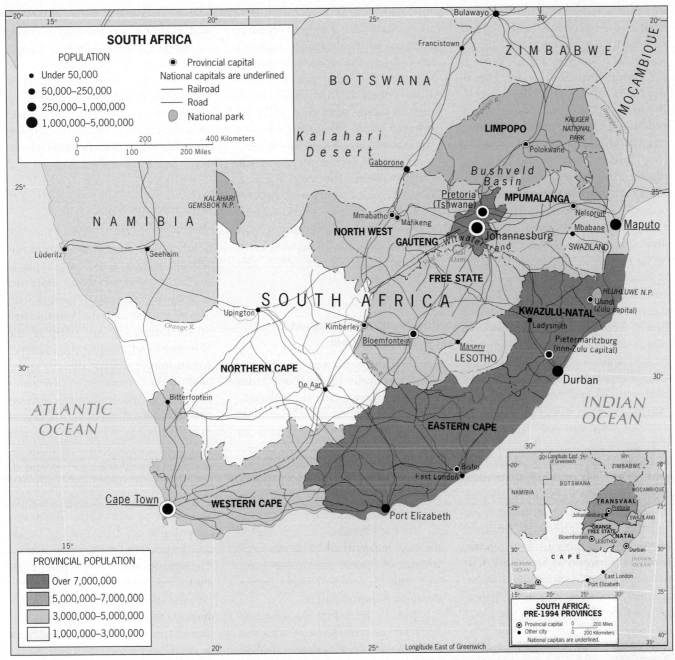

FIGURE 6B-3

© H. J. de Blij, P. O. Muller, and John Wiley & Sons, Inc.

Kwazulu-Natal border and below the Great Escarpment. The Tswana still occupy ancestral lands along the Botswana border. Cape Town remains the core area of the Coloured population; Durban still has the strongest South Asian imprint. Travel through South Africa, and you will recognize the diversity of rural cultural landscapes as they change from Swazi to Ndebele to Venda.

This longstanding regionalism was among the factors that led the Afri-kaner government to institute its **separate development** scheme, which was linked to **apartheid**, but it could not hold back the tide of urbanization that intermingled the populations. Millions of workers, job-seekers, and illegal migrants converged on the cities, creating vast shantytowns on their margins. In the legal African townships such as Johannesburg's Soweto (SOuth WEstern TOwnships) and in the squatter settlements the anti-apartheid movement burgeoned, and the

strength of the African National Congress (ANC) movement grew, laying the groundwork for its eventual dominance in South Africa's political arena.

Political Change

Largely through the unmatched statesmanship of Nelson Mandela, South Africa did not undergo a massive revolution with great upheaval. Instead it is today governed by a multiracial party, the erstwhile architects of apartheid are in parliament as a legitimate opposition, and its economy is the largest and healthiest in the realm. Check Appendix B and you will see that, on the African mainland, South Africa's per-capita GNI is in a class by itself. By some calculations, South Africa produces 45 percent of all of Subsaharan Africa's GDP.

Together with political change came a change in administrative structure of the country; in fact, this change occurred before the historic elections of 1994. Before 1994 South Africa was divided into four provinces: the Cape, by far the largest and centered on the legislative capital of Cape Town; Natal, anchored by the port city of Durban; the Transvaal (meaning across the Vaal River), focused on the administrative capital of Pretoria and including the great Johannesburg metropolis; and the Orange Free State, the old Boer stronghold with Bloemfontein as its headquarters (see inset map, Fig. 6B-3).

This structure was replaced by nine provinces (Fig. 6B-3), creating a federal arrangement in which each province has its own administration while being represented in the central government. The boundaries of Natal and the Orange Free State remained essentially unchanged, but Natal's name was changed to Kwazulu-Natal, recognizing the Zulu presence there, and the Orange Free State became simply the Free State. But the Cape Province was divided into four new provinces, one of them overlapping into the Transvaal, and the rest of the Transvaal was divided into three, one of which is Gauteng, which includes the Johannesburg-Witwatersrand-Pretoria megalopolis.

A New Era Dawns

In June 1999, ANC leader Thabo Mbeki, who had served as President Mandela's deputy, became the country's second popularly elected president. In other African states, the succession from heroic founder-of-the-nation to political inheritor-of-the-presidency often has not gone well,

but South Africa's new constitution proved its worth. The new president faced several disadvantages: unavoidable comparisons to the incomparable Mandela; the rising tide of AIDS, which Mbeki controversially attributed to causes other than HIV; his awkward stance toward the chaos in Zimbabwe. Nevertheless, Mbeki's economic policies, social programs, and foreign initiatives (The Congo, Rwanda) have served his country well. In 2007 Jacob Zuma succeeded Mbeki as leader of the ANC and, following a brief constitutional crisis, Zuma was elected president in 2009—the RSA's first leader of Zulu ancestry.

How the Economy Evolved: Diamonds and Gold

Ever since diamonds were discovered at Kimberley in the 1860s, South Africa has been synonymous with minerals. The Kimberley finds, made in a remote corner of the Orange Free State, set into motion a new economic geography. Rail lines were laid from the coast to the diamond capital even as fortune seekers, capitalists, and tens of thousands of African workers, many from as far afield as Moçambique, streamed to the site. One of the capitalists was Cecil Rhodes, of Rhodes Scholarship fame, who used his fortune to help Britain dominate Southern Africa.

Just 25 years after the diamond finds, prospectors discovered what was long to be the world's greatest goldfield on a ridge called the Witwatersrand (Fig. 6B-3). Johannesburg became the gold capital of the world, and a new and even larger stream of foreigners arrived, along with a huge influx of African workers. Cheap labor enlarged the profits. Johannesburg grew explosively, satellite towns developed, and black townships mushroomed.

During the twentieth century, South Africa proved to be richer than had been foreseen. Additional goldfields were discovered in the Orange Free State. Coal and iron ore were found in abundance, which gave rise to a major iron and steel industry. Other

metallic minerals, including chromium and platinum, yielded large revenues on world markets. Asbestos, manganese, copper, nickel, antimony, and tin were mined and sold; a thriving metallurgical industry developed in South Africa itself. Capital flowed into the country, white immigration grew, farms and ranches were laid out, and markets multiplied.

Infrastructural Gains During Apartheid

South Africa's cities grew enormously. Johannesburg was no longer just a mining town: it became an industrial complex and a financial center as well. The old Boer capital, Pretoria, just 50 kilometers (30 mi) north of the Witwatersrand, became the country's administrative center during apartheid. In the Orange Free State, major industrial growth (including oil-from-coal technology) matched the expansion of mining. While the core area developed megalopolitan characteristics, most coastal cities also expanded. Durban's port served not only the Witwatersrand but a wider regional hinterland as well. Cape Town was becoming South Africa's second-largest city; its port, industries, and productive agricultural hinterland gave it primacy over a wide area.

The labor force for all this development, from mines to railroads and from farms to highways, came from the African peoples of South Africa (and, indeed, from beyond its borders as well). Even during apartheid, workers from Moçambique, Zimbabwe, Botswana, and the other neighboring countries sought jobs in South Africa while millions of South African villagers also left their homes for the towns and cities. In the process they built the best infrastructure of any African country. But apartheid ruined South Africa's prospects. The cost of the separate development program was astronomical. Social unrest during the decade preceding the end of apartheid created a vast educational gap among youngsters. International sanctions

The leafy suburbs and high-rise CBD of Johannesburg and the shantytown housing of a still poverty-stricken section of crowded Soweto, one of its satellite towns, show the jarring social contrasts in South Africa's ever-more-unequal society. Future political stability will depend on the government's ability to improve living conditions in the poorest areas, rural as well as urban. Although significant advancement has been made in housing, water supply, and electricity hookups, the enormity of the problem, and constraints ranging from political realities to available resources, slow the pace of progress in a country that is in a race against time. (Left) © Martin Harvey/Gallo Images/Getty Images, Inc. (Right) © Sergio Pitamitz/Robert Harding World Imagery/© Corbis

against the race-obsessed regime damaged the economy.

The Economy Today

In many respects South Africa is the most important country in Subsaharan Africa, and the entire realm's fortunes are bound up with it. No African country attracts more foreign investment or foreign workers. None has the universities, hospitals, and research facilities. No other has the military forces capable of intervening in African trouble spots. Few have the free press, effective trade unions, independent courts, or financial institutions to match South Africa's. And with a population now passing 50 million (79 percent black, just under 10 percent white, 9 percent Coloured, 2.6 percent Asian), South Africa has a large, multiracial, and growing middle class.

Despite its successes, South Africa continues to face long-range problems. The economy continues to depend far too much on the export of minerals and metals (diamonds, platinum, gold, iron, and steel) at a time when this dependence entails risks at home and abroad. At home, those union-friendly labor laws, including rising wages, are making mining less profitable. Abroad, commodity prices are unreliable. In 1970 South Africa produced nearly 70 percent of the world's gold; today it produces less than 15 percent. And the manufacturing sector of the economy remains too weak. Even while the black middle class is expanding, unemployment among blacks may be as high as 50 percent—and the gap between rich and poor is growing. Conflicts between locals and immigrants are on the rise. Land reform, an urgent matter in a country where land alienation reached huge proportions, is too slow in the view of many. Add to this the scourge of AIDS and the government's failure to address this crisis effectively, and South Africa's luster is overshadowed by serious problems.

The Middle Tier

Between South Africa's northern border and the region's northern limit lie two groups of states: those with borders adjoining South Africa and those beyond.

Five countries form the ***Middle Tier***, and all border on South Africa: Zimbabwe, Namibia, Botswana, and the ministates of Lesotho and Swaziland (Fig. 6B-2). As the map shows, four of these five are landlocked. Diamond-exporting (and upper-middle-income) **Botswana** occupies the heart of the Kalahari Desert and surrounding steppe; despite its lucrative diamonds, the majority of its 1.8 million inhabitants are subsistence farmers. In 2010 Botswana remained the most severely AIDS-afflicted country in all of tropical Africa. **Lesotho** and **Swaziland**, both traditional kingdoms, depend very heavily on remittances from their workers in South African mines, fields, and factories.

The most important state in the Middle Tier undoubtedly is **Zimbabwe**

(13.8 million), landlocked but well endowed with mineral and agricultural resources. Zimbabwe (the country is named after ancient stone ruins in its interior) is mostly an elevated plateau between the Zambezi and Limpopo rivers, with the desert to the west and the Great Escarpment to the east. Its core area is defined by the mineral-rich Great Dyke and its environs, extending southwest from the vicinity of the capital, Harare, to the country's second city, Bulawayo. Copper, asbestos, and chromium (of which Zimbabwe is one of the world's leading sources) are among its major mineral exports, but Zimbabwe is not just an ore-exporting country. Farms are capable of producing maize (corn), tobacco, tea, sugar, cotton, and other crops. Two nations form most of Zimbabwe's population: the Shona (82 percent) and the Ndebele (14 percent). A tiny minority of whites owned the best farmland and organized the agricultural economy. Mounting environmental and economic problems beginning in the 1980s led to rising social and political tensions. President Mugabe and his dominant party allowed white farms to be invaded by squatters who sometimes killed the owners; corruption rose and human rights were severely curbed. Moreover, this country is not able to feed itself due to mismanagement of the agricultural sector. The economy is in ruins, and the population is declining as people flee the chaos. Once-promising Zimbabwe is the tragedy of the region today.

Southern Africa's youngest independent state, **Namibia** (2.1 million), is a former German colony with a territorial peculiarity: the so-called Caprivi Strip linking it to the right bank of the Zambezi River (Fig. 6B-2), another consequence of colonial partitioning. Administered by South Africa from 1919 to 1990, Namibia is named after one of the world's driest deserts (the Namib, which lines its coast). This state is about as large as Texas and Oklahoma, but only its far north receives enough moisture to permit subsistence farming, which is why most of the people live close to the Angolan border. Mining in the Tsumeb area and ranching in the vast steppe country of the south form the leading commercial activities. The capital, Windhoek, is centrally situated opposite Walvis Bay, the main port. German influence still lingers in what used to be called South West Africa, as does an Afrikaner presence from the apartheid period. Although orderly land reform is underway, unresolved issues remain, and much of the population still lives in poverty.

The Northern Tier

In the four countries that extend across the *Northern Tier* of Southern Africa—Angola, Zambia, Malawi, and Moçambique—problems abound. **Angola** (17.7 million), formerly a Portuguese colony, together with its **3 exclave** (outlier) of Cabinda had a thriving economy based on a wide range of mineral and agricultural exports at the time of independence in 1975. But then Angola fell victim to the Cold War, with northern peoples choosing a communist course and southerners falling under the sway of a rebel movement backed by South Africa and the United States. The results included a devastated infrastructure, idle farms, looting of diamonds, hundreds of thousands of casualties, and millions of landmines that continue to kill and maim. But Angola's oil wealth (the country ranks second in Subsaharan Africa in oil production) yields about U.S. $5 billion per year, and the return of stability is attracting investors to begin rebuilding the ruined country. In the exclave of Cabinda, however, oil-rich Angola now confronts a secessionist movement.

On the opposite coast, the other major former Portuguese dependency, **Moçambique** (21.3 million), fared poorly in a different way. Without Angola's mineral base and with limited commercial agriculture, Moçambique's chief asset was its relative location. Its two major ports, Maputo and Beira, handled large volumes of exports and imports for South Africa, Zimbabwe, and Zambia. But upon independence Moçambique also chose a Marxist course with unfortunate economic and political consequences. Here, too, a rebel movement supported by South Africa caused civil conflict, created famines, and generated a stream of more than a million refugees toward Malawi. Railroad and port facilities lay idle, and Moçambique at one time was ranked by the United Nations as the world's poorest country. In recent years, the port traffic has been somewhat revived, and Moçambique and South Africa are working on a joint Maputo Development Corridor (Fig. 6B-2), but it will take generations for Moçambique to recover.

Landlocked **Zambia** (12.7 million), the product of British colonialism, shares the mineral riches of the Copperbelt with The Congo's Katanga Province. Not only have commodity prices on which Zambia depends fluctuated wildly, but Zambia's outlets—Lobito in Angola as well as Beira in Moçambique—and the railways leading there were made inoperative by Cold War conflicts. In recent years, China has taken an interest in Zambia's minerals, and the Chinese are investing in railroad repairs as well as mining operations.

Neighboring **Malawi** (14.5 million) has an almost totally agricultural economic base and is frequently challenged by environmental degradation. But as the 2000s ended, things looked more positive as some bumper harvests were produced and Malawi was able to export its surplus to Zimbabwe.

EAST AFRICA

To the east of the chain of Great Lakes that marks the eastern border of The Congo (Lakes Albert, Edward, Kivu, and Tanganyika), the land rises from the Congo Basin to the East African Plateau. Hills and valleys, fertile soils, and copious rains mark this transition in Rwanda and Burundi. Eastward the rainforest disappears and the open savanna cloaks the countryside. Great volcanoes tower above a rift-valley-dissected highland. At the heart of the region lies Lake Victoria. In the north the surface rises above 3300 meters (10,000 ft), and so deep are the trenches cut by faults and rivers there that the land was called, appropriately, Abyssinia (now Ethiopia).

Five countries, in addition to the highland component of Ethiopia, form this East African region: Kenya, Tanzania, Uganda, Rwanda, and Burundi (Fig. 6B-4). Here the Bantu peoples that make up most of the population met Nilotic peoples from the north.

Kenya (40.2 million) is neither the largest nor the most populous country in East Africa, but over the past half-century it has been the dominant state in the region. Its skyscrapered capital at the heart of its core area, Nairobi, home to 3.5 million, is the region's largest city and hub for many activities; its port, Mombasa, is the region's busiest.

After independence, Kenya chose the capitalist path of development, aligning itself with Western interests. Without major known mineral deposits, Kenya depended on exports of coffee and tea as well as other food products, and on a tourist industry based on its magnificent landscapes. Tourism became its largest single earner of foreign exchange, and Kenya prospered, apparently proving the wisdom of its economic planners.

But serious problems arose. Kenya during the 1980s had the highest rate of population increase in the world, and population pressure on farmlands and on the fringes of the wildlife reserves mounted. Poaching became widespread, and tourism declined. During the late 1990s, El Niño-induced rains affected parts of Kenya, causing landslides and washing away large segments of the crucial

Kenya has become one of the world's major exporters of flowers, but not without controversy. Here, workers are picking roses in a huge, 2-hectare (5-acre) greenhouse, part of an industry that currently concentrates around Lake Naivasha in the Eastern Rift Valley. European companies have set up these enterprises, buying large swaths of land and attracting workers far and wide. Locals see wildlife and waters threatened, but the government appreciates the revenues from an industry that, in 2009, accounted for about 20 percent of Kenya's agricultural export income. This is a revealing core-periphery issue: companies growing and exporting the flowers to European markets use pesticides not tolerated in Europe itself, so that the ecological damage prevented by regulations in the global core falls on the periphery, where rules are less stringent and the environment suffers. The burgeoning flower industry around Lake Naivasha also has created social problems as thousands of workers, attracted by rumored job opportunities, have converged on the area and find themselves living in squalid conditions. © Marta Nascimento/REA/Redux

highway between Nairobi and Mombasa. This disaster was followed by a severe drought lasting several years, bringing famine to the interior. Meanwhile, government corruption siphoned off funds that should have been invested. Democratic principles were violated, and relationships with Western allies were strained. The AIDS epidemic brought another setback. And then Kenya sustained damage from terrorist attacks in Nairobi and Mombasa, further impacting the tourist industry.

Early in the twenty-first century Kenya was stable and relatively prosperous. However, geography, history, and politics have placed the Kikuyu (22 percent of the population) in a position of power, although there are other major peoples (see Fig. 6B-4) and several smaller groups. The Luhya,

Luo, Kalenjin, and Kamba together constitute about 50 percent of the population, and on the territorial margins of the country there are peoples such as the Maasai, Turkana, Boran, and Galla. Unfortunately, the contested outcome of the 2007 presidential election—the opposition felt cheated by allegedly fraudulent results keeping the Kikuyu president in office—uncorked what had been longstanding tensions, and violence erupted. Kenya's continuing challenge is to overcome this still-simmering crisis and sustain a political system that will ensure democracy and represent the interests of Kenya's disparate peoples.

Tanzania (a hybrid name derived from **Tan**ganyika plus **Zan**zibar) is the biggest and most populous East African country (42.1 million). Its

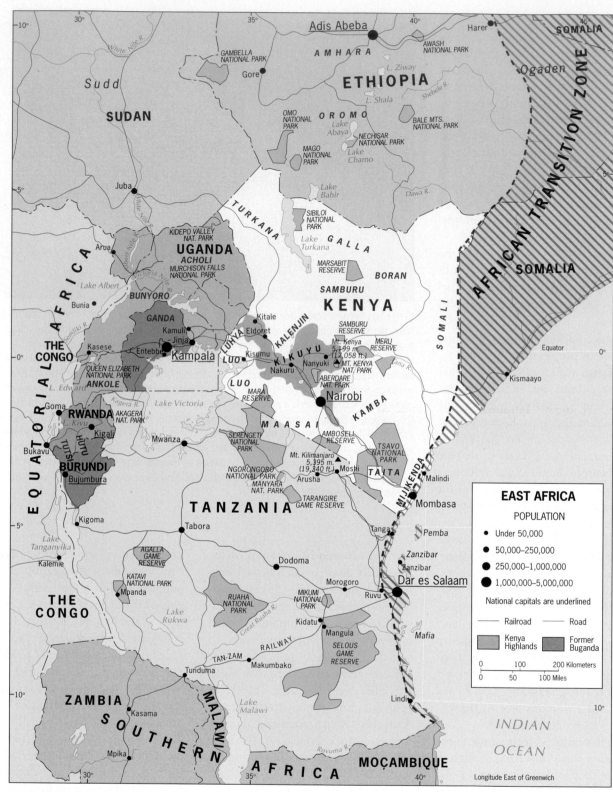

FIGURE 6B-4

© H. J. de Blij, P. O. Muller, and John Wiley & Sons, Inc.

total area exceeds that of the other four countries combined. Tanzania has been described as a country without a core because its clusters of population and the zones of productive capacity lie dispersed—mostly on its margins on the east coast (where the capital, Dar es Salaam, is located), near the shores of Lake Victoria in the northwest, near Lake Tanganyika in the far west, and near Lake Malawi in the interior south. This is in sharp contrast to Kenya, which has a well-defined core area in the Kenya Highlands (centered on Nairobi in the heart of the country). Moreover, Tanzania is a country of many peoples, none numerous enough to dominate the state. About 100 ethnic groups coexist; one-third of the population, mainly on the coast, are Muslims.

After receiving its independence Tanzania embarked on a socialist path toward development, including an extensive but poorly planned farm collectivization program. The tourist industry declined sharply, and Tanzania became one of the world's poorest countries. But Tanzania did achieve remarkable political stability and a degree of democracy that none of the other East African states attained. Since 1990, the government has changed course, but the AIDS crisis, problems with the Zanzibar merger, and involvement in the troubles of neighboring countries, including Rwanda, have been costly. Yet today, Tanzania's prospects are improving as the tourist industry has rebounded and political stability continues.

The country known today as **Uganda** (31.1 million) contained this region's most important African state when the British colonialists arrived: the Kingdom of Buganda, peopled by the Ganda, located on the northwest shore of Lake Victoria (Fig. 6B-4; also see Fig. 6A-5). The British established their headquarters near the Ganda capital of Kampala and utilized the Ganda to control Uganda through indirect rule. Thus the Ganda became the dominant nation in multicultural Uganda, and when the British left, they bequeathed Uganda a complicated federal system designed to perpetuate Ganda supremacy.

The system failed, bringing to power one of Africa's most brutal dictators, Idi Amin. Uganda had a strong economy based on coffee, cotton, and other farm exports, and on copper mining; its Asian minority of about 75,000 dominated local commerce. Amin ousted all the Asians, exterminated his opponents, and destroyed the economy. In addition, the AIDS epidemic struck Uganda severely. Recovery has been slow since Amin's expulsion, but Uganda has advanced toward more representative government and has made significant gains in the struggle against AIDS. Unfortunately, Ugandan forces played a less constructive role in the conflicts involving Rwanda and The Congo.

As the map shows, Uganda is a **4 landlocked state** and depends on Kenya for an outlet to the ocean. Its relative location adjacent to unstable Sudan, Rwanda, and The Congo constitutes a formidable challenge. And in the far north, a rebel group called the Lord's Resistance Army has been terrorizing the local Acholi people for decades.

Rwanda and **Burundi** would seem to occupy Tanzania's northwest corner, and indeed they were part of the German colonial domain conquered before World War I. But during that war Belgian forces attacked the Germans from their Congo bases and were awarded these territories when the conflict ended in 1918. Local peoples had no power to stop this imperial land-grab.

Rwanda (10.1 million) and Burundi (9.4 million), Africa's most densely populated countries, are physiographically part of East Africa, but their cultural geography is linked to the north and west. Here, Tutsi pastoralists from the north subjugated Hutu farmers, who had themselves made serfs of the local Twa (pygmy) population, setting up a conflict that was originally ethnic but became cultural. Certain Hutu were able to advance in the Tutsi-dominated society, becoming to some extent converted to Tutsi ways, leaving subsistence farming behind, and rising in the social hierarchy. These so-called moderate Hutu were—and are—frequently targeted by other Hutus, who resent their position in society. This longstanding discord, worsened by colonial policies, had repeatedly devastated both countries and in 1994 resulted in the horrific Rwanda genocide. Unfortunately, this conflict continues to simmer and has spilled over into The Congo. As many as 4 million people have perished as Hutu, Tutsi, Ugandan, and Congolese rebel forces have fought for control over areas of the eastern Congo. Only massive international intervention could stabilize the situation, but the world has turned a blind eye to the region's woes—again.

The highland zone of **Ethiopia** also forms part of East Africa. Adis Abeba, the historic capital, was the headquarters of a Coptic Christian, Amharic empire that held its own against the colonial intrusion except for a brief period from 1935 to 1941, when the Italians defeated it. Indeed, the Ethiopians, based in their mountain fortress (Adis Abeba lies more than 2400 meters/8000 ft above sea level), became colonizers themselves, taking control of much of the Islamic part of the African Horn to the east.

Ethiopia's natural outlets are toward the Gulf of Adan and the Red Sea, but its government was forced to yield independence to Eritrea (part of the African Transition Zone) and the country is now effectively landlocked. A bitter border war between the two countries, starting in 1998, cost some 100,000 lives. Physiographically and culturally, however, Ethiopia is part of East Africa, and the Amhara and Oromo peoples are neither Arabized nor Muslim: they are Africans. However, Figure 6B-4 shows that functional linkages between Ethiopia and East Africa remain weak, but this is likely to change in the future.

Madagascar

Off Africa's east coast lies the world's fourth-largest island, Madagascar (Fig. 6B-5). About 2000 years ago the first human settlers arrived here—not from Africa but from distant Southeast Asia. A powerful Malay kingdom of the Merina flourished in the highlands, whose language, Malagasy, became Madagascar's tongue (Fig. 6A-9).

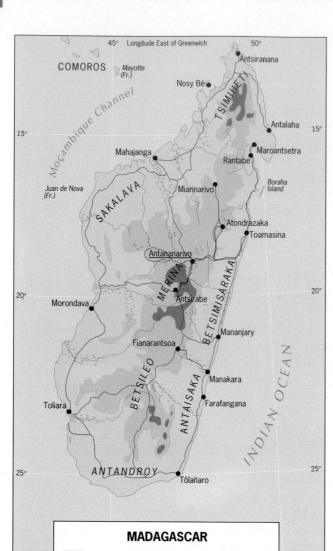

FIGURE 6B-5

© H. J. de Blij, P. O. Muller, and John Wiley & Sons, Inc.

The Malay immigrants eventually brought Africans to their island, but today the Merina (ca. 5 million) and the Betsimisaraka (ca. 3 million) remain the largest of about 20 ethnic groups in the population of 20 million. After successfully resisting colonial invasion, the locals eventually yielded to France, and French became the *lingua franca* of the educated elite.

Madagascar is not part of either East Africa or Southern Africa. Its human as well as wildlife population is unique; its cultural landscapes still carry Southeast Asian imprints. Here the people eat rice grown on terraced paddies, not corn. Population pressure is shrinking forests and animal refuges, the economy is weak, political stability is elusive, and poverty permeates in one of the world's most scenic outposts.

EQUATORIAL AFRICA

The term *equatorial* is not just locational but also environmental. The equator bisects Africa, but only the western two-thirds of central Africa features the conditions associated with the low-elevation tropics: intense heat, high rainfall and extreme humidity, little seasonal variation, rainforest and monsoon-forest vegetation, and enormous biodiversity. To the east, beyond the Western Rift Valley, elevations rise, and cooler, more seasonal climatic regimes prevail. As a result, we recognize two regions in these lowest latitudes: Equatorial Africa to the west and East Africa to the east.

Equatorial Africa is physiographically dominated by the bowl-shaped Congo Basin. The Adamawa Highlands separate this region from West Africa; rising elevations and climatic change mark its southern limits (see the ***Cwa*** boundary in Fig. G-7). Its political geography consists of eight states, of which The Congo (formerly Zaïre) is by far the largest in both territory and population (Fig. 6B-6).

Five of the other seven states— Gabon, Cameroon, Congo, Equatorial Guinea, and São Tomé and Príncipe—all have coastlines on the Atlantic Ocean. The Central African Republic and Chad, the south of which is part of this region, are landlocked. In certain respects, the physical and human characteristics of Equatorial Africa extend even into southern Sudan. This vast and complex region is in many ways the most troubled region in the entire Subsaharan Africa realm.

The Congo

As the map shows, The Congo (known [officially] as the Democratic Republic of the Congo [DRC]) has only a tiny window (37 kilometers/23 mi) on the Atlantic Ocean, barely enough to accommodate the wide mouth of the Congo River. Oceangoing ships can reach the port of Matadi, inland from which falls and rapids make it necessary to move goods by road or rail to the capital, Kinshasa. Lack of navigability also necessitates transshipment between Kisangani and Ubundu, and at Kindu. Follow the railroad south from Kindu, and you reach another narrow corridor of territory at the city of Lubumbashi. That vital part of Katanga Province contains most of The Congo's major mineral resources, including copper and cobalt.

With a territory not much smaller than the United States east of the Mississippi, a population of just over 70 million, a rich and varied mineral base, and much good agricultural land, The Congo would seem to have all the ingredients needed to lead this region and, indeed, Africa. But powerful centrifugal forces, arising from its physiography and cultural geography, pull The Congo apart. The immense forested heart of this basin-shaped country creates communication and transportation barriers between east and west, north and south. Many of The Congo's productive areas lie along its periphery, separated by enormous distances. These areas tend to look across the border, to one or more of The Congo's nine neighbors, for outlets, markets, and often ethnic kinship as well.

Crisis in the Interior

The Congo's civil wars of the 1990s started in one such neighbor, Rwanda, and spilled over into what was then still known as Zaïre. Rwanda has for centuries been the scene of conflict between sedentary Hutu farmers and invading Tutsi pastoralists. Colonial

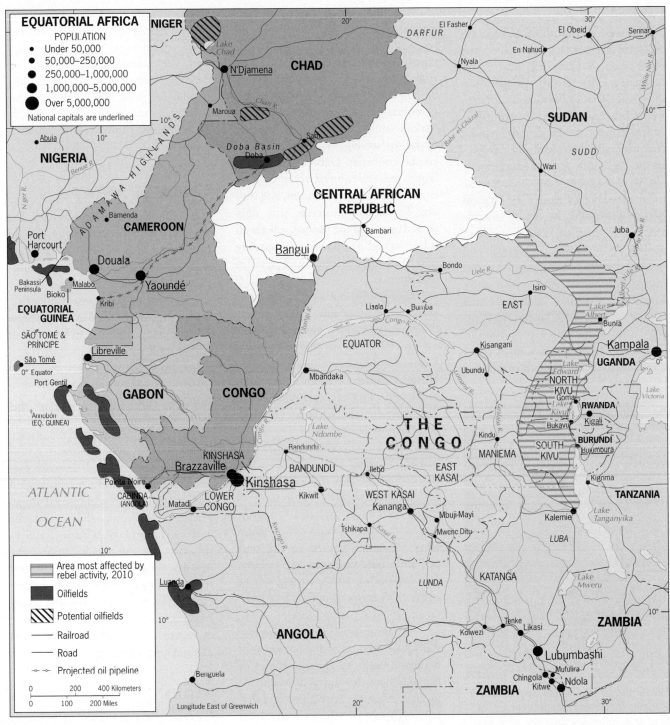

EQUATORIAL AFRICA
POPULATION
- Under 50,000
- 50,000–250,000
- 250,000–1,000,000
- 1,000,000–5,000,000
- Over 5,000,000

National capitals are underlined

Area most affected by rebel activity, 2010

Oilfields

Potential oilfields

Railroad

Road

Projected oil pipeline

0   200   400 Kilometers
0   100   200 Miles

Longitude East of Greenwich

FIGURE 6B-6

© H. J. de Blij, P. O. Muller, and John Wiley & Sons, Inc.

borders and practices worsened the situation, and after independence a series of terrible crises followed. In the mid-1990s the latest of these crises generated one of the largest refugee streams ever seen in the world, and that conflict engulfed eastern (and later northern and western) Congo. The death toll will never be known, but estimates range from 5 to 6 million, a calamity that was not enough to propel the international community into concerted peacemaking action. By 2004 a combination of power transfer in the capital, Kinshasa, negotiation among the rebel groups and the African states involved in the conflict, UN assistance, and general exhaustion had produced a semblance of stability in all but some eastern areas of The Congo (Fig. 6B-6). But in 2007 civil unrest erupted again and continues intermittently, adding to the ever-rising human toll.

Across the River

To the west and north of The Congo and Ubangi rivers lie Equatorial Africa's other seven countries (Fig. 6B-6). Two of them are landlocked. **Chad**, straddling the African Transition Zone as well as the regional boundary with West Africa, is one of Africa's most remote countries, although recent oil discoveries in the south and assistance from China in exploiting those reserves are today changing its status. The **Central African Republic**, chronically unstable and poverty-stricken, never was able to convert its agricultural potential and mineral resources (diamonds, uranium) into real progress. And one country consists of two small, densely forested volcanic islands: **São Tomé and Príncipe**, a ministate with a population of only 220,000 whose economy is being transformed by recent oil discoveries.

Ghana's coastal town of Elmina may be too small to show up on our regional map, but this was the site of the first European settlement in West Africa. It was a key node in Portuguese maritime trade and later became notorious during the slave trade era. Today it is a fishing port of about 20,000, but the industry is not doing well. The local fishers must compete with the larger foreign trawlers that harvest the waters off the coast of West Africa, and cold-storage facilities remain minimal, so that almost all of the local catch is smoked and sold on the local market. This keeps prices low, so that Elmina's economy has changed very little even after the government in 1983 introduced economy-boosting policies. Talk to the locals (fishing and related work are the livelihoods of over half the population), and you realize that it was the capital of Accra that benefited most from the new economic rules, not places like this.
© Richard J. Grant

The four coastal states present a different picture. All four possess oil reserves and share the Congo Basin's equatorial forests; petroleum and timber, therefore, rank prominently among their exports. In **Gabon**, this combination has produced Equatorial Africa's only upper-middle-income economy (see Fig. G-10). Of the four coastal countries, Gabon also has the largest proven mineral resources, including manganese, uranium, and iron ore. Its capital, Libreville (the only coastal capital in the region), reflects all this in its high-rise downtown, bustling port, and rapidly expanding squatter settlements.

Cameroon, not as well endowed with oil or other raw materials, has the region's strongest agricultural sector by virtue of its higher-latitude location and higher-relief topography. Western Cameroon is one of the more developed parts of Equatorial Africa and includes the capital, Yaoundé, and the port of Douala.

With five neighbors, **Congo** (the "other" Congo) could be a major transit hub for this region, especially for The Congo (DRC) if it recovers from civil war. Its capital, Brazzaville, lies across the Congo River from Kinshasa and is linked to the port of Pointe Noire by road and railway. But devastating power struggles have negated Congo's geographic advantages.

As Figure 6B-6 shows, **Equatorial Guinea** consists of a rectangle of mainland territory and the island of Bioko, where the capital of Malabo is located. A former Spanish colony that remained one of Africa's least-developed territories, Equatorial Guinea, too, has been affected by the oil business in this area. Petroleum products now dominate its exports, but, as in so many other oil-rich countries, this bounty has not significantly raised incomes for most of the people.

One other territory would seem to be a part of Equatorial Africa: **Cabinda**, wedged between the two Congos just to the north of the Congo

River's mouth. But Cabinda is one of those colonial legacies on the African map—it belonged to the Portuguese and was administered as part of Angola. Today it is an exclave of independent Angola, and a valuable one: it contains major oil reserves. But, as noted earlier, problems of secession loom here.

WEST AFRICA

West Africa extends south from the margins of the Sahara to the Gulf of Guinea coast and from Lake Chad west to Senegal (Fig. 6B-7). Politically, the broadest definition of this region includes all those states that lie to the south of Western Sahara, Algeria, and Libya and to the west of Chad (itself sometimes included) and Cameroon. Within West Africa, a rough division can be made between the large, mostly steppe-and-desert states that extend across the southern Sahara (Chad included) and the smaller, better-watered coastal states.

Enormous cultural diversity predated the colonial era in this region and continues to be exhibited there. France and Britain dominated the colonial map of West Africa, and to this day the interaction among the region's states is limited. But West Africa's cultural vitality, historic legacies, populous cities, crowded countrysides, and bustling markets combine to create a regional imprint that is distinct and pervasive.

Nigeria

Nigeria, the region's cornerstone, is home to just over 155 million people, by far the largest population of any African country. When Nigeria achieved full independence from Britain in 1960, its new government was faced with the daunting task of administering a European political creation containing three major nations and nearly 250 other peoples ranging from several million to a few thousand in number.

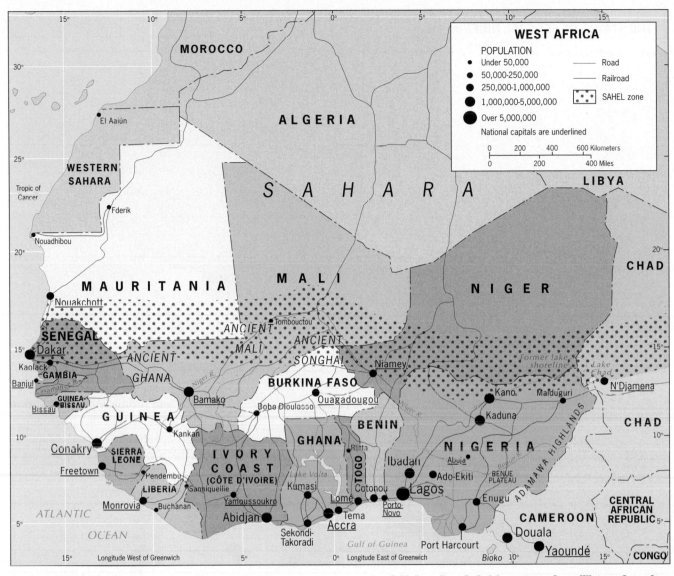

FIGURE 6B-7

© H. J. de Blij, P. O. Muller, and John Wiley & Sons, Inc.

For reasons obvious from the map (Fig. 6B-8), Britain's colonial imprint always was stronger in the two southern regions than in the north. Christianity became the dominant faith in the south, and southerners, especially the Yoruba, took a lead role in the transition from colony to independent state. The choice of Lagos, the port of the Yoruba-dominated southwest, as the capital of a federal Nigeria (and not one of the cities in the more populous north) reflected British desires for the country's future. A three-region federation, two of which lay in the south, would ensure the primacy of the non-Islamic part of the state. But this framework did not last long. In 1967, the Ibo-dominated Eastern Region declared its independence as the Republic of Biafra, leading to a three-year civil war at a cost of 1 million lives. Since then, Nigeria's federal system has been modified repeatedly; today there are 36 States, and the capital has been moved from Lagos to more-centrally-located Abuja (Fig. 6B-8).

Fateful Oil

Large oilfields were discovered beneath the Niger Delta during the 1950s, when Nigeria's agricultural sector produced most of its exports (peanuts, palm oil, cocoa, cotton) and farming still had priority in national and State development plans. Soon, revenues from oil production dwarfed all other sources, bringing the country a brief period of prosperity and promise. But before long Nigeria's petroleum riches brought more bust than boom. Misguided development plans now focused on grand, ill-founded industrial schemes and expensive luxuries such as a national airline; the continuing mainstay of the vast majority of Nigerians, agriculture, fell into neglect. Worse, poor management, corruption, outright theft of oil revenues during military misrule, and excessive borrowing against future oil income led to economic disaster. The country's

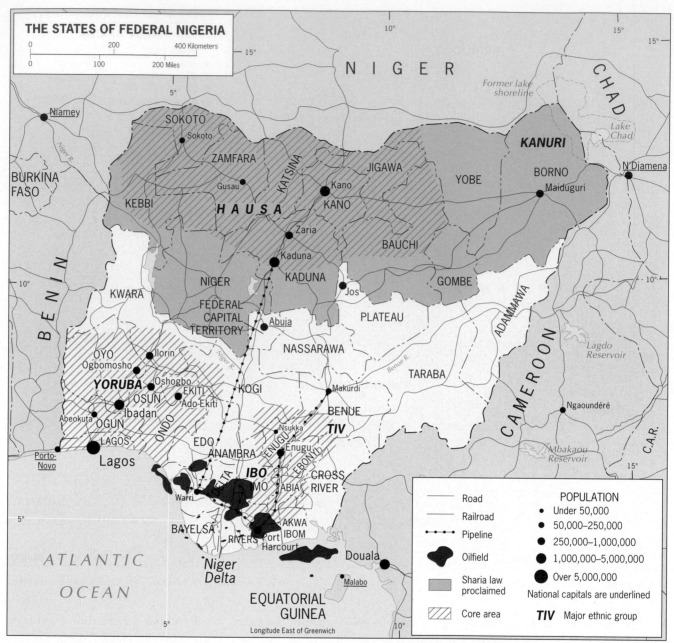

FIGURE 6B-8

© H. J. de Blij, P. O. Muller, and John Wiley & Sons, Inc.

infrastructure collapsed. In the cities, basic services broke down. In the rural areas, clinics, schools, water supplies, and roads to markets crumbled. In the area of the Niger Delta, local people beneath whose land the oil was being exploited demanded a share of the revenues and reparations for ecological damage; the military regime under General Abacha responded by arresting and executing nine of their leaders. On global indices of national well-being, Nigeria sank to the lowest rungs even as its production ranked it

as high as the world's tenth-largest oil producer, with the United States its chief customer.

Islam Ascendant

In 1999, Nigeria's hopes were raised when, for the first time since 1983, a democratically elected president was sworn into office. But Nigeria's problems (now also including a deepening AIDS crisis) worsened when northern States, beginning with Zamfara, proclaimed **5** **Sharia** (strict Islamic) **law.**

When Kaduna State followed suit, riots between Christians and Muslims devastated the venerable capital city of Kaduna. There, and in 11 other northern States (Fig. 6B-8), the imposition of Sharia law led to the departure of thousands of Christians, intensifying the cultural fault line that threatens the cohesion of the country. Although Kaduna State temporarily repealed its decision, the Islamic revivalism now exhibited in the north raises the prospect that Nigeria, West Africa's cornerstone and one of this realm's most important states, may succumb to devolutionary forces arising from its location in the African Transition Zone. For Africa, that would be a calamity.

Coast and Interior

Nigeria is one of 17 states (counting Chad and offshore Cape Verde) that form the region of West Africa. Four of these countries, comprising a huge territory under desert and steppe environments but containing small populations (Fig. 6B-7), are landlocked: Burkina Faso, Niger, Mali, and Chad (the latter two are discussed in Chapter 7B).

Burkina Faso, with the majority of its population Muslim, is fully within the Islamic world. It is a poor landlocked state with undeveloped reserves of gold; its economy relies on exports of cotton. Among the coastal states, **Ghana**, once known as the Gold Coast, was the first West African state to achieve independence, with a democratic government and a sound economy based on cocoa exports. A pair of grandiose post-independence schemes can be seen on the map: the port of Tema, intended to serve a vast West African hinterland, and Lake Volta, which resulted from the region's largest dam project. Mismanagement contributed to Ghana's economic collapse, but in the 1990s, a military regime was replaced by a stable and democratic government and recovery began. **Ivory Coast** (officially Côte d'Ivoire) translated three decades of autocratic but stable rule into economic progress, but excesses by the country's president-for-life cost it dearly. One of these excesses involved the construction of a Roman Catholic basilica to rival St. Peter's in Rome in the president's home village, Yamoussoukro, also designated to replace Abidjan as the country's new capital. At the turn of this century, the political succession became entangled in a north-south, Muslim-Christian schism that threatened the country's stability. Democratic **Senegal**, on the far west coast, demonstrates what stability can achieve: without oil, diamonds, or other valuable income sources and with an overwhelmingly subsistence-farming population, Senegal nevertheless managed to achieve some of the region's highest GNI levels. Almost 95 percent Muslim and dominated by the Wolof, the ethnic group concentrated in the capital of Dakar, and with continuing close ties to France, Senegal has even been able to overcome both a failed effort to unify with its English-speaking enclave, **Gambia**, and a secession movement in its southwestern Casamance district.

Other parts of West Africa have been afflicted by civil war and horror. **Liberia**, founded in 1822 by freed American slaves who returned to Africa with the help of American colonization societies, was governed by Americo-Liberians for six generations and sold rubber and iron ore abroad.

FROM THE FIELD NOTES...

© H. J. de Blij

"Wherever you go in West Africa, you are struck by the bright colors, especially in the markets. And Freetown, Sierra Leone on this Friday seems to be one huge market, the streets lined with stalls, young and old crowding intersections in huge numbers. Most of the goods were for local trade; compared to what you see in East Africa, where tourism plays a bigger role, these markets still have a strong domestic atmosphere. Music plays everywhere, in shops, out in the open; and the whole city seems suffused with the aromas of spicy foods."

www.conceptcaching.com

But a military coup in 1980 ended that era, and full-scale civil war beginning in 1989 embroiled virtually every ethnic group in the country. About a quarter of a million people, one-tenth of the population, perished; hundreds of thousands of others fled as refugees, many to Sierra Leone, which also was originally founded as a haven for freed slaves, in this case by the British in 1787. **Sierra Leone** went the all-too-familiar route from self-governing Commonwealth member to republic to one-party state to military dictatorship. In the 1990s, worse was to follow: a rebel movement, funded by diamond sales, inflicted dreadful punishment on the local population. Early in the twenty-first century a remarkable turnabout has occurred as free elections were held and the country is stable and rebuilding. The same cannot yet be said of **Guinea**, where dictatorial mismanagement and a violent power struggle in 2009–2010 have ruined economic opportunities in both agriculture and mining.

We should remember, however, that all these conflicts we have mentioned are not representative of vast and populous West Africa. Tens of millions of farmers and herders who manage to cope with fast-changing environments in this zone between desert and ocean live remote from the newsmaking conflicts along the well-watered coast. Time-honored systems continue to serve: for example, the local village markets that drive the traditional economy. Visit the countryside, and you will find that some village markets are not open every day but, rather, every three or four days. Such a system ensures that all villages get a share in the exchange network. These **7** **periodic markets** represent one of the many traditions enduring here, even as the cities beckon the farmers and burst at the seams. This region's great challenges are economic survival and nation-building through political stability under some of the most challenging circumstances in the world.

THE AFRICAN TRANSITION ZONE

As Figures 6B-9 and 6B-1 show, the African Transition Zone is unlike the four regions just discussed. This is where Subsaharan Africa's cultures intersect with the world of Islam, and it is Islam's gateway into the Subsaharan African realm. It encompasses some entire countries that are also part of the formal regions on the map (such as Senegal and Niger), it divides others into Muslim and non-Muslim sectors (such as Chad, Sudan, and Nigeria), and, in Africa's Horn, it comprises countries that do not form parts of other regions (Eritrea, Djibouti, Somalia). As elsewhere in the world where geographic realms meet or overlap, complications mark the African Transition Zone. In some areas, the transition from Muslim to non-Muslim society is gradual, as it remains in the northern parts of Ivory Coast and Nigeria despite the recent cultural problems erupting in these important states. Elsewhere the break is sharp, as it is in eastern Ethiopia and Sudan where a traditional border between Christian/animist-African cultures and Muslim-African communities is abrupt. The southern border of the African Transition Zone, the religious frontier sometimes referred to as Africa's **8** **Islamic Front**, is therefore neither static nor everywhere the same.

Conflict marks much of the African Islamic Front (mapped in Fig. 6B-9): long-term in Sudan, where a

To the people living in Nigeria's "middle belt," the Islamic Front is much more than a geographic concept: it is like living in an earthquake zone where things can fall apart in a moment. In Figure 6B-8, you can see Plateau State bordering Islamic-law Kaduna and Bauchi States, with a land extension northward separating these two. In that land extension is where Jos, the State capital, is situated, and in that city Muslim and Christian tensions are never far from the surface. In late November 2008 local elections led to rumors that the mainly Christian Peoples Democratic Party had won over the mainly Muslim All Nigerian Peoples Party. Riots broke out, and in a few days hundreds died and much property destruction took place. Mosques and churches, as well as more than 3000 shops in one market alone, were destroyed before the government could restore order. This photo, taken in early December, shows the Nigerian military separating an advancing group of protesters from their targets by barricading a road in central Jos while local elders try to calm the people. It took the national army and State law enforcement several more days to control the violence, but it will not be the last time such strife will erupt along Nigeria's Islamic Front. © AFP/Getty Images, Inc.

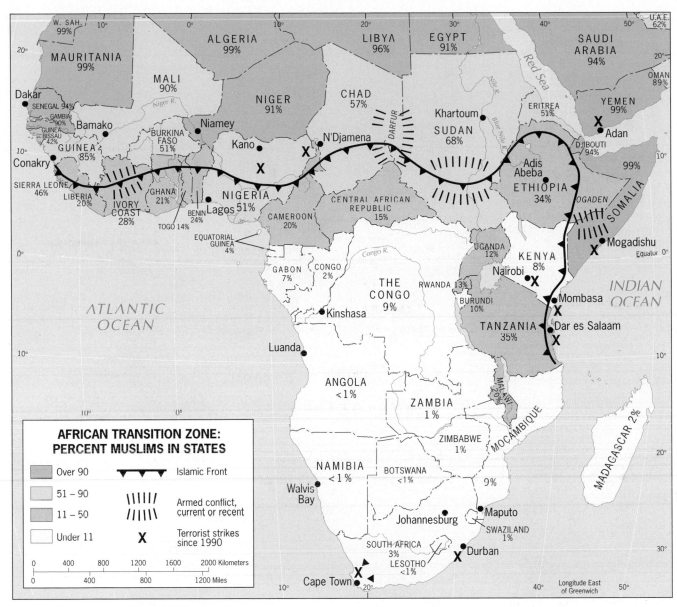

FIGURE 6B-9

© H. J. de Blij, P. O. Muller, and John Wiley & Sons, Inc.

30-year war for independence by non-Islamic African rebels in the south against the Arabized, Muslim north has cost millions of lives; intermittent in Nigeria, where Islamic revivalism is currently clouding the country's prospects; and recent in Ivory Coast, where political rivalries spurred religious strife in a once stable country. Perhaps the most conflict-prone part of the African Transition Zone, however, lies in the east, involving not only Sudan but also the historic Christian state of Ethiopia, nearly encircled by dominantly Muslim societies.

The Horn of Africa

Africa's Horn (Fig. 6B-10) is an especially volatile subregion of the African Transition Zone. In landlocked **Ethiopia**, more than one-third of the population of just over 83 million is Muslim, virtually all of it concentrated in the east. The costly conflict with neighboring **Eritrea**, which is about 50 percent Muslim, has damaged the economies of both countries (that issue was not over religion but over boundary definition). The mini-state of **Djibouti**, about 95 percent Muslim, has taken on added impor-

tance since it lies directly across from Yemen and overlooks the narrow entry to the Red Sea, the Bab el Mandeb Strait, a ***choke point*** in international commerce. But the key component in the eastern sector of the African Transition Zone is **Somalia**, where 9.5 million people, virtually all Muslim, live in a harsh desert that forces cross-border migration into Ethiopia's Ogaden area in pursuit of seasonal pastures. As many as 3 to 5 million Somalis live permanently on the Ethiopian side of the border, but this is not the only division the Somali "nation" faces. In reality, the Somali

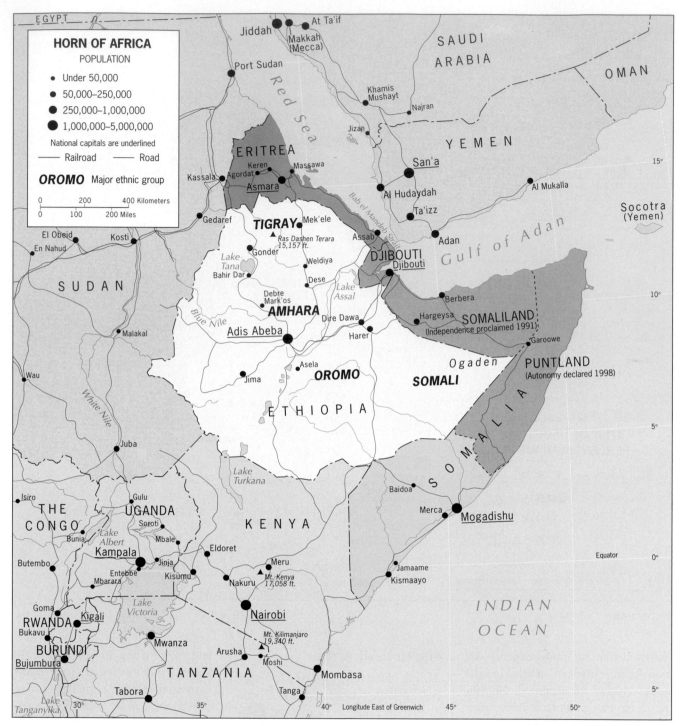

FIGURE 6B-10

© H. J. de Blij, P. O. Muller, and John Wiley & Sons, Inc.

people is an assemblage of five major ethnic groups fragmented into hundreds of clans engaged in an endless contest for power as well as survival.

In the early 2000s, Somalia's condition as a **9 failed state** led to the country's fragmentation into three parts (Fig. 6B-10). In the north, the sector called *Somaliland* proclaimed

independence in the 1990s and remains by far the most stable of the three. It functions essentially as an African state, although the international community has not recognized it as such. In the east, a conclave of local chiefs declared their territory to be separate from the rest of Somalia, calling it *Puntland*, and asserted an

unspecified degree of autonomy. In the south, where the official capital, Mogadishu, is located on the Indian Ocean coast, local secular warlords and Islamic militias continued their struggle for supremacy, the warlords supported by U.S. funding. In 2006, Islamic militias stormed into the capital and took control, ousting the warlords

This boat photographed near Hobyo, Somalia is one of about 100 speedboats operated by the so-called Central Regional Coast Guard, a pirate group formed in 2005 with about 400 men in its ranks today. As recently as 2008, this organization attacked 29 ships in the waters off eastern Africa, eventually extorting a total of $U.S. 10 million from their owners. More recently, stronger protection of commercial shipping in the area has reduced the pirates' success rate, and some pirates have been killed by defensive actions. But the threat continues. © Veronique de Viguerie/Getty Images, Inc.

and proclaiming their determination to create an Islamic state. The continuing combination of a failed state, Islamic fundamentalists in control, and al-Qaeda in the mix makes this part of the African Transition Zone a persistent focus of international concern.

As this chapter has underscored, Africa is a continent of infinite social diversity, and Subsaharan Africa is a realm of matchless cultural history. Relatively recent environmental change has widened the Sahara, which separates north from south. The push of Islam turned the north to Mecca and the south to Christian proselytism. Millions of Africans were carried away in bondage; the cruelties of colonialism subjugated those who were left

behind into a political straitjacket forged in Berlin. Africa's equatorial heart is an incubator for debilitating diseases second to none in the world, but Africa's health problems have, until recently, not been a top priority in the medical world. When the richer countries finally abandoned their African empires, African peoples were left

a task of reconstruction in which they received little help and found their products uncompetitive on markets where rich-country producers basked in subsidies. We said at the beginning of Chapter 6A that Africa's time and turn will come again, but it will take far more than is being done today to rectify half a millennium's malefactions.

POINTS TO PONDER

- Land reform in South Africa is lagging dangerously.
- According to the latest data, South Africa is now the most unequal society in the world.
- Rwanda's government plans to give every child aged 9–12 a virtually unbreakable laptop with the capacity to download free educational software and electronic books.
- In 2010, Angola was China's top oil supplier, having surpassed both Saudi Arabia and Iran.

Minaret among the derricks as mosque and oil wells vie for space on Azerbaijan's Caspian shore.
© Sergei Supinsky/epa/© Corbis

7A

NORTH AFRICA/SOUTHWEST ASIA: DEFINING THE REALM

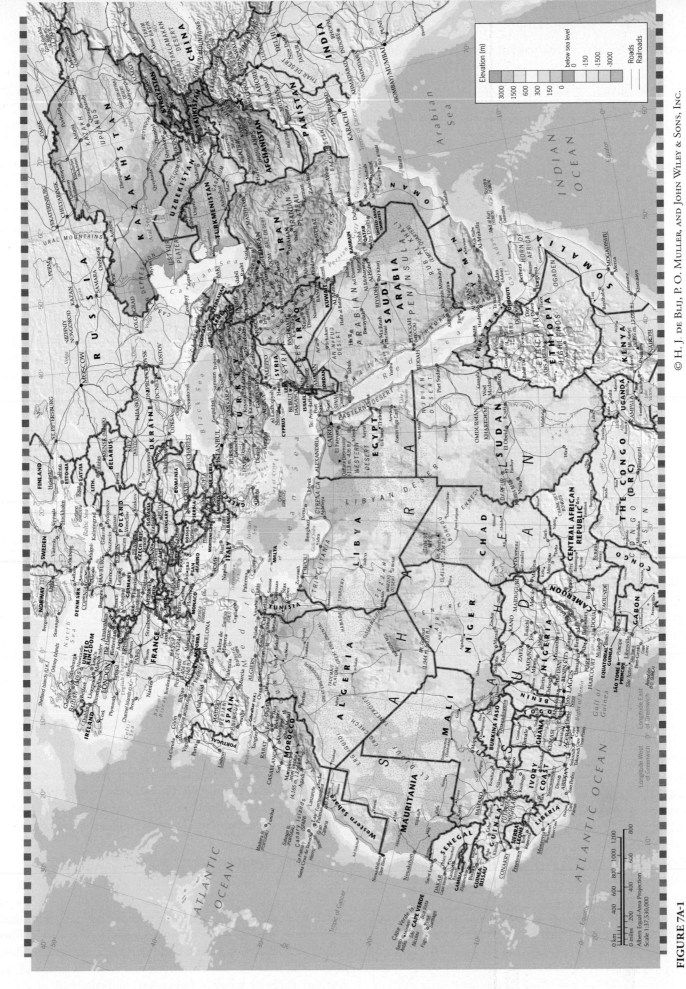

FIGURE 7A-1

© H. J. de Blij, P. O. Muller, and John Wiley & Sons, Inc.

From Morocco on the shores of the Atlantic to the mountains of Afghanistan, and from the Horn of Africa to the steppes of inner Asia, lies a vast geographic realm of enormous cultural complexity. It stands at the crossroads where Europe, Asia, and Africa meet, and it is part of all three continents (Fig. 7A-1). Throughout history, its influences have radiated to these continents and to practically every other part of the world as well. This is one of humankind's primary source areas. On the Mesopotamian Plain between the Tigris and Euphrates rivers (in modern-day Iraq) and on the banks of the Egyptian Nile arose several of the world's earliest civilizations. In its soils, plants were domesticated that are now grown from the Americas to Australia. Along its paths walked prophets whose religious teachings are still followed by hundreds of millions. And early in the second decade of the twenty-first century, the heart of this realm is beset by some of the most bitter and dangerous conflicts in the world.

NAMING THIS PIVOTAL REALM

As you may conclude from its long and somewhat cumbersome name—North Africa/Southwest Asia—this is a sprawling and geographically complex realm. In our era of high-speed communication you will sometimes see it referred to as *NASWA*, the first letters of its regional deployment. And it is tempting to refer to it in other kinds of geographic shorthand, based on some of its dominant features. But such generalizations can be very misleading, and some examples follow.

A "Dry World"?

This realm is, for instance, often called the Dry World, containing as it does the vast Sahara as well as the Arabian Desert. But most of the realm's people live where there is water—along the Nile River, along the hilly Mediterranean coastal strip or *tell* (meaning mound in Arabic) of northwesternmost Africa, along the Asian eastern and northeastern shores of the Mediterranean Sea, in the Tigris-Euphrates Basin, in far-flung desert oases, and along the lower mountain slopes of Iran south of the Caspian Sea as well as Turkestan to the northeast. Figure 7A-2 reflects this close relationship between population distribution and available water. Many of this realm's prominent geographic features—river valleys and deltas, moist coastlines, well-watered mountain basins—are virtually defined by the clusters you see on this map.

So we know this world realm as a place where water is almost always at a premium, where consumers (such as Israelis and Palestinians) compete for limited supplies, where peasants often struggle to make soil and moisture yield a small harvest, where nomadic peoples and their livestock still migrate across dust-blown flatlands, where oases are green islands in the desert that sustain local farming, supply weary travelers, and support trade across vast expanses of aridity.

But this is also the land of the Nile, the lifeline of Egypt, the crop-covered *tell* of coastal Algeria, the verdant shores of western Turkey, and the meltwater-fed valleys of Central Asia. Compare Figure 7A-1 to Figure G-7, and the dominance of *B* climates becomes evident, underscoring just how water-dependent this realm's clustered population is.

Is *This* the "Middle East"?

This realm is commonly referred to as the Middle East. That must sound odd to someone in, say, India, who might think of a Middle West rather than a Middle East! The name, of course, reflects the biases of its source: the Western world, which saw a Near East in Turkey; a Middle East in Egypt, Arabia, and Iraq; and a Far East in China and Japan. Still, the term has taken hold, and it can be seen and heard in everyday usage by scholars, journalists, and members of the United

MAJOR GEOGRAPHIC QUALITIES

NORTH AFRICA/SOUTHWEST ASIA

1. North Africa and Southwest Asia were the scene of several of the world's great ancient civilizations, based in its river valleys and basins.

2. From this realm's culture hearths diffused ideas, innovations, and technologies that changed the world.

3. The North Africa/Southwest Asia realm is the source of three world religions: Judaism, Christianity, and Islam.

4. Islam, the most recent of the major religions to arise in this realm, transformed, unified, and energized a vast domain extending from Europe to Southeast Asia and from Russia to East Africa.

5. Drought and unreliable precipitation dominate natural environments in this realm. Population clusters exist where water supply is adequate to marginal.

6. Certain countries of this realm have enormous reserves of oil and natural gas, creating great wealth for some but doing little to raise the living standards of the majority.

7. The boundaries of the North Africa/Southwest Asia realm consist of volatile transition zones in several places in Africa and Asia.

8. Conflict over water sources and supplies is a constant threat in this realm, where population growth rates are high by world standards.

9. The Middle East, as a region, lies at the heart of this realm; and Israel lies at the center of the Middle East conflict.

10. Religious, ethnic, and cultural discord often cause instability and strife in this realm.

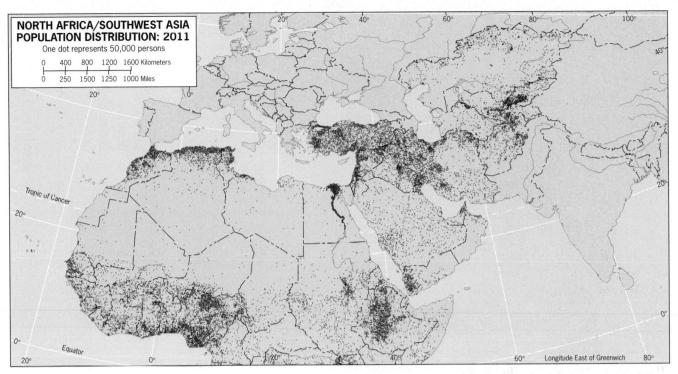

FIGURE 7A-2

© H. J. de Blij, P. O. Muller, and John Wiley & Sons, Inc.

Nations. Even so, we feel that it should only be applied to one of the regions of this sprawling realm, not to the realm as a whole.

An "Arab World"?

Another shorthand label often used for North Africa/Southwest Asia is the Arab World. This term implies a uniformity that does not actually exist. First, the name *Arab* is applied loosely to the peoples of this area who speak Arabic and related languages, but ethnologists normally restrict it to certain occupants of the Arabian Peninsula—the Arab "source." In any case, Turks are not Arabs, and neither are most Iranians or Israelis. Moreover, although the Arabic language prevails from Mauritania in the west across all of North Africa to the Arabian Peninsula, Syria, and Iraq in the east, it is not spoken in other parts of this realm. In Turkey, for example, Turkish is the major language, and it has Ural-Altaic rather than Arabic's Semitic or Hamitic roots; the Iranian language belongs to the Indo-European linguistic family (see Fig. G-9). Other Arab World languages that have separate ethnological identities are spoken by the Jews of Israel, the Tuareg people of the Sahara, the Berbers of northwestern Africa, and the peoples of the wide transition zone between North Africa and Subsaharan Africa to the south.

An "Islamic World"?

Lastly, a name given to this realm is the World of Islam. The prophet Muhammad (Mohammed) was born in Arabia in AD 571, and in the centuries after his death in 632, Islam spread into Africa, Asia, and Europe. This was the age of Arab conquest and expansion. Their armies penetrated Southern Europe, their caravans crossed the deserts, and their ships plied the coasts of Asia and Africa. Along these routes they carried the Muslim (Islamic) faith, converting the ruling classes of the states of the West African savanna, threatening the Christian stronghold in the highlands of Ethiopia, penetrating the deserts of inner Asia, and pushing into India and even the island extremities of Southeast Asia. Today, the Islamic faith extends far beyond the limits of the realm under discussion (Fig. 7A-3). Nor is the World of Islam entirely Muslim. Judaism, Christianity (notably in Egypt and Lebanon), and other faiths survive in the heartland of the Islamic World. So this connotation is not satisfactory either.

HEARTHS OF CULTURE

This geographic realm occupies a pivotal part of the world: here Africa, the source of humanity, meets Eurasia, crucible of human cultures. Two million years ago, the ancestors of our species walked from East Africa into North Africa and Arabia and spread all across Asia. Less than one hundred thousand years ago, *Homo sapiens* crossed these lands on their way to Europe, Australia, and, eventually, the Americas. Ten thousand years ago, human communities in what we now call the Middle East began to domesticate plants and animals, learned to irrigate their fields, enlarged their settlements into towns, and formed the earliest states. The world's dominant monotheistic religions originated here,

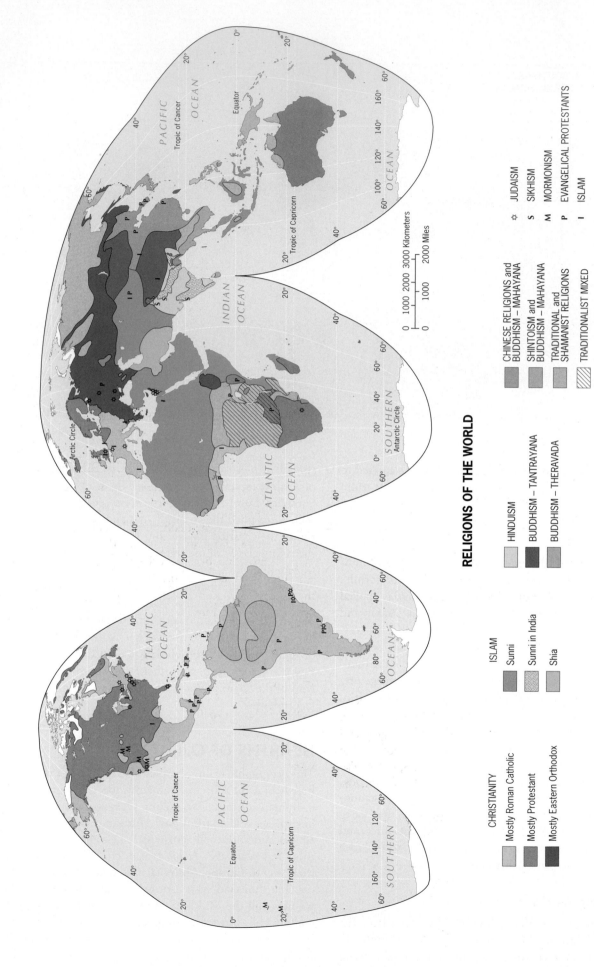

RELIGIONS OF THE WORLD

CHRISTIANITY

Mostly Roman Catholic

Mostly Protestant

Mostly Eastern Orthodox

ISLAM

Sunni

Sunni in India

Shia

HINDUISM

BUDDHISM – TANTRAYANA

BUDDHISM – THERAVADA

CHINESE RELIGIONS and
BUDDHISM – MAHAYANA

SHINTOISM and
BUDDHISM – MAHAYANA

TRADITIONAL and
SHAMANIST RELIGIONS

TRADITIONALIST MIXED

☆ JUDAISM

S SIKHISM

M MORMONISM

P EVANGELICAL PROTESTANTS

I ISLAM

0 1000 2000 3000 Kilometers

0 1000 2000 Miles

FIGURE 7A-3

© H. J. de Blij, P. O. Muller, and John Wiley & Sons, Inc.

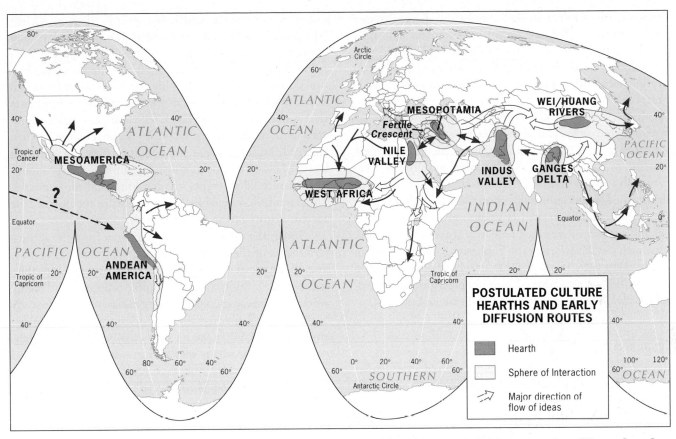

FIGURE 7A-4

© H. J. DE BLIJ, P. O. MULLER, AND JOHN WILEY & SONS, INC.

first Judaism, then Christianity, and most recently the heart of the realm was stirred and mobilized by the teachings of Muhammad and the Quran (Koran), and Islam began. Today this realm is a cauldron of religious and political activity and turmoil, weakened by conflict but empowered and enriched by oil in some places, plagued by poverty in many others.

Dimensions of Culture

In the introductory chapter, we discussed the concept of culture and its regional expression in the cultural landscape. **1 Cultural geography**, we noted, is a wide-ranging and comprehensive field that studies spatial aspects of human cultures, focusing not only on cultural landscapes but also on **2 culture hearths**—the crucibles of civilization, the sources of ideas, innovations, and ideologies that changed regions and realms. Those ideas and innovations spread far and wide through a set of processes that we study under the rubric of **3 cultural diffusion**. Because we understand these processes better today, we can reconstruct many of the ancient routes by which the knowledge and achievements of culture hearths spread—that is, *diffused*—to other areas.

Another aspect of cultural geography, also relevant in the context of the North Africa/Southwest Asia realm, is the study of **4 cultural landscapes** that a dominant

culture creates. Human cultures exist in long-term accommodation with (and adaptation to) their natural environments, exploiting opportunities that these environments present and coping with the extremes they can impose. In the process, they fuse their physical landscape and cultural landscape into an interacting unity.

Rivers and Communities

In the basins of the major rivers of this realm (the Tigris-Euphrates system of modern-day Turkey, Syria, and Iraq as well as the Nile of Egypt) lay two of the world's earliest culture hearths (Fig. 7A-4). Mesopotamia (land amidst the rivers) had fertile alluvial soils, abundant sunshine, ample water, and animals and plants that could be domesticated. Here, in the Tigris-Euphrates lowland between the Persian Gulf and the uplands of present-day Turkey, arose one of humanity's first culture hearths, a cluster of communities that grew into larger societies and, eventually, into the world's first states. (Early state development probably was going on simultaneously in East Asia's river basins as well.)

Mesopotamia

Mesopotamians were innovative farmers who domesticated crops such as wheat and knew when to sow and harvest various crops, water their fields, and store their surpluses. Their knowledge diffused to villages near and far, and a *Fertile Crescent*, a region of significant agricultural productivity, evolved, extending from Mesopotamia across southeastern Turkey into Syria and the eastern Mediterranean coast beyond.

Irrigation was the key to prosperity and power in Mesopotamia, and urbanization was its reward. Among many settlements in the Fertile Crescent, some thrived, grew, enlarged their hinterlands, and diversified socially and occupationally; others failed. What determined success? One theory, the **5 hydraulic civilization theory**, holds that cities that could control irrigated farming over large hinterlands held power over others, used food as a weapon, and thrived. One such city, Babylon on the Euphrates River, endured for nearly 4000 years (from 4100 BC). A busy port, its walled and fortified center endowed with temples, towers, and palaces, Babylon for a time was the world's largest city.

Egypt and the Nile

Egypt's cultural evolution may have started even earlier than Mesopotamia's, and its focus lay upstream from (south of) the Nile Delta and downstream from (north of) the first of the Nile's series of rapids, or cataracts. This part of the Nile Valley is surrounded by inhospitable desert, and unlike Mesopotamia (which lay open to all comers), the Nile provided a natural fortress here. The ancient Egyptians converted their security into progress. The Nile was their highway of trade and interaction; it also supported agriculture through irrigation. The Nile's cyclical ebb and flow was much more predictable than that of the Tigris-Euphrates river system. By the time Egypt fell victim to outside invaders (about 1700 BC), a full-scale urban civilization had emerged. Ancient Egypt's artist-engineers left a magnificent legacy of massive stone monuments, some of them containing treasure-filled crypts of god-kings called Pharaohs. These tombs have enabled archeologists to reconstruct the ancient history of this culture hearth.

Today, the world continues to benefit from the accomplishments of the ancient Mesopotamians and Egyptians. They domesticated cereals (wheat, rye, barley), vegetables (peas, beans), fruits (grapes, apples, peaches), and many animals (horses, pigs, sheep). They also advanced the study of the calendar, mathematics, astronomy, government, engineering, metallurgy, and a host of other skills and technologies. In time, many of their innovations were adopted and then modified by other cultures in the Old World and eventually in the New World as well. Europe was the greatest beneficiary of these legacies of Mesopotamia and ancient Egypt, whose achievements constituted the foundations of Western civilization.

 FROM THE FIELD NOTES...

© Jan Nijman

"Flying north over central Egypt from Luxor's ruins toward Cairo along the Nile's ribbon of green, I pondered the famous remark by the ancient Greek historian Herodotus that "Egypt is the gift of the River Nile." The Nile brings water from the interior of Africa that crosses the vast northern deserts and is crucially important to a country with as little rainfall as Egypt. About 95 percent of the Egyptian population lives within 20 kilometers (12 mi) of the river's banks. Almost all of Egypt's agriculture has always depended on irrigation, which is how this narrow strip of desert turned fertile. Notice the razor-sharp boundaries between irrigated and non-irrigated land. Without a doubt, the Nile was the lifeline that supported one of the ancient world's major civilizations."

www.conceptcaching.com

Decline and Decay

As Figure G-7 reminds us, many of the early cities of this culture hearth lay in what is today desert territory. Assuming that there were no good reasons to build large settlements in the middle of deserts, we may hypothesize that **6** **climate change** sweeping over this region, not a monopoly over irrigation techniques, gave certain cities in the ancient Fertile Crescent an advantage over others. Climate change, associated with shifting environmental zones after the final Pleistocene glacial retreat, may have destroyed the last of the old civilizations. Perhaps overpopulation and human destruction of the natural vegetation contributed to the process. Indeed, some cultural geographers suggest that the momentous innovations in agricultural planning and irrigation technology were not "taught" by the seasonal flooding of the rivers but were forced on the inhabitants as they tried to survive changing environmental conditions.

The scenario is not difficult to imagine. As outlying areas began to fall dry and farmlands were destroyed, people congregated in the already crowded river valleys—and made every effort to increase the productivity of the land that could still be watered. Eventually overpopulation, destruction of the watershed, and perhaps reduced rainfall in the rivers' headwater areas dealt the final blow. Towns were abandoned to the encroaching desert; irrigation canals filled with drifting sand; croplands dried up. Those who could migrated to areas that were still reputed to be productive. Others stayed, their numbers dwindling, increasingly reduced to subsistence.

As old societies disintegrated, power emerged elsewhere. First the Persians, then the Greeks, and later the Romans imposed their imperial designs on the tenuous lands and disconnected peoples of North Africa/Southwest Asia. Roman technicians converted North Africa's farmlands into irrigated plantations whose products went by the boatload to Roman Mediterranean shores. Thousands of people were carried off as slaves to the cities of the new conquerors. Egypt was quickly colonized, as was the area we now call the Middle East. One region that lay distant, and therefore remote from these invasions, was the Arabian Peninsula, where no major culture hearth or large cities had emerged and where the turmoil had not affected Arab settlements and nomadic routes.

STAGE FOR ISLAM

In a remote place on the Arabian Peninsula, where the foreign invasions of the Middle East had had little effect on the Arab communities, an event occurred early in the seventh century that was to change history and affect the destinies of people in many parts of the world. In a town called Mecca (Makkah), about 70 kilometers (45 mi) from the Red Sea coast in the Jabal Mountains, a man named Muhammad in the year AD 611 began to receive revelations from Allah (God). Muhammad (571–632) was then in his early forties. Convinced after some initial self-doubt that he was indeed chosen to be a prophet, Muhammad committed his life to fulfilling the divine commands he believed he had received. Arab society was in social and cultural disarray, but Muhammad forcefully taught Allah's lessons and began to transform his culture. His personal power soon attracted enemies, and in 622 he fled from Mecca to the safer haven of Medina (Al Madinah), where he continued his work until his death. This moment, the *hejira* (meaning migration), marks the starting date of the Muslim era, Year 1 on Islam's calendar. Mecca, of course, later became Islam's holiest place.

The Faith

The precepts of Islam in many ways constituted a revision and embellishment of Judaic and Christian beliefs and traditions. All of these faiths have but one god, who occasionally communicates with humankind through prophets; Islam acknowledges that Moses and Jesus were such prophets but considers Muhammad to be the final and greatest prophet. What is earthly and worldly is profane; only Allah is pure. Allah's will is absolute; Allah is omnipotent and omniscient. All humans live in a world that Allah created for their use but only to await a final judgment day.

Islam brought to the Arab World not only the unifying religious faith it had lacked but also a new set of values, a new way of life, a new individual and collective dignity. Islam dictated observance of the Five Pillars: (1) repeated expressions of the basic creed, (2) daily prayer, (3) a month each year of daytime fasting (Ramadan), (4) the giving of alms, and (5) at least one pilgrimage to Mecca in each Muslim's lifetime (the *hajj*). Islam prescribed and proscribed in other spheres of life as well. It forbade alcohol, smoking, and gambling. It tolerated polygamy, although it acknowledged the virtues of monogamy. Mosques appeared in Arab settlements, not only for the (Friday) sabbath prayer, but also as social gathering places to knit communities closer together. Mecca became the spiritual center for a divided, widely dispersed people for whom a collective focus was something new.

The Arab-Islamic Empire

Muhammad provided such a powerful stimulus that Arab society was mobilized almost overnight. The prophet died in 632, but his faith and fame spread like wildfire. Arab armies carrying the banner of Islam formed, invaded, conquered, and converted wherever they went. As Figure 7A-5 shows, by AD 700 Islam had reached far into North Africa, into Transcaucasia, and into most of Southwest Asia. In the centuries that followed, it penetrated southern and

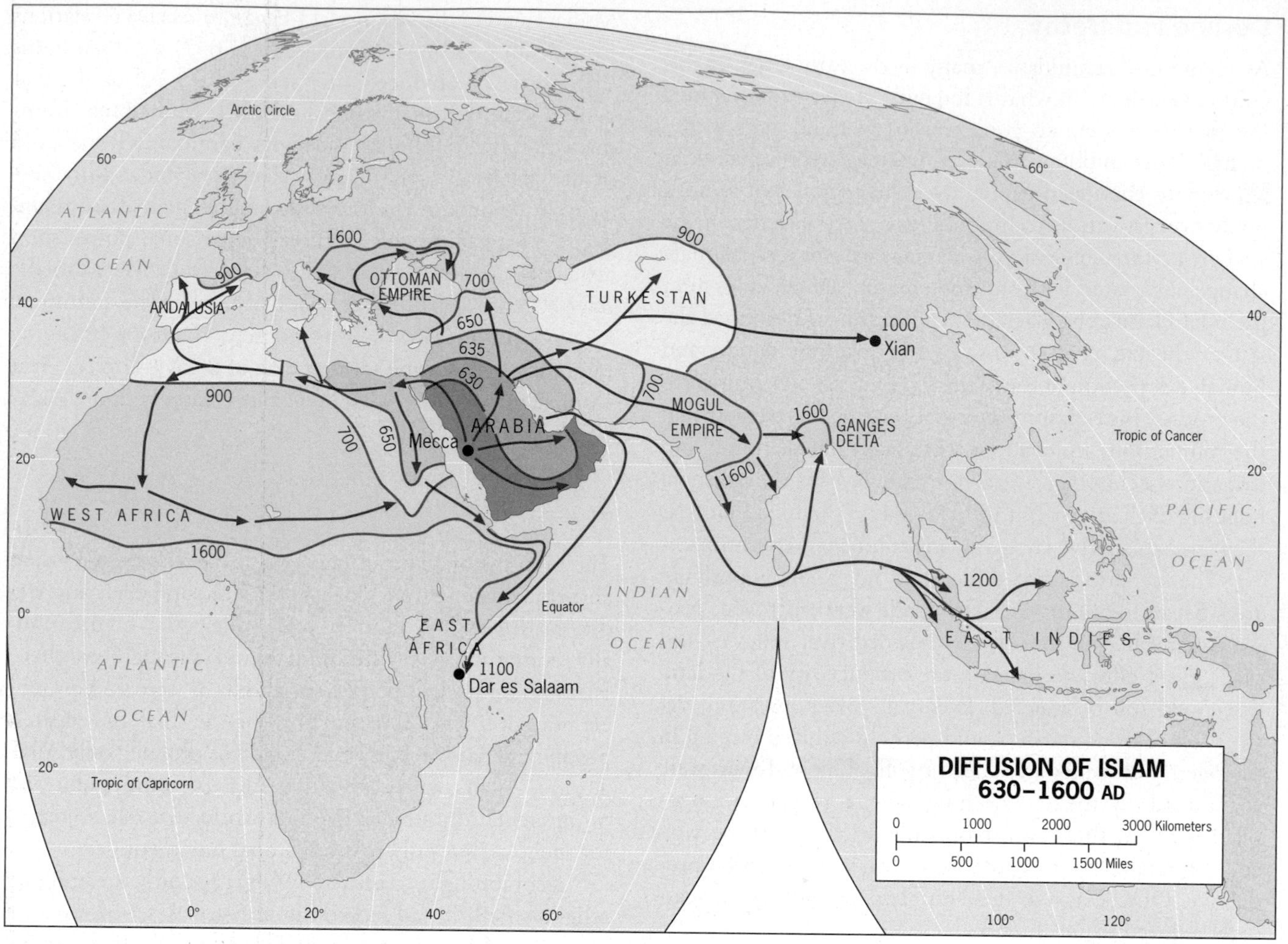

FIGURE 7A-5

© H. J. de Blij, P. O. Muller, and John Wiley & Sons, Inc.

eastern Europe, Central Asia's Turkestan, West Africa, East Africa, and South and Southeast Asia, even reaching China by AD 1000.

Routes of Diffusion

The spread of Islam provides a good illustration of a series of processes known as **7** **spatial diffusion** that focus on the way ideas, inventions, and cultural practices propagate through a population in space and time. In 1952, the Swedish geographer named Torsten Hägerstrand published his fundamental study of spatial diffusion entitled *The Propagation of Diffusion Waves*. He reported that diffusion takes place in two forms: **8** **expansion diffusion**, when propagation waves originate in a strong and durable source area and spread outward, affecting an ever larger region and population; and **9** **relocation diffusion**, in which migrants carry an innovation, idea, or (for example) a virus from the source to distant locations and it diffuses from there. Islam spread mostly through expansion diffusion (the worldwide spread of AIDS is a case of relocation diffusion).

Islam on the March

Both expansion and relocation diffusion include several types of processes. The spread of Islam, as Figure 7A-5 shows, initially proceeded via a form of expansion diffusion called **10** **contagious diffusion** as the faith moved from village to village across the Arabian Peninsula, North Africa, and Southwest Asia. But Islam got powerful boosts when kings, chiefs, and other high officials were converted, who in turn propagated the faith downward through their bureaucracies to far-flung subjects. This form of expansion diffusion, called **11** **hierarchical diffusion**, served Islam well.

But the map leaves no doubt that Islam later spread by relocation diffusion as well, notably to the Ganges Delta in South Asia, to present-day Indonesia, and to East Africa. That process continues to this day, but the heart of Islam remains in Southwest Asia. There, Islam became the cornerstone of an Arab Empire with Medina as its first capital. As the empire grew by expansion diffusion, its headquarters were relocated from Medina to Damascus (in present-day Syria) and later to Baghdad on the Tigris River (Iraq). And it prospered. In architecture, mathematics, and science, the

© H. J. de Blij

"Seville was a Muslim capital of Iberia, and walking the streets of the old city center is an adventure in cultural and historical geography. Seville fell to the Muslim invaders in 711, and the Muslims built a large mosque in the heart of town. Later the Christians, who ousted the Muslims in 1248, destroyed most of the mosque and built a huge cathedral on the site, but they saved the ornate minaret, made it the bell tower, and called it the Giralda (left). But the finest Islamic legacy in Seville surely is the Alcazar Palace (above), begun in the late twelfth century and embellished and finished by the Christians after their victory. Its arched façades and intricately carved walls remain a monument to the Muslim architects and artists who designed and created them."

www.conceptcaching.com

Arabs overshadowed their European contemporaries. The Arabs established institutions of higher learning in many cities including Cairo, Baghdad, and Toledo (Spain), and their distinctive cultural landscapes united their far-flung domain. Non-Arab societies in the path of the Muslim drive were not only Islamized, but also Arabized, adopting other Arab traditions as well. Islam had spawned a culture; it still lies at the heart of that culture today.

As we noted, Islam's expansion eventually was checked in Europe, Russia, and elsewhere. But a map showing the total area under Muslim sway in Eurasia and Africa reveals the enormous dimensions of the domain affected by **12 Islamization** at one time or another (Fig. 7A-6). Islam continues to expand, now mainly by relocation diffusion.

There are Islamic communities in cities as widely scattered as Vienna, Singapore, and Cape Town, South Africa; Islam is also growing rapidly in the United States. With more than 1.6 billion adherents today (23 percent of humankind), Islam is a vigorous and burgeoning cultural force around the world.

ISLAM DIVIDED

For all its vigor and success, Islam is still fragmented into sects. The earliest and most consequential division arose after Muhammad's death. Who should be his legitimate successor? Some believed that only a blood relative should follow the prophet as leader of Islam. Others, a majority,

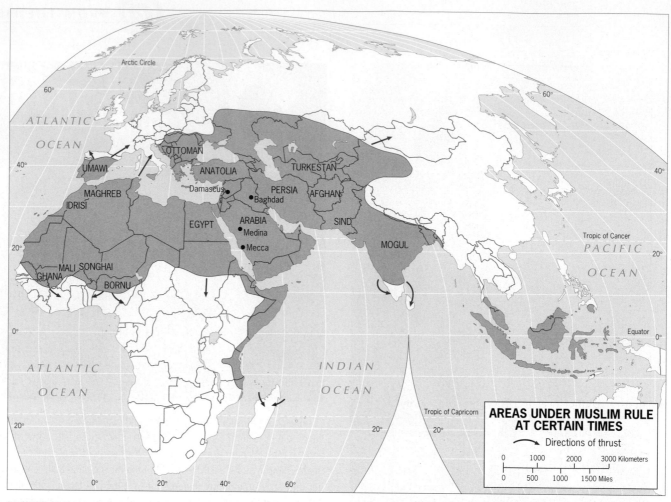

FIGURE 7A-6

© H. J. de Blij, P. O. Muller, and John Wiley & Sons, Inc.

felt that any devout follower of Muhammad was qualified to take over. The first chosen successor was the father of Muhammad's wife (and thus not a blood relative). But this did not satisfy those who wanted to see a man named Ali, a cousin of Muhammad, made *caliph* (successor). When Ali's turn came, his followers, the ***Shi'ites***, proclaimed that Muhammad finally had a legitimate successor. This offended the ***Sunnis***, those who did not see a blood relationship as necessary for the succession. From the beginning of this disagreement, the numbers of Muslims who took the Sunni side far exceeded those who regarded themselves as Shia (followers) of Ali. The great expansion of Islam was largely propelled by the Sunnis; the Shi'ites survived as minorities scattered throughout the realm. Today, about 85 percent of all Muslims are Sunnis, but the Shia–Sunni schism is often at the root of many conflicts in this realm.

The Strength of Shi'ism

But the Shi'ites vigorously promoted their version of the faith. In the early sixteenth century their work paid off: the royal house of Persia (modern-day Iran) made Shi'ism the only legal religion throughout its vast empire. That domain extended from Persia into lower Mesopotamia (modern Iraq), into Azerbaijan, and into western Afghanistan and Pakistan. As the map of religions (Fig. 7A-3) shows, this created for Shi'ism a large culture region and gave the faith unprecedented strength. Iran remains the bastion of Shi'ism in the realm today, and the appeal of Shi'ism continues to radiate into neighboring countries and even farther afield.

During the late twentieth century, Shi'ism gained unprecedented influence in the realm. In its heartland, Iran, a *shah* (king) tried to secularize the country and to limit the power of the *imams* (mosque officials); he provoked a revolution that cost him the throne and made Iran a Shi'ite Islamic republic. Before long, Iran was at war with neighboring, Sunni-ruled Iraq, and Shi'ite parties and communities elsewhere were invigorated by the newfound power of Shi'ism. From Arabia to Africa's northwestern corner, Sunni-ruled countries warily watched their Shi'ite minorities, newly imbued with religious fervor. Mecca, the sacred place for both Sunnis and Shi'ites, became a battleground during the week of the annual pilgrimage, and for a time the (Sunni) Saudi Arabian government denied entry to Shi'ite pilgrims. That schism has healed somewhat, but intra-Islamic sectarian differences run deep, as the ongoing conflict in Iraq demonstrates.

Smaller Islamic sects further diversify this realm's religious landscape. Some of these sects play a disproportionately large role in the societies and countries of which they are a part, as we will see in Chapter 7B.

Religious Revivalism in the Realm

Another cause of intra-Islamic strife lies in the resurgence of religious fundamentalism, or as Muslims refer to it, **13 religious revivalism**. In the 1970s, the imams in Shi'ite Iran wanted to reverse the shah's moves toward liberalization and secularization: they wanted to (and did) recast society in traditional, revivalist Islamic molds. An *ayatollah* (leader under Allah) replaced the shah in 1979; Islamic rules and punishments were instituted. Urban women, many of whom had been considerably liberated and educated during the shah's regime, were forced to return to more traditional Islamic roles. Vestiges of Westernization, encouraged by the shah, disappeared. Under these new conditions, even the war against Iraq (1980–1990), which began as a conflict over territory, became a holy war that cost more than a million lives.

Islamic revivalist fundamentalism did not rise in Iran alone, nor was it confined to Shi'ite communities. Many Muslims—Sunnis as well as Shi'ites—in all parts of the realm disapproved of the erosion of traditional Islamic values, the corruption of society by European colonialists and later by Western modernizers, and the declining power of the faith in the secular state. As long as economic times were good, such dissatisfaction remained submerged. But when jobs were lost and incomes declined, a return to fundamental Islamic ways became more appealing.

Muslim Against Muslim

This set Muslim against Muslim in all the regions of the realm. Revivalists fired the faith with a new militancy, challenging the status quo from Afghanistan to Algeria. The militants forced their governments to ban "blasphemous" books, to resegregate the sexes in schools, to enforce traditional dress codes, to legitimize religious-political parties, and to heed the wishes of the *mullahs* (teachers of Islamic ways). Militant Muslims proclaimed that democracy inherited from colonialists and adopted by Arab nationalists was incompatible with the rules of the Quran (Koran).

The rift between moderate Muslims and militant revivalists began to spill over into other parts of the world decades ago, from Pakistan to the African Transition Zone and from the Caucasus to the Philippines. Even while Islamic countries such as Iran and Algeria struggled to overcome the instability and damage arising from their internal strife, militant activists planned and carried out attacks against the allies of the moderates—oil-guzzling Western countries guilty of military intervention, propping up unrepresentative regimes, mistreatment of immigrant Muslims, and cultural corruption. Before the worst of these attacks (as of mid-2010) destroyed the World Trade Center in New York and severely damaged the Pentagon outside Washington, D.C., the French had foiled a similar assault on the Eiffel Tower in Paris, the Russians had suffered hundreds of casualties in Moscow and Transcaucasia, and U.S. targets were hit in the Middle East, the Arabian Peninsula, East Africa, and elsewhere.

Back to Basics

Although it has been argued that these attacks do not represent Islam as a religion in conflict with the West, they do have a religious context. During the 1990s, following the Cold War defeat of Soviet forces in Afghanistan and the subsequent abandonment of that country by the United States and its allies, a radical organization named *al-Qaeda* emerged, dedicated to punishing the perceived enemies of Islamic peoples by the most effective means at its disposal: terrorist attacks. However, the leader and chief financier of this movement, Usama bin Laden, had a bigger agenda. His ultimate goal was to overthrow the regime of rich princes ruling his home country, Saudi Arabia, accusing them of collusion with the infidel United States, of apostasy (faithlessness), and worse. Bin Laden wanted to make Saudi Arabia, birthplace of Muhammad and guarantor of holy Mecca, a true Islamic state.

Bin Laden was not the first Saudi to wish to return his country to its Islamic roots. During the eighteenth century an Islamic theologian named Muhammad ibn Abd al-Wahhab (1703–1792) founded a movement to bring Islamic society on the peninsula back to the most fundamental, traditional, and puritanical form of the faith. So strict were his teachings that he was expelled from his hometown in 1744—a fateful exile because he moved to a place that happened to be the headquarters of Ibn Saud, ruler over a sizeable swath of the Arabian Peninsula. Ibn Saud formed an alliance with al-Wahhab, who had inherited a fortune when his wife died; between them, they set out on a campaign of conquest that made the Saudi dynasty the rulers of the region and **14 Wahhabism** the dominant form of Islam. Even the relatively liberal Ottoman sultans could not suppress either the Saudis or the Wahhabis, and in 1932, when the modern Kingdom of Saudi Arabia was established, Wahhabism became its official and dominant form of Islam.

Despite its reference to the name of the founder, Wahhabism today is a term used mostly by non-Muslims and outsiders. Followers of Wahhabi doctrines call themselves *Muwahhidun* (Unitarians) to signify the strict and fundamental nature of their beliefs. These are the doctrines

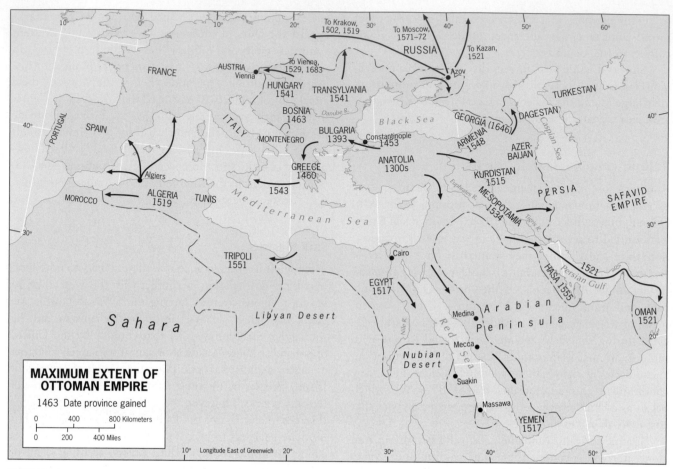

FIGURE 7A-7

© H. J. de Blij, P. O. Muller, and John Wiley & Sons, Inc.

Usama bin Laden accuses the Saudi princes of having violated, and these are the harsh teachings so often cited in the Western media as emanating from the pulpits of Saudi Arabia's revivalist mosques.

Islam and Other Religions

Two additional major faiths had their sources in the *Levant* (the area extending from Greece eastward around the Mediterranean coast to northern Egypt), Judaism and Christianity, and both are older than Islam. Islam's rise submerged many smaller Jewish communities, but the Christians, not the Jews, waged centuries of holy war against the Muslims, seeking, through the Crusades, not only to drive Islam back but to reestablish Christian communities where they had dominated before the Islamic expansion. The aftermath of that campaign still marks the realm's cultural landscape today. A substantial Christian minority (about 40 percent of the population) remains in Lebanon, and Christian minorities also survive in Israel, Syria, Egypt, and Jordan. Strained relations between the long-dominant Christian minority in Lebanon and the Muslim majority (itself divided into five sects) contributed to the disastrous armed

struggle that engulfed the country in the 1970s and 1980s. It reemerged in 2006, and Lebanon has once again fallen into turmoil.

By far the most intense conflict in modern times has pitted the Jewish state, Israel, against its Islamic neighbors near and far. Israel's United Nations-sponsored creation in 1948 precipitated more than six decades of intermittent strife and attempts at mediation; it also caused friction among Islamic states in the region. Jerusalem—holy city for Judaism, Christianity, and Islam—lies in the crucible of this confrontation.

The Ottoman Aftermath

It is ironic that Islam's last great advance into Europe eventually led to European occupation of Islam's very heartland. The Ottomans (named after their leader, Osman I), based in what is today Turkey, conquered Constantinople (now Istanbul) in 1453 and pushed into southeastern Europe. Soon Ottoman forces were on the doorstep of Vienna; they also invaded Persia, Mesopotamia, and North Africa (Fig. 7A-7). At its height, the Ottoman Empire under Suleyman the Magnificent (ruled 1522–1560) was the most powerful

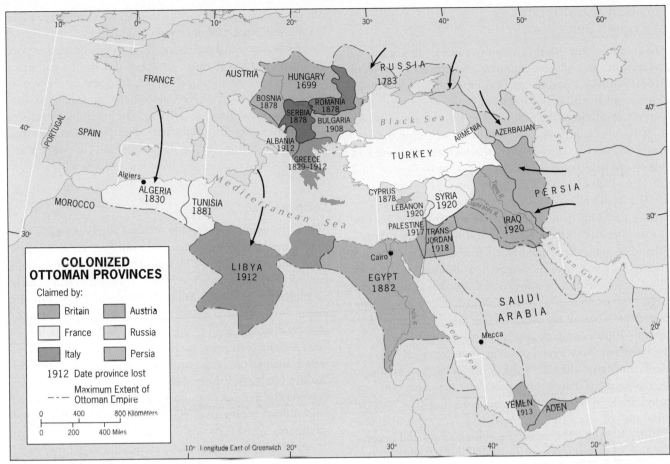

FIGURE 7A-8

© H. J. DE BLIJ, P. O. MULLER, AND JOHN WILEY & SONS, INC.

state in western Eurasia. As Figure 7A-7 shows, its armies even advanced toward Moscow, Kazan, and Krakow from a base at Azov (near present-day Rostov in Russia). The Turks also launched marine attacks on Sicily, Spain, and France.

The Ottoman Empire survived for more than four centuries (it ended in 1923), but it lost territory as time went on, first to the Hungarians, then to the Russians, and later to the Greeks and Serbs until, after World War I, the European powers took over its provinces and made them colonies—colonies we now know by the names of Syria, Iraq, Lebanon, and Yemen (Fig. 7A-8). As the map shows, the French and the British took large possessions; even the Italians annexed part of the Ottoman domain.

The boundary framework that the colonial powers created to delimit their holdings was not satisfactory. As Figure 7A-2 illustrates, this realm's population of nearly 600 million is clustered, fragmented, and strung out in river valleys, coastal zones, and crowded oases. The colonial powers laid out long stretches of boundary as ruler-straight lines across uninhabited territory; they saw no need to adjust these boundaries to cultural or physical features in the landscape. Other boundaries, even some in desert zones, were poorly defined and never marked on the ground. One product of

this process was Iraq, delimited by (then) British Colonial Secretary Winston Churchill in 1921 as a kingdom with Baghdad as its capital; another was Jordan, centered on Amman. Later, when the colonies had become independent states, such boundaries led to quarrels, even armed conflicts, among neighboring Muslim states.

THE POWER AND PERIL OF OIL

Travel through the cities and towns of North Africa and Southwest Asia, talk to students, shopkeepers, taxi drivers, and migrants, and you will hear the same refrain: "Leave us in peace, let us do things our traditional way. Our problems—with each other, with the world—seem always to result from outside interference. The stronger countries of the world exploit our weaknesses and magnify our quarrels. We want to be left alone."

That wish might be closer to fulfillment were it not for two relatively recent events: the creation of the state of Israel and the discovery of some of the world's largest oil reserves. We will discuss the evolution of Israel in Chapter 7B. Here we focus on the realm's most valuable export product: oil.

Location and Dimensions of Known Reserves

About 30 of the world's countries have significant oil reserves. But five of these countries, all located in the North Africa/Southwest Asia realm, in combination possess larger reserves (approximately 55 percent of the world total) than the rest combined. The estimated reserves of this Big Five in 2010, averaged from several sources including the governments of the countries themselves and reported in billions of barrels, were as follows: (1) Saudi Arabia, 267; (2) Iran, 138; (3) Iraq, 115; (4) Kuwait, 104; and (5) the United Arab Emirates, 98. The next-ranking country, Venezuela, had known reserves just below 90 billion barrels; Russia about 65; Libya, Nigeria, and Kazakhstan between 30 and 45; and the United States just above 20. No other country's known reserves surpassed 16 billion barrels as of 2010. It should also be noted that Canada may possess the largest oil reserves of all—but most of this supply is contained in Alberta's *tar sands* (as we saw in Chapter 3A), which requires complex and expensive recovery and refining processes. Yet with oil prices at unprecedented high levels, Canada's production is likely to grow. And finally, as we reported in Chapter 5B, spectacular new discoveries are being made in Brazil's offshore oilfields; if as-yet-unproven estimates of these massive reserves are confirmed, that country is poised to surpass the United States and become a major oil exporter before the end of this decade.

In general terms, oil (and associated natural gas) exists in this realm in three discontinuous zones (Fig. 7A-9). The most productive of these zones extends from the southern and southeastern part of the Arabian Peninsula northwestward around the rim of the Persian Gulf, reaching into Iran and continuing northward into Iraq, Syria, and southeastern Turkey, where it peters out. The second zone lies across North Africa and extends from north-central Algeria eastward across northern Libya to Egypt's Sinai Peninsula, where it ends. The third zone begins on the margins of the realm in eastern Azerbaijan, continues eastward under the Caspian Sea into Turkmenistan and Kazakhstan, and also reaches into Uzbekistan, Tajikistan, Kyrgyzstan, and Afghanistan.

Producers and Consumers

Saudi Arabia is the world's largest oil exporter, but in recent years Russia has risen to second place. As we noted in Chapters 2A and 2B, oil and natural gas are by far Russia's most valuable commodities, and the Russian state direly needs the revenues they produce. Thus Russia, with modest (though expanding) known reserves, is vigorously exporting to international markets and is now a leading force in the global energy picture. In combination, however, production by the countries of the North Africa/Southwest Asia realm far exceeds that from all other sources. The United States,

not surprisingly, is the world's leading importer even while it consumes almost all of its own production.

As Figure G-10 indicates, the production and export of oil and gas has elevated several of this realm's countries into the higher-income categories. But petroleum wealth also has enmeshed these Islamic societies and their governments in global strategic affairs. When regional conflicts create instability in producing countries that have the potential to disrupt supply lines, powerful consumers are tempted to intervene—and have done so.

Colonial Legacy

When the colonial powers laid down the boundaries that partitioned this realm among themselves, no one knew about the riches that lay beneath the ground. A few wells had been drilled, and production in Iran had begun as early as 1908 and in Egypt's Sinai Peninsula in 1913. But the major discoveries came later, in some cases after the colonial powers had already withdrawn. Some of the newly independent countries, such as Libya, Iraq, and Kuwait, found themselves with wealth undreamed of when the Turkish Ottoman Empire collapsed. As Figure 7A-9 shows, however, others were less fortunate. A few countries had (and still have) potential. The smaller, weaker emirates and sheikdoms on the Arabian Peninsula always feared that powerful neighbors would try to annex them (Kuwait faced this prospect in 1990 when Iraq invaded it). The unevenly distributed oil wealth, therefore, created another source of division and distrust among Islamic neighbors.

A Foreign Invasion

The oil-rich countries of the realm found themselves with a coveted energy source but without the skills, capital, or equipment to exploit it. These had to come from the Western world and entailed what many tradition-bound Muslims feared most: a strong foreign presence on Islamic soil, foreign intervention in political as well as economic affairs, and penetration of Islamic societies by the vulgarities of Western ways.

Many Muslims' worst fears came true in Iran during the 1950s, when the government of Iran tried to control British oil exploitation centered on Abadan and the Persian Gulf. The imperious ways of the British, the luxurious life of the European expatriates compared to the abject poverty of Iranian workers, and the unfair terms of the concession under which the oil was exported all contributed to the rise of nationalist sentiment in Iran, whose leaders appealed to the United States for help in negotiating a better deal with the British. American President Harry Truman had some sympathy for the Iranians, but his successor Dwight Eisenhower worried about Iran turning toward communism. Capitalizing on political disputes within Iran, CIA operatives engineered the overthrow of elected Premier Mohammad

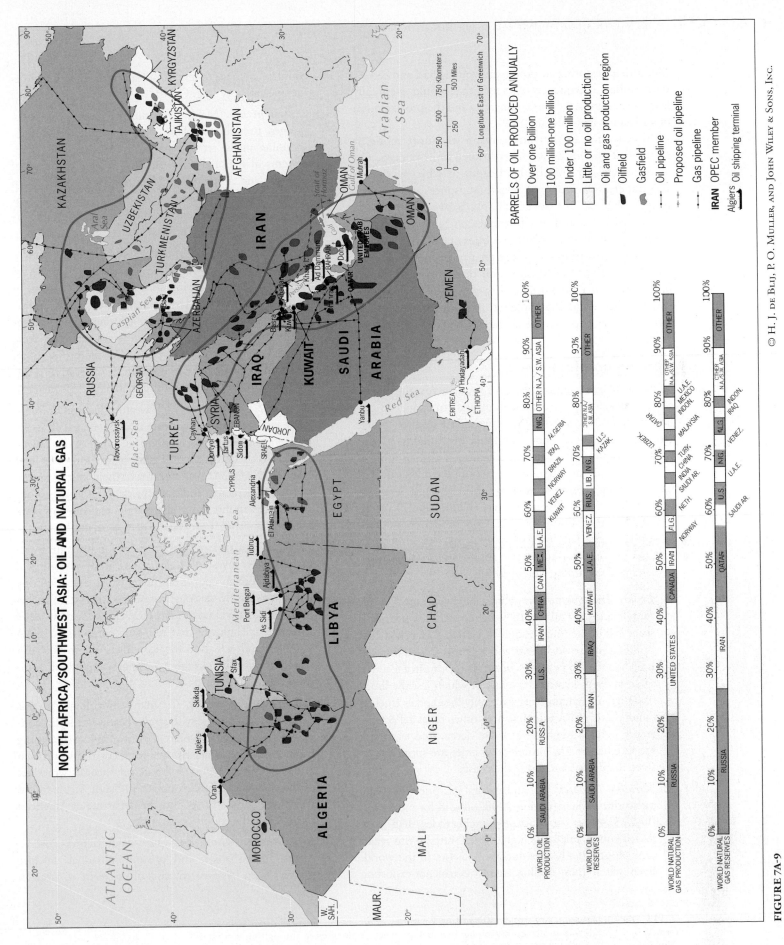

NORTH AFRICA/SOUTHWEST ASIA: OIL AND NATURAL GAS

BARRELS OF OIL PRODUCED ANNUALLY

- Over one billion
- 100 million–one billion
- Under 100 million
- Little or no oil production
- Oil and gas production region
- Oilfield
- Gasfield
- Oil pipeline
- Proposed oil pipeline
- Gas pipeline
- **IRAN** OPEC member
- Algiers Oil shipping terminal

FIGURE 7A-9

© H. J. de Blij, P. O. Muller, and John Wiley & Sons, Inc.

Mosaddeq, restoring to power the young shah and setting into motion a sequence of events that would eventually lead to the demise of Iran's monarchy and the proclamation of an Islamic Republic in 1979.

Even when foreign intervention was more subtle, it intensified clashes between traditional and modern. Oil revenues created cultural landscapes in which gleaming skyscrapers towered over ornate mosques and historic neighborhoods. The social chasms between rich, well-connected elites and less fortunate citizens bred resentment. Involvement of Western armies in regional conflicts produced the unthinkable: foreign armies on Arabia's hallowed soil. To many Muslims, all this violated the most basic tenets of their faith, and some retaliated with violence of their own. That violence, in the form we define as terrorism, has become one of the challenges of our time.

Oil's Geographic Impact

As is true of every natural resource found across the planet, oil has completely transformed cultural landscapes in certain parts of North Africa/Southwest Asia and left others virtually unchanged. For hundreds of millions of this realm's inhabitants, a patch of tillable soil and a source of water still mean more to daily life than all the oil in OPEC.* Countrysides in Arab as well as non-Arab regions in this part of the world continue to carry the imprints of centuries of cultural tradition, not decades of oil-driven modernization. Nevertheless, oil and natural gas—their location, production, transportation, and sale—have produced massive changes, including the following:

1. *Urban Transformation.* Undoubtedly the most visible manifestation of oil wealth is the modernization (and, as some call it, the "futurization") of cities. By far the tallest building in the world rises above Dubai in a state called the United Arab Emirates, but it is only one in a forest of glass-encased skyscrapers, many of which test the limits of design and engineering. Not only the capitals (such as Saudi Arabia's Riyadh) but also other cities reflect the riches oil has brought—yes, even Mecca. And, as we will see in Chapter 7B, entirely new cities are springing up from the deserts and along the coasts.

2. *Variable Incomes.* Oil and natural gas prices fluctuate on world markets (during 2008, oil sold for more than [U.S.] $140 per barrel before plunging to less than $40). When energy prices are high, several states in this realm rank among the highest-income societies in the world. Even when they decline, many petroleum-exporting countries manage to remain at least in the upper-middle-income category (Fig. G-10).

3. *Infrastructure.* Massive spending on airports, seaports, bridges, tunnels, four-lane highways, public buildings, shopping malls, recreational facilities, and other components of national infrastructures creates an image of comfort and affluence quite unlike that prevailing in countries without major oil or gas revenues. Saudi Arabia is now engaged in a vast modernization project extending from coast to coast.

4. *Industrialization.* A number of far-sighted governments among those with oil wealth, realizing that reserves will not last forever, are investing some of their income in industries that will outlast the era of massive oil exporting. Petrochemical manufacturing using domestic supplies, aluminum, steel, and fertilizers are among these industries, although others potentially more promising, for example in high-technology fields, are not yet a significant part of this important initiative.

5. *Regional Disparities.* Oil wealth, like other high-value resources, tends to create strong regional contrasts both within and between countries. Saudi Arabia's ultra-modern east coast is a world apart from most of its interior, which remains a land of desert, oasis, vast distances, isolated settlements, and slow change. This is not unique to countries in this realm: sharp regional disparities mark oil-rich countries from Algeria to Azerbaijan.

6. *Foreign Investment.* Governments and private entrepreneurs have invested massive quantities of oil-generated wealth in foreign countries, buying stock and acquiring prestigious hotels, famous stores, and other high-cachet properties. These investments have created a network of international linkages that not only connects NASWA states and individuals to the economies of non-NASWA countries but also links them to the growing Islamic communities in those states.

7. *Foreign Involvement.* To many inhabitants of this realm, especially those with strong Islamic-revivalist convictions, the inevitable presence of foreigners (including businesspeople, politicians, architects, engineers, and even armed forces) on Islamic soil is an unwelcome byproduct of the energy era. Public opinion in Saudi Arabia forced its ruling regime to negotiate the departure of American troops from the country's territory.

8. *Intra-Realm Migration.* Oil wealth allows governments, industrialists, and private individuals to hire workers from less-favored parts of the realm to work in

---

*OPEC, the Organization of Petroleum Exporting Countries, is the international oil cartel (syndicate) formed by 12 producing countries to promote their common economic interests through the setting of joint pricing policies and the limitation of market options for consumers. Its 8 NASWA members are Algeria, Iran, Iraq, Kuwait, Libya, Qatar, Saudi Arabia, and the United Arab Emirates.

the oilfields, ports, and in other mainly menial capacities. This has brought many Shi'ites to the countries of the eastern zone of the Arabian Peninsula, where they now form significant sectors of national populations; hundreds of thousands of Palestinian Arabs also sought temporary work in that region's industries. In 2010 it was estimated that Saudi Arabia, with a population of just under 30 million, hosted more than 5 million foreign workers from within the realm.

9. *Migration from Other Realms.* The labor market in such places as Dubai and Abu Dhabi in the United Arab Emirates has also attracted workers from beyond this realm's borders during recent periods of rapid growth. Wages in such countries as Pakistan, India, Sri Lanka, and Bangladesh are even lower than those paid by the building industry or private employers in oil-boom-driven NASWA states. These workers serve mainly as domestics, gardeners, trash collectors, and the like; at the beginning of 2009, about 80 percent of the total population of Dubai consisted of foreign employees. Over the past few years, working conditions and wage issues also created protests that exposed the harsh circumstances faced by many guest workers throughout this realm.

10. *Diffusion of Revivalism.* Oil and gas revenues are used by Islamist regimes to support Islamic communities as well as their mosques and cultural centers throughout the world. No NASWA state has spent more money on such causes than Saudi Arabia, and thousands of mosques from England to Indonesia prosper as a result. This example of relocation diffusion creates nodes of recruitment for the faith—and ensures the dissemination of revivalist principles.

Oil brought this realm into contact with the outside world in ways unforeseen just a century ago. Oil has strengthened and empowered some of its peoples, but has dislocated and imperiled others. It has truly been a double-edged sword. So before we turn to the regions of this realm in Chapter 7B, we should remind ourselves that the great majority of NASWA's inhabitants are not, in their daily lives, directly affected by the changes the energy era has brought. Most Moroccans, Tunisians, Egyptians, Jordanians, Yemenis, Turks, and many millions of others—Kurds, Palestinians, Berbers, Tuaregs—make ends meet by trading or farming or working at jobs their parents and grandpar-

ents performed. Take the case of Iran, which as a country earns about two-thirds of its income from oil and natural gas. Only one-half of 1 percent of that country's workforce (just over 100,000 out of more than 20 million workers) earn a salary from energy or energy-related work. By far the largest number in any single occupation are the farmers (5 million). And for all their oil, Iranians in 2007 earned a mere (U.S.) $10,800 per person—less than the average Turk and one-third of the average earned per capita in Singapore. As we will discover, it is cultural geography rather than economic geography that dominates the regionalization of the North Africa/Southwest Asia realm.

The nearly 30 states of the North African/Southwest Asia geographic realm display an often stunning degree of variety and diversity. Oil and natural gas have generated wealth, but also severe inequality and disparity—not only *within* the countries with large reserves, but also *between* those that have and those that have not. Ultramodern city skylines on the Arabian Peninsula stand in sharp contrast to Nile Valley villages virtually unchanged for millennia. Superhighways cross deserts still traversed by nomadic traders riding camels. Social geographies vary as greatly as economic ones: more than 80 percent of women are literate in Lebanon, just 40 percent in Yemen. Population growth in oil-poor Tunisia is well below the world average; in oil-rich Saudi Arabia it is twice the global rate. Representative government is making gains in Turkey, losing ground in Iran. If we could devise a measure of national individuality, this might prove to be the most diverse realm of all—and not even the momentous spread of Islam created a cloak of cultural conformity. Interposed east-west between China and Europe, and north-south between Russia and Africa, this crossroads of history remains a geographic kaleidoscope.

POINTS TO PONDER

- The accelerating diffusion of Islam is changing the cultural geography of the margins of this realm.
- By some measures this realm displays the world's strongest social contrasts.
- Central Asia is witnessing an Islamic revival, but old Soviet ways persist. Dictatorial rule and suppression of dissent continue.
- Water-supply crises are widespread and frequent in this driest of the world's most populous geographic realms.

Jerusalem's urban landscape is rife with cultural contrasts—and is also home to the holy ground that means so much to Jews, Christians, and Muslims. © H. J. de Blij

7B

NORTH AFRICA/SOUTHWEST ASIA: REGIONS OF THE REALM

≡ Egypt and the Lower Nile Basin
The Maghreb and Its Neighbors
The Middle East
The Arabian Peninsula
The Empire States
Turkestan

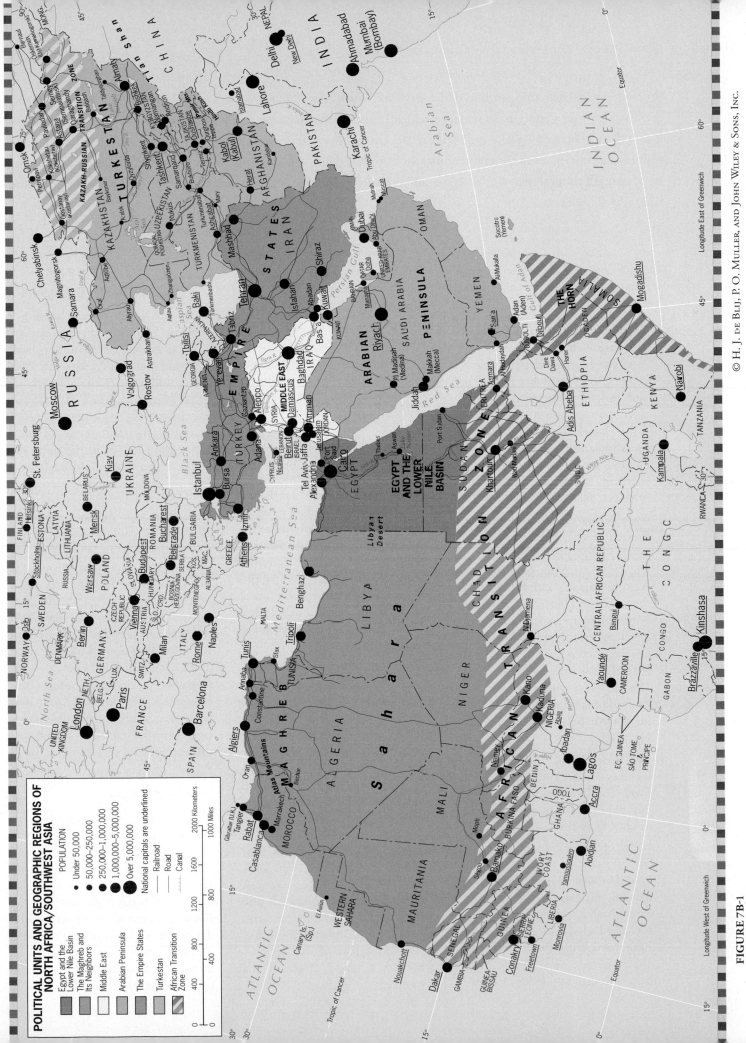

POLITICAL UNITS AND GEOGRAPHIC REGIONS OF NORTH AFRICA/SOUTHWEST ASIA

POPULATION
- · Under 50,000
- • 50,000–250,000
- ● 250,000–1,000,000
- ⬤ 1,000,000–5,000,000
- ⬤ Over 5,000,000

National capitals are underlined

- Railroad
- Road
- Canal

Egypt and the Lower Nile Basin
The Maghreb and Its Neighbors
Middle East
Arabian Peninsula
The Empire States
Turkestan
African Transition Zone

0   400   800   1200   1600   2000 Kilometers
0   200   400   600   800   1000 Miles

FIGURE 7B-1

© H. J. de Blij, P. O. Muller, and John Wiley & Sons, Inc.

dentifying and delimiting regions in the sprawling North Africa/Southwest Asia realm is quite a challenge. Population clusters are widely scattered in some areas, highly concentrated in others. Cultural transitions and landscapes—internal as well as peripheral—make it difficult to discern a regional framework. This, furthermore, is a highly changeable realm and always has been. Several centuries ago it extended into eastern Europe; now it reaches into Central Asia, where an Islamic **1** cultural revival (the regeneration of a long-dormant culture through internal renewal and external infusion) is well under way.

The following are the regional components of this far-flung realm today (Fig. 7B-1):

1. *Egypt and the Lower Nile Basin.* This region in many ways constitutes the heart of the realm as a whole. Egypt (together with Iran and Turkey) is one of the realm's three most populous countries. It is the historic focus of this part of the world and a major political and cultural force. It shares with its southern neighbor, Sudan, the waters of the lower Nile River.

2. *The Maghreb and its Neighbors.* Western North Africa (the *Maghreb*) and the areas that border it also form a region, consisting of Algeria, Tunisia, and Morocco at the center, and Libya, Chad, Niger, Mali, Burkina Faso, and Mauritania along the broad periphery. The last five of these countries are located astride or adjacent to the broad African Transition Zone where the Arab-Islamic realm of North Africa merges into Subsaharan Africa (see Fig. 6B-9).

3. *The Middle East.* This key region includes Israel, Jordan, Lebanon, Syria, and Iraq. In effect, it is the crescent-like zone of countries that extends from the eastern Mediterranean coast to the head of the Persian Gulf.

4. *The Arabian Peninsula.* Dominated by the enormous territory of Saudi Arabia, the Arabian Peninsula also includes the United Arab Emirates (UAE), Kuwait, Bahrain, Qatar, Oman, and Yemen. Here lies the source and focus of Islam,

the holy city of Mecca; here, too, lie many of the world's greatest oil deposits.

5. *The Empire States.* We refer to this region as the Empire States because two of the realm's giants, both the centers of major historic empires, dominate its geography. Turkey, the heart of the Ottoman Empire, now is the realm's most secular state and aspires to joining the European Union. Shi'ite Iran, the core of the erstwhile Persian Empire, has become an Islamic republic. To the north, Azerbaijan, once a part of the Persian Empire, lies in the turbulent, Muslim-infused Transcaucasian Transition Zone. To the south, the island of Cyprus is divided between Turkish and Greek spheres, the Turkish sector being another remnant of imperial times.

6. *Turkestan.* Turkish influence ranged far and wide in southwestern Asia, and following the Soviet Union's collapse in 1991 that influence turned out to be durable and strong. In the five former Soviet Central Asian republics the strength and potency of Islam vary, and their post-Soviet governments deal warily (sometimes forcefully) with Islamic revivalists. Russian influence continues, democratic institutions remain weak, and Western (chiefly American) involvement has increased since the campaign against terrorism began and Afghanistan—which also forms part of this region—became a major target of that campaign.

EGYPT AND THE LOWER NILE BASIN

Egypt occupies a pivotal location in the heart of a realm that extends over 9600 kilometers (6000 mi) longitudinally and some 6400 kilometers (4000 mi) latitudinally. At the northern end of the Nile and of the Red Sea, at the eastern end of the Mediterranean Sea, in the northeastern corner of Africa across from Turkey to the north and Saudi Arabia to the east, adjacent to Israel, to Islamic Sudan, and to Libya, Egypt lies in the crucible of this realm. Because it owns the Sinai Peninsula, Egypt, alone among states on the African continent, has a foothold in Asia. This foothold gives it a coast overlooking the strategic Gulf of Aqaba (the northeasternmost arm of the Red Sea). Egypt also controls the Suez Canal, the vital link between the Indian and Atlantic oceans and lifeline of Europe. The capital, Cairo (Al Qahira), is the realm's largest city and a leading center of Islamic civilization. We hardly need to further justify Egypt's designation (together with northern Sudan) as a discrete region.

Gift of the Nile

Egypt's Nile is the aggregate of two great branches upstream: the White Nile, which originates in the streams that feed Lake Victoria in East Africa, and the Blue Nile, whose source lies in Lake Tana in Ethiopia's highlands. The two Niles converge at Khartoum in modern-day Sudan. About 95 percent of Egypt's 78 million people live within 20 kilometers (12 mi) of the great river's banks or in its delta (Figs. 7B-2, 7A-2).

Before dams were constructed on the Nile, the ancient Egyptians used *basin irrigation*, building fields with earthen ridges and trapping the annual floodwaters with their fertile silt, to grow their crops. That practice continued for thousands of years until, during the nineteenth century, the construction of permanent dams made it possible to irrigate Egypt's farmlands year round. These dams, with locks for navigation, controlled the Nile's annual flood, expanded the country's cultivable area, and allowed the farmers to harvest more than one crop per year on the same field. In a single century, all of Egypt's farmland was brought under *perennial irrigation*. The largest of these dams, the Aswan High Dam, began operating in 1968, controlling water and sediments as well as bringing reliability to what had been an uncertain flood regime. It created Lake Nasser, which extends into Sudan, where 50,000 people had to be resettled to make way for it. The Aswan High Dam increased Egypt's irrigable land by nearly 50 percent and today provides the country with about 40 percent of its electricity.

Valley and Delta

Egypt's elongated oasis along the Nile, just 5 to 25 kilometers (3 to 15 mi) wide, broadens north of Cairo across a delta anchored in the west by the great city of Alexandria and in the east by Port Said, gateway to the Suez Canal. The delta contains extensive farmlands, but it is a troubled area today. The ever more intensive use of the Nile's water and silt upstream is depriving the delta of much needed replenishment. And the low-lying delta is at risk due to geological subsidence and sea-level rise as a consequence of global warming, increasing the danger of salt-water intrusion from the Mediterranean Sea that can severely damage soils.

Egypt's millions of subsistence farmers, the *fellaheen*, struggle daily to make their living off the land, as did the peasants of the Egypt of five millennia ago. Rural landscapes seem barely to have changed; ancient tools are still used, and dwellings remain rudimentary. Poverty, disease, high infant mortality rates, and low incomes prevail. The Egyptian government is embarked on grandiose plans to expand its irrigated acreage, but little of this has come to pass.

Subregions of Egypt

Egypt has six subregions, mapped in Figure 7B-2. Most Egyptians live and work in Lower (i.e., northern) and Middle Egypt, subregions ① and ②, the country's core area anchored by Cairo and flanked by the leading port and second manufacturing center, Alexandria. The economy has benefited from further oil discoveries in the Sinai (⑥) and in the Western Desert (④), so that Egypt now is self-sufficient and even exports some petroleum products. Cotton and textiles are the other major source of external income, but the important tourist industry has repeatedly been damaged by Islamic extremists. As the population continues to grow, the gap separating food supply and demand widens, and Egypt today must import grain. Since the late 1970s, Egypt has been a major recipient of U.S. foreign aid.

Egypt today is at a crossroads in more ways than one. Its planners know that reducing the rather high birth rate would improve the demographic situation, but revivalist Muslims object to any programs that limit family size. Its accommodation with Israel helps ensure foreign aid but divides the people. Its government faces a fundamentalist challenge as well as a rising demand for more democracy. Egypt's future, in this crucial corner of the realm, is clouded.

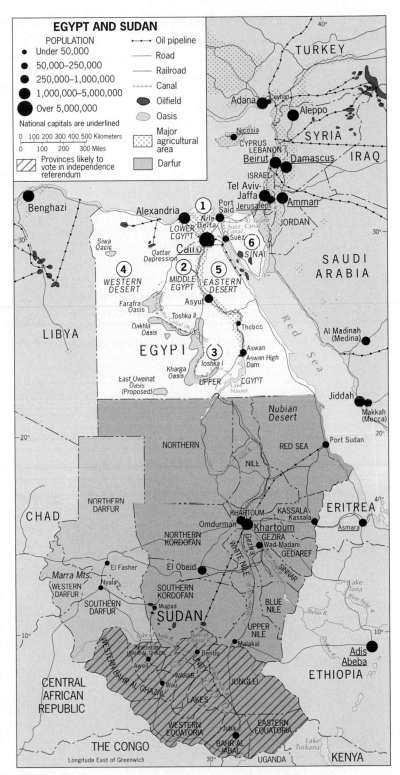

FIGURE 7B-2   © H. J. DE BLIJ, P. O. MULLER, AND JOHN WILEY & SONS, INC.

Divided Sudan

As Figure 7B-1 shows, Egypt is flanked by two countries that have posed challenges to its leadership: Sudan to the south and Libya to the west. Sudan, more than twice as large as Egypt and with 41 million people, lies centered on the confluence of the White Nile (from Uganda) and the Blue Nile (from Ethiopia). Here the twin capital, Khartoum–Omdurman, anchors a sizeable agricultural area where cotton was planted during colonial times. The British colonial administration combined northern Sudan, which was Arabized and Islamized, with a large area to the south, which was African and where many villagers had been Christianized. After the British departed, the regime in Khartoum sought to impose its Islamic rule on this portion of the African Transition Zone that formed southern Sudan, and a bitter civil war ensued. The cost in human lives and dislocation over the better part of the past three decades is incalculable.

Sudan has a 500-kilometer (300-mi) coastline on the Red Sea, where Port Sudan lies almost directly across from Jiddah and Makkah (Mecca) in Saudi Arabia. The country's economy for many years was typical of the energy-poor global periphery, exchanging sheep, cotton, and sugar for oil, with Saudi Arabia the main trading partner. The civil war in the African Transition Zone impoverished the Islamic regime in Khartoum, and per-capita income in Sudan was one of the world's lowest. All that changed in the 1990s.

Oil in the Desert

During the 1980s it became clear that significant oil reserves lie in Sudan, and during the 1990s discoveries were made in Kordofan Province and elsewhere, including major deposits in the embattled south (Fig. 7B-2). This complicated the political as well as the economic situation, and in short order Sudan became an exporter instead of an importer of oil. Foreign companies, particularly from China, which was willing to overlook human rights violations by the Khartoum regime, soon arrived to explore and exploit. The government then saw its income rise, allowing it to purchase more weapons. Local peoples living on top of or near the newly found deposits were forcibly relocated. Ultramodern buildings in Khartoum reflected the newfound wealth for the few.

Leaders in the South saw oil as a ticket to self-sufficiency. The 2005 peace agreement that had set a timetable for possible independence had stipulated that the South would be receiving half the revenues, but the North kept the records and was not trustworthy. Southerners were divided over the independence issue—would it be better to be part of prosperous Sudan or to go it alone and become a landlocked state? The issue is not resolved, but the South will be able to vote on independence in 2011. Southerners talk of a pipeline to the Indian Ocean through Kenya to avoid dependence on Sudan, but nothing has come of it yet. Sudan's oil bonanza will have politico-geographical consequences either way.

Tragedy in Darfur

Even as the issue of Southern independence heated up, a terrible crisis engulfed Sudan's western provinces of Darfur (Fig. 7B-2), where the local people, the Fur, have for centuries lived as Arabized pastoralists in the north and as settled farmers in the south. It is not clear how the conflict began, but a number of officials of

The "international community" has been slow to respond to the crisis in Darfur, arguing over the question of whether the assault by Arabized Muslim militias on sedentary African farmers in this western province of Sudan technically constitutes genocide, and relying on an undermanned, poorly equipped, and inadequately mandated African Union force to stop the violence. The result can be seen in this photograph of the northern Darfur village of Bandago, where virtually every dwelling has been burned by *janjaweed* militias empowered by Khartoum's Islamic regime. By the middle of 2010, more than 2.6 million people had been driven from their homes, leaving behind charred remains like this; as many as 400,000 had died; and this part of the African Transition Zone faced famine and dislocation for many years to come, whatever the result of the power struggles afflicting them. © AP/Wide World Photos the Khartoum regime suspected the southern Fur of sympathizing with rebels in the far South who were opposed to Islamic rule. In 2003 northern pastoralist militias, encouraged by the regime, rode into the villages and fields of the southern Fur, burning homes, destroying crops, and killing thousands. As many as 2.6 million people were driven from their homes, and a combination of violence and resultant disease took a toll that, by 2010, was approaching 400,000.

THE MAGHREB AND ITS NEIGHBORS

The countries of northwestern Africa are collectively called the **Maghreb**, but the Arab name for them is more elaborate than that: *Jezira-al-Maghreb*, or "Isle of the West," in recognition of the great Atlas Mountain range rising like a huge island from the Mediterranean Sea to the north and the sandy flatlands of the immense Sahara to the south.

The countries of the Maghreb (sometimes spelled *Maghrib*) include Morocco, last of the North African kingdoms; Algeria, a secular republic beset by the religious-political problems we noted earlier; and Tunisia, smallest and most Westernized of the three (Fig. 7B-3). Neighboring Libya, facing the Mediterranean between the Maghreb and Egypt, is unlike any other North African country: an oil-rich desert state whose population is almost entirely clustered in settlements along the coast.

Atlas Mountains

Whereas Egypt is the gift of the Nile, the Atlas Mountains facilitate the settled Maghreb. These high ranges wrest from the rising air enough orographic rainfall to sustain life in the intervening valleys, where good soils support productive farming. From the vicinity of Algiers eastward along the coast into Tunisia, annual rainfall

averages more than 75 centimeters (30 in), a total more than three times as high as that recorded for Alexandria in Egypt's delta. Even 240 kilometers (150 mi) inland, the slopes of the Atlas still receive over 25 centimeters (10 in) of rainfall. The effect of the topography can even be read on the world climate map (Fig. G-7): where the highlands of the Atlas terminate, aridity immediately begins.

The Atlas Mountains are aligned southwest-northeast and begin within Morocco as the High Atlas, with elevations close to 4000 meters (13,000 ft). Eastward, two major ranges dominate the landscapes of Algeria proper: the Tell Atlas to the north, facing the Mediterranean, and the Saharan Atlas to the south, overlooking the great desert. Between these two mountain chains, with each consisting of several parallel ranges and foothills, lies a series of intermontane basins markedly drier than the northward-facing slopes of the Tell Atlas. Within these

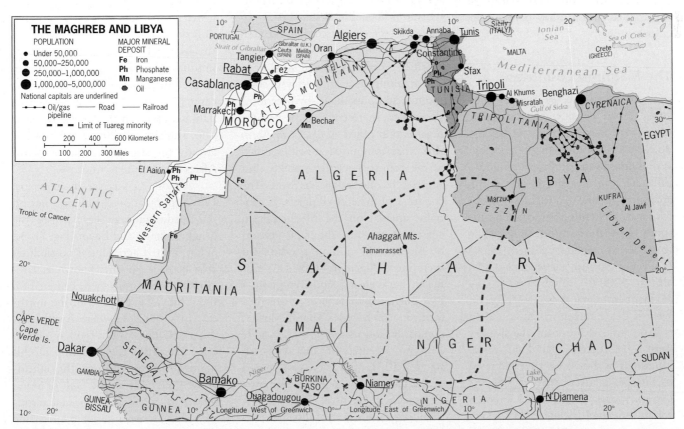

FIGURE 7B-3 © H. J. de Blij, P. O. Muller, and John Wiley & Sons, Inc.

© H. J. de Blij

"The *medina* (old city) in Marrakech, Morocco is lively, and the central square is crowded with locals and visitors alike. But the neighborhoods and streets and alleys beyond the tourist destinations reveal a different aspect of this sprawling city where modernization and tradition coexist. You'll look in vain here for signs in English—or, for that matter, French . . ."

www.conceptcaching.com

valleys, the **2** rain shadow effect of the Tell Atlas is reflected not only in the steppe-like natural vegetation but also in land-use patterns: pastoralism replaces cultivation, and stands of short grass and bushes blanket the countryside.

The Maghreb Countries

Between desert and sea, the Maghreb states display considerable geographic diversity. **Morocco**, a conservative kingdom in a revolution-marked region, is tradition-bound and weak economically. Its core area lies in the north, anchored by four major cities; but the Moroccans' attention is focused on the south, where the government is seeking to absorb its neighbor, Western Sahara (a former Spanish dependency with almost 500,000 inhabitants, many of them immigrants from Morocco). Even if this campaign is successful, it will do little to improve the lives of most of Morocco's 32 million people. Hundreds of thousands have migrated to Europe, many of them via Spain's two tiny exclaves on the Moroccan coast, Ceuta and Melilla, or via rickety boats to the Canary Islands (Fig. 7B-3).

Algeria, France's former colony whose agricultural potential drew more than a million European settlers, now has an economy based primarily on its substantial oil and gas reserves. Its bitter war of liberation (1954–1962) was followed by recurrent conflict between Islamists and military-backed secular authorities, costing over 100,000 lives and resulting in a long-delayed compromise still punctuated by occasional skirmishes. In recent years a terrorist organization called *Al-Qaeda in North Africa* has committed atrocities aimed at the country's relative stability, but Algeria's reconciliation has held. The country's primate city, Algiers, is centrally situated on the Mediterranean coast and contains about 10 percent of Algeria's 36 million people. But even more Algerians—officially estimated between 1.5 and 2 million—have migrated to France, where they form one of Europe's largest Muslim population sectors.

The smallest of the Maghreb states, **Tunisia**, lies at the eastern end of the region. As Appendix B shows, Tunisia in many ways outranks surrounding countries. Stability, national cohesion, and the maximization of

limited natural resources (no oil reserves here comparable to neighboring Algeria or Libya) have yielded a higher GNI, a higher urbanization level, higher social indicators, and a lower growth rate (among its population of 11 million) than elsewhere in the Maghreb. Most of the country's productive capacity lies in the north, in the hinterland of its large, historic capital, Tunis. As the authoritarian national government slowly loosens its grip, Tunisia's relations with Europe—already the strongest in all of North Africa—continue to improve.

Almost rectangular in its shape, **Libya** (population 7 million), lies on the Mediterranean Sea between the Maghreb states and Egypt (Fig. 7B-3). What limited agricultural possibilities exist lie in the district known as Tripolitania in the northwest, centered on the capital, Tripoli, and in the northeast in Cyrenaica, where Benghazi is the urban focus. But it is oil, not farming, that drives the economy of this highly urbanized country. The oilfields lie well inland from the Gulf of Sidra, linked by pipelines to coastal terminals. Libya's two interior corners, the desert Fezzan district in the southwest and the Kufra oasis in the southeast, are connected to the coast by two-lane roads subject to sandstorms.

Adjoining Saharan Africa

As Figure 7B-3 shows, the Maghreb countries and Libya adjoin several desert-dominated states in the African Transition Zone. These are African countries with strong Islamic imprints, but as Figure 6B-9 shows, they lie north of the Islamic Front with only one exception, Chad. Vast and sparsely populated **Mauritania** is the most Islamized of these countries, with about half its population of less than 4 million concentrated in and near the capital, Nouakchott, base of a small fishing fleet. Neighboring **Mali** relies on the waters of the upper Niger River, on which its capital of Bamako lies. A multiparty, democratic state, Mali is also multicultural, with a sizeable minority adhering to traditional beliefs and a small Christian minority. To the east of Mali lies **Niger**, which shares boundaries with Algeria and Libya; like Mali, this is one of the world's least urbanized countries (17 percent). Unlike Mali, however, Niger has only a short stretch of the middle course of the Niger River, where its capital, Niamey, is located. And **Chad**, directly south of Libya, has the lowest percentage of Muslim inhabitants but the strongest division between Islamized north and Christian/animist south. As Figure 7B-1 reminds us, the capital, N'Djamena, is located on the southern edge of the African Transition Zone, right next to the Islamic Front.

≡ THE MIDDLE EAST: CRUCIBLE OF CONFLICT

The regional term **Middle East**, we noted earlier, is not satisfactory, but it is so common and generally used that avoiding it creates more problems than it solves. It originated when Europe was the world's dominant realm and when places were *near*, *middle*, and *far*

from Europe: hence a Near East (Turkey), a Far East (China, Japan, Korea, and other countries of East Asia), and a Middle East (Egypt, Arabia, Iraq). If you check definitions used in the past, you will see that these terms were applied inconsistently: Syria, Lebanon, Palestine, and even Jordan sometimes were included in the Near East, and Persia and Afghanistan in the Middle East.

Today, the geographic designation *Middle East* has a more specific meaning. And at least half of it has merit: this region, more than any other, lies at the middle of the far-flung Islamic realm (Fig. 7B-1). To the north and east of it, respectively, lie Turkey and Iran, with Muslim Turkestan beyond the latter. To the south lies the Arabian Peninsula. And to the west lie the Mediterranean Sea and Egypt, and the rest of North Africa. This, then, is the pivotal region of the realm, its very heart.

Five countries form the Middle East (Fig. 7B-4): Iraq, largest in population and territorial size, facing the Persian Gulf; Syria, next in both categories and fronting the Mediterranean; Jordan, linked by the narrow Gulf of Aqaba to the Red Sea; Lebanon, whose survival as a unified state has come into question; and Israel, Jewish nation in the crucible of the Muslim world. Because of the extraordinary importance of this region in world affairs, we focus in some detail on issues of cultural, economic, and political geography in the discussion that follows.

Iraq's Enduring Importance

Figure 7B-4 explains, even at a glance, why Iraq is pivotal among the states of the Middle East. With about 60 percent of the region's total area, more than 40 percent of its predominantly Arab population, and most of its valuable energy and agricultural resources, Iraq is key to the Middle East's fortunes. Iraq also is heir to the early Mesopotamian states and empires that emerged in the basin of its twin rivers,

the Tigris-Euphrates, and the country is studded with matchless archeological sites and museum collections.

Much of this heritage was disastrously damaged in 2003 when the United States led a military invasion of Iraq in the aftermath of the 9/11 terrorist attacks in America. The map indicates why this attack had (and continues to have) critical consequences for the region and beyond: Iraq has six neighbors, all of them affected in serious ways. To the west lie Syria, Iraq's ally during dictator Saddam Hussein's despotic rule, and Jordan, whose leaders chose Iraq's side during an early 1990s conflict with southern neighbor Kuwait. To the south, Iraq possesses a lengthy border with Saudi Arabia. And to the north, Iraq has borders with Iran, against which it fought a bitter war during the 1980s, and with Turkey, source of its crucial river lifelines.

But perhaps the most obvious feature in Figure 7B-4 is the lavender-colored, striped zone that not only covers northeastern Iraq but also large parts of Turkey and Iran as well as smaller parts of other countries. This is the area where the majority of the inhabitants are not Arabs, but Kurds, a people who have been discriminated against by all the countries in which they live. About 5 million Kurds inhabit the hilly north of Iraq; adjacent southeastern Turkey is the traditional home of 15 million more, and another 8 million live in western Iran. Together, they form one of the world's largest **3** **stateless nations**, divided and often exploited by those who rule over them. The American invasion of Iraq gave the Kurds something they had not experienced anywhere else in their domain: a chance at regional self-government and a large measure of cultural freedom.

With so many land neighbors, it is no surprise that Iraq is nearly landlocked. Figure 7B-4 shows how short and congested its single outlet to the Persian Gulf is; it is for this reason that Saddam tried to conquer and annex neighboring Kuwait in 1990.

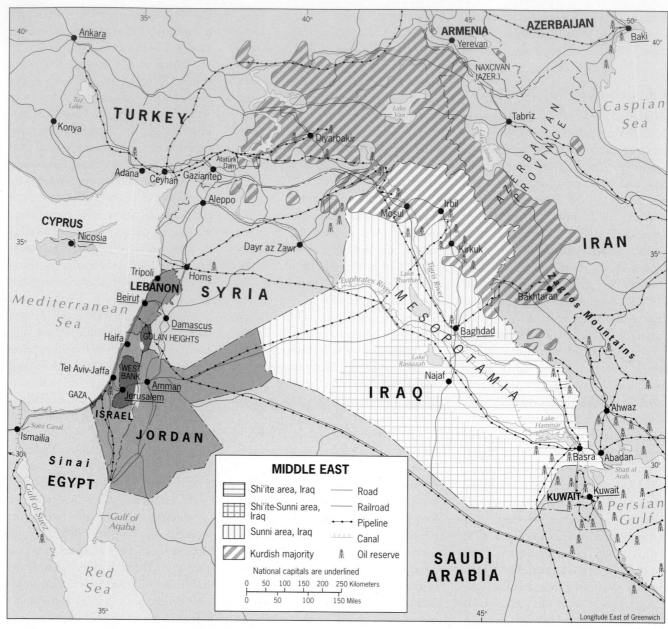

FIGURE 7B-4

© H. J. DE BLIJ, P. O. MULLER, AND JOHN WILEY & SONS, INC.

As a result, a network of pipelines across Iraq's neighbors must carry much of Iraq's oil export volume to coastal terminals in other countries—another reason these countries have an interest in what happens to Iraq in the years ahead.

Discordant Political and Cultural Geographies

Let us now turn to the larger-scale map of Iraq (Fig. 7B-5), which shows how Iraq is divided into three cultural domains in which religion, tradition, and custom form the basis of its political geography. The largest population sector (within the national total of 31 million) is that of the 20 million Shi'ites in the southeast, whose religious affinities are with neighboring Iran; the next largest is the Sunni minority of about 6 million in the north and west, which ruled the country before the 2003 invasion; and the smallest of Iraq's major components is that of the Kurds, roughly 5 million in their northeastern stronghold. Even at this scale, however, the map is a generalization of a considerably more intricate mosaic. The borders between the major cultural domains in reality are transition zones, most of them without sharp definition (see the Kurd-Sunni frontier). Also not shown at this scale are smaller minorities of Turkmen, Assyrians, and others clustering in the north. And most important of all, the population of Baghdad, the capital, has grown larger than either the Sunni or the Kurdish domains, a

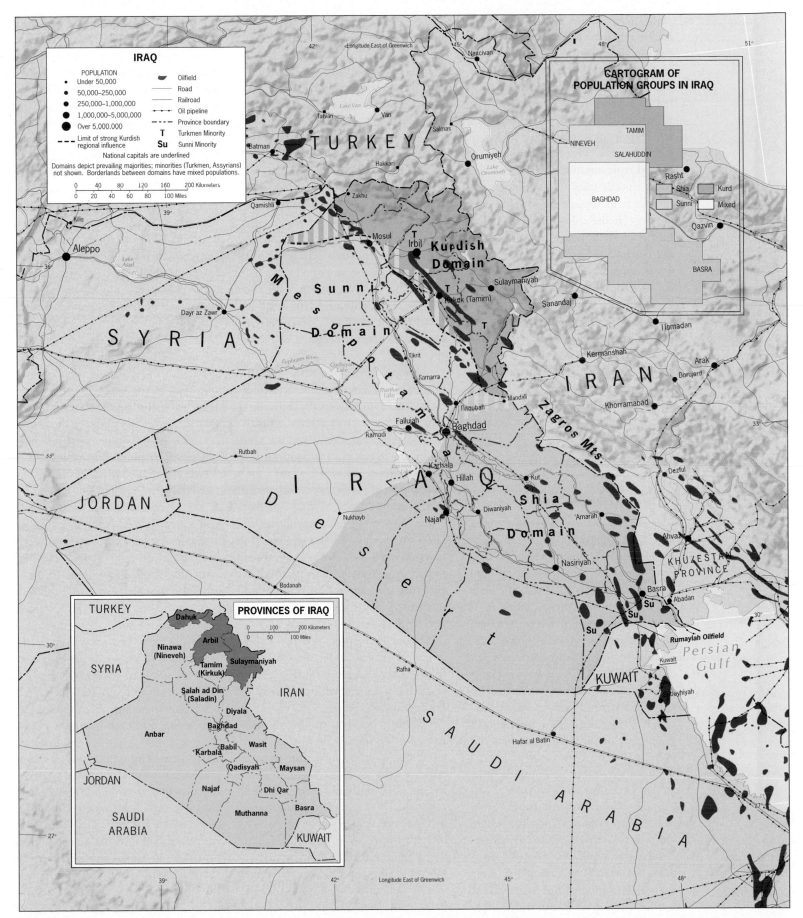

FIGURE 7B-5

microcosm of the divided state (see the upper inset map in Fig. 7B-5).

The multipurpose war in Iraq unleashed terrible sectarian violence and enabled al-Qaeda terrorists, who had not previously been active here, to worsen the security situation. Even so, Iraq's people repeatedly voted in elections that advanced the country toward representative government, and some middle-class Iraqis (among the hundreds of thousands who fled the country in the mid-2000s) began to return to their homeland. By 2010, even as suicide bombers continued to attack targets from religious pilgrims to government facilities, American combat forces were scheduled to withdraw from Iraqi soil. But it remained uncertain whether Iraq would be able to achieve the structural stability and political democracy envisaged by those who initiated a conflict that claimed an estimated 100,000 Iraqi lives and caused the death of some 4500 allied soldiers.

The Region's Other Muslim-Dominated States

Like Iraq before 2003, **Syria** is ruled by a minority. Although Syria's population of 21 million is about 75 percent Sunni Muslim, the ruling elite comes from a smaller Islamic sect based there, the Alawites. Leaders of this powerful minority have retained control over the country for decades, at times by ruthless suppression of dissent. In 2000, president-for-life Hafez al-Assad died and was succeeded by his son, Bashar, signaling a continuation of the political status quo. For 25 years, part of this status quo was the occupation and control of neighboring Lebanon, but in 2005 this came to an end as the Syrians withdrew.

Like Lebanon and Israel, Syria has a Mediterranean coastline where crops can be raised without irrigation. Behind this densely populated coastal zone, Syria has a much larger interior than its neighbors, but its areas of productive capacity are widely dis-

persed. Damascus, in the southwest corner of the country, was built on an oasis and is considered to be the world's oldest continuously inhabited city. It is now the capital of Syria, with a population of close to 3 million.

The far northwest is anchored by Aleppo, the focus of cotton- and wheat-growing areas in the shadow of the Turkish border. Here the Orontes River is the chief source of irrigation water, but in the eastern part of the country the Euphrates Valley is the crucial lifeline. It is in Syria's interest to develop its northeastern provinces, and recent discoveries of oil there will contribute to this development. But the war in neighboring Iraq has triggered problems along Syria's eastern border, which will delay its plans for this area.

Jordan, Syria's southern neighbor, is a classic case (and victim) of changing relative location. This desert kingdom was a product of the Ottoman collapse, but when Israel was created in 1948 it lost its window on the Mediterranean Sea as well as the (now Israeli) port of Haifa, previously in the British-administered Mandate of Palestine. Following its independence in 1946 with about 400,000 inhabitants, newly created Israel soon bequeathed Jordan some 500,000 West Bank Palestinians and, later, a huge inflow of refugees. Today Palestinians outnumber original residents by more than two to one in the population of 6-plus million. It may be said that the 47-year rule by King Hussein, ending in 1999, was the key centripetal force that held the country together.

With a poverty-stricken capital, Amman, lacking in oil reserves, and possessing only a small and remote outlet to the Gulf of Aqaba, Jordan has survived with U.S., British, and other aid. It lost its West Bank territory in the 1967 war with Israel, including its sector of Jerusalem (then the kingdom's second-largest city). No third country has a greater stake in a settlement between Israel and the Palestinians than does Jordan.

The map suggests that **Lebanon** has significant geographic advantages in this region: a lengthy coastline on the Mediterranean Sea; a well-situated capital, Beirut, on its shoreline; oil terminals along its coast; and a major capital (Syria's Damascus) in its hinterland. The map at the scale of Figure 7B-4 cannot reveal yet another asset: the fertile, agriculturally productive Bekaa Valley in the eastern interior.

French colonialism following the Ottoman period created a territory, which in 1930 was about equally Muslim and Christian. Beirut had become known as the Paris of the Middle East. After independence in 1946, Lebanon functioned as a democracy and did well economically as the region's leading banking and commercial center. But the Muslim sector of the population grew much faster than the Christian one, and in the late 1950s the Arabs launched their first rebellion against the established order. Following the first of several waves of influx by Palestinian refugees, Lebanon fell apart in 1975 in a civil war that wrecked Beirut, devastated the economy, and left the country at the mercy of Syrian forces, which had entered to take control.

For so small a country (just over 4 million, including more than 400,000 naturalized and non-citizen Palestinians), Lebanon is permeated with religious and ethnic factions and is prone to sectarian breakdown. The Lebanese were unhappy with what they felt to be a Syrian occupation, yet the occupation continued for several decades. In the meantime, an Iran-sponsored terrorist movement, Hizbollah, came into being, and even became a political force. The Syrians were ousted in 2005 under United Nations auspices, and a stable new Lebanese government coalesced around a power-sharing agreement. However, the peace lasted only a few years. In 2006 a Hizbollah kidnapping of Israeli soldiers provoked an Israeli attack, shattering the reconstructed physical and political infrastructure. Lebanon's situation today remains precarious.

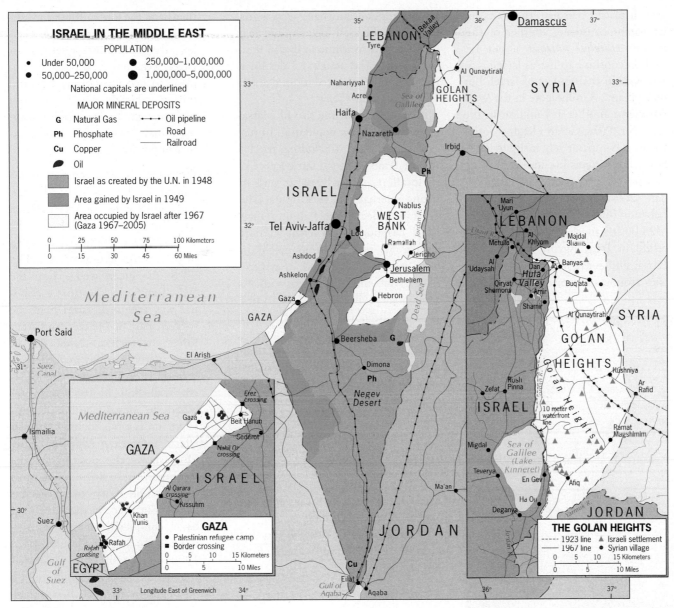

FIGURE 7B-6

© H. J. de Blij, P. O. Muller, and John Wiley & Sons, Inc.

Israel and the Palestinian Territories

Israel, the Jewish state, lies in the heart of the Arab World (Fig. 7B-1). Since 1948, when Israel was created as a homeland for the Jewish people on the recommendation of a United Nations commission, the Arab-Israeli conflict has overshadowed all else in the region.

Figure 7B-6 helps us understand the complex issues involved here. In 1946 the British, who had administered this area in post-Ottoman times, granted independence to what was then called Transjordan, the kingdom east of the Jordan River. In 1948, the orange-colored area became the UN-sponsored state of Israel—including, of course, land that had long belonged to Arabs in this territory called Palestine.

As soon as Israel proclaimed its independence, neighboring Arab states attacked it. Israel, however, not only held its own but pushed the Arab forces back beyond its borders, gaining the green areas shown in Figure 7B-6. Meanwhile, Transjordanian armies crossed the Jordan River and annexed the yellow-colored area named the West Bank (of the Jordan River), including part of the city of Jerusalem. The king called his newly enlarged country Jordan.

More conflict followed. In 1967 a six-day-long war produced a major Israeli victory: Israel took the Golan Heights from Syria, the West Bank from Jordan, and the Gaza Strip from Egypt, and conquered the entire Sinai Peninsula all the way to Egypt's Suez Canal. In later peace agreements, Israel returned the Sinai but not the Gaza Strip to Egypt.

All this strife produced a huge outflow of Palestinian Arab refugees and displaced persons. The Palestinian Arabs constitute another of this region's **stateless nations**; about 1.3 million continue to live as Israeli citizens within the borders of Israel, but more than 2.3 million are in the West Bank and another 1.4 million in the Gaza Strip (the Golan Heights population is comparatively insignificant). Some Palestinians live in neighboring and nearby countries, including Jordan (2.7 million), Syria (435,000), Lebanon (405,000), and Saudi Arabia (325,000); another 200,000 or so reside in Iraq, Egypt, Kuwait, and Libya; and some live in other countries around the world, including the

Israel's Security Barrier takes several forms—a concrete wall, an iron fence—and its effect also has multiple dimensions. This section of the barrier separates the village of Abu Dis on the outskirts of Jerusalem (left) from the West Bank (right). While Israel defines this demarcation as a security issue, the Arabs, citing its location inside Palestinian territory, argue that it is motivated primarily by politics.
© AP/Wide World Photos

United States. Many have been assimilated into the local societies, but tens of thousands continue to live in refugee camps. The stateless Palestinian population is estimated to total about 10 million in 2011.

Israel is about the size of Massachusetts and has a population of just under 8 million (including its 1.3 million Arab citizens), but because of its location amid adversaries and its strong international links, these data do not reflect Israel's importance. Israel has built a powerful military even as its regional Arab neighbors have grown stronger and its policies arouse Arab and Muslim passions. As a democracy with strong Western ties, Israel has been a recipient of massive U.S. financial aid, and U.S. foreign policy has been to seek an accommodation between Jews and Palestinians, as well as between Israel and its Arab neighbors. Recently, Washington initiated a push toward a two-state solution that would create a Palestinian state alongside Israel, but as always there are thorny issues to be dealt with and a need for political will and accommodation on both sides that does not appear to be forthcoming.

The geographic obstacles to this kind of an accommodation include the following:

1. **The West Bank.** Even after its capture by Israel in 1967, the West Bank might have become a Palestinian homeland (and possibly a state), but Jewish immigration to the area made such a future difficult. In 1977 only 5000 Jews lived in the West Bank; by 2011 there were over 500,000, making up more than 18 percent of the population and creating a seemingly inextricable jigsaw of Jewish and Arab settlements (Fig. 7B-7), making the two-state solution difficult.

2. **The Golan Heights.** The inset map on the right in Figure 7B-6 suggests how difficult the Golan Heights issue is. The Heights overlook a large area of northern Israel, and

they flank the Jordan River and crucial Lake Kinneret (the Sea of Galilee), the main water reservoir for Israel. Relations with Syria are not likely to become normalized until the Golan Heights are returned, but in democratic Israel the political climate may make ceding this territory impossible.

3. **Jerusalem.** The United Nations intended Tel Aviv to be Israel's capital, and Jerusalem an international city. But the Arab attack and the 1948–1949 war allowed Israel to drive toward Jerusalem (see Fig. 7B-6, the green wedge into the West Bank). By the time a ceasefire was arranged, Israel held the western part of the city, and Arab forces the eastern sector. But in this eastern sector lay major Jewish historic sites, including the Western Wall. Still, in 1950 Israel declared the western sector of Jerusalem its capital, making this, in effect, a **forward capital.** Figure 7B-8 shows the position of the (black) armistice line, leaving most of the Old City in Jordanian hands. But then, in the 1967 war, Israel conquered all of the West Bank, including East Jerusalem; in 1980 the Jewish state reaffirmed Jerusalem's status as capital, calling on all nations to move their embassies from Tel Aviv. Meanwhile, the government redrew the map of the ancient city, building Jewish settlements in a ring around East Jerusalem that would terminate the old distinction between Jewish west and Arab east. This enraged Palestinian leaders, who still view Jerusalem as the eventual headquarters of a hoped-for Palestinian state

4. **The Security Fence.** In response to the infiltration of suicide terrorists, Israel has now walled off most of the West Bank along the Security Barrier border shown in Figure 7B-7 and the photo at left. The new wall does not conform to the de facto postwar boundary of the West Bank,

cutting off an additional 10 percent of its territory and in effect annexing it (and its inhabitants) to Israel. This reinforcement of the West Bank border imposes a hardship on Palestinians, for some of whom it runs between their homes and their farm fields. Palestinians demand that the wall be taken down; Israelis reply that the Palestinian Authority must control the terrorists.

5. *The Gaza Strip*. Israel in 2005 decided to withdraw from the Gaza Strip, removing all Jewish settlements and yielding control of the area to the (then) Palestinian Authority (Fig. 7B-6, left inset map). A power struggle ensued between the two main Palestinian political factions, Fatah and the more militant Hamas, and Hamas won a disputed 2006 election. As rockets smuggled in from Egypt through tunnels dug beneath the Gaza-Egyptian border fell on Israel causing some deaths, injuries, and damage, Israel warned the Hamas regime of retaliation. Hamas authorities either would not or could not stop the rockets fired from Gaza, and in 2009 Israel launched a massive attack, killing about 1300 Gaza residents including many innocent civilians and inflicting heavy infrastructure damage, including a UN facility. Condemnation of Israel's assault as disproportionate was worldwide, but a majority of Israeli citizens approved of it.

Israel lies in the path of a fast-moving geopolitical storm; the issues raised above are only part of the overall problem (others involve water rights, compensation for expropriated land, Arabs' "right of return" to prerefugee abodes, and the form a Palestinian state would take). Meanwhile, an Iranian president called for Israel's being "wiped off the map" even as his country appears to be on its way to nuclear weapons capacity and already has the means to deliver them. In this

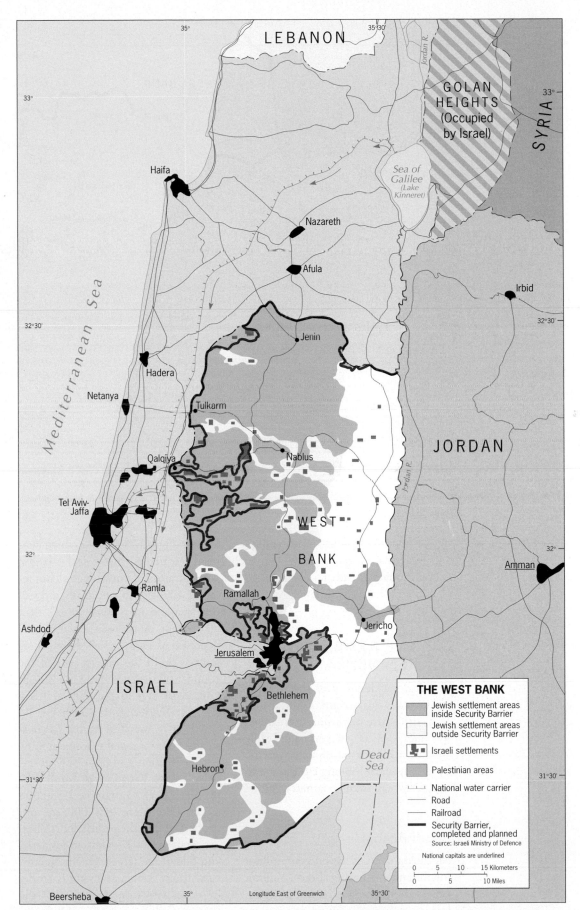

FIGURE 7B-7

© H. J. de Blij, P. O. Muller, and John Wiley & Sons, Inc.

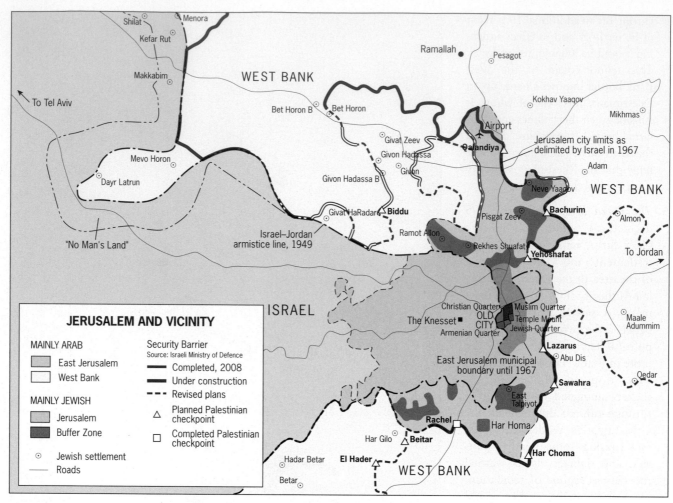

FIGURE 7B-8

age of nuclear, chemical, and biological weapons, Israel's search for an accommodation with its Arab neighbors is a race against time.

THE ARABIAN PENINSULA

The regional identity of the Arabian Peninsula is clear: south of Jordan and Iraq, the entire peninsula is encircled by water. This is a region of old-style sheikdoms made fabulously wealthy by oil, modern-looking emirates, and the site where Islam originated.

Saudi Arabia

Saudi Arabia itself has only 30 million inhabitants (including about 7 million expatriate workers) in its vast territory, but we can see the kingdom's impor-

tance in Figure 7A-9: the Arabian Peninsula contains the world's largest concentration of known petroleum reserves. Saudi Arabia occupies most of this area and may possess more than 20 percent of the world's liquid oil deposits. As Figure 7B-9 shows, these reserves lie in the eastern part of the country, particularly along the Persian Gulf coast and in the Rub al Khali (Empty Quarter) to the south.

As a region, the Arabian Peninsula is environmentally dominated by a desert habitat and politically dominated by the Kingdom of Saudi Arabia (Fig. 7B-9). Containing 2,150,000 square kilometers (830,000 sq mi), Saudi Arabia is the realm's fourth biggest state; only Kazakhstan, Algeria, and Sudan are larger. On the peninsula, Saudi Arabia's neighbors (moving clockwise from the head of the Persian Gulf) are Kuwait, Bahrain,

Qatar, the United Arab Emirates, the Sultanate of Oman, and the Republic of Yemen. Together, these countries on the eastern and southern fringes of the peninsula contain more than 35 million inhabitants; the largest by far is Yemen, with 24 million.

Figure 7B-9 reveals that most of the economic activities in Saudi Arabia are concentrated in a wide belt across the "waist" of the peninsula, from the oil boomtown of Dhahran on the Persian Gulf through the national capital of Riyadh in the interior to the Mecca–Medina area near the Red Sea. A modern transportation and communications network has recently been completed, but in the more remote zones of the interior Bedouin nomads still ply their ancient caravan routes across the vast deserts. For decades, Saudi Arabia's royal families were virtually the sole beneficiaries of

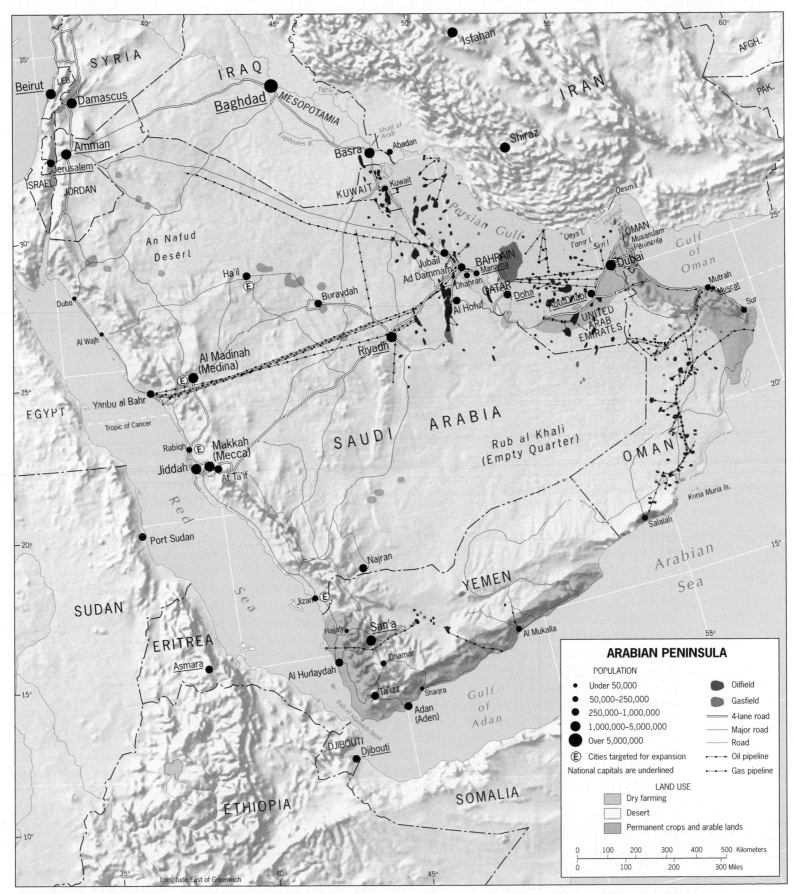

SYRIA

IRAQ

IRAN

AFGH.

PAK.

Isfahan

Beirut
LEB.
Damascus

Baghdad   MESOPOTAMIA

Shiraz

Amman

Basra   Abadan

Jerusalem

ISRAEL

JORDAN

KUWAIT

Kuwait

Qesm I.

An Nafud
Desert

Ha'il

Buraydah

Jubail

Ad Dammam

BAHRAIN
Manama

Dhahran

QATAR
Doha

Abu Dhabi

UNITED
ARAB
EMIRATES

OMAN

Musandam
Peninsula

Gulf
of
Oman

Dubai

Mutrah
Muscat

Sur

Duba

Al Wajh

Al Madinah
(Medina)

Riyadh

Al Hofuf

Yanbu al Bahr

Tropic of Cancer

EGYPT

SAUDI   ARABIA

Rub al Khali
(Empty Quarter)

OMAN

Rabigh

Makkah
(Mecca)

Jiddah

At Ta'if

Red

Sea

Khuria Muria Is.

Salalah

Port Sudan

Najran

Arabian

Sea

SUDAN

Jizan

YEMEN

Al Mukalla

Hajjah

San'a

Dhamar

Shaqra

ERITREA

Asmara

Al Hudaydah

Ta'izz

Gulf
of
Aden

Adan
(Aden)

DJIBOUTI   Djibouti

ETHIOPIA

SOMALIA

ARABIAN PENINSULA

POPULATION

- Under 50,000
- 50,000–250,000
- 250,000–1,000,000
- 1,000,000–5,000,000
- Over 5,000,000

Ⓔ Cities targeted for expansion

National capitals are underlined

Oilfield

Gasfield

4-lane road

Major road

Road

Oil pipeline

Gas pipeline

LAND USE

Dry farming

Desert

Permanent crops and arable lands

0   100   200   300   400   500   Kilometers

0   100   200   300 Miles

Longitude East of Greenwich

FIGURE 7B-9

© H. J. de Blij, P. O. Muller, and John Wiley & Sons, Inc.

their country's wealth, and there was hardly any impact on the lives of villagers and nomads. When the oil boom arrived in the 1950s, foreign laborers were brought in (today there are about 7 million, many of them Shi'ite) to work in the oilfields, ports, factories, and as domestics. The east boomed, but the rest of the country has lagged behind.

Disparities and Uncertainties

Efforts to reduce Saudi Arabia's regional economic disparities have been impeded by the enormous cost of bringing water from deep-seated sources to the desert surface to stimulate agriculture, by the sheer size of the country, and by the high rate of population growth (2.6 percent). Still, housing, health care, and education have seen major improvement, and industrialization also has been stimulated in such places as Jubail on the Persian Gulf and Yanbu, near Medina, on the Red Sea.

Saudi Arabia's conservative monarchism, official friendship with the West, and social contrasts resulting from its economic growth have raised political opposition, for which no adequate channels exist. Relations with the United States were affected by the role of Saudi suicide terrorists in 9/11, Saudi financial support for Wahhabist extremists, and additional issues arising in the wake of the events of 2001. More recently, the Saudis have resisted increasing oil production despite record prices, testing their friendship with the United States. Reliance on oil income, though currently quite profitable, is economically risky in the long run because the country's social benefits tend to become dependent on these revenues.

On the Periphery

Five of Saudi Arabia's six neighbors on the Arabian Peninsula face the Persian and Oman gulfs (Fig. 7B-7) and are monarchies in the Islamic tradition. All five also derive substantial revenues from oil. Their populations range from 0.8 to 4.6 million in addition to hundreds of thousands of foreign workers; these are not strong or influential states. Nonetheless, they do display considerable geographic diversity. *Kuwait*, at the head of the Persian Gulf, almost cuts Iraq off from the open sea, an issue the 1990–1991 Gulf War did not settle. *Bahrain* is an island-state, a tiny territory with dwindling oil reserves. It has become an important banking center for the region and has some of the most progressive social policies in this part of the world. Its nearly 800,000 people are evenly split between Sunnis and Shi'ites; nearly two-thirds of its labor force is foreign. Neighboring *Qatar* consists of a peninsula jutting out into the Persian Gulf. Its oil reserves are also dwindling, but it has capitalized on its increasing natural gas reserves.

The *United Arab Emirates (UAE)*, a federation of seven emirates, faces the Persian Gulf between Qatar and Oman. A reigning sheik is the absolute monarch in each of the emirates, and the seven sheiks together form the Supreme Council of Rulers. In terms of oil revenues, however, there is no equality: two emirates—Abu Dhabi and Dubai—have most of the reserves. Dubai (Dubayy) has converted its favorable geographic location together with its oil revenues into a booming economy—recently tempered by global recession—in which tourism, international transit functions, international higher education, banking, and trade are symbolized by ultramodern architecture, engineering feats (including the world's tallest building), and state-of-the-art entertainment complexes.

The eastern corner of the Arabian Peninsula is occupied by the Sultanate of *Oman*, another absolute monarchy, centered on the capital, Muscat. Figure 7B-9 shows that Oman consists of two parts: the large eastern corner of the peninsula and a small but critical cape to the north, the Musandam Peninsula, which protrudes into the Persian Gulf to form a nar-row channel or **4** **choke point**—the Hormuz Strait (Iran lies on the opposite shore). Tankers leaving the other Gulf states must negotiate this narrow, twisting channel at slow speed, and during politically tense times warships have had to protect them. Iran's claim to several small islands near the Strait that are owned by the UAE is a potential source of dispute.

This brings us to what is, in so many ways, Saudi Arabia's most substantial peninsular neighbor: *Yemen*. The boundary between Saudi Arabia and Yemen has only recently been satisfactorily delimited in an area where there may be substantial oil reserves. Another boundary, between former North Yemen and South Yemen, was erased in 1990, when the two countries joined to form the present state. San'a, formerly the capital of North Yemen, retained that status; Adan (Aden), the only major port along a lengthy stretch of the peninsula, anchors the south.

Yemen, with a population of 23.6 million—55 percent Sunni and 45 percent Shi'ite—initially made significant progress toward representative government and political stability, but centrifugal forces and terrorist activity caused later setbacks. A Shi'ite rebellion against the central government has destabilized the north even as a secessionist movement flares in the south, where oilfields encourage such aspirations. In 2000, a terrorist attack on a U.S. warship docked in Adan killed 17 sailors, nearly sank the vessel, and ushered in a series of subsequent attacks on foreigners, government facilities, and other targets for which an organization called *Al-Qaeda of the Arabian Peninsula* claimed responsibility. Al-Qaeda's intentions to widen its terrorist reach beyond Yemen were confirmed in 2009 by its attempt to instruct and equip a Nigerian citizen for the purpose of downing an American jetliner.

Figure 7B-9 underscores the geographic attractions Yemen has for al-Qaeda. Not only does Yemen lie at the back door of Saudi Arabia, sworn enemy of Usama bin Laden, whose

© H. J. de Blij

"The port of Mutrah, Oman, like the capital of Muscat nearby, lies wedged between water and rock, the former encroaching by erosion, the latter crumbling as a result of tectonic plate movement. From across the bay one can see how limited Mutrah's living space is, and one of the dangers here is the frequent falling and downhill sliding of large pieces of rock. It took about five hours to walk from Mutrah to Muscat; it was extremely hot under the desert sun but the cultural landscape was fascinating. Oil also drives Oman's economy, but here you do not find the total transformation seen in Kuwait or Dubai. Townscapes (as in Mutrah) retain their Arab-Islamic qualities; modern highways, hotels, and residential areas have been built, but not at the cost of the older and the traditional. Oman's authoritarian government is slowly opening the country to the outside world after long-term isolation."

www.conceptcaching.com

family lived here before he was born; its southern tip overlooks one of the world's busiest choke points at the entrance to the Red Sea (the Bab-el-Mandeb Strait or "Gate of Grief"), where ships converge and risk capture by pirates. Across the Gulf of Aden lies another terrorist haven, Somalia, and across the Red Sea lies Eritrea, a logical target for al-Qaeda's further expansion. From Yemen, al-Qaeda's militants can stoke strife along Africa's Islamic Front (see Fig. 6B-9) even as they plot their global campaigns. Yemen's government searches for ways to regain stability, resist secession, and inhibit terrorist activity, but its own resources are limited and it cannot accept major Western assistance without endangering its legitimacy among its own people.

THE EMPIRE STATES

Two major states, both with imperial histories, dominate the region that lies immediately to the north of the Middle East and Persian Gulf (Fig. 7B-1), where Arab ethnicity gives way but Islamic culture endures—Turkey and Iran. Although they share a short border and are both Muslim countries, they display significant differences as well. Even their versions of Islam are different: Turkey is an officially secular but dominantly Sunni state, whereas Iran is the heartland of Shi'ism. Turkey's leaders have worked to establish satisfactory relations with Israel; Iran's president promised to "wipe Israel off the map." In recent years, Turkey has been trying to negotiate entry into the European Union. Iran's goals have been quite different as that country has defied international efforts to constrain its nuclear ambitions. Two smaller countries are inextricably bound up with these two regional powers: island *Cyprus* to the southwest, still divided today between Turks and Greeks in a way that threatens Turkey's European ambitions (discussed in Chapter 1B); and oil-rich, Caspian Sea-bordering *Azerbaijan* to the northwest, with ethnic and religious ties to Iran but economic links elsewhere (see Chapter 2B).

Turkey

As Figure 7B-10 indicates, Turkey is a mountainous country of generally moderate relief; it also tends to exhibit considerable environmental diversity, ranging from steppe to highland (see Fig. G-7). On central Turkey's dry Anatolian Plateau, villages are small, and subsistence farmers grow cereals and raise livestock. Coastal plains are not large, but they are productive as well as densely populated. Textiles (from home-grown cotton) and farm products dominate the export economy, but Turkey also has substantial mineral reserves, some oil in the southeast, massive dam-building projects on the Tigris and Euphrates rivers, and a small steel industry based on domestic raw materials.

In Chapter 7A we chronicled the historical geography of the Ottoman

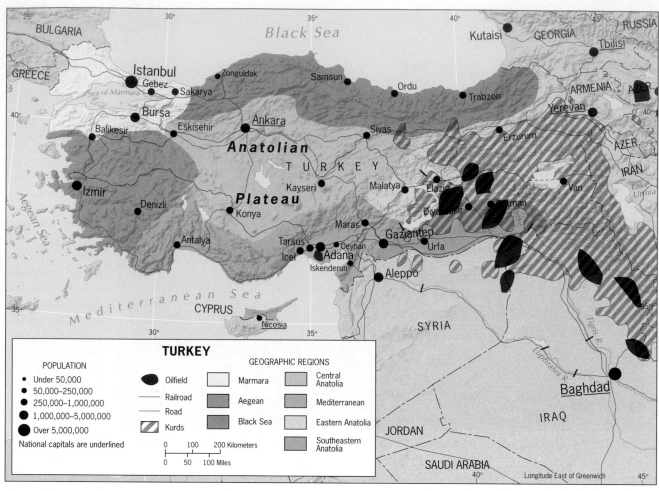

FIGURE 7B-10

© H. J. DE BLIJ, P. O. MULLER, AND JOHN WILEY & SONS, INC.

Empire, its expansion, cultural domination, and collapse. By the beginning of the twentieth century, the country we now know as Turkey lay at the center of this decaying and corrupt state, ripe for revolution and renewal. This occurred in the 1920s and thrust into prominence a leader who became known as the father of modern Turkey: Mustafa Kemal, known after 1933 as Atatürk, meaning Father of the Turks.

Capitals New and Old

The ancient capital of Turkey was Constantinople (now Istanbul), located on the Bosphorus, part of the strategic straits connecting the Black and Mediterranean seas. But the struggle for Turkey's survival had been waged from the heart of the country, the Anatolian

Plateau, and it was here that Atatürk decided to place his seat of government (Fig. 7B-10). Ankara, the new capital, possessed certain advantages: it would remind the Turks that they were (as Atatürk always said) Anatolians; it lay nearer the center of the country than Istanbul; and it could therefore act as a stronger unifier. Istanbul lies on the threshold of Europe, with the minarets and mosques of this largest and most varied Turkish city rising above a townscape that resembles those in eastern Europe (see photo next page).

Even though Atatürk moved the capital eastward and inward, his orientation was westward and outward. To implement his plans for Turkey's modernization, he initiated reforms in almost every sphere of life within the country. Islam, formerly the state religion, lost its official status, and

Turkey became a secular state whose army ensured that the Islamists would not take over again. The state took over most of the religious schools that had controlled education. The Roman alphabet replaced the Arabic. A modified Western code supplemented Islamic law. Symbols of old—growing beards, wearing the fez—were prohibited. Monogamy was made law, and the emancipation of women was begun. The new government emphasized Turkey's separateness from the Arab World, and in many ways it has remained aloof from the affairs that engage other Islamic states.

Turkey and Its Neighbors

From before Atatürk's time, Turkey has had a history of mistreating minorities. Soon after the outbreak of

© H. J. de Blij

"Having spent the night in Istanbul's Beyoğlu District on the European side, I walked across the Atatürk (Unkapani) Bridge to the Eminönü District on the south side of the Golden Horn, intending to make it to the upscale Fatih District later. But getting a clear view of the Hagia Sofia 'Museum' was not easy from this angle, amid the congested buildup in this hilly area. In a line of taxis at a taxi stand one had a sign saying 'Speak English' so I asked the driver about getting to a good vantage point, and he showed me on my map how to find a lookout point on the terrace of a local restaurant. From there, I could see the bridge I had crossed to my left, the Hagia Sofia Church with its later minarets in the distance, the Sultan Haseki Hürrem Baths to the right, and in the foreground a simple, local mosque. 'You should know the rules about mosques,' he said. 'The number of minarets reveals the importance of any mosque. One minaret, and it's local, neighborhood. Two, and it's likely to be more prosperous, perhaps the work of a very successful citizen or community. Any more than two, and the builders have to have not only religious sanction but also state permission to build. And if they get that, they go for four.' Thousands of mosques, many of them architectural treasures, grace the urban landscape of Istanbul, whose population today, according to local geographers, is 21 million."

the First World War, the (pre-Atatürk) regime decided to expel all the Armenians, concentrated in the country's northeast. Nearly 2 million Turkish Armenians were uprooted and brutally forced out; an estimated 600,000 died in a campaign that still arouses anti-Turkish emotions among Armenians today. In modern times, Turkey also has been criticized for its treatment of its large and regionally concentrated Kurdish population. About one-fifth of Turkey's population of 77 million is Kurdish, and successive Turkish governments have mishandled relationships with this minority nation, even prohibiting the use of Kurdish speech and music in public places during one especially repressive period. The historic Kurdish homeland lies in the southeast of Turkey, centered on Diyarbakir, but millions of Kurds have moved to the shantytowns around Istanbul—and to jobs in the countries of the European

Union. With Kurdish nationalism rising across the border in embattled Iraq, Turkey has responded in two principal ways: by suppressing Kurdish insurgent movements both in Turkey itself and across the border in Iraq, and by awarding more rights and freedoms to those Kurds not involved in armed resistance. Thus Turkey improved its human rights record, a key stumbling block in its aspirations to join the European Union. Simultaneously, Ankara made tentative but significant moves to rehabilitate its relations with Armenia, addressing an erosive problem of long standing.

Turkey and the EU

With its diversified economy, secular government, and independent posture in international affairs, Turkey is the key, predominantly Islamic state potentially eligible for admission into

the European Union (Albania and Kosovo may precede it). Turkey's soccer team already plays in the European championships, and for a long time Turkey's westward orientation, membership in NATO, its many migrants, and its solid economic ties to that realm made this a distinct possibility. Already in a customs union with the EU since 1995, Turkey in 1999 was formally declared an accession candidate. Formal negotiations toward admission began in 2005.

However, there are various major obstacles in the way of this becoming a reality. The Cyprus and Kurdish issues have both negatively affected Turkey's prospects for joining the EU. Admitting Turkey would put Europe's boundary on the doorstep of a dangerous neighborhood, which is a concern to some, though a positive aspect for others. Another consideration is Turkey's human rights record, which remains inadequate. In particular,

Turkey's numerous "honor killings," usually of girls and women by men, arouse European anger—not only because they take place (heinous murders occur in non-Islamic societies as well) but because they constitute fully half of all murders committed in the country—and because they continue to be tolerated through ineffective gender-biased law enforcement and insufficient public condemnation. In February 2010, a teenage girl was buried alive by male members of her family for the "crime" of unsupervised talking to boys in her neighborhood; the incident sparked outrage in the European media and rekindled opposition to Turkey's candidacy for EU membership.

Nonetheless, Turkey has made some progress. The state is now a functioning democracy, and its secular values are constitutionally protected so that even an Islamic party majority in government does not portend radicalization even as it may encourage symbols of piety. For example, after the largest Islamic party won a majority in Turkey's 2007 elections, relations with Israel continued even as the new government lifted the long-existing ban on women wearing head scarves. As noted earlier, the new government also continued to seek accommodation with its large Kurdish minority as well as with neighboring (Christian) Armenia.

But while Turkey might be closing in on European Union targets and requirements, enthusiasm among Turks for joining the EU is diminishing. Many Turks take offense at the unrestrained public debate concerning their candidacy going on in Europe, often including criticisms of Turkish customs and traditions not only in Turkey itself but also in Turkish communities already established in Europe (shocking "honor killings" also take place in these neighborhoods). And even if the EU eventually does extend an invitation to Turkey, it is by no means certain that a future Turkish government would accept it.

Iran

Long known as *Persia*, Iran, Turkey's neighbor to the east, also has a history of imperial conquest. In 1971, the then-reigning shah and his family celebrated the 2500th anniversary of Persia's first monarchy with unmatched royal splendor. But by 1979, revolution had engulfed Iran, and Shi'ite fundamentalists drove from power the shah who had ascended his throne through American intervention. The monarchy was replaced by an Islamic republic, and a frightful wave of retribution followed.

A Crucial Location and Dangerous Terrain

As Figure 7B-1 shows, Iran occupies a critical area in this turbulent realm. It controls the entire corridor between the Caspian Sea and the Persian Gulf. To the west it adjoins Turkey and Iraq, both historic enemies. To the north (west of the Caspian Sea) Iran borders both Azerbaijan and Armenia, where once again Islam confronts Christianity. To the east Iran meets Afghanistan and Pakistan, and to the northeast lies volatile Turkmenistan.

Iran, as Figure 7B-11 demonstrates, is a country of mountains and deserts. The heart of the country is an upland, the Iranian Plateau, that lies surrounded by even higher mountains, including the Zagros in the west, the Elburz in the north along the Caspian Sea coast, and the mountains of Khurasan to the northeast. This mountainous topography signals danger: here the Eurasian and Arabian tectonic plates converge, causing major and often devastating earthquakes. The Iranian Plateau therefore is actually a huge highland basin marked by salt flats and wide expanses of sand and rock. The highlands wrest some moisture from the air, but elsewhere only oases interrupt the arid monotony—oases that for countless centuries have been stops on the area's caravan routes.

City and Countryside

In ancient times, Persepolis in southern Iran (located near the modern city of Shiraz) was the focus of a powerful Persian kingdom, a city dependent on **5** **qanats**, underground tunnels carrying water from moist mountain slopes to dry flatland sites many miles away. Today, Iran's population of 74 million is 67 percent urban, and the capital, Tehran, lies far to the north, on the southern slopes of the Elburz Mountains. This mushrooming metropolis of more than 8 million, lying at the heart of modern Iran's core area, still depends in part on the same kinds of qanats that sustained Persepolis more than 2000 years ago. As such, Tehran symbolizes the internal contradictions of Iran: a country in which modernization has taken hold in the cities, but little has changed in the vast countryside, where the mullahs led their peasant followers in a revolution that overthrew a monarchy and installed a theocracy.

Energy and Conflict

As Figure 7A-9 shows, Iran possesses the realm's second-largest concentration of oil reserves, which amount to more than half of those in Saudi Arabia. Petroleum and natural gas production provide about 90 percent of the country's income. The reserves lie in a zone along the southwestern periphery of Iran's territory, and Abadan became its petroleum capital near the head of the Persian Gulf. But Iran is a large and populous country, and the wealth oil generated could not transform it in the way the last shah intended, a transformation that had it occurred might have staved off the revolution. Modernization remained but a veneer: in the villages away from Tehran's polluted air, the holy men continued to dominate the lives of ordinary Iranians. As elsewhere in the Islamic world, urbanites, villagers, and nomads remained enmeshed in a web of production and profiteering, serfdom, and indebtedness that always

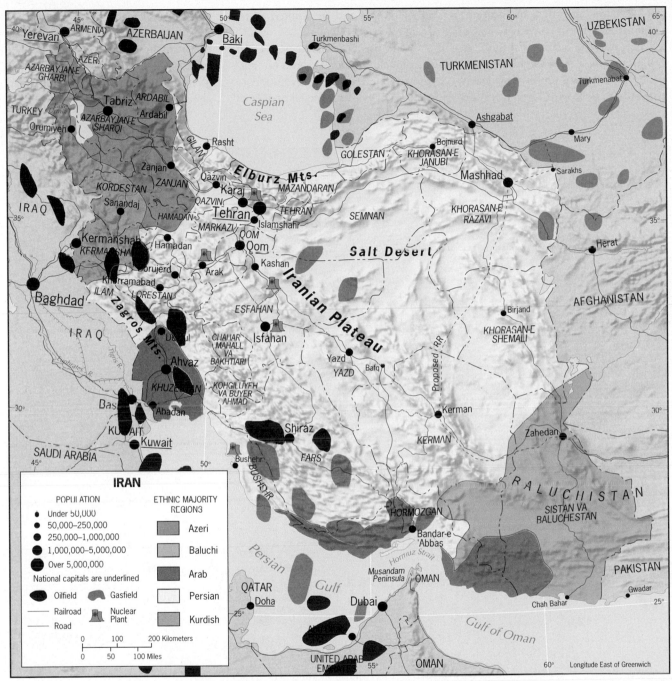

FIGURE 7B-11

© H. J. de Blij, P. O. Muller, and John Wiley & Sons, Inc.

characterized traditional society here. The revolution swept this system away, but it did not improve the lot of Iran's millions. A devastating war with Iraq (1980–1990), into which Iran ruthlessly poured hundreds of thousands of its young men, sapped both the coffers and energies of the state. When it was over, Iran was left poorer, weaker, and aimless, its revolution spent on unproductive pursuits.

In the early years of the twenty-first century, evidence abounded that the people of Iran remained divided between conservatives determined to protect the power of the mullahs and reformers seeking to modernize and liberalize Iranian society. In Iran's 2005 election, many reformist candidates were disqualified by religious conservatives who hold ultimate authority, and a president who had proved himself to be mildly reformist was succeeded by a far more extreme figure, Mahmoud Ahmadinejad, who, shortly after taking office, proclaimed that the state of Israel should be "wiped off the map." In 2009, Ahmadinejad's reelection elicited charges of fraud and produced widespread and deadly riots.

Meanwhile, Iran was pursuing a course toward nuclear-power-generating capability with implications that trouble many countries. In addition, the Iranians want to construct a natural gas pipeline across Pakistan to India, making India highly dependent on the Iranian supply. The deal is currently stalled, with Iran's price too high for India and the United States trying to cancel the plan.

Iran's rise as a nuclear power and its regional ambitions should not surprise anyone. Although revolutionary Iran disavowed the imperial designs of its Persian predecessors, this does not mean that its national interests now end at its borders. Tehran has a major stake in developments in Iraq and Afghanistan, both neighbors; it has an already-nuclear and unstable Pakistan

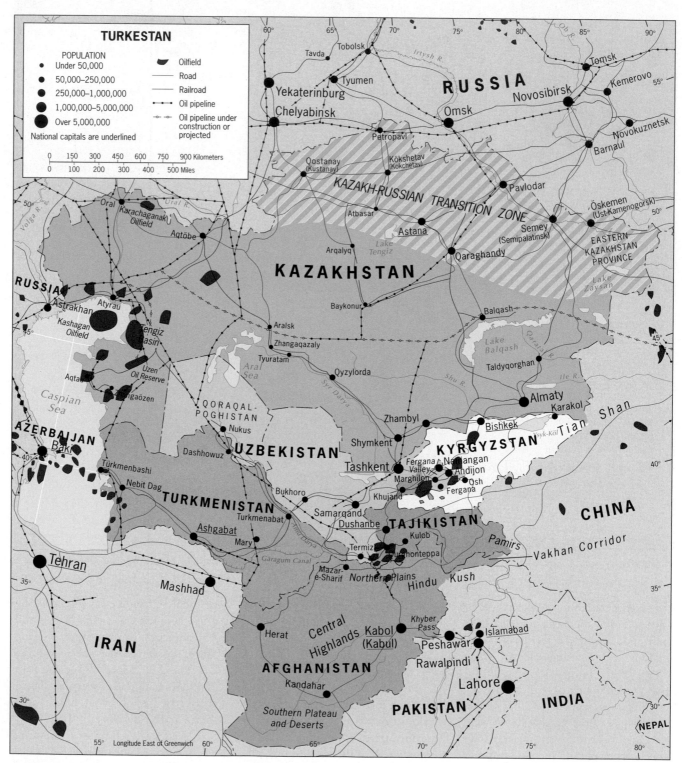

FIGURE 7B-12

© H. J. de Blij, P. O. Muller, and John Wiley & Sons, Inc.

on its eastern border; it has a strong interest in the fate of Shi'ite majorities and minorities elsewhere in the realm; it has long been an avowed enemy of Israel and a strong supporter of Palestinian causes; and it has funded organizations labeled terrorist whose actions have even reached across the Atlantic. Iran's revolution ousted the shah and terminated a monarchy, but it did not extinguish all ties to an imperial past.

TURKESTAN: THE SIX STATES OF CENTRAL ASIA

For centuries Turkish (Turkic) peoples spread outward across a vast Central Asian domain that extended from Mongolia and Siberia to the Black Sea. Propelled by population growth and energized by Islam, they penetrated Iran, defeated the Byzantine Empire, and colonized much of eastern Europe. Eventually, their power declined as Mongols, Chinese, and then Russians invaded their strongholds. But these conquerors could not expunge them, as the names on the modern map prove (Fig. 7B-12). The latest conquerors, the Chinese and particularly the Russian czars and their communist successors, created five Soviet Socialist Republics named after the majority peoples within their borders. Thus the Kazakhs, Turkmen, Kyrgyz, and other Turkic peoples retained some geographic identity in what was then Soviet Central Asia.

Central Asia—*Turkestan*—is a still-changing region. In some areas, the cultural landscapes of neighboring realms extend into it, for example, in northern (Russian) Kazakhstan. In other areas, Turkestan extends into adjoining realms, as in the Xinjiang Uyghur Autonomous Region of westernmost China. Some areas once penetrated by Turkic peoples are no longer dominated by them—for instance, Afghanistan. And certain peoples now living in Turkestan are not of Turkic ancestry, notably the Tajiks. This is a fractious region in sometimes turbulent transition.

As we define it, Turkestan includes six states: (1) *Kazakhstan*, territorially larger than the other five combined but situated astride a major ethnic transition zone; (2) *Turkmenistan*, relatively closed and dictatorial, with important frontage on the Caspian Sea and bordering both Iran and Afghanistan; (3) *Uzbekistan*, the most populous state and situated at the heart of the region; (4) *Kyrgyzstan*, wedged between powerful neighbors and chronically unstable; (5) *Tajikistan*, regionally and culturally divided as well as strife-torn; and (6) *Afghanistan*, engulfed in almost continuous war since it was invaded by Soviet forces in 1979.

The States of Former Soviet Central Asia

During their hegemony over Central Asia, the Soviets tried to suppress Islam and install secular regimes (this was their objective when they invaded Afghanistan as well), but today Islam's cultural revival is one of the defining qualities of this region. From Almaty to Samarqand, mosques are being repaired and revived, and Islamic dress again is part of the cultural landscape. National leaders have made high-profile visits to Mecca; most are sworn into office touching the Quran. All of Central Asia's countries now observe Islamic holidays. In other ways, too, Turkestan reflects the norms of the North Africa/Southwest Asia realm: in its dry-world environments and tight clustering of its population, its mountain-fed waterways irrigating farms and fields, its sectarian conflicts, its oil-based economies. It also is a region where democratic government remains an elusive goal.

Turkestan's intricate ethnolinguistic mosaic is vividly represented in Figure 7B-13. Although states in this region tend to be named after their largest ethnic population sectors, political boundaries do not match the distribution of cultures. Perhaps the most complex cultural fabric is that of Afghanistan, populated by Pushtuns (occasionally

spelled Pashtuns), Tajiks, Hazaras, Uzbeks, Turkmen, Baluchis, and other, smaller groups. Add to this the region's rugged relief, vast deserts, and limited surface communications, and you can see why national integration is a particularly difficult challenge here.

Kazakhstan is the region's giant and borders two even greater territorial titans: Russia and China (Fig. 7B-12). During the Soviet period, northernmost Kazakhstan was heavily Russified and become, in effect, part of Russia's Eastern Frontier (see Fig. 2B-1). Rail and road links crossed the area mapped as the Kazakh–Russian Transition Zone, connecting the north to Russia. The Soviets made Almaty, in the heart of the Kazakh domain, the territory's capital. Today the Kazakhs are in control, and they in turn have moved the capital to Astana, right in the heart of the Transition Zone, where almost 4.5 million Russians (who constitute 27 percent of the total national population of 16 million) still live. Clearly, Astana is another **6** forward capital.

Figure 7B-12 reveals Kazakhstan's situation as a corridor between the huge oil reserves of the Caspian Basin and China. Oil and gas pipelines nearing completion across Kazakhstan will eliminate, or greatly reduce, China's dependence on oil carried by tankers along distant international sea lanes.

Uzbekistan occupies the heart of Turkestan and borders every other state in it. Uzbeks not only make up 80 percent of the population of 28 million, but also form substantial minorities in several neighboring states. The capital, Tashkent, lies in the eastern core area of the country, where most of the people live in towns and farm villages, and the crowded Fergana Valley is the focus. In the far west lies the shrunken Aral Sea, whose feeder streams were diverted into cotton fields and croplands during the Soviet occupation; heavy use of pesticides contaminated the groundwater, and countless thousands of local people continue to suffer severe medical problems as a result.

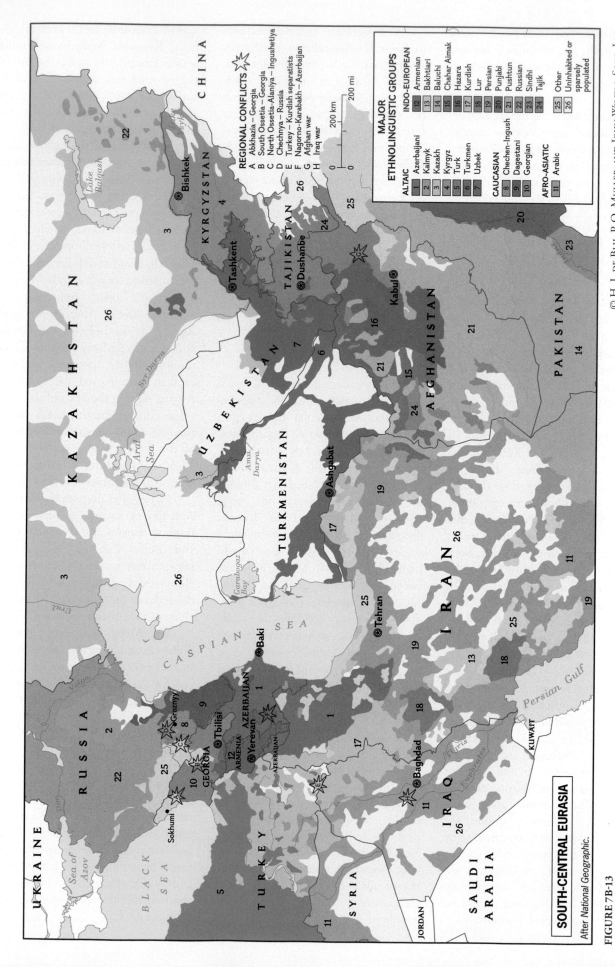

SOUTH-CENTRAL EURASIA

After National Geographic.

FIGURE 7B-13

MAJOR ETHNOLINGUISTIC GROUPS

ALTAIC
1 Azerbaijani
2 Kalmyk
3 Kazakh
4 Kyrgyz
5 Turk
6 Turkmen
7 Uzbek

CAUCASIAN
8 Chechen–Ingush
9 Dagestani
10 Georgian

AFRO-ASIATIC
11 Arabic

INDO-EUROPEAN
12 Armenian
13 Bakhtiari
14 Baluchi
15 Chahar Aimak
16 Hazara
17 Kurdish
18 Lur
19 Persian
20 Punjabi
21 Pushtun
22 Russian
23 Sindhi
24 Tajik

25 Other
26 Uninhabited or sparsely populated

REGIONAL CONFLICTS
A Abkhazia – Georgia
B South Ossetia – Georgia
C North Ossetia-Alaniya – Ingushetiya
D Chechnya – Russia
E Turkey – Kurdish separatists
F Nagorno–Karabakh – Azerbaijan
G Afghan war
H Iraq war

0   200 km
0   200 mi

Turkmenistan, the autocratic desert republic that extends all the way from the Caspian Sea to the borders of Afghanistan, has a population of little more than 5 million, of which more than three-quarters are Turkmen. During the Soviet era, the communist planners began work on a massive project: the Garagum (Kara Kum) Canal designed to bring water from Turkestan's eastern mountains into the heart of the desert. Today the canal is 1100 kilometers (700 mi) long, and it has enabled the cultivation of some 1.2 million hectares (3 million acres) of cotton, vegetables, and fruits. The plan is to extend the canal all the way to the Caspian Sea, but meanwhile Turkestan has hopes of greatly expanding its oil and gas output from Caspian coastal and off-shore reserves. But look again at Figure 7B-12: Turkmenistan's relative location is clearly not advantageous for export routes.

Kyrgyzstan's topography and political geography are reminiscent of the Caucasus. Mapped in yellow in Figure 7B-12, Kyrgyzstan lies intertwined with Uzbekistan and Tajikistan to the point of having exclaves and enclaves along its borders. The Kyrgyz, for whom the Soviets established this "republic," constitute less than two-thirds of the population of about 5.4 million. Uzbeks and other minorities form a complex cultural geography; mountains and valleys isolate communities and complicate nation-building. The agricultural economy is quite weak, consisting of pastoralism in the mountains and farming in the valleys. About 70 percent of the people profess allegiance to Islam, and Wahhabism has gained a strong foothold here. The town of Osh is often referred to as the headquarters of the movement in Turkestan.

Tajikistan's mountainous scenery is even more spectacular than Kyrgyzstan's, and here, too, topography is a barrier to the integration of a multicultural society. The Tajiks, who constitute about 65 percent of the population of

If you were asked where this photo might have been taken, Russia would seem to be a good guess, or perhaps Belarus (although the English word "bar" might not be expected there). Certainly the bleak apartment building in the background seems to be of Soviet socialist vintage. But this is the entrance to a casino on the main shopping street of Almaty, the largest city and former capital of Kazakhstan. The Soviets and their Kazakh allies administered Kazakhstan from Almaty (then known as Alma-Ata) for some 70 years, and the Russian imprint remains strong in this city. But, as Figure 7B-12 shows, the Russian imprint is even stronger in the country's north, and in 1998 the president and his party, over the objections of the mainly Kazakh opposition, moved the national capital from scenic and warm Almaty to drab and cold Astana. Yet Almaty retains its attractions: more than 40 casinos as well as countless bars and clubs coexist with Sunni Islam, Kazakhstan's leading religion. © Abbie Trayler-Smith/Panos Pictures

7.6 million, are ethnically Persian (Iranian), not Turkic, and speak a language related to Persian (an Indo-European tongue). Most Tajiks, despite their Persian affinities, are Sunni Muslims, not Shi'ites. Small though Tajikistan is, regionalism plagues the state: the government in Dushanbe is often at odds with the barely connected northern area (see Fig. 7B-12), a hotbed not only of Islamic revivalism but also of anti-Tajik, Uzbek activism.

Fractious Afghanistan

Afghanistan, the southernmost country in this region, exists because the British and Russians, competing for hegemony in this area during the nineteenth century, agreed to tolerate it as a cushion, or **7** **buffer state**, between them. This is how Afghanistan acquired the narrow extension

leading from the main territory eastward to the Chinese border—the Vakhan Corridor (Fig. 7B-14). As the colonialists delimited it, Afghanistan adjoined the domains of the Turkmen, Uzbeks, and Tajiks to the north, Persia (now Iran) to the west, and the western flank of British India (now Pakistan) to the east.

Landlocked and Fractured

Geography and history seem to have conspired to divide Afghanistan. As Figure 7B-14 shows, the towering Hindu Kush range dominates the center of the country, creating three broad environmental zones: the relatively well-watered and fertile northern plains and basins; the rugged, earthquake-prone central highlands; and the desert-dominated southern plateaus. Kabol (Kabul), the capital,

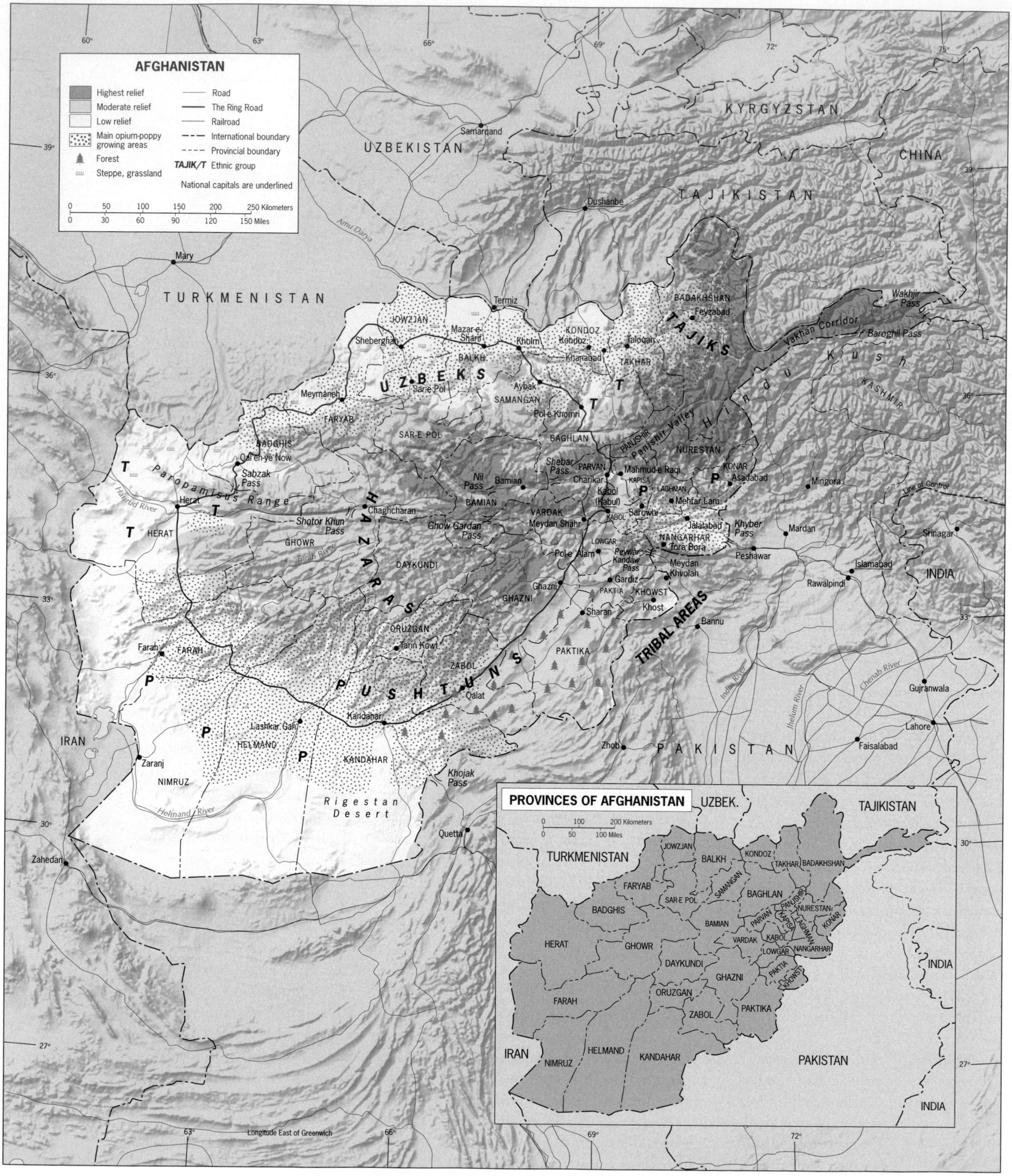

AFGHANISTAN

Highest relief	Road
Moderate relief	The Ring Road
Low relief	Railroad
Main opium-poppy growing areas	International boundary
Forest	Provincial boundary
Steppe, grassland	*TAJIK/T* Ethnic group

National capitals are underlined

| 0 | 50 | 100 | 150 | 200 | 250 Kilometers |
| 0 | 30 | 60 | 90 | 120 | 150 Miles |

PROVINCES OF AFGHANISTAN

| 0 | 100 | 200 Kilometers |
| 0 | 50 | 100 Miles |

FIGURE 7B-14

© H. J. de Blij, P. O. Muller, and John Wiley & Sons, Inc.

lies on the southeastern slope of the Hindu Kush, linked by narrow passes to the northern plains and by the famous Khyber Pass to Pakistan.

Across this variegated landscape moved countless peoples: Greeks, Turks, Arabs, Mongols, and others. Some settled here, their descendants today speaking Persian, Turkic, and other languages. Others left archeological remains or no trace at all. The present population of Afghanistan (just under 35 million) has no ethnic majority. This is a country of minorities in which the Pushtuns of the east are the most numerous but make up barely 40 percent of the total. The second-largest minority are the Tajiks (ca. 25 percent), a world away across the Hindu Kush, concentrated in the zone near Afghanistan's border with Tajikistan. The Hazaras of the central highlands and the south, the Uzbeks and Turkmen in the northern border areas, the Baluchis of the southern deserts, and other, smaller groups scattered across this country create one of the world's most complex cultural mosaics. Two major languages, Pushtun and Dari (the local variant of Persian), plus several others create a veritable Tower of Babel here.

Costs of Conflict

Episodes of conflict have marked the history of Afghanistan, but none was as costly as its involvement in the Cold War (1945–1990). Following the Soviet intervention of 1979, the United States supported the Muslim opposition, the *Mujahideen* (meaning strugglers), with modern weapons and money, and the Soviets were forced to withdraw. Soon the factions that had been united during the anti-Soviet campaign were in conflict, delaying the return of some 4 million refugees who had fled to Pakistan and Iran. The situation resembled the pre-Soviet past: a feudal country with a weak and ineffectual government in Kabol.

In 1994, what at first seemed to be just another warring faction appeared on the scene: the so-called **8** **Taliban** ("seekers" or students of religion) from religious schools in Pakistan. Their avowed aim was to end Afghanistan's chronic factionalism and endemic corruption by instituting strict Islamic law. Popular support in the war-weary country, especially among the Pushtuns, led to a series of successes, and by 1996 the Taliban had taken Kabol.

The Taliban's imposition of Islamic law was so strict and severe that Islamic as well as non-Islamic countries objected. Restrictions on the activities of women ended their professional education, employment, and freedom of movement, and had a devastating impact on children as well. Public amputations and stonings enforced the Taliban's code. In the process, Afghanistan became a haven for groups of revolutionaries whose goals went far beyond those of the Taliban: they plotted attacks on Western interests throughout the realm and threatened Arab regimes they deemed compliant with Western priorities. Taliban-ruled, cave-riddled, remote and isolated Afghanistan was an ideal locale for these outlaws. Already in possession of arms and ammunition (Soviet as well as American) left over from the Cold War, they also benefited from Afghanistan's huge, illicit opium trade. It is estimated that in a typical year Afghanistan produces over 90 percent of the world's opium. Much of the revenue found its way into the coffers of the conspirators.

Failed State, Terrorist Base

With such resources, the militants were able to organize and launch attacks against several Western targets in the realm and elsewhere, but events took a fateful turn in 1996 that would empower them as never before. During the conflict with the Soviets, the Mujahideen cause had been supported not only by the United States but also by a devout and revivalist Saudi Muslim named Usama bin (son of) Laden. The child of a construction billionaire who had more than 50 children including 22 sons, Usama graduated from a university in Jiddah in 1979 and headed for Afghanistan with an inherited fortune estimated at about U.S. $300 million to help the anti-Soviet campaign. Well connected in Saudi Arabia and now a pivotal figure in Afghanistan, bin Laden saw his fame as well as his war chest grow as the Soviet intervention collapsed at the end of the 1980s. Following the withdrawal of the communist forces, he returned to Saudi Arabia, where he denounced his government for allowing U.S. troops on Saudi soil during the 1990–1991 Gulf War. The Saudi regime responded by stripping him of his citizenship and expelling him, and bin Laden fled to Sudan. There he set up several legitimate businesses to facilitate his now-global financial transactions, but he also established terrorist training camps. Under much international pressure, the Khartoum regime ousted him in 1996, and bin Laden returned to a country he knew would welcome him again: Afghanistan.

Bin Laden's fateful return to Afghanistan coincided with the Taliban's conquest of Kabol, and now he helped its forces push northward into the fertile and productive northern plains. Meanwhile, a terrorist organization named **9** **al-Qaeda** took root in the country, a global network that would further the aims of the revolutionaries once loosely allied. Afghanistan became al-Qaeda's headquarters, and bin Laden its director; its exploits were funded by numerous Islamic sources and ranged from terrorist training in local camps to lethal attacks on American targets, including a warship in Yemen's port of Adan and two U.S. embassies in East Africa.

On February 26, 1993, terrorists exploded a massive car bomb in the basement garage of the World Trade Center in New York, but their objective—to topple the 110-story tower—failed. Even as those responsible went on trial, al-Qaeda's leaders were planning the suicide attacks of September 11, 2001 that destroyed the buildings, killed thousands, caused billions of dollars in physical damage, and inflicted an untold amount of psychological damage. Several weeks later, United States and British forces, with the acquiescence of Pakistan, attacked both the Taliban regime and the al-Qaeda infrastructure in Afghanistan. Proof of bin Laden's and al-Qaeda's complicity in the September 11 assault was found in videotape and documentary form.

The Taliban and al-Qaeda leaderships may not have counted on Pakistan's compliance with Western demands, but they surely knew what the consequences of 9/11 would be for Afghanistan and its people. Once again a foreign power would invade the country and set its political course, the associated upheaval enabling rapacious warlords to exploit the peoples of border provinces far from the capital.

Clouded Future

In the immediate aftermath of al-Qaeda's September 11, 2001 attacks, the American response engendered reasons for optimism: the Taliban were quickly defeated, a more representative government was forming, warlords were co-opted or sidelined, Pushtun and other refugees were streaming back into Afghanistan, and life in Kabol and other cities and towns returned to some semblance of normal. A brilliant U.S. ambassador (Zalmay Khalilzad), born in Afghanistan and speaker of the key local languages, and the skillful military commander of American forces in the country (Gen. David Barno), combined their talents to achieve significant progress against daunting odds. And in 2004, Afghanistan held elections that produced a representative government headed by President Hamid Karzai.

But already the seeds of disaster had been sown: the invasion of Iraq in 2003 had begun to divert attention and resources from Afghanistan to that ill-fated campaign. Soon, both Khalilzad and Barno were gone. In a pattern recurring throughout Afghanistan's history, opponents of modernization in any form resumed their attacks on foreign forces and local facilities (such as reopened girls' schools). From their mountain hideaways in the cave-riddled Afghanistan-Pakistan border zone that had protected Usama bin Laden from capture, Taliban warriors not only challenged Western armies in Afghanistan but also staged a massive attack in northern Pakistan, where they briefly held a large area and conducted their hallmark public executions, destroyed schools, kidnapped prominent citizens, and terrorized shop-

Fearsome-looking Taliban fighters in an "undisclosed location" in Afghanistan brandish their weapons and practice their threatening glares. In mid-2010 the international effort to rescue Afghanistan from the Taliban scourge was in trouble as public support for the campaign dwindled in Afghanistan itself as well as in the United States and other countries contributing to it.
© Reuters/© Corbis

keepers before Pakistani forces beat them back.

In 2010, Afghanistan seemed to have regained its prominence among American priorities: even as the war in Iraq was winding down, the U.S. government approved a significant increase in combat troops and support personnel to augment its inadequate forces. But the moment of opportunity may have come and gone. Afghanistan's physical geography, social complexity, experience with foreigners and resultant warrior culture, Islamic intensity and traditional male dominance, endemic corruption, and highly limited subsistence opportunities combine to create an arena in which no military power has ever prevailed over the long term. Even as Usama bin Laden taunts the targets of his terrorism from al-Qaeda's highland sanctuary, Afghanistan has lost its primacy as a terrorist haven: al-Qaeda had gone global—from Pakistan's Waziristan to Yemen to North Africa and beyond. The question in post-9/11 America is how long the U.S. public will support an increasingly costly campaign whose prospect of success is negligible.

Even as United States armed forces withdraw from Iraq, an even greater challenge looms: the pacification and stabilization of Afghanistan and the defeat, in neighboring Pakistan as well as in Afghanistan, of the Taliban movement. This challenge is most urgent (and difficult) in Afghanistan's eastern and southern provinces, where high relief, distance, and poor surface communications favor the Taliban. This image shows U.S. soldiers on patrol near a bridge at Nishagam in Konar province (see Fig. 7B-14), not far from the border with the Federally Administered Tribal Areas of Pakistan, a Taliban stronghold. © Liu Jin/AFP/Getty Images, Inc.

This chapter opened by emphasizing the diversity and variety among cultures, countries, and regions in the North Africa/Southwest Asia realm. Our regional survey then began with a discussion of Egypt and its ancient heritage of life concentrated in one river valley and delta, and has just ended with an account of an Afghanistan fractured by history and geography as few countries are. It has chronicled geographies of isolation and intervention, of resistance to change and the welcoming of modernization. Here lie the sources of much of what is called "Western" civilization. Here were born the two largest belief systems (in terms of adherents) on the planet. Beneath these lands lie the energy resources that propelled the world's modernization. And today this realm, in its regions and on its margins, is roiled by dangerous conflicts on whose resolution the future of the world depends.

POINTS TO PONDER

- As the second decade of the twenty-first century opened, Saudi Arabia was exporting more oil to China than the United States.

- The tallest building in the world, the half-mile-high Burj Dubai (Arabic for Dubai Tower) opened in January 2010. You can see 95 kilometers (60 mi) from its observation deck on the 124th floor.

- Half of all murders committed annually in Turkey are "honor killings" by family members, mainly of girls and women.

- "Waging war against God" is a capital offense in Iran.

Taj Mahal surrounded by the cityscape of Agra, India. © Jan Nijman

IN THIS CHAPTER

The monsoon is still the key to India's fortunes
South Asia as a birthplace of religions
Two nuclear powers quarrel over Kashmir
The ever-present threat of terrorism
Backward agriculture and cutting-edge IT
South Asia's missing girls

CONCEPTS, IDEAS, AND TERMS

8A

SOUTH ASIA: DEFINING THE REALM

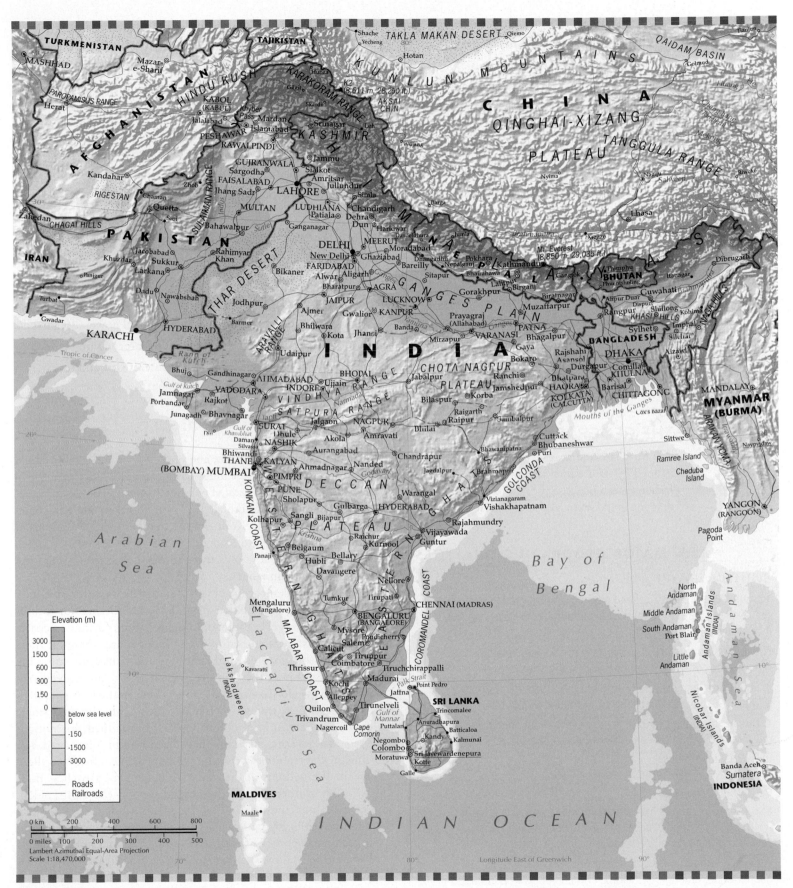

FIGURE 8A-1

© H. J. de Blij, P. O. Muller, and John Wiley & Sons, Inc.

outh Asia is a realm of almost magical geographic names: Mount Everest, Kashmir, the Khyber Pass, the Ganges River. There was a time when this realm was legendary and prized. Remember that it was "India" and its fabled wealth that the European explorers were after, from Vasco da Gama to Columbus to Magellan. Before them, the fourteenth-century North African geographer Ibn Batutta had traveled to South Asia overland, and his writings about its riches were met with astonishment and even disbelief. From the sixteenth century onward, European trading companies derived enormous profits from commerce in the realm.

However, by the late nineteenth century South Asia seemed to have become remote from the affairs of the world—hungry, weak, exploited, an epitome of the global periphery. Even after independence in 1947, India and the other countries of the realm long remained among the world's poorest. For decades, population growth outstripped economic expansion.

Today, for a number of reasons, South Asia commands the world's attention once again. It is poised to become the most populous realm on Earth in 2011 (see the Data Table in Appendix B). Two of its states, India and Pakistan, often find themselves in conflict and both are nuclear powers. In the remote mountain hideaways of Pakistan, a terrorist organization's leaders planned attacks that changed the skyline of New York and altered the battleground of Iraq. In the ports of India, a growing navy reflects that country's determination to become more than a regional power. Meanwhile, outsourcing by U.S. companies to India has become a hot topic and India's spectacular rise in information technology has changed the industry. Our daily lives will increasingly be affected by what happens in this crowded and restive part of the world.

THE GEOGRAPHIC PANORAMA

The Eurasian landmass incorporates all or part of six of the world's geographic realms, and of these six none is more clearly defined by nature than the one we call South Asia. Figure 8A-1 shows us why: the huge triangle that divides the northern Indian Ocean between the Arabian Sea and the Bay of Bengal is so sharply demarcated by mountain walls and desert wastes that you could take a pen and mark its boundary, from the Naga Hills in the east through the Great Himalaya and Karakoram in the north to the Hindu Kush and the Iran-bordering wastelands of Baluchistan in the west. Note how short the distances are over which the green of habitable lowlands turns to the dark brown of massive, snowcapped mountain ranges.

South Asia's kaleidoscope of cultures may be the most diverse in the world, proving that neither formidable mountains nor forbidding deserts could prevent foreign influences from further diversifying an already variegated realm. We will encounter many of these influences in this chapter, but South Asia also had one unifying force of sorts: the British Empire, which in its late-nineteenth-century heyday held sway over all of it. When, in the mid-twentieth century, the British wanted to transfer their authority to a single regional government, local objections swiftly nullified this notion. That regional government would have been Hindu-dominated, but Muslims in the realm's eastern and western flanks refused, as did a pair of small kingdoms in the mountainous north, as well as Buddhists in the southern island then called Ceylon (now Sri Lanka). Negotiations and compromises produced partition and the political boundaries seen in Figure 8A-1. As a result, India, the realm's giant, is flanked by six countries (Pakistan, Nepal, Bhutan, Bangladesh, Sri Lanka, and the Maldives) as well as a remaining disputed territory, Kashmir.

Since Islam is Pakistan's official religion (India has none) and that faith is a key criterion in defining the realm we designated as North

MAJOR GEOGRAPHIC QUALITIES

SOUTH ASIA

1. South Asia is clearly defined physiographically, and much of the realm's boundary is marked by mountains, deserts, and the Indian Ocean.

2. South Asia's great rivers, especially the Ganges, have for tens of thousands of years supported huge population clusters.

3. South Asia, and especially northern India, was the birthplace of major religions that include Hinduism and Buddhism.

4. Due to the realm's natural boundaries, foreign influences in pre-modern South Asia came mainly via a narrow passage in the northwest (Khyber Pass).

5. South Asia covers just over 3 percent of the Earth's land area but contains nearly 23 percent of the world's human population.

6. South Asia's annual monsoon continues to dominate life for hundreds of millions of subsistence and commercial farmers. Failure of the monsoon cycle spells economic crisis.

7. Certain remote areas in the realm's northern mountain perimeter are a dangerous source of friction between India, Pakistan, and China.

Africa/Southwest Asia, should Pakistan be included in the latter? The answer lies in several aspects of Pakistan's historical geography. One criterion is ethnic continuity, which links Pakistan to India rather than to Afghanistan or Iran. Another factor involves language: although Urdu is Pakistan's official language, English is the *lingua franca*, as it is in India. Still another factor, of course, is Pakistan's evolution as part of the British South Asian Empire. Furthermore, the boundary between Pakistan and India does not signify the eastern frontier of Islam in Asia. As we shall see, more than 170 million of India's nearly 1.2 billion citizens are Muslims (which is just about as many as there are in all of Pakistan), and millions live very close to the Indian side of the border whose creation cost so many lives in 1947. And not only are Pakistan and India linked in the cultural-historical arena: they are locked in a deadly and dangerous embrace in embattled Kashmir.

The close integration of Pakistan with South Asia will not surprise you after you study the realm's physiography in Figure 8A-1: the natural boundary in this part of the realm lies west of the Indus River, not east. Pakistan today remains part of a realm that changes not in the Punjab, but at the Khyber Pass, the highland gateway to Afghanistan.

SOUTH ASIA'S PHYSIOGRAPHY

From snowcapped peaks to tropical forests and from bone-dry deserts to lush farmlands, this part of the world presents a virtually endless array of ecologies and environments, a diversity that is matched by its cultural mosaic. The broad outlines of this realm's physiography are best understood against the backdrop of its fascinating geologic past.

A Tectonic Encounter

As Figure 8A-2 shows, the spectacular relief in the north of this realm resulted initially from the collision of two of the Earth's great tectonic plates (see Fig. G-4). About 10 million years ago, after a long geologic journey following the breakup of the supercontinent Pangaea (Fig. 6A-3), the Indian plate encountered Eurasia. In this slow-motion, accordion-like collision, parts of the crust were pushed upward, thereby creating the mighty Himalaya Mountains. This process is still going on—at the rate of 5 millimeters (0.2 in) a year. One major outcome was that the northern margins of the South Asian realm were thrust upward to elevations where permanent snow and ice make the landscape look polar. The swing of the seasons melts enough of this snow in spring and summer to sustain the great rivers below, providing water for farmlands that support hundreds of millions of people. The Ganges, Indus, and Brahmaputra all have their origins in the Himalayas. Only south of the Ganges Basin does the massive plateau begin that

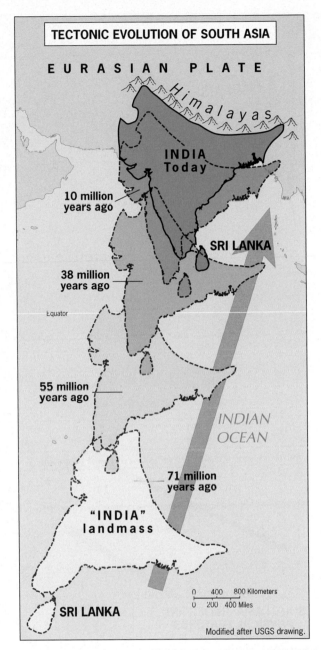

FIGURE 8A-2   © H. J. de Blij, P. O. Muller, and John Wiley & Sons, Inc.

marks the much older geologic centerpiece of the Indian Plate as it drifted northeast toward Eurasia.

The Monsoon

Physical geography, therefore, is crucial here in South Asia—but not just on and below the ground. What happens in the atmosphere is critical as well. The name "South Asia" is almost synonymous with the term **1** **monsoon** because the annual rains that come with its onset, usually in June, are indispensable to subsistence as well as commercial agriculture in the realm's key country, India.

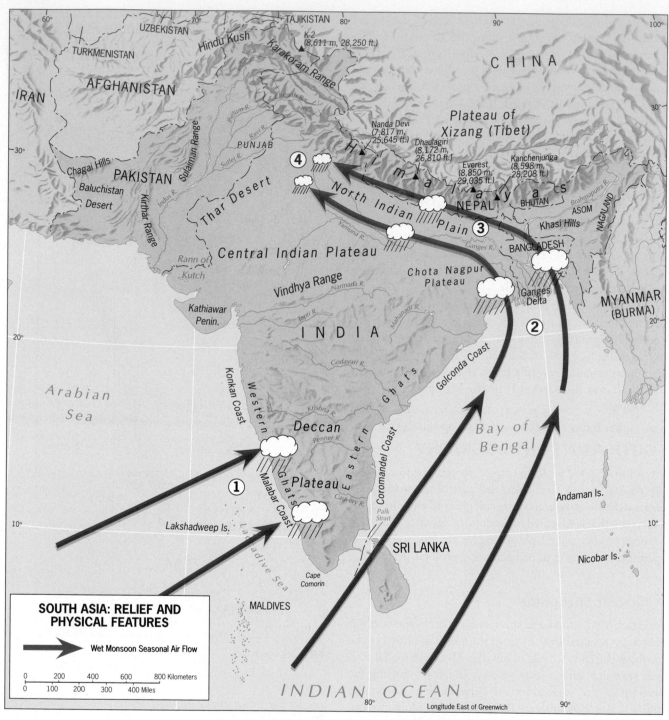

FIGURE 8A-3

© H. J. DE BLIJ, P. O. MULLER, AND JOHN WILEY & SONS, INC.

Figure 8A-3 shows how the monsoon works. As the South Asian landmass heats up during the spring, a huge low-pressure system forms above it. This low-pressure system begins to draw in vast volumes of air from over the ocean onto the land. When the inflow of moist oceanic air reaches critical mass in early June, the ***wet monsoon*** has arrived. It may rain for 60 days or more. The countryside turns green, the paddies fill, and another dry season's dust and dirt are washed away. The region is reborn (see photo pair next page). Some of the moisture-laden air is forced upward against the Western Ghats ① , cooling as it rises and condensing large amounts of rainfall. Other streams of air flow across the Bay of Bengal and get caught up in the convection over northeastern India and Bangladesh ② . Seemingly endless rain now inundates a much larger area, including the entire North Indian Plain. The mountain wall of the Himalayas stops the air from spreading and the rain from dissipating ③ . Thus the airflow is steered westward, drying out as it advances toward Pakistan ④ .

After persisting for weeks, the system finally breaks down and the wet monsoon gives way to periodic rains and, eventually, another dry season. Then the anxious wait begins for the next year's monsoon, for without it India would face disaster. In much of rural India, life can hang by a meteorological thread.

Physiographic Regions

Figure 8A-3 underscores South Asia's overall division into three zones: northern mountains, southern plateaus, and, in between, a wide crescent of river lowlands.

The **northern mountains** extend from the Hindu Kush and Karakoram ranges in the northwest through the Himalayas in the center (Everest, the world's tallest peak, lies on the Nepal-China border) to the ranges of Bhutan and the Indian State of Arunachal Pradesh in the east. Dry and barren in the west on the Afghanistan border, the ranges become green and tree-studded in Kashmir, forested in the lower-lying sections of Nepal, and even more densely vegetated in Arunachal Pradesh. Transitional foothills, with many deeply eroded valleys cut by rushing meltwater, lead to the river basins below.

The belt of **river lowlands** extends eastward from Pakistan's lower Indus Valley (the area known as Sindh) through India's wide Gangetic Plain and then on across the great double delta of the Ganges and Brahmaputra in Bangladesh (Fig. 8A-1). In the east, this physiographic region is often called the North Indian Plain. To the west lies the lowland of the Indus River, which rises in Tibet, crosses Kashmir, and then bends southward to receive its major tributaries from the Punjab ("Land of Five Rivers").

Peninsular India is mainly plateau country, dominated by the massive **Deccan**, a tableland built of basalt that poured out when India separated from Africa during the breakup of Pangaea. The Deccan (meaning "South") tilts to the east, so that its highest areas are in the west and the major rivers flow into the Bay of Bengal. North of the Deccan lie two other plateaus, the Central Indian Plateau to the west and the Chota Nagpur Plateau to the east (Fig. 8A-3). On the map, also note the Eastern and Western Ghats: "ghat" means step, and it connotes the descent from Deccan plateau elevations to the narrow coastal plains below. Onshore winds of the annual wet monsoon bring ample rain to the Western Ghats. As a result, here lies one of India's most productive farming areas and one of southern India's largest population concentrations.

BIRTHPLACE OF CIVILIZATIONS

South Asia has a special place in the historical geography of humankind. First, recent research indicates that the Ganges Basin was a crucial stop on the great out-of-Africa migration that carried our species from the shores of the Red Sea

The arrival of the annual rains of the wet monsoon transforms the Indian countryside. By the end of May, the paddies lie parched and brown, dust chokes the air, and it seems that nothing will revive the land. Then the rains begin, and blankets of dust turn into layers of mud. Soon the first patches of green appear on the soil, and by the time the monsoon ends all is green. The photograph on the left, taken just before the onset of the wet monsoon in the State of Goa, shows the paddies before the rains begin; three months later the countryside looks as on the right. © Steve McCurry/Magnum Photos, Inc.

to the coast of Australia. Here, perhaps as far back as 70,000 years ago, environmental conditions enabled groups of hunter-gatherers to stabilize and expand. From the Indian subcontinent, they went on to Southeast Asia and Australia, and, scholars conclude, emigrated westward to the Middle East and Europe as well. The Ganges Basin may have been a crucible of humanity, but that was only the beginning. Subsequently, the great river basins of South Asia gave rise to two separate world-class culture hearths (see Fig. 7A-4), and continue to nurture the second-largest population concentration on Earth (Fig. G-8).

Indus Valley Civilization

Whatever the distant past, we know that a complex and technologically advanced civilization had emerged in the Indus Valley by about 2500 BC, simultaneous with other Bronze Age "urban revolutions" in Egypt and Mesopotamia. The Indus Valley civilization was centered on two major cities, Harappa and Mohenjo-Daro, which may have been capitals during different periods of its history (Fig. 8A-4); in addition, there were more than 100 smaller urban settlements. The locals apparently called their state **Sindhu**, and both **Indus** (for the river) and **India** (for the later state) may derive from this name. Although the influence of this civilization extended as far east as present-day Delhi, it did not last because of environmental change and, perhaps,

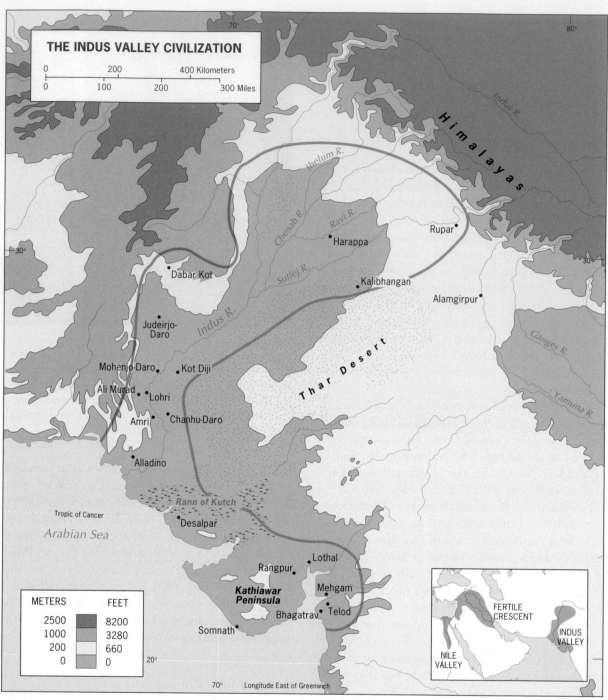

FIGURE 8A-4

© H. J. de Blij, P. O. Muller, and John Wiley & Sons, Inc.

because the political center of gravity shifted southeastward into the Ganges Basin.

Aryans and the Origins of Hinduism

Around 1500 BC, northern India was invaded by the ***Aryans*** (peoples speaking Indo-European languages based in what is today Iran). As the Iron Age dawned in India, acculturated Aryans began the process of welding the Ganges Basin's isolated tribes and villages into a new organized system, and urbanization made a comeback. The Aryans brought their language (Sanskrit, related to Old Persian) and a new social order to the vast riverine flatlands of northern India. Their settlement here was also accompanied by the emergence of a religious belief system, ***Vedism***. Out of the texts of Vedism and local creeds there arose a new religion—***Hinduism***—and with it a new way of life.

It is thought that the arrival and accommodation of Aryans in this new society demanded a system of **2 social stratification** that would solidify the powerful position of

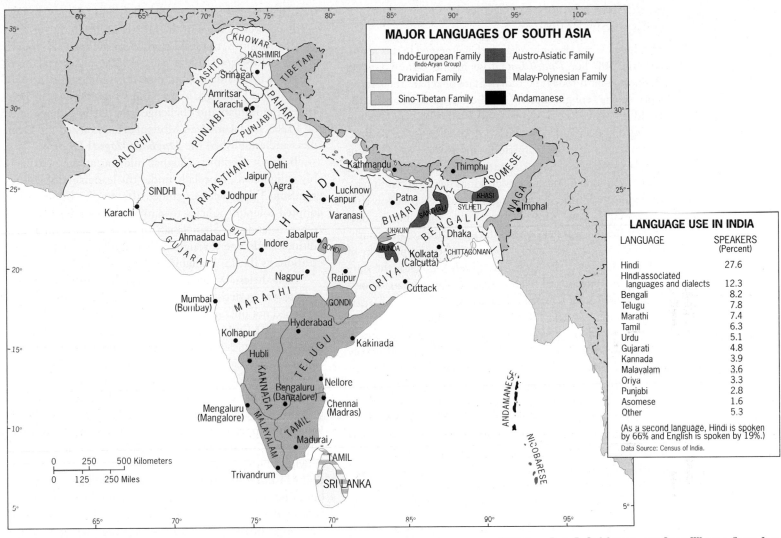

MAJOR LANGUAGES OF SOUTH ASIA

☐ Indo-European Family (Indo-Aryan Group)	■ Austro-Asiatic Family	
▨ Dravidian Family	▨ Malay-Polynesian Family	
☐ Sino-Tibetan Family	■ Andamanese	

LANGUAGE USE IN INDIA

LANGUAGE	SPEAKERS (Percent)
Hindi	27.6
Hindi-associated languages and dialects	12.3
Bengali	8.2
Telugu	7.8
Marathi	7.4
Tamil	6.3
Urdu	5.1
Gujarati	4.8
Kannada	3.9
Malayalam	3.6
Oriya	3.3
Punjabi	2.8
Asomese	1.6
Other	5.3

(As a second language, Hindi is spoken by 66% and English is spoken by 19%.)
Data Source: Census of India.

FIGURE 8A-5

© H. J. DE BLIJ, P. O. MULLER, AND JOHN WILEY & SONS, INC.

the Aryans and be legitimized through religion. Starting about 3500 years ago, a combination of regional integration, the organization of villages into controlled networks, and the emergence of numerous small city-states produced a hierarchy of power among the people, a ranking from the very powerful (Brahmins—highest-order priests) to the weakest. Hinduism's class-based ***caste system*** is highly controversial in the West (and among groups of Indians too) because of its rigidity and the ways in which it justifies structural inequality. Those in the lowest castes, deemed to be there because they deserved it given their past lives, are worst off, without hope of advancement and at the mercy of those higher up on the social ladder.

What is really most striking about Hindu civilization is that, after more than three millennia, it is still around and functioning with more or less the same beliefs, the same pantheon of gods, the same caste system, the same temples, and the same religious practices. You can visit Varanasi or any other holy place on the Ganges River today and witness *poojas* (prayers, offerings) that were performed in similar

fashion more than 3000 years ago. Nowhere in the world has a major culture been sustained this long. It helps explain why the South Asian realm has such a powerful spiritual appeal to many Westerners, even if they are dismissive of the caste system.

Although Hinduism spread across South Asia and even reached the Southeast Asian realm (especially Cambodia and Indonesia), Indo-European languages never took hold in the southern part of the subcontinent. As Figure 8A-5 shows, **3 Indo-European languages** (several of which are rooted in Sanskrit) dominate the western and northern parts of the realm, whereas the southern languages belong to the **4 Dravidian** family—languages that were indigenous to the realm even before the arrival of the Aryans. But these are not fossil languages: they remain vibrant today and have long literary histories. Telugu, Tamil, Kanarese (Kannada), and Malayalam are spoken by some 250 million people. In India's northern and northeastern fringes, Sino-Tibetan languages dominate, and smaller pockets of Austro-Asiatic speakers can be found in eastern India and neighboring Bangladesh.

Buddhism and Other Indigenous Religions

Hinduism is not the only religion that emerged in this realm. Around 500 BC, **Buddhism** arose in the eastern Ganges Basin in what is today the Indian State of Bihar. The famous story of the "enlightenment" of the Prince Siddhartha (the Buddha) took place in the town of Bodh Gaya, and his following soon expanded in all directions. The appeal of Buddhism was (and is) especially strong among lower-caste Hindus and large numbers have converted through the ages. Even some well-known ancient Hindu kings were known to have turned to Buddhism, prompting their subjects to follow suit. Interestingly, Buddhism emerged inside India but its ultimate influence was felt beyond the realm in East and Southeast Asia. Today, less than 1 percent of the population of India is Buddhist (81 percent are Hindu), but it is the state religion in Bhutan and a large majority (more than 70 percent) of Sri Lankans are Buddhist as well.

Another (smaller) indigenous religion that evolved alongside Hinduism since ancient times is **Jainism**, which is often described as a more purist, principled, and deeply spiritual form of Hinduism. It is especially well known for its uncompromising stand on nonviolence and vegetarianism. Jains account for under 1 percent of the population in India. Finally, we should take note of **Sikhism** as one of the realm's indigenous religions, a blend of sorts of Islamic and Hindu beliefs. This religion, practiced by about 2 percent of the population, is much younger, of course. It emerged around AD 1500, a few centuries after Islam became a dominating force in much of the South Asian realm.

FOREIGN INVADERS

The Reach of Islam

In the late tenth century, Islam came rolling like a giant tide across South Asia, spreading across Persia and Afghanistan, through the high mountain passes, into the Indus Valley, across the Punjab, and into the Ganges Basin, converting virtually everybody in the Indus Valley and foreshadowing the emergence, many centuries later, of the Islamic Republic of Pakistan. By the early thirteenth century, the Muslims had established the long-surviving and powerful **Delhi Sultanate**, which expanded across much of the northern tier of the peninsula. The Muslims also came by sea, arriving at the Ganges-Brahmaputra Delta and spreading their faith from the east as well as the west, in the process laying the foundation of today's predominantly Islamic state of Bangladesh.

Why did millions of Hindus convert to Islam? It was a combination of compulsion and attraction. The princes of India's local states faced the choice of cooperation or annihilation, and thus the elites of the native states found it prudent to convert. And, as was the case with Buddhism, Islam was a welcome alternative for Hindus of lower caste.

Thus Islam was the faith of the rulers and of the disadvantaged, a powerful force in the heartland of Hinduism.

The Muslims, not unlike other conquerors, also fought among themselves. In the early 1500s, a descendant of Genghis Khan named Babur placed his forces in control of Kabol (Kabul) in Afghanistan, and from that base he penetrated the Punjab and challenged the Delhi Sultanate. In the 1520s, his Islamicized Mongol armies ousted the Delhi rulers and established the **Mughal (Mogul) Empire**.

By most accounts, Mughal rule was at times remarkably enlightened, especially under the leadership of Babur's grandson Akbar, who expanded the empire by force but adopted tolerant policies toward Hindus under his sway; Akbar's grandson, Shah Jahan, made his mark on India's cultural landscape through such magnificent architectural creations as the Taj Mahal in the city of Agra.

Nonetheless, by the early eighteenth century the Mughal Empire was in decline. Maratha, a Hindu state in the west, expanded not only into the peninsular south but also northward toward Delhi, capturing the allegiance of local rulers and weakening Islam's hold. Fractured India now lay open to still another foreign intrusion, this time from Europe.

Reflecting on more than seven centuries of Islamic rule in South Asia, it is remarkable that Islam never achieved proportional dominance over the realm as a whole. While Pakistan is more than 96 percent Muslim and Bangladesh 83 percent, India—where the Delhi Sultanate and the Mughal Empire were centered—remains less than 15 percent Muslim today. Islam may have arrived like a giant tide, but Hinduism stayed afloat and outlasted the invasion. The accompanying caste system has many critics and understandably so, but it is hard to deny the ingenuity of the Hindu religion as it withstood first the Islamic invasions and then the European onslaught that culminated in the incorporation of the entire realm into the British Empire.

The European Intrusion

By the middle of the eighteenth century, London had taken over much of the trade in South Asia. British power was imposed through the East India Company (EIC), which represented the empire but whose main purpose was economic control. The British took advantage of the weakened and fragmented power of the Mughals and followed a strategy commonly known as "indirect rule." They left local rulers in place as long as they extracted the desired trading arrangements. The EIC not only controlled trade with Europe in spices, cotton, and silk goods, but also India's longstanding commerce with Southeast Asia, which until then was in the hands of Indian, Arab, and Chinese merchants. This system worked well (for the British) for almost a century, but by then political developments and heightened tensions were making it inevitable that the British government itself would need to take over from the EIC and assume direct responsibility. Thus "East India" became part of the British

© H. J. de Blij

"More than a half-century after the end of British rule, the centers of India's great cities continue to be dominated by the Victorian-Gothic buildings the colonizers constructed here. Here is evidence of a previous era of globalization, when European imprints transformed urban landscapes. Walking the streets of Mumbai (the British called it Bombay) you can turn a corner and be forgiven for mistaking the scene for London, double-deckered buses and all. One of the British planners' major achievements was the construction of a nationwide railroad system, and railway stations were given great prominence in the urban architecture. I had walked up Naoroji Road, having learned to dodge the wild traffic around the circles in the Fort area, and watched the throngs passing through Victoria (now Chhatrapati Shivaji) Station. Inside, the facility is badly worn, but the trains continue to run, bulging with passengers hanging out of doors and windows."

www.conceptcaching.com

colonial empire in 1857 (a period of rule, or *raj*, that would endure for the next 90 years), and Queen Victoria officially became its empress 20 years later.

Colonial Transformation

British colonialism in South Asia coincided with the Industrial Revolution in Europe, and the impact of Britain on the realm must be understood in that context. South Asia became, in large part, a supplier of raw materials needed to keep the factories going in Manchester, Birmingham, and other industrial centers in Britain. For instance, when the supply of cotton from the American South came to a halt during the U.S. Civil War in the early 1860s, the British quickly encouraged (and enforced) cotton production in what is today western India.

When the British took power in South Asia, this was a realm with already considerable industrial development (notably in metal goods and textiles) and an active trade with both Southwest and Southeast Asia. The colonialists saw this as competition, and soon India was exporting raw materials and importing manufactured products—from Europe, of course. Local industries declined, and Indian merchants lost their markets.

Colonialism did produce assets for India. The country was bequeathed one of the most extensive transport networks of the colonial era, particularly the railroad system—even though the network focused on interior-to-seaport linkages

rather than fully interconnecting the various parts of the country. British engineers laid out irrigation canals through which millions of hectares of land were brought into cultivation. Coastal settlements that had been founded by Britain developed into major cities and bustling ports, led by Bombay (now Mumbai), Calcutta (now Kolkata), and Madras (now Chennai). These three cities still rank among India's largest urban centers, and their cityscapes bear the unmistakable imprint of colonialism.

British rule also produced a new elite among the South Asian natives. They had access to education and schools that combined English and Indian traditions, and their Westernization was reinforced through university education in Britain. This elite drew from Hindu and Muslim communities, and it was to play a major part in the rising demands for self-rule and independence. These demands started to gather momentum in the early twentieth century and could no longer be denied when World War II came to an end in the mid-1940s.

THE GEOPOLITICS OF MODERN SOUTH ASIA

Partition and Independence

Even before the British government decided to yield to demands for independence, it was clear that British India

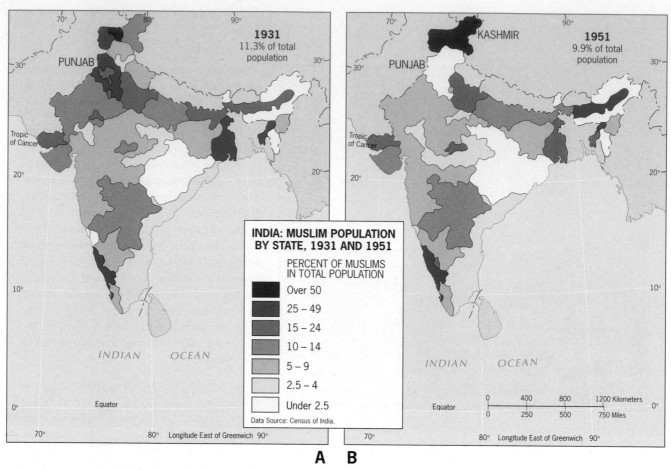

FIGURE 8A-6

© H. J. de Blij, P. O. Muller, and John Wiley & Sons, Inc.

would not survive the coming of self-rule as a single political entity. As early as the 1930s, Muslim activists were promoting the idea of a separate state. As the colony moved toward independence, a major political crisis developed that eventually resulted in the separation of India and Pakistan. But **5** **partition** was no simple matter. True, Muslims were in the majority in the western and eastern sectors of British India, but smaller Islamic clusters were scattered throughout the realm. Furthermore, the new boundaries between Hindu and Muslim communities had to be drawn right through areas where both sides coexisted—thereby displacing millions.

The consequences of this migration for the social geography of India were especially far reaching. Comparing the country's 1931 and 1951 distributions of Muslims in Figure 8A-6, you can see the impact on the Indian Punjab and in what is today the State of Rajasthan. (Since Kashmir was mapped as three entities before the partition and as one afterward, the change there represents an administrative, not a major numerical, alteration.) Even in the east a Muslim exodus occurred, as reflected on the map by the State of West Bengal, adjacent to Bangladesh, where the Islamic component in the population declined substantially.

The world has seen many **6** **refugee** migrations but none involving so many people in so short a time as the one

resulting from British India's partition (which occurred on Independence Day, August 15, 1947). Scholars who study the refugee phenomenon differentiate between "forced" and "voluntary" migrations, but as this case underscores, it is not always possible to separate the two. Many Muslims, believing they had no choice, feared for their future in the new India and joined the stampede. Others had the means and the ability to make a decision to stay or leave, but even these better-off migrants undoubtedly sensed a threat.

The great majority of Hindus who lived on the "wrong" side of the border moved as well. The Hindu component of present-day Pakistan may have neared 16 percent in 1947 but is only about 1 percent today; in Bangladesh, which was named East Pakistan at the time of partition, it declined from 30 percent to 16 percent today. Partition therefore created a new cultural and geopolitical landscape in South Asia.

India–Pakistan

From the moment of their separate creation, India and Pakistan have had a tenuous relationship. Upon independence, present-day Pakistan was united with present-day Bangladesh, and the two countries were respectively called West Pakistan and East Pakistan. As we have noted, the basis for this scheme was Islam: in both Pakistan and Bangladesh, Islam is the state religion. Between the two Islamic wings of Pakistan lay Hindu India. But there was little else to unify the Muslim easterners and westerners, and their union lasted less than 25 years. In 1971, a costly war of secession, in which India supported East Pakistan, led to the collapse of this unusual arrangement. East Pakistan, upon its "second independence" in 1971, took the name Bangladesh; and since there was no longer any need for a "West" Pakistan, that qualifier was dropped and the name Pakistan remained on the map.

India's encouragement of independence for Bangladesh emphasized the continuing tension between Pakistan and India, which had already led to war in 1965, to further conflict during the 1970s over Jammu and Kashmir, and to periodic flare-ups over other issues. During the Cold War, India tilted toward Moscow, while Pakistan found favor in Washington because of its strategic location adjacent to Afghanistan. Armed conflict between the two South Asian countries seemed to be a regional matter—until the early 1990s, when their arms race took on ominous nuclear proportions. Since then, the specter of nuclear war has hung over the conflicts that continue to embroil Pakistan and India, a concern not just for the South Asian realm but for the world as a whole. No longer merely a decolonized, divided, and disadvantaged country trying to survive, Pakistan has taken a crucial place in the political geography of a geographic realm in turbulent transition.

The relationship between India and Pakistan is especially sensitive because there are still so many Muslims in India. Massive as the 1947 refugee movement was, it left far more Muslims in India than those who had departed. From about 360 million before partition, the number of Muslims in India declined sharply, but it remained a huge minority, one that was growing rapidly to boot. By 2011, it surpassed 170 million, more than 14 percent of the total population—the largest cultural minority in the world and almost as large as Pakistan's entire population (181 million).

What this means is that a sizeable portion of India's population has "natural" sympathies vis-à-vis Pakistan. Their presence works at times as a brake on hawkish Indian policies toward Islamabad. On the other hand, conflict with Pakistan can have detrimental effects on Hindu-Muslim relations inside India, and over the years this has led to communal violence and deadly clashes. Moreover, this issue is further complicated today by the alleged role of Indian Muslims in terrorist activities in India, orchestrated from Pakistan.

Contested Kashmir

When Pakistan became an independent state following the partition of British India in 1947, its capital was Karachi on the south coast, near the western end of the Indus Delta.

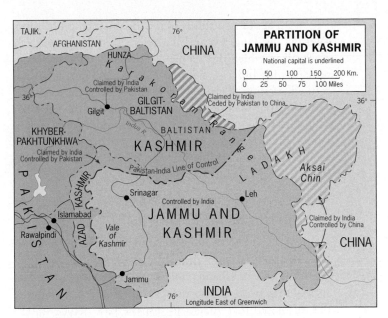

FIGURE 8A-7   © H. J. de Blij, P. O. Muller, and John Wiley & Sons, Inc.

As the map shows, however, the present capital is Islamabad. By moving the capital from the "safe" coast to the embattled interior, and by placing it on the doorstep of the contested territory of Kashmir, Pakistan announced its intent to stake a claim to its northern frontiers. And by naming the city Islamabad, Pakistan proclaimed its Muslim foundation here in the face of the Hindu challenge. This politico-geographical usage of a national capital can be assertive, and Islamabad exemplifies the principle of the **7** **forward capital**.

Kashmir is a territory of high mountains surrounded by Pakistan, India, China, and, along more than 50 kilometers (30 mi) in the far north, Afghanistan (Fig. 8A-7). Although known simply as Kashmir, the area actually consists of several political divisions, including the Indian State properly referred to as Jammu and Kashmir, a major bone of contention between India and Pakistan.

When partition took place in 1947, the existing States of British India were asked to decide whether they wanted to be incorporated into India or Pakistan. In most of the States, the local ruler made this decision, but Kashmir was an unusual case. It had about 5 million inhabitants at the time, nearly three-quarters of them Muslims, but the maharajah of Kashmir himself was a Hindu. When he decided not to join Pakistan and instead aimed to retain autonomous status, this was answered with a Muslim uprising supported by Pakistan. The maharajah, in turn, called for help from India. After more than a year's fighting and through the intervention of the United Nations, a cease-fire line left most of Jammu and Kashmir (including nearly four-fifths of the territory's population) in Indian hands. Eventually, this line—now known as the Line of Control—began to appear on maps as the final boundary settlement, and Indian governments have proposed that it be so recognized.

With about two-thirds of the inhabitants still Muslim, Pakistan has long demanded a referendum in Jammu and Kashmir in which the people can decide for themselves to remain with India or become part of Pakistan. India has refused, arguing that there is a place for Muslims in secular India but not for Hindus in the Islamic Republic of Pakistan. Given the specter of terrorism in India and the dangerous precedent of a concession given India's enormous ethnic and regional diversity, the Kashmir conflict is unlikely to be resolved for years to come.

The Specter of Terrorism

Comparatively successful as the integration of India's Muslim communities into the fabric of the Indian state has been, the risk of Islamic violence, directed against Indian society in general, is also rising. A brief review of some of the major terrorist attacks in just the past decade underscores the severity of the situation. In 2001, Islamic terrorists with roots in Pakistan assaulted the Indian Parliament in New Delhi and brought the two countries to the brink of war. A series of bombs set off in different parts of Mumbai in 2003 killed more than 100 people and injured many more. Three years later, a series of simultaneous, coordinated attacks struck several trains along a commuter line in Mumbai, killing about 200. The most daring attack to date followed in 2008, when terrorists targeted Mumbai's most upscale, Westernized hotels. Nearly 200 people died and hundreds were wounded. Live pictures of smoke billowing from the famous Taj Mahal Hotel in southern Mumbai were seen around the world. Finally, in early 2010, a bombing in Pune, another major city not far from Mumbai, claimed 11 more lives.

The group responsible for the two most recent attacks was *Lashkar-e-Taiba* (the Party of the Righteous), a Pakistan-based organization that among other things aims to return Kashmir to Islamic rule. These events have serious implications for India, where the overwhelming majority of Muslim citizens have remained uninvolved in extremist causes. It seems that a small number of Indian Muslims are joining local terrorist cells with links to Pakistan and perhaps other Islamic countries. Their terrorist acts lead to investigations that offend ordinary and peaceful Indian Muslims, radicalizing a number of them and expanding the market for Islamic militancy. It is still too early to gauge the potential impact of this development on a country that has long and justly prided itself on its multicultural democracy, but the portents for India's political, social, and economic geography are obviously serious.

In the meantime, Pakistan's northwestern frontier is effectively managed by Taliban forces that provide shelter and training grounds to the al-Qaeda terrorist network. This border zone with Afghanistan is well beyond the control of the Pakistani government. U.S. efforts to defeat the Taliban in Afghanistan continue, but are thwarted by the Taliban's ability to move back and forth across the border. Thus the United States is increasing pressure on Islamabad to confront the Taliban on the Pakistani side of the border in this remote mountain refuge.

It is a delicate geopolitical chess game. Pakistan is careful not to alienate its Islamic base even if it despises the northern extremists, and it is fearful of an economically stronger India and its growing intimacy with the United States. India is deeply concerned about Pakistan's role in terrorism on Indian soil and, even worse, the possibility of fundamentalists taking control of Pakistan's nuclear capabilities; at the same time, it must also guard against increased tensions between Hindus and Muslims inside India. India is also impatient with American reluctance to choose its side in the Kashmir conflict. The United States, in turn, is sympathetic to the world's biggest democracy but needs Pakistan to be an ally in the global counterterrorism campaign. Each of these parties is walking a tightrope, where the slightest mistake could have deadly consequences.

Chinese Claims

An overview of this realm's geopolitical landscape would not be complete without noting the powerful presence of China, South Asia's northeastern neighbor, and that country's territorial claims. As Figure 8A-7 shows, the northeastern extension of Jammu and Kashmir is claimed by China. This conflict has been fairly quiet in recent years, but officially neither China nor India show the slightest sign of conceding. Another, historically more volatile dispute lies in India's far northeast, where China claims the bulk of the territory of the poetically named Indian State of Arunachal Pradesh (land of the dawn-lit mountains). Here India borders Tibet, and back in 1914, long before the Chinese took control, independent Tibet agreed to this boundary. Tibet (now called Xizang by the Chinese) figures prominently in this conflict, especially because the exiled Dalai Lama has called India his second home since the 1960s. The brief China-Indian war of 1962 was fought over this border, but the matter of Tibet was never far removed. Today, this issue remains unresolved.

EMERGING MARKETS AND FRAGMENTED MODERNIZATION

In recent years, optimistic television news reports and articles in the popular press have been proclaiming a new era for South Asia, marked by rising growth rates for the realm's national economies, rewards from globalization and modernization, and increasing integration into the global economy. India, obviously the key to the realm, has even been described as "India Shining" during this wave of enthusiasm.

And indeed, a combination of circumstances, ranging from America's involvement with Pakistan in the campaign against terrorism to the real estate and stock market booms in India, suggest that a new era has arrived. But consider this: more than two-thirds of India's nearly 1.2 billion people continue to live in poverty-stricken rural areas, their villages and lives virtually untouched by what is happening in the cities (where, by the way, tens of millions of urban dwellers inhabit some of the world's poorest slums). Fully one-third of Pakistan's population lives in abject poverty; female literacy is below 30 percent. Half of the people of Bangladesh, and nearly half of those in the realm as a whole, live on the equivalent of one U.S. dollar per day or less. It is estimated that half the children in South Asia are malnourished and underweight, a majority of them girls—this at a time when the world is able to provide adequate calories for all its inhabitants, if not adequately balanced daily meals. It still remains to be seen if the benefits of newfound economic growth can be spread around widely enough for the good of South Asia's masses.

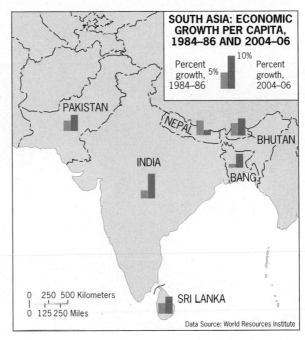

FIGURE 8A-8 © H. J. de Blij, P. O. Muller, and John Wiley & Sons, Inc.

Economic Liberalization

Most countries in this realm have liberalized their economies since the late 1980s, as part of a worldwide turn toward **8** neoliberalism. This involves privatization of state-run companies, lowering of international trade tariffs, reduction of government subsidies, cutting of corporate taxes, and overall deregulation to spur business activity. It was an important change from previous times in which markets were tightly controlled by large central governments that since independence had espoused ideologies opposed to unfettered capitalism. This change of direction was unavoidable. The ineffective policies of the past, continuing grinding poverty, and near fiscal bankruptcy demanded support from the International Monetary Fund; the IMF, in turn, demanded structural reforms.

The results of these reforms have been especially noticeable in India, Bangladesh, and Pakistan, where economic growth rates climbed to levels never seen before (Fig. 8A-8). Most of this growth is in manufacturing, services, finance, and, in India, information technology (IT). A more open economy has attracted increased foreign investment, and during the past two decades a new (urban) middle class has emerged. This new middle class may account for only 25 percent or so of the population, but in South Asia that translates into fully 300 million people—a huge new consumer market for an array of products ranging from cars to personal computers.

Nevertheless, that still leaves a staggering 1 billion South Asians who have *not* acquired middle-class status, for whom little has changed, who are overwhelmingly dependent on agriculture, and who are not likely to log on to the new information economy anytime soon.

The Significance of Agriculture

Across South Asia, more than half of the entire workforce is employed in agriculture, ranging from about 40 percent in Pakistan to about 80 percent in Nepal. But overall productivity is low, and the contribution from agriculture to the overall economy is only around 20 percent. Incomes in rural areas are much lower than in major cities, and the same is true for the standard of living. About two-thirds of South Asia's population is rural, and even those who do not work in agriculture tend to rely on it indirectly.

Millions of lives every year depend on a good harvest. As Figure 8A-3 shows, the wet monsoon brings life-giving rains to the southwestern (Malabar) coast, and a second branch from the Bay of Bengal turns and sweeps across northern India toward Pakistan, losing strength (and moisture) as it proceeds. This means that amply watered eastern India and Bangladesh as well as the southwestern coastal strip grow rice as their staple crop, but drier northwestern India and Pakistan raise wheat.

Farmers' fortunes tend to vary with geography, and this can be illustrated in the situation on either side of India's Western Ghat Mountains. The monsoon rains are generally plentiful along the western slopes of the mountain range, from the southern tip of the subcontinent up to northern Maharashtra. Here, you can see the hillside vegetation assume all shades of fresh green come the month of June, a sure sign that the harvest will be bountiful. But on the eastern (rain shadow) side, and further into the interior of the Deccan Plateau, it is a different story. The rains don't come as often and don't last as long. Farming becomes a gamble with nature, and life becomes precarious. Many of

© Jan Nijman

"In the remote, rural areas of western India's Maharashtra State, life has not changed much for centuries. On the drier interior side of the Western Ghats, the wet monsoon is less predictable and so is the harvest. Farms are small and worked by extended families, mostly without any machinery. Millet is the main crop in these parts. Here, a grandfather pumps irrigation water from a well with the use of a bullock while the rest of the family does the more exacting work in the surrounding fields. Just as the bullock goes around and around, day after day, life tends to repeat itself in a cyclical pattern for generation after generation in India's myriad small villages."

www.conceptcaching.com

the farmers here are members of lower castes, landless, indebted, and have the hardest time making ends meet. Almost every year, Maharashtra's inland districts report several thousand (!) farmer suicides as desperately poor peasants end their lives because they can no longer provide for their families.

It is clear that the majority of people in this realm depend on agriculture and that governments must aim their economic policies to improve agricultural productivity to raise the standard of living in rural areas. But they have a long way to go. The demands on governments are many, and they often seem distracted by economic sectors that can make a faster and greater monetary contribution, such as manufacturing, financial services, and IT—the kinds of economic activity that take place in the big cities, far from the impoverished countryside.

SOUTH ASIA'S POPULATION GEOGRAPHY

Given its enormous human content, the South Asian realm's areal size is relatively quite small. It totals less than two-fifths the size of equally populous East Asia. Comparing the world's two giants shows that China's area is almost three times as large as India's. The total population of Subsaharan Africa is less than half of South Asia's, in an area almost five times as large. Adjectives such as "teeming,"

"crowded," and "jammed" are often used to describe the realm's habitable living space, and with good reason. South Asia's intricate cultural mosaic is tightly packed, with only the deserts in the west and the mountain fringe in the north displaying large empty spaces (Fig. 8A-9); in fact, you can even see the outlines of the densely populated river basins in the dot pattern of the population distribution map.

But what is the significance of the realm's enormous population size and of its density, and how are South Asia's population dynamics related to issues of development? The field of **9** **population geography** focuses on the characteristics, distribution, growth, and other aspects of spatial demography in a country, region, or realm as this relates to soils, climates, land ownership, social conditions, economic development, and other factors. In the South Asian context, it is useful to concentrate on four demographic dimensions: the role of density, the demographic transition, age distributions and economics, and the gender bias in birth rates. As we shall see, population issues are often more complex than they first appear.

Population Density and the Question of Overpopulation

10 **Population density** measures the number of people per unit area (such as a square kilometer or square mile) in a country, province, or, as is the case in Figure 8A-9, an entire

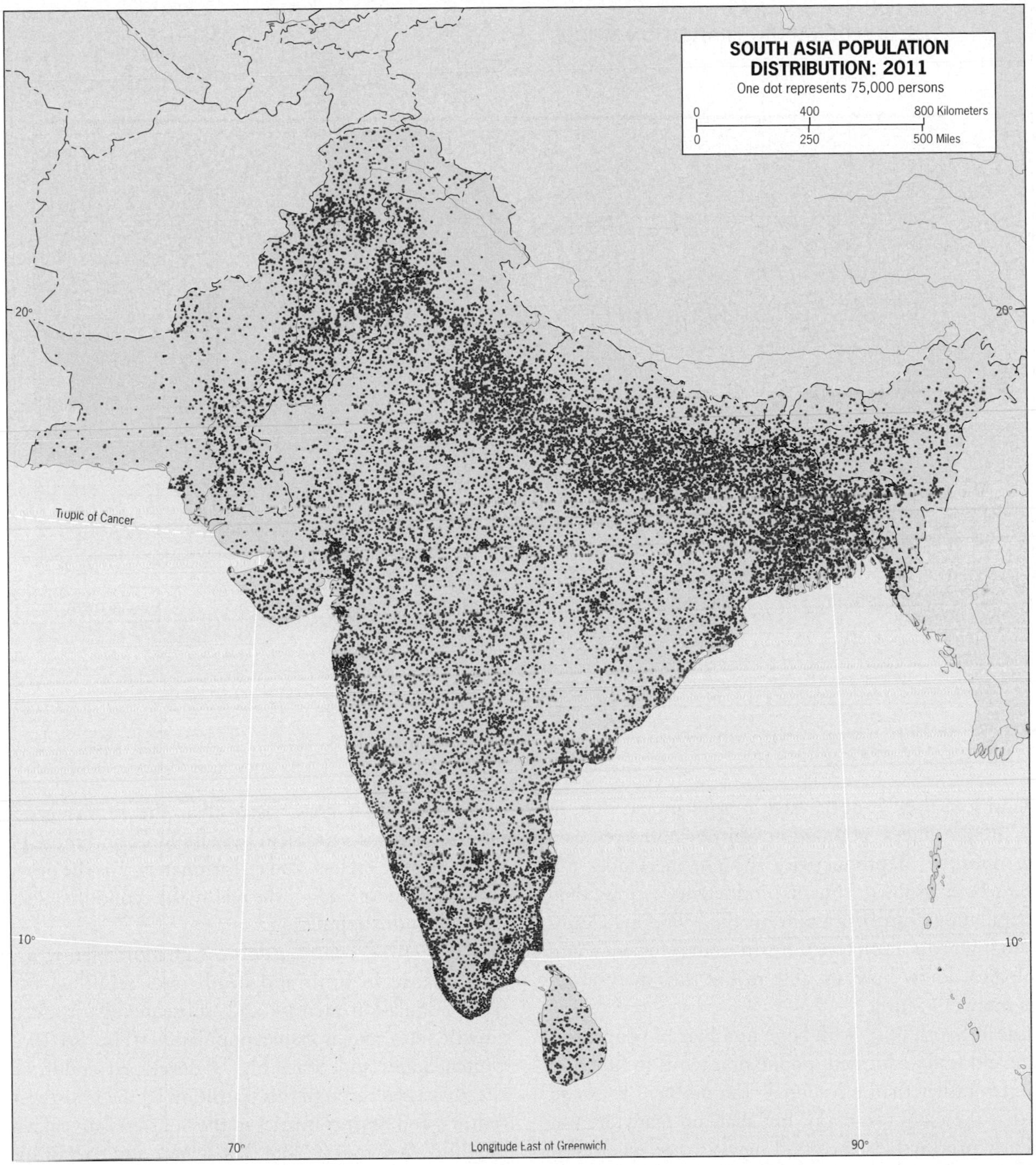

FIGURE 8A-9

© H. J. de Blij, P. O. Muller, and John Wiley & Sons, Inc.

realm. We distinguish between two types of measures. ***Arithmetic density*** is simply the number of people per area, usually a country. **11** **Physiologic density** is a more meaningful measure because it takes into account only land that is arable, that can be used for food production. Please take a careful look at the data displayed for South Asia in Appendix B, and you will see that, for example in Pakistan, the two measures are quite different due to that country's large deserts and inhospitable mountain ranges.

Until recently, South Asia's persistent poverty was often related to its enormous and rapidly growing popula-

tion and its high population densities. The idea was that there were simply "too many mouths to feed": the realm was "overpopulated." The notion of ***overpopulation*** can be compelling and seems to make sense at an intuitive level because every country or region can be thought of having a limited "carrying capacity."

But things are more complex than that. If you look again at Appendix B, you will find that some countries with high densities, such as the Netherlands or Japan, are doing very well, and it is not necessarily because they have such impressive natural resources (they do not). The point is that

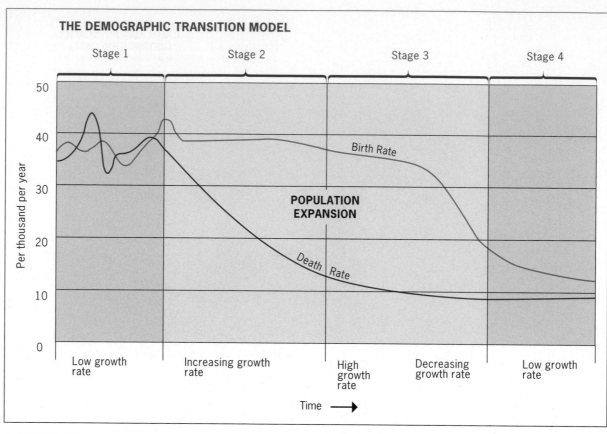

FIGURE 8A-10

© H. J. de Blij, P. O. Muller, and John Wiley & Sons, Inc.

high density in itself is not always a problem and that, in certain circumstances, population can be considered as a **human resource**. If productivity is high, there does not appear to be a problem, but if productivity is low, then large populations can be a drain on the economy. Countries with high education levels, institutional efficiency, and technological know-how are able to use their natural resources more efficiently.

Thus in South Asia, with large numbers of people still illiterate and undereducated, population tends to function as a *burden* rather than a resource. The problem is not so much that there are too many, but that too many are not sufficiently productive. The good news is that, as a result of higher economic growth rates stemming from economic reforms, there is more money to invest in education. The bad news so far is that not enough of the money is actually spent that way.

The Demographic Transition

The relevance of population issues to development go far beyond density, which is really just a snapshot of the population pattern at a given time. It gets more interesting—and more complicated—when we relate population change to economic trends. For example, for a considerable time South

Asia's population grew faster than the realm's economy. Clearly that was a problem because more and more people had to survive on less. Today, fortunately, it is the other way around: in most parts of the realm, the economy is growing faster than the population.

The term **12 demographic transition** refers to a structural change in birth and death rates resulting, first, in rapid population increase and, subsequently, in declining growth rates and a stable population (Fig. 8A-10). The United States and other highly developed countries had already passed through this transition by the mid-twentieth century, and most countries in the South Asian realm are in the third stage today. Note that stage 2 and part of stage 3, with high birth rates and low death rates (due to medical advances), entail a population expansion. In South Asia, this expansion occurred from the 1950s through the 1970s.

The key issue, of course, is for the birth rates to come down so that growth rates will drop and the population will stabilize. This is happening today, but the process is not yet complete. Figure 8A-11 shows how **13 fertility rates** (the number of births per woman) have dropped across the realm over the past quarter-century. Only Sri Lanka seems to have completed the transition, although Bhutan is now very close. Elsewhere, fertility rates are still too high (India by itself has been adding about 15 million people *per year*

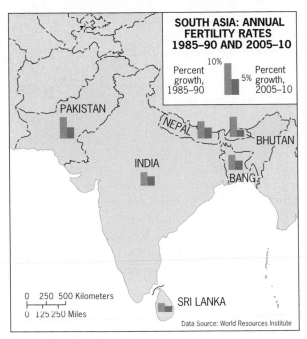

FIGURE 8A-11 © H. J. de Blij, P. O. Muller, and John Wiley & Sons, Inc.

during the past decade), but at least they are moving in the right direction.

Demographic Burdens

The immediate significance of demography to economics lies in what we call the **14** **demographic burden**. This term refers to the proportion of the population that is either too old or too young to be productive and that must be cared for by the productive population. Typically, the most productive population in developing countries is represented in the age cohorts from 20 to 50 years. A country with low death rates and high birth rates will have a relatively large share of old and young people, and thus a large demographic burden. Obviously, the way to reduce this burden is to lower birth rates.

Let us now turn to examine Figure 8A-12 and compare today's **15** **population pyramids** (diagrams showing the age–sex structure) for India and China. The latter has been more successful in curtailing births since 1980, so China now faces a smaller demographic burden than India.

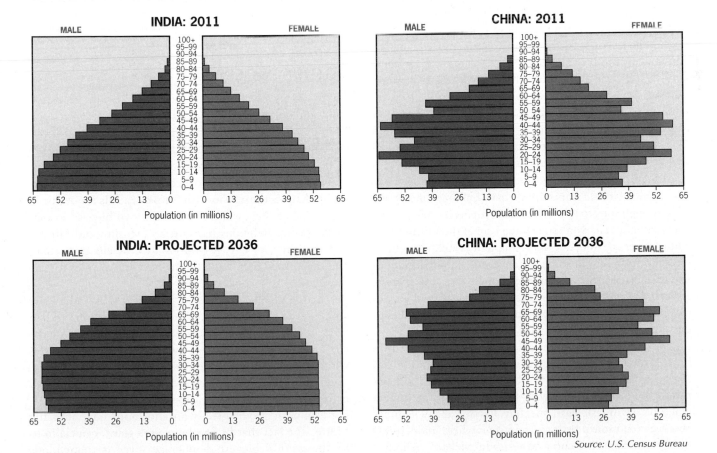

FIGURE 8A-12 © H. J. de Blij, P. O. Muller, and John Wiley & Sons, Inc.

But, interestingly, what is advantageous today can become a disadvantage tomorrow. Look at Figure 8A-12 again and see what the population profiles will be like 25 years from now. Assuming that India will be able to further reduce its birth rate in the coming years, its demographic burden will be less than China's one generation from now. In China, today's productive cohort will have moved on to old age, thereby adding to its demographic burden. It is another reason for India-boosters to be optimistic about the future. Yet some other major hurdles remain, and some forms of birth control are as morally reprehensible as they are economically counterproductive.

The Missing Girls

Issues of family planning and birth control shed an interesting light on the "fragmented modernization" of South Asia. As we saw, the realm finds itself in an advanced stage of the demographic transition wherein birth rates have started to come down. But take a close look at India's population pyramid for 2011 and note that among young children the males far outnumber the girls. In fact, males outnumber females well into middle age.

Traditionally, boys are valued more than girls because they are thought to be more productive income-earners, because they are entitled to land and inheritance, and because they do not require a dowry at the time of marriage. When a couple gets married (often arranged, and at a young age), the bride comes into the care of the groom's family, where she also contributes her work in and around the house. For this, the bride's family must provide a dowry that can impose a major expense on her parents. For these reasons, the birth of a boy is a greater cause for celebration than that of a girl. "Raising a daughter," as one saying goes, "is like watering your neighbor's garden."

One reason for the high fertility rates in the past was that families would continue to have children until there were enough sons to take care of the parents in their old age (the girls, after all, would be taking care of their future husband's parents). When a poor couple repeatedly produces girls and not boys, in some instances the family decides to end the life of the newborn daughter. It is this ***female infanticide*** or *gendercide* that causes the unnatural gender bias in South Asia's population profiles (the same applies to parts of China).

But why, as the economic situation has improved, as birth rates have come down, and as modernization has started to set in, do we still observe this skewed **16** sex ratio? The answer is that with fewer children, the importance of having at least one boy has for many families become even more pressing. And here is where "modernization" throws another curve ball: newly available technologies of ultrasound scanning and rising incomes (i.e., the growing affordability of a scan) have induced many families to determine the gender of the unborn child and decide on abortion if the child is female. Thus in recent years the sex ratio has become more, not less, skewed, and the most extreme ratios are now found in some of the most developed parts of the realm, such as the Indian States of Punjab and Haryana.

In the long run, of course, this leads to a shortage of females, which becomes particularly apparent at marriage age. In some areas, families now face a problem in finding a bride for their sons, and this "bachelor angst," in turn, is leading to a change in attitudes. Look at India's population pyramid in 2036 and note that over the next 25 years the sex ratio is expected to become less skewed. It is another example of the dynamic nature of demography—and yet another reason to be hopeful about better times to come for South Asia.

Notwithstanding the serious matter of the "missing girls," it is difficult to make any generalizations about gender relations across this populous realm, in part because of religious and regional diversity as well as rural-urban differences. These are in many respects male-dominated societies, especially at a young age. But it is useful to remember that Pakistan, India, Sri Lanka, and Bangladesh have all had female prime ministers who held their countries' most powerful political office. That has yet to happen in the United States.

FUTURE PROSPECTS

South Asia is a realm in transition—politically, economically, and demographically. It is a realm that seems clearly bounded by nature, yet it is vitally linked to Southwest Asia and, increasingly, the entire world. It is also a realm that is at times difficult to read. India–Pakistan tensions continue to be a cause for concern, and the specter of terrorism haunts those who wish for peace. This is not just in the hands of the governments of the two biggest nations in the realm. Religious movements (Muslim and Hindu) and the way they engage politics are crucially important, and the United States and China have major roles to play as well.

Economically, there is no question that India's rise will increasingly demand the world's attention. Indian transnational corporations will continue to penetrate the global economy, and the growing Indian middle class with its appetite for consumption will increasingly draw interest from producers around the world. That English is the subcontinent's *lingua franca* and that IT is a leading economic sector give it an enormous edge into the future. The fact that India, the realm's giant, can claim to be the world's biggest democracy gives it tremendous credibility.

India's leading high-tech center is Bengaluru (formerly Bangalore: another of India's recent place-name changes), and the Indian corporation Infosys Technologies, based here in this far southern city, is the country's leading software exporter. This Silicon Valley-like campus in suburban Electronics City is the heart of the company's booming outsourcing business (Bank of America and Citigroup are among its American customers). While some skilled workers have emigrated to Europe, Australia, and the U.S., where their skills earn them more than at home, many more choose to stay here and benefit from the higher local income the IT manufacturers pay. Thus Bengaluru attracts India's best and brightest, whose children attend the city's superior schools. And this is a pleasant place to live: at 900 meters (3000 ft) above sea level, even the summer weather is tolerable. Bengaluru, however, is growing so rapidly that its infrastructure cannot keep up, so that congestion, traffic jams, and commuting times are all growing as well. Sound familiar? © Deepak G. Pawar/The India Today Group/GettyImages, Inc.

When, during the next several decades, South Asia passes through the demographic transition; when it keeps the peace; when it continues its leading role in the global IT sector; when its economic growth is used to educate and empower the masses; and when the reorganization of agriculture allows more productive and prosperous lives—and these are all real possibilities—then this populous and wondrous South Asian realm may yet turn out to be the biggest story of the twenty-first century.

POINTS TO PONDER

- The South Asian realm contains three of the world's mightiest rivers and the world's largest human concentration.
- After more than 60 years, the conflict over Kashmir is still unresolved.
- The most skewed sex ratios occur in the most prosperous parts of the realm.
- Islamic terrorism is the single greatest threat to South Asia's future.

Pilgrims bathing in the sacred Ganges at Varanasi, India's holiest city.
© David Zimmerman/Masterfile

8B

SOUTH ASIA: REGIONS OF THE REALM

Pakistan
India
Bangladesh
Mountainous North
Southern Islands

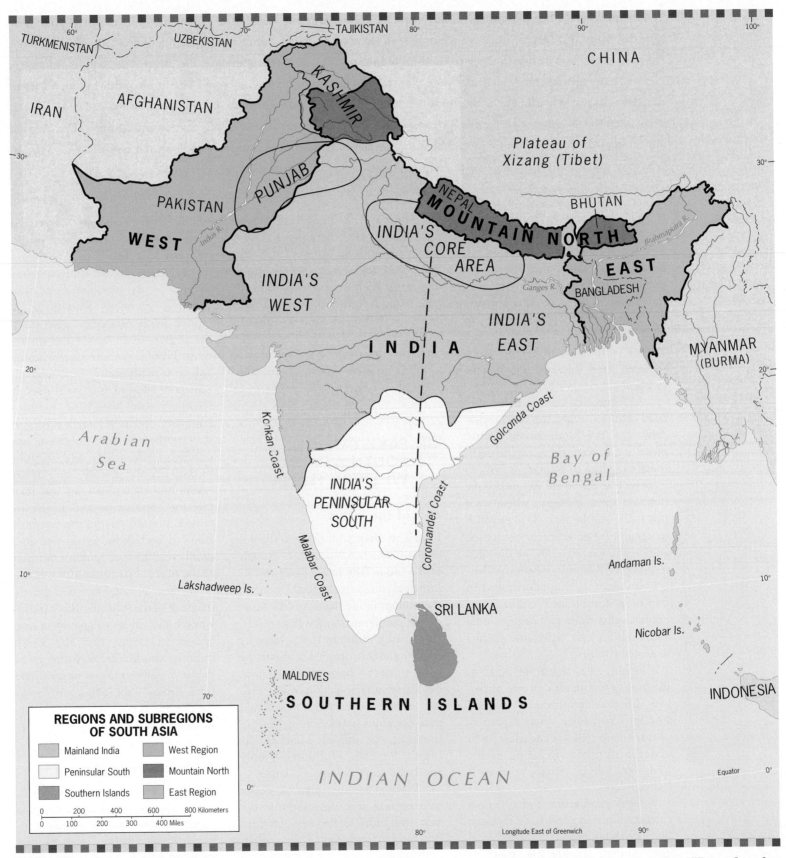

© H. J. de Blij, P. O. Muller, and John Wiley & Sons, Inc.

REGIONS AND SUBREGIONS OF SOUTH ASIA

- Mainland India
- Peninsular South
- Southern Islands
- West Region
- Mountain North
- East Region

0   200   400   600   800 Kilometers

0   100   200   300   400 Miles

India, the cornerstone of South Asia, has a core area centered in the wide basin of the Ganges River, the historic heart of this realm (Fig. 8B-1). India is a region as well as a state, with key subregions based on cultural and other criteria to be discussed later in this chapter. To the west lies Islamic Pakistan, whose lifeline, the Indus River and its tributaries, creates a core area in the Punjab. But note that the Punjab subregion extends across the border with India. What you see here is one aftermath of an enormous human tragedy. When independence approached, Muslim leaders in the west demanded the creation of an Islamic state. A boundary drawn hastily in 1947 between a future Muslim West Pakistan and a dominantly Hindu India precipitated the sudden migration of millions of people in the border area—Muslims westward and Hindus eastward—with countless casualties and innumerable human tragedies.

In South Asia's east, the British had earlier drawn a border between (Hindu) Bengal and (dominantly Muslim) East Bengal, and that border became the boundary of what became known as East Pakistan, the political connection based upon shared Islamic values. In 1971, however, East Pakistan severed its link with West Pakistan (which renamed itself Pakistan) and became independent Bangladesh.

As Figure 8B-1 shows, three separated entities make up the region defined as the Mountainous North. From east to west, these are the traditional kingdom of Bhutan, as isolated a country as the world has today; the troubled state of Nepal, where royal rule failed but the struggle to establish representative government continues; and, at the western end, the disputed territory of Kashmir, where the British boundary-making effort failed to resolve a complicated cultural and political situation, and where India and Pakistan have repeatedly clashed in armed conflict.

Finally, the region mapped as the Southern Islands consists of two very different countries: dominantly Buddhist Sri Lanka, where a quarter-century-long civil war ended in 2009, and the ministate Republic of the Maldives, whose official religion is Islam.

Physiographically, India mainly consists of plateau-and-basin country, but its neighbors to the north, Nepal and Bhutan, straddle the southern slopes of the mighty Himalaya range and its foothills. The waters of the Indian Ocean separate India from neighbors Sri Lanka and the Maldives in the south. Thus there are good geographic reasons for using the political framework to designate the regions of this realm. Culture and nature constitute formidable regionalizing factors here, even within India itself. The cultural forces that broke British India apart at the time of independence continue to exert enormous power in still-evolving South Asia.

PAKISTAN: ON SOUTH ASIA'S WESTERN FLANK

Gift of the Indus

If, as is so often said, Egypt is the gift of the Nile, then Pakistan is the gift of the Indus. The Indus River and its principal tributary, the Sutlej, nourish the ribbons of life that form the heart of this populous country (Fig. 8B-2). Territorially, Pakistan is not large by Asian standards; its area is about the same as that of Texas plus Louisiana. But Pakistan's population of 181 million makes it one of the world's ten most populous states. Among Muslim countries (its official name is the Islamic Republic of Pakistan), only Southeast Asia's Indonesia is larger.

Pakistan lies like a gigantic wedge between Iran and Afghanistan to the west and India to the east. Here in Pakistan lay South Asia's earliest urban civilizations, whose innovations radiated southeastward into the triangular peninsula. Here lies South Asia's Muslim frontier, contiguous to the great Islamic realm to the west and irrevocably linked to the enormous Muslim minority to its east.

Pakistan's cultural landscapes bear witness to its transitional location. Teeming, disorderly Karachi is the typical South Asian city; as in India, the largest urban center lies on the coast. Historic, architecturally Islamic Lahore is reminiscent of the scholarly centers of Muslim Southwest Asia. In Pakistan's east, the 1947 partition boundary divides a Punjab subregion that is otherwise continuous—a land of villages, wheatfields, and irrigation ditches. In the northwest, Pakistan resembles Afghanistan in its huge migrant populations and its mountainous frontier. And in the far north, Pakistan and India are locked in a deadly conflict over Kashmir. A legacy of the times of partition, this territory is claimed by Pakistan because the majority of the inhabitants is Muslim, while India refuses to give up control because it claims that the Hindu minority would have no future inside Pakistan. This issue has plagued relations between the two countries for decades and is not likely to be resolved in the foreseeable future.

A Hard Place to Govern

At independence (West) Pakistan had a bounded national territory, a capital, a cultural core, and a population—but few centripetal forces to bind state and nation. The disparate subregions of Pakistan shared the Islamic faith and an aversion for Hindu India, but little else. Karachi and the coastal

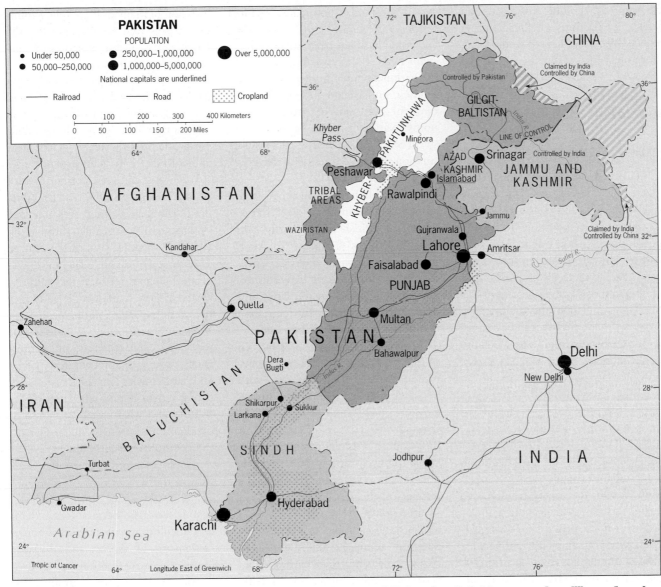

FIGURE 8B-2

© H. J. de Blij, P. O. Muller, and John Wiley & Sons, Inc.

south, the desert of Baluchistan, the city of Lahore and the Punjab, the rugged northwest along Afghanistan's border, and the mountainous far north remain worlds apart. Urdu is the official language, and English is still the *lingua franca* of the elite. Yet several other major languages prevail in different areas (see Fig. 8A-5), and ways of life vary enormously.

Successive Pakistani governments, civilian as well as military, turned to Islam to provide the common bond that history and geography had denied the nation. In the process, Pakistan became one of the world's most theocratic states. But even Islam itself is not unified in restive Pakistan. About 77 percent of the people are Sunni Muslims, and the Shia minority numbers approximately 20 percent. Sunni fanatics intermittently attack Shi'ites, leading to retaliation and establishing grounds for subsequent revenge.

To govern so diverse and fractious a country would challenge any system, and so far Pakistan has failed the test. Democratically elected governments have repeatedly squandered their opportunities, only to be overthrown by military coups. Pakistan's recent economic boom has not filtered down to the poor; literacy rates are not rising; health conditions are not improving significantly; national institutions are weak (for instance, there are only about 2 million registered taxpayers in a country of more than 180 million); and one consequence of the global antiterrorism campaign is that Pakistanis, who used to be overwhelmingly secular in their political choices, are now increasingly joining Islamic parties.

Meanwhile, too little is being done to confront a growing water-supply crisis, an insurgency festers in Baluchistan, the army is incapable of establishing control over mountainous Waziristan (where al-Qaeda and Taliban groups maintain hideouts), the issue of Kashmir costs Pakistan dearly, and relations with neighboring India (which if satisfactory would bring enormous benefits) remain conflicted.

Subregions of Pakistan

Punjab

Pakistan's core area is the Punjab (Fig. 8B-2), the Muslim heartland across which the post-independence boundary between Pakistan and India was superimposed. (As a result, India also has a region called Punjab, sometimes spelled Panjab there.) Pakistan's Punjab is home to more than 55 percent of the country's population. In the triangle formed by the Indus River and its tributary, the Sutlej, live almost 100 million people. Punjabi is the language here, and wheat farming is the mainstay.

Three cities anchor this core area: Lahore, the outstanding center of Islamic culture in the realm; Faisalabad; and Multan. Lahore, now home to more than 7 million people, lies close to the India–Pakistan border. Founded around 2000 years ago, Lahore was situated favorably to become a great Muslim center during the Mughal period, when the Punjab was the main corridor into India. After partition in 1947, the city received hundreds of thousands of refugees and grew rapidly. Lahore did lose its eastern hinterland, but its new role in independent Pakistan sustained its growth.

Punjab's relationship with Pakistan's other three provinces is one of the country's weak points. Both the governments and residents of those provinces feel uneasy about the dominance of Punjab, the populous, powerful core of the country from which most of the army is drawn.

Sindh

The lower Indus River is the key to life in Sindh (Fig. 8B-2), but the Punjab controls the waters upstream, which is one of the issues dividing Pakistan.

When the Punjab-dominated regime proposes to build dams across the Indus and its major tributaries, Sindhis (who make up almost one-fourth of the national population) are reminded of their underrepresentation in government and talk of greater autonomy. Nationalist, anti-Pakistan rage swept Sindh following the 2007 assassination of Sindh's presidential candidate, Benazir Bhutto.

The ribbon of fertile, irrigated, alluvial land along the lower Indus, where the British laid out irrigation systems, makes Sindh a Pakistani breadbasket for wheat and rice. Commercially, cotton is king here, supplying textile factories in the cities and towns (textiles account for more than half of Pakistan's exports by value).

But the dominant presence in southern Sindh is the chaotic, crime-ridden megacity of Karachi, with its stock market, dangerous streets, crowded beaches, and poverty-stricken shantytowns, a place of searing contrasts under a broiling sun. Karachi grew explosively during and after partition in 1947, when refugee Muhajirs from across the new border with India streamed into this urban area, setting off riots and gang battles whose aftermath still simmers. With little effective law enforcement, Karachi has today become a hotbed of terrorist activity, but somehow the city still functions as Pakistan's (and Afghanistan's) major maritime outlet and seat of Sindh's provincial government.

North West Frontier

As Figure 8B-2 shows, the North West Frontier Province lies wedged between the powerful Punjab to the east and troubled Afghanistan to the west, with the territory long known as the Tribal Areas intervening in the south. The Tribal Areas have had a certain degree of autonomy ever since British times and the Pakistani government's reach in these parts is very limited. The province's mountainous physical geography reflects its remoteness and the

isolation of many of its people. Mountain passes lead to Afghanistan; the Khyber Pass, already noted as the historic route of invaders, is legendary (see photo next page). Coming from Afghanistan, the Khyber's road takes you directly to the provincial capital, Peshawar, which lies in a broad, alluvium-filled, fertile valley where wheat and corn drape the countryside.

During the Soviet occupation of Afghanistan in the 1980s and later during the Taliban regime, several million Pushtun refugees streamed through the Khyber and other passes into refugee camps in this area. Following the ouster of the Taliban from Afghanistan in late 2001, the great majority of Pushtuns returned home. The aptly named North West Frontier Province remains a conservative, deeply religious, militant territory, where Islamic political parties and movements are proportionately stronger than in any other part of the country and where the obstruction of national policies (including antiterrorist operations) is a common goal.

Pakistan's northwest frontier is critical to the U.S. war in Afghanistan, as well as the wider campaign against Islamic terrorism because it is extremely difficult to monitor the cross-border movements of Taliban and al-Qaeda forces. The area is a hotbed of terrorist activity and training camps. The United States constantly prods Pakistan to try to secure the province militarily, but it is not clear to what extent Pakistan is willing—or able—to comply. This does not stop America from flexing its own muscles: news reports in April 2010 indicated that high-intensity bombings by CIA-operated drones (small unmanned planes) had killed dozens of militants in North Waziristan (Fig. 8B-2). However, the precarious position of this U.S. (and Pakistani) enterprise became clear within hours as the Taliban swiftly retaliated with a brazen attack on the American consulate in Peshawar (more than two dozen people were killed or wounded by a powerful bomb followed by a barrage of rockets).

FROM THE FIELD NOTES...

© Barbara A. Weightman

"Driving through the Khyber Pass that links Afghanistan and Pakistan was a riveting experience. From the dry and dusty foothills on the Afghanistan side, with numerous ruins marking historic battles, I traversed the rugged Hindu Kush via multiple hairpin turns and tunnels into Pakistan. This is one of the most strategic passes in the world, but for invading armies it was no easy passage: defensive forces had the advantage in these treacherous valleys. But now the roads and tunnels facilitate the movement of refugees, drugs, and arms, and they are used by militant separatist forces from both sides of the border. Pakistan's North West Frontier Province, and especially the city of Peshawar, are hotbeds of these activities."

www.conceptcaching.com

Baluchistan

As Figure 8B-2 indicates, Baluchistan is by far the largest of Pakistan's four provinces, accounting for (not including Kashmir) nearly half of the national territory but inhabited only by an estimated 8.5 million people or barely 4 percent of the country's population. For a sense of its terrain, take a look at Figure 8A-1; much of this vast territory is desert, with mostly barren mountains that wrest some moisture from the air only in the northeast. Sheep raising is the leading livelihood here, and wool the primary export. In the far north, Baluchistan abuts the Tribal Areas and Afghanistan. The provincial capital, Quetta, lies in this zone. This province could easily be called the "South West Frontier Province."

However, Baluchistan is not unimportant. Beneath its parched surface lie possibly substantial reserves of oil as well as coal, and already the province produces most of Pakistan's natural gas. The construction of a major port and energy terminal is under way at Gwadar on Baluchistan's southwestern coast. The port will serve Baluchistan but also become a transshipment port for oil and gas from Iran and the Caspian Basin, destined for markets in East Asia. China has dispatched engineers to plan and build this port with a loan to Pakistan.

But Baluchis complain that they were left out of the process. Note that 90 percent of local residents have no energy supply themselves. And there are other persistent problems. Even though there is groundwater in several localities, 8 out of 10 Baluchis do not have access to clean water. For decades, short-lived local rebellions have signaled dissatisfaction with the government, but more recently the Baluchistan Liberation Army (BLA) has instigated a more serious and more durable insurgency. Indeed, three Chinese engineers were killed and over a dozen wounded in a recent terrorist attack at Gwadar.

Pakistan's Prospects

Pakistan is a country of enormous cultural contrasts, where the modern and medieval exist side by side, where you hear residents of one province call those of another (but not themselves) "Pakistanis," and where a sense of nationhood is still elusive among many people whose loyalties to their family, clan, and village are stronger.

Pakistan has experienced so many cycles of progress and failure that confidence is in short supply. And yet there are areas of progress against all odds: a combination of expanded irrigation and Green Revolution farming techniques has allowed Pakistan during the past decade to export some rice (although wheat imports continue), and the country's manufacturing and services sectors have shown substantial growth. Exports include not only cotton-based textiles but also carpets, tapestries, and leather goods. Domestic manufacturing remains limited, but Pakistan has built its own steel mill near Karachi. Meanwhile, the authorities struggle to control Pakistan's growing production and trade in opium and hashish. Neighboring Afghanistan remains the world's dominant source of this illegal commerce, and the spillover effect continues; but Pakistan has many remote corners where poppy fields yield high returns and trade routes are well established. In this as in so many other spheres, Pakistan displays the contradictory symptoms of a state in transition.

This western flank of South Asia is the realm's most critical region, today more so than ever before. Islamic Pakistan's coherence and stability have become crucial at a time when the global struggle against terrorism is entangling its leaders with Western power and priorities. The role of Pakistan's government in this struggle is disputed and resented by some of its own people, who express their distaste by voting for militant Islamic parties, voicing support for the resurgent Taliban movement, and giving refuge to fugitive al-Qaeda operatives in their homes. Militancy and instability are no longer confined to the northwest or the far north: in early 2010, two suicide bombings in Lahore took more than 70 lives; the government blamed al-Qaeda and the Taliban. The future of Pakistan hangs in the balance, and with it the stability of South Asia.

INDIA: GIANT OF THE REALM

If you have been reading the press and watching television over the past few years, you have seen the increasing attention being paid to India—not just in North America but around the world. Examples of news coverage include the alleged "strategic partnership" between the United States and India, the outsourcing of American jobs to India, the rapid rise of a new Indian middle class, and the emergence of large Indian companies that are making their presence felt around the world. Undoubtedly, India is on the move—but will it really become the next economic superpower?

Certainly India has the dimensions to make the world take notice of it. Not only does it occupy three-quarters of the great land triangle of South Asia: India also is poised to overtake China to become the most populous country on Earth before the middle of this century. Already, India is the world's biggest democracy, a federation of 28 States and several additional Territories with a population nearing 1.2 billion.

That India has endured as a unified state is a politico-geographical miracle. The country contains a cultural mosaic of immense ethnic, religious, linguistic, and economic diversity and contrast; it is truly a state of many nations. Upon independence in 1947, India adopted a democratic, secular, federal system of government, giving regions and peoples some autonomy and identity, and allowing others to aspire to such status.

India's pluralist and democratic complexion cannot be fully understood without considering the huge influence of Mahatma Gandhi, the great spiritual and political leader who played a central role in the gaining of independence from the British. More than anyone else, he symbolized tolerance, reason, nonviolence, and perseverance—the basic principles on which the Indian freedom struggle and the new state were founded. Jawaharlal Nehru, India's first prime minister, also exerted enormous influence through his strong, unshakable beliefs in democracy and secular government. Interestingly, these beliefs and principles can be said to be firmly embedded in Hindu culture. After all, Hinduism is in some ways an extremely open, diverse, and introspective religion and way of life. Indeed, in his book *The Argumentative Indian*, Harvard's Nobel laureate economist Amartya Sen emphasizes this very idea—that Indian civilization,

In the spring of 2009, the Pakistani government began a major military offensive against the Taliban in an area known as the Swat Valley, part of the Malakand Division of the North West Frontier Province located to the north of the city of Peshawar. The Taliban had established a virtual insurgent state in this far northern area of the country, and their armed incursions southward had overrun the regional center of Mingora and were threatening the divisional headquarters of Mardan (see Fig. 8A-1). The offensive drove the Taliban back, but at a high cost to the civilian population. By mid-2009, a humanitarian crisis of massive proportions had dislocated as many as 2 million civilians, many of them surviving in refugee camps such as this at Swabi, where 18,000 people depend on help from the Red Cross/Red Crescent (ICRC).
© Jeroen Oerlemans/Panos Pictures

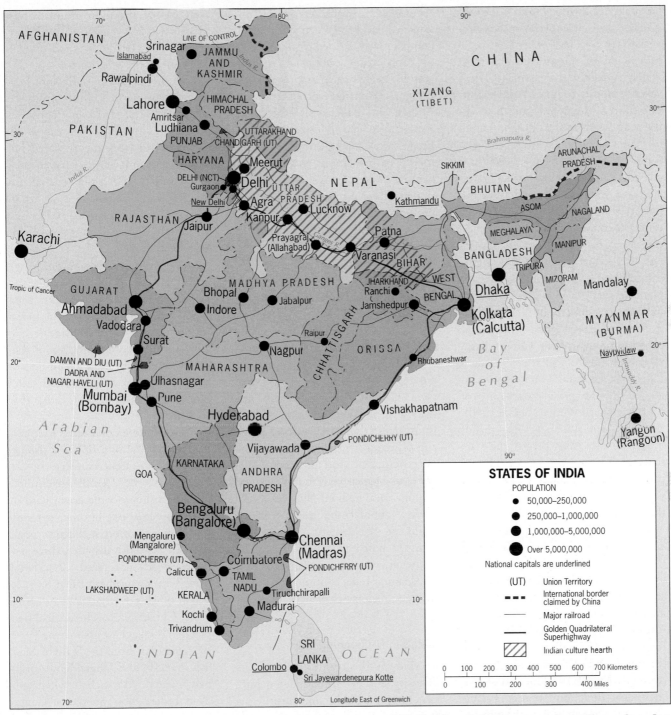

FIGURE 8B-3

© H. J. de Blij, P. O. Muller, and John Wiley & Sons, Inc.

despite its spiritual tendencies, has an inherent penchant for democracy.

Political Geography

A Federation of States and Peoples

The map of India's political geography shows a federation of 28 States, 6 Union Territories (UTs), and 1 National Capital Territory (NCT) (Fig. 8B-3). The federal government retains direct authority over the UTs, all of which are small in both territory and population. The NCT, however, includes Delhi and the adjoining capital, New Delhi, and has more than 17 million inhabitants.

This form of political spatial organization is mainly the product of India's restructuring following independence from Britain. Its State boundaries reflect the broad outlines of the country's cultural mosaic: as far as possible the system recognizes languages, religions, and cultural traditions. Indians speak 14 major and numerous minor languages, and while Hindi is the official language, it is by no means universal (see Fig. 8A-5). At the time of independence, Hindi was the most widely spoken native language, but it was the mother tongue for only about one-third of the population. It is telling that the famous Midnight Speech of Prime Minister Nehru at the moment of independence in 1947 was given in English—Hindi or any other language would have been divisive and politically impossible. English soon became the country's *lingua franca*.

As Figure 8B-3 shows, the territorially largest States lie in the heart of the country and on the great southward-pointing peninsula. Uttar Pradesh (197 million, according to the 2010 population estimate) and Bihar (96 million) constitute much of the Ganges Basin and also form the core area of modern India. Maharashtra (111 million), anchored by the great coastal city of Mumbai (called Bombay before 1996), also has a population larger than that of most countries. West Bengal, the State that adjoins Bangladesh, contains 89 million residents, about 16 million of whom live in its urban focus, Kolkata (known as Calcutta before 2000).

These are staggering numbers, and they do not decline much toward the south. Southern India consists of four States linked by a discrete history and by their distinct Dravidian languages. Facing the Bay of Bengal are Andhra Pradesh (84 million) and Tamil Nadu (67 million), both part of the hinterland of the city of Chennai (formerly Madras) and located on the coast near their joint border. Facing the Arabian Sea are Karnataka (almost 59 million) and Kerala (34 million).

As Figure 8B-3 indicates, India's smaller States lie mainly in the northeast, on the far side of Bangladesh, and in the northwest, toward Jammu and Kashmir. North of Delhi, physical and cultural landscapes change from the flatlands of the Ganges to the hills and mountains of spurs of the Himalayas. In the State of Himachal Pradesh, forests cover the hillslopes and high relief

reduces living space; less than 7 million people live here, many in small, comparatively isolated clusters.

The map becomes even more complicated in the distant northeast, beyond the narrow corridor between Bhutan and Bangladesh. The dominant State here is Asom (Assam), home to just over 30 million, famed for its tea plantations and important because its petroleum and gas production amounts to more than 40 percent of India's total. In the Brahmaputra Valley, Asom resembles the India of the Ganges. But in almost all directions from Asom, things change.

North of Asom, in sparsely populated Arunachal Pradesh (1.3 million), we are in the Himalayan offshoots once again. To the east, in Nagaland (2.2 million), Manipur (2.4 million), and Mizoram (1 million), lie the forested and terraced hillslopes that separate India from Myanmar (Burma). This is an area of numerous ethnic groups (more than a dozen in Nagaland alone) and of frequent rebellion against Delhi's government. And to the south, the States of Meghalaya (2.6 million) and Tripura (3.5 million), hilly and still wooded, border the teeming floodplains of Bangladesh. Here in the country's northeast, where peoples are restive and where population growth is still soaring, India confronts one of its strongest regional challenges.

India's Ever-Changing Map

In view of the country's enormous diversity and its federal democratic structure, it should not surprise us that India's political map is the product of endless compromise. The newly devised framework in 1947, based on the major regional languages, proved to be unsatisfactory to many communities in India. In the first place, many more languages are in use than the 14 that had been officially recognized. Demands for additional States soon arose. As early as 1960, the State of Bombay was divided into two language-based States, Gujarat

and Maharashtra. In 1966, Hindu-dominated Haryana was carved out of Punjab to give Sikhs (and Punjabi speakers) more self-control.

As noted above, India's northeast harbors numerous ethnic groups in a varied, forest-clad topography. The Naga, a cluster of peoples whose domain had been incorporated into Asom (Assam) State, rebelled soon after India's independence. A protracted war brought federal troops into the area; after a truce and lengthy negotiations, Nagaland was proclaimed a State in 1961. That led the way for other politico-geographical changes in India's problematic northeastern wing. In the State of Manipur, separatist groups have challenged Indian authority for more than three decades (few tourists come here, more than 1000 miles from New Delhi and on the doorstep of Myanmar). In 2009 alone, this conflict cost more than 400 lives.

Devolutionary pressures have continued throughout India's existence as an independent country. In certain cases, the federal government and the military come down hard on the insurgents, but in other cases the government is more inclined to negotiate. In 2000, for instance, three new States were recognized: Jharkhand, carved out of southern Bihar State on behalf of 18 poverty-stricken districts there; Chhattisgarh, where tribal peoples had been agitating since the 1930s for separation from the State of Madhya Pradesh; and Uttarakhand (originally named Uttaranchal), which split from India's most populous, Ganges Basin, core-area State of Uttar Pradesh on the basis of its highland character and lifeways (Fig. 8B-3).

As recently as the end of December 2009, India's central government appointed a committee to consider the creation of a new State of Telangana out of the northwestern sector of Andhra Pradesh. Telangana is an inland area that sees itself as marginalized by coastal elites. The proposed State counts about 31 million people and covers approximately one-third of Andhra Pradesh. The issue played out through 2010, was fiercely contested, and remained unresolved at press time.

Over the past several years there have been troubling signs of still another challenge to India's federal government: communist- (avowedly Maoist-) inspired rebellions that seem to be transforming into a coordinated revolutionary campaign. It is known in India as the **Naxalite** movement, named after a town in the State of West Bengal where it was founded in the 1960s.

Mainly active in India's poorest and most disaffected States—such as Bihar, Jharkhand, and even Andhra Pradesh—the Naxalites appeal to the poor and other minorities, whose plight is blamed on India's elites and neoliberal economic policies. They remain active in one-third of India's more than 600 districts (the administrative unit below the State) and maintain a violent presence in about 90 districts. They blow up railroad tracks, attack police stations, and intimidate villagers. In early 2010, they killed 24 police officers in West Bengal, murdered 12 villagers in Bihar, and ambushed and killed 76 Indian soldiers in the eastern State of Chhattisgarh. The Indian government has declared the Naxalites to be the country's greatest internal security threat. Stability is key to India's economic prospects, and this is what the Naxalites try to undermine.

Communal Tensions

The term **1 communal tension** (communal violence) refers to the several different kinds of conflicts that recur among India's highly diverse sociocultural groups. Most commonly, these conflicts have a base in (politicized) religion, but they can also be caste-based.

The Sikhs

The Sikhs (the word means "disciples") adhere to a religion that was created about five centuries ago to unite

Although the Maoist Naxalite rebels denied involvement, Indian authorities blamed their leadership for planning and executing the carefully timed destruction of two trains in May 2010, their most audacious terrorist act through that date. In this dramatic photo, the cars of the rust-colored train, struck by an explosion on the track about 150 kilometers (90 mi) west of Kolkata on the way to Mumbai, lie crushed against those of the oncoming blue train, which could not stop to avert even greater carnage. More than 150 passengers were killed and many more injured. The State of West Bengal, where this tragedy happened, is a hotbed of Naxalite anti-government activity. © AP/Wide World Photos

warring Hindus and Muslims into a single faith. They rejected what they considered to be the negative aspects of Hinduism and Islam, and Sikhism gained millions of followers in the Punjab and adjacent areas. During the colonial period, many Sikhs supported the British administration of India, and by doing so they won the respect and trust of the colonialists, who employed tens of thousands of Sikhs as soldiers and policemen. By 1947, there was a large Sikh middle class in the Punjab. Today, they still exert a strong influence over Indian affairs, far in excess of the less than 2 percent of the population (about 23 million) they constitute.

For a time after India's independence, the Sikhs created India's most serious separatist problem, demanding the formation of an autonomous State they wanted to call *Khalistan*. The government sought to defuse the situation in 1966 by dividing the

original State of Punjab into a Sikh-dominated northwest—which retained the name Punjab—and a Hindu-dominated southeast (Haryana; see Fig. 8B-3). But the conflict intensified again in the 1970s when Prime Minister Indira Gandhi declared a "national emergency" and imposed strong centralized rule across the federation. Tensions culminated in an attack by federal military forces on Sikh rebels holed up in the Golden Temple in the city of Amritsar, Sikhdom's holiest site. In the wake of this attack, Mrs. Gandhi was assassinated by her own Sikh bodyguards in 1984 and more violence followed. Since the 1980s, both sides have sought to gain better understanding.

The Muslims

When India became a sovereign state and the great population shifts across its borders had ended, the country

© Jan Nijman, Photo by Zach Woodward

"It is a Friday afternoon as we make our way through a main thoroughfare in one of Mumbai's biggest slums. Dharavi 'houses' approximately half a million people on less than 2 square kilometers! About 20 percent are Muslims and the overwhelming majority of all others are dalits. Increasingly, the Muslims live spatially segregated in their own tightly-knit neighborhoods. There are several mosques but there is little space inside them. So the men, taking a break from work nearby, place their mats alongside the road and prepare for prayer outdoors. Having a faith is important to people living and working in Dharavi's filthy and impoverished environs, especially for the Muslim minority."

www.conceptcaching.com

was left with a Muslim minority of about 35 million widely dispersed throughout the country. By 1991, that minority had grown to nearly 99 million, representing 11.7 percent of the total population. The current Muslim population is estimated to be 171 million, about 14.5 percent, and clearly it continues to grow much faster than the Indian population in general. Muslims are an actual majority in Jammu and Kashmir, Asom, and West Bengal, while their largest absolute numbers are in the bigger states of Uttar Pradesh, Bihar, and Maharashtra. India's great geographic advantage is that the Muslim minority is not regionally concentrated, avoiding what otherwise could be a dangerous secession movement.

The position of Muslims in Indian society is not easy to separate from India's relations with Pakistan, from the issue of Kashmir, and from acts of Islamic terrorism that have haunted India in recent years (e.g., attacks on the Parliament buildings in 2001 and on Mumbai's upscale hotels in 2008). At present, India

holds the unenviable distinction of being the most frequently targeted country in the world for terrorist attacks. As a result, relations between Muslims and Hindus are worsening. Concern over linkages between foreign (read Pakistani) terrorists and local Muslim subversives is changing Hindu attitudes and tactics, a topic elaborated in the next section.

The most serious threat to Muslim integration into Indian society arises from their comparatively low level of education and poor economic standing. Official statistics show that less than 4 percent of Muslims nationwide have completed secondary school. And according to one recent government report, Muslims have now become as poor as the members of India's lowest-ranked Hindu castes. Even as small cadres of Muslim university graduates have become successful professionals and have increased their representation among high-profile politicians, film stars, and sports icons, the vast majority appear to be stuck near the bottom of the social ladder.

Hindutva

Hindu extremism or fundamentalism may seem like a contradiction in terms, but a movement has come to the fore in recent decades that seeks to remake India as a society in which Hindu principles prevail. **2** *Hindutva*, or Hinduness, is variously expressed as Hindu nationalism, Hindu heritage, or Hindu patriotism. It has been the guiding agenda for the Bharatiya Janata Party (BJP), a powerful force in national politics and such bigger States as Maharashtra, Gujarat, and Madhya Pradesh. *Hindutva* fanatics want to impose a Hindu curriculum on schools, change the flexible family law in ways that would make it unacceptable to Muslims, inhibit the activities of non-Hindu religious proselytizers, and forge an India in which non-Hindus are essentially outsiders.

This naturally worries Muslims as well as other minorities, but it also concerns those who understand that India's secularism—its separation of religion and state—is indispensable to the survival of democracy. Moderate Hindus and non-Hindus in India oppose such notions, which are as divisive as any India has faced. It is easy to see how Hindu hardliners and Islamic fundamentalists can fuel each other's agendas and maintain a vicious cycle of conflict.

Recent State and national elections have seen a decline in the fortunes of the BJP and, more importantly, an internal party struggle between moderates and *hindutva* hardliners in which the former have seemed to prevail. Overall, Indian voters have not rushed to embrace this radicalization of Hinduism, a confirmation of the continuing robustness and vitality of India's democracy.

The Persistence of Caste

Hinduism's benign and admirable properties combine with a system of social stratification that is generally derided in the West (and by a large number of Indians as well). Under Hindu dogma, castes are fixed layers in society whose ranks are based on ancestries, family ties, and occupations. The **3 caste system** has its origins in the early social divisions into priests and warriors, merchants and farmers, and it is also thought to have had a racial basis (the Sanskrit term for caste, *varna*, means color). More specifically, caste became associated with specific professions and, over the centuries, its complexity grew until India had thousands of castes, some of them with a few hundred members and others containing millions.

Hindus believe in reincarnation, and a person is thought to be born into a particular caste based on his or her actions in a previous existence. Hence it would not be appropriate to counter such ordained caste assignments by permitting movement (or even contact) from a lower caste to a higher one. Persons of a particular caste could perform only certain jobs, wear only certain clothes, and worship only in prescribed ways at par-

ticular places. They or their children could not eat, play, or even walk with people of a higher social status.

The *untouchables* occupying the lowest tier of all were the most debased, wretched members of this rigidly structured social system. Indeed, the term "untouchable" has such negative connotations that it has been replaced several times. Mahatma Gandhi, a powerful critic of the system, introduced the term *harijans*, meaning children of God. But more recently this label was also thought to have a condescending connotation, and it gave way to the term ***dalits*** (the oppressed), indicating a greater sense of awareness and assertiveness among them.

In the isolated villages deep in the countryside, dalits still suffer from severe discrimination and harsh treatment by higher castes. Children often are made to sit on the floor of their classroom (if they go to school at all); dalits are not allowed to draw water from the village well because they might pollute it; they must take off their shoes, if they own any, when they pass higher-caste houses; and they cannot take jobs in professions other than the one they were born into (sweeping, cleaning, and the like).

Today, dalits are estimated to constitute more than 15 percent of all Hindus while Brahmins, the highest caste, account for a comparable share. The rest of the population occupies a wide range of in-between castes. The Indian government officially abolished castes at independence, but the system has proven hard to dismantle. Many dalits have chosen to convert to other religions, most notably Buddhism. It is estimated that nearly 90 percent of Buddhists in India were originally dalits, and it appears that even after converting they do not lose this stigma.

Successive Indian governments have now introduced an elaborate system of affirmative action on behalf of *Scheduled Castes* (the official government label for dalits). This effort has

had more effect in urban than in rural areas of India. In any case, dalits now have reserved for them places in the schools, a fixed percentage of State and federal government jobs, and a quota of seats in national and State legislatures. These jobs are often highly desirable because they tend to be white-collar and much better paid than those generally available to dalits.

Because of the great number of minorities in India, affirmative action schemes have become increasingly complex over time. In addition to the existing quota of 15 percent of federal jobs for dalits, the central government proposed in 2010 to reserve another 10 percent of public-sector jobs for Muslims and an additional 5 percent for other religious minorities such as Sikhs. In another proposal that is presently being debated, one-third of all Parliament seats would go to women. This was protested by some Muslim organizations that feared there would not be enough eligible Muslim women, and they subsequently demanded a sub-quota for them. In the world's biggest and most diverse democracy, politics can be a highly complicated matter!

Just how far India's political pendulum can swing was shown in provincial legislative elections in Uttar Pradesh State in 2007, where a dalit party won an absolute majority and where a woman named Mayawati Kumari was the first dalit to become a chief minister of one of India's major States. It was a stunning victory that made headlines throughout the country and shook up the political establishment, revealing the growing power of the lowest castes and marking a turning point for India's representative government. But most lower-caste Indians are still faced with very limited opportunities, widespread discrimination, and abject poverty. In traditional India, caste provided stability and continuity; in contemporary India, it constitutes an often painful and difficult legacy.

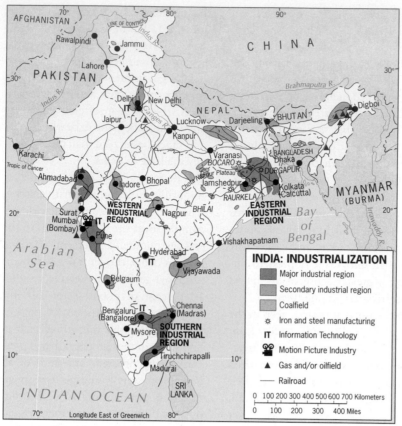

FIGURE 8B-4   © H. J. de Blij, P. O. Muller, and John Wiley & Sons, Inc.

Economic Geography

East and West

The most commonly cited, and most clearly evident, regional division of India is between north and south. The north is India's heartland, the south its Dravidian appendage; the north speaks Hindi as its *lingua franca*, the south prefers English over Hindi; the north is bustling and testy, the south seems slower and less agitated.

But there is another, as yet less obvious, but potentially more significant divide across India. In Figure 8B-4, draw a line from Lucknow, on the Ganges River, south to Chennai, near the northern tip of Tamil Nadu State (cf. Fig. 8B-1). To the west of this line, India is showing signs of economic progress, the kind of productive activity that has brought Pacific Rim countries such as Thailand and Indonesia a new life. To the east, India has more in common with the less-promising countries that also face the Bay of Bengal:

Bangladesh and Myanmar (Burma). The manufacturing map may seem to suggest that much of India's industrial strength lies near Kolkata, but the heavy industries built here by the state in the 1950s are now outdated, uncompetitive, and in steady decline. The State of Bihar represents the stagnation that afflicts much of India east of our line: by several measures it ranks among the poorest of the 28 States.

Compare this to western India. The State of Maharashtra, the hinterland of Mumbai, leads India in many categories, and Mumbai leads Maharashtra. Many smaller, private industries have emerged here, manufacturing goods ranging from umbrellas to satellite dishes and from toys to textiles. Across the Arabian Sea lie the oil-rich economies of the Persian Gulf. Hundreds of thousands of workers from western India have found jobs there, sending money back to families from Punjab to Kerala. More importantly, many have used their foreign incomes

to establish service industries back home. Outward-looking western India, in sharp contrast to the inward-looking east, has begun to establish other ties to the outside world. Satellite links have propelled Bengaluru (formerly Bangalore) to become the center of a growing software-producing complex that reaches world markets. Maharashtra's economic success also has spilled over into Gujarat, its northern neighbor, and even landlocked Rajasthan (the next State to the north beyond Gujarat) is experiencing the beginnings of what, by Indian standards, is a boom.

Life is in the Harvest

Agriculture provides more jobs in India than any other economic sector, and India's fortunes (and misfortunes) remain strongly tied to farming. More than two-thirds of the population still lives on (and from) the land, spread in and around the country's 600,000 villages. There, traditional farming methods persist, yields per hectare remain among the world's lowest, and hunger and malnutrition still afflict millions even as grain surpluses accrue. The relatively few areas of modernization, as in the wheatlands of the Punjab, are islands in a sea of stagnation. Land reform has essentially failed; roughly one-quarter of India's entire cultivated area, including much of the best land, is still owned by less than 5 percent of the country's landholders, who have much political influence and obstruct redistribution. Perhaps half of all rural families own less than 1 hectare (2.5 acres) or no land at all; and an estimated 175 million live and work as tenants, always uncertain of their fate.

As everywhere, agriculture depends strongly on physiography and climate. In India, the monsoon is absolutely critical, its abundance and timing a reliable predictor of the harvest. Figure 8B-5 maps India's agricultural geography and reflects the rainfall patterns and monsoonal cycle depicted in Figure 8A-3. Rice dominates along the

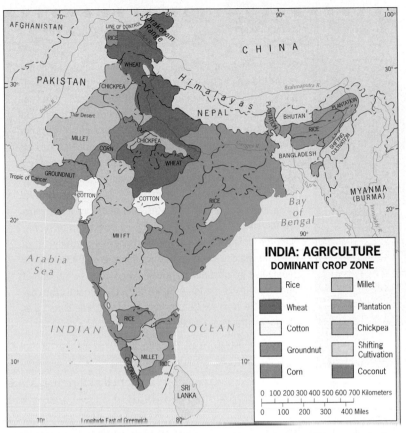

FIGURE 8B-5   © H. J. DE BLIJ, P. O. MULLER, AND JOHN WILEY & SONS, INC.

Arabian Sea-facing southwestern coast (on the rainy side of the Western Ghats) and in the monsoon-drenched peninsular northeast; where drier conditions develop, wheat and other grains prevail. Even though the country has a good variety of crops, yields are far too low as a result of inefficient land ownership and small farmers' lack of access to inputs such as fertilizer, irrigation equipment, and machinery.

If there is any to sell, getting produce to market is yet another struggle for millions of farmers. In 2010, almost half of India's 600,000 villages could not be accessed by truck or car, and in this era of modern transportation animal-drawn carts still far outnumber motor vehicles nationwide.

Manufacturing and Information Technology

India's industrial sectors have significantly improved over the past two decades, but the pace of change is still too slow to meet the country's needs. The map of manufacturing geography (Fig. 8B-4) is a legacy of colonial times, with coastal Mumbai, Kolkata, and Chennai anchoring major industrial zones, and textiles—the entry-level industry of disadvantaged countries—dominating the manufacturing scene. Other industries, such as steel, machinery, and building materials, have become much stronger, and some of the companies involved have even become major global players. For example, Tata Steel and Mittal Steel (now officially called ArcelorMittal), rank among the half-dozen biggest steel producers in the world.

These changes were made possible through major policy changes in the early 1990s, when India liberalized its economy after four decades of stifling overregulation and inefficiency. Along with many other developing countries, India went down the path of neoliberalism: trade barriers came down, state companies were privatized, and foreign investment was encouraged. Although it is not yet clear what the impact will be on India's massive working class, there is no doubt that many companies became far more competitive and profitable. This, in turn, increased the government's tax revenues, allowing it to invest in, among other things, infrastructural projects. Since 2000, India's economy has grown faster than ever before. Much of this growth is related to the information technology sector.

India's information technology (IT) industries, clustered in and around Bengaluru (Bangalore), Hyderabad, Mumbai, and New Delhi, draw much international attention, in part because of the large-scale outsourcing of U.S. jobs to India. Leading Indian software companies such as Infosys, Wipro, and Satyam are now household names all over urban India. The growth of software and IT services has been spectacular and now accounts for about 8 percent of GDP and no less than one-quarter of merchandise exports. But this sector employs only about 2 percent of the workforce, with at least ten times as many people holding government jobs!

What India needs far more, in terms of employment, are manufacturing industries competitively selling goods on world markets, putting tens of millions to work, and transforming the economy. Clearly, India is very different from China. The two may have comparably-sized populations, but the Chinese economy employs 12 times as many workers in manufacturing and it has a much larger urban middle class. India, on the other hand, has shown spectacular growth but too much of it is confined to IT, and too much benefits only a small, highly educated, urban elite.

Urbanization

India is famous for its great and teeming cities, but the country is not yet an urbanized society. Only 28 percent of the population currently lives in cities and towns—compared to an average of about 80 percent across the developed

© Jan Nijman

"With the help of a real estate agent, I toured some of the new residential developments in India's biggest city, Mumbai. The new urban middle class wants better and more spacious housing, but given the lack of space in the city it has to be a high-rise far away from the center. From a new construction site I looked out onto the northeastern edge of the city, marked by an already completed condominium complex. These homes are of a much better quality than the average dwelling in Mumbai, containing most of the luxuries we take for granted in the United States. Two-bedroom apartments sell for about U.S.$75,000 and up—a huge sum of money even for India's upwardly mobile classes, and totally out of reach for the city's masses. Half of Mumbai's 20-plus million people live in substandard housing, and many in the kind of slums you see in the foreground, nestled up against the coveted middle-class residences. As the new middle class takes off, the contrasts are getting sharper."

www.conceptcaching.com

world. But while rural-urban migration has long stabilized at a low level in the West, the rate of urbanization in India is much higher. People by the hundreds of thousands are arriving in the already overcrowded cities, swelling urban India by about 3 percent annually—twice as fast as the overall population growth. Not only do the cities attract as they do everywhere, but many villagers are also driven off the land by desperate conditions in the countryside. As villagers manage to establish themselves in Mumbai or Kolkata or Chennai, they help their relatives and friends to join them in squatter settlements that often are populated by newcomers from the same locality, bringing their language and customs with them and cushioning the stress of the move.

As a result, India's cities display staggering social contrasts. Squatter shacks without any amenities crowd cheek-by-jowl against the walls of modern apartment high-rises and condominiums. Hundreds of thousands of homeless roam the streets

and sleep in parks, under bridges, and even on sidewalks. As crowding intensifies, social stresses multiply.

Figure 8B-3 displays the distribution of major urban centers in India. Except for Delhi–New Delhi, the largest cities have coastal locations: Kolkata dominates the east, Mumbai the west, and Chennai the south. The overriding influence of these coastal cities is a colonial legacy. But urbanization also has expanded in the interior, notably in the core area. In 2010, even though less than 30 percent of the population was urban, India had 43 metropolitan areas with populations exceeding 1 million.

When you arrive in any Indian city, you are struck by the number of small shops everywhere—tiny businesses wedged into every available space in virtually every nonpublic building along every street. Even the upper, walk-up floors are occupied by shops, their advertising signs suspended from windows and balconies. According to a study by the Indian government's Department of Consumer Affairs, India has the highest

density of retail outlets of any country in the world: a total of 15 million shops (compared to well below 1 million in the United States, where the marketplace is 13 times richer). After farming, the retail sector is India's largest provider of jobs.

What keeps all these small stores in business? Most of them earn very little and can afford to stay open only because they are part of what economic geographers call the **4** informal sector—they are essentially unregistered, pay no rent and probably no taxes, use family labor, have been handed down through generations, and keep going because India's economic geography, bound by longstanding protective government regulation, has been very slow to change. When you are in India you may wonder how so many shops can survive, but the answer is in the throngs of people on the sidewalks (spilling over into the clogged traffic in the streets). Not many of these people are wealthy, but all of them need basic goods and some can afford small luxuries, and so the shops tend to be busy all day.

This bustling retail scene notwithstanding, the newest statistics confirm that change is under way, especially in the cities. India's middle class, now estimated to include some 300 million people (equal to the entire U.S. population), is expanding rapidly, and something very new is beginning to appear in the urban landscape: shopping malls. As recently as 2000, this vast country containing more than one-sixth of humankind did not have a single shopping center! But by 2005 the 100th had opened its doors, and by late 2009 more than 500 were in operation. You will see American fast-food restaurants among the establishments in these malls as well as the brand names of numerous other foreign companies, proving that globalization has already breached India's walls. Nevertheless, if 300 million middle-class Indians can afford to visit a mall, more than 700 million Indians cannot, and the gap is widening by the day. Can India's economic transformation be accomplished without severe social dislocation?

Infrastructural Challenges

Now, as the second decade of the twenty-first century opens, India is constructing a nationwide four-lane superhighway that is going to link the four anchors of the urban system (Delhi, Mumbai, Chennai, and Kolkata) and in the process connect 15 other major cities along this national route called the *Golden Quadrilateral* (Fig. 8B-3). The impacts of this project are multiple: it is expanding urban hinterlands, commuters are using it to travel farther to work than ever before, many once-remote villages now have a link to markets, and it is accelerating the rural-to-urban migration flow that will transform India in this century.

Improving India's infrastructure, however, is not enough to overcome all serious impediments to the country's economic advancement. Take another look at Figure 8B-3 and note that the Golden Quadrilateral crosses numerous State boundaries. In the United States, we are used to seeing thousands of trucks on the interstate highways, crossing from one State into another without slowing down; there are truck-weighing stations along these expressways, but in the newest ones all the truck has to do is slow down enough for electronic surveillance to record its passing.

In India, truck drivers face a very different experience. It can take as long as nine days, including more than 30 hours waiting at State-border checkpoints and tollbooths, for a loaded truck to travel from Kolkata to Mumbai via Chennai, or less than the distance from Los Angeles to New Orleans on Interstate-10. Drivers are subject to daunting piles of paperwork and repeated demands for bribes, and when mechanical problems occur it may take days to get a truck back on the road. It will take more than expanding India's highway network to improve the circulation the country so desperately needs.

The Energy Problem

If you have a friend in (or from) India, you know someone who is familiar with power outages. They are a way of life in India, where electricity demand routinely exceeds the available supply, where governments cannot bring themselves to require customers to pay for the actual cost of the power they consume, and where power grids, generating equipment, and other associated infrastructure are in bad shape. On top of all this, keep in mind that hundreds of millions of villagers still have no electrical supply at all.

And yet electrical power is key to India's modernization. Already, some foreign companies doing business in India are importing their own generators, and others are discouraged from investing in factories and other facilities because the power supply is so unreliable. Most of India's electricity is generated in thermal plants burning coal, oil, or natural gas; about 25 percent comes from hydroelectric sources, and about 3 percent from nuclear plants. The problems are many: India has substantial coal deposits, but the railroads cannot handle the transport to power plants. So India must import coal, but the ports do not have the required capacity. And India possesses only limited oil and natural gas reserves. Add to this an increasingly inadequate national power-supply grid in a country with a still-exploding population and even faster-growing demand, and you have trouble.

A key remedy, of course, lies in increased oil and gas imports, but here India runs into geopolitical problems. While the American government was eager to give India leeway in the nuclear arena, the United States made it clear that it does not like India's plan to buy Iranian natural gas via a pipeline across Pakistan. India's other options lie in interior Asia, but those sources are more distant and pipeline construction would involve further diplomatic complications. Once again, the other alternative—importing oil and gas via tankers and ports—is constrained by inadequate infrastructure. Energy will remain a problem for some time to come.

India's Prospects

So what lies in store for India? Some economic geographers suggest that India might leapfrog China and move quickly from an "underdeveloped" to a "postindustrial" information-based economy. Certainly India has the requisite intellectual clout. But to reach India's hundreds of millions of potential wage-earners, India also needs a vigorous expansion of its secondary industries, those that make goods (other than textiles) and sell them at home and abroad. Here's how it may happen: when an economy churns along the way China's has over the past three decades, labor and production costs tend to rise. That causes manufacturers to look for places where cheaper labor will reduce such costs. India, with its long history of local manufacturing, huge domestic market, and vast reservoir of capable labor, would then take its turn on the world stage.

As always in India, cultural issues also intrude when it comes to economic change. In 2008, a division of one of India's largest companies, Tata Motors, introduced a tiny automobile (the Nano) that would sell for little more than U.S. $2000. Experts internationally praised this vehicle as a marvel of Indian ingenuity and engineering, and when Tata proposed to build the factory to produce it in one of India's poorest States, West Bengal, it planned to boost the job market where this seemed to be most needed. But Tata's planners failed to reckon with the farmers in West Bengal and their political party. First, urged on by political leaders, these farmers refused to negotiate over the price of their land. Next, the company faced rallies and blockades as it tried to persuade locals that the thousands of prospective jobs would more than compensate for the approximately 400 hectares (1000 acres) of land it sought. Then, suddenly, the company withdrew its plans and began considering offers from other States. But the implications were clear: if the Tata Group, an Indian corporation, could not build a factory in an Indian State, would foreign investors looking for cheap labor want to risk similar problems?

Nonetheless, India's prospects are brightening, if not yet in the dramatic terms used in popular-media hype. A middle class of more than 300 million demands goods that range from cell phones (in recent years, 2.5 million new subscribers were signing up every month) to motor bikes (10,000 per day were being sold). Even if Tata's mini-car was not going to be built in West Bengal State, Chennai was already becoming an Indian automobile center: Korean and German cars are coming off assembly lines not only to be marketed in India but also to be sent to foreign consumers.

India's economy today ranks as the world's sixth-largest, and many

economists suggest that, whatever happens, India by 2020 may be in third place. Over the past several years, the economy has grown by an average of 7 percent—less than China's, but far ahead of growth rates in most other parts of the world. As we will see in Chapters 9A and 9B, China's dramatic growth resulted from decisions at the top, a transformation planned and implemented in controlled detail. In India the economy is growing from the bottom up, with all the traditional chaos that makes India a country like no other. As with China's provinces, some of India's States will advance ahead of others, and India's already incredible socioeconomic contrasts will intensify. But over time, India could achieve what China hitherto has not: an economic and cultural geography of consensus.

≡ BANGLADESH: CHALLENGES OLD AND NEW

On the map of South Asia, Bangladesh looks like another State of India: the country occupies the area of the **5 double delta** of India's great Ganges and Brahmaputra rivers, and India almost completely surrounds it on its landward side (Fig. 8B-6). But Bangladesh is an independent country, born in 1971 after its brief war for independence against Pakistan, with a territory about the size of Wisconsin. Today it remains one of the poorest and least developed countries on Earth, with a population of 153 million that is growing at an annual rate of 1.7 percent.

A Vulnerable Territory

Not only is Bangladesh a poor country; it also is highly susceptible to damage from **6 natural hazards**. During the twentieth century, eight of the ten deadliest natural disasters in the entire world struck this single country. The most recent of these was

in 1991 when a cyclone (as hurricanes are called in this part of the world) killed more than 150,000 people. Smaller cyclones occur every year multiple times and almost routinely kill anywhere between dozens and thousands of people.

The reasons for Bangladesh's vulnerability can be deduced from Figures 8B-6 and 8A-1. Southern Bangladesh consists of the deltaic plain of the combined Ganges–Brahmaputra river system, integrating fertile alluvial (river-deposited silt) soils that attract farmers with the low elevations that endanger them when water rises. The shape of the Bay of Bengal forms a funnel that sends cyclones and their storm surges of wind-whipped water barreling into the delta coast. Without money to build seawalls, floodgates, elevated shelters in sufficient numbers, or adequate escape routes, hundreds of thousands of people are at continuous risk, with deadly consequences.

Limits to Opportunity

Bangladesh remains a nation of subsistence farmers. Only about one-fourth of the population lives in urban settlements, and more than half of the workforce is engaged in agriculture. Dhaka, the megacity capital, and the cities of Chittagong, Rangpur, Khulna, and Rajshahi are the only urban centers of consequence. Moreover, Bangladesh has one of the highest **7 physiologic densities** in the world (1679 people per square kilometer/4349 per square mile), and only higher-yielding varieties of rice and the introduction of wheat in the crop rotation (where climate permits) have improved diets and food security. But diets remain poorly balanced and, overall, the country's nutrition level is unsatisfactory.

Bangladesh's birth rate has come down substantially in the past few decades but is still too high. At the same time, per-capita incomes have risen notably but remain well below

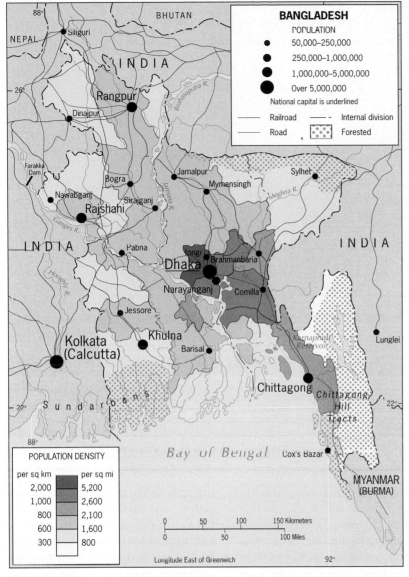

FIGURE 8B-6   © H. J. DE BLIJ, P. O. MULLER, AND JOHN WILEY & SONS, INC.

THE MOUNTAINOUS NORTH

As Figure 8B-1 shows, a tier of land-locked countries and territories lies across the mountainous zone that walls India off from China. One of them, Kashmir, is in a state of near-war. Another, Sikkim, was absorbed by India in 1975 and made into one of its federal States. But Nepal and Bhutan retain their independence.

Nepal

Nepal, northeast of India's Hindu core, has a population of 28 million and is the size of Illinois. It has three geographic zones (Fig. 8B-7): a southern, subtropical, fertile lowland called the Terai; a central belt of Himalayan foothills with swiftly flowing streams and deep valleys; and the spectacular high Himalayas themselves (topped by Mount Everest) in the north. The capital, Kathmandu, lies in the east-central part of the country in an open valley of the central hill zone.

Nepal, birthplace of Buddha, is materially poor but culturally rich. The Nepalese are a people of many origins, including India, Tibet, and interior Asia. About 85 percent are Hindu, and Hinduism is the country's official religion; but Nepal's Hinduism is a unique blend of Hindu and Buddhist ideals. Thousands of temples and pagodas ranging from the simple to the ornate grace the cultural landscape, especially in the valley of Kathmandu, the country's core area. Although over a dozen languages are spoken, 90 percent of the people also speak Nepali, a language related to Indian Hindi.

Nepal is a troubled country suffering from severe underdevelopment, with the lowest GNI in the realm. It is the only country in the realm with a declining income per capita over the past two decades. It also faces strong centrifugal social and political forces. Environmental degradation, crowded farmlands and soil erosion, as well as

those in India or Pakistan. The troubled textile industry provides most of the country's foreign revenues, along with remittances from Bangladeshis working abroad, but the once-thriving jute industry continues its decline. The discovery of a natural gas reserve is now the subject of a national debate: home consumption or money-making export?

Bangladesh is a dominantly Muslim society, though not one dominated by revivalists (for example, 30 seats in the national legislature are reserved for women). But Bangladesh's politics are chaotic, patronage-driven, corrupt, and seemingly beyond redemp-

tion. Foreign aid, essential to the country's survival, pours in, but the country remains mired among the world's lowest-income societies. Its relations with neighboring India have at times been strained over water resources (India controls the Ganges, which is Bangladesh's lifeline), cross-border migration (one-sixth of the population is Hindu), and transit between parts of India across Bangladesh's north (refer to Fig. 8B-4 to see the reason). Bangladesh faces an uphill battle to improve people's livelihoods. For most Bangladeshis, survival is the leading industry and all else is luxury.

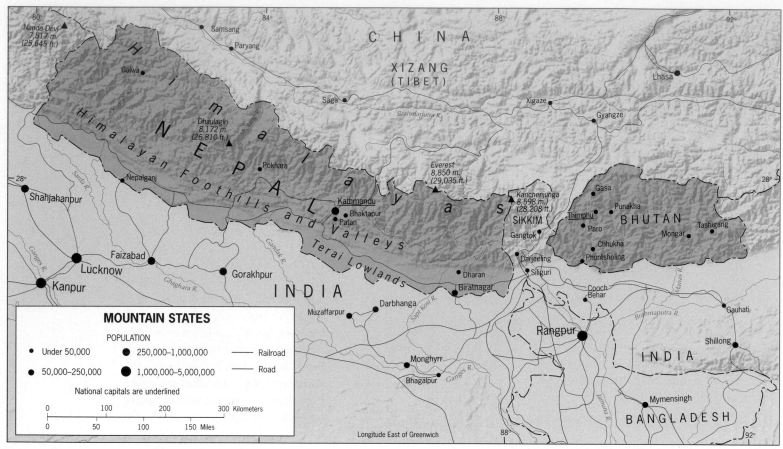

FIGURE 8B-7

© H. J. de Blij, P. O. Muller, and John Wiley & Sons, Inc.

deforestation scar the countryside. The Himalayan peaks form a world-renowned tourist attraction, but tourist spending in Nepal, always relatively modest, has been cut back because of the recurrent disorder associated with a Maoist-communist insurrection and the collapse of the country's monarchy.

Nepal's cultural and political geographies have long been in turmoil. Strong regionalism divides the country both north-south and east-west. The southern Terai with its tropical lowlands is a world apart from the Himalayan foothills in the central interior, and the peoples of the west have origins and traditions different from those in the east. An absolute (Hindu) monarchy held Nepal together until 1990, when antigovernment demonstrations and resulting casualties persuaded the king to lift the ban on political parties. But chaos soon resumed, the king was assassi-

nated, and by 2002 Maoist rebels controlled nearly half of the country, turning Nepal into a *failed state*. The ruling king (the brother of the former king) was effectively forced from office in 2006, the insurgents were brought into the political system, and in the following year a new constitution formalized the abolition of the monarchy and the reinvention of Nepal as a secular federal republic. Peaceful elections in 2008 confirmed the capacity of Nepal's new system to accommodate the still-powerful centrifugal forces in the country, but it remains to be seen if the new government can deliver on the demands of the country's highly diverse communities and constituencies.

Bhutan

Mountainous Bhutan, with a population of 740,000, lies wedged between India and China's Tibet (Fig. 8B-7),

the only other buffer between Asia's giants. In landlocked, fortress-like Bhutan, time seems to have stood still. Bhutan long was a constitutional monarchy ruled by a king whose absolute power was unquestioned by his subjects. But in 2007 the reigning king, who had recently succeeded his father, and perhaps with an eye on what had happened to the monarchy in nearby Nepal, decided to order his subjects to vote for a political party in a newly created democracy. Thus in 2008 Bhutan went from an absolute monarchy to a multiparty democracy on the orders of its monarch, and Thimphu, the capital, became the seat of a new National Assembly.

In the mountainous countryside, the symbols of Buddhism, the state religion, dominate the cultural landscape, and the government, in its development policies, stresses the importance of spiritual fulfillment alongside the satisfaction of material needs. Still, there is social tension through the presence of a Nepalese (Hindu) minority; and a Bhutanese refugee population remains housed in camps in eastern Nepal. Add to this a still-unresolved boundary issue with China and the possibility of rising material expectations, and newly democratic Bhutan faces some important challenges.

Forestry, hydroelectric power, and tourism all have much potential here, and Bhutan has considerable mineral resources. But isolation and inaccessibility preserve traditional ways of life in this mountainous ***buffer state***.

THE SOUTHERN ISLANDS

As Figure 8B-1 shows, South Asia's subcontinental landmass is flanked by several sets of islands: Sri Lanka just off the southern tip of India, the Maldives in the Indian Ocean to the southwest, and the Andaman Islands (which belong to India) marking the eastern edge of the Bay of Bengal.

The Maldives

The Maldives consists of more than a thousand tiny islands whose combined area is just 300 square kilometers (115 sq mi) and whose highest elevation is less than 2 meters (6 ft) above sea level. Its population of just over 300,000 from Dravidian and Sri Lankan sources is now virtually 100 percent Muslim, with one-quarter of it concentrated on the capital island named Maale. The Maldives might be unremarkable, except that, as the table in Appendix B shows, this country has the realm's highest GNI per capita. The locals have translated their palm-studded, beach-fringed islands into a tourist mecca that attracts tens of thousands of mainly European visitors annually. The central question is whether the Maldives' economy can withstand the global downturn that started in 2008, because tourism is a vulnerable industry.

The Maldives' low elevation has been singled out repeatedly in assessments of the future impact of global warming, which would produce rising sea levels. Those risks became sudden reality on December 26, 2004 when an Indian Ocean tsunami generated off Indonesia swept over the islands, killing more than 100 residents as well as tourists and destroying resort facilities along the shore as well as inland.

The Maldives may continue as the richest country per capita in South Asia, but there are concerns beyond those of low relief and global warming. A democratic election in 2008 was a first, but whether representative government can survive economic hard times is an open question. In 2009, the economy had contracted due to a decrease in tourists and diminished fish exports, and the end of the crisis was not yet in sight in mid-2010.

Sri Lanka: Paradise Lost and Regained?

Sri Lanka (known as Ceylon before 1972), the compact, pear-shaped island located just 35 kilometers (22 mi) across the Palk Strait from southernmost India, became independent from Britain in 1948 (Fig. 8B-8). There were good reasons to create a separate sovereignty for Sri Lanka. This is neither a Hindu nor a Muslim country: the majority of its population of just over 20 million people—some 70 percent—are Buddhists. Furthermore, unlike Pakistan or India, plantations dominate Sri Lanka, and commercial farming still is the mainstay of the agricultural economy.

The great majority of Sri Lanka's population is descended from migrants who came to this island from northwestern India beginning about 2500 years ago. Those migrants introduced the advanced culture of their source area, building towns and irrigation systems and bringing Buddhism. Today, their descendants, known as the Sinhalese, speak a language (Sinhala) that

belongs to the Indo-European language family of northern India.

The Dravidians who lived on the mainland, just across the Palk Strait, came much later. During the nineteenth century the British brought hundreds of thousands of Tamils to labor on their tea plantations, and soon a small minority became a substantial component of Ceylonese society. The Tamils brought their Dravidian tongue to the island and also introduced their Hindu faith. At the time of independence, they constituted more than 15 percent of the population; today they total about 18 percent.

When Ceylon became independent, it was one of the great hopes of the postcolonial world. The country had a sound economy plus a democratic government, and it was renowned for its tropical beauty. Its reputation soared when a massive effort succeeded in eradicating malaria and when family-planning campaigns reduced population growth while the rest of the realm was experiencing a population explosion. Rivers from the cooler, forested interior highlands fed the paddies that provided ample rice; crops from the moist southwest paid the bills, and the capital, Colombo, grew to reflect the optimism that prevailed.

In the midst of this glowing scenario, the seeds of disaster were already being sown. Sri Lanka's Tamil minority soon began proclaiming its sense of exclusion, demanding better treatment from the Sinhalese majority. Although the government recognized Tamil as a "national language" in 1978, sporadic violence marked the Tamil campaign, and in 1983 full-scale civil war broke out. Now many in the Tamil community demanded a separate Tamil state to encompass the north and east of the country (see Fig. 8B-8 inset map), and a rebel army calling itself the Liberation Tigers of Tamil Eelam (LTTE) confronted Sri Lanka's national forces.

The Tamil Tigers first took the Jaffna Peninsula in the far north (Fig. 8B-8), next they extended their control over the Vanni Region just to its

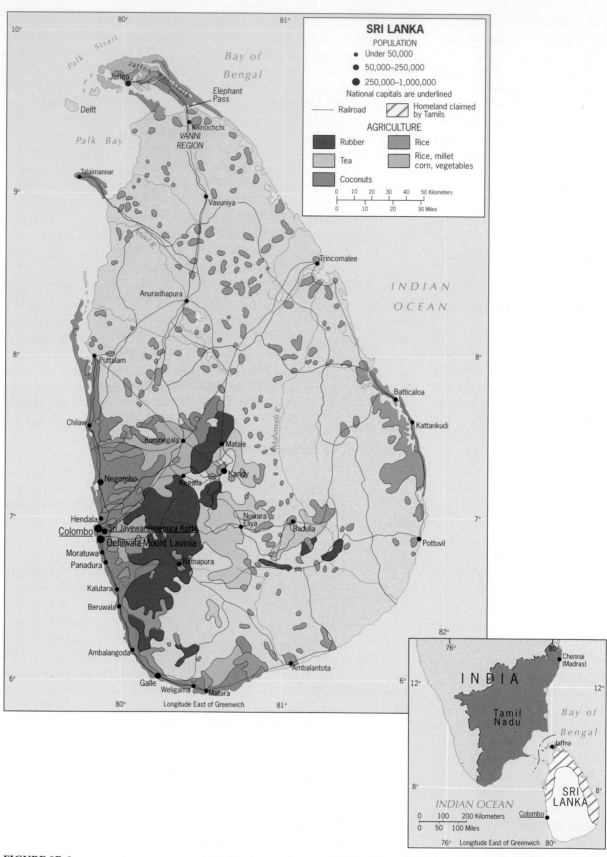

FIGURE 8B-8

© H. J. de Blij, P. O. Muller, and John Wiley & Sons, Inc.

south, and soon they started publishing maps that revealed their ultimate demand: the entire northern and eastern periphery of Sri Lanka would become the secessionist state of *Eelam* (whose most recent rendering is indicated by the striped zone in the inset map). Not only had an **8 insurgent state** (a concept discussed in Chapter 5B) formed in northeastern Sri Lanka: the "Tigers" backed up these demands through terrorist attacks in the capital, Colombo, and unleashed combat in many parts of the state they were seeking to establish.

Eventually the town of Kilinochchi, on the mainland near the entry to the Jaffna Peninsula, became the LTTE "capital," with all the hallmarks of an insurgent state. The Tigers collected taxes, administered banks and issued currency, organized police and court systems, established schools, and even ran hotels and guest houses for visitors. By 2002, the government of Sri Lanka had agreed to a cease-fire, and negotiations leading to the country's partitioning—at least from the LTTE perspective—began.

But in 2007 the Colombo government decided to embark on the counteroffensive that marks the third and decisive stage in the success or failure of an insurgent state. After five years of intermittent fighting, cease-fires, and fruitless negotiations, it concluded that the only way to settle the conflict was through an all-out military campaign. On January 2, 2009, President Mahinda Rajapaksa announced the capture of Kilinochchi and proclaimed that the threat of partition in Sri Lanka had ended.

In the spring of 2009, the final battle of the counteroffensive took place in the Vanni Region amid terrible losses among civilians, who were used as human shields by the doomed LTTE even as government forces impeded their escape. The leader of the insurgency was killed, and the Tamil Tigers' insurgent state disappeared from the map. An international outcry over the tactics of the Sri Lankan military was

This is the tragic end of an insurgent state overpowered by a national counteroffensive. Here in northeastern Sri Lanka, the Tamil Tigers made their last stand before their defeat by the Sri Lankan military in May 2009 (this photo was taken exactly one year later). Thousands of civilians, used as human shields in the final stage of the battle, died or were wounded; as many as 200,000 were displaced by the fighting. In the aftermath of this civil war, the Sinhalese-dominated government must find ways to reintegrate the Tamils and others swept up in the insurgents' campaign, in order to avoid a return to the social conditions that gave rise to the insurgency in the first place. © STR/Reuters/© Corbis

followed by expressions of concern over the future of Sinhalese-Tamil relations in this still-fractured country.

Reconciliation is urgently needed, but even though several million Tamils had never wanted anything to do with an armed insurrection, the Colombo government seemed to focus more on retribution than pacification. More than three decades ago the LTTE insurgency arose in the first place because of Hindu-Tamil anger over the government's Buddhist-Sinhalese cultural dominance and social policies. Therefore, only more even-handed management of national affairs can stave off the resumption of the cycle that transformed South Asia's great success story into a dreadful tragedy whose after-effects will be felt for decades. To the contrary, Rajapaksa's first act following his reelection to a second term in early 2010 was to have his main challenger incarcerated on charges of conspiracy. Divided Sri Lanka still awaits a turn toward peaceful cooperation.

POINTS TO PONDER

- The United States needs Pakistan in the global campaign against terrorism, but Pakistan's government must be careful not to (further) alienate its overwhelmingly Muslim population.

- In 2009, for the first time in India's history, the value of manufacturing output overtook that of farming.

- If India goes ahead with reserving 33 percent of its Parliament seats for women, it will have almost twice as much female representation as the U.S. Congress.

- Nepal, struggling to emerge from failed-state status, is the only country in South Asia in which per-capita incomes have declined in recent years.

The Forbidden City side of the Gate of Heavenly Peace on Beijing's Tiananmen Square. © H. J. de Blij.

IN THIS CHAPTER

The realm's complex political structure

Mainland-offshore connections across the millennia

East Asia's fascinating physical stage

The astonishing speed of China's economic transformation

The ethnic kaleidoscope: Challenges and opportunities

The realm's increasingly stressed resource base

CONCEPTS, IDEAS, AND TERMS

State formation	1
Dynasty	2
Natural resources	3

9A

EAST ASIA: DEFINING THE REALM

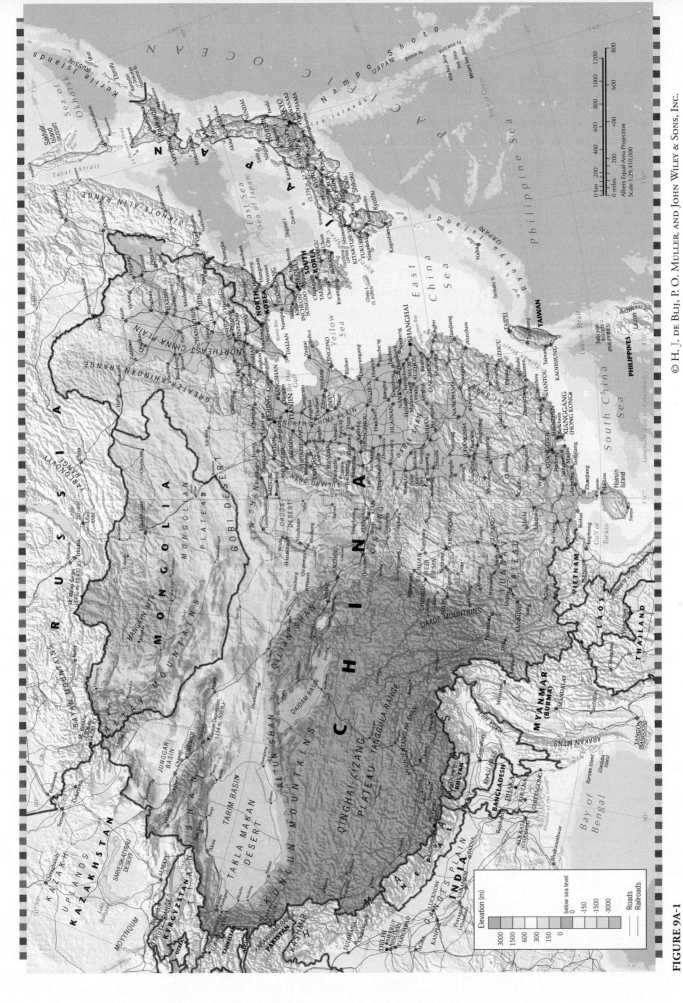

FIGURE 9A-1

© H. J. de Blij, P. O. Muller, and John Wiley & Sons, Inc.

East Asia is a geographic realm like no other. At its heart lies the world's most populous country, the product of what may be the world's oldest continuous civilization. On its Pacific mainland shores an economic transformation is in progress that has no parallel in world history. Its offshore islands witnessed the first use of atomic weapons on civilian populations, the end of the Second World War that had begun in Europe, and the postwar emergence of one of the world's most powerful economies. Few lives in this world were left unaffected, directly or indirectly, by the momentous events that occurred in East Asia over the past two generations. Just look around you. Japanese-designed automobiles, Chinese-made furniture, South Korean motorbikes, Taiwanese computers—from toys to textiles and from hardware to software, East Asian products fill streets and stores, homes and hotels.

It has all happened with astonishing speed. Some of us can remember when plastic toys from Japan and poor-grade textiles from Hong Kong were all you expected from this part of the world. If you wanted a good typewriter or a fine sweater, you looked for German or British trademarks. If you bought a car, it would be American or maybe European. You were no more likely to find anything useful made in China than you were to buy anything from Russia.

But then things changed, and fast. Japan led the way, turning its World War II defeat into postwar economic triumph. By the early 1970s, Japan's economic growth—compared to China's seemingly total stagnation—seemed to justify Japan's recognition as a discrete geographic realm, an economic engine for the world and a strong competitor for America's top-ranking economy. But then Hong Kong, South Korea, and autonomous Taiwan began to show the world what other peoples in East Asia were capable of. When the ruler of communist China, the realm's giant, opened the door to an American president in 1972, a pivotal moment in global history (and geography) had arrived. Before the end of the century, China had taken its place as modern East Asia's core.

THE GEOGRAPHIC PANORAMA

As Figure 9A-1 shows, the East Asian geographic realm forms a roughly triangular wedge between the vast expanses of Russia to the north and the populous countries of South and Southeast Asia to the south, its border often marked by high mountain ranges or remote deserts. The dark brown on the map designates the highest mountains and plateaus, which create a vast arc north of the Himalayas before bending southward and becoming lower (and shaded tan) toward Myanmar, Laos, and Vietnam in Southeast Asia. Here in the southwest, where Tibet is located, mountains and plateaus alike are covered by permanent ice and snow, the ranges crumpled up like the folds of an accordion. Three major rivers, their valleys parallel for hundreds of kilometers, reveal the orientation of this high-relief topography. Northward, note how rapidly the mountains give way to broad, flat deserts whose names appear prominently: the Takla Makan in the far west, the Gobi where China meets Mongolia, the Ordos in the embrace of what looks like a huge meander of a river we will learn more about later, the Huang He (Yellow River).

One basin, lodged between high mountains to its west and lower ones elsewhere, and of special interest, is the Sichuan (Red) Basin. It looks rather small on the map, but is home to more than 100 million people and, we will find, is important for other reasons as well. In this country of mountains and deserts, living space is at a premium. And speaking of living space, the green areas on the map, which have the

MAJOR GEOGRAPHIC QUALITIES

EAST ASIA

1. East Asia is encircled by snowcapped mountains, vast deserts, cold climates, and Pacific waters.

2. East Asia was one of the world's earliest culture hearths, and China is one of the world's oldest continuous civilizations.

3. East Asia is the second most populous geographic realm, but its population remains strongly concentrated in its eastern regions.

4. China, the world's largest state demographically, is the current rendition of an empire that has expanded and contracted, fragmented, and unified many times during its long existence.

5. China today remains a strongly rural society, and its vast river basins feed hundreds of millions in a historic pattern that continues today.

6. China's sparsely peopled western regions are strategically important to the state, but they lie exposed to minority pressures and Islamic influences.

7. Along China's Pacific frontage an economic transformation is taking place, affecting all the coastal provinces and creating an emerging Pacific Rim region.

8. Increasing regional disparities and fast-changing cultural landscapes are straining East Asian societies.

9. Japan, one of the economic giants of the East Asian realm, has a history of colonial expansion and wartime conduct that still affects international relations here.

10. East Asia may witness the rise of the world's next superpower as China's economic and military strength and influence grow—and if China avoids the devolutionary forces that fractured the Soviet Union.

11. The political geography of East Asia contains a number of flashpoints that can generate conflict, including Taiwan, North Korea, and several island groups in the realm's seas.

lowest relief and (often) the most fertile soils, are home to the vast majority of this realm's population. Here the great rivers that come from the melting ice and snow in the interior highlands have been depositing their sediment for eons, and when humans domesticated plants and started to grow crops, this was the place to be. That was thousands of years ago—perhaps as long as 10,000 years—and ever since, this has been the largest cluster of humanity on Earth.

But the East Asian realm is not confined to the mainland of mountains and river basins. You can imagine how its offshore islands were populated: the Korean Peninsula seems to form part of a bridge pointing to the southernmost island of what is today Japan, and from there, it seems likely that the early migrants moved farther north until they reached Hokkaido. In warmer times, they may have ventured even beyond, onto the Kurile Islands. And in the south, Taiwan lies even closer to mainland China than Japan does to Korea, whereas tropical Hainan Island, the realm's southernmost point, is almost—but not quite—connected to the small peninsula that reaches toward it.

In terms of total land area, though, East Asia is mostly mainland—but the islands and their peoples have played huge roles in forging this realm's regional geography. And the waters between mainland and islands (the Taiwan Strait, the South China Sea, the East China Sea, the Yellow Sea, the Korea Strait, and others) also figure prominently in the geographic saga of this realm. Today, the Japanese and the Chinese are arguing over the ownership of small islands in these waters, small islands with large oil reserves and fishing grounds both sides claim. So the map is more complicated than that wedge-shaped triangle in Figure 9A-1 at first suggests.

POLITICAL GEOGRAPHY

It is all too easy to refer to China when you mean East Asia, because China is the dominant country in East Asia, contains more than 85 percent of the realm's population, and has taken an ever-more prominent role on the world stage. But there are five other political entities on East Asia's map: Japan, South Korea, North Korea, Mongolia, and Taiwan. Note that we refer here to *political entities* rather than *states*. In this realm, the distinction is important: Taiwan refers to itself as the Republic of China (ROC), but it is not recognized as a sovereign state by most members of the international community. The communist administration in Beijing, capital of the People's Republic of China (PRC), regards Taiwan as part of China and as a temporarily wayward province. And North Korea is viewed by many members of the international community as a *failed state*, whose (communist) regime has one of the world's most dreadful human rights records; North Korea is not a fully functional member of the United Nations.

Nevertheless, China is now the realm's dominant entity, even if its overall economy in some respects still

On April 14, 2010 a devastating earthquake struck Jiegu Town in Qinghai Province, in an area where Tibetans form the ethnic majority. By the time this photo was taken more than a week later, Chinese authorities were urging Tibetan monks, who had rushed in to help, to leave the area because they were "causing trouble" and "spreading false information." The blue huts you see here were erected as part of the relief effort mounted by the Chinese authorities. © AFP/Getty Images, Inc.

ranks behind Japan's. It is important to keep in mind, as you read this chapter, that parts of what we map today as regional components of China were not part of China in the past, and that other areas, outside China today, are regarded by many Chinese as Chinese property (for example, a large sector of the Indian State of Arunachal Pradesh, a section of Kashmir, and portions of the Russian Far East). As we will see, China's imperial past saw the state expand and shrink and expand again, leaving unfinished business on land as well as at sea. On such issues, Chinese emotions can run very deep.

And for good reason. Six hundred years ago China already was the largest nation on the planet, its history spanning thousands of years, its unmatched fleets exploring lands and peoples in South Asia and East Africa, its technologies unequaled. But even the Chinese could not curb the growing presence of Europe's colonial powers, and eventually the British, French, Russians, and Germans accomplished the unthinkable, taking control over most of the Chinese state. When the Japanese emulated the Europeans and seized their own colonial empire, much of it at China's expense, humiliation was complete—and never forgotten. To this day, the sense that China's borders (including those with Russia and India) were imposed by outsiders is never far from the surface.

ENVIRONMENT AND POPULATION

To understand the complex physical geography of East Asia shown in Figure 9A-1, it is useful to refer back to Figures G-4 and G-5 in the introductory chapter. The high, snowcapped

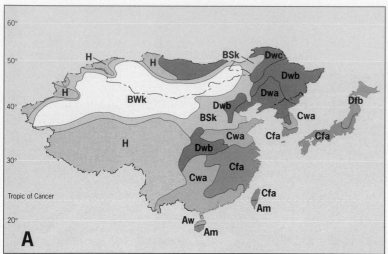

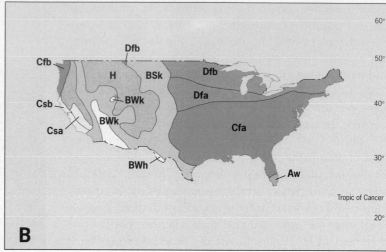

CLIMATES OF EAST ASIA AND THE CONTERMINOUS UNITED STATES
After Köppen-Geiger

A HUMID EQUATORIAL CLIMATE	B DRY CLIMATE	C HUMID TEMPERATE CLIMATE	D HUMID COLD CLIMATE	H HIGHLAND CLIMATE
Am Short dry season	BS Semiarid	Cf No dry season	Df No dry season	H Unclassified highlands
Aw Dry winter	BW Arid	Cw Dry winter	Dw Dry winter	
	h=hot k=cold	Cs Dry summer a=hot summer b=cool summer c=short, cool summer		

FIGURE 9A-2

© H. J. de Blij, P. O. Muller, and John Wiley & Sons, Inc.

mountain ranges of the realm's southwestern interior result from the gigantic collision of the Indian and Eurasian tectonic plates (Fig. 8A-2), pushing the Earth's crust upward and creating not only the mountain ranges of which the Himalaya is the most famous, but also the dome-like Qinghai-Xizang (Tibetan) Plateau. In Figure G-5, note the high incidence of earthquakes associated with this collision, converging on a narrow zone of instability that crosses southwestern China and leads into Southeast Asia. The devastating May 12, 2008 earthquake in Sichuan Province that killed more than 70,000 originated in this danger zone and is only one in an endless series that will continue to take its toll (see photo previous page). From this map it is also obvious that the Pacific Ring of Fire, with its lethal combination of volcanism and earthquakes, endangers Japan far more than it does China—but, as we will note later, even eastern China is not free of earthquake risk.

Looking at the map of East Asian climates (Fig. 9A-2, left map), we should not be surprised that the western and northern sectors of this realm are dominated by conditions that do not favor substantial population clusters. Permanent snow and ice cover much of the area mapped as *H* (highland climes) including Tibet (Xizang) and Qinghai. Northward, the *B* climates (desert and steppe) prevail because this vast expanse lies about as far from maritime influences (and air masses) as you can get in Asia. Mongolia is one of the driest countries in the world, but even here—

and even in frosty Tibet—there are places where people manage to eke out a living. But the map leaves no doubt as to why the great majority of East Asians live in the eastern part of this realm.

When we compare the climates prevailing in the East Asian realm to those familiar to us in North America (Fig. 9A-2, right map), it is immediately obvious that the *C* or humid-temperate climates are more extensive in the United States than in East Asia. Note especially the comparative location of the milder *Cfa* climate, which in the United States extends beyond 40 degrees north latitude, up to New England, but which in China yields to colder *D* climates at a latitude equivalent to Virginia's. Thus the capital, Beijing, has the warm summers but long, bitterly cold winters characteristic of *D* climates. Take a closer look at Figure 9A-2 and it is clear that both the Korean Peninsula and the Japanese islands lie astride this transition from *C* to *D* climates. South Korea is significantly milder and moister than North Korea, and Japan's largest island is temperate in the south but cold in the north. Not surprisingly, the least densely populated island of Japan is northernmost Hokkaido, where the climate is rather like that of northern Wisconsin.

Comparing just the United States and China, note that while *C* or *D* climates prevail over more than half of the United States, these climates predominate over less than one-third of China—even though the United States

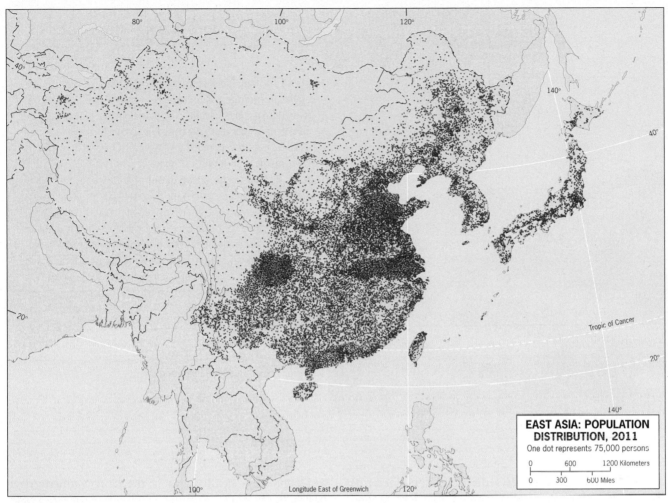

FIGURE 9A-3

© H. J. de Blij, P. O. Muller, and John Wiley & Sons, Inc.

has only 308 million people compared to China's 1.34 billion. That is what makes the population distribution map (Fig. 9A-3) so noteworthy: the overwhelming majority of East Asia's people reside in the easternmost one-third of the realm's territory, creating the most densely settled large population cluster in Earth and depending largely on the limited area colored green in Figure 9A-1.

Several times in earlier chapters of this book we have noted the capacity of humans to live under virtually all environmental circumstances; technological developments enable the survival of year-round communities in Antarctica, on oil platforms at sea, in the driest of deserts, and on the highest of plateaus. But the world population distribution map (Fig. G-8) still reminds us how we got our modern start—through farming and herding. The fertile river basins and coastal plains of East Asia supported ever-larger farming populations, whose descendants still live on that same land: for all its current industrialization and urbanization, more than half the population of China remains rural to this day. Environment, in the form of elevation, relief, water supply, soil fertility, and climate played the crucial

role in the evolution of the population distribution displayed in Figure 9A-3—and will continue to do so for centuries to come.

The Great Rivers

Compared to other world geographic realms, East Asia has relatively few political entities and only a small number of prominent identities. Not counting Taiwan, there are just five states (Europe has more than 40). The Korean Peninsula is the only significant one of its kind. Japan is the only offshore country, and along with its many small islands the only ones that really matter are the four major ones. And for all its physiographic diversity, gigantic China is sometimes called the product of four great river systems and their basins, deltas, and estuaries.

These rivers and their tributaries are visible, but do not stand out as clearly in Figure 9A-1 as they do on the physiographic map (Fig. 9A-4). So important are they as designators of the realm's regional geography that we should acquaint ourselves with them here. Of the four, the two in

FROM THE FIELD NOTES...

© H. J. de Blij

"From the train, traveling from Beijing to Xian, we had a memorable view of the Loess Plateau. Loess is a fine-grained dust formed from rocks pulverized by glacial action, blown away by persistent winds, and deposited in sometimes well-defined locales. Loess covers much of the North China Plain, which is what makes it so fertile, but there it is not as thick as in the Loess Plateau in the middle basin of the Huang He (Yellow River). Here the loess averages 75 meters (250 ft) in thickness and in places reaches as much as 180 meters (600 ft). Because loess has some very distinctive physical properties, it tends to create unusual landscapes. Through a complicated physical process following deposition, loess develops the capacity to stand upright in walls and columns, and resists collapse when it is excavated. As a result the landscape looks terraced: streams cut deep and steep-sided valleys. The Loess Plateau is a physiographic region, but it is a cultural region too. Hundreds of thousands of people have literally dug their homes into vertical faces of the kind shown here, creating cave-like, multi-room dwellings with wooden exterior doors."

www.conceptcaching.com

the middle are in many ways the most important: the **Huang He** (Yellow River) that makes a huge loop around the Ordos Desert and then flows into the Bohai Gulf, and the **Yangzi** River, probably the most famous river in China's historical geography, called the **Chang Jiang** (Long River) upstream.

As the map shows, the Yellow River and its tributaries form the sources of water vital to the historic core area of China, the North China Plain, where the capital, Beijing, is located. The Yangzi River is the major artery of the Lower Chang Basin, and at its mouth lies China's largest city, Shanghai. Both the Huang and the Yangzi rivers originate in the snowy mountains of the Qinghai-Xizang Plateau, a reminder that these remote environments are critical to hundreds of millions of people thousands of kilometers away.

The other two rivers have much shorter courses, but the one in the south, the **Pearl** River, called the **Xi** River upstream, forms an estuary that has become China's (and East Asia's) greatest hub of globalization, an ongoing saga we will discuss in Chapter 9B. This is where you find Hong Kong and, right next to it, the fastest-growing major city in the history of humanity. Finally, the northernmost of China's four major rivers, the **Liao** River, originates near the margin of the Gobi Desert and then forms an elbow across the Northeast China Plain to reach the Bohai Gulf flowing southward. Here the climate is colder, low relief scarcer,

and population smaller than in the more southerly river basins, but mining and industry create opportunities lacking for agriculture. As is obvious, each of East Asia's river-based clusters has its own combination of problems and potentials.

UNFOLDING THE CULTURAL MAP

The story of hominin settlement in East Asia goes back hundreds of thousands of years, perhaps as much as a million years. Many archeological sites in mainland Eurasia have yielded evidence of *Homo erectus*, "Upright Man," including perhaps the most famous of all: Peking Man. In a cave near the Chinese capital of Beijing (then called Peking by Westerners), an archeologist in 1927 found a single tooth he recognized as representing a hominin. Later excavations proved him correct as parts of more than three skeletons were unearthed confirming that Peking Man and his contemporaries, hundreds of thousands of years ago, made stone and bone tools, had a communal culture, controlled the use of fire, and cooked their meat.

But when *Homo sapiens*, modern humans, appeared in East Asia between 60,000 and 40,000 years ago, the hominins could not compete with their larger-brained, better-organized competitors, and they soon vanished from the landscape. But initially the new invaders lived pretty much

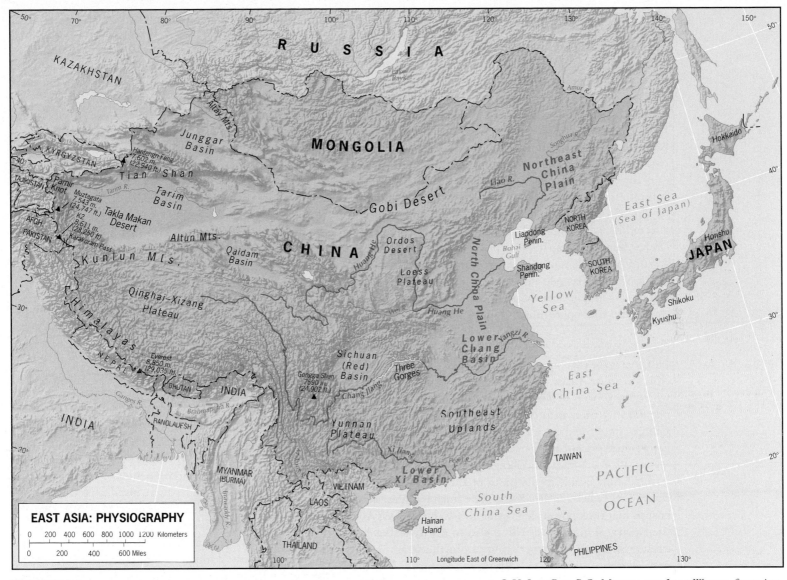

FIGURE 9A-4

© H. J. de Blij, P. O. Muller, and John Wiley & Sons, Inc.

the same way their predecessors did: they hunted, gathered edibles, and, importantly, they became ever better at fishing, which drew them to East Asia's coasts and islands. *Homo erectus* communities appear not to have crossed the waters onto nearby islands, even those they could see from land; it fell to *Homo sapiens* to take this momentous step. Although the evidence is still being assembled, it appears that modern humans crossed the Korea Strait to Japan at about the same time other migrants were entering North America via Alaska, between 14,000 and 13,000 years ago.

Whatever the timing, which is still uncertain, the first humans to cross into Japan appear to have been the Jomon, a people of as-yet-undetermined geographic or ethnic origins but who migrated northward throughout the Japanese archipelago. Later invaders, following the same route, overpowered the Jomon, whose last surviving descendants, the Ainu, now number a mere 20,000 and live in northernmost Hokkaido.

Thus began a sequence of events that was to continue for thousands of years: the diffusion of peoples, innovations, and ideas from mainland East Asia via the Korean Peninsula into what is today Japan. Initially, the new arrivals subsisted the same way their predecessors did: from hunting, gathering, and fishing. But momentous developments on the mainland, in what we call today the Lower Chang Basin and the North China Plain, brought a new era of farming and communal life that radiated into Korea and Japan and made these areas an integral part of the East Asian geographic sphere.

THE VIEW FROM EAST ASIA: STATES AND DYNASTIES

We in the Western world often take it for granted that the pivotal events that led to the taming of animals and the selective farming of plants, the herding of livestock and the harvesting of crops, the storage of produce and the growth of villages into towns, all began in what we now call the Middle East. The story of rivers ebbing and flowing, the "lessons" of natural irrigation, the planned harvesting of grains, the rise of cities, and the organization of the first states is the story of "Western" civilization, the saga of Mesopotamia and ancient Egypt, the Tigris and Euphrates, and the Nile. From the Fertile Crescent, we were taught, these revolutionary changes diffused to other parts of Eurasia and then to the rest of the world.

The Chinese have long taken a different view. After the communists took control of China in 1949, students were taught that China's evolution was entirely domestic, not affected by what might have been happening in "the Far West." But following the end of Chairman Mao's rule in 1976, China's economic opening was paralleled by an academic revival that permitted archeological research to go wherever it might lead. Many Chinese scholars had been waiting for that opportunity.

The results were staggering, and even today new discoveries continue to change the picture of China's—and East Asia's—early regional development. It was already clear that East Asia was one of the few places in the world where the process of **1** state formation had occurred independently thousands of years ago, and that the modern Chinese state could trace its roots to that ancient time, long before there was a Greece or a Rome. For a very long time, it was assumed that the Chinese state was born on the North China Plain, in the basin of the Yellow River. All early states (wherever they arose) were ruled by elites, small groups that took power in growing towns, but in East Asia their political history was chronicled in **2** dynasties because a succession of rulers came from the same line of male descent, sometimes enduring for centuries. And the earliest dynasty about which much was known with certainty was the Shang Dynasty, which lasted from around 1766 BC until about 1080 BC.

But then the discoveries began to overturn the established knowledge. First, evidence for an even older dynasty, the Xia Dynasty, was found in the area where the Yellow River is joined by its Wei tributary, and the remains of its capital gave it the name *Erlitou* culture, dated from circa 2200 BC to 1770 BC. And then the whole notion that the cradle of Chinese civilization lay in this middle Yellow River Basin was overturned by the discovery of even more ancient and complex cultures in other areas, hundreds of kilometers away. Now we know about such cultures as the Yangshao (6000 BC), Hongshan (4500 BC), and Liangzhu (3500 BC), and others that long predated even the Xia city builders. And it is clear that such crops as millet were domesticated in China nearly 10,000 years ago, rice about 9000 years ago, and wheat some 5000 years ago. The early Chinese did not need Middle Eastern guidance when it came to farming, urbanization, or state formation.

Geographically, the most consequential result of these discoveries is the knowledge that East Asia, as a realm, was not the product of one dominant culture that emerged on the North China Plain and spread outward, but was forged from numerous cultures that flourished in several areas including the Lower Chang Basin to the south and even beyond. During this Neolithic ("new stone") period, these regional cultures specialized in skills and arts that diffused far afield. Now we know that certain industries important in East Asia to this day are of early Neolithic vintage: two 8000-year-old pots in the shape of a silkworm cocoon were found in Hebei Province near Beijing. Don't be surprised when, visiting China, you hear students suggest that maybe the inventions we Westerners think diffused from west to east instead spread from east to west!

Beyond the River Basins

Eventually the most powerful dynasties did establish themselves on the North China Plain, but their influence was not confined to present-day China. Even as East Asia's cultures matured and dynasties endured for centuries, their domains expanded and their innovations reached ever-more-distant communities. The Korean Peninsula, where tradition has it that the first king ascended the throne more than 4000 years ago and where the modern period is said to date from 57 BC, witnessed growing Chinese influence; in the late seventh century AD the Chinese even helped one Korean kingdom defeat two competitors. Later the Chinese helped the Koreans repel Japanese invaders, but it was not until China's final dynasty, the Qing (Manchu), that the Chinese forced the Koreans to acknowledge China's regional supremacy. That authority, as it turned out, would not be permanent: during much of the nineteenth century the Korean Peninsula was in turmoil as European, Chinese, and Japanese interests competed for influence, a contest that ended in 1895 when the Japanese ousted the Chinese and made Korea their colony.

In Japan, where local tradition places the country's beginnings in 660 BC, the ruling elites did not confront a Chinese presence but borrowed heavily from Chinese culture, including town and city plans, building styles, legal models, and even writing systems. Thus the cultural geography of the East Asian realm acquired a common base that, despite long-term Korean independence and Japanese modifications and adaptations to the Chinese norms they had borrowed, acquired an overall Chinese denomination. Buddhism matured in China and diffused to Japan; Confucianism infused Korean kingdoms. The realm was being defined in numerous ways.

Peoples in the Realm's Core and Periphery

If we were studying the East Asian realm six centuries ago, during China's Ming Dynasty, we would have had no doubt

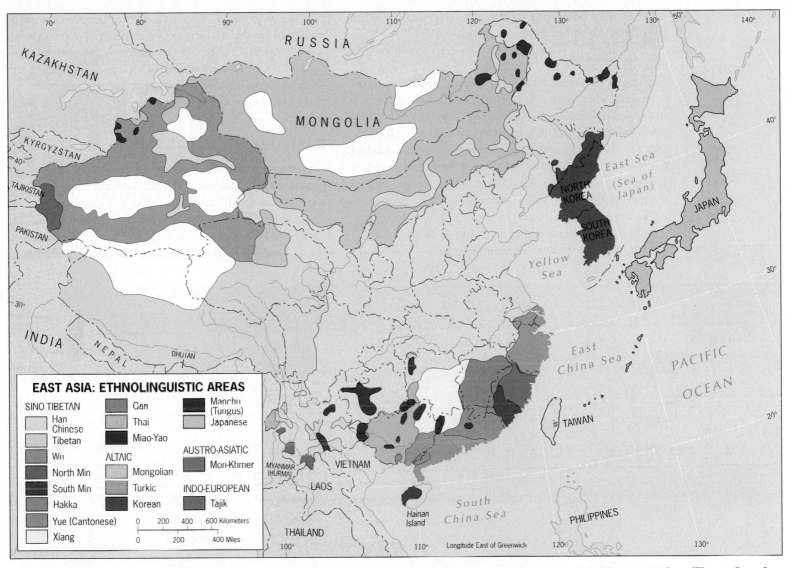

FIGURE 9A-5

© H. J. de Blij, P. O. Muller, and John Wiley & Sons, Inc.

as to the location of core and periphery. China was the core. Encircling the Chinese state in the periphery were not only the Koreans and the Japanese but also the Mongols and Tatars to the north and northwest; the Kazakhs, Kyrgyz, Tajiks, and Uyghurs to the west; the Tibetans, Nepalese, and others to the southwest; and peoples too numerous to identify individually to the south, including majorities (Burmans, Thais, Vietnamese) and minorities in states that were still forming in South and Southeast Asia. Like all empires, the China of the Ming emperors expanded and contracted over time, but during the next (and last) dynasty, that of the Qing, many of the states and peoples just mentioned fell under Chinese rule. But it was the emperors' last hurrah: European and Japanese imperialists challenged Qing rule, took control over most of the state's core area, ousted the Chinese from much of the periphery, and left the country in chaos, bringing about the end of five millennia of dynastic rule a century ago in 1911.

But by that time, the ethnic geography of East Asia had already become very complex. Even after losing control over peoples from Korea to Vietnam and from Mongolia to Burma (now called Myanmar), the Chinese state of the twentieth century still governed numerous minorities. East Asia today, therefore, remains a mosaic of ethnicities and languages (Fig. 9A-5).

We will examine the implications of Figure 9A-5 more closely in Chapter 9B, but for now we should note that this map highlights the cultural diversity of a realm in which China dominates numerically, but where infusions from elsewhere are evident. We already know of the Mongols, the Muslim Uyghurs, and the Buddhist Tibetans; but the most varied and most numerous minority groups inhabit the southeastern corner of this realm, from the island of Hainan to the mouth of the Yangzi River. For example, the Yue language, middle green on the map (it used to be called Cantonese), is the common language in the crucial Pearl

River Estuary. Many of the other minority tongues shown have links to Southeast Asian languages.

And don't be misled by what the map seems to show about Han Chinese (or "Mandarin"). True, the elites and the highly educated speak standard Mandarin, with those in the south somewhat less true to the Beijing-area version than those in the north. But ordinary people in villages a few kilometers apart may not be able to understand each other at all. What they are able to do is read the characters in which standard Chinese (locally called *Putonghua*) is written. And so, wherever you are in China, when you watch an anchorperson read the news on television, you will see a ticker scroll below showing in characters what is being said on-screen. Viewers in Sichuan may not understand the newsreader in Beijing, but still will be able to get the news by reading the tape.

The ethnolinguistic mosaic of the East Asian realm is paralleled by regional variations in belief systems that we will discuss in Chapter 9B, because beliefs—their origins, diffusion, and current regional expression—all reflect the diverse ways cultural and political geographies are interconnected in this populous realm.

RESOURCES OF EAST ASIA

Given that the East Asian geographic realm contains nearly one-quarter of the entire human population on Planet Earth, it is not difficult to imagine the magnitude of the demand for **3** **natural resources** here. While China was moribund under its early communist administration, and before postwar Japan embarked on its headlong rush to

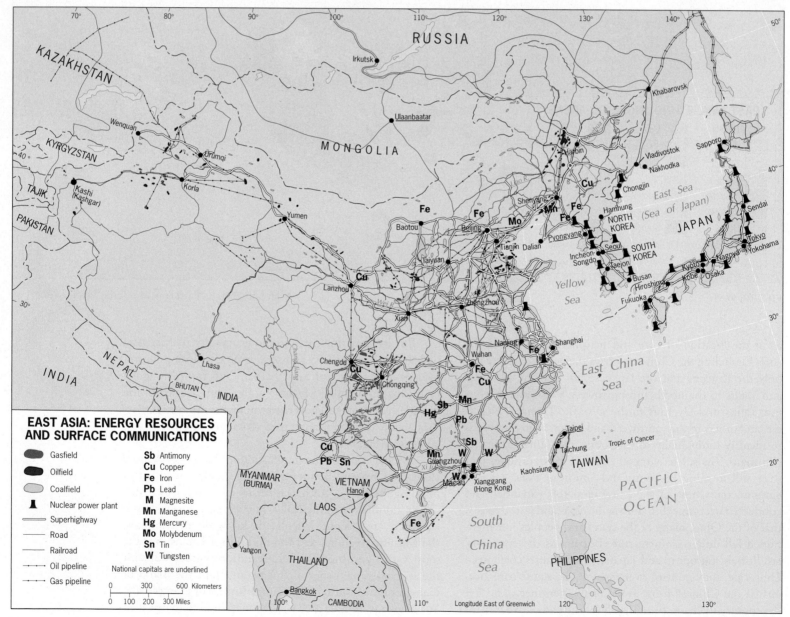

EAST ASIA: ENERGY RESOURCES AND SURFACE COMMUNICATIONS

- Gasfield
- Oilfield
- Coalfield
- Nuclear power plant
- Superhighway
- Road
- Railroad
- Oil pipeline
- Gas pipeline

- **Sb** Antimony
- **Cu** Copper
- **Fe** Iron
- **Pb** Lead
- **M** Magnesite
- **Mn** Manganese
- **Hg** Mercury
- **Mo** Molybdenum
- **Sn** Tin
- **W** Tungsten

National capitals are underlined

0   300   600 Kilometers
0 100 200 300 Miles

FIGURE 9A-6

© H. J. de Blij, P. O. Muller, and John Wiley & Sons, Inc.

world economic power, East Asia's requirements remained modest by global standards. But then Japan's economic success, followed by China's sudden adoption of market economics, created needs unimagined and unprecedented.

Japan, about the size of Montana but with a population of more than 100 million, showed what lay ahead. With few domestic resources to support large-scale manufacturing, the Japanese set up global networks through which moved commodities ranging from oil and natural gas to iron ore and chemicals. Urbanizing and modernizing populations demand ever more consumer goods, and Japanese products poured onto domestic as well as foreign markets. Japan became Australia's top-ranked customer for commodities. Japanese-owned fleets of freighters plied the oceans.

When China took off in the 1980s, economic geographers cast a wary eye on the geologic map. Until then, China's biggest resource had been its fertile, river-deposited alluvium (silt): despite the communist regime's best efforts, state-run industries planned to satisfy domestic needs were no match for globalizing Japan. Most Chinese were farmers. Staving off famines was a never-ending preoccupation. But when China opened its doors to the world, and its cities burgeoned even as its industries proliferated, its needs—for oil, gas, metals, food, electricity, water—multiplied. Before long, China had replaced Japan as Australia's number-one customer. And Chinese manufacturers and suppliers looked for commodities from Indonesia to Iraq and from Tanzania to Brazil. If China holds an advantage, it is in the so-called rare earth elements not commonly known, such as thulium (used in lasers), praseodymium (aircraft parts), lanthanium (electric automobiles), and promethium (X-ray equipment). By some measures, East Asia contains nearly 95 percent of the known deposits of these minerals, which are increasingly utilized in missile technology and "green" energy applications.

But China needs the world because East Asia's storehouse of other known resources is not encouraging (Fig. 9A-6). Coal reserves are widespread and can satisfy the expanding coal-fired energy system (at the cost of thousands of miners' lives every year). But oil reserves, while also widespread, tend to be modest and diminishing; those in northeastern and far-western China are the largest, although exploration continues offshore. Nothing in China compares with the massive iron ore deposits and alloy-metal sources available to Russia when it industrialized to meet the Nazi challenge; and nothing in China compares with the oil and gas reserves of the Middle East and Arabian Peninsula. Hence China competes with Japan for energy pipelines from Russia and with other sources for its industrial raw materials. So in just a few short years,

In China's headlong rush to become a global economic power, environmental protection has always taken a back seat. This scene of industrial production in Chongqing, the Sichuan Basin's largest city, shows the consequences. © Yann Layma/The Image Bank/Getty Images, Inc.

China has become the world's biggest customer as well as a massive exporter.

In Chapter 9B we examine the consequences. East Asia's ascent to the globalizing world's principal stage has put it in the forefront of regional development, but at a high environmental cost in terms of atmospheric pollution and water contamination (see photo above). The East Asian geographic realm is in the midst of an industrial, social, and political revolution the outcome of which is far from clear.

POINTS TO PONDER

- India is expressing concern over China's initiatives on its perimeter as the Chinese build ports, transport routes, and mining operations in countries that surround India.

- The Japanese island of Okinawa is the scene of a dispute with the United States over regional security responsibilities. Key issues: the relocation of an American military base and the number of U.S. soldiers Tokyo wants Okinawa to host.

- Japan and the (then) Soviet Union never signed a peace treaty after World War II, and Russia continues to occupy a group of islands just northeast of Japan.

- China in 2010 objected vigorously to the sale of military aircraft by the United States to Taiwan because China views Taiwan as a temporarily wayward province to be returned to Chinese control.

The razor-sharp "urban front" of Wenzhou, China rolls relentlessly forward. © Mark Leong/Redux Pictures

IN THIS CHAPTER

Evolving China: From ancient roots to modern transformation
Minority problems roil China
China's global economic challenge
Rebounding Japan: Still the realm's powerhouse
How geography could help reunite the Korean Peninsula
Signs of progress over the issue of Taiwan

CONCEPTS, IDEAS, AND TERMS

Sinicization	1
Hanification	2
Extraterritoriality	3
Special Economic Zone (SEZ)	4
Geography of development	5
Overseas Chinese	6
Buffer state	7
Economic tiger	8
Regional complementarity	9
State capitalism	10
Modernization	11
Areal functional organization	12

9B

EAST ASIA: REGIONS OF THE REALM

Han China
China's Outer Periphery
Mongolia
The Korean Peninsula
Japan
Taiwan

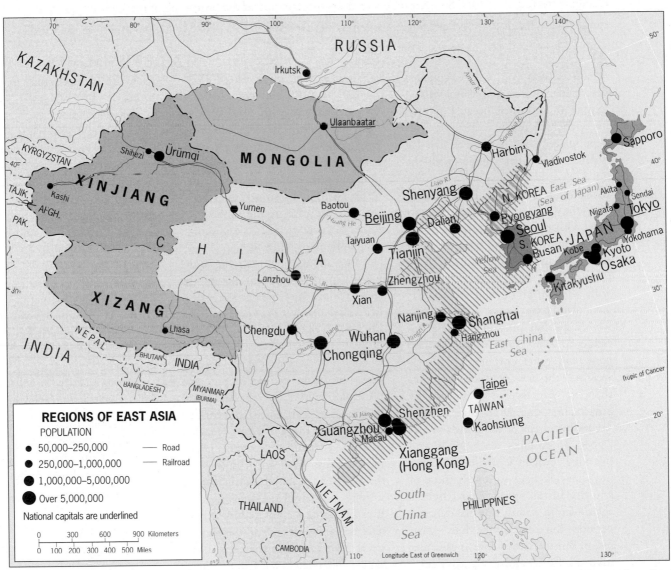

REGIONS OF EAST ASIA
POPULATION
- ● 50,000–250,000
- ● 250,000–1,000,000
- ● 1,000,000–5,000,000
- ● Over 5,000,000

—— Road
—— Railroad

National capitals are underlined

0   300   600   900 Kilometers
0 100 200 300 400 500 Miles

FIGURE 9B-1

© H. J. de Blij, P. O. Muller, and John Wiley & Sons, Inc.

In Chapter 9A we defined the boundaries of the East Asian geographic realm, using its political-geographical layout as a frame of reference. But when it comes to the geographic regions of this physiographically and culturally diverse realm, things get rather more complicated. One problem is the speed with which East Asia is changing. In the introductory chapter we emphasized that regions are always subject to change, but East Asia over the past half-century has been a special case. The spectacular takeoff and subsequent faltering of Japan; the meteoric rise of eastern China; the transformation of South Korea from moribund dictatorship to burgeoning (and high-income) democracy—these are just some of the changes affecting East Asia. So what follows reflects the situation at the beginning of our century's second decade, and we cannot even imagine what the regional geography of this realm will look like at the beginning of the third.

A half-century ago, some geographers described East Asia as a realm with an "empty heart"—not because China, its once and future core area, was depopulated, but because China lay isolated, with little contact with the outside world and less influence over its Eurasian neighbors. Imagine this: after the Chinese communists quarreled with their Soviet advisors in the 1950s and threw them out, just a few *dozen* foreigners were left in all of China! What little contact there was with the world beyond its borders went via Hong Kong. China's cities were huge but showed little sign of modernization. The old, colonial-era bank and other corporate buildings were still the newest structures (other than the gray, faceless tenements the Soviets introduced) in their CBDs. Today, hundreds of thousands of foreigners do business, teach, study, and live long-term in Chinese cities that now have skylines to match those of any city in the world. And China is again—as it was during its dynastic period—the core of the East Asian realm. But this time China is asserting itself as a future global superpower.

You can see these developments reflected in Figure 9B-1. Note that we map a subregion (in a striped pattern) as the Formative Pacific Rim Region; today we could call this the heart of East Asia's core, the focus of China's modern transformation. But we do not yet know how this success story will spread westward. At present, the contrasts between the increasingly affluent Pacific Rim and the much poorer and still-rural west are a matter

of growing concern to the Chinese themselves. Will the core fracture? Can China bridge the gap between coast and interior? In realms discussed previously we have seen the phenomenon of internal and external peripheries (most notably in the Russian realm north and south of the Caucasus). In the case of the China that forms the core of East Asia, a similar geography is evolving.

The colonial powers employed the name China Proper to denote the "real" China, the China of the people who call themselves **the people of Han**. They do so because the Han Dynasty (206 BC to AD 220) is seen as the time when China became the real China, when the state adopted Confucian principles, culture flourished as never before, cities burgeoned (present-day Xian was the "Rome of China"), armies conquered distant lands, trading goods moved across the heart of Eurasia via the Silk Road to and from the Roman Empire, and population grew steadily. Ever since, the majority Chinese have referred to themselves and their country as *Han China*. In the eyes of foreigners, Han China was China Proper.

Han China on the map corresponds approximately to the large ethnolinguistic area mapped as Han Chinese in Figure 9A-5 and to the beige-colored zone labeled China Proper in Figure 9B-1. If Han China is the core, then the heart of that core lies in the east where the great cities of Beijing, Shanghai, Hong Kong, and Guangzhou are situated. But to the west, in that vast territory centered on Lanzhou, is quite another Han China.

Here the Han are rural, poor, and far removed—geographically as well as functionally—from the booming east. This is Han China's inner periphery. Beyond, as Figure 9B-1 shows, lies an outer periphery, that includes blue-colored Tibet (which the Chinese call Xizang) and Xinjiang (colored orange) under Chinese control, and Mongolia (green) outside of it. And in the east, the periphery looks quite different. Here lie North Korea, where China has significant influence; South Korea, now a strong trading partner of China; and Japan. At present, Taiwan also forms part of this periphery, but as we shall see, further change may be in the offing there too.

To summarize, we will discuss East Asia's regions in the following sequence:

1. **Han China**
   The Pacific Rim: Heart of the Core Area
   China's Inner Periphery

2. **China's Outer Periphery**
   Xizang (Tibet)
   Xinjiang

3. **Mongolia**

4. **The Korean Peninsula**

5. **Japan**

6. **Taiwan**

HAN CHINA

Citizens of the People's Republic of China (PRC) have no doubt: not only does their country still have the largest population of any state in the world ("still" because India is closing in), but theirs is the oldest continuous

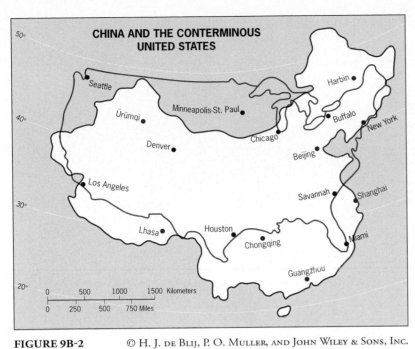

CHINA AND THE CONTERMINOUS UNITED STATES

FIGURE 9B-2   © H. J. de Blij, P. O. Muller, and John Wiley & Sons, Inc.

civilization on the planet. Yet China's territory does not match its demography. For all its 1.34 billion people, China is not a great deal larger than the conterminous United States (Fig. 9B-2). Distances in China are comparable to those in the Lower 48. It is about the same distance from Beijing to Shanghai as it is from New York to Chicago. And as the map shows, east-west distances also are similar. Latitudinally, though, China extends farther north as well as south of the contiguous United States, creating a greater range of natural environments. In the far northeast, China comes close to Siberian cold. The south has Caribbean warmth—with typhoons (hurricanes) to match.

Evolving China

China's great antiquity, as we noted earlier, is chronicled in dynasties, the earliest of which are still shrouded in mystery. But, as with any long-existing state, some of China's dynastic rulers proved more productive than others. Some dynasties bequeathed long-term geographic legacies to the state. Others molded the Chinese nation. The Han Dynasty did both, which is why

we speak to this day of the Chinese as the People of Han.

But the Zhou Dynasty, centuries before the Han, witnessed the arrival of Buddhism in China, saw Confucius walk the pathways of the north, started the building of the Great Wall, and, in another context, spread the use of chopsticks for eating. Confucianism was to become China's guiding philosophy for two millennia. Chopsticks are still in use.

More than a thousand years after the Han, something happened that the Russians would understand. The Mongols drove the local rulers from power and took over the state. But instead of imprinting Mongol rules on their Chinese subjects, the Mongols underwent **1** **Sinicization** or, as it is sometimes said, **2** **Hanification**—that is, they adopted many of the ways of the People of Han. Hostility between the Mongols and the Chinese never diminished, however. Marco Polo saw it first-hand during his early medieval travels to "Cathay" and reported on it when he returned to Europe.

Mongol (Yuan Dynasty) rule ended in 1368, and it is amazing that, in the five and a half centuries dynastic China was subsequently to endure,

only two dynasties were to rule: the Ming (1368 to 1644) and the Qing (1644 until its collapse in 1911). The Ming rulers started out triumphantly: they annexed North Korea, Mongolia, even Myanmar (Burma). They dispatched huge fleets into Pacific and Indian Ocean waters to explore the wider world. But it all came crashing down under conditions we talk a great deal about today: climate change. A major environmental shift called the *Little Ice Age* that had struck Europe in the previous century now afflicted Ming China. Wheatfields lay barren. With 100 million mouths to feed, the emperors worried about revolution. They burned the fleets, built barges, and extended the Grand Canal from the Lower Chang Basin to transport rice to the hungry north. The Ming Dynasty never recovered.

By the mid-1600s, the Manchus, a northern people with historic Mongol-Tatar links, saw their opportunity and took control in Beijing. The People of Han are still shaking their heads about it—a people numbering about 1 million managed to grab the reins of power over a nation of several hundred million. But it happened again: the invaders adopted Chinese ways, kept the Ming systems of administration and education, called themselves Qing, and set about expanding the empire. It was, as we noted earlier, the emperors' last hurrah. The empire was never larger (Fig. 9B-3), but it was also weak. Our map shows the maximum dimensions the Qing Empire acquired as well as the colonial domains the Europeans, Russians, and Japanese appropriated. Five (or more) millennia of dynastic rule over China ended in war, revolution, and final collapse in 1911, only a century ago.

China in Disarray

Figure 9B-3 provides but a mere hint of China's disarray. Chinese emperors (and their subjects) had come to regard the state as invincible, but the colonial powers demonstrated otherwise.

FIGURE 9B-3

© H. J. de Blij, P. O. Muller, and John Wiley & Sons, Inc.

Economically, the colonialists imported cheap manufactures into China, whose handicraft industries succumbed in the face of unbeatable competition. In addition, British merchants imported large quantities of opium from British India into China, and soon this addictive intoxicant was destroying the very fabric of Chinese cultural life. When the Qing government tried to resist this influx, the First Opium War (1839–1842) proved disastrous. The victorious British forced China's rulers to acquiesce to opium importation; the breakdown of Chinese sovereignty was under way. When the Chinese tried that approach again, the Second Opium War 15 years later led to the Chinese being forced to allow the cultivation of the opium poppy in China itself. Chinese society was disintegrating under a narcotic tide that proved uncontrollable.

Meanwhile, the outlines of Figure 9B-3 were in the making. The Beijing court was forced to grant concessions and leases to foreign merchants; China ceded Hong Kong to the British even as British ships sailed up the Yangzi to consolidate a huge sphere of influence in China's midsection. Portugal took Macau, Germany established itself on the Shandong Peninsula, the French encroached on China from their Asian colonies farther south, and the Russians entered the Northeast, then still called *Manchuria*. The Japanese then invaded Korea, annexed the Ryukyu Islands (they still possess them, including Okinawa, today), and colonized Formosa (now called Taiwan) in 1895.

All over China, under the doctrine of **3** extraterritoriality, the British (and other colonial powers) created *concessions* in which Europeans (as well as Russians and Japanese) were immune from local Chinese law. Some of these urban enclaves, business as well as residential, were even off-limits to Chinese citizens. In many other buildings, parks, and additional amenities, Chinese found themselves unable to enter without permission from foreigners. This contributed to the loss of face that helped provoke the so-called Boxer Rebellion of 1900, when bands of revolutionaries roamed cities and countryside, attacking and killing not only foreigners but also their Chinese collaborators. It was a portentous uprising, put down with great loss of life by a multinational force consisting of British, Russian, French, Italian, German, Japanese, and American soldiers. China had no friends to call upon.

Toward a New China

Even as China's public order continued to disintegrate, a better-organized revolutionary movement arose, the so-called Nationalist movement under a prominent leader named Sun Yat-sen. Blaming the sclerotic dynastic regime for China's troubles, these Nationalists in 1911 mounted a fierce attack on the emperor's garrisons all over China. Within a few months the 267-year-old Qing dynasty was overthrown, and with it the system that had survived for thousands of years at China's helm.

The Nationalists, however, faced insurmountable problems in their own efforts to impose a new order on chaotic China. They did negotiate an end to the extraterritorial treaties, and, because of their well-orchestrated military campaign, they did aquire greater legitimacy than the previous regime had when confronting colonial interests. But the Nationalist government set up its base in the southern city of Canton (now Guangzhou), leaving another would-be government to attempt to rule from Peking (Beijing), the old imperial headquarters. Meanwhile, a group of intellectuals in Shanghai founded the Chinese Communist Party; one of its prominent members was a young man named Mao Zedong.

During the chaotic 1920s, the Nationalists and the Communist Party at first cooperated, with the remaining foreign presence their joint target. After Sun Yat-sen's death in 1925, Chiang Kai-shek became the Nationalists' leader, and by 1927 the foreigners were on the run, escaping by boat and train or falling victim to rampaging Nationalist forces. But soon the Nationalists began purging communists even as they pursued foreigners, and in 1928, when Chiang established his Nationalist capital in the city of Nanjing on the banks of the Yangzi, it appeared that the Nationalists would emerge victorious from their campaigns. They had driven the communists ever deeper into the interior, and by 1933 the Nationalist armies were on the verge of encircling the last communist stronghold in the area around Ruijin in Jiangxi Province. This led to a momentous event in Chinese history: the ***Long March***. Nearly 100,000 people—soldiers, peasants, leaders—marched westward from Ruijin in 1934, a communist column that included Mao Zedong and Zhou Enlai. Nationalist forces rained attack after attack on the marchers, and of the original 100,000, about three-quarters were eliminated. But new sympathizers joined along the way (see the route marked in Fig. 9B-3), and the 20,000

survivors found a refuge in the remote, mountainous interior of Shaanxi Province, 3200 kilometers (2000 miles) away. There, they prepared for a renewed campaign that would eventually bring them to power.

Japan in China

Although many foreigners fled China during the 1920s and 1930s, others seized the opportunity presented by the contest between the Nationalists and communists. The Japanese took control over the Northeast, and when the Nationalists proved unable to dislodge them, they set up a puppet state there, appointed a Manchu ruler to represent them, and called their possession Manchukuo.

The inevitable full-scale war between the Chinese and the Japanese broke out in 1937, with the Nationalists bearing the brunt of it (which gave the communists further opportunity to regroup). The gray boundary in Figure 9B-3 shows how much of China the Japanese conquered. The Nationalists moved their capital inland to Chongqing, and the communists controlled the territory centered on Yanan to the north. China had been broken into three pieces.

The Japanese committed unspeakable atrocities during their campaign in China. Millions of Chinese citizens were shot, burned, drowned, subjected to gruesome chemical and biological experiments, and otherwise wantonly victimized. Years later, when China's economic reforms of the 1980s and 1990s led to a renewed Japanese presence in China, the Chinese public and its leaders called for Japan to acknowledge and apologize for these wartime abuses. The unqualified apology the Chinese desire, in word and deed, has not been forthcoming. Some Japanese history textbooks still do not acknowledge what happened to the satisfaction of the Chinese, and surveys indicate strong public sentiment on this still-sensitive issue.

The Rise of Communist China

After the U.S.-led Western powers defeated Japan in 1945, the civil war in China quickly resumed. The United States, hoping for a stable and friendly government in China, attempted to mediate the conflict but at the same time recognized the Nationalists as the legitimate government. The United States also aided the Nationalists militarily, destroying any chance of genuine and impartial mediation. By 1948, it was clear that Mao Zedong's well-organized militias would defeat Chiang Kai-shek. Chiang kept moving his capital—back to Guangzhou, seat of Sun Yat-sen's first Nationalist government, then again to Chongqing. In the late summer of 1949, after a series of disastrous defeats in which hundreds of thousands of Nationalist forces were killed, the remnants of Chiang's faction gathered Chinese treasures and valuables and fled to the island of Taiwan. There, they acquired control of the government and proclaimed their own Republic of China. Meanwhile, on October 1, 1949, standing in front of the assembled masses at the Gate of Heavenly Peace in Beijing's Tiananmen Square, Mao Zedong proclaimed the birth of the People's Republic of China.

China under Communist Rule

After more than six decades of communist rule, China today is a society transformed. It has been said that the year 1949 actually marked the beginning of a new dynasty not so different from the old, an autocratic system that dictated from the top. In that view, Mao Zedong simply bore the mantle of his dynastic predecessors. Only the family lineage had fallen away; now it would be communist "comrades" succeeding each other.

And certainly some of China's old traditions continued during the communist era, but in many other ways Chinese society was totally overhauled. Benevolent or otherwise, the dynastic rulers of old China headed a country in which—for all its splendor, strength, and cultural richness—the fate of landless people and of serfs often was indescribably miserable; in which floods, famines, and diseases could decimate the populations of entire regions without any help from the state; in which local lords could (and often did) repress the people with impunity; in which children were sold and brides were purchased. The European intrusion made things even worse, bringing slums, starvation, and deprivation to the millions who had moved to the cities.

The communist regime, dictatorial though it was, attacked China's weaknesses on many fronts, mobilizing virtually every able-bodied citizen in the process. Land was confiscated from the wealthy; farms were collectivized; dams and levees were built with the hands of thousands; the threat of hunger for millions receded; health conditions improved; child labor was reduced. But China's communist planners also made terrible mistakes. The ***Great Leap Forward***, requiring the reorganization of the peasantry into communal brigade teams to speed up industrialization and make farming more productive, had the opposite effect and so disrupted agriculture that between 20 and 30 million people died of starvation between 1958, when the program was implemented, and 1962, when it was abandoned.

Mao ruled China from 1949 to 1976, long enough to leave lasting imprints on the state. Another of his communist dictums had to do with population. Like the Soviets (and influenced by a horde of Soviet advisors and planners), Mao refused to impose or even recommend any population policy, arguing that such a policy would represent a capitalist plot to constrain China's human resources. As a result, China's population mushroomed explosively during his rule.

Yet another costly episode of Mao's rule was the so-called ***Great Proletarian Cultural Revolution***, launched by Mao Zedong during his final decade in power (1966–1976). Fearful that Maoist communism was becoming contaminated by Soviet "deviationism" and concerned about his own stature as its revolutionary architect, Mao unleashed a campaign against what he viewed as emerging elitism in society. He mobilized young people living in cities and towns into cadres known as Red Guards and ordered them to attack "bourgeois" elements throughout China, criticize Communist Party officials, and root out "opponents" of the system. He shut down all of China's schools, persecuted untrustworthy intellectuals, and encouraged the Red Guards to engage in what he called a renewed revolutionary experience. The results were disastrous: Red Guard factions took to fighting among themselves, and anarchy, terror, and economic paralysis followed. Thousands of China's leading intellectuals died, moderate leaders were purged, and teachers, elderly citizens, and older revolutionaries were tortured to make them confess to crimes they did not commit. As the economy suffered, food and industrial production diminished. Violence and famine killed as many as 30 million people as the Cultural Revolution spun out of control. One of those who survived was a Communist Party leader who had himself been purged and thereafter been reinstated—Deng Xiaoping. Deng was destined to lead the country in the post-Mao period of economic transformation.

Reorganizing Nation and State

Imagine the challenge the communist administration of China faced when it took power in 1949, and try to envision the dilemmas that arose when China's economy took off not long after the disintegration of Mao's regime in 1976. Here was a country of more than a billion people crowded into the eastern third of it, with far-flung minorities inhabiting remote peripheries and with boundaries in contention with neighbors. Here were remnant foreign dependencies and a large, nearby island ruled by rearguard Nationalists. Moreover, the administrative divisions inherited from earlier regimes were anything but functional.

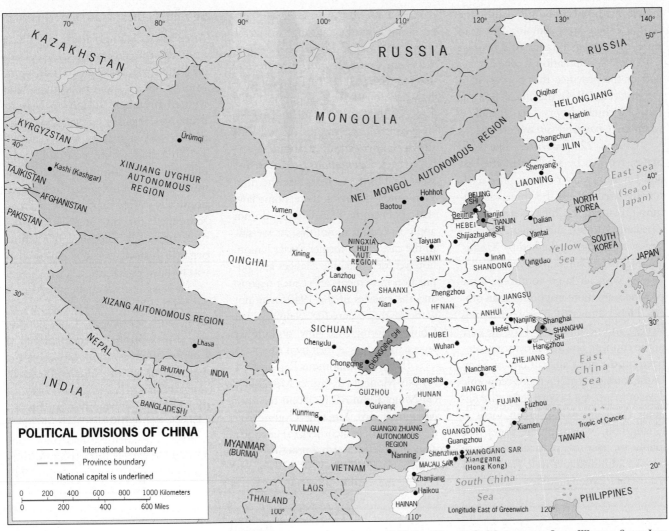

FIGURE 9B-4

© H. J. de Blij, P. O. Muller, and John Wiley & Sons, Inc.

First Mao Zedong's regime, and then the modernizers who followed Mao, brought order to the chaotic map of China and, in the modernizers' case, adjusted it to the changing needs of the state. In Mao's time, China was organized into the capital and 22 provinces; but in Deng's time, special adjustments were needed for administrative purposes. For example, in 1997 China took over the British dependency of Hong Kong; but Hong Kong could not simply be made China's 23rd province. So the administration established a new category in the administrative hierarchy, which was later awarded to former-Portuguese Macau as well, and—perhaps—to Taiwan sometime in the future. Also, some cities in China (other than the capital of Beijing) grew more important than others

in the scheme of things, and it was decided to recognize those cities with the special municipal status of *Shi.*

Political and Administrative Divisions

Before we investigate the emerging human geography of contemporary China, let us acquaint ourselves with the country's political and administrative framework (Fig. 9B-4). For administrative purposes, China is divided into the following units:

4 Central-government-controlled Municipalities (*Shi's*)

22 Provinces

2 Special Administrative Regions (SARs)

5 Autonomous Regions (ARs)

The four central-government-controlled ***Municipalities*** (or ***Shi's***) are the capital, Beijing; the nearby port city of Tianjin; China's largest metropolis, Shanghai; and the Chang River port of Chongqing in the interior. These *Shi's* form the cores of China's most populous and important subregions, and direct control over them from the capital entrenches the central government's power.

We should keep in mind that the administrative map of China continues to change—and to pose problems for geographers. The city of Chongqing was made a *shi* in 1996, and its municipal area was enlarged to incorporate not only the central urban area but an enormous hinterland covering all of eastern Sichuan Province. As a result, the urban population of Chongqing is officially 30 million, making this the world's largest metropolis—but in truth, the central urban area contains less than seven million inhabitants. And because Chongqing's population is officially *not* part of the province that borders it to the west (Fig. 9B-4), the official population of Sichuan dropped by 30 million when the *Chongqing Shi* was created.

China's 22 **Provinces**, like U.S. States, tend to be smallest in the east and largest toward the west. The territorially smallest are the three easternmost provinces on China's coastal bulge: Zhejiang, Jiangsu, and Fujian. The two largest are Qinghai, flanked by Xizang, and Sichuan, China's Midwest.

As with all large countries, some provinces are more important than others. Hebei Province nearly surrounds Beijing and occupies much of the core of the country. Shaanxi Province is centered on the great ancient city of Xian. In the southeast, momentous economic developments are occurring in Guangdong Province, whose urban focus is Guangzhou. When, in the pages that follow, we refer to a particular province or other administrative unit, Figure 9B-4 is a useful locational guide.

In 1997, the British dependency of Hong Kong (Xianggang) was taken over by China and became the country's inaugural **Special Administrative Region (SAR)**. In 1999, Portugal similarly transferred Macau, opposite Hong Kong on the Pearl River estuary, to Chinese control, creating the second SAR under Beijing's administration.

The 5 **Autonomous Regions (ARs)** were established to recognize the non-Han minorities living there. Some laws that apply to the Han Chinese do not apply to certain minorities. As we saw in the case of the former Soviet Union, however, demographic changes and population movements affect such regions, and the policies of the 1940s may not work in the twenty-first century. Han Chinese immigrants now outnumber several minorities in their own ARs.

The five Autonomous Regions are: (1) Nei Mongol AR (Inner Mongolia); (2) Ningxia Hui AR (adjacent to Inner Mongolia); (3) Xinjiang Uyghur AR (China's northwest corner); (4) Guangxi Zhuang AR (far south, bordering Vietnam); and (5) Xizang AR (Tibet, anchoring the southwest).

Seizing the Post-Mao Opportunities

After Mao Zedong's death in 1976, a political struggle among China's leaders pitted the so-called purists (communist bosses and intellectuals who wanted to adhere strictly to Marxist principles) against those who called themselves pragmatists, who envisaged a China that would continue to be politically and administratively communist but would open itself economically to free-market forces. Deng Xiaoping and his pragmatist allies soon prevailed, and in 1979 China began a historic transformation that was to have global impact.

It was no small challenge. Almost everything in China was old, inefficient, and poorly run. Consider the four subregions that were identified in Chapter 9A, which are based on the river basins of the Huang (Yellow), Yangzi/Chang, Pearl/Xi, and Liao. In the Northeast (formerly Manchuria), relatively well endowed in coal and metal reserves, the Japanese had built substantial industrial infrastructure during their colonial occupation (in 1949

@ **FROM THE FIELD NOTES...**

© H. J. de Blij

"We flew over the North China Plain for an hour, and one could not miss the remarkably regular spacing of the villages on this seemingly table-flat surface. Many of these villages became communes during the communist reorganization of China's social order. The elongated buildings (many added since the 1950s) were each designed to contain one or more extended families. After the demise of Mao and his clique, the collectivization effort was reversed and the emphasis returned to the small nuclear family. In this village and others we overflew, the long sides of the buildings almost always lay east-west, providing maximum sun exposure to windows and doors, and enhancing summer ventilation from longitudinal breezes. The more distant view looks hazy, because the North China Plain during the spring is wafted by loess-carrying breezes from the northwest. On the ground, a yellow-gray hue shrouds the atmosphere until a passing weather front temporarily clears the air."

www.conceptcaching.com

© H. J. de Blij

"Shanghai in April 2009 showed no sign of any economic slowdown and gave much evidence of the environmental consequences. From my vantage point on the top floor of the Jin Mao Tower in Pudong District, buildings beyond the Huangpu River disappeared in a dense smog on a hot spring day that locals described to me as "pretty clear." China's prodigious economic growth has made the People's Republic the world's largest polluter of the atmosphere, surpassing the United States. Some projections suggest that, by 2020, China may be emitting twice as much pollution as America, but since China has more than four times the population, the U.S. would still lead in terms of pollutants per capita. That's just one reason the United States shouldn't invoke the "China excuse" to weaken its own efforts to reduce pollution from motor vehicles, factories, and other sources. "Enjoy this view of the Oriental Pearl Television Tower," Li Sheng told me. "We're building the Shanghai Tower right down there where you see the construction site, and it'll be in the way. And it'll be half as tall again as your Empire State Building." Well, let's hope the view from the top will be clearer."

www.conceptcaching.com

the Northeast contained half of China's railroad mileage). Climate, as we noted, limits agriculture here. The communist regime had made the industrial redevelopment of the Northeast a priority, and cities and towns grew exponentially. But even though the Northeast eventually produced more than one-quarter of China's industrial output, its state-run factories were inefficient, workers' perks were unaffordable in the long run, and the cost of products was, to put it mildly, uncompetitive. There was no way the Northeast would be an asset in an economically open China.

Southward, the great North China Plain—scene of those early dynasties as well as the site of the capital and the original core area of expanding China—had its own problems: several hundred million people living on overcrowded farmlands where communist planners had expropriated land and established collectives at a terrible cost

(especially during the so-called Great Leap Forward) but where major efforts directed at flood control, expanded irrigation, improved fertilization, and better farm equipment had positive results. Yet look again at Figure 9B-4: provinces whose names we rarely hear (Hebei, Henan, and Shangdong) each have more people than France, Britain, or Italy. How could those populous, mainly rural provinces, periodically in the grip of famine, become winners in the great game of globalization?

Even during the (1966–1976) Cultural Revolution, however, the Lower Chang Basin—the subregion anchored by Shanghai and networked by the Yangzi and its tributaries—retained some of its historic identity and energy. Drab, teeming, and decrepit though it had become, Shanghai had not completely lost its intellectual vigor, artistic individuality, risk-taking entrepreneurs, or even its opponents of communist dogma. From day one, Deng and his

pragmatists in Beijing saw in Shanghai what they could not yet foresee in their own backyard. Here was a place near the Pacific coast and the mouth of the realm's greatest river, loaded with talent and accustomed to taking chances. And in its vast hinterland lay more than flood-prone wheatfields: the Sichuan Basin alone had a population (at the time) of more than 100 million growing everything from rice to tea and from fruits to spices.

If China's planners needed encouragement, all they had to do was look to the southernmost subregion of China's east: the Pearl River Estuary, outlet of the Xi River. There was Hong Kong, about as vivid a contrast between success and failure as China could present. When Deng and his pragmatists took charge, Hong Kong was a thriving port city importing raw materials by the shipload and disgorging manufactured products that sold on markets all over the world. Make no mistake: Hong

FIGURE 9B-5   © H. J. de Blij, P. O. Muller, and John Wiley & Sons, Inc.

Kong may have been a British colony, but it was Chinese managers and Chinese workers who propelled its economy. If you were a visitor to Hong Kong in the 1970s, they took you to a place on a hillside from where you could peer across the fortified border into "Red China"—and observe some villages with duck ponds and paddies and wooden fishing boats. You didn't need an interpreter to understand what you were seeing.

New Economic Zones on the Pacific Rim

But no one would have believed what was about to happen. Many economic geographers and others studying the fast-changing situation assumed that Deng and his advisors would go slow, opening China's door just slightly to avoid the political pressures that would surely come with market economics. Deng, however, thought otherwise. By setting new economic rules that would initially apply only to coastal cities, areas, and provinces, the anticipated economic changes would at first impact only China's Pacific Rim, leaving most of the rest of the country comparatively unaffected. Figure 9B-5 shows how it would work.

The government introduced a system of **4** **Special Economic Zones (SEZs)**, so-called **Open Cities**, and **Open Coastal Areas**, which would attract technologies and investments

from abroad and transform the economic geography of eastern China. In these economic zones, investors are offered numerous incentives. Taxes are low. Import and export regulations are eased. Land leases are simplified. The hiring of labor under contract is allowed. Products made in the economic zones may be sold on foreign markets and, under some restrictions, in China as well. Even Taiwanese enterprises may operate here. And profits earned are allowed to be sent back to the investors' home countries.

When Deng's government made the decisions that would reorient China's economic geography, location was a prime consideration. Beijing wanted China to participate in the global market economy, but it also wanted to cause as little impact on interior China as possible—at least in the first stages. The obvious answer was to position the Special Economic Zones along the coast. Initially, four SEZs were established in 1980, all with particular locational properties (Fig. 9B-5):

1. **Shenzhen**, adjacent to then-booming British Hong Kong on the Pearl River Estuary in Guangdong Province

2. **Zhuhai**, across from then still-Portuguese Macau, also on the Pearl River Estuary in Guangdong Province

3. **Shantou**, opposite southern Taiwan, a colonial treaty port, also in Guangdong Province, source of many expatriate Chinese now living in Thailand

4. **Xiamen**, on the Taiwan Strait, also a colonial treaty port (then known as Amoy in the local dialect), in Fujian Province, source of many Chinese now based in Singapore, Indonesia, and Malaysia

In 1988 and 1990, respectively, two additional SEZs were proclaimed:

5. **Hainan Island**, declared an SEZ in its entirety, its potential success linked to its location near Southeast Asia

6. *Pudong*, across the Huangpu River from Shanghai, China's largest city, different from other SEZs because it was a giant state-financed project designed to attract large multinational companies.

And the process continues, with the newest SEZ authorized in 2006:

7. *Binhai New Area*, the coastal zone of the northern port city of Tianjin, a long-established Open City with considerable foreign investment, now elevated to SEZ status, projected to outperform even Shanghai–Pudong and Shenzhen itself.

The grand design of China's economic planners, therefore, was to stimulate economic growth in the coastal provinces and to capitalize (as it were) on the exchange opportunities created by location; the availability of funding; the proximity of foreign investors in Southeast Asia, Taiwan, Japan, and, importantly, still-British Hong Kong; the presence of cheap labor; and the promise of world markets eager for low cost Chinese products.

Remember that hillside view from Hong Kong in the 1970s? In those days the name *Shenzhen* was hardly known. But during the following decade, those rice paddies and duck ponds gave way to the fastest-growing city in the history of humanity—from 20,000 in the late 1970s to 8 million just 30 years later. At the same time, the entire Pearl River Estuary was rapidly evolving into a massive manufacturing complex (Fig. 9B-6). What made it happen? When China opened the Shenzhen SEZ, hundreds of companies moved their factories from neighboring Hong Kong to Shenzhen, capitalizing on lower wages and fewer taxes, less environmental regulation, and weaker official oversight (Marx would surely have turned in his grave). China, meanwhile, built business-friendly infrastructure; expanded seaport facilities, roads, railroads, and airports; and constructed apartment

From open fields and rural villages, the Binhai New Area (BNA) on the coast at Tianjin, only 120 kilometers (75 mi) southeast of Beijing, is rising fast from its start in 2009. Official proclamations declare that this will become still another international free trade zone on China's Pacific Rim, favored by its proximity to the capital and designed to compete with Pudong (Shanghai) and Shenzhen bordering Hong Kong. © Lo Mak/Redlink/©Corbis

buildings to accommodate hundreds of thousands of workers. Soon Shenzhen was the shining example of Deng's Open China policy, to be emulated by the other SEZs from Zhuhai (next to Macau) to Pudong (Shanghai) to Binhai (Tianjin).

Geography of Development

Did all this hurt Hong Kong's economy? In fact, it ushered in a new and even more prosperous era for the (then still) British colony. Hong Kong now performed less manufacturing, but far more financing, banking, fiscal management, and other service-oriented functions, requiring the greater skills already proven to prevail among its population of 7-plus million during earlier economic transitions. Such transitions are the subject of a field called the **5** **geography of development**, which focuses on raw-material distributions, cultural traditions, historical factors (such as the lingering effects of colonialism), environmental issues, and the role of forces of loca-

tion in the transitions that national and regional economies undergo.

Before the current wave of globalization began to affect economic development processes virtually everywhere on Earth, economists and economic geographers tried to identify the factors that led some societies (but not others) to follow a particular path from traditional subsistence to mass production and consumption. One of these scholars, Walt Rostow, proposed a sequential model that focused on cultural "flexibility," the presence of "progressive" leadership, the acceptance of and public support for modernization, the resulting "takeoff," the achievement of "maturity" (marked by product specialization and foreign trade), and the ultimate "mass consumption" stage, by which time most workers would be employed in service and service-related industries. Certainly you could find real-life examples of this theoretical model (Hong Kong is one), however it took insufficient account of evolving core-periphery contrasts within regions or even individual countries, and did not anticipate or account for

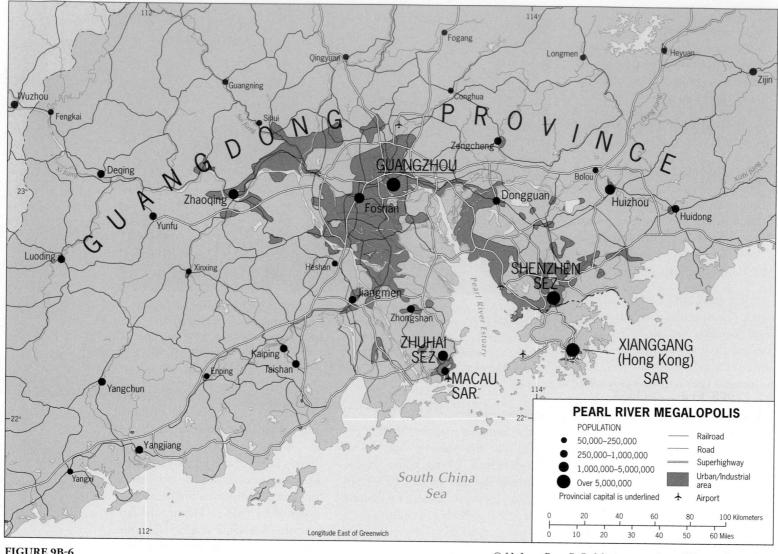

FIGURE 9B-6

© H. J. de Blij, P. O. Muller, and John Wiley & Sons, Inc.

globalization forces that sometimes propelled certain parts of countries rapidly ahead while leaving others barely affected.

Take the case of China. Its post-Mao leadership was reformist, but not progressive (representative government is not seen as essential to development; and a number of Chinese policymakers say that "democracy constrains dynamism"). Deng and his pragmatists answered the question "How can we use our resources best in our effort to speed economic development?" by deliberately favoring the best-located (that is, coastal) provinces and cities, by opening its Pacific Rim to market forces, by exploiting its huge working class, and by encouraging millions of

6 **Overseas Chinese**—most of whom had left China from those very coastal provinces to settle (and often thrive) in other countries—to invest their money in their ancestral homeland.

Development in the current era has therefore taken on a new meaning. Contrary to Rostow's multistage model of adaptation and consensus, development shaped by contemporary conditions of globalization has frequently become a winner-take-all form of opportunism, often with little constraint and with much exploitation of those who find themselves in the path of "progress." Scholars studying this phenomenon use the adjective **neoliberal** to describe both its "new" and its "unconstrained" features.

China's Inner Periphery

One result of China's opening its doors to globalization's forces was that the coastal provinces and cities boomed and disparities between coast and interior rapidly widened. Few maps could illustrate this core-periphery contrast more dramatically than Figure 9B-7, which three-dimensionally displays the relative gross domestic product (GDP) contributed by each province to the national income. The higher the "terrain," the greater the income; the Pacific Rim, even the more slowly developing lower Northeast, looks like a wall between coastal core and interior periphery.

As Figure 9A-5 reminds us, both the core and inner periphery are peopled dominantly by Han Chinese.

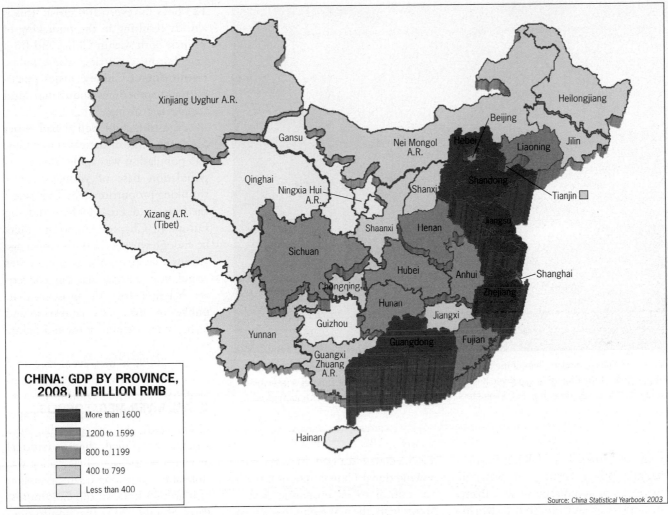

CHINA: GDP BY PROVINCE, 2008, IN BILLION RMB

More than 1600

1200 to 1599

800 to 1199

400 to 799

Less than 400

Source: *China Statistical Yearbook 2003*

FIGURE 9B-7

© H. J. de Blij, P. O. Muller, and John Wiley & Sons, Inc.

Although there are minorities in the periphery, the great majority of the people in the western provinces are Han Chinese, and hundreds of millions of them continue to subsist on less than three American dollars a day. China's Pacific Rim transformation drew many millions more to the SEZ factories as well as urban and regional construction sites in what has been the largest internal migration in human history, but those who stayed behind found themselves remote from the opportunities brought by globalization—and often exploited by local mayors, party bosses, and others when it came to land, property, or produce markets. It does not make the international news, but China's inner periphery is rife with public protests in the streets and villages. It is a dangerous situation,

and the Chinese communist regime is showing an increasing awareness of the need for mitigation.

One way Beijing's planners are responding to this uneven development is by spreading the privileges enjoyed by the coastal SEZs into the urban centers of the interior. With workers' hourly salaries rising in the coastal provinces, some corporations are finding it profitable to move their production facilities into the periphery. To encourage them to do so, there are now new economic zones in such deep-interior cities as Kunming (Yunnan Province) and Lanzhou (Gansu Province). Another is the construction of massive public works in the interior. For example, the notion that the mighty Yangzi/Chang River could become a development corridor from

coast to interior was given substance by the construction of the gigantic hydroelectric Three Gorges Dam on the Chang River and the elevation of the upriver city of Chongqing to the rank of *Shi*, the only interior city so recognized. Other enormous projects are in the planning stage, including the diversion of part of the upper course of the Chang across the north to supply water to the upper Huang (Yellow) River and particularly the North China Plain in its lower basin. In addition, China is building more than 110,000 kilometers (70,000 mi) of four-lane highways to improve surface links between coast and interior, and hundreds of airports are being expanded or newly constructed.

Despite all this and more, China's Pacific Rim continues to race ahead

Three Gorges Dam is emblematic of China's modern "era of the megaproject." The dam wall rises 180 meters (600 ft) above the now-inundated valley floor of the Chang/Yangzi River. It is 2 kilometers (1.2 mi) wide and creates a reservoir—largest of its kind in the world—that extends more than 600 kilometers (400 mi) upstream. © Wen Zhenxiao/XinHua/XinhuaPress/©Corbis

while the interior is falling further behind. When farmers in Shaanxi Province are summarily ousted from the lands they cultivate (but to which they do not have title and which they cannot use to secure a bank loan) because some corporation with connections to the regime wants to build a factory, it is all too clear that China needs another transformation to achieve the consensus Rostow saw as crucial to sustained development.

Population Issues

China has been the world's most populous state for many centuries, and during the population explosion of the twentieth century it also had one of the fastest-growing populations on the planet. During Mao's rule, when China still was a dominantly agricultural society, families were encouraged to have numerous children, and China was growing at a rate approaching 3 percent per year. That was the policy inherited by Deng and his

reformers who realized, however, that slowing down China's rate of growth was critical to its economic future. Accordingly, the new regime embarked on a vigorous population-control program that imposed on Han Chinese families (but not minorities) a one-child limit. Enforced by sometimes draconian means, this policy had the desired effect: by the mid-1980s China's population was growing at 1.2 percent, and by 2010 the rate was 0.5 percent, well below half the rate for the world as a whole (1.3 percent).

The policy had its desired economic effect, but its other results were less salutary. China's is a dominantly patriarchal society, where male children are much preferred; rates of female fetus abortion, infanticide, and abandonment skyrocketed, resulting in a gender imbalance that raises concern for the future. China's administration estimates that, if the policy is not modified, the society will be short some 30 million brides (the reported gender imbalance in births today is

123 boys for every 100 girls). This is already resulting in the trafficking of women both within China and from neighboring countries, where public resentment of Chinese males pursuing (and sometimes abducting) local females is growing.

The one-child policy had other outcomes. One of these was that China's population was aging; that is, its population base of youngsters was shrinking proportionately. That raised the fears we discussed when studying Europe in Chapter 1A: would there be enough workers in the younger age groups to support the growing older population? For China, this concern was heightened by the prospect that, unlike in European countries and Japan, China would grow old before it grew rich.

In 2009, a rare event occurred in the PRC: official toleration of public debate regarding Politburo policy. In the English-language newspaper *China Daily* and other media, brief articles mentioned that a national debate over the one-child policy was "tolerable," signaling open discussion of it at high levels of the Communist Party. Should China abandon its one-child policy, the world will need to revise its population projections for the twenty-first century.

CHINA'S OUTER PERIPHERY

Historically, empires have expanded and contracted, acquiring and losing territory as well as subjects, and leaving their imprint where they once were. Russians in Central Asia, British in East Africa, French in Indochina, Japanese in Taiwan all came and went, leaving behind languages, religions, infrastructures, and traditions.

And so it was with China. Figure 9B-3 reminds us just how vast China's Qing Dynasty empire was and how much it lost—but Figure 9B-1 reveals that China still is an empire today. Its beyond-the-Han domain includes a pair of large Autonomous Regions and a

© Jan Nijman

"Tibet is one of the most colorful and spiritual places on Earth, high in the Himalayas with its steely blue skies and dry, crisp air. Since 1950, Tibet has been controlled by the People's Republic of China. The Chinese military seem omnipresent and it is hard to escape a sense of forceful occupation, especially in the capital city, Lhasa. In late 2009 I explored the busy streets around the famous Jokham Temple, a major pilgrimage destination for Tibetan Buddhists. There, too, the Chinese were conspicuously present, with observation posts perched atop buildings as if to remind ordinary Tibetans of the political order of the day."

www.conceptcaching.com

third one rapidly being integrated into Han China itself. Two other, smaller Autonomous Regions in the inner periphery are similarly misnamed (see Fig. 9B-4). There is nothing "autonomous" about China's minority entities: just ask the Tibetans of Xizang and the Uyghurs of Xinjiang.

Xizang (Tibet)

Tibet (called *Xizang* by the Han) is the icebound heartland of Tibetan-Buddhist culture that extends into Qinghai Province and, importantly, into a corner of northeastern India in the Indian State of Arunachal Pradesh. Tibet shook off Chinese domination at the end of the nineteenth century, but in 1950, almost immediately after the communist regime took control in Beijing, it ordered the Red Army into Xizang to recapture the territory. Tibetan society had been organized around the fortress-like monasteries of Buddhist monks who paid allegiance to their supreme leader, the Dalai Lama.

The Chinese wanted to modernize this feudal system, but the Tibetans clung to their traditions, and in 1959

the army crushed an uprising. With the Dalai Lama now exiled and the people powerless, the Chinese destroyed much of Tibet's cultural heritage, looting its religious treasures and works of art. Under Deng's administration, part of this heritage was returned to Xizang, but Beijing also encouraged Han Chinese to move to Tibet and built "the world's highest railroad" (opened in 2006) across the Qinghai-Xizang Plateau to the capital of Lhasa. While the Dalai Lama travels the world making the 3 million Tibetans' case and asking not for independence but for genuine autonomy, Hanification continues, resulting in another uprising in 2008 when the world's eyes were focused on the imminent Beijing Olympics.

Xinjiang

Xinjiang is the westernmost outpost of China's modern empire and is even larger than Xizang. With 22 million people, about half of them Han Chinese, the Xinjiang-Uyghur AR is even more important than its Tibetan, Buddhist neighbor. The maps show why: here China meets the peoples of Turkestan and the faith of Islam; here

China has significant energy reserves; and here, in the remote, clear-sky, desert-dominated far west, China built its original space program. And the map shows something else: only one country lies between this western outpost and the much larger oil and gas reserves of the Caspian Sea Basin: Kazakhstan (Fig. 9B-1). As we noted in Chapter 7B, the pipelines are already in place and additional ones are now under construction (see Fig. 7B-12).

But Xinjiang's Autonomous Region is no more autonomous than Tibet. During the Qing Dynasty the Muslim peoples here—including Uyghurs, Kazakhs, Kyrgyz, Tajiks, and others—fell under Chinese control; but even when the communist regime took over in Beijing, fewer than one in 20 inhabitants of Xinjiang was Han Chinese. Today, only six decades later, nearly 50 percent are Han, and Xinjiang has become a two-tiered society. The modern, regimented, and urbanized component is centered on the revitalized capital of Ürümqi and the nearby model city of Shihezi, located in the northern, Junggar Basin; the traditional, religious, and rural component is anchored by the historic town of

Kashgar (Kashi on Chinese maps) in the western corner of the southern, Tarim Basin (see Fig. 9B-1). Where the Uyghurs, who form the majority among the local peoples, meet the Han, as happens in the factories of Ürümqi and the streets of Kashi, the results are sometimes tragic. In 2009, a skirmish between Uyghur workers and Han managers at a plant on the outskirts of Ürümqi spun out of control and led to deadly rioting in which some 200 Han Chinese citizens were killed. In 2008, prior to the Beijing Olympics, local extremists bombed a police patrol in Kashi. For years, Uyghur militants have pursued an intermittent and fruitless campaign of violence against Han-Chinese domination.

Both Xizang and Xinjiang exhibit the properties of peripheries: local cultures and traditional economies are overpowered by national interests and global systems, widening disparities not only between core and periphery but also between and among societies within the periphery itself. When, as in China's case, the political system affords inadequate opportunity for the expression of grievances and the representation of local interests, the symptoms of marginalization become entrenched.

Inner Mongolia

The third large Autonomous Region shown in Figure 9B-4 is Inner Mongolia, originally established to acknowledge the presence within China of a substantial population with ancestral affinities to the people of next-door, present-day Mongolia. As Figure 9B-3 reminds us, Mongolia, too, once was a part of the Qing Dynasty empire, but out of the collapse of that empire a century ago and the ensuing chaos in Han China as well as the rise of the Soviet Union, a Mongolian "People's Republic" came into being. As the modern border between China and Mongolia, however, does not reflect the domain of the Mongolian people, hence the Nei Mongol Autonomous Region.

Proximity to Han China, good connections to the Chinese core area, the modest size of the Mongol population, massive immigration by Han Chinese, and economic integration have essentially voided the identity of this AR and, for all intents and purposes, have made it a sector (albeit an economically lagging one) of "China Proper." Nothing at all comparable to Tibetan or Uyghur separatism occurs here; traditional Mongol ways of life have been largely submerged under Hanification. A cultural landscape of sedentary farms and modernizing towns has replaced the pastoral, nomadic lifestyle still surviving on the vast plains of Mongolia itself.

We now turn to the East Asia lying beyond China's borders. Some of what we encounter will be reminiscent of China's inner periphery, but elsewhere the mainland and island entities of this realm display all the properties of core areas—not just the East Asian core, but the global core (see Fig. G-11). Here, in fact, was where the East Asian transformation began that in short order reached the PRC and changed the world.

MONGOLIA

That East Asian transformation, however, has not reached Mongolia. This immense, landlocked, isolated country wedged between China and Russia, with an area larger than Alaska but a sparse population of under 3 million, suggests a steppe- and desert-dominated vacuum between two of the world's most powerful countries (Fig. 9B-1).

This used to be the domain of a powerful people who swept westward to challenge the Russians and southward to rule China, but today it is a weak and vulnerable country whose 800,000 herders and their millions of sheep follow nomadic tracks along the fenceless fringes of the vast Gobi Desert, where Siberian cold periodically causes severe human and livestock losses. During the Soviet era, the loca-

tion of the capital of Ulaanbaatar symbolized Mongolia's security against Chinese encroachment, but now Chinese investment and involvement are growing. As a historic **7** **buffer state** within interior Asia, Mongolia has few options.

THE KOREAN PENINSULA

If you were able to cross the tightly-sealed border between China and North Korea, it would seem (as in the case of Mongolia) that you have once again gone from core to periphery. No evidence here of the economic success story we tend to associate with the geographic term "Korea." But if you could travel freely to the other border between North Korea and South Korea and cross it, you would have no doubt. The Korean Peninsula was and continues to be the scene of what economic geographers refer to as an economic miracle. Once a sclerotic dictatorship with a moribund economy, South Korea now has a dynamic economy typical of one of the burgeoning **8** **economic tigers** of the realm.

The Korean Peninsula is not that large: its total territory is about the size of Idaho, and much of it (especially the north) is mountainous and rugged. Nonetheless, its population stands at 73 million (Idaho's is only 1.6 million). The map (Fig. 9B-8) suggests that the Korean Peninsula is very nearly a bridge connecting East Asia's mainland to the islands of Japan, and in ancient times, when sea levels were lower, the link was even closer. As a result, Korea has long been the stage for the contest between Chinese and Japanese cultural influences. More recently it became the scene of devastating great-power conflict between communist and anticommunist armies. The cease-fire line you see on the map marks the armistice that ended the Korean War of 1950–1953. But unlike the Berlin Wall and the

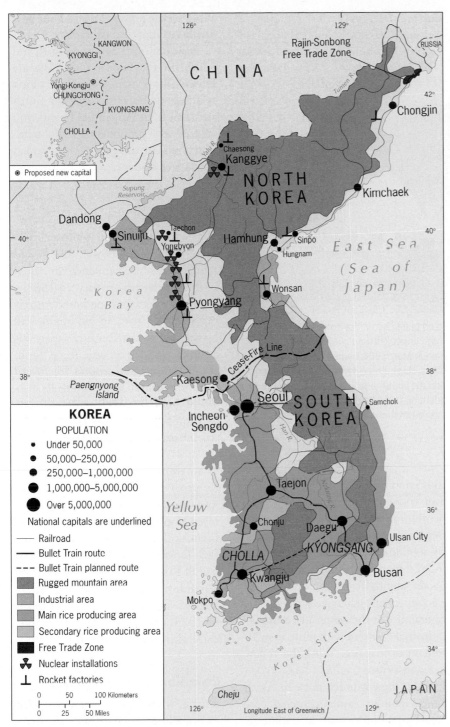

FIGURE 9B-8   © H. J. de Blij, P. O. Muller, and John Wiley & Sons, Inc.

"administrative" purposes. The territory north of the 38th parallel was placed under the control of Soviet forces; south of this line, the United States was in control. In 1950, communist forces from northern Korea invaded the south in a forced-unification drive, unleashing a devastating conflict that swept southward across the peninsula, back northward across the 38th parallel, and then drew in Chinese armed forces who pushed the front southward again; finally, in 1953, the Korean War ended at the Cease-Fire Line not far from where the 38th parallel had marked the original boundary (Fig. 9B-8). Ever since, a demilitarized zone (DMZ) has more or less hermetically sealed North from South. And ever since, North and South Korea have grown apart, still in danger of renewed conflict.

North Korea

Six decades of communist rule have turned North Korea (24 million) into one of the poorest, hungriest, and most regimented nations on Earth. A regime that imprisons its citizens for the slightest infractions, operates a gulag reputedly worse than that of the Soviets, starves its people as punishment, and isolates its subjects from the rest of the world nevertheless finds sufficient resources to achieve missile technology, nuclear capability, as well as associated weaponry with which to blackmail its neighbors and infect the world. Refugees' and escapees' stories tell of poverty and misery, but China, North Korea's ideological ally, has sealed its lengthy border in the interest of "stability" rather than pressuring the rulers in the capital of Pyongyang to reconsider their policies. Recent tentative moves toward change—a few cross-border visits by South Koreans to families in the North, some tourists allowed into the North, several (failed) industrial cooperatives, an American symphony orchestra visiting the capital—have essentially gone nowhere.

Iron Curtain, it is still there. Thus North and South Korea still await reunification.

In truth, the Koreans have been divided, partitioned, colonized, and occupied many times during their history. Even when outsiders were not involved, their indigenous kingdoms struggled for supremacy. During the Qing Dynasty, Chinese emperors intervened at will. As China fell apart, the Japanese conquered the peninsula and annexed it as their colony in 1910. When Japan was defeated at the end of World War II in 1945, the Allied powers divided Korea for

As Figure 9B-8 shows, the Cease-Fire Line marking the DMZ that forms the land boundary on the Korean Peninsula extends westward into the Yellow Sea, making Paengnyong Island part of South Korea. Offshore waters have long been a source of dispute between the two Koreas, with a dangerous escalation in March 2010, when a South Korean warship suffered an explosion, broke in half, and sank with the loss of 46 sailors. North Korea denied all involvement, but recovery of the two major parts of the ship (the stern is raised in this dramatic photo) enabled experts to reconstruct what happened. A submarine-launched torpedo, not an on-board accident or a wayward mine, destroyed the vessel in what many South Koreans saw as an act of war. Once again, state terrorism brought the Korean Peninsula perilously close to open conflict. © AFP/Getty Images, Inc.

South Korea

The tragedy of the Korean Peninsula is especially painful when put in geographic context. As all environmental maps indicate, North Korea and South Korea actually need each other: the North has raw materials required by industries in the South and South Korea produces food needed in the North. This is a condition known as **9 regional complementarity**, but in Korea the political situation prevents it from becoming operational to mutual advantage.

The DMZ, therefore, marks an ultimate core-periphery border at this peninsular scale. South Korea (49 million) emerged from the ravages of the Korean War as an unstable, dictatorial, politically and economically corrupt nation, its initial **10 state capitalism** propelled by powerful industrial conglomerates in cahoots with the politicians. But then democracy took hold, corruption was confronted (if not altogether tamed), and

the economy took off. South Korea, ever more connected to the world, became an early economic tiger on the Pacific Rim, the world's largest shipbuilding nation, a major automobile manufacturer, a producer of goods ranging from iron and steel to electronics and chemicals. Seoul, the capital, lies at the center of an urban-industrial complex with more than 10 million inhabitants facing the Yellow Sea across from China.

But there is much more to South Korea than its capital-centered core area. The country's historic, traditional regions still figure in its contemporary economic geography (Fig. 9B-8). In the southeast, *Kyongsang*, located on the Korea Strait opposite Japan's main island and centered on the city of Busan, has become South Korea's second-ranking manufacturing complex (the first Hyundai cars were made at Ulsan). And in the far southwest lies *Cholla*, anchored by the city of Kwangju, an area historically margin-

alized by the powers in Seoul but now finally coming into its own.

The contrasts between North and South on the Korean Peninsula are startling—and dangerous. Take a look at the comparative statistics for North and South in Appendix B and you will see why proponents of globalization often cite this part of the world as supportive evidence.

JAPAN

When we assess China's prospects of becoming a superpower, we should remember what happened in Japan in the nineteenth century. In 1868, a group of reform-minded modernizers seized power from an old guard, and by the end of that century Japan was a military and economic force. From the factories in and around Tokyo and from urban-industrial complexes elsewhere poured a stream of weapons and equipment that the Japanese used to embark on colonial expansion. By the mid-1930s, Japan lay at the center of an empire that included all of the Korean Peninsula, the whole of China's Northeast (which the Japanese called Manchukuo), the chain of Ryukyu Islands off the central coast of China, and Taiwan, as well as the southern half of Russia's Sakhalin Island (called Karafuto). Not even a disastrous earthquake, which destroyed much of Tokyo in 1923 and killed 143,000 people, could slow the Japanese drive.

Colonial Wars and Recovery

During World War II Japan expanded its domain farther than the architects of the 1868 modernization could ever have anticipated. By early December 1941, Japan had conquered extensive parts of China Proper, all of French Indochina to the south, and most of the small islands lying in the western Pacific. Then, on December 7, 1941, Japanese-built aircraft carriers moved Tokyo's warplanes within striking distance of Hawai'i, and the surprise attack on Pearl Harbor reinforced

© H. J. de Blij

"The city of Kyoto, chronologically Japan's second capital (after Nara; before Tokyo), is the country's principal center of culture and religion, education and the arts. Tree-lined streets lead past hundreds of Buddhist temples; tranquil gardens provide solace from the bustle of the city. I rode the bullet train from Tokyo and spent my first day following a walking route recommended by a colleague, but got only part of the way because I felt compelled to enter so many of the temple grounds and gardens. And not only Buddhism, but also Shinto makes its mark on the cultural landscape. I passed under a *torii*—a gateway usually formed by two wooden posts topped by two horizontal beams turned up at their ends—which signals that you have left the secular and entered the sacred, and found this beautiful Shinto shrine with its orange trim and olive-green glazed tiles."

www.conceptcaching.com

Japan's confidence in its war machine. Soon the Japanese overran the Philippines, the Netherlands East Indies (now Indonesia), Thailand, and British Burma and Malaya, and drove a wide corridor through the heart of China to the border with Vietnam.

A few years later, Japan's expansionist era was over. Its armies had been driven from virtually all its possessions, and when American nuclear bombs devastated two Japanese cities in 1945, the country lay in ruins. But once again, aided this time by an enlightened U.S. postwar administration, Japan surmounted disaster.

Japan's economic recovery and its rise to the status of world economic superpower was the success story of the second half of the twentieth century. Japan lost a war and an empire, but it scored many economic victories in a new global arena. Japan became an industrial giant, a technological pacesetter, a fully urbanized society, a political power, and a highly affluent nation. Cities everywhere have reliable Japanese cars on their streets;

tourists across the globe snap pictures with Japanese cameras; laboratories the world over use Japanese optical equipment. From microwave ovens to DVDs, from oceangoing ships to plasma TVs, Japanese-designed products flood world markets.

Japan's brief colonial adventure helped lay the groundwork for other economic successes along the western Pacific Rim. The Japanese ruthlessly exploited the natural and human resources of Korea and Taiwan, but they also installed a new economic order there. After World War II, this infrastructure facilitated an economic transition—and soon made both Taiwan and South Korea competitors in world markets.

British Tutelage

After the modernizers took control of Japan in 1868—an event known as the ***Meiji Restoration*** (the return of "enlightened rule" centered on the Emperor Meiji)—they turned to Britain for guidance in reforming their

nation and its economy. During the decades that followed, the British advised the Japanese on the layout of cities and the construction of a railroad network, the location of industrial plants, and the organization of education. The British influence still is visible in the Japanese cultural landscape: the Japanese, like the British, drive on the left side of the road. Consider how this affects the effort to open the Japanese market to U.S.-made automobiles!

The Japanese reformers of the late nineteenth century undoubtedly saw many geographic similarities between Britain and Japan. At that time, most of what mattered in Japan was concentrated on the country's largest island, Honshu (literally, *mainland*). The ancient capital, Kyoto, lay in the interior, but the modernizers wanted a coastal, outward-looking headquarters. So they chose the town of Edo, situated on a large bay where Honshu's eastern coastline bends sharply (Fig. 9B-9). They renamed the place ***Tokyo*** (meaning eastern capital), and

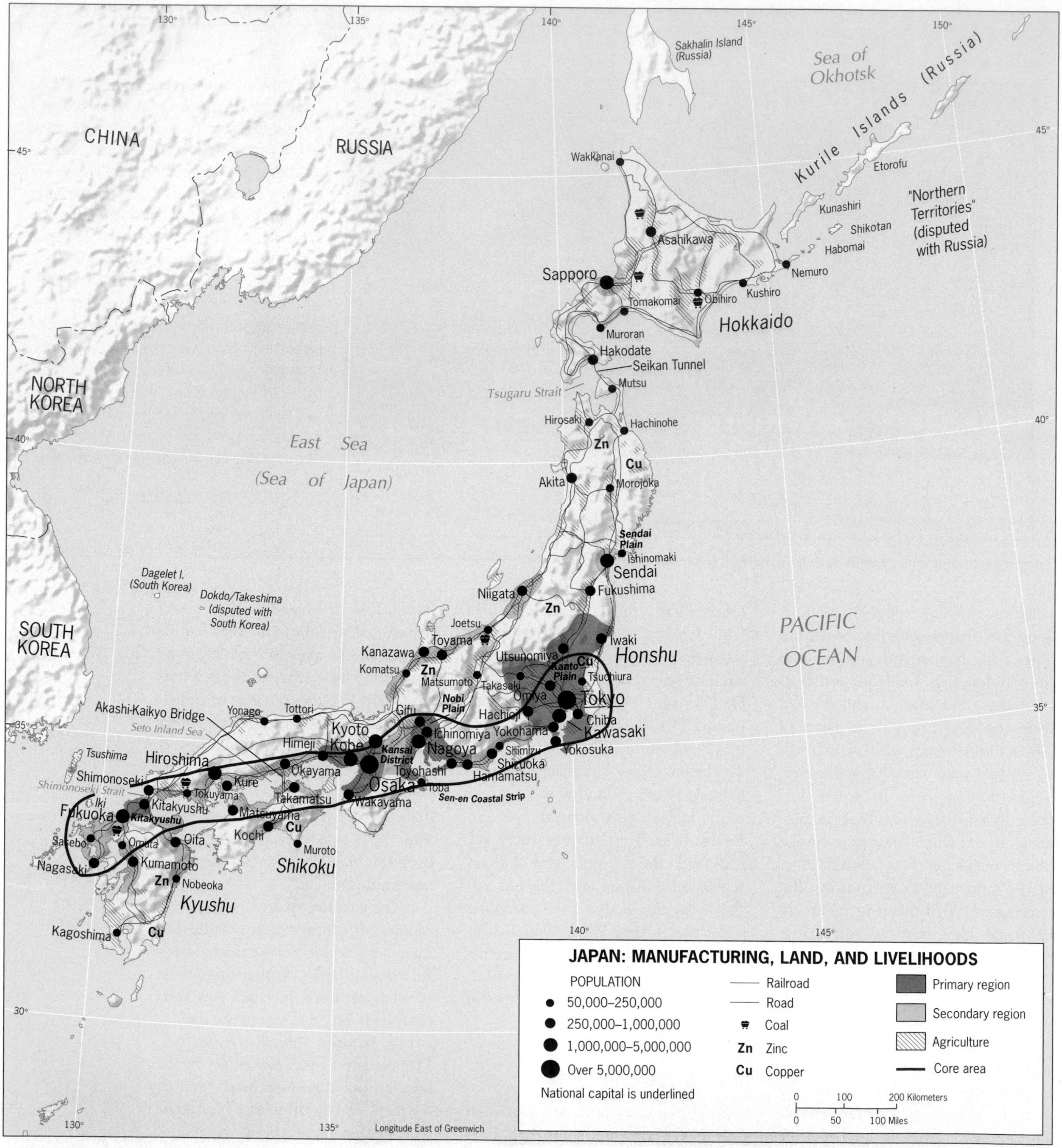

JAPAN: MANUFACTURING, LAND, AND LIVELIHOODS

POPULATION
- 50,000–250,000
- 250,000–1,000,000
- 1,000,000–5,000,000
- Over 5,000,000

National capital is underlined

— Railroad
— Road
Coal
Zn Zinc
Cu Copper

Primary region
Secondary region
Agriculture
— Core area

0   100   200 Kilometers
0   50   100 Miles

FIGURE 9B-9

© H. J. de Blij, P. O. Muller, and John Wiley & Sons, Inc.

little more than a century later it was the largest urban agglomeration on Earth. Honshu's coasts were also near mainland Asia, where raw materials and potential markets for Japanese products could be found. The notion of a greater Japanese empire followed naturally from the British example.

Spatial Contrasts and Constraints

In other ways, the British and Japanese archipelagoes, at opposite ends of the Eurasian landmass, differ considerably. In total area, Japan is larger. In addition to Honshu, Japan has three other large islands—Hokkaido to the north and Shikoku and Kyushu to the south—as well as numerous small islands and islets (Fig. 9B-9). Much of this territory is mountainous and steep-sloped, geologically young, earthquake-prone, and studded with volcanoes. Britain has lower relief, is older geologically, does not suffer from severe earthquakes, and has no active volcanoes. And in terms of the raw materials for industry, Britain was much better endowed than Japan. Self-sufficiency in iron ore and high-quality coal gave Britain a head start that lasted a century.

Japan's high-relief topography also constrained its economic development. Except for the ancient capital of Kyoto, all of Japan's major cities are perched along the coast, and virtually all lie partly on artificial land claimed from the sea. Sailing into Kobe harbor near Osaka (the country's second-largest city), one passes artificial islands designed for high-volume shipping and connected to the mainland by automatic space-age trains. Enter Tokyo Bay, and the refineries and factories to the east and west stand on huge expanses of landfill that have pushed the bay's shoreline outward. With just under 128 million people, four-fifths of whom reside in towns and cities, Japan uses its habitable living space very intensively—and expands it wherever possible.

As Figure 9B-9 shows, farmland in Japan is both limited and regionally fragmented. Urban sprawl has invaded much of the cultivable land. In the hinterland of Tokyo lies the Kanto Plain; around Osaka, the Kansai District; and surrounding Nagoya, the Nobi Plain—each a major farming zone under relentless urban pressure. All three of these plains lie within Japan's fragmented but well-defined core area (delimited by the red line on the map), the heart of Japan's prodigious manufacturing complex.

Modernization, Japanese Style

The reformers who set Japan on a new course in 1868 set in motion a process of **11** modernization, but in so doing they managed to build on, not replace, Japanese cultural traditions. We in the Western world tend to equate modernization with Westernization: urbanization, the spread of transport and communications facilities, the establishment of a market (money) economy, the breakdown of local traditional communities, the proliferation of formal schooling, and the acceptance and adoption of foreign innovations. In the non-Western world, the process is often viewed differently. There, modernization is seen as an outgrowth of colonialism, the perpetuation of a system of wealth accumulation introduced by foreigners driven by greed. In this view, the local elites who replaced the colonizers in the newly independent states merely continue the disruption of traditional societies, rather than truly modernizing them. Traditional societies, they argue, can be modernized without being Westernized.

In this context, Japan's modernization is unique. Having long resisted foreign intrusion, the Japanese did not achieve the transformation of their society by importing a Trojan horse; it was done by Japanese planners, building on the existing Japanese infrastructure, to fulfill Japanese objectives. Certainly, Japan imported foreign technologies and adopted innovations from the British and others, but the Japan that was built, a unique combination of modern and traditional elements, was basically an indigenous achievement.

The Role of Relative Location

Japan's changing fortunes during the past century reveal the influence of *relative location* in the country's development. When the Meiji Restoration took place, Britain, located on the other side of the Eurasian landmass, lay at the center of a global empire. The colonization and Europeanization of the world were in full swing. The United States was still a developing country, and the Pacific Ocean was an avenue for European imperial competition. Japan, even while it was conquering and consolidating its first East Asian colonies (the Ryukyus, Taiwan, Korea), lay remote from the mainstream of global change.

Then Japan became embroiled in World War II and dealt severe blows to the European colonial armies in Asia. The Europeans never recovered: the French lost Indochina, and the Dutch were forced to abandon their East Indies (now Indonesia). When the war ended, Japan was defeated and devastated, but at the same time the Japanese had done much to diminish the European presence in the Pacific Basin. Moreover, the global situation had changed dramatically. The United States, Japan's trans-Pacific neighbor, had become the world's wealthiest and most powerful, whereas Britain and its global empire were fading. Suddenly Japan was no longer remote from the mainstream of global action: now the Pacific was becoming an avenue to the world's richest markets. And Japan's relative location—its situation relative to the economic and political foci of the world—had changed. Therein lay much of the opportunity that the Japanese seized after the postwar rebuilding of their country.

Spatial Organization

As important as relative location is spatial organization. Imagine 128 million people crowded into a territory the size of Montana (population: 990,000), most of it mountainous, subject to frequent earthquakes and volcanism, with no domestic oilfields, little coal, few raw materials for industry, and very little level land for farming. If Japan today were an underdeveloped country in need of food relief and foreign aid, there would be abundant explanations for its condition, including overpopulation, inefficient farming, and energy shortages.

True, only an estimated 18 percent of Japan's national territory is designated as habitable. And Japan's large population is crowded into some very big cities. Moreover, Japan's agriculture is not especially efficient. But Japan defeated these odds by calling on old Japanese virtues: organizational efficacy, massive productivity, dedication to quality, and adherence to common goals. Even before the Meiji Restoration, Japan was a tightly organized country of some 30 million citizens.

All this proved invaluable to the modernizers when they set Japan on its new course. The country's industrial growth could be based on the urban and manufacturing development that was already taking place. As noted earlier, Japan does not possess major domestic raw-material sources, so no substantial internal reorganization was necessary. However, some cities were better situated than others relative to those limited local resources and, more important, external sources of raw materials. As Japan's regional organization took shape, a hierarchy of cities developed; Tokyo took and kept the lead, but other cities rapidly developed into industrial centers.

Areal Functional Organization

This process was governed by a geographic principle known as **12 areal functional organization**, a set of five

Tokyo, at the center of one of the largest metropolises in the world, continues to change. Land-filling and bridge-building in the bay continue; skyscrapers sprout amid low-rise neighborhoods in this earthquake-prone area; traffic congestion worsens. The red-painted Tokyo Tower, a beacon in this part of the city, was modeled on the Eiffel Tower in Paris but, as a billboard at its base announces, is an improvement over the original: lighter steel, greater strength, less weight. Tokyo Bay, part of which can be seen from this vantage point, was the scene of one of history's most costly environmental disasters during the earthquake of 1923. A giant tsunami (seismic sea wave) swept up the bay from the epicenter even as landfills liquefied and buildings sank into the mud. The death toll exceeded 140,000; a much larger population than lived in Tokyo in 1923 is now at risk of a repeat. © Yann Arthus-Bertrand/Photo Researchers, Inc.

interrelated tenets that help explain the evolution of regional organization, not only in Japan but across the world. Human activity has a *spatial focus*. It is *concentrated* in some locale, whether a farm or a factory or a store. Every one of these establishments occupies a particular *location*; no two of them can occupy exactly the same spot on the Earth's surface (even in high-rises there is a vertical form of absolute location). Nor can any human activity proceed in total isolation, so *interconnections* develop among these various establishments. This *system* of interconnections grows more complex as human capacities and demands expand. Each system (for example, farmers sending crops to market and buying equipment at service centers) forms a unit of areal functional organization.

In the introductory chapter, we referred to *functional regions* as sys-

tems of spatial organization; we can map units of areal functional organization as regions. These regions evolve because of so-called creative imaginations, in which people apply their cultural experience and technological know-how to organize and rearrange their living space. Finally, we can recognize levels of development in areal functional organization, a ranking of places and regions based on the type, extent, and intensity of exchange. Those levels are *subsistence, transitional*, and *exchange*.

Coastal Development

Japan's level of development, exchange, is the highest of the three categories. It is reflected in the organizational map (Fig. 9B-9). This map also tells us a great deal about the nature of Japan's exchange economy, its external orientation, and its dependence on foreign trade. All of the country's primary and secondary regions lie on the coast.

Dominant among these regions is the **Kanto Plain** (Fig. 9B-9), the heart of Japan's core area, which is focused on the Tokyo urban area and contains about one-third of the country's population. Among its advantages are an unusually extensive area of low relief, a fine natural harbor at Yokohama, a relatively mild and moist climate, and a central location with respect to the country as a whole. Its principal disadvantage lies in its vulnerability to earthquakes. The Kanto Plain and its Tokyo-centered metropolis (population 26.7 million) lie at the convergence of three tectonic plates, and Toyko has a centuries-long history of devastating earthquakes that have struck the region, on average, about every 70 years since 1633 (Figs. G-4, G-5).

The second-ranking economic region in Japan is called the **Kansai District** (Fig. 9B-9), and it contains the Osaka-Kobe-Kyoto triangle and is located at the east end of the Inland Sea between Honshu and Shikoku. Osaka and Kobe are major industrial centers and busy ports, but the Kansai District also yields large harvests of rice, Japan's staple food. Between the Kanto Plain and the Kansai District lies the **Nobi Plain** (Fig. 9B-9), where Nagoya is the key city. And, as the map shows, the Japanese core area is anchored in the west by the conurbation centered on **Kitakyushu**, situated not on Honshu Island but in the northwestern corner of Kyushu, Japan's southernmost major island. This five-city conurbation, of which Nagasaki is a part, continues to grow, favored by its location relative to Japan's Pacific Rim neighbors.

Japan Local and Global

In 2010, technical analyses and media reports focused on a momentous transition: Japan's national economy, once a powerful second only to that of the United States with no other rivals in sight, was about to be overtaken by that of China. There was a time (between the 1960s and the mid-1990s) when Japan's global reach was so dominant that a number of geographers designated Japan as a discrete geographic realm. Today, Japan does not even rank first in its own neighborhood.

Japan's two-decade-long economic slowdown had many causes ranging from government mismanagement and inefficiency to growing competition (such as from South Korean cars and Taiwanese high-technology products). Globalizing Japan, linked to many other economies, suffered when cyclical downturns affected its partners. Japanese corporations and workers, long accustomed to relationships that guaranteed lifelong employment and comfortable retirement, found that global competition made such commitments unaffordable. And during the economic boom, Japan's busy investors had bought up all kinds of properties, from landmark buildings in New York and movie studios in Hollywood to department stores in Europe and hotels in Hawai'i. When the values of those assets plummeted, economic damage resulted. For the first time in decades, Japan suffered a crisis of confidence and, more tangibly, something unfamiliar to the burgeoning Japan of the late twentieth century: rising unemployment.

Make no mistake: Japan still has a large and strong economy, but its future is clouded for several reasons, of which global competition for markets and raw materials are just two. Another problem lies in Japan's rapidly aging society, projected to decline from 128 million today to barely 95 million in 2050 and only about 65 million by the end of this century. Ethnically homogeneous Japan has historically resisted immigration, and when the government tried to recruit ethnic Japanese living in Brazil (and elsewhere) to return home, the experiment was not particularly successful. A shrinking base of qualified workers threatens the government's capacity to sustain social programs for all.

Still another set of problems has to do with Japan's international relations. Japan never signed a peace treaty with the (then) Soviet Union after the Second World War because the Russians had occupied and refused to return four small island groups in the Kurile chain northeast of Hokkaido (Fig. 9B-9). Failed negotiations for the return of these "Northern Territories," as they are called by the Japanese, have cost Japan the opportunity to play a key role in the economic development of the Russian Far East, where crucial energy as well as mineral resources abound.

Furthermore, relations with both South and North Korea are troubled—with the South over ownership of a small island group in the East Sea (Sea of Japan) and over memories of Japanese misconduct during World War II, and with the North over the kidnapping of Japanese citizens by North Korean agents. Moreover, Japan has obvious reasons to feel particularly threatened by North Korea's nuclear weapons development. And add to all this a number of lingering issues with China over Japanese actions during the colonial and wartime periods, and it is clear that Japan has to contend with discomforts other than just economic ones.

Ever since the end of the Second World War in 1945, Japan has adhered to a constitution that essentially forbids its rearmament and to a relationship with the United States that has involved the stationing of tens of thousands of American armed forces on Japanese soil. But Japan's 2009 election campaign brought to the fore an unusually forceful reappraisal of these issues, and public opinion has been shifting toward a stronger military posture and the ouster of U.S. troops. One result of the international community's failure to constrain North Korea's nuclear aspirations may be Japan's military revival. Another consequence may be a diminution of America's capacity to maintain the stability of a potentially volatile region.

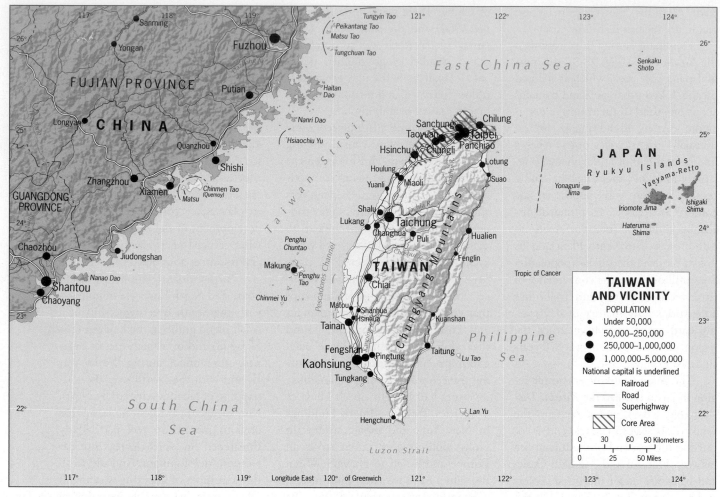

FIGURE 9B-10

© H. J. de Blij, P. O. Muller, and John Wiley & Sons, Inc.

≡ TAIWAN

Mention the island of Taiwan in China, and you are likely to be greeted with a frown and a headshake. Taiwan, your host may tell you, is a problem foreigners do not understand. Virtually all of the 23 million people of Taiwan are Chinese. Taiwan was part of China during the Qing Dynasty. Taiwan was stolen from China by Japanese imperialists in 1895, when it was known as Formosa. Then, when communists and "nationalists" were fighting each other for control of mainland China right after World War II, and the communists were winning, the Nationalists in 1949 fled by plane and boat to Taiwan, where they overpowered the locals. Even as Mao Zedong was proclaiming the People's Republic of China (PRC) in Beijing, the loser,

Chiang Kai-shek, named his regime in Taiwan's capital, Taipei, the Republic of China (ROC)—and told the world that the ROC was China's "legitimate" government.

Of course, the PRC never agreed to this, but the ROC had powerful friends, including America. Chiang Kai-shek's regime was installed in the United Nations in China's seat. Washington sent massive aid to support the island's economic recovery and weapons to ensure its security. While the PRC languished under communist rule, Taiwan (the name commonly used for the ROC) grew economically. And over time, Taiwan's political system matured into a functioning (if turbulent) democracy. Chiang's followers were even voted out of office, but not by a party that wanted to reunite Taiwan with the

PRC. What the Taiwanese wanted was economic prosperity, and they got it. Taiwan now became another of East Asia's economic tigers. Well before the PRC's Pacific Rim transformation started, Taiwan was exporting personal computers, telecommunications equipment, and precision electronic instruments from factories that dotted its western coastal zone from Taipei and Hsinchu in the north to Kaohsiung in the south (Fig. 9B-10).

But the ROC could not match its economic prowess with political progress. The PRC was adamant: things might be going well in Taiwan, but Taiwan was still a "wayward province" that must be reunited with the motherland. And when U.S. President Nixon arrived in Beijing in 1972 for a visit that was to change the world, Taiwan was a bargaining chip. Soon, the ROC's United Nations delegation was sent away and representatives of Beijing were seated in its place. Many countries around the world that had recognized Taiwan as the legitimate heir to China's leadership suddenly changed their tune. Beijing's leaders set about trying to isolate the ROC, and to a large extent they succeeded. When Deng Xiaoping and his pragmatists set up China's coastal SEZs, some of them right across from Taiwan, it looked like Taiwan's days as a political entity and as an economic power were numbered.

Geography, however, was to intervene. With billions of dollars in reserves and with good connections to Overseas Chinese in Southeast Asia, Taiwan had some strong cards to play—and the Beijing regime could not afford to deny Taiwanese companies permission to exploit SEZ opportunities. And so, via the "back door" of Hong Kong, Taiwanese entrepreneurs built thousands of factories in mainland China, many of them located directly across the Taiwan Strait. Taiwanese businesspeople now funneled hundreds of millions of dollars into China's Pacific Rim development boom. Today, even in difficult economic times, per-capita annual income in Taiwan exceeds U.S. $16,000—triple that of China and far more than that of several European countries. Few

beneficiaries would want to see their economic well-being imperiled by political adventures.

Entity and Identity

When we introduced the East Asian realm, we referred to its political components as including a failed state and a non-state—the latter the "entity" of Taiwan. Unfinished geopolitical business of this kind entails risks, but there are signs that all parties want to find a solution to Taiwan's unresolved status. Given the likelihood that the PRC will never allow Taiwan to attain independence, and the growing prospect that the United States, Taiwan's chief guarantor, could not secure Taiwan against Chinese armed intervention, many Taiwanese as well as their allies and adversaries are seeking a long-

range, negotiated solution. Among the options is one that appears to be working in Hong Kong, where the principle called **One Nation, Two Systems** has functioned better than many observers anticipated. As a Special Administrative Region (SAR), Hong Kong's economy has thrived, and freedoms not available in the Shi's, Provinces, or Autonomous Regions continue to prevail. In time, the Taiwanese may negotiate a status, internationally as well as domestically, that benefits all concerned.

In sum, the East Asian realm is as fraught with risk as it is filled with promise. As China's power rises, Japan's (comparatively) wanes, and America's role changes. East Asia is headed for a transition that will substantially define the route to the New World Order to which we alluded at the beginning of this book.

POINTS TO PONDER

- In the two decades between 1990 and 2010, China's economy vaulted from tenth place in the world to second.
- Construction began in 2010 on a 29-kilometer (18-mi) bridge-tunnel that will link Hong Kong to Macau and Zhuhai across the mouth of the Pearl River, part of China's ever-expanding infrastructure.
- At the end of 2009, the world's fastest train service began on dedicated tracks between Beijing and Shanghai. It now takes less than three hours to travel the 1068 kilometers (664 mi).
- Today an average Japanese citizen eats about 60 kilograms (132 lbs) of rice annually, only about half the amount consumed 50 years ago.

Buddhist-inspired Angkor Wat, Cambodia, the realm's most famous relic. © Jan Nijman

IN THIS CHAPTER

Natural hazards abound: A dangerous part of the world
An intricate ethnic and cultural mosaic
Infusion of religions from near and far
Overseas Chinese and their economic power
Political boundaries and their genetic classification
State territorial morphology: Why the shape of a country matters

CONCEPTS, IDEAS, AND TERMS

10A

SOUTHEAST ASIA: DEFINING THE REALM

FIGURE 10A-1

© H. J. de Blij, P. O. Muller, and John Wiley & Sons, Inc.

Southeast Asia is a realm of peninsulas and islands, a corner of Asia bounded by India on the northwest and China on the northeast (Fig. 10A-1). Its western coasts are washed by the Indian Ocean, and to the east stretches the vast Pacific. From all these directions, Southeast Asia has been penetrated by outside forces. From India came traders; from China, settlers; from across the Indian Ocean, Arabs to engage in commerce and Europeans to build empires; and from across the Pacific, Americans. Southeast Asia has been the scene of countless contests for power and primacy—the competitors have come from near and far.

Southeast Asia's geography in some ways resembles that of eastern Europe. It is a mosaic of smaller countries on the periphery of two of the world's largest states. It has been a **1 buffer zone** between powerful adversaries. It is a **2 shatter belt** in which stresses and pressures from without and within have fractured the political geography. Like eastern Europe, Southeast Asia exhibits great cultural diversity. This is a realm of hundreds of cultures and ethnicities, numerous languages and dialects, global as well as local religions, and diverse national economies ranging from high- to low-income.

MAJOR GEOGRAPHIC QUALITIES

SOUTHEAST ASIA

1. Southeast Asia extends from the peninsular mainland to the archipelagos offshore. Because Indonesia controls part of New Guinea, its functional region reaches into the neighboring Pacific geographic realm.

2. Southeast Asia, like eastern Europe, has been a shatter belt between powerful adversaries and has a fractured cultural and political geography shaped by foreign intervention.

3. Southeast Asia's physiography is dominated by high relief, crustal instability marked by volcanic activity and earthquakes, and tropical climates.

4. A majority of Southeast Asia's more than 600 million people live on the islands of just two countries: Indonesia, with the world's fourth-largest population, and the Philippines. The rate of population increase in the Insular region of Southeast Asia exceeds that of the Mainland region.

5. Although the overwhelming majority of Southeast Asians have the same ancestry, cultural divisions and local traditions abound, which the realm's divisive physiography sustains.

6. The legacies of powerful foreign influences, Asian as well as non-Asian, continue to affect the cultural landscapes of Southeast Asia.

7. Southeast Asia's political geography exhibits a variety of boundary types and several categories of state territorial morphology.

8. The Mekong River, Southeast Asia's Danube, has its source in China and borders or crosses five Southeast Asian countries, sustaining tens of millions of farmers, fishing people, and boat owners.

9. The realm's giant in terms of territory as well as population, Indonesia, has not asserted itself as the dominant state because of mismanagement and corruption; but Indonesia has enormous potential.

A GEOGRAPHIC OVERVIEW

Figure 10A-1 displays the relative location and dimensions of the Southeast Asian geographic realm, an assemblage of 11 countries situated on the Asian mainland and on thousands of islands, large and small, extending from Sumatera* in the west to New Guinea in the east, and from Luzon in the north to Timor in the south. We will become familiar with only the largest and most populous of these islands, but sailing the seas of this part of the world you would pass dozens of smaller ones every day, and if you could stop you would find even the tiniest populated islands to have their own character resulting from the cultural sources of their residents, the environmental challenges they face and the opportunities they exploit, their modes of dress and the structure of their dwellings, even the vivid colors with which they often decorate their boats.

The giant of this realm, in terms of both area and population, is the far-flung archipelago of Indonesia, labelled appropriately on the map by the largest letters. In the east, the state of Indonesia extends beyond the Southeast Asian realm into the Pacific Realm because it controls the western half of an island—New Guinea—whose indigenous peoples are not Southeast Asian. We focus on this unusual situation later (and on New Guinea as a whole in Chapter 12), but note it here because unusual borders and divided islands are a hallmark of this realm.

On the Asian mainland, Southeast Asia is bordered by India to the northwest and by China to the northeast, both sources of immigrants, cultural infusions, economic initiatives, and other relationships evident in the realm's cultural landscapes. Also originating there are rivers that play a major role in the lives of many millions of people south of the realm border. As we will find, the core areas of several of Southeast Asia's most populous countries lie in the basins of these major rivers.

Before we get started, it is helpful to get acquainted with the key states of this realm, some of which will already be familiar. North of Indonesia lies the Philippines, well known to Americans because this

---

*As in Africa and South Asia, names and spellings have changed with independence. In this chapter we will use contemporary spellings, except when we refer to the colonial period. Thus Indonesia's four major islands are Jawa, Sumatera, Kalimantan (the Indonesian part of Borneo), and Sulawesi. The Dutch called them Java, Sumatra, Borneo, and Celebes, respectively.

was once an American colony and is still a source of many immigrants to the United States. Also north of Indonesia is Malaysia, easy to find on the map because its core area lies on the long peninsula that almost connects mainland Asia with Indonesia. At the tip of that peninsula lies another famous geographic locality: Singapore, the great economic success story of Southeast Asia and a "global city" by any measure.

On the Asian mainland, the name Vietnam still resonates in America, which fought a bitter and costly war there in the 1960s and 1970s. As the map shows, Vietnam looks like a sliver of land extending from its border with China to the delta of the greatest of all Southeast Asian rivers, the Mekong. Looking westward, neighboring Laos and Cambodia may not be so familiar, but next comes centrally positioned Thailand, the dream (and reality) of millions of tourists and one of the world's most interesting countries. And then, on Southeast Asia's western margin, we come to Myanmar, one of the planet's most tragic places, where human potential and talent as well as natural endowment and opportunity were ruined and wasted by power struggles and the force of arms, resulting in a poverty-stricken state where time seems to have stood still.

SOUTHEAST ASIA'S PHYSICAL GEOGRAPHY

In certain ways, Southeast Asia is reminiscent of Middle America, a fractured realm of islands and peninsulas flanking a populous mainland studded with high mountains and deep valleys. It is useful to look back at Figures G-4 and G-5 to see why: both Southeast Asia and Middle America are dangerous places, where the Earth's crust is unstable as tectonic plates are in collision, earthquakes are a constant threat, volcanic eruptions take their toll, tropical cyclones lash sea and land, and floods, landslides, and other natural hazards make life riskier than in most other parts of the world.

In terms of human geography, Southeast Asia is part of the Pacific Rim. But in physiographic terms, it forms part of the Pacific Ring of Fire and all the hazards this designation brings with it. As recently as 2004 an undersea earthquake off westernmost Indonesia caused a **3** tsunami in the Indian Ocean that killed more than 300,000 people along coastlines from Sumatera to Somalia. This was only the latest in an endless string of natural disasters originating in Southeast Asia whose effects were felt far beyond the realm's borders. In 1883, the Krakatau volcano between Sumatera and Jawa exploded with a death toll estimated at 30,000. In 1815, the Tambora volcano in the chain of islands east of Jawa known as the Lesser Sunda Islands blew up, darkening skies throughout the world and affecting climates around the planet (the year that followed is still known as the "year without a summer" when crops failed, economies faltered, and people went hungry as far away as

When you take the road from Semarang on Jawa's north coast toward Yogyakarta to the south, via Borobudur, menacing volcanic Mount Merapi is a constant companion, visible almost all the way. People in the surrounding countryside as well as nearby towns live with constant risk from eruptions and associated earthquakes in this especially unstable, Indonesian part of the Pacific Ring of Fire. © Dean Conger/©Corbis

Egypt, New England, and France). Research on what may have been the most calamitous of all such eruptions suggests that, about 73,000 years ago, the Toba volcano on Sumatera exploded with such force that its ash and soot not only darkened skies and affected weather, but changed global climate for perhaps as long as 20 years and threatened the very survival of the human population, then still small in number and widely dispersed. Some scientists postulate that this eruption caused such widespread casualties that human genetic diversity was significantly diminished.

It therefore goes without saying that high relief dominates Southeast Asia, from the Arakan Mountains in western Myanmar to the glaciers (yes, glaciers!) of New Guinea. Take a close look at Figure 10A-1 and you can see how many elevations approach or exceed 3000 meters (10,000 ft), and, using the elevation guide in the upper right-hand corner, how mountainous and hilly much of the realm's topography is. Lengthy ranges form the backbones not only of islands such as Sulawesi and Sumatera, but also of the Malay Peninsula, most of Vietnam, and of the border zone between Thailand and Myanmar. In Figure G-5 it is possible to trace these volcano-studded mountain ranges by their earthquake epicenters and volcanic records.

Exceptional Borneo

But among the islands there is one significant exception. The bulky island named Borneo in Figure 10A-1 has high elevations (Mount Kinabalu in the north reaches 4101 meters [13,455 ft]), yet has no volcanoes and negligible earth tremors.

This island has been called a stable "mini-continent" amid a mass of volcanic activity, a slab of ancient crust that long ago was pushed high above sea level by tectonic forces and was subsequently eroded into its present landscapes. Borneo's soils are not nearly as fertile as those of the volcanic islands, so that an equatorial rainforest developed here that long survived the human population explosion, giving sanctuary to countless plant and animal species including Southeast Asia's great ape, the Orangutan. That era is now ending as human encroachment on Borneo's tropical habitat is accelerating, logging is destroying the remaining forest, and roads and farms are penetrating its interior.

The remnants of Borneo's equatorial rainforest form part of a much larger stand of rainforest that once extended over far more of this realm than is the case today. Other than Borneo, eastern Sumatera still has some limited expanses of it, including a few Orangutan sanctuaries. So does less populous and more remote New Guinea, which was never reached by the Orangutans. As we noted in the introductory chapter, equatorial rainforests still stand today in three major areas of the world: Equatorial Africa, Amazonian South America, and here in Southeast Asia. The combination of climatic conditions that sustains these forests—warm temperatures and consistent, year-round rainfall—produces the biologically richest and ecologically most complex vegetation regions on Earth. A huge number and variety of trees and other plants grow in very close proximity, vying for space and sunlight both horizontally and vertically. But despite all this luxuriant growth, the soils on which rainforests grow are not rich in nutrients. The greatest concentration of nutrients is in the decaying vegetation on the forest floor, from which the new generation of plants derive their sustenance. Local people who have inhabited the forest for a long time have developed ways to use plants, animals, and soils to make their living, but when farmers remove the trees wanting to use the soil on the assumption that it will support their crops, failure usually results. Nevertheless, population pressure everywhere in the tropics continues to shrink what remains of the world's rainforests.

Relative Location and Biodiversity

Look again at Figure 10A-1, but in a more general way. The Malay Peninsula, adjacent Sumatera, Jawa, and the Lesser Sunda Islands east of Jawa seem to form a series of stepping stones toward New Guinea, and in the lower right-hand corner of the map you can see Australia's Cape York Peninsula seeming to reach toward New Guinea. What would happen, you might ask, if sea level were to drop and those narrow bodies of water between the peninsulas and the islands dried up?

That is precisely what occurred—not just once, but repeatedly in the geologic history of Southeast Asia. Long

ago, the Orangutans whose descendants remain in Indonesia today were able to migrate from the mainland into warmer equatorial latitudes because sea levels dropped during periods of glaciation, and islands separated by water today were temporarily connected by land bridges. More recently, early human arrivals here also got assistance from nature: Australia's Aborigines probably managed to cross what remained of the deeper trenches between islands by building rafts, but their epic journey was facilitated by land bridges as well.

Time and again, therefore, Southeast Asia was a receptacle for migrating species. Combine this with the realm's tropical environments, and it is no surprise that it is known for its **4** biodiversity. Scientists estimate that fully 10 percent of the Earth's plant and animal species are found in this comparatively small realm. Biogeographers can trace the progress of many of these species from the mainland across the archipelago toward Australia, but some were stopped by deeper trenches that remained filled with water even when sea levels dropped. An especially deep trench lies in the Lombok Strait between the islands of Bali (next to Jawa) and Lombok (Fig. 10A-1), a biogeographical boundary first recognized by the naturalist Alfred Russel Wallace, a contemporary of Charles Darwin.

As we shall presently see, Southeast Asia's biodiversity had a fateful impact on its historical geography. Among the realm's specialized plants, the spices attracted outsiders from India, China, and Europe, with consequences still visible on the map today.

Four Major Rivers

Water is the essence of life, and among world realms Southeast Asia is comparatively well endowed with moisture (Fig. G-7). Ample, sometimes even excessive, rainfall fills the rice-growing paddies of Indonesia and the Philippines. On the Southeast Asian mainland, where annual rainfall averages are somewhat lower and the precipitation is more seasonal, major rivers and their tributaries fill irrigation channels and form fertile deltas. The map of population distribution clearly highlights this spatial relationship between rivers and people expressed in distinct coastal clustering (Fig. 10A-2).

As Figure 10A-1 reminds us, rivers that are crucial to life in certain realms sometimes have their sources in neighboring realms, a geographic issue that can lead to serious regional discord. What right does an "upstream" state in one realm have to dam, or otherwise interfere with, a river that is vital to a state (or states) downstream? In Southeast Asia's case, two major rivers originate in China, but two others are internal. The two that flow from China are the *Mekong*, which crosses the mainland from north to south, and the *Red* River that reaches the sea in northern Vietnam. The

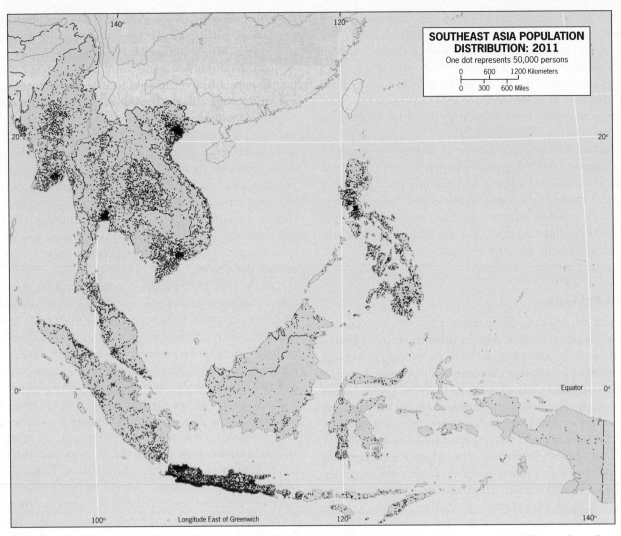

FIGURE 10A-2   © H. J. de Blij, P. O. Muller, and John Wiley & Sons, Inc.

local rivers are the ***Irrawaddy***, the lifeline of Myanmar, and the ***Chao Phraya***, the key artery of Thailand.

The Mighty Mekong

From its source among the snowy highlands of China's Qinghai-Xizang (Tibetan) Plateau, the Mekong River rushes and flows some 4200 kilometers (2600 mi) to its delta in southernmost Vietnam. This "Danube of Southeast Asia" crosses or borders five of the realm's countries, supporting rice farmers and fishing people, forming a transportation route where roads are few, and providing electricity from dams upstream. Tens of millions of people depend on the waters of the Mekong, from subsistence farmers in Cambodia to apartment dwellers in China. The Mekong Delta in southern Vietnam is one of the realm's most densely populated areas and produces enormous harvests of rice.

But problems loom. China is building a series of dams across the Lancang (as the Mekong is called there) to supply surrounding Yunnan Province with electricity. Although such hydroelectric dams should not interfere with water flow, countries downstream worry that a severe dry spell in the interior would impel the Chinese to slow the river's flow to keep their reservoirs full. Cambodia is especially concerned over the future of the Tonlé Sap, a large natural lake filled by the Mekong (see Fig. 10B-2). In Vietnam, farmers worry about salt water invading the delta's paddies should the Mekong's level drop. And the Chinese may not be the only dam builders in the future: Thailand has expressed an interest in building a dam on the Thai-Laos border where it is defined by the Mekong.

In such situations, the upstream states have an advantage over those downstream. Several international organizations have been formed to coordinate development in the Mekong Basin, including the Mekong River Commission

(MRC), founded more than 50 years ago. China has offered to sell electricity from its dams to Thailand, Laos, and Myanmar. Coordinated efforts to reduce deforestation in the Mekong's drainage basin have had some effect. After consultations with the MRC, Australia built a bridge linking Laos and Thailand. There is even a plan to make the Mekong navigable from Yunnan to the coast, creating an alternative outlet for interior China.

Sail the Mekong today, however, and you are struck by the slowness of development along this artery. Wooden boats, thatch-roofed villages, and teeming paddies mark a river still crossed by antiquated ferries and flanked by few towns. Of modern infrastructure, one sees little. And yet the Mekong and its basin form the lifeline of mainland Southeast Asia's dominantly rural societies.

Rivers and States

Whereas the Mekong River affects the lives of peoples in several Southeast Asian states, the other three major Southeast Asian rivers are essentially local. As the map shows, the Red River forms the focus for the heavily populated Tonkin Plain in northern Vietnam where the capital, Hanoi, lies on its banks. In Thailand, the relatively short but crucial Chao Phraya, on which Bangkok lies, is just one of a series of channels in that river's delta, formed by numerous streams from the country's interior. And Myanmar's Irrawaddy River crosses that country from north to south, its valley one of the world's leading rice-producing areas and its largest city, Yangon, at the corner of its delta.

POPULATION GEOGRAPHY

Examine the map of Southeast Asia's population distribution (Fig. 10A-2), and you are immediately struck by the huge concentration of people on a relatively small island in Indonesia—a cluster larger than any other in the realm, the four mainland river deltas included. The island is Jawa, and its population of more than 140 million is not only larger than any other, but also Jawa alone has more people than any of the countries in the entire realm except Indonesia, of which it is a part. And even there, it contains more than half the country's population.

This is even more remarkable because Indonesia is not yet a highly urbanized country. In 2010, more than 50 percent of all Indonesians still lived in rural areas, and although Jawa, as we shall see in Chapter 10B, is the most highly urbanized island of Indonesia, tens of millions of people live off the land even there. What makes all this possible is a combination of fertile volcanic soil, ample water, and extremely warm temperatures that enable Jawa's farmers to grow three crops of rice in a single paddy during a single year, helping feed a national population still growing faster than the global average. But make no mistake: the darkest red clusters on Jawa also signify fast-growing urban areas, where Indonesians are building a new economy with global linkages.

Within Indonesia, the contrasts between Jawa and the four other major islands—Sumatera, Borneo (Kalimantan), Sulawesi, and particularly Indonesian New Guinea—reflect the core-periphery relationship between the two sectors of this country. Such contrasts, less sharply defined, also mark other Southeast Asian countries, as the map suggests: the primate cities here (Bangkok, Manila, Yangon, Kuala Lumpur) are particularly dominant. Vietnam has two such anchors, representing respectively its historic northern (Hanoi) and southern (Ho Chi Minh City) core areas. You do not need the names on the map to recognize these huge cities or their dominance.

As the table in Appendix B indicates and the map confirms, Southeast Asia's states are not, by world standards, especially populous. Indonesia today is the world's fourth-ranking country in terms of population (247 million), but no other Southeast Asian country has even half of Indonesia's numbers. Three mainland countries contain between 50 and 100 million people. But take Laos, quite a large country territorially (about the size of the United Kingdom) and note that its population barely exceeds 6 million; similarly, Cambodia, half the size of Germany, has about 15 million. In part, such modest numbers on the mainland reflect natural conditions less favorable to farming than those prevailing on volcanic soils or in fertile river basins, but more generally this realm did not grow as explosively during the last century as did neighboring realms. Indeed, Appendix B indicates that several countries in Southeast Asia today are growing more slowly than the global average of 1.3 percent annually.

The Ethnic Mosaic

Southeast Asia's peoples come from a common stock just as (Caucasian) Europeans do, but this has not prevented the emergence of regionally or locally discrete ethnic or cultural groups. Figure 10A-3 displays the broad distribution of ethnolinguistic groups in the realm, but be aware that this is a generalization. At the scale of this map, numerous small groups cannot be depicted.

Figure 10A-3 shows the rough spatial coincidence, on the mainland, between major ethnic group and modern political state. The Burman dominate in the country formerly called Burma (now Myanmar); the Thai occupy the state once known as Siam (now Thailand); the Khmer form the nation of Cambodia and extend northward into Laos; and the Vietnamese inhabit the long strip of territory facing the South China Sea.

Territorially, by far the largest population shown in Figure 10A-3 is classified as Indonesian, the inhabitants

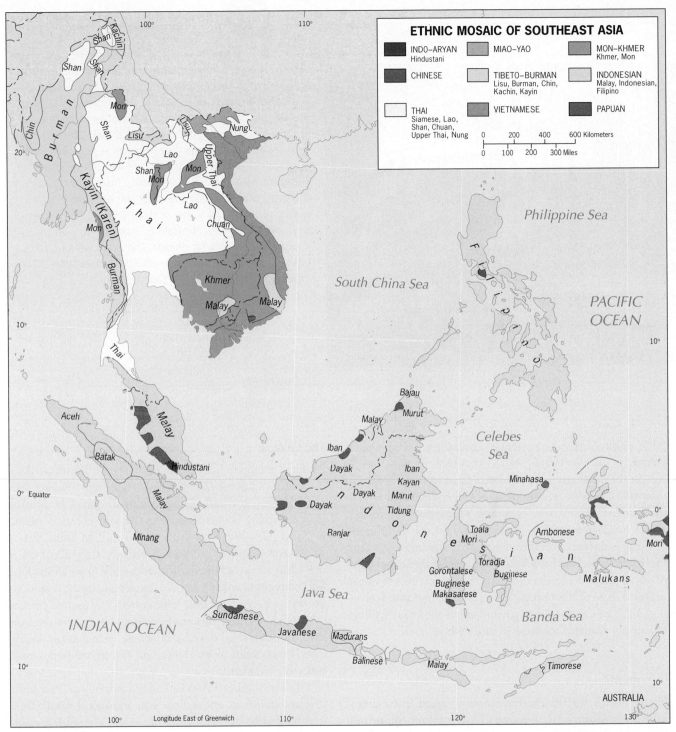

ETHNIC MOSAIC OF SOUTHEAST ASIA

■ INDO-ARYAN Hindustani	MIAO-YAO	MON-KHMER Khmer, Mon
■ CHINESE	TIBETO-BURMAN Lisu, Burman, Chin, Kachin, Kayin	INDONESIAN Malay, Indonesian, Filipino
THAI Siamese, Lao, Shan, Chuan, Upper Thai, Nung	VIETNAMESE	PAPUAN

0   200   400   600 Kilometers

0   100   200   300 Miles

FIGURE 10A-3

© H. J. de Blij, P. O. Muller, and John Wiley & Sons, Inc.

of the great island chain that extends from Sumatera west of the Malay Peninsula to the Malukus (Moluccas) in the east and from the Lesser Sunda Islands in the south to the Philippines in the north. Collectively, all these peoples shown on the map—the Filipinos, Malays, and Indonesians—are known as Indonesians, but they have been divided by history and politics. And note as well that the Indonesians in Indonesia itself include Javanese,

Madurese, Sundanese, Balinese, and other major groups; hundreds of smaller ones cannot be mapped at this scale. In the Philippines, too, island isolation and contrasting ways of life are reflected in the cultural mosaic. Also part of this Indonesian ethnic-cultural complex are the Malays, whose heartland lies on the Malay Peninsula but who form minorities in other areas as well. Like most Indonesians, the Malays are Muslims, although Islam is a more powerful

© H. J. de Blij

"Like most major Southeast Asian cities, Bangkok's urban area includes a large and prosperous Chinese sector. No less than 14 percent of Thailand's population of 67 million is of Chinese ancestry, and the great majority of Chinese live in the cities. In Thailand, this large non-Thai population is well integrated into local society, and intermarriage is common. Still, Bangkok's "Chinatown" is a distinct and discrete part of the great city. There is no mistaking Chinatown's limits: Thai commercial signs change to Chinese, goods offered for sale also change (Chinatown contains a large cluster of shops selling gold, for example), and the urban atmosphere, from street markets to bookshops, is dominantly Chinese. This is a boisterous, noisy, energetic part of multicultural Bangkok, a vivid reminder of the Chinese commercial success in Southeast Asia."

www.conceptcaching.com

force within Malay society than it generally is in Indonesian culture.

In the northern part of the mainland region, numerous minorities inhabit remote parts of the countries in which the Burman (Burmese), Thai, and Vietnamese dominate. Those minorities tend to occupy areas on the peripheries of their countries, where the terrain is mountainous and the forest is dense, and where the governments of their national states do not have complete control. This remoteness and sense of detachment give rise to aspirations of secession, or at least resistance to governmental efforts to establish full authority, often resulting in bitter ethnic conflict.

Immigrants

Figure 10A-3 further reminds us that, again like eastern Europe, Southeast Asia is home to major ethnic minorities from outside the realm. On the Malay Peninsula, note the South Asian (Hindustani) cluster. Such Hindu communities with Indian ancestries exist in many parts of the peninsula, but in the southwest they form the majority in a small area. In Singapore, too, South Asians form a significant minority. These communities arose during the European colonial period, but South Asians had arrived in this realm many centuries earlier, propagating Buddhism and leaving architectural and cultural imprints on places as far away as Jawa and Bali.

The Chinese

By far the largest immigrant minority in Southeast Asia, however, is Chinese. The Chinese began arriving here during the Ming and early Qing (Manchu) dynasties, but the largest exodus occurred during the late colonial period (1870–1940), when as many as 20 million immigrated. The European powers at first encouraged this influx, using the Chinese in administration and trade. But soon these **5** **Overseas Chinese** began to congregate in the major cities, where they established Chinatowns and gained control over much of the commerce (see photo above). By the time the Europeans tried to reduce Chinese immigration, World War II was about to break out and the colonial era would end soon afterward.

Today Southeast Asia is home to as many as 30 million Overseas Chinese, more than half the world total. Their lives have often been difficult. The Japanese relentlessly persecuted those Chinese who lived in Malaya during World War II. Later, during the 1960s, Chinese in Indonesia were accused of communist sympathies, and hundreds of thousands were killed. More recently, in the late 1990s, Indonesian mobs again attacked Chinese and their property, this time because of their relative wealth and because many Chinese had become Christians during the colonial era and were now targeted by Islamic throngs. Resentment still continues, expressed by episodic flare-ups against Chinese in various parts of Southeast Asia.

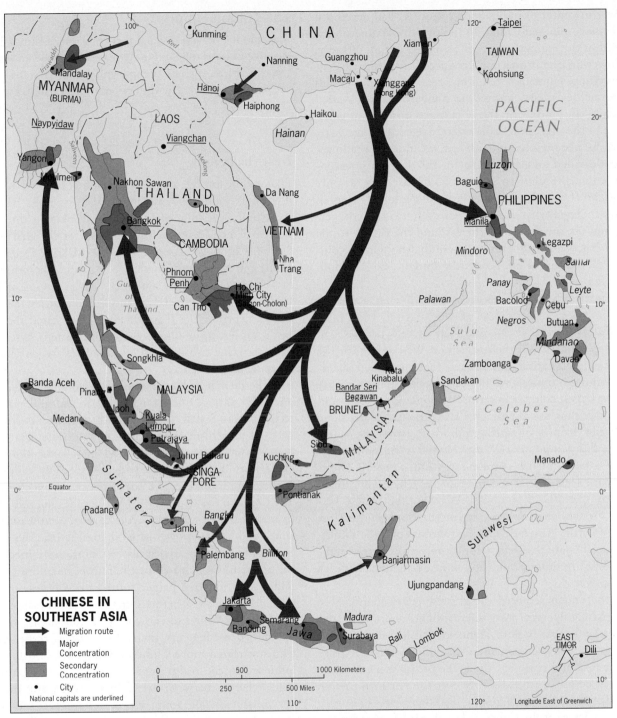

FIGURE 10A-4

© H. J. de Blij, P. O. Muller, and John Wiley & Sons, Inc.

Figure 10A-4 shows the migration routes and current concentrations of Chinese in Southeast Asia. Most originated in southern China's Fujian and Guangdong provinces, and a large number invested much of their wealth back in China when it opened up to foreign businesses three decades ago. Clearly, the Overseas Chinese of Southeast Asia have played a significant role in shaping the economic miracle of the Pacific Rim.

COLONIALISM'S HERITAGE: HOW THE POLITICAL MAP EVOLVED

When the European colonizers arrived in Southeast Asia, they encountered a patchwork of kingdoms, principalities, sultanates, and other traditional political entities whose leaders they tried to co-opt, overpower, or otherwise fold into their imperial schemes. There was no single, powerful center of indigenous culture as had developed in Han-dominated China. In the river basins and on the plains of the mainland, as well as on the islands offshore, a flowering of cultures had produced a diversity of societies whose languages, religions, arts, music, foods, and other achievements formed an almost infinitely varied mosaic—but none of those cultures rose to imperial power. The European colonizers forged empires here, often by playing one state off against another; the Europeans divided and ruled. Out of this foreign intervention came the modern map of Southeast Asia, as only Thailand (formerly Siam) survived the colonial era as an independent entity. Thailand was useful to two competing powers, the French to the east and the British to the west. It was a convenient buffer, and although the colonists carved pieces off Thailand's domain, the kingdom endured.

Indeed, the Europeans accomplished what local powers could not: the formation of comparatively large, multicultural states that encompassed diverse peoples and societies and welded them together. Were it not for the colonial intervention, it is unlikely that the 17,000 islands of far-flung Indonesia would today constitute the world's fourth-largest country in terms of population. Nor would the nine sultanates of Malaysia have been united, let alone with the peoples of northern Borneo across the South China Sea. For good or ill, the colonial intrusion consolidated a realm of few culture cores and numerous ministates into less than a dozen countries. The leading colonial competitors in Southeast Asia were the Dutch, British, French, and Spanish (with the Spanish later replaced by the Americans in their stronghold, the Philippines). The Japanese had colonial objectives here as well, but these aspirations came and went during the course of World War II.

The Dutch acquired the greatest prize: control over the vast archipelago now called Indonesia (formerly the Netherlands East Indies). France established itself on the eastern flank of the mainland, controlling all territory east of Thailand and south of China. The British conquered the Malay Peninsula, gained power over the northern part of the island of Borneo, and established themselves in Burma (Myanmar) as well. Other colonial powers also gained footholds, but not for long. The exception was Portugal, which held on to its eastern half of the island of Timor (Indonesia) until well after the Dutch had been ousted from their East Indies.

Figure 10A-5 shows the colonial framework in the late nineteenth century, before the United States assumed control over the Philippines in 1898. Note that while Thailand survived as an independent state, it lost territory to the British in Malaya and Burma and to the French in Cambodia and Laos.

The Colonial Imprint

The colonial powers divided their possessions into administrative units as they did in Africa and elsewhere. Some of these political entities became independent states when the colonial powers withdrew or were ousted by force (Fig. 10A-5).

French Indochina

France, one of the mainland's leading colonial powers, divided its Southeast Asian empire into five units. Three of these units lay along the east coast: Tonkin in the north next to China, centered on the basin of the Red River; Cochin China in the south, with the Mekong Delta as its focus; and in between these two, Annam. The other two French territories were Cambodia, facing the Gulf of Thailand, and Laos, landlocked in the interior. Out of these five French dependencies there emerged three states: the three east-coast territories ultimately became a single state, Vietnam; the other two—Cambodia and Laos—each achieved separate independence.

The French had a name for their empire: *Indochina*. The *Indo* part of Indochina refers to cultural imprints received from South Asia: the Hindu presence; the importance of Buddhism, which came to Southeast Asia via Sri Lanka (Ceylon) and its seafaring merchants; the influences of Indian architecture and art (especially sculpture), writing and literature, and social structures and patterns. The *China* in the name Indochina signifies the role of the Chinese here. Chinese emperors coveted Southeast Asian lands, and China's power penetrated deep into the realm. Social and political upheavals in China, combined with the opportunities created by the European colonists, sent millions of Sinicized people southward. Chinese traders, pilgrims, seafarers, fishermen, and many others sailed from southeastern China to the coasts of Southeast Asia and established settlements there. Over time, those settlements attracted more Chinese immigrants, and Chinese influence in the realm grew (Fig. 10A-4). Not surprisingly, relations between the Chinese settlers and the earlier inhabitants of Southeast Asia have at times been strained, even violent. The Chinese presence in Southeast Asia is long-term, but the invasion has continued into modern times.

The name *Indochina* can only refer to a part of mainland Southeast Asia, however; it cannot be used to refer to the realm as a whole. Although the *Indo* segment of this regional name can be taken to also refer to the Buddhist influences that dominate here, it makes no reference to the momentous arrival of Islam, introduced by Arab seafarers in the twelfth and thirteenth centuries, and destined to transform the cultural geography of this realm.

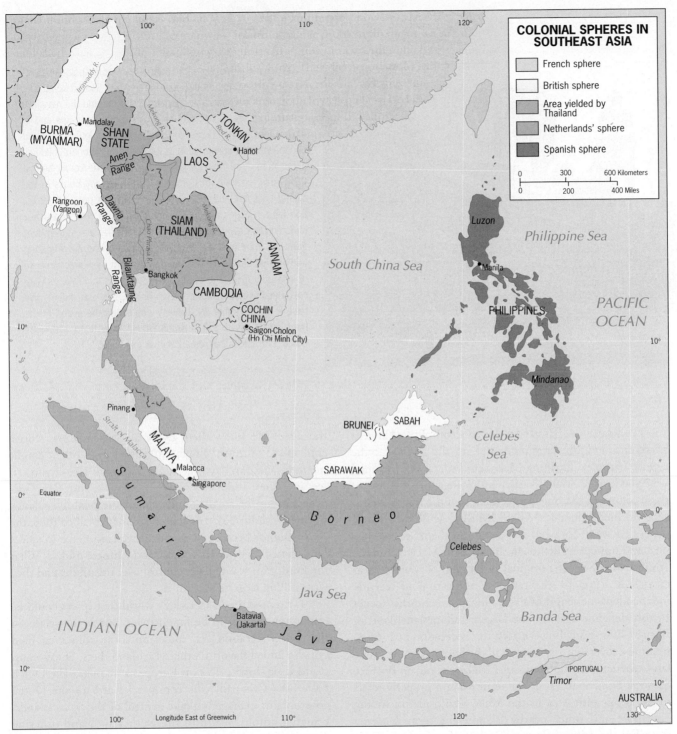

FIGURE 10A-5

British Imperialism

The British ruled a pair of major entities in Southeast Asia (Burma and Malaya) in addition to a large part of northern Borneo and many small islands in the South China Sea. Burma was attached to Britain's Indian Empire; from 1886 until 1937, it was governed from distant New Delhi. But when British India became independent in 1947 and split into several countries, Burma was not part of the grand design that created West and East Pakistan (the latter now Bangladesh), Ceylon (now Sri Lanka), and India. Instead, Burma (today called Myanmar) was given the status of a sovereign republic in 1948.

FROM THE FIELD NOTES...

© H. J. de Blij

"Walking along Tuanku Abdul Rahman Street in Kuala Lumpur, I had just passed the ultramodern Sultan Abdul Samad skyscraper when this remarkable view appeared: the old and the new in a country that seems to have few postcolonial hang-ups and in which Islam and democracy coexist. The British colonists designed and effected the construction of the Moorish-Victorian buildings in the foreground (now the City Hall and Supreme Court); behind them rises the Bank of Commerce, one of many banks in the capital. Look left, and you see the Bank of Islam, not a contradiction here in economically diversified Malaysia. And just a few hundred yards away stands St. Mary's Cathedral, across the street from still another bank. Several members of the congregation told me that the church was thriving and that there was no sense of insecurity here. 'This is Malaysia, sir,' I was told. 'We're Muslims, Buddhists, Christians. We're Malays, Chinese, Indians. We have to live together. By the way, don't miss the action at the Hard Rock Café on Sultan Ismail Street.' Now there, I thought, was a contradiction as remarkable as this scene."

www.conceptcaching.com

In Malaya, the British developed a complicated system of colonies and protectorates that eventually gave rise to the equally complex, far-flung Malaysian Federation. Included were the former Straits Settlements (Singapore was one of these colonies), the nine protectorates on the Malay Peninsula (former sultanates of the Muslim era), the British dependencies of Sarawak and Sabah on the island of Borneo, and numerous islands in the Strait of Malacca and the South China Sea. The original Federation of Malaysia was created in 1963 by the political unification of recently independent mainland Malaya, Singapore, and the former British dependencies on the largely Indonesian island of Borneo. Singapore, however, left the Federation in 1965 to become a sovereign city-state, and the remaining units were later restructured into peninsular Malaysia and, on Borneo, Sarawak and Sabah. Thus the term *Malaya* properly refers to the geographic area of the Malay Peninsula, including Singapore and other nearby islands; the term *Malaysia* identifies the politico-geographical entity of which Kuala Lumpur is the capital city.

Netherlands "East Indies"

Following in the wake of the Portuguese, the first European colonizers in this realm, the Dutch came in search of what Southeast Asia had to offer—and now one aspect of this realm's biodiversity had fateful consequences. Among the plants domesticated by the local people on the islands of present-day Indonesia was a group collectively known as

the *spices*. We know them today as black pepper, cloves, cinnamon, nutmeg, ginger, turmeric, and other condiments essential to flavorful meals. In Figure 10A-1 you can see, in eastern Indonesia between Sulawesi and New Guinea, a group of small islands called the Maluku Islands (formerly Moluccas). The Dutch colonizers called these the **Spice Islands** because of the lucrative commerce in spices long carried on by Arab, Indian, and Chinese traders. What the Europeans wanted was control over this trade, and they were willing to go to war for it.

It may seem odd in today's world that spices could be important enough to be fought over, but at that time, in the pre-refrigeration era, spices not only conserved food but also added flavor to otherwise bland diets. Spices commanded sky-high prices on European markets. The Dutch East India Company, the commercial arm of the Dutch government, managed to take control of the Spice Islands, bringing untold wealth to the Netherlands and ushering in a period of enrichment through colonial exploitation known as the country's Golden Age.

From the mid-seventeenth to the late eighteenth century, the Dutch could develop their East Indies sphere of influence almost without challenge, for the British and French were preoccupied with the Indian subcontinent. By playing the princes of Indonesia's states against one another in the search for economic concessions and political influence, by placing the Chinese in positions of responsibility, by imposing systems of forced labor in areas directly under its control, and by rearranging land ownership and power structures, the Company had a disastrous effect on the Indonesian societies it subjugated.

Java (Jawa), the most populous and productive island, became the focus of Dutch administration; from its capital at Batavia (now Jakarta), the Dutch East India Company extended its sphere of influence into Sumatra (Sumatera), Celebes (Sulawesi), much of Borneo (Kalimantan), and the smaller islands of the East Indies. This was not accomplished overnight, and the struggle for territorial control was carried on long after the Company had yielded its administration to the Netherlands government. Dutch colonialism therefore threw a girdle around Indonesia's more than 17,000 islands, paving the way for the creation of the realm's largest and most populous state (home to a quarter-billion today).

From Spain to the United States

In the colonial tutelage of Southeast Asia, the Philippines, long under Spanish domination, had a unique experience. As early as 1571, the islands north of Indonesia were under Spain's control (they were named for Spain's King Philip II). Spanish rule began when Islam was penetrating the southern Philippines via northern Borneo. The Spaniards spread their Roman Catholic faith with great zeal, and between them the soldiers and priests consolidated Hispanic dominance over the predominantly Malay population. Manila, founded in 1571, became a profitable waystation on the route between southern China and western Mexico (Acapulco was the main trans-Pacific destination for the galleons leaving Manila's port). There was much profit to be made, but the indigenous people shared little in it. Great landholdings were awarded to loyal Spanish civil servants and to men of the church. Oppression eventually yielded revolution, and Spain was confronted with a major uprising in the Philippines when the Spanish-American War broke out elsewhere in 1898.

As part of the settlement of that war, the United States replaced Spain in Manila as colonial proprietor. That was not the end of the revolution, however. The Filipinos now took up arms against their new foreign ruler, and not until 1905, after terrible losses of life, did American forces manage to pacify their new dominion. Subsequently, U.S. administration in the Philippines was more progressive than Spain's had been. In 1934, Congress passed the Philippine Independence Law, providing for a ten-year transition to sovereignty. But before independence could be arranged, World War II intervened. In 1941, Japan conquered the islands, temporarily ousting the Americans; U.S. forces returned in 1944 and, with widespread Filipino support, defeated the Japanese in 1945. The agenda for independence was resumed, and in 1946 the sovereign Republic of the Philippines was proclaimed.

Today, all of Southeast Asia's states are independent, but centuries of colonial rule have left strong cultural imprints. In their urban landscapes, their education systems, and countless other ways, this realm still carries the legacy of its colonial past.

STATES AND BOUNDARIES

It is tempting to view the state as a living thing, an organism: after all, it has a heart (the capital), arteries (roads, railways), circulation (automobiles, trains, planes), lungs (parks, forests)—and in that analogy, its boundaries represent skin. This last comparison is not as far-fetched as it might seem. Although we tend to think of boundaries as lines on the map or fences on the ground, the legal definition of a boundary goes much farther than that. In fact, boundaries are actually invisible vertical planes extending above the ground into the air and below the ground into soil and rock (or water). Where these planes intersect with the ground, they form lines—the lines on the map.

As we shall soon see, boundaries give states their shape or form: some are bulky, others elongated and thin; many are instantly recognizable on a map even without their names (for example, Italy and its boot, Egypt and its corner, Afghanistan with its panhandle). And if people can be thin-skinned, so can states—about their borders. Countries can be very territorial even about the smallest parcel of land.

A useful way to think about boundaries is to regard them as contracts between states. Such contracts take the form of treaties in which the boundary is described in words, much like a legal survey of a private property. This often-elaborate description is called the **definition** of the boundary, but like all contracts, the terms of that definition may be debatable, sometimes giving rise to disputes. Putting the terms of the treaty on a detailed, large-scale map is the job of cartographers, whose **delimitation** of the treaty language is what we observe in atlases—and sometimes in international courts. And at times, national governments decide that they must emphasize their boundaries in practice to protect their interests, constructing walls, fences, or other kinds of artificial barriers. Through such boundary **demarcation**, states try to enhance their security (Israel), deter illegal in-migration (United States), or control out-migration (the Berlin Wall between 1961 and 1989).

Classifying Boundaries

Boundaries have many functions, and to understand this (and to learn why some boundaries tend to produce more trouble than others) it is helpful to view them in categorical

© Barbara A. Weightman

"I stood on the Laotian side of the great Mekong River which, during the dry season, did not look so great! On the opposite side was Thailand, and it was rather easy for people to cross here at this time of the year. But, the locals told me, it is quite another story in the wet season. Then the river inundates the rocks and banks you see here, it rushes past, and makes crossing difficult and even dangerous. The buildings where the canoes are docked are built on floats, and rise and fall with the seasons. The physiographic-political boundary between Thailand and Laos lies in the middle of the valley we see here."

www.conceptcaching.com

perspective. Some boundaries conform to elongated features in the natural landscape (mountain ranges, rivers) and have **physiographic** origins. Others coincide with historic breaks or transitions in the cultural landscape and are sometimes referred to as *anthropogeographic* or more recently as ***ethnocultural*** boundaries. And as any world political map shows, many boundaries are simply straight lines, defined by endpoint coordinates with no reference to physical or cultural landscape features; these ***geometric*** boundaries often lead to problems when valuable natural resources are found to lie across them.

In general, the colonial powers and their successor governments defined the boundaries of Southeast Asia more judiciously than was the case in several other now-postcolonial areas of the world, notably Africa, the Arabian Peninsula, and Turkestan. The colonial powers that established the original treaties tried to define boundaries to lie in remote and/or sparsely peopled areas: for example, across interior Borneo. Nonetheless, certain Southeast Asian boundaries have triggered problems, among them the geometric boundary between Papua, the portion of New Guinea ruled by Indonesia, and the country of Papua New Guinea, which occupies the eastern part of the island. The artificiality of this boundary is resulting in rising secessionist feelings among the population of Indonesian Papua.

Even on a small-scale map of the kind we use in this chapter, we can categorize the boundaries in this realm. A comparison between Figures 10A-1 and 10A-3 reveals that the boundary between Thailand and Myanmar over long stretches is anthropogeographic (ethnocultural), notably where the name *Kayin* (*Karen*), the Myanmar minority, appears in Figure 10A-3. And Figure 10A-1 shows that a large segment of the Vietnam–Laos boundary is physiographic-political, coinciding with the Annamite Cordillera (Highlands).

Boundaries in Changing Times

A number of the world's boundaries are centuries—even many centuries—old, while others are of recent, even postcolonial, origin. Another way of interpreting their functions involves a look at their evolution as part of the cultural landscape they partition. Political geographers have been doing this for a long time; a pioneer in this field was Richard Hartshorne (1899–1992), who argued that boundaries should be assigned to one of four categories according to their ***genetic classification***. As it happens, you can find examples of all four of these boundary types in Southeast Asia.

Certain boundaries, Hartshorne reasoned, were defined and delimited before the present-day human landscape developed. In Figure 10A-6 (upper-left map), the border between Malaysia and Indonesia across the island of Borneo is an example of the first boundary type, an **6 antecedent boundary**. Most of this border passes through very sparsely inhabited tropical rainforest, and the break in settlement can even be detected on the realm's population map (Fig. 10A-2).

GENETIC POLITICAL BOUNDARY TYPES

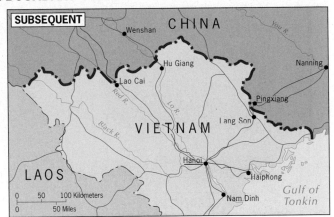

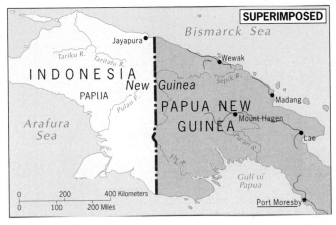

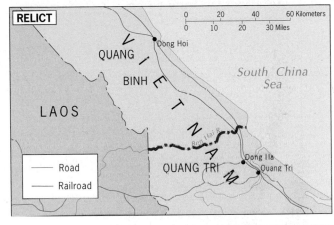

FIGURE 10A-6

© H. J. de Blij, P. O. Muller, and John Wiley & Sons, Inc.

A second category of boundaries evolved as the cultural landscape of an area took shape and became part of the ongoing process of accommodation between several states. These **7** **subsequent boundaries** are represented in Southeast Asia by the map in the upper right of Figure 10A-6, which shows in some detail the border between Vietnam and China. This border is the result of a long process of adjustment and modification, the end of which may not yet have come.

The third category involves boundaries drawn forcibly across a unified or at least homogeneous cultural landscape. The colonial powers did this when they divided the island of New Guinea by delimiting a boundary in a nearly straight line (curved in only one place to allow for a bend in the Fly River), as shown in the lower-left map of Figure 10A-6. The **8** **superimposed boundary** they delimited gave the Netherlands the western half of New Guinea. When Indonesia became independent in 1949, the Dutch did not yield their part of New Guinea, which is peopled mostly by ethnic Papuans, not Indonesians. In 1962, the Indonesians invaded the territory by force of arms, and in 1969 the United Nations recog-

nized its authority there. This made the colonial, superimposed boundary the eastern border of Indonesia and had the effect of projecting Indonesia from the Southeast Asian realm into the neighboring Pacific realm. Geographically, all of New Guinea forms part of the Pacific Realm.

The fourth genetic boundary type is the so-called **9** **relict boundary**—a border that has ceased to function but whose imprints (and sometimes influence) are still evident in the cultural landscape. The boundary between former North and South Vietnam (Fig. 10A-6, lower-right map) is a classic example: once demarcated militarily, it has had relict status since 1976 following the reunification of Vietnam in the aftermath of the Indochina War (1964–1975).

Southeast Asia's boundaries have colonial origins, but they have continued to influence the course of events in postcolonial times. Take one instance: the physiographic boundary that separates the main island of Singapore from the rest of the Malay Peninsula, the Johor Strait (see Fig. 10B-6). That physiographic-political boundary facilitated, perhaps crucially, Singapore's secession from the state of Malaysia in 1965. Without it, Malaysia might have been

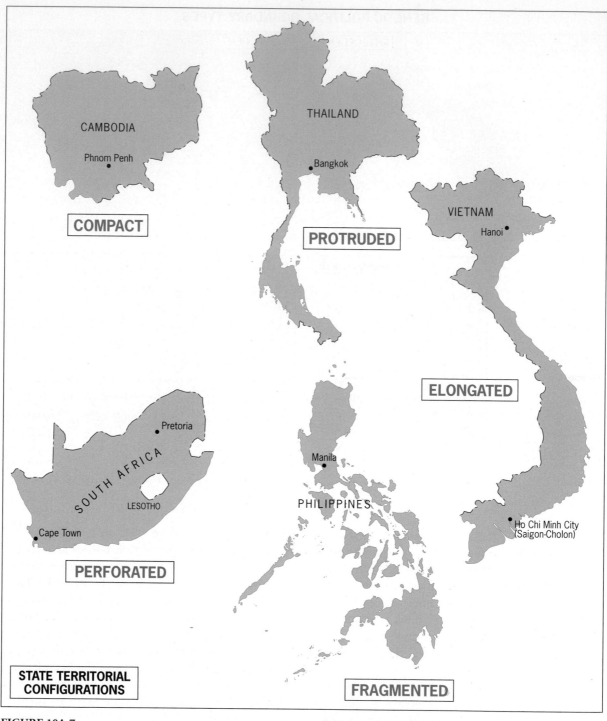

FIGURE 10A-7

© H. J. de Blij, P. O. Muller, and John Wiley & Sons, Inc.

persuaded to stop the separation process; at the very least, territorial issues would have arisen to slow the sequence of events. As it was, no land boundary needed to be defined: the Johor Strait demarcated Singapore and left no question as to its limits*

State Territorial Morphology

Boundaries define and delimit states; they also create the mosaic of often interlocking territories that give individual countries their shape. This shape or ***territorial morphology*** can affect a state's condition, even its survival. Viet-

---

*Except one: a tiny island at the eastern entrance to the Strait named Pedra Blanca (as Singapore calls it) or Pulau Batu Putih (the Malaysian version), which is still disputed today.

nam's extreme elongation has influenced its existence since time immemorial. And, as we will note in Chapter 10B, Indonesia has tried to redress its fragmented nature (thousands of islands) by promoting unity through the "transmigration" of residents of Jawa from the most populous island to many of the others.

Political geographers identify five dominant state territorial configurations, all of which we have encountered in our world regional survey but which we have not categorized until now. All but one of these shapes is represented in Southeast Asia, and Figure 10A-7 provides the terminology and examples:

- **10 Compact states** have territories shaped somewhere between round and rectangular, without major indentations. This encloses a maximum amount of territory within a minimum length of boundary. Southeast Asian example: Cambodia.

- **11 Protruded states** (sometimes called *extended*) have a substantial, usually compact territory from which extends a peninsular corridor that may be landlocked or coastal. Southeast Asian examples: Thailand and Myanmar.

- **12 Elongated states** (also called *attenuated*) have territorial dimensions in which the length is at least six times the average width, creating a state that lies astride environmental or cultural transitions. Southeast Asian example: Vietnam.

- **13 Fragmented states** consist of two or more territorial units separated by foreign territory or by water. Subtypes include mainland-mainland, mainland-island, and island-island. Southeast Asian examples: Malaysia, Indonesia, the Philippines, and East Timor.

- **14 Perforated states** completely surround the territory of other states, so that they have a "hole" in them. No Southeast Asian example; the most illustrative current case is South Africa, perforated by Maryland-sized Lesotho.

In Chapter 10B, we will have frequent occasion to refer to the shapes of Southeast Asia's states. For so comparatively small a realm with so few countries, Southeast Asia displays a considerable variety of state morphologies. But one point of caution: states' territorial morphologies do not determine their viability, cohesion, unity, or lack thereof; they can, however, influence these qualities. Cambodia's compactness has not ameliorated its divisive political geography, for instance. As we will find in our survey of the realm's regional geography, though, shape plays a key role in the still-evolving political and economic geography of Southeast Asia.

POINTS TO PONDER

- Islam, Buddhism, Christianity, and Hinduism all have a significant presence in Southeast Asia. Islamic extremists are in conflict with those of other faiths (as well as Muslim moderates) in several parts of the realm.

- Population pressure and economic activity are decimating one of the world's last great stands of equatorial rainforest, including unmatched wildlife refuges.

- Chinese minorities in Southeast Asia control disproportionate chunks of national economies.

- With its thousands of islands large and small adjoining a peninsular mainland, Southeast Asia has one of the world's most complicated maritime-boundary frameworks.

The stupas at Jawa's Borobudur, one of the world's greatest Buddhist monuments. © H. J. de Blij

IN THIS CHAPTER

Modernizing Vietnam's north to catch up to the booming south
Thailand's most perfect of kings and his protest-plagued kingdom
As Myanmar suffers, it becomes increasingly important to China
Running out of space in Singapore
Democratic Indonesia takes a leading role
Awaiting Pacific Rim development in the lagging Philippines

CONCEPTS, IDEAS, AND TERMS

10B

SOUTHEAST ASIA: REGIONS OF THE REALM

≡ Mainland Southeast Asia
Insular Southeast Asia

FIGURE 10B-1

© H. J. de Blij, P. O. Muller, and John Wiley & Sons, Inc.

The regional geography of Southeast Asia evolved from the empires of competing colonial powers whose domination ensured compliance with their imperial designs, but great credit goes to the postcolonial governments of this realm which confronted strong centrifugal forces and generally succeeded in maintaining the stability of this regional framework. Neither outsiders (American intervention in the 1960s was aimed at dividing Vietnam into a communist North and a noncommunist South) nor locals (such as insurgents in Malaysia and secessionists in Indonesia) could fracture multicultural states, whose components did not start their independent existence with much in common culturally or economically. There were times when it seemed that Malaysia might break up, but only Singapore left that federation, a negotiated secession without violence. Sprawling Indonesia dealt with separatists by force as well as through persuasion and accommodation. The troubled birth of East Timor, the realm's youngest state, is the exception, not the rule. Meanwhile, Southeast Asia's governments have become more representative (with the tragic exception of Myanmar); its economies have matured and, in the majority of cases, prospered; and the realm's vital statistics (see Appendix B) reflect its overall development.

So much of the Southeast Asian realm consists of the waters between and among its land areas that, on the globe, it seems to be much larger territorially than it really is. Indeed, the realm's mainland component is actually smaller than its island entities combined: the mainland (including the peninsular part of Malaysia) consists of about 2,100,000 square kilometers (800,000 sq mi) whereas the islands, even without including Indonesian Papua (which belongs to the Pacific Realm), account for some 2,400,000 square kilometers (930,000 sq mi)—a ratio of just over 46 percent on the mainland to slightly less than 54 percent on the islands. But, as we will see, Malaysia overall exhibits stronger island than mainland properties, tilting the balance even more toward offshore Southeast Asia.

With 11 countries ranging from ministates Singapore and Brunei to regional giants Indonesia and Myanmar, Southeast Asia's states fall into every World Bank income category (see Fig. G-10). Singapore got rich from trade; Brunei from oil. Myanmar stayed poor when it did not have to because mismanagement prevented it from realizing its potential. Environmental, locational, and administrative problems plagued, and then restrained, Laos and Cambodia. But Malaysia forged ahead, Thailand was upward bound until recent political

setbacks, and conservatively communist Vietnam is beginning to loosen its economic shackles. There was a time when this realm, in overall terms, was on a par with Subsaharan Africa. No longer.

Southeast Asia's first-order geographic regionalization recognizes the mainland-island dichotomy that is obvious from any map of this realm. But things are somewhat more complicated than that because the southern part of the Malay Peninsula, occupied by the states of Malaysia and Singapore, exhibits physiographic, historical, and cultural circumstances to justify its inclusion in the insular (island) rather than the mainland region. Using the political framework as our grid, we can map the broadest regional division of Southeast Asia in Figure 10B-1:

- **Mainland region:** Vietnam, Cambodia, Laos, Thailand, Myanmar (Burma)
- **Insular region:** Malaysia, Singapore, Indonesia, Brunei, the Philippines, East Timor (Timor-Leste)

Special note should also be taken of the discordance between the political grid and the cultural-geographic reality on the eastern periphery of the Insular region. Here, Indonesia controls the western half of the island of New Guinea, but all of New Guinea belongs in the neighboring Pacific Realm.

▤ MAINLAND SOUTHEAST ASIA

Five countries form the Mainland region of Southeast Asia (Fig. 10B-1): the three remnants of Indochina—Vietnam, Cambodia, and Laos—in the east; Thailand in the center; and Myanmar (formerly Burma) in the west. One religion, Buddhism, dominates cultural landscapes, but this is a multicultural, multiethnic region. Although still one of the less urbanized regions of the world, the Mainland contains several major cities and the pace of urbanization is increasing. And as Figure 10B-1 shows, two countries (Vietnam and Myanmar) possess dual core areas.

We approach this region from the east, beginning our survey in Indochina where the United States fought and lost a disastrous war that ended in 1975, but whose impact on America continues to be felt today. After the Indochina War started formally in 1964 (U.S. involvement in Vietnam actually began earlier), some scholars warned that the conflict might spill over from Vietnam into Laos and Cambodia, and then into Thailand, Malaysia, and even Myanmar. This view was based on the **1 domino theory**, which holds that destabilization and conflict from any cause in one country can result in the collapse of order in one or more neighboring countries, triggering a chain of events that can affect a series of contiguous states in a region.

Whether history will prove these scholars right or wrong continues to be debated. Cambodia and Laos were indeed disastrously affected as the original Vietnam War became the wider Indochina War, but not the other "dominoes" (in fact, the last Laotian Hmong refugees who had been driven from their homes and who had found refuge in Thailand were forcibly ousted in late 2009). To some observers, this invalidated the domino theory, which was grounded in the capitalist–communist struggle of the twentieth century. But communist insurgency was (and is) only one way a state may be destabilized with dire consequences for its neighbors. Powerful states that intervene in weaker countries, as the United States did in Iraq in 2003 and Russia in Georgia in 2008, do so at the risk of destabilizing those countries' neighbors by setting in motion arms races and by sowing the seeds of future regional conflict. Ethnic-cultural strife in one state can also engulf neighboring countries, as happened when Rwanda's civil wars spread into the neighboring Congo (where a government fell as a result), and also involved Burundi and Uganda. In our crowded world, potential dominoes abound.

Vietnam

Vietnam (population: 88 million) still carries the scars of the Indochina War, although the vast majority of Vietnamese have no personal memory of that terrible conflict. The more immediate concerns in Vietnam today are to reconnect the country with the outside world and to integrate its 2000-kilometer (1200-mi) strip of highly elongated territory through better infrastructure (Fig. 10B-2). But it is an emerging economy that is now making considerable progress. Poverty has dropped dramatically in the past decade as Vietnam has become a player in the global economy. Today it is regarded

as a new "hot" location for assembly plants, which are attracted by low wages (lower than in China) and economic development zones.

French Legacy

The French colonialists recognized that Vietnam, whose average width is under 240 kilometers (150 mi), was not a homogeneous colony, so they divided it into three units: (1) **Tonkin**, land of the Red River Basin and Delta and centered on Hanoi in the north; (2) **Cochin China**, domain of the Mekong Delta and centered on Saigon in the south; and (3) **Annam**, focused on the ancient city of Hué in the middle. Today, the Vietnamese prefer to use *Bac Bo, Nam Bo,* and *Trung Bo,* respectively, to designate these areas.

The Vietnamese (or Annamese, also Annamites, after their cultural heartland) speak the same language, although the northerners can easily be distinguished from southerners by their accent. As elsewhere in their colonial empire, the French made their language the *lingua franca* of Indochina, but their tenure was cut short by the Japanese, who invaded Vietnam in 1940. During the Japanese occupation, Vietnamese nationalism became a powerful force, and after the Japanese defeat in 1945, the French could not regain control. In 1954, the French suffered a disastrous final trouncing on the battlefield at Dien Bien Phu in the far northwest and were ousted from the country.

North and South

But even after its forces routed the colonizers, Vietnam did not become a unified state. Separate regimes took control: a communist one in Hanoi and a noncommunist counterpart in Saigon. Vietnam's pronounced elongation had made things difficult for the French; now it played its role during the postcolonial period. Note, in Figure 10B-2, that Vietnam is widest in the north and south, with a narrow "waist" in its middle zone. North and

South Vietnam were worlds apart, and those worlds were represented in Hanoi by communism and in Saigon by anticommunism. For more than a decade the United States tried to prop up the Saigon regime that controlled the south, but the communists prevailed, and like China, Vietnam still has a communist government today. As many as 2 million Vietnamese refugees set out on often-flimsy boats onto the South China Sea; of those who survived, a majority settled in the United States.

Today, contrasts between north and south continue, although they are diminishing. The capital, Hanoi, long lagged behind bustling Saigon, but today its skyline reflects modernization and the links to its port of Haiphong are vastly improved. With nearly 5 million residents, Hanoi anchors the northern (Tonkin Plain) core area of Vietnam, the lower basin of the Red River (its agricultural hinterland). On rural roads, goods are still moved by human- or animal-drawn cart, but the roads themselves are being improved. The south, however, is experiencing even more profound change, with significant economic growth as the government slowly, and on its own terms, is opening up the country to the global economy and in the process is lifting many out of poverty.

Vietnam in Transition

In the first decades of communist rule, private enterprise was abolished and farmers were compelled to join collectives on the communist Chinese model, resulting in near-famine conditions in the early 1980s. This coincided with China's economic opening, and Hanoi's leaders were quick to follow suit, though more timidly. By the mid-1990s, free enterprise was encouraged; farmers were allowed to cultivate for profit, and foreign investment was welcomed, though under far more restrictive terms than in China. The government has continued to provide adequate social services with health

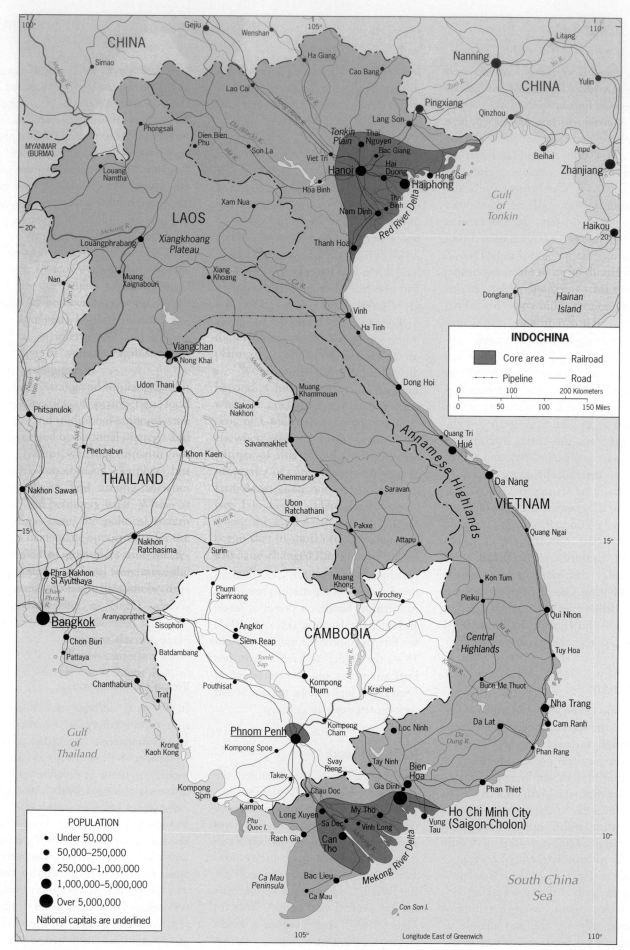

FIGURE 10B-2

© H. J. de Blij, P. O. Muller, and John Wiley & Sons, Inc.

© H. J. de Blij

"Although modernization has changed the skylines of Vietnam's major cities (Saigon more so than Hanoi), most of the country's urban areas look and function much as they have for generations. Main arteries throb with commerce and traffic (still mostly bicycles and mopeds); side streets are quieter and more residential. No high-rises here, but the problems of dense urban populations are nevertheless evident. Every time I turned into a side street, the need for street cleaning and refuse collection was obvious, and drains were clogged more often than not. "We haven't allowed our older neighborhoods to be destroyed like they have in China," said my colleague from Hanoi University, "but we also don't have an economy growing so fast that we can afford to provide the services these people need." Certainly the Pacific Rim contrast between burgeoning coastal China and slower-growing Vietnam—a matter of communist-government policy—is evident in their cities. "As you see, nothing much has changed here," she said. "Not much evidence of globalization. But we're basically self-sufficient, nobody goes hungry, and the gap between the richest and the poorest here in Vietnam is a fraction of what it is now in China. We decided to keep control." You would find more agreement on these points here in Hanoi than in Saigon, but the Vietnamese state is stable and progressing."

www.conceptcaching.com

indicators better than those in China and with secondary school attendance rising rapidly.

One focus of Vietnam in recent years has been on agricultural production. In 2002 the government announced that Vietnam, long a rice producer, had become the world's largest exporter of coffee (it ranks second today), a lucrative global commodity. That claim was perhaps a bit premature because Vietnam is still far behind top ranking Brazil, but it is now the second-largest producer of bulk coffee in the world. However, because mass-produced bulk coffee is grown plantation-style in the sun, the expansion of this commodity into the interior highlands has resulted in deforestation. This activity is having a significant environmental impact, leading to worsening annual floods that bedevil the once-stable countryside.

Vietnam has a relatively fast-growing national economy even in these difficult economic times. Ho Chi Minh City (Saigon's official name since 1976) is booming again and now has a Special Economic Zone based on the Chinese model, as well as the New Saigon business and residential district. Saigon and environs now contribute 25 percent of Vietnam's industrial output and one-third of its tax revenues. In 2007, Vietnam was admitted to the World Trade Organization, and more such changes are yet to come. Still, Hanoi's communist planners have kept tighter economic control over the country than their Chinese counterparts and are determined not to allow runaway capitalism of the Chinese variety—nor the wealth disparities that come with it.

Cambodia

Compact Cambodia is heir to the ancient Khmer Empire whose capital was Angkor and whose legacy is a vast landscape of imposing monuments including the great Buddhist-inspired temple complex, Angkor Wat. Today, 90 percent of Cambodia's more than 15 million inhabitants are ethnic Khmers, with the remainder divided between Vietnamese and Chinese. The present capital, Phnom Penh, lies on the Mekong River, which crosses Cambodia before it enters and forms its huge delta in Vietnam (Fig. 10B-2).

Geographically, Cambodia enjoys a number of advantages; compact states (see Fig. 10A-7) enclose a maximum amount of territory within a minimum of boundary, and cultural homogeneity tends to diminish centrifugal forces. But neither spatial morphology nor homogeneous ethnicity could withstand the impact of the Indochina War, which led to communist revolution and the systematic murder of as many as 2 million Cambodians by the Maoist terror group, the *Khmer*

Rouge. Once self-sufficient and a food exporter, Cambodia today must import food. Corrupt and violent politics, chronic instability, and rural dislocation make this one of Southeast Asia's poorest countries. Clearly, its postwar trauma continues.

Laos

Landlocked Laos has no fewer than five neighbors, one of which is East Asia's giant, China (Fig. 10B-2). The Mekong River forms a long stretch of its western boundary, and the important sensitive border with Vietnam to the east lies in mountainous terrain. With just slightly more than 6 million people (over half of them ethnic Lao, related to the Thai of Thailand), Laos lies surrounded by comparatively powerful states. The country has no railroads, just a few miles of paved roads, and very little industry; it is only 27 percent urbanized (the

capital, Viangchan, lies on the Mekong and has an oil pipeline to Vietnam's coast).

Laos has long included one small corner of the Golden Triangle of opium-poppy-cultivation fame, but under international (especially U.S. and European) pressure, the communist regime has forced the mainly hill-tribe people who produce most of it to abandon their crops. Because this was the only way they could make a living, these farmers were sent to resettlement villages in the lowlands, where they contracted malaria and other diseases not prevalent in their upland domain and were subject to cultural disintegration. The reward for the authorities was the continuation of foreign aid; the cost to the powerless hill people (especially the women) is incalculable. Here is an outstanding example of the power of the globalizing core reaching into the weakest of peripheries.

Thailand

In virtually every way, Thailand is the leading state of the Mainland region. In contrast to its neighbors, Thailand has been a strong participant in the Pacific Rim's economic development. Its capital, Bangkok (population: 7 million), is the largest urban center in the Mainland region and one of the world's most prominent primate cities. The country's population, 67 million in 2010, is growing at the slowest rate in this geographic realm. Over the past few decades, only political instability and uncertainty have inhibited economic progress. Thailand is a constitutional monarchy with an elected parliament; its progress toward a stable democracy has experienced a history of setbacks. The latest of these began in 2006, when the armed forces ousted a controversial but popular prime minister, precipitating a power struggle that has severely impacted the national economy, notably the vitally important tourist industry that is still languishing years later. Mass demonstrations immobilized cities, closed down airports, and paralyzed surface transport at enormous cost to industry and commerce.

Thailand has a compact heartland in which lie the core area, capital, and leading zones of productive capacity, while a 1000-kilometer (600-mi) **2 protrusion** (a corridor of land extending away from the rest of the state), in places less than 32 kilometers (20 mi) wide, extends southward to the border with Malaysia (Fig. 10B-3). The boundary that defines this protrusion runs down the length of the upper Malay Peninsula to the Kra Isthmus, where neighboring Myanmar peters out and Thailand fronts the Andaman Sea (an arm of the Indian Ocean) as well as the Gulf of Thailand.

The Restive Peninsular South

In the entire country, no place lies farther from the capital than the southern end of this tenuous protrusion. Consider this spatial situation in the context of Figure 10A-3. The Malay

During the spring of 2010 a political crisis in Thailand turned into a social disaster when rebels supporting an ousted prime minister battled supporters of a discredited government in the streets of Bangkok, causing dozens of deaths, hundreds of injuries, extensive property damage, and economic havoc. By the time the army and police overpowered the "red shirts," parts of this globalized city looked like a war zone; as defeat loomed, the rebels torched one of the world's largest shopping malls, the Central World. © AFP/Getty Images, Inc.

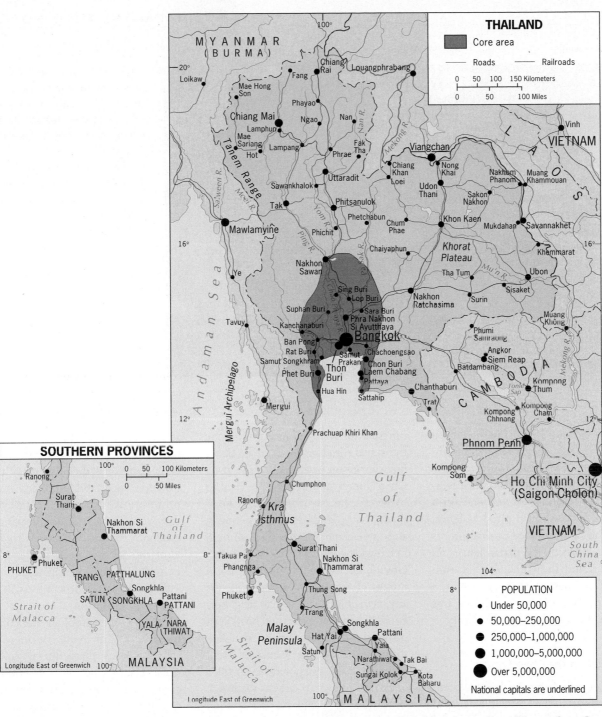

FIGURE 10B-3

© H. J. de Blij, P. O. Muller, and John Wiley & Sons, Inc.

ethnic population group extends from Malaysia more than 300 kilometers (200 mi) into Thai territory. In Thailand's five southernmost provinces (Fig. 10B-3, inset map), 85 percent of the inhabitants are Muslims (the figure for Thailand as a whole is less than 5 percent). The political border between Thailand and Malaysia is porous; in fact, you can cross it at will almost anywhere along the Kolok River by canoe, and inland it is a matter of walking a forest trail. For more than a century, the southern provinces have had closer ties with Malaysia across the border than with Bangkok 1000 kilometers (600 mi) away, and the Thai government has not tried to restrict movement or impose onerous rules on the Muslim population here.

But in the new era of Islamic terrorism, and in view of rising violence in this southern frontier, the south is now coming to national attention. The Pattani United Liberation Organization, named after the functional capital of the Muslim south, was active in

© H. J. de Blij

"Little did I realize, as I walked the central business district of Yangon (Myanmar) in mid-April 2008, that these streets would be turned into chaos by Cyclone Nargis just days later. This is Sule Pagoda Road, a few blocks north of the famed Octagonal Pagoda, the golden landmark in the center of town. Many signs in English remind you that this was once a British colony."

www.conceptcaching.com

the 1960s and 1970s but had been dormant since. Then a series of bombings at a Pattani hotel, several schools, a Chinese shrine, and a Buddhist temple in 2001 and 2002 ushered in a new era of Muslim-inspired violence. Thailand's pronounced protrusion, its porous border, its dependence on tourism, and the vulnerable location of its Andaman-coast tourist facilities combine to raise concern over this distant part of its national territory.

Bangkok on the Chao Phraya

As Figure 10B-1 shows, Thailand occupies the heart of the Mainland region of Southeast Asia. Even though Thailand has no Red, Mekong, or Irrawaddy Delta, its central lowland is watered by a set of streams that flow off the northern highlands and the Khorat Plateau to the east. One of these streams, the Chao Phraya, is the Rhine of Thailand. From the head of the Gulf of Thailand to Nakhon Sawan, this river is a highway of traffic. Barge trains loaded with rice head for the coast, ferry boats head upstream, freighters transport tin and

tungsten (of which Thailand is among the world's leading producers). Bangkok sprawls on both sides of the lower Chao Phraya floodplain, here flanked by skyscrapers, pagodas, factories, boatsheds, ferry landings, luxury hotels, and myriad modest dwellings all built in crowded confusion on very swampy ground. On the right bank of the Chao Phraya, Bangkok's west side, lie the city's remaining *klong* (canal) neighborhoods, where waterways and boats constitute the transport system. Bangkok still is known as the "Venice of Asia," although many *klongs* have been filled in and paved over to serve as roadways. Unfortunately Bangkok's location on this soggy terrain has compounded its notorious traffic problems as a desperately needed underground transit system cannot be constructed.

Myanmar

Thailand's western neighbor, Myanmar (referred to as Burma in anti-regime circles), is one of the world's poorest and most isolated countries where, it seems, time has stood still

for centuries. Long languishing under one of the world's most corrupt and brutal military dictatorships, it has potential as an oil producer, but its promise must await the regime change that would make multinationals more welcome. Myanmar shares with Thailand the narrow neck of the Malay Peninsula—but look again at Figure 10B-3 and note the contrast in surface communications.

Myanmar's problems are complicated by a shift in the Burmese core area that took place during colonial times. Prior to the colonial period, the core of embryonic Burma lay in the so-called dry zone between the Arakan Mountains (in the far west) and the Shan Plateau, which covers the country's triangular easternmost extension (Fig. 10B-4). The urban focus of the state was Mandalay, which had a central situation and relative proximity to the non-Burmese highlands all around. Then the British developed the agricultural potential of the Irrawaddy Delta farther south, and Rangoon (now called Yangon) became the hub of the colony. The Irrawaddy

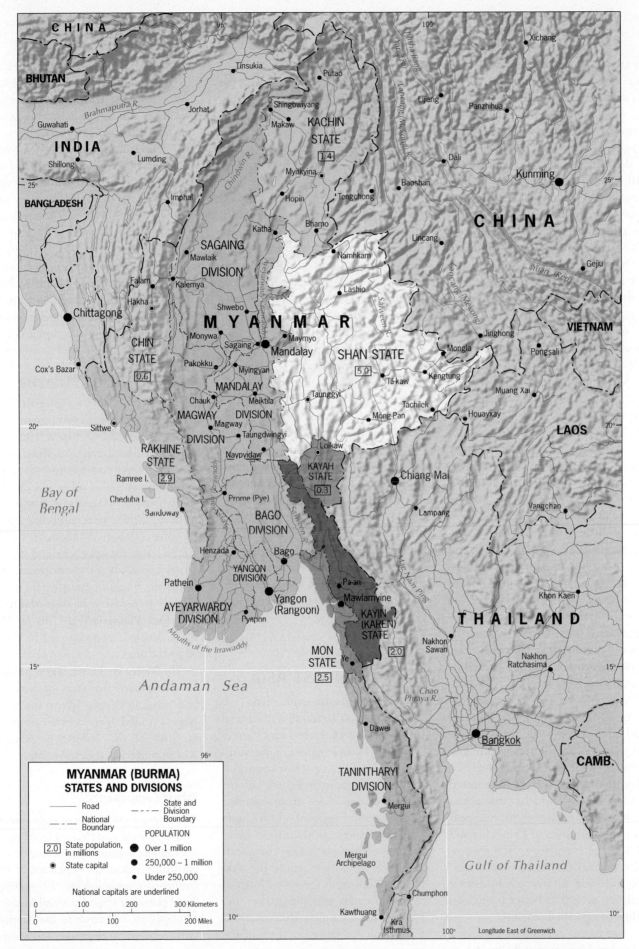

MYANMAR (BURMA)
STATES AND DIVISIONS

—— Road	‑‑‑ State and Division Boundary
—·—·— National Boundary	

POPULATION

2.0 State population, in millions

● State capital

● Over 1 million

● 250,000 – 1 million

• Under 250,000

National capitals are underlined

0   100   200   300 Kilometers

0   100   200 Miles

FIGURE 10B-4

© H. J. de Blij, P. O. Muller, and John Wiley & Sons, Inc.

waterway links the old and the new core areas, but the center of gravity has shifted to the south.

As Figure 10A-3 indicates, the peripheral peoples (11 minorities) of Myanmar occupy a significant part of the state. A closer look at Figure 10B-4 shows that their domains hold the status of State, of which there are seven; the Burman-dominated areas (68 percent of a national population of just over 50 million) are designated as Divisions. The Shan of the northeast and far north, who are related to the neighboring Thai, account for about 9 percent of the population, or 4.5 million. The Kayin (Karen), who constitute more than 7 percent (3.5 million), live in the neck of Myanmar's protrusion and have proclaimed their desire to create an autonomous territory within a federal Myanmar. The Mon (2.3 percent, or 1.2 million) were in what is today Myanmar long before the Burmans and introduced Buddhism to the area; they want the return of ancestral lands from which they were ousted. Although the powerful military has dealt such aspirations a series of setbacks, centrifugal forces continue to bedevil the central regime. The military government's response has been to exert power by all available means rather than to accommodate these forces, even to the point of stifling political discourse—let alone opposition—among the Burmans themselves.

In 2007, the generals moved the capital 300 kilometers (200 mi) northward from Yangon to a newly constructed national-government center—ostensibly a city, but it can barely be called that as it is poorly connected to anywhere else and has few urban amenities. This center is located next to the interior town of Pyinmana and is known as Naypyidaw. The government asserts that its more central location will yield greater administrative efficiency as well as better protection against (unspecified) enemies.

Two recent events have decidedly shaken up Myanmar and exposed its failings in governance to the world in a way that its secretive junta does not like. In late 2007, a sizeable uprising of the people was led by Buddhist monks, and the world held its breath to see if this most repressive of states would finally open up. But that was not to be: the junta yet once again clamped down, proclaiming that dissent would not be tolerated. Then in the spring of 2008, a powerful tropical cyclone slammed into the Irrawaddy Delta, killing at least 130,000 and displacing 2 to 3 million people. The world watched as the junta initially did not acknowledge the severity of the situation and only slowly mobilized relief for those affected. At first, foreign relief workers were prevented from bringing assistance that the government itself could not provide, and only weeks later was limited entry permitted for workers and supplies.

Myanmar today ranks among the world's poorest, least developed, and most corrupt countries—not because of its intrinsic indigence but because a rapacious military regime, tolerated and even commercially engaged by the international community, represses its peoples, crushes their aspirations, and demonstrates the liabilities of life in the global periphery.

▤ INSULAR SOUTHEAST ASIA

On the peninsulas and islands of Southeast Asia's southern and eastern periphery lie 6 of the realm's 11 states (Fig. 10B-1). Few regions in the world contain so diverse a set of countries. Malaysia, the former British colony, consists of two major areas separated by hundreds of miles of South China Sea. The realm's southernmost state, Indonesia, sprawls across thousands of islands from Sumatera in the west to New Guinea in the east. North of the Indonesian archipelago lies the Philippines, a country that once was a U.S. colony. These are three of the most severely fragmented states on

Earth, and each has faced the challenges that such politico-geographical division brings. This Insular region of Southeast Asia also contains two small but significant sovereign entities: a city-state and a sultanate. The city-state is Singapore, once a part of Malaysia (and one instance in which internal centrifugal forces were too great to be overcome). The sultanate is Brunei, an oil-rich Muslim territory on the island of Borneo that seems to be transplanted from the Persian Gulf. In addition, a third small entity, East Timor, only achieved independence in 2002. Few parts of the world are more varied or interesting geographically.

Mainland-Island Malaysia

The state of Malaysia represents one of the three categories of fragmented states discussed in Chapter 10A: the mainland-island type, in which one part of the national territory lies on a continent and the other on one or more islands. Malaysia is a colonial political artifice that combines a pair of quite disparate components into a single state: the southern end of the Malay Peninsula and the northern sector of the island of Borneo. These are known, respectively, as **West Malaysia** and **East Malaysia** (Fig. 10B-1). The name *Malaysia* came into use in 1963, when the original Federation of Malaya, on the Malay Peninsula, was expanded to incorporate the areas of Sarawak and Sabah in Borneo. When the name Malaya is used, it refers to the peninsular part of the Federation, whereas Malaysia refers to the total entity.

Ethnic Components

The Malays of the peninsula, traditionally a rural people, displaced older aboriginal communities there and now constitute just over 50 percent of the country's population of 29 million. They possess a strong cultural identity expressed in adherence to the Muslim faith, a common language, and a sense of territoriality that arises from their perceived Malayan origins and their collective view of Chinese, Indian, European, and other foreign intruders.

The Chinese came to the Malay Peninsula and to northern Borneo in substantial numbers during the colonial period, and today they constitute 24 percent of Malaysia's population (they are the largest single group in Sarawak). Hindu South Asians were in this area long before the Europeans, and for that matter before the Arabs and Islam arrived on these shores; today they still form a substantial minority of 7 percent of the population, clustered, like the Chinese, on the western side of the peninsula.

West Malaysia:
The Dominant Peninsula

The populous peninsular part of Malaysia remains the country's dominant sector with 11 of its 13 States and 80 percent of its population. Here the Malay-dominated government has strictly controlled economic and social policies while pushing the country's modernization. During the Asian economic boom of the 1990s, Malaysia's government embraced the notion of symbols: the capital, Kuala Lumpur, was endowed with (at that time) the world's tallest building; a space-age airport outpaced Malaysia's needs; a high-tech administrative capital was built at Putrajaya; and a nearby development was called Cyberjaya—all part of a so-called *Multimedia Supercorridor* to anchor Malaysia's core area (Fig. 10B-5).

The chief architect of this program was Malaysia's long-term and autocratic head of state, Mahathir bin Mohamad, leader of the Malay-dominated party that forms the majority in government. Mahathir had the support not only of the great majority of the country's Malays, but also the important and influential ethnic Chinese minority, which saw him as the only acceptable alternative to the more fundamentalist Islamic party challenging his rule. But Malaysia's headlong rush to modernize

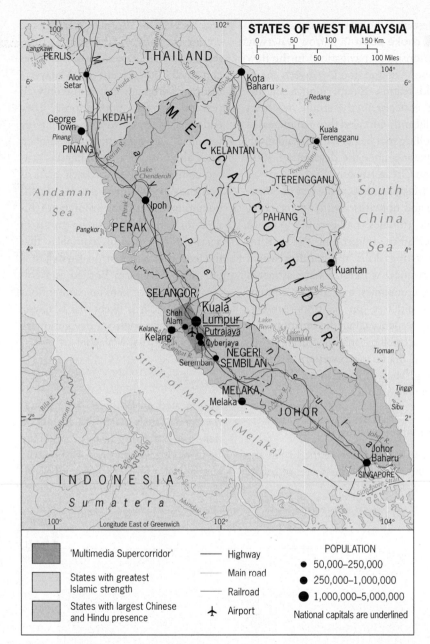

FIGURE 10B-5   © H. J. de Blij, P. O. Muller, and John Wiley & Sons, Inc.

caused a backlash among more conservative Muslims, which in 2001 led to Islamist victories in two States, tin-producing Kelantan and energy-rich but socially poor Terengganu. As the fundamentalist governments in those eastern States imposed strict religious laws, Malaysians talked of two corridors marking their country: the Multimedia Supercorridor in the west and the Mecca Corridor in the east (Fig. 10B-5). But after Mahathir's resignation in 2003 and the appointment of Abdullah Badawi as his more moderate

successor, Islamist fervor in the Mecca Corridor, which had featured calls for a *jihad* in Malaysia, began to wane.

The Malay Peninsula's primacy began long ago, during colonial times, when the British created a substantial economy based on rubber plantations, palm-oil extraction, and mining (tin, bauxite, copper, iron). Singapore, at the southern end of it, was the most prized possession until Singapore seceded from the Malaysian Federation in 1965. The Strait of Malacca (Melaka) to the west continues to be one of the world's busiest and most strategic waterways, a **3** **choke point** (a narrow waterway that constrains navigation) in the flow of resources and goods between major world realms. Piracy, long a significant concern in the strait, has been significantly curtailed through increased patrolling that began with the aid given to the area after the devastating 2004 Indian Ocean tsunami.

Malaysia, despite the loss of Singapore and notwithstanding its recurrent ethnic troubles, has become a major player on the burgeoning Pacific Rim. The strong skills and modest wages of the local workforce have attracted many companies, and the government has capitalized on its opportunities, such as encouraging the creation of a high-technology manufacturing complex on the far northwestern island of Pinang, where Chinese outnumber Malays by two to one.

East Malaysian Borneo

The decision to combine the 11 Sultanates of Malaya with the States of Sabah and Sarawak on Borneo, creating the country now called Malaysia, had far-reaching consequences. Those two States constitute 60 percent of Malaysia's territory (although they house only 20 percent of the population). They endowed Malaysia with major energy resources and huge stands of timber. They also complicated Malaysia's ethnic complexion because each State is home to more than two dozen indigenous groups (in fact, the immigrant Chinese form the largest single group in Sarawak). These locals complain that the federal government in Kuala Lumpur treats Malaysian Borneo as a colony, and politics here are contentious and fractious. It is likely that Malaysia will eventually confront devolutionary forces here in East Malaysia.

Brunei

Also located on Borneo—where Sarawak and Sabah meet—is Brunei, a rich, oil-exporting Islamic sultanate far from the Persian Gulf. Brunei, the remnant of a former Islamic kingdom

that once controlled all of Borneo and areas beyond, came under British control and was granted independence in 1984. Just slightly larger than Delaware and with a population of 430,000, the sultanate is a mere ministate—except for the discovery of oil in 1929 and natural gas in 1965, which made this one of Southeast Asia's two richest countries (Brunei and Singapore continue to vie for the realm's highest per-capita GNI). And there are indications that further discoveries will be made in the offshore zone owned by Brunei. The Sultan of Brunei rules as an absolute monarch; his palace in the capital, Bandar Seri Begawan, is reputed to be the world's largest. He will have no difficulty finding customers for his oil in energy-poor eastern Asia.

Singapore

In 1965, a fateful event occurred in Southeast Asia. Singapore, the crown jewel of British colonialism in this realm, seceded from the recently independent (1963) Malaysian Federation

 FROM THE FIELD NOTES...

© H. J. de Blij

"Only when you rent a boat and see the reclamation work up close in the Jurong Islands (just off Singapore's southwestern shore) do you realize how massive a project this is to expand the city-state's space. Massive amounts of sand and earth are being repositioned to close channels, connect islands, and provide more space, primarily for industrial expansion. But this goes on in several other places along the Singaporean shore, including nearby Sentosa where locals and visitors go in ever-growing numbers for relaxation and entertainment, and where space is also at a premium."

www.conceptcaching.com

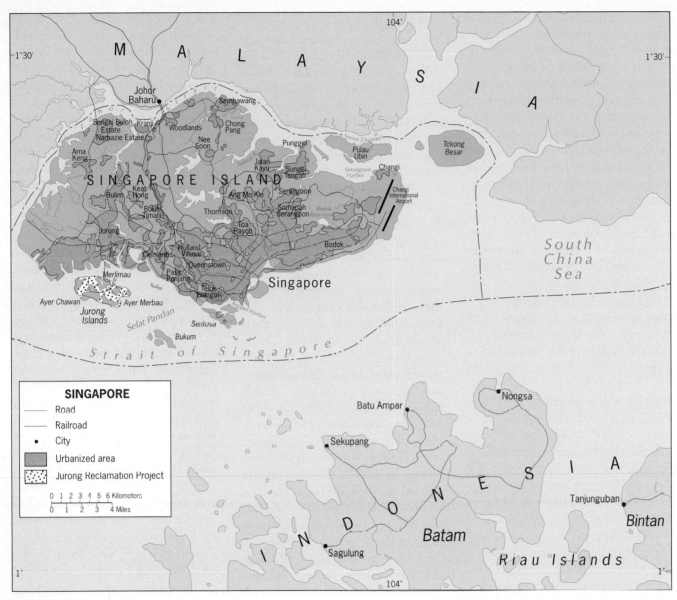

FIGURE 10B-6 © H. J. de Blij, P. O. Muller, and John Wiley & Sons, Inc.

and became a sovereign state, albeit a ministate (Fig. 10B-6). With its magnificent relative location, its multiethnic and well-educated population, and its firm government, Singapore then overcame the severe limitations of space and the absence of raw materials to become one of the leading economic tigers of the Pacific Rim.

With a mere 619 square kilometers (239 sq mi) of territory, space is at a premium in Singapore, and this is a constant worry for the government. Singapore's only local spatial advantage over Hong Kong is that its small territory is less fragmented (there are just a few small islands in addition to the compact main island). With a population of just under 5 million and an expanding economy, Singapore must develop space-conserving high-tech and service industries.

Benefiting from its relative location, the old port of Singapore had become one of the world's busiest (by numbers of ships served) even before its independence. It thrived as an **4** *entrepôt* between the Malay Peninsula, Southeast Asia, Japan, and other emerging economic powers on the Pacific Rim and beyond. Crude oil from Southeast Asia still is unloaded and refined at Singapore, then shipped to East Asian destinations. Raw rubber from the adjacent peninsula and from Indonesia's nearby island of Sumatera is shipped to Japan, the United States, China, and other countries. Timber from Malaysia, rice, spices, and other foodstuffs are processed and forwarded via Singapore. In return, automobiles, machinery, and equipment are imported into Southeast Asia and distributed through Singapore.

But that is the old pattern. Singapore's leaders today are redirecting the city-state's economy and moving toward an ever more technologically sophisticated industrial base. In Singapore the essentially one-party government tightly controls business as well as other aspects of life and borders on authoritarian. Some newspapers and magazines have been banned for criticizing the regime, and there are even fines for chewing gum, eating on the subway, failing to flush a public toilet, and wearing hair deemed longer than appropriate. However, its overall success following secession has tended to keep the critics quiet: while GNI per capita from 1965 to 2003 multiplied by a factor of more than 15—to almost U.S. $30,000—that of neighboring Malaysia reached just $10,320. Among other things, Singapore became (and for several years remained) the world's largest producer of disk drives for personal computers.

To accomplish its revival, Singapore has moved in several directions. First, it is concentrating on three growth areas: information technology, automation, and biotechnology. Second, it espouses notions of a **5 Growth Triangle** involving Singapore's developing neighbors, Malaysia and Indonesia; those two countries would supply the raw materials and cheap labor, and Singapore the capital and technical know-how. Third, Singapore opened its doors to capitalists of Chinese ancestry who left Hong Kong when China took it over and who wanted to relocate their enterprises in Singapore. The population in Singapore is 77 percent Chinese, 14 percent Malay, and 8 percent South Asian. The government is Chinese-dominated, and its policies have served to sustain Chinese control. Indeed, China itself often cites Singapore's combination of authoritarianism and economic success as proving that communism and market economies can coexist.

Indonesia

The fourth-largest country in the world in terms of human numbers is also the globe's most expansive **6 archipelago.** Spread across a chain of more than 17,000 mostly volcanic islands, Indonesia's nearly 250 million people live both separated and clustered—separated by water and clustered on islands large and small.

The complicated map of Indonesia requires close study (Fig. 10B-7). Five large islands dominate the archipelago territorially, but one of these, easternmost New Guinea, is not part of the Indonesian culture sphere, although its western half is under Indonesian control. The other four major islands are collectively known as the Greater Sunda islands: *Jawa* (Java),* smallest but by far the most populous and important; *Sumatera* (Sumatra) in the west, directly across the Strait of Malacca from Malaysia; *Kalimantan*, the southern, Indonesian sector of large, compact, mini-continent Borneo; and wishbone-shaped, *Sulawesi* (Celebes) to the east. Extending eastward from Jawa are the Lesser Sunda Islands, including Bali and, near the eastern end, Timor. Another important island chain that lies within Indonesia is the Maluku (Molucca) Islands, between Sulawesi and New Guinea. The central water body of Indonesia is the Java Sea.

The Major Islands

Indonesia is a Dutch colonial creation, and Jawa was chosen as its colonial headquarters with Batavia (now Jakarta) as the capital. Today, **Jawa** remains the core of Indonesia. With about 140 million inhabitants, Jawa is one of the world's most densely peopled places (see Figs. G-8 and 10A-2) as well as one of the most agriculturally productive, with its rice paddies

rising up on the flanks of active and soil-fertilizing volcanoes. This combination of extreme population density on top of a tectonically active crustal zone has the potential for disaster. You may recall the 2004 tsunami off the coast of Sumatera, but two years later there was a significant earthquake centered just 25 kilometers (15 mi) from the southern-coast city of Yogyakarta. Moreover, Mount Merapi, also near Yogyakarta, was threatening to erupt at the time.

Jawa also is the most highly urbanized part of a country in which more than 50 percent of the people still live off the land; the Pacific Rim boom of the 1990s had a decisive impact here. The megacity of Jakarta (10.1 million), on the northwestern coast, became the heart of a larger conurbation now known as *Jabotabek*, consisting of *Ja*karta as well as *Bo*gor, *Ta*ngerang, and *Bek*asi. During the two decades from 1990 to 2010 the population of this megalopolis grew from 15 to 28 million. Already, Jabotabek is home to more than 10 percent of Indonesia's entire population and nearly 25 percent of its urban population. Thousands of factories, their owners taking advantage of low prevailing wages, were built in this area, straining its infrastructure and overburdening the port of Jakarta. On an average day, hundreds of ships lie at anchor, awaiting docking space to offload raw materials and take on finished products.

As has always been the case in Indonesia, Jawa is where the power lies. As a cultural group (though itself heterogeneous), the Jawanese constitute about 45 percent of the country's population. In the middle of the island, a politically powerful sultanate centers on Yogyakarta; Jawa also is the main base for the two national Islamic movements, one comparatively

---

*As in Africa and South Asia, names and spellings have changed with independence. In this chapter we will use contemporary spellings, except when we refer to the colonial period. Thus Indonesia's four major islands are Jawa, Sumatera, Kalimantan (the Indonesian part of Borneo), and Sulawesi. The Dutch called them Java, Sumatra, Borneo, and Celebes, respectively.

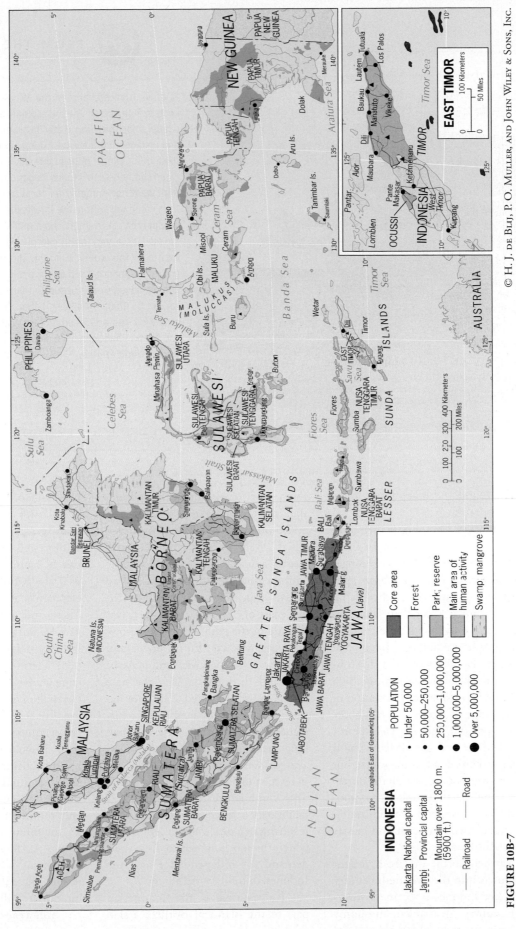

FIGURE 10B-7

© H. J. de Blij, P. O. Muller, and John Wiley & Sons, Inc.

© H. J. de Blij

"I drove from Manado on the Minahasa Peninsula in north-eastern Sulawesi to see the ecological crisis at Lake Tondano, where a fast-growing water hyacinth is clogging the water and endangering the local fishing industry. On the way, in the town of Tomolon, I noticed this side street lined with prefabricated stilt houses in various stages of completion. These, I was told, were not primarily for local sale. They were assembled from wood taken from the forests of Sulawesi's northern peninsula, then taken apart again and shipped from Manado to Japan. "It's a very profitable business for us," the foreman told me. "The wood is nearby, the labor is cheap, and the market in Japan is insatiable. We sell as many as we can build, and we haven't even begun to try marketing these houses in Taiwan or China." At least, I thought, this wood was being converted into a finished product, unlike the mounds of logs and planks I had seen piled up in the ports of Borneo awaiting shipment to East Asia."

www.conceptcaching.com

moderate and the other increasingly fundamentalist.

Sumatera, Indonesia's western-most island, forms the western shore of the busy Strait of Malacca; Singapore lies across the Strait from approximately the middle of the island. Although much larger than Jawa, Sumatera has only about one-third as many people (49 million). In colonial times the island became a base for rubber and palm-oil plantations; its high relief makes possible the cultivation of a wide range of crops, and neighboring Bangka and Belitung yield petroleum and natural gas. Palembang is the key urban center in the south, but current attention focuses on the north. There, the Batak people accommodated colonialism and Westernization and made Medan one of Indonesia's Pacific Rim boom cities.

Farther north, the Aceh fought the Dutch into the twentieth century and, after Indonesia became a sovereign country, demanded autonomy and even outright independence for

their State. Rebels fought the Indonesian military to a costly stalemate, and thousands died in the conflict—which would probably continue today but for a dramatic turn of events. The seafloor epicenter of the 2004 Indian Ocean tsunami lay near the far northern coast of Sumatera, and Aceh was directly in the path of the most powerful ocean waves. Entire towns and roads were swept away and tens of thousands died; Banda Aceh, the State capital, was devastated. The international relief effort opened Aceh to foreigners in ways the Indonesians had long prevented, and the Indonesian army as well as the rebels were engaged in rescue missions rather than warfare. This combination of circumstances facilitated a truce and negotiations resulting in a peace agreement under which the rebels agreed to drop their demand for independence and the Indonesian military withdrew.

Kalimantan is the Indonesian part of the island of Borneo, a slab of the Earth's crystalline crust whose

backbone of tall mountains is of erosional, not volcanic, origin. Larger than Texas, Borneo has a deep, densely rainforested interior that is a last refuge for some 35,000 Orangutans as well as dwindling numbers of Asian elephants, rhinoceroses, and tigers. These and a number of indigenous peoples survive even as loggers and farmers undermine their habitat. Borneo's Pleistocene heritage sustains a comparatively small human population (13 million on the Indonesian side, just over five percent of the country's population) on poor soils. Indigenous peoples, principally the Dayak clans, have traditionally had less impact on the natural environment than the Indonesian and Chinese immigrants and multinational corporations that log the forests and clear woodland for farms. As Figure 10B-7 shows, the only towns of any size in Kalimantan lie on or near the coast; routes into the interior are still few and far between.

Sulawesi consists of a set of intersecting, volcanic mountain ranges rising above sea level; the 800-kilometer (500-mi) Minahasa Peninsula, propelled by volcanic action, is still building itself into the Celebes Sea. This northern peninsula, a favorite of the Dutch colonizers, remains the most developed part of an otherwise rugged and remote island, with Manado its relatively prosperous urban focus. Seven major ethnic groups inhabit the valleys and basins between the mountains, but the population of about 19 million also includes many immigrants from Jawa, especially in and around the southern center of Ujungpandang. Subsistence farming is the leading mode of life, although logging, some mining, and fishing augment the economy. Clashes between Muslims and Christians occur intermittently in remote areas.

Papua, the Indonesian name for the western part of the island of New Guinea, has become an issue in Indonesian politics. Bordered on the east by a classic superimposed geometric boundary (Figs. 10B-7; 10A-6, lower left map), it was taken over by Indonesia from the Dutch in 1969. Papua contains about 22 percent of Indonesia's territory, but its population is under 3 million—just 1 percent of the national total. The indigenous inhabitants of this territory, which is in effect a colony, are Papuan, most living in remote reaches of this mountainous and densely forested island. Papua is economically important to Indonesia, for it contains what is reputed to be the world's richest gold mine and its second-largest open-pit copper mine, but political consciousness has now reached the Papuans. The Free Papua Movement has become increasingly active, holding small rallies in the capital, Jayapura, displaying a Papuan flag, and demanding recognition.

Diversity in Unity

Indonesia's survival as a unified state is as remarkable as India's and Nigeria's. With more than 300 discrete ethnic clusters, over 250 languages, and just about every religion practiced on Earth (although Islam dominates), actual and potential centrifugal forces are powerful here. Wide waters and high mountains perpetuate both cultural distinctions and differences. Indonesia's national motto is *bhinneka tunggal ika*: diversity in unity.

What Indonesia has achieved is etched against the country's continuing cultural heterogeneity. There are dozens of distinct aboriginal cultures; virtually every coastal community has its own roots and traditions. And the majority, the rice-growing Indonesians, include not only the numerous Jawanese—who have their own cultural identity—but also the Sundanese (who constitute 14 percent of Indonesia's population), the Madurese (8 percent), and others. Perhaps the best impression of this cultural diversity comes from the string of islands that extends eastward from Jawa to Timor, the Lesser Sunda Islands (Fig. 10B-7). The rice-growers of Bali adhere to a modified version of Hinduism, giving the island a unique cultural atmosphere; the population of Lombok is mainly Muslim, with some Balinese Hinduism; Sumbawa is a Muslim community; Flores is mostly Roman Catholic. In western Timor, Protestant groups dominate; in the now-independent east, where the Portuguese ruled, Roman Catholicism prevails. Nevertheless, Indonesia nominally is the world's largest Muslim country: overall, 86 percent of the people adhere to Islam, and in the cities the silver domes of neighborhood mosques rise densely above the townscape. Although until recently Indonesian Islam has been relatively moderate, lately more overt Islamization has been on the rise, with new laws banning public displays of affection and limiting the types of clothes women may wear in public.

And Indonesia has hardly been immune from terrorism. Serious terrorist attacks have occurred in the tourist areas of Bali and in central Jakarta, mounted by a terrorist organization named *Jemaah Islamiyah (JI)*, apparently headquartered in eastern Jawa. In addition to Indonesians, many Australians, for whom Indonesia, and especially Bali, is a popular tourist destination, have lost their lives in these attacks.

Transmigration and the Outer Islands

As noted above Indonesia's population of almost 250 million makes it the world's fourth most populous country, but Jawa, we noted earlier, contains approximately 55 percent of it. With about 140 million people on an island the size of Louisiana, population pressure is enormous here. Moreover, Indonesia's annual rate of population growth is above the realm's average of 1.3 percent. To deal with this problem, and at the same time to strengthen the core's power over outlying areas, the Indonesian government long pursued a policy known as **7** transmigration (known as *transmigrasi* in Indonesian), inducing many from the densely populated inner islands (e.g., Jawa, Bali, Madura) to relocate to the sparsely inhabited outer islands (e.g., Sumatera, Kalimantan, Sulawesi). During the last few decades of the twentieth century, millions moved to peripheral locales as part of the government-sponsored program.

It was the Dutch colonialists who started the Transmigration Program, but when Indonesia achieved independence President Sukarno abandoned it. Then, in 1974, President Suharto revived it, and over the next 25 years as many as 8 million Jawanese were relocated to other islands. But surveys showed that half of these migrants experienced a decline in their standard of living; many were reduced to bare subsistence on tropical-forest land taken away from its indigenous inhabitants that turned out to be unsuitable for the farming methods used by the settlers. Cultural conflict, ecological havoc, and deforestation led

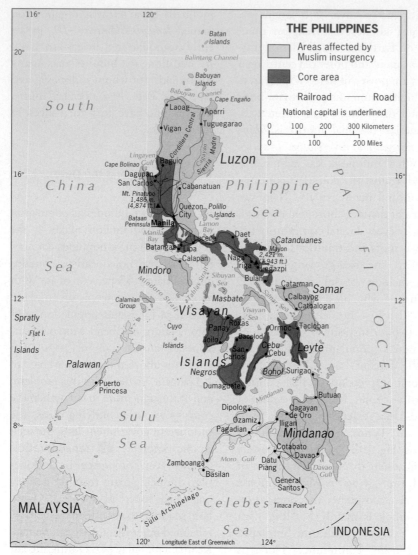

FIGURE 10B-8 © H. J. de Blij, P. O. Muller, and John Wiley & Sons, Inc.

Indeed, the sea may come to mean more to East Timor than mere connectivity. Offshore to the southeast, beneath the waters of the Timor Sea between East Timor and Australia, lie valuable oil and gas reserves (partially mapped in the Fig. 10B-7 inset) whose ownership will depend on how the new country's maritime boundaries (a concept discussed in Chapter 12) are to be drawn. After some difficult, lengthy negotiations, East Timor and Australia agreed in 2005 to defer the maritime-boundary issue for 50 years while East Timor receives a satisfactory share of the revenues from the known oil-yielding reserves. If its continued political turmoil can be calmed, this will go a long way toward providing East Timor with the resources it needs to sustain itself.

The Philippines

North of Indonesia, across the South China Sea from Vietnam, and south of Taiwan lies an archipelago of more than 7000 islands (only about 460 of them larger than 1 square kilometer [0.4 sq mi] in area) inhabited by 94 million people. The islands of the Philippines can be viewed as three groups: (1) Luzon, largest of all, and Mindoro in the north; (2) the Visayan group in the center; and (3) Mindanao, second-largest, located in the south (Fig. 10B-8). Southwest of Mindanao lies a small group of islands, the Sulu Archipelago, nearest to Malaysian Borneo, where Muslim-based insurgencies have kept this area in turmoil.

Few of the generalizations we have been able to make for Southeast Asia could apply in the Philippines without qualification. The country's location relative to the mainstream of change in this part of the world has had much to do with this situation. The islands, inhabited by peoples of Malay ancestry with Indonesian strains, shared with much of the rest of Southeast Asia an early period of Hindu cultural influence, which was strongest in the south and southwest and diminished northward.

the Indonesian government to cancel the program in 2001, but the damage it did will long outlast this ill-fated initiative.

East Timor

The easternmost of the Lesser Sunda Islands is Timor, the eastern part of which was a Portuguese colony, overrun by Indonesia in 1975 and annexed in 1976, which became the scene of a bitter struggle for independence. When the people of East Timor in 1999 were allowed to express their views on independence, this Connecticut-sized territory with only about 800,000 inhabitants (now up to 1.2 million) took the first steps toward statehood.

East Timor (officially known as *Timor-Leste*) became an independent state in 2002. But as the inset map in Figure 10B-7 shows, its sovereignty creates a number of politico-geographical complications. The political entity of East Timor consists of a main territory, where the capital named Dili is located, and a small ***exclave*** to the west on the north coast of (Indonesian) western Timor called Ocussi (spellings vary, and it is sometimes mapped as Ocussi-Ambeno or Ambeno Province). Even though there is a road from the exclave to the main territory, relations between East Timor and Indonesia may not make this link viable, so that the two parts of the new state have to be connected by boat traffic.

Next came the Chinese invasion, felt more strongly on the largest island of Luzon in the northern part of the Philippine archipelago. Islam's arrival was delayed somewhat by the position of the Philippines well to the east of the mainland and to the north of the Indonesian islands. The few southern Muslim beachheads were soon overwhelmed by the Spanish invasion during the sixteenth century. Today the Philippines, adjacent to the world's largest Muslim state (Indonesia), is 83 percent Roman Catholic, 8 percent Protestant, and only 5 percent Muslim.

Muslim Insurgency

The Philippines' small Muslim population, concentrated in the southeastern flank of the archipelago (Fig. 10B-8), has long decried its marginalization in this predominantly Christian country. Over the past generation a half-dozen Muslim organizations, including the Moro (Moor) National Liberation Front and the Moro Islamic Liberation Front, have promoted the Muslim cause with tactics ranging from peaceful negotiation with the government to violent insurgency. Densely forested Basilan Island became the base for an especially extreme group, Abu Sayyaf, which received support through the al-Qaeda network. International counterterror operations were extended into this part of the Philippines when American troops joined Filipino forces in pursuit. The Muslim challenge is siphoning off a disproportionate share of Manila's operating budget.

People and Culture

Out of the Philippines melting pot, where Malay, Arab, Chinese, Japanese, Spanish, and American elements have met and mixed, has emerged the distinctive Filipino culture. It is not a homogeneous or unified culture, as is reflected by the nearly 90 Malay languages in use in the islands, but it is in many ways unique. At independence in 1946, the largest of the Malay languages, Tagalog (also called Pilipino), became the country's official language. But English is widely learned as a second language, and a Tagalog-English hybrid, Taglish, is increasingly heard today. The Chinese component of the population is small (less than two percent) but dominant in local business.

The Philippines population, concentrated where the good farmlands lie, is densest in three general areas: (1) the northwestern and south-central part of Luzon; (2) the southeastern extension of Luzon; and (3) the islands of the Visayan Sea between Luzon and Mindanao (see Fig. 10A-2). Luzon is the site of the capital, Manila-Quezon City (11.7 million—one-eighth of the entire national population), a megacity facing the South China Sea. Alluvial as well as volcanic soils, together with ample moisture in this tropical environment, produce self-sufficiency in rice and other staples and make the Philippines a net exporter of farm products despite a fairly high population growth rate of 2.1 percent.

Perhaps more than any other people, Filipinos take jobs in foreign countries in massive numbers, proving their capacity to succeed in jobs they cannot find at home. The global merchant marine would not exist without Filipino sailors, and Filipina nurses and domestic workers can be found from Dubai to Dubuque. Funds sent home to family members by the emigrants make the Philippines a world leader in monetary inflow known as **8 remittances**.

Prospects

The Philippines seems to get little mention in discussions of Asia-Pacific developments, even though it would seem to be well positioned to share in the Pacific Rim's economic growth. It has participated in offshore manufacturing in its numerous Export Processing Zones (which are quite similar to the Mexican *maquiladoras* discussed in Chapter 4B), but the necessary trade linkages have not been created and sustained. Also, governmental mismanagement as well as political instability have slowed the country's participation, but during the 1990s the situation improved. Despite a series of jarring events—the ouster of U.S. military bases, the damaging eruption of a volcano near the capital, the violence of Muslim insurgents, and a dispute over the nearby Spratly Islands in the South China Sea—the Philippines made substantial economic progress during that decade. Its electronics and textile industries (mostly in the Manila hinterland) expanded continuously, and additional foreign investment arrived. But agriculture continues to dominate the Philippines' economy, unemployment remains high, further land reform is badly needed, and social restructuring (reducing the controlling influence of a comparatively small group of families over national affairs) must occur. The country now is a lower-middle-income economy, and given a longer period of political stability and success in reducing the population growth rate, it will rise to the next level and finally take its place among the Pacific Rim's growth centers.

POINTS TO PONDER

- Islamic-Buddhist conflict in southernmost Thailand and Islamic-Christian strife in the southern Philippines are troubling two of this realm's most populous countries.
- A series of attacks by Muslim fanatics on Christian churches in Malaysia in 2010 threatened the delicate interfaith balance of this dominantly Islamic country.
- Under the constitution of the state of Malaysia, a person of Malay ancestry must adhere to the Muslim faith.
- A combination of representative government, social stability, and economic growth is propelling Indonesia to the forefront of Southeast Asian affairs.

Eastern edge of Australia's Outback, where distances are measured in hundreds of kilometers. © Paul Dymond/Alamy

IN THIS CHAPTER

Australia's amazing biogeography
Australia's changing population
Aboriginal claims to land and resources
China covets Australian commodities
Foreign policy dilemmas downunder
New Zealand's matchless physiography

CONCEPTS, IDEAS, AND TERMS

11

THE AUSTRAL REALM

 Australia
New Zealand

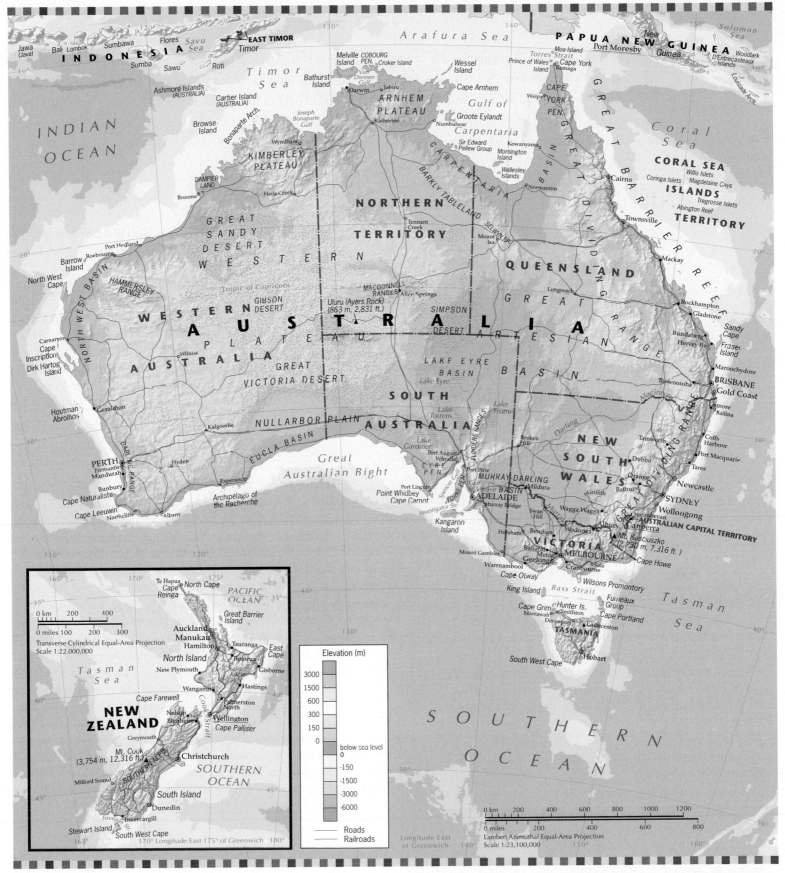

FIGURE 11-1

© H. J. de Blij, P. O. Muller, and John Wiley & Sons, Inc.

The Austral Realm is geographically unique (Fig. 11-1). It is the only geographic realm that lies entirely in the Southern Hemisphere. It is also the only realm that has no land link of any kind to a neighboring realm and is thus completely surrounded by ocean and sea. It is second only to the Pacific as the world's least populous realm. Appropriately, its name refers to its location (**1** *austral* means south)—a location far from the sources of its dominant cultural heritage but close to its newfound economic partners on the Asian Pacific Rim.

DEFINING THE REALM

Two countries constitute the Austral Realm: Australia, in every way the dominant one, and New Zealand, physiographically more varied but demographically much smaller than its giant partner (Fig. 11-2). Between them lies the Tasman Sea. To the west lies the Indian Ocean, to the east the Pacific, and to the south the frigid Southern Ocean.

This southern realm is at a crossroads. On the doorstep of populous Asia, its Anglo-European legacies are now infused by many other cultural strains. Polynesian Maori in New Zealand and Aboriginal communities in Australia are demanding greater rights and more acknowledgment of their cultural heritage. Pacific Rim markets are buying large quantities of raw materials. Japanese and other Asian tourists fill hotels and resorts. Queensland's tropical Gold Coast resembles Honolulu's Waikiki. The streets of Sydney and Melbourne display a multicultural panorama unimagined just two generations ago. All these changes have stirred political debate. Issues ranging from immigration quotas to indigenous land rights dominate, exposing social fault lines (city versus Outback in Australia; North versus South Island in New Zealand). Aborigines and Maori were first in this realm, then the Europeans arrived, and now Asians are a significant economic and cultural element.

LAND AND ENVIRONMENT

Physiographic contrasts between massive, compact Australia and elongated, fragmented New Zealand are related to their locations with respect to the Earth's tectonic plates (consult Fig. G-4). Australia, with some of the geologically most ancient rocks on the planet, lies at the center of its own plate, the Australian Plate. New Zealand, younger and less stable, lies at the convulsive convergence of the Australian and Pacific plates. Earthquakes are rare in Australia, and volcanic eruptions are unknown; New Zealand has plenty of both. This locational contrast is also reflected by differences in relief (Fig. 11-1). Australia's highest relief occurs in what Australians call the Great Dividing Range, the mountains that line the east coast from the Cape York Peninsula to southern Victoria, with an outlier in Tasmania. The highest point along these old, now eroding mountains is Mount Kosciusko, 2230 meters (7316 ft) tall. In New Zealand, entire ranges are higher than this—Mount Cook, for example, reaches 3754 meters (12,316 ft).

West of Australia's Great Dividing Range, the physical landscape generally has low relief, with some local exceptions such as the Macdonnell Ranges near the center; plateaus and plains dominate (Fig. 11-1). The Great Artesian Basin is a key physiographic region, providing underground water sources in what is otherwise desert country; to the south lies the continent's predominant river system, the Murray-Darling. The area mapped as *Western Plateau and Margins* in the lower map of Figure 11-3 contains much of Australia's mineral wealth.

Climates

Figure G-7 reveals the effects of latitudinal position and interior isolation on Australia's climatology. In this respect, Australia is far more varied than New Zealand, its climates ranging from tropical in the far north, where rainforests flourish, to Mediterranean in parts of the south. The interior is dominated by desert and steppe conditions, the

 MAJOR GEOGRAPHIC QUALITIES

THE AUSTRAL REALM

1. Australia and New Zealand constitute a geographic realm by virtue of territorial dimensions, relative location, and dominant cultural landscape.

2. Despite their inclusion in a single geographic realm, Australia and New Zealand differ physiographically. Australia has a vast, dry, low-relief interior; New Zealand is mountainous and has a temperate climate.

3. Australia and New Zealand are marked by peripheral development—Australia because of its aridity, New Zealand because of its topography.

4. The populations of Australia and New Zealand are not only peripherally distributed but also highly clustered in urban centers.

5. The economic geography of Australia and New Zealand is dominated by the export of livestock products and specialty goods such as wine. Australia also has significant wheat production and mining of various resources.

6. Australia and New Zealand are being integrated into the economic framework of the Asian Pacific Rim, principally as suppliers of raw materials.

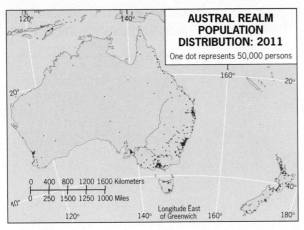

AUSTRAL REALM POPULATION DISTRIBUTION: 2011
One dot represents 50,000 persons

FIGURE 11-2

© H. J. de Blij, P. O. Muller, and John Wiley & Sons, Inc.

steppes providing the grasslands that sustain tens of millions of livestock. Only in the east does Australia have an area of humid temperate climate, and here lies most of the country's economic core area. New Zealand, by contrast, is totally under the influence of the Southern and Pacific oceans, creating moderate, moist conditions, temperate in the north and colder in the south.

The Southern Ocean

Twice now we have referred to the **2** **Southern Ocean**, but try to find this ocean on maps and globes published by famous cartographic organizations such as the National Geographic Society and Rand McNally. From their maps you would conclude that the Atlantic, Pacific, and Indian oceans reach all the way to the shores of Antarctica. Australians and New Zealanders know better. They experience the frigid waters and persistent winds of this great weather-maker on a daily basis.

For us geographers, it is a good exercise to turn the globe upside down now and then. After all, the usual orientation is quite arbitrary. Modern mapmaking started in the Northern Hemisphere, and the cartographers put their hemisphere on top and the other at the bottom. That is now the norm, and it can distort our view of the world. In bookstores in the Southern Hemisphere, you see upside down maps showing Australia and Argentina at the top, and Europe and Canada at the bottom. But this matter has a serious side. A reverse view of the globe shows us how vast the ocean encircling Antarctica is. The Southern Ocean may be remote, but its existence is real.

Where do the northward limits of the Southern Ocean lie? This ocean is bounded not by land but by a marine transition called the **3** **Subtropical Convergence**. Here the cold, extremely dense waters of the Southern Ocean meet the warmer waters of the Atlantic, Pacific, and Indian oceans. It is quite sharply defined by changes in temperature, chemistry, salinity, and marine fauna. Flying over it, you can actually observe it in the changing colors of the water: the Antarctic side is a deep gray, the northern side a greenish blue.

Although the Subtropical Convergence moves seasonally, its position does not vary far from latitude 40° South, which also is the approximate northern limit of Antarctic icebergs. Defined this way, the great Southern Ocean is a huge body of water that moves clockwise (from west to east) around Antarctica, which is why we also call it the **4** **West Wind Drift**.

Biogeography

One of this realm's defining characteristics is its wildlife. Australia is the land of kangaroos and koalas, wallabies and wombats, possums and platypuses. These and numerous other *marsupials* (animals whose young are born very early in their development and then are carried in an abdominal pouch) owe their survival to Australia's early isolation during the breakup of Gondwana (see Fig. 6A-3). Before more advanced mammals could enter Australia and replace the marsupials, as happened in other parts of the world, the landmass was separated from Antarctica and India, and today it contains the world's largest assemblage of marsupial fauna.

Australia's vegetation has distinctive qualities as well, notably the hundreds of species of eucalyptus trees native to this geographic realm. Many other plants form part of Australia's unique flora, some with unusual adaptation to the high temperatures and low humidity that characterize much of the continent.

The study of fauna and flora in spatial perspective combines the disciplines of biology and geography in a field known as **5** **biogeography**, and Australia is a giant laboratory for biogeographers. In the Introduction, we noted that several of the world's climatic zones are named after the vegetation that marks them: tropical savanna, steppe, tundra. When climate, soil, vegetation, and animal life reach a long-term, stable adjustment, vegetation forms the most visible element of this ecosystem.

Biogeographers are especially interested in the distribution of plant and animal species, as well as in the relationships between plant and animal communities and their natural environments. (The study of plant life is called *phytogeography*; the study of animal life is called *zoogeography*.) These scholars seek to explain the distributions the map reveals. In 1876 one of the founders of biogeography, Alfred Russel Wallace, published a book entitled *The Geographical Distribution of Animals* in which he fired the first shot in a long debate: where does the zoogeographic boundary of Australia's fauna lie? Wallace's fieldwork in the area revealed that Australian forms exist not only in Australia itself but also in New Guinea and in some islands to the northwest. As already introduced in Chapter 10, Wallace

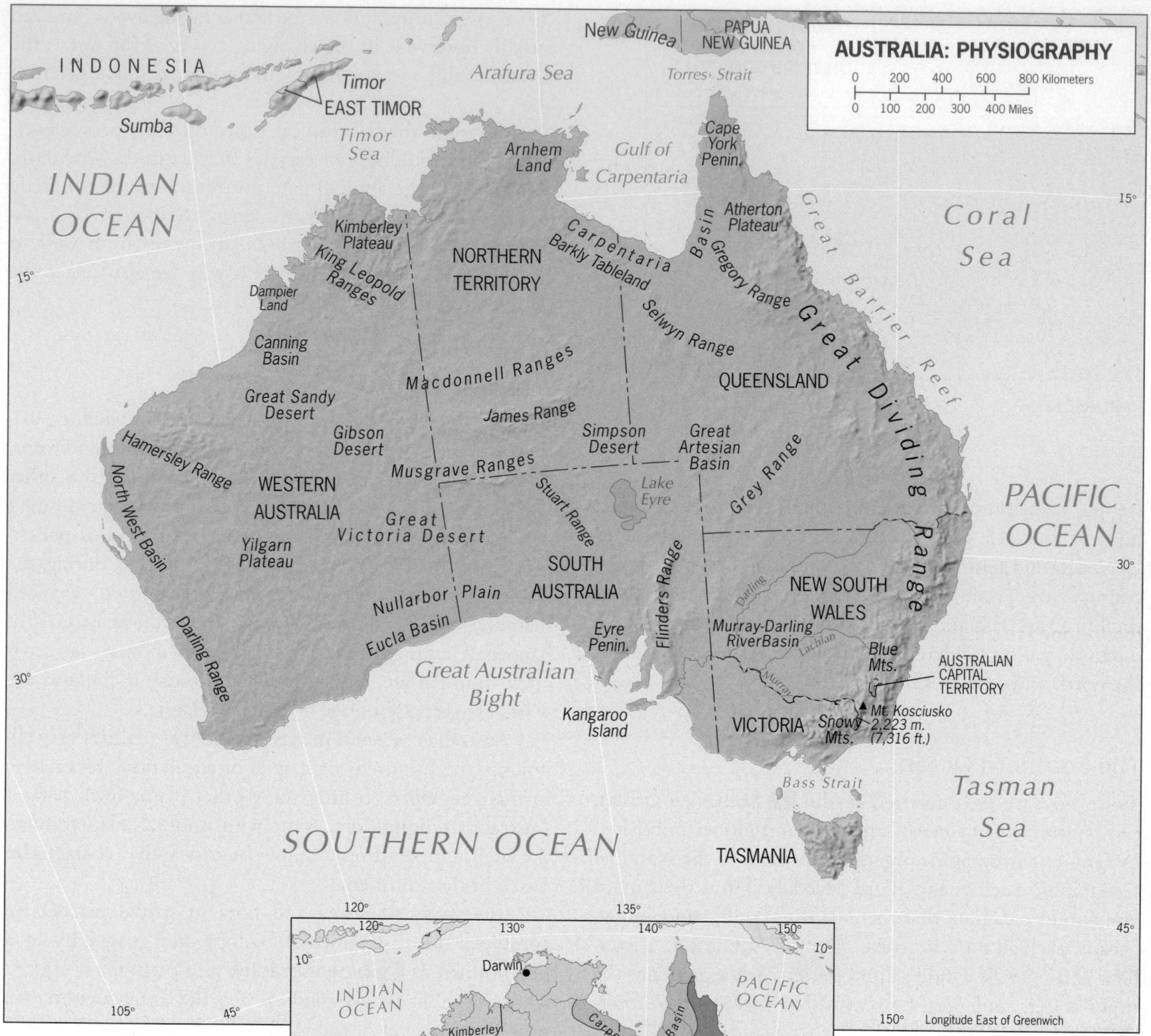
AUSTRALIA: PHYSIOGRAPHY
0 200 400 600 800 Kilometers
0 100 200 300 400 Miles
INDONESIA
Timor
EAST TIMOR
Sumba
New Guinea
PAPUA NEW GUINEA
Arafura Sea
Torres Strait
Timor Sea
INDIAN OCEAN
Arnhem Land
Gulf of Carpentaria
Cape York Penin.
Atherton Plateau
Coral Sea
Kimberley Plateau
King Leopold Ranges
Dampier Land
NORTHERN TERRITORY
Carpentaria Basin
Barkly Tableland
Gregory Range
Great Barrier Reef
Selwyn Range
Canning Basin
Great Dividing Range
Macdonnell Ranges
QUEENSLAND
Great Sandy Desert
James Range
Gibson Desert
Simpson Desert
Great Artesian Basin
Hamersley Range
WESTERN AUSTRALIA
Musgrave Ranges
Lake Eyre
Grey Range
PACIFIC OCEAN
North West Basin
Stuart Range
Great Victoria Desert
Yilgarn Plateau
SOUTH AUSTRALIA
Flinders Range
Darling
NEW SOUTH WALES
Nullarbor Plain
Eucla Basin
Eyre Penin.
Murray-Darling RiverBasin
Lachlan
Blue Mts.
AUSTRALIAN CAPITAL TERRITORY
Darling Range
Great Australian Bight
Murray
Mt. Kosciusko 2,223 m. (7,316 ft.)
Kangaroo Island
VICTORIA
Snowy Mts.
Bass Strait
Tasman Sea
SOUTHERN OCEAN
TASMANIA
15°
30°
45°
105°
120°
135°
150°
Longitude East of Greenwich

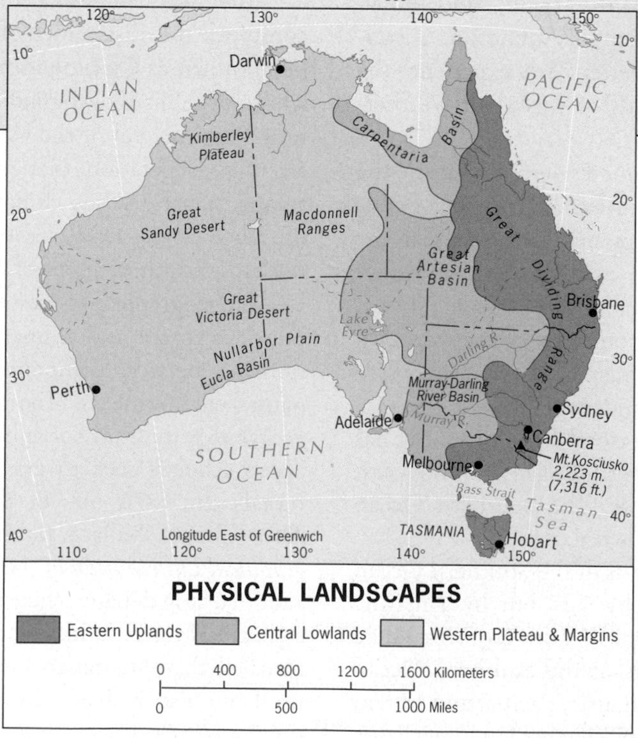
120°
130°
140°
150°
10°
20°
30°
40°
Darwin
INDIAN OCEAN
PACIFIC OCEAN
Kimberley Plateau
Carpentaria Basin
Great Sandy Desert
Macdonnell Ranges
Great Dividing Range
Great Artesian Basin
Brisbane
Great Victoria Desert
Lake Eyre
Darling R.
Nullarbor Plain
Eucla Basin
Murray-Darling River Basin
Perth
Sydney
Adelaide
Murray R.
Canberra
SOUTHERN OCEAN
Melbourne
Mt.Kosciusko 2,223 m. (7,316 ft.)
Bass Strait
Tasman Sea
TASMANIA
Hobart
Longitude East of Greenwich
110°
PHYSICAL LANDSCAPES
Eastern Uplands
Central Lowlands
Western Plateau & Margins
0 400 800 1200 1600 Kilometers
0 500 1000 Miles

FIGURE 11-3

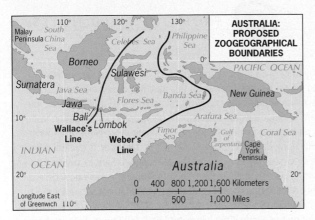

FIGURE 11-4 © H. J. de Blij, P. O. Muller, and John Wiley & Sons, Inc.

proposed that the faunal boundary should lie between Borneo and Sulawesi, and just east of Bali (Fig. 11-4).

6 **Wallace's Line** soon was challenged by other researchers, who found species Wallace had missed and who visited islands Wallace had not. There was no question that Australia's zoogeographic realm ended somewhere in the Indonesian archipelago, but where? Western Indonesia was the habitat of non-marsupial animals such as tigers, rhinoceroses, and elephants, as well as primates; New Guinea clearly was part of the realm of the marsupials. How far had the more advanced mammals progressed eastward along the island stepping stones toward New Guinea? The zoogeographer Max Weber found evidence that led

him to postulate his own **Weber's Line**, which, as Figure 11-4 shows, lay very close to New Guinea.

The Human Impact

Not all research in zoogeography or phytogeography deals with such large questions. Much of it focuses on the relationships between particular species and their habitats—that is, the environment they normally occupy and of which they form a part. Such environments change, and these changes can spell disaster for the species. Climate change forces species to advance or retreat and is a topic often investigated by biogeographers. But it is not always clear whether environmental change or human intervention caused the problem. In Australia, the arrival of the **7** **Aboriginal population** (about 50,000 years ago) appears to have caused an ecosystem collapse because there is no evidence of significant climate change at the time. In an article in *Science*, Gifford H. Miller and his colleagues report that widespread burning of the existing forest, shrub, and grassland vegetation across Australia appears to have led to the spread of desert scrub and to the rapid extinction of most of the continent's large mammals soon after the human invasion occurred. The species that survived faced a second crisis when European colonizers introduced their livestock, leading to the further destruction of remaining wildlife habitats. Survivors include marsupials such as the koala bear and the wombat, but the list of extinctions is far longer.

REGIONS OF THE REALM

Australia is the dominant component of the Austral Realm, a continent-scale country in a size category that also includes China, Canada, the United States, and Brazil. For two reasons, however, Australia has fewer regional divisions than do the aforementioned countries: the comparatively uncomplicated physiography of Australia and its diminutive human numbers. Our discussion, therefore, uses the core-periphery concept as a basis for investigating Australia and focuses on New Zealand as a region by itself.

AUSTRALIA

On January 1, 2001, Australia celebrated its 100th birthday as a state, the Commonwealth of Australia, recognizing (still) the British monarch as the head of state and entering its second century with a strong economy, stable political framework, high standard of living for most of its people, and favorable prospects ahead. Positioned on the Pacific Rim, nine-tenths as large as the 48 contiguous U.S. States, well endowed with farmlands and vast pastures, rivers, underground water, minerals, and energy resources, served

by good natural harbors, and populated by 21.6 million mostly well-educated people, Australia is one of the most fortunate countries in the world.

Sharing the Bounty

Not everyone in Australia shares adequately in all this good fortune, however, and the less advantaged made their voices heard during the celebrations. The country's indigenous (Aboriginal) population, though a small minority today of about 550,000, remains disproportionately disadvantaged in almost every way, from lower life expectancies to higher unemployment than average, from lower high school graduation rates to higher imprisonment ratios. But the nation is now embarked on a campaign to address these ills, with a formal apology issued by the government in 2008; its conciliatory actions range from enhanced social services for Aboriginals to favorable court decisions involving Aboriginal land claims.

When Australia was born as a federal state, its per-capita GNI was the highest in the world. Australia's bounty fueled

© H. J. de Blij

"My most vivid memory from my first visit to Alice Springs is spotting vineyards and a winery in this parched, desert environment as the plane approached the airport. I asked a taxi driver to take me there, and got a lesson in economic geography. Drip irrigation from an underground water supply made viticulture possible; the tourist industry made it profitable. None of this, however, is evident from the view seen here: a spur of the Macdonnell Ranges overlooks a town of bare essentials under the hot sun of the Australian desert. What Alice Springs has is centrality: it is the largest settlement in a vast area Australians often call "the centre." Not far from the mid-point on the nearly 3200-kilometer (2000-mi) Stuart Highway from Darwin on the Northern Territory's north coast to Adelaide on the Southern Ocean, Alice Springs also was the northern terminus of the Central Australian Railway (before the line was extended north to Darwin in 2003), seen in the middle distance. The shipping of cattle and minerals is a major industry here. You need a sense of humor to live here, and the locals have it: the town actually lies on a river, the intermittent Todd River. An annual boat race is held, and in the absence of water the racers carry their boats along the dry river bed. No exploration of Alice Springs would be complete without a visit to the base of the Royal Flying Doctor Service, which brings medical help to outlying villages and homesteads."

www.conceptcaching.com

a huge flow of exports to Europe, and Australians prospered. That golden age could not last forever, and eventually the country's share of international trade declined. Still, Australia today ranks among the top 15 countries in the world in terms of GNI, and for the vast majority of Australians life is comfortable. In terms of the indicators of development discussed in Chapter 9B, Australia is far ahead of all its western Pacific Rim competitors except Japan. As Australians celebrated their first century, they were, on average, earning far more than Thais, Malaysians, or Koreans. In terms of consumption of energy per person, number of automobiles and kilometers of roads, levels of health, and literacy, Australia is a high-income developed country.

Distance

Australians often talk about distance. One of their leading historians, Geoffrey Blainey, labeled it a tyranny—an imposed remoteness from without and a divisive part of life within. Even today, Australia is far from nearly everywhere on Earth. A trans-Pacific jet flight from Los Angeles to Sydney takes about 14 hours nonstop and is correspondingly expensive. Freighters carrying products to European markets take ten days to two weeks to get there. Inside Australia, distances also are of continental proportions, and Australians pay the price—literally. Until some upstart private airlines started a price war, Australians paid more per mile for their domestic flights than air passengers anywhere else in the world.

But distance also was an ally, permitting Australians to ignore the obvious. Australia was a British progeny, a European outpost. Once you had arrived as an immigrant from Britain or Ireland, there were a wide range of environments, magnificent scenery, vast open spaces, and seemingly limitless opportunities. When the Japanese Empire expanded, Australia's remoteness saved the day. When immigration became an issue, Australia in its comfortable isolation could adopt an all-white admission policy that was not officially terminated until 1976. When boat people by the hundreds of thousands fled Vietnam in the mid-1970s aftermath of the Indochina War, almost none reached Australian shores.

Immigrants

Today Australia is changing and rapidly so. Immigration policy now focuses on the would-be immigrants' qualifications, skills, financial status, age, and facility with the English language. With regard to skills, high-technology specialists, financial experts, and medical personnel are especially welcome. Relatives of earlier immigrants, as well as a quota of genuine asylum-seekers, also are admitted. In

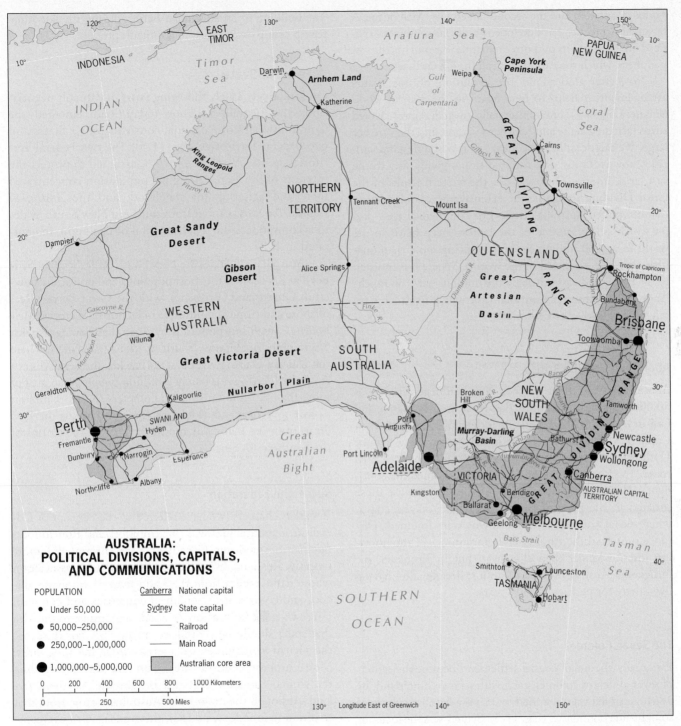

AUSTRALIA: POLITICAL DIVISIONS, CAPITALS, AND COMMUNICATIONS

POPULATION

- Under 50,000
- 50,000–250,000
- 250,000–1,000,000
- 1,000,000–5,000,000

Canberra National capital
Sydney State capital
—— Railroad
—— Main Road

■ Australian core area

0 200 400 600 800 1000 Kilometers
0 250 500 Miles

FIGURE 11-5

© H. J. DE BLIJ, P. O. MULLER, AND JOHN WILEY & SONS, INC.

recent years, total immigration has been between 120,000 and 180,000 annually, which keeps the country's population growing. According to the latest data, Australia's declining natural rate of increase today stands at 0.7 percent. The country is fast becoming a truly multicultural society. In Sydney, for instance, one in five residents is now of Asian ancestry. Overall, a quarter of Australia's 2010 population is foreign-born, and another quarter consists of first-generation Australians.

Core and Periphery

As Figure 11-2 shows, Australia is a large landmass, but its population is heavily concentrated in a core area that lies in the east and southeast, most of which faces the Pacific Ocean (here named the Tasman Sea between Australia and New Zealand). Figure 11-5 above shows that this crescent-like Australian heartland extends from north of the city of Brisbane to the vicinity of Adelaide and includes the largest city, Sydney, the capital, Canberra, and the second-largest city, Melbourne. A secondary core area has developed in the far southwest, centered on Perth and its outport, Fremantle. Beyond lies the vast periphery, which the Australians call the **8 Outback**.

To better understand the evolution of this spatial arrangement, it helps to refer again to the map of world climates (Fig. G-7). Environmentally, Australia's most favored strips face the Pacific and Southern oceans, and they are not large. We can describe the country as a coastal rimland with cities, towns, farms, and forested slopes giving way to the vast, arid, interior Outback. On the western flanks of the Great Dividing Range lie the extensive grassland pastures that catapulted Australia into its first commercial age—and on which still graze one of the largest sheep herds in the world (over 160 million sheep, producing more than one-fifth of all the wool sold in the world). Where it is moister, to the north and east, cattle by the millions graze on ranchlands. This is frontier Australia, over which livestock have ranged for nearly two centuries.

Aborigines and the British

Aboriginal Australians reached this landmass as long as 50,000 years ago, crossed the Bass Strait into Tasmania, and had developed a patchwork of indigenous cultures when Captain Arthur Phillip sailed into what is today Sydney Harbor (1788) to establish the beginnings of modern Australia. The Europeanization of Australia doomed the continent's Aboriginal societies. The first to suffer were those situated in the path of British settlement on the coasts, where penal colonies and free towns were founded. Distance protected the Aboriginal communities of the northern interior longer than elsewhere; but in Tasmania, the indigenous Australians died off in just decades after having lived there for perhaps 45,000 years.

The Seven Colonies

Eventually, the major coastal settlements became the centers of seven different colonies, each with its own hinterland; by 1861, Australia was delimited by its now-familiar pattern of straight-line boundaries (Fig. 11-5). Sydney was the focus for New South Wales; Melbourne, Sydney's rival, anchored Victoria. Adelaide was the heart of South Australia, and Perth lay at the core of Western Australia. Brisbane was the nucleus of Queensland, and Hobart was the seat of government in Tasmania. The largest clusters of surviving Aboriginal people were in the so-called Northern Territory, with Darwin, on Australia's tropical north coast, its colonial city. Notwithstanding their shared cultural heritage, the Australian colonies were at odds not only with London over colonial policies but also with each other over economic and political issues.

The building of an Australian nation during the late nineteenth century was a slow and difficult process.

A Federal State

On January 1, 1901, following years of difficult negotiations, the Australia we know today finally emerged: the Commonwealth of Australia, consisting of six States and two Federal Territories (Fig. 11-5). The two Federal Territories are the *Northern Territory*, assigned to protect the interests of the large Aboriginal population concentrated there and agitating for statehood, and the *Australian Capital Territory*, carved from southern New South Wales to accommodate the federal capital of Canberra, inaugurated in 1927.

Australia's six States are New South Wales (capital Sydney), at 7.0 million the most populous and politically powerful; Queensland (Brisbane), with the Great Barrier Reef offshore and tropical rainforests in its north; Victoria (Melbourne), small but populous by Australian standards with 5.3 million inhabitants; South Australia (Adelaide), where the Murray-Darling river system reaches the sea; Western Australia (Perth) with barely 2 million people in an area of more than 2.5 million square kilometers (nearly 1 million sq mi); and Tasmania (Hobart), the island across the Bass Strait from the mainland and in the path of the storms of the Southern Ocean.

Successful Federation

In earlier chapters, we have referred to the concept of federalism, a political idea with ancient Greek and Roman roots familiar to Americans, Canadians, South Asians, and, more recently, Russians. It is a notion of communal association whose name comes from the Latin *foederis*, implying alliance and coexistence, a union of consensus and common interest—a **9 federation**. It stands in contrast to the idea that states should be centralized, or unitary. For this, too, the ancient Romans devised a term: *unitas*, meaning unity. Most European countries are **10 unitary states**, including the United Kingdom of Great Britain and Northern Ireland. Although the majority of Australians came from that tradition (a kingdom, no less), they managed to overcome their differences and establish a Commonwealth that was, in effect, a federation of States with different viewpoints, economies, and objectives, separated by enormous distances along the rim of an island continent.

An Urban Culture

Despite the vast open spaces and romantic notions of frontier and Outback, Australia is an urban country, with 87 percent of all Australians living in cities and towns. As

© H. J. de Blij

"See a scene like this, and you realize why Australians refer to their land as "the lucky country." Australia's periphery has much magnificent scenery ranging from spectacular cliffs to dune-lined beaches, but nothing matches Sydney Harbor on a sunny, breezy day when sailboats by the hundreds emerge from coves and inlets and, if you are fortunate enough to be on the water yourself, every turn around a headland presents still another memorable view. Ask the captains of ocean liners plying the world what port is their favorite, and most will point to this magnificent estuary with its narrow entrance and secluded bays as the grandest of all."

www.conceptcaching.com

noted earlier, the core of Australia lies in the southeast of the continent (Fig. 11-5). Yet for all its vastness and youth, Australia has developed a remarkable cultural identity, a sameness of urban and rural landscapes that persists from one end of the continent to the other. Sydney, often called the New York of Australia, lies on a spectacular estuarine site, its compact, high-rise central business district overlooking a port bustling with ferry and freighter traffic. Sydney is a vast, sprawling metropolis of 4.4 million, with multiple outlying centers studding its far-flung suburbs; brash modernity and reserved British ways somehow blend here. Melbourne (3.8 million), sometimes regarded as the Boston of Australia, prides itself on its more interesting architecture and more cultured ways. Brisbane, the capital of Queensland, which also anchors Australia's Gold Coast and adjoins the Great Barrier Reef, is the Miami of Australia; unlike Miami, however, its residents can find nearby relief from the summer heat in the mountains of its immediate hinterland (as well as at its beaches). Perth, Australia's San Diego, is one of the world's most isolated cities, separated from its nearest Australian neighbor by two-thirds of a continent and from Southeast Asia and Africa by thousands of kilometers of ocean.

And yet, each of these cities—as well as the capitals of South Australia (Adelaide), Tasmania (Hobart), and, to a lesser extent, the Northern Territory (Darwin)—exhibits an Australian character of unmistakable quality. Life is orderly and unhurried. Streets are clean, slums are few, graffiti rarely seen. By American and even European standards, violent crime (though rising) is uncommon. Standards of public transport, city schools, and health-care provision are high. Spacious parks, pleasing waterfronts, and plentiful sunshine make Australia's urban life more acceptable than that almost anywhere else in the world. Critics of Australia's way of life say that this very pleasant state of affairs has persuaded Australians that hard work is not really necessary. But a few days' experience in the commercial centers of the major cities contradicts that assertion: the pace of life is quickening. The country's cultural geography evolved as that of a European outpost, prosperous and secure in its isolation. Now Australia is reinventing itself as a major component in an Australo-Asian chain, a Pacific partner in a transformed regional economic geography.

Economic Geography

Because of its early history as a treasure trove of raw materials, Australia to this day is seen worldwide as a country whose economy depends primarily on its exportable natural resources. In fact, however, Australia's economy depends overwhelmingly on services, not commodity exports. Tourism alone contributes around 5 percent—about the same as the value of mineral exports. The action is in those bustling coastal cities far more than in the Outback.

When Australia became established as a state, however, it needed goods from overseas, and here the tyranny of distance played a key role. Imports from Britain (and later the United States) were expensive mainly because of transport costs. This

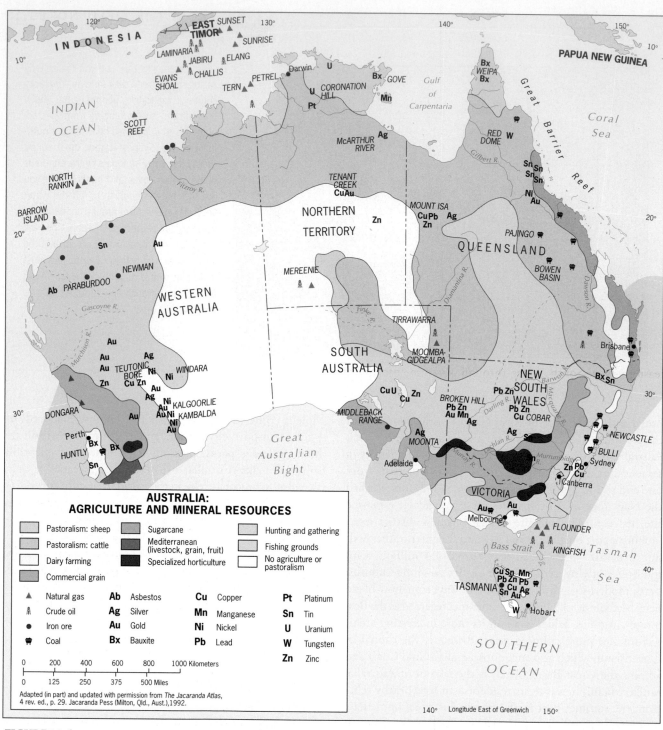

AUSTRALIA:
AGRICULTURE AND MINERAL RESOURCES

Pastoralism: sheep
Pastoralism: cattle
Dairy farming
Commercial grain

Sugarcane
Mediterranean
(livestock, grain, fruit)
Specialized horticulture

Hunting and gathering
Fishing grounds
No agriculture or
pastoralism

▲ Natural gas   **Ab** Asbestos   **Cu** Copper   **Pt** Platinum
🛢 Crude oil   **Ag** Silver   **Mn** Manganese   **Sn** Tin
● Iron ore   **Au** Gold   **Ni** Nickel   **U** Uranium
⛏ Coal   **Bx** Bauxite   **Pb** Lead   **W** Tungsten
                                                 Zn Zinc

0   200   400   600   800   1000 Kilometers
0 125 250 375 500 Miles

Adapted (in part) and updated with permission from *The Jacaranda Atlas*,
4 rev. ed., p. 29. Jacaranda Pess (Milton, Qld., Aust.),1992.

FIGURE 11-6

© H. J. de Blij, P. O. Muller, and John Wiley & Sons, Inc.

encouraged local entrepreneurs to set up their own industries to produce such goods more cheaply. Economic geographers call such industries **11 import-substitution industries,** and this is how local industrialization got its start.

When the prices of foreign goods became lower because transportation was more efficient and therefore cheaper, local businesses demanded protection from the colonial governments, and high tariffs were erected against imported goods. Local products now could continue to be made inefficiently because their market was guaranteed. If Japan could not afford this, how could Australia? We can see the answer on the map (Fig. 11-6). Even before federation in 1901, all the colonies could export valuable minerals whose earnings shored up those inefficient, uncompetitive local industries. By the time the colonies unified, pastoral industries were also contributing income. So the miners and the farmers paid for those imports Australians could not produce themselves, plus the products made in the cities. No wonder the cities grew: here were secure manufacturing jobs, jobs in state-run service enterprises, and jobs in the growing government bureaucracy. When we noted earlier that Australians once had the highest per-capita GNI in the world, this was initially achieved in the mines and on the farms, not in the cities.

Agricultural Abundance

Australia has material assets of which other countries on the Pacific Rim can only dream. In agriculture, sheep-raising was the earliest commercial venture, but it was the technology of refrigeration that brought world markets within reach of Australian beef producers. Wool, meat, and wheat have long been the country's big three income earners; Figure 11-6 displays the immense pastures in the east, north, and west that constitute the ranges of Australia's huge herds. The zone of commercial grain farming forms a broad crescent extending from northeastern New South Wales through Victoria into South Australia, and covers much of the hinterland of Perth. Keep in mind the scale of this map: Australia is only slightly smaller than the 48 contiguous States of the United States! Commercial grain farming in Australia is big business. As the climatic map (Fig. G-7) would suggest, sugarcane grows along most of the warm, humid coastal strip of Queensland, and Mediterranean crops (including grapes for Australia's highly successful and expanding wine industry) cluster in the hinterlands of Adelaide and Perth. Mixed horticulture concentrates in the basin of the Murray River system, including rice, grapes, and citrus fruits, all under irrigation. And, as elsewhere in the world, dairying has developed near the large metropolitan areas. With its considerable range of environments, Australia yields a diversity of crops.

Mineral Wealth

Australia's mineral resources, as Figure 11-6 shows, also are diverse. Major gold discoveries in Victoria and New South Wales produced a ten-year gold rush starting in 1851 and ushered in a new economic era. By the middle years of that decade, Australia was producing 40 percent of the world's gold. Subsequently, the search for more gold led to the discoveries of additional minerals. New finds are still being made today, and even oil and natural gas have been found both inland and offshore (see the symbols in Fig. 11-6 in the Bass Strait between Tasmania and the mainland, and off the northwestern coast of Western Australia). Coal is mined at many locations, notably in the east near Sydney and Brisbane but also in Western Australia and even in Tasmania; before coal prices fell, this was a valuable export.

About 320 kilometers (200 mi) south-southwest of Port Hedland in the Pilbara region of Western Australia, Mount Whaleback near the town of Newman is one of the world's largest sources of high-grade iron ore. China's insatiable demand for this ore has generated here the planet's largest "open-cut" iron mine, a huge and growing gash in the natural landscape—just one such impact resulting from Australia's role as raw-material supplier to industrializing economies. © Photoshot Holdings Ltd./Alamy

Substantial deposits of metallic and nonmetallic minerals abound—from the complex at Broken Hill and the mix of minerals at Mount Isa to the huge nickel deposits at Kalgoorlie and Kambalda, the copper of Tasmania, the tungsten and bauxite of northern Queensland, and the asbestos of Western Australia. A closer look at the map reveals the wide distribution of iron ore (the red dots), and for this raw material as for many others, Japan was Australia's best customer for many years. Today, China has become Australia's leading customer.

Manufacturing's Limits

Australian manufacturing, as we noted earlier, remains oriented to domestic markets. Australian automobiles, electronic equipment, and cameras are clearly not challenging the Pacific Rim's economic tigers for a place on world markets—not yet, at any rate. Australian manufacturing is diversified, producing some machinery and equipment made of locally produced steel as well as textiles, chemicals, paper, and many other items. These industries cluster in and near the major urban areas where the markets are. The domestic market in Australia is not large, but it remains relatively affluent. This makes it attractive to foreign producers, and Australia's shops are full of high-priced goods from Japan, South Korea, Taiwan, the U.S., and Europe. Indeed, despite its long-term protectionist practices, Australia still does not produce many goods that could be manufactured at home. Overall, the economy continues to display symptoms of a still-developing rather than a fully developed country.

Australia's Future

The Commonwealth of Australia is changing, but its neighbors in Southeast and South Asia are changing even faster. Australia's European bonds are weakening as its Asian ties are strengthening and it plays a bigger role in the Pacific Rim. Australia does face some challenges at home as well. These include: (1) Aboriginal claims, (2) concerns over immigration, (3) environmental degradation, and (4) issues involving Australia's status and regional role.

Aboriginal Issues

For several decades the Aboriginal issues focused on two questions: the formal admission by the government and majority of mistreatment of the Aboriginal minority with official apologies and reparations, and land ownership. The first question was resolved in 2008 when newly elected (now former) Prime Minister Kevin Rudd offered a formal apology for the historic mistreatment of the Aborigines. The second question has major geographic implications. Although making up only 2 percent of the total population, the Aboriginal population of 550,000 (including many of mixed ancestry) has been gaining influence in national affairs, and in the 1980s Aboriginal leaders began a campaign to obstruct exploration on what they designated as

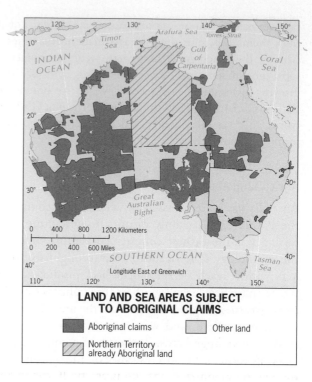

FIGURE 11-7

© H. J. DE BLIJ, P. O. MULLER, AND JOHN WILEY & SONS, INC.

The name Cabramatta conjures up varied reactions among Australians. During the 1950s and 1960s, many immigrants from southern Europe settled in this western suburb of Sydney, attracted by affordable housing. In the 1970s and 1980s, Southeast Asians arrived here in large numbers, and during this period Cabramatta, in the eyes of many, became synonymous with gang violence and drug dealing. More recently, however, Cabramatta's ethnic diversity has come to be viewed in a more favorable light, and is now seen as the "multicultural capital of Australia," a tourist attraction and proof of Australia's capacity to accommodate non-Europeans. Meanwhile, Cabramatta has been spruced up with Oriental motifs of various kinds. This "Freedom Gate" in the Vietnamese community is flanked by a Ming horse and a replica of a Forbidden City lion—all reflecting better times for an old gateway for immigrants as well as refugees. © Trip/Eric Smith.

ancestral and sacred lands. When the Europeans arrived as colonists they had invoked the doctrine of *terra nullis* (land owned by no one) and occupied land that had been used communally by the indigenous people. The situation is similar to what happened in the Americas when Amerindian ideas of land ownership conflicted with the private ownership ideas of the Europeans. Until 1992, Australians had taken it for granted that Aboriginals had no right to land ownership, but in that year the Australian High Court made the first of a series of rulings in favor of Aboriginal claimants. A subsequent court decision implied that vast areas (probably as much as 78 percent of the whole continent) could potentially be subject to Aboriginal claims (Fig. 11-7). Today the **12** **Aboriginal land issue** remains mainly (though not exclusively) an Outback issue, but it has the potential to overwhelm Australia's court system and to inhibit economic growth.

Immigration Issues

The immigration issue is older than Australia itself. Fifty years ago, when Australia had less than half the population it has today, 95 percent of the people were of European ancestry, and more than three-quarters of them came from the British Isles. Eugenic (race-specific) immigration policies maintained this situation until the 1970s. Today, the picture is dramatically different: of 21.6 million Australians, only about one-third have British-Irish origins, and Asian immigrants outnumber both European immigrants and the natural increase each year. During the early 1990s, about 150,000 legal immigrants arrived in Australia annually, most from Hong Kong, Vietnam, China, the Philippines, India, and Sri Lanka. Annual immigration quotas have since been reduced, most recently to 80,000, but Asian immigrants continue to outnumber those from Western sources. This situation upsets some in the long-dominant Anglo-Australian society, and restrictions on foreign ownership of Australian real estate reflect fears of what extremist groups call the Asianization of the country. But the obvious success of many Asian settlers in Australia evinces the opportunities still available in this free and open society. Sydney, the main recipient of the Asian influx, has become a mosaic of ethnic neighborhoods, some of which have gone through periods of gang violence and drug dealing but have stabilized and even prospered over time (see photo at left). Still, multiculturalism will remain a long-term challenge for Australia in the twenty-first century.

Environmental Issues

13 **Environmental degradation** is practically synonymous with Australia. First the Aborigines, then the Europeans and their livestock inflicted heavy damage on Australia's natural environments and ecologies. Great stands of magnificent forest were destroyed. In Western Australia, centuries-old trees were simply ringed and left to die so that the sun could penetrate through their leafless crowns to nurture the grass below for pasture for introduced livestock. In island Tasmania, where Australia's native eucalyptus tree reaches its greatest dimensions (comparable to North American redwood stands), tens of thousands of hectares of this irreplaceable treasure have been lost to chain saws and pulp

mills. Many of Australia's unique marsupial species have been driven to extinction, and many more are endangered or threatened. "Never have so few people wreaked so much havoc on the ecology of so large an area in so short a time," observed a geographer in Australia recently. But awareness of this environmental degradation is growing. In Tasmania, the Green environmentalist political party has now become a force in State affairs, and its activism has slowed deforestation, dam-building, and other development projects. Still, many Australians fear the environmentalist movement as an obstacle to economic growth at a time when the economy needs stimulation. This, too, is an issue for the future.

Issues of Status and Role

Several issues involving Australia's status at home, relations with neighbors, and position in the world are also stirring up national debate. A persistent domestic question is whether Australia should become a republic, ending the status of the British monarch as the head of state, or whether it should continue its status quo in the British Commonwealth.

Relations with neighboring *Indonesia* and *East Timor* have become more complex. For many years, Australia had what may be called a special relationship with Indonesia, whose help it needs in curbing illegal seaborne immigration. It was also profitable for Australia to counter international (UN) opinion and recognize Indonesia's 1976 annexation of Portuguese East Timor, for in doing so Australia could deal directly with Jakarta for the oil reserves under the Timor Sea (see Fig. 10B-7). Thus Australia gave neither recognition nor support to the rebel movement that fought for independence in East Timor. But this story had a surprisingly happy ending. When the East Timorese

During the first decade of the twenty-first century Australia lay in the grip of its worst drought on record, with calamitous consequences ranging from deadly forest fires to parched farmlands. One of Australia's most profitable agricultural industries, winemaking, suffered severely as many winegrowers were driven off their land. The photo on the left shows a once-mature vineyard near Yarrawonga, on the New South Wales-Victoria border north of Melbourne along the Murray River, desiccated and abandoned during the summer of 2009. When the rains finally came, they caused disastrous floods, washing away topsoil and eroding the heat-baked countryside. The photo on the right, taken in January 2010, shows the results at Coonamble, New South Wales, along a tributary of the Darling River. (Left) © Global Warming Images/Alamy (Right) © Photo by James Croucher/Newspix/Getty Images, Inc.

campaign for independence succeeded in 1999 and Indonesian troops began an orgy of murder and destruction, Australia sent an effective peacekeeping force and spearheaded the United Nations effort to stabilize the situation. Today, a new chapter has opened in Australia's relations with these northern neighbors.

Australia also has a long-term relationship with *Papua New Guinea (PNG)*, as is noted in Chapter 12. This association, too, has gone through difficult times. In recent years, the inhabitants of Papua New Guinea have strongly resisted privatization, World Bank involvement, and globalization generally. Australia assists PNG in several spheres, but its motives are sometimes questioned. In 2001, the construction of a projected gas pipeline from PNG to the Australian State of Queensland precipitated fighting among tribespeople over land rights, resulting in dozens of casualties, and Australian public opinion reflected doubts concerning the appropriateness of this venture.

When political violence and chaos overtook the *Solomon Islands* east of Papua New Guinea in 2003, Australian forces intervened: a failing state in Australia's neighborhood could become a base for terrorist activity.

Immediately after the 9/11 attack on U.S. targets in New York and Washington, Australian leaders expressed strong support for the American campaign against terrorism, but they made it a point also to assure the Indonesian government that the War on Terror was not a war on Islam. Although the great majority of Australians supported this stance and troops were sent to Iraq, their will was severely tested in 2002 when a terrorist attack on a nightclub on the Indonesian island of Bali killed 88 vacationing Australians—and again in 2005 when another attack in Bali was followed by the apprehension of a terrorist cell preparing for an assault within Australia itself. The elections of 2007 brought in a new Labor government with a different orientation, especially toward the United States, and Australia's troops were pulled out of Iraq.

Territorial dimensions, relative location, and raw-material wealth have helped determine Australia's place in the world and, more specifically, on the Pacific Rim. Australia's population, still barely 20 million in the second decade of the twenty-first century, is smaller than Malaysia's and only slightly larger than that of the Caribbean island of Hispaniola. However, Australia's importance in the international community far exceeds its human numbers.

▤ NEW ZEALAND

Fifteen hundred miles east-southeast of Australia, in the Pacific Ocean across the Tasman Sea, lies New Zealand, also known as *Aotearoa* in Maori (meaning "land of the long white cloud"). In an earlier age, New Zealand would have been part of the Pacific geographic realm because its population was all Maori, a people with Polynesian roots.

But New Zealand, like Australia, was invaded and occupied by Europeans. Today, its population of 4.4 million is about 70 percent European, and the Maori form a substantial minority of about 660,000, with many of mixed Euro-Polynesian ancestry (including Pacific Islanders).

New Zealand consists of two large mountainous islands and many scattered smaller islands (Fig. 11-8). The two large islands, with the South Island somewhat larger than the North Island, look diminutive in the great Pacific Ocean, but together they are larger than Britain. In contrast to Australia, the two main islands are mostly mountainous or hilly, with several peaks rising far higher than any on the Australian landmass. The South Island has a spectacular snowcapped range appropriately called the Southern Alps, with numerous peaks reaching beyond 3300 meters (10,000 ft). The smaller North Island has proportionately more land under low relief, but it also has an area of central highlands along whose lower slopes lie the pastures of New Zealand's chief dairying district. Hence, while Australia's land lies relatively low in elevation and exhibits much low relief, New Zealand's is on the average high and is dominated by rugged terrain.

Human Spatial Organization

The most promising areas for habitation, therefore, are the lower-lying slopes and lowland fringes on both islands. On the North Island, the largest urban area, Auckland, occupies a comparatively low-lying peninsula. On the South Island, the largest lowland is the agricultural Canterbury Plain, centered on Christchurch. What makes these lower areas so attractive, apart from their availability as cropland, is their magnificent pastures. The range of soils and pasture plants allows both summer and winter grazing. Moreover, the Canterbury Plain, the chief farming region, also produces a wide variety of vegetables, cereals, and fruits. About half of all New Zealand is pasture land, and much of the farming provides fodder for the pastoral industry. Sixty million sheep and eight million cattle dominate these livestock-raising activities, with wool, meat, and dairy products providing about two-thirds of the islands' export revenues.

Despite their contrasts in size, shape, physiography, and history, New Zealand and Australia share a number of characteristics. Apart from their joint British heritage, they share a sizeable pastoral economy with growth in specialty goods such as wines, a small local market, the problem of great distances to world markets, and a desire to stimulate (through protection) domestic manufacturing. The very high degree of urbanization in New Zealand (86 percent of the total population) once again resembles Australia: substantial employment in city-based industries, mostly the processing and packing of livestock and farm products, as well as government jobs.

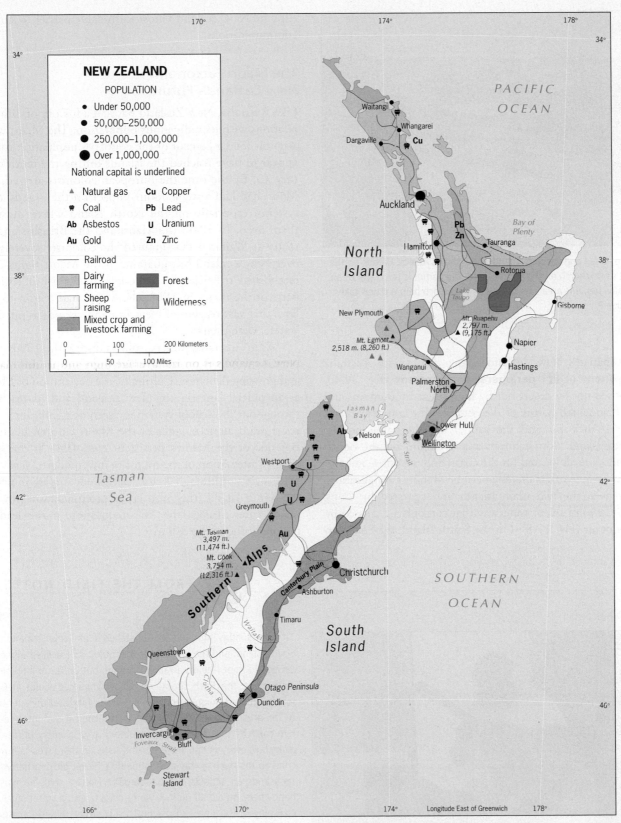

NEW ZEALAND

POPULATION
- Under 50,000
- 50,000–250,000
- 250,000–1,000,000
- Over 1,000,000

National capital is underlined

▲ Natural gas   **Cu** Copper
🪣 Coal   **Pb** Lead
Ab Asbestos   **U** Uranium
Au Gold   **Zn** Zinc

— Railroad

Dairy farming

Sheep raising

Mixed crop and livestock farming

Forest

Wilderness

0   100   200 Kilometers
0 50 100 Miles

PACIFIC OCEAN

Waitangi
Whangarei
Dargaville
Cu
Auckland
Bay of Plenty
Hamilton
Pb Zn
Tauranga
North Island
Rotorua
Gisborne
Lake Taupo
New Plymouth
Mt. Ruapehu 2,797 m. (9,175 ft.)
Mt. Egmont 2,518 m. (8,260 ft.)
Napier
Wanganui
Hastings
Palmerston North
Lower Hutt
Wellington

Tasman Bay
Ab
Nelson
Cook Strait

Tasman Sea

Westport
U
U
U
Greymouth
Mt. Tasman 3,497 m. (11,474 ft.)
Mt. Cook 3,754 m. (12,316 ft.)
Au
Southern Alps
Canterbury Plain
Christchurch
Ashburton
Waitaki R.
Timaru
Queenstown
Clutha R.
South Island
Otago Peninsula
Dunedin
Invercargill
Bluff
Foveaux Strait
Stewart Island

SOUTHERN OCEAN

166°   170°   174° Longitude East of Greenwich 178°

FIGURE 11-8

© H. J. de Blij, P. O. Muller, and John Wiley & Sons, Inc.

New Zealand's landscapes are varied and spectacular. This view shows Akaroa Harbor, on the Banks Peninsula just south of Christchurch on the South Island. The harbor is the crater of an ancient volcano filled in by seawater. This area was settled by the French before it was taken over by the British. © A. WinklerPrins

Spatially, New Zealand further shares with Australia its pattern of **14 peripheral development** (Fig. 11-2), imposed not by desert but by high rugged mountains and the fragmented nature of the country. The country's major cities—Auckland and the capital of Wellington on the North Island; Christchurch and Dunedin on the South Island—are all located on the coast, and the entire rail and highway network is thereby peripheral in its configuration. Moreover, the two main islands are separated by Cook Strait, a wind-swept waterway that can only be crossed by ferry or air (Fig. 11-8). On the South Island, the Southern

Alps are New Zealand's most formidable barrier to surface communications.

The Maori Factor and New Zealand's Future

Like Australia, New Zealand has had a history of difficult relations with its indigenous population. The Maori, who account for 15 percent of the country's population today, appear to have reached the islands during the tenth century AD. By the time the European colonists arrived, the Maori had had a tremendous impact on the islands' ecosystems, especially on the North Island where most of them lived. In 1840, the Maori and the British signed a treaty at Waitangi that granted the colonists sovereignty over New Zealand but guaranteed the Maori rights over established tribal lands. Although the British abrogated parts of the treaty in 1862, the Maori had reason to believe that vast reaches of New Zealand, as well as offshore waters, were theirs.

As in Australia, judicial rulings during the 1990s supported the Maori position, which led to expanded claims and growing demands. Culturally, the declaration of Maori as an official language in New Zealand and its teaching throughout the school system are seen as significant progress toward an acceptance of the Maori cultural heritage. But one of the Maori's persistent complaints is the slow pace of integration of the minority into modern New Zealand society. Although Maori claims encompass much of the South Island, they also cover prominent sites in the major cities. Today, the Maori question is the leading national issue in the country.

 FROM THE FIELD NOTES...

© H. J. de Blij

"It was Sunday morning in Christchurch on New Zealand's South Island, and the city center was quiet. As I walked along Linwood Street I heard a familiar sound, but on an unfamiliar instrument: the Bach sonata for unaccompanied violin in G minor—played magnificently on a guitar. I followed the sound to the artist, a Maori musician of such technical and interpretive capacity that there was something new in every phrase, every line, every tempo. I was his only listener; there were a few coins in his open guitar case. Shouldn't he be playing before thousands, in schools, maybe abroad? No, he said, he was happy here, he did all right. A world-class talent, a street musician playing Bach on a Christchurch side street, where tourists from around the world were his main source of income. Talk about globalization."

www.conceptcaching.com

The Green Factor

New Zealand is well known for its progressive politics and high quality of life. Among the factors that contribute to superior environmental progressiveness is its status as one of the leading "green" societies in the world, with a long-active Green Party and an established program of environmental conservation. Although the Maori and then the European colonists degraded the New Zealand landscape, their descendants have been exemplary in observing environmentally-friendly and sustainable policies.

Environmental scientists recently ranked New Zealand as number one (the United States was twenty-eighth) in a report that examined a range of environmental indices such as clean water, air pollution, renewable energy, and biodiversity conservation. Approximately 30 percent of its land area is protected from development, and in 2007 New Zealand declared that it would become the first carbon-neutral country in the world. Already over 70 percent of its energy comes from renewable sources (hydro and geothermal), but it aspires to a goal of 90 percent by 2025. This is far beyond what any country has set out to achieve. It is also a nuclear-free country that does not permit even visiting naval vessels with nuclear capacities to stop at its ports. The country has also established Environmental Courts that hear cases involving environmental management decisions. With its green initiatives, New Zealand demonstrates that a country, albeit one with a small population, can work to improve the environment if it possesses the appropriate political will at the highest level.

Dominant cultural heritage and prevailing cultural landscape form two criteria on which the delimitation of the Austral Realm is based. But in both Australia and New Zealand, the cultural mosaic is changing, and the convergence with neighboring realms is proceeding.

POINTS TO PONDER

- Australia's population is growing rapidly, with immigration (nearly 250,000 per year) exceeding natural increase (about 150,000 annually).
- Australia is on course to become the world's second-largest exporter of liquefied natural gas by 2020.
- Australia has suffered through a 13-year drought with only minor relief, depleting irrigation waters, drying up rivers, and devastating agricultural areas.
- Over the past decade, Australians have killed about 50 million kangaroos. Almost all the meat has gone into pet food.

One of the islands of Kiribati, near where the equator and International Date Line intersect—as well as the realm's three regions. © H. J. de Blij

IN THIS CHAPTER

Water, water everywhere—but who owns it?
Islands high and low
The conundrum of divided Papua
Colonists and foreigners: Still there, still dominant
The partitioning of Antarctica
Geopolitics in the warming Arctic Basin

CONCEPTS, IDEAS, AND TERMS

Marine geography	1
Territorial sea	2
High seas	3
Continental shelf	4
Exclusive Economic Zone (EEZ)	5
Maritime boundary	6
Median-line boundary	7
High-island cultures	8
Low-island cultures	9
Antarctic Treaty	10

12

PACIFIC REALM AND POLAR FUTURES

Melanesia
Micronesia
Polynesia

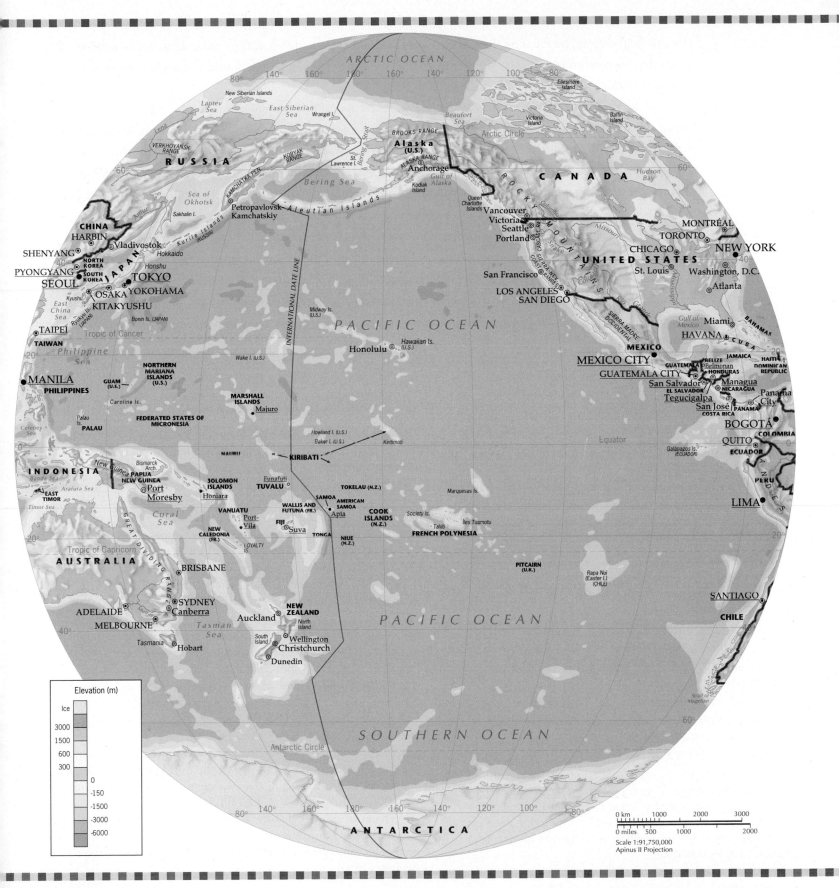

Elevation (m)

Ice
3000
1500
600
300
0
-150
-1500
-3000
-6000

0 km 1000 2000 3000

0 miles 500 1000 2000

Scale 1:91,750,000
Apinus II Projection

FIGURE 12-1

© H. J. de Blij, P. O. Muller, and John Wiley & Sons, Inc.

In this final chapter, we focus on three of the world's vast expanses so different from those discussed in previous chapters that we need an entirely new perspective. Water and ice, not land and soil, dominate the physiography. Environmental changes going on here affect the entire planet.

The largest of these expanses, the *Pacific Realm*, covers almost half the globe from the Bering Sea to the Southern Ocean and from Ecuador to Indonesia. The Pacific Ocean, larger than all the world's land areas combined, is studded with tens of thousands of islands of which New Guinea is by far the largest. The two polar areas to which it is linked could hardly be more different: the *Arctic* is water, frozen at the surface, so that you can reach the North Pole in a submarine; the *Antarctic* is land, weighed down by the planet's largest accumulation of permanent ice. The surface journey to the South Pole, first accomplished exactly a century ago, has cost many an explorer his life.

Climate change is altering the geography everywhere, as is enhanced technology of mineral (especially oil and gas) exploration and exploitation. Rising sea levels would imperil thousands of populated low-elevation islands. Areas of ocean floor once beyond the scope of human intervention are now coming within reach. Melting ice is clearing new maritime shipping routes. But, as we will see, international rules to govern who owns what in these remote frontiers are subject to dispute. When land boundaries jut outward into the sea, it depends on who draws the maps. We should be aware of the potential for competition, even conflict.

 ## MAJOR GEOGRAPHIC QUALITIES

THE PACIFIC REALM

1. The Pacific Realm's total area is the largest of all geographic realms. Its land area, however, is the smallest, as is its population.

2. The island of New Guinea, with 9.3 million people, alone contains over 75 percent of the Pacific Realm's population.

3. The Pacific Realm, with its wide expanses of water and numerous islands, has been strongly affected by United Nations Law of the Sea provisions regarding states' rights over economic assets in their adjacent waters.

4. The highly fragmented Pacific Realm consists of three regions: Melanesia (including New Guinea), Micronesia, and Polynesia.

5. Melanesia forms the link between Papuan and Melanesian cultures in the Pacific.

6. The Pacific Realm's islands and cultures may be divided into volcanic high-island cultures and coral-based low-island cultures.

7. In Micronesia, U.S. influence has been particularly strong and continues to affect local societies.

8. In Polynesia indigenous culture exhibits remarkable consistency and uniformity throughout the region, its enormous dimensions and dispersal notwithstanding. Yet at the same time, local cultures are nearly everywhere severely strained by external influences. In Hawai'i, as in New Zealand, indigenous culture has been largely submerged by Westernization.

DEFINING THE REALM

Our survey begins with the Pacific geographic realm that covers nearly an entire hemisphere of this world, the one commonly called the Sea Hemisphere (Fig. 12-1). This Sea Hemisphere meets the Russian and North American realms in the far north and merges into the Southern Ocean in the south. Despite the preponderance of water, this fragmented, culturally complex realm does possess regional identities. It includes the Hawaiian Islands, Tahiti, Tonga, and Samoa—fabled names in a world apart.

In terms of modern cultural and political geography, Indonesia and the Philippines are not part of the Pacific Realm, although Indonesia's political system reaches into it; nor are Australia and New Zealand part of it. Before the European invasion and colonization, Australia would have been included because of its Aboriginal population as well as New Zealand because of its Maori population's Polynesian affinities. But the Europeanization of their countries has engulfed indigenous Australians and Maori New Zealanders, and the regional geography of Australia and New Zealand today is decidedly not Pacific. In New Guinea, on the other hand, Pacific peoples remain numerically and culturally the dominant element.

In examining the Pacific Realm, it is important to keep in mind the dimensional contrasts this part of the world presents. The realm is enormous, but its total land area is a mere 975,000 square kilometers (377,000 sq mi), about the size of Texas and New Mexico, and over 90 percent of this lies in New Guinea.* The population is so widely scattered that a distribution map would be impractical, and it only totals a bit more than 12 million (about as many people as there are in Beijing, China).

---

*The figures in the Data Table in Appendix B do not match these totals because only the political entity of Papua New Guinea is listed, not the Indonesian Province of Papua that occupies the western part of the island. Here, as Figure 10B-1 shows, the political and the realm boundaries do not coincide.

COLONIZATION AND INDEPENDENCE

The Pacific islands were colonized by the French, British, and Americans; an indigenous Polynesian kingdom in the Hawaiian Islands was annexed by the United States and is now the fiftieth State. Still, today, the map is an assemblage of independent and colonial territories (Fig. 12-1). Paris controls New Caledonia and French Polynesia. The United States administers Guam and American Samoa, the Line Islands, Wake Island, Midway Islands, and several smaller islands; the United States also has special relationships with other territories, former dependencies that are now nominally independent. The British, through New Zealand, have responsibility for the Pitcairn group of islands, and New Zealand administers and supports the Cook, Tokelau, and Niue Islands. Easter Island, that storied speck of land in the southeastern Pacific, is part of Chile. Indonesia rules Papua, the western half of the island of New Guinea.

Other island groups have now become independent states. The largest are Fiji, once a British dependency, the Solomon Islands (also formerly British), and Vanuatu (until 1980 ruled jointly by France and Britain). Also on the current map, however, are such microstates as Tuvalu, Kiribati, Nauru, and Palau. Foreign aid is crucial to the survival of most of these countries. Tuvalu, for example, has a total area of around 25 square kilometers (10 sq mi), a population of some 10,000, and a per-capita GNI of about U.S. $1500, derived from fishing, copra sales (coconut meat used to make oil), and some tourism. But what really keeps Tuvalu going is an international trust fund set up by Australia, New Zealand, the United Kingdom, Japan, and South Korea. Annual grants from that fund, as well as money sent back to families by workers who have left for New Zealand and elsewhere, allow Tuvalu to survive.

THE PACIFIC REALM AND ITS MARINE GEOGRAPHY

Certain land areas may not be part of the Pacific Realm (we mentioned the Philippines and New Zealand), but the Pacific Ocean extends from the shores of North and South America to mainland East and Southeast Asia and from the Bering Sea to the Subtropical Convergence. This means that several seas, including the Sea of Japan (East Sea), the East China Sea, and the South China Sea, are part of the Pacific Ocean. As we will see below, this relationship matters. Pacific coastal countries, large and small, mainland and island, compete for jurisdiction over the waters that bound them.

The Pacific Realm and its ocean, therefore, form an ideal place to focus on **1 marine geography**. This field encompasses a variety of approaches to the study of oceans and seas; some marine geographers focus on the biogeography of coral reefs, others on the geomorphology of beaches, and still others on the movement of currents and drifts. A particularly interesting branch of marine geography has to do with the definition and delimitation of political boundaries at sea. Here geography meets political science and maritime law.

The State at Sea

Littoral (coastal) states do not end where atlas maps suggest they do. States have claimed various forms of jurisdiction over coastal waters for centuries, closing off bays and estuaries and ordering foreign fishing fleets to stay away from nearby fishing grounds. Thereby arose the notion of the **2 territorial sea**, where all the rights of a coastal state would prevail. Beyond lay the **3 high seas**, free, open, and unfettered by national interests.

It was in the interest of colonizing, mercantile states to keep territorial seas narrow and high seas wide, thus interfering as little as possible with their commercial fleets. In the seventeenth and eighteenth centuries the territorial sea was 3, 4, or at most 6 nautical miles wide, and the colonizing powers claimed the identical widths for their colonies (1 nautical mile − 1.85 kilometers or 1.15 statute miles).*

During the twentieth century, these constraints weakened. States without trading fleets saw no reason to limit their territorial seas. States with nearby fishing grounds traditionally exploited by their own fleets wanted to keep the increasing number of foreign trawlers away. States with shallow **4 continental shelves**, offshore continuations of coastal plains, wished to control the resources on and below the seafloor, made more accessible by improved technology. States also disagreed on the methods by which offshore boundaries, whatever their width, should be defined. Early efforts by the League of Nations (the precursor to the UN) during the 1920s to resolve these issues met with only partial success, mainly in the technical area of boundary delimitation.

Scramble for the Oceans

In 1945, the United States helped precipitate what has become known as the scramble for the oceans. President Harry S Truman issued a proclamation that claimed U.S. jurisdiction and control over all the resources "in and on" the continental shelf down to its margin, around 100 fathoms (600 ft or 183 m) deep. In some areas, the shallow continental shelf of the United States extends more than 300 kilometers (200 mi) offshore, and Washington did not

---

*Here and in the rest of this section we will give all distances involving maritime boundaries in nautical miles only. Throughout the modern history of maritime law, the latter unit has been the only one used for this type of boundary-making. Any distance stated in nautical miles can be converted to its metric equivalent by multiplying it by 1.85. The table in Appendix A will aid in this.

want foreign countries drilling for oil just beyond the 3-mile territorial sea.

Few observers foresaw the impact the Truman Proclamation would have, not only on U.S. waters but on the oceans everywhere, including the Pacific. It set off a rush of other claims. In 1952, a group of South American countries, some with little continental shelf to claim, issued the Declaration of Santiago, claiming exclusive fishing rights up to a distance of 200 nautical miles off their coasts. Meanwhile, as part of the Cold War competition, the Soviet Union urged its allies to claim a 12-mile territorial sea.

UNCLOS Intervention

At this point the United Nations intervened, and a series of UNCLOS (United Nations Conference on the Law of the Sea) meetings began. These meetings addressed issues ranging from the closure of bays to the width and delimitation of the territorial sea, and after three decades of negotiations they achieved a convention that changed the political and economic geography of the world's oceans forever. Among its key provisions were the authorization of a 12-mile territorial sea for all countries and the establishment of a 200-mile (230-statute-mi/370 km)-wide **5 Exclusive Economic Zone (EEZ)** over which a coastal state would have total economic rights. Resources in and under this EEZ (fish, oil, minerals) belong to the coastal state, which could either exploit them or lease, sell, or share them as it saw fit. We already mentioned these zones in Chapter 2A in regard to the new competition for the Arctic Ocean, but as you will see EEZs take on even greater importance in this Pacific Realm.

These new provisions had a far-reaching impact on the world's oceans and seas (Fig. 12-2), especially the Pacific. Unlike the Atlantic Ocean, the Pacific is studded with islands large and small, and a microstate consisting of one small island suddenly acquired an EEZ covering 166,000 square nautical miles. European colonial powers still holding minor Pacific possessions (most notably France) saw their maritime jurisdictions vastly expanded. Small low-income archipelagos could now bargain with large, rich fishing nations over fishing rights in their EEZs. And for all the UNCLOS Convention's provisions for the "right of innocent passage" of shipping through EEZs and via narrow straits, the world's high seas have obviously been diminished.

Maritime Boundaries

The extension of the territorial sea to 12 nautical miles and the EEZ to an additional 188 nautical miles created new **6 maritime boundary** problems. Waters less than 24 nautical miles wide separate many countries all over the world, so that **7 median lines**, equidistant from opposite shores, have been delimited to establish their territorial seas. And even more countries lie closer than 400 nautical miles apart, requiring further maritime-boundary delimitation to determine their EEZs. In such maritime regions as the North Sea, the Caribbean Sea, and the Japan, East China, and South China seas (and recently the Arctic Ocean), a maze of maritime boundaries emerged, some of them subject to dispute. Political changes on land can also lead to significant modifications at sea.

A case in point involves newly independent East Timor and its neighbor across the Timor Sea, Australia. Australia had divided the waters and seafloor of the Timor Sea with Indonesia while recognizing Indonesia's 1976 annexation of East Timor. When East Timor achieved independence in 2002, the so-called *Timor Gap* became an issue: where was the median line that would divide Timorese and Australian claims to the oil and gas reserves in this zone? The Australians argued that since the line defined with Indonesia predated East Timor's independence, it should continue in effect, giving Australia the bulk of the energy resources. But UNCLOS regulations, to which Australia subscribed, required a redelimitation, giving East Timor a much larger share. This led Australia to withdraw from UNCLOS in 2002 so that it would not be bound by its rules. After difficult negotiations, the Australians in 2005 offered East Timor a half share of the energy revenues from the natural gas reserve that was in dispute, in addition to income already flowing from another major gasfield in the so-called Joint Petroleum Development Area. In return, East Timor's government agreed to defer the maritime-boundary issue for 50 years, leaving a future generation to solve this question. It was estimated that, over the next 30 years, East Timor might receive U.S. $13 billion from these sources, and, depending on the life and capacity of the reserves, perhaps even more. This will go a long way toward meeting the young state's most urgent needs.

EEZ Implications

The UNCLOS provisions created opportunities for some states to expand their spheres of influence. Wider territorial-sea and EEZ allocations raised the stakes: claiming an island now entailed potential control over a huge maritime area. In Chapter 2A we highlighted the need to carve up the Arctic as the ice melts and resources beckon. In Chapters 9B and 10B, we referred to several island disputes off mainland East Asia, which involve Japan and Russia, Japan and South Korea, Japan and China, China and Vietnam, and China and the Philippines. Ownership of many islands there is uncertain, and small specks of island territory have become large stakes in the scramble for the oceans. For example, in the case of the Spratly Islands (Fig. 10B-1), six countries claim ownership, including both Taiwan and China. China's island claims in the South China Sea support Beijing's contention that this body of water is part and parcel of the Chinese state—a position that worries other states with coasts facing it.

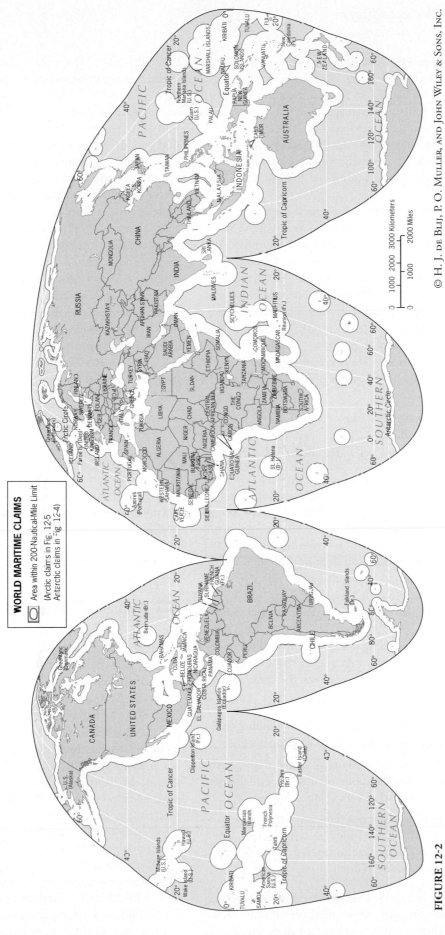

WORLD MARITIME CLAIMS

Area within 200-Nautical-Mile Limit

(Arctic claims in Fig. 12-5
Antarctic claims in Fig. 12-4)

FIGURE 12-2

© H. J. de Blij, P. O. Muller, and John Wiley & Sons, Inc.

Figure 12-2 reveals what EEZ regulations have meant to Pacific Realm countries such as Tuvalu, Kiribati, and Fiji, with nearly circular EEZs now surrounding the clusters of islands in this vast ocean space. Japan, Taiwan, and other fishing nations have purchased fishing rights in these EEZs from the island governments. Nonetheless, violations of EEZ rights do occur; recently, Vanuatu and the Philippines were at odds over unauthorized Filipino fishing in Vanuatu's EEZ.

The process of boundary delimitation continues. In an earlier edition of this book, we included a map of the South China Sea and nearby waters off East and Southeast Asia showing the median-line boundaries delimited according to UNCLOS specifications. But that map changed when coastal states engaged in bilateral and multilateral negotiations—and argued over island ownership with major boundary implications. Indeed, the Pacific (and world) map of maritime boundaries remains a work in progress.

This point is illustrated by a development that occurred late in 1999, when the implications of provisions in Article 76 of the 1982 UNCLOS Convention were reexamined. Whenever a continental shelf extends beyond the 200-mile limit of the EEZ, the Convention apparently allows a coastal state to claim that extension as a natural prolongation of its landmass. Although there is as yet no regulation that would also allow the state to extend its EEZ to the edge of the continental shelf, it is not difficult to foresee such an amendment to the Convention. As it stands, states are now delimiting their proposed natural prolongations under a current deadline, which puts poorer states at a disadvantage since the required marine surveys are very expensive. Lately, this issue has become quite heated, especially in the Arctic Ocean as countries scramble to claim prolongations. A recent article in *Science* reported that the big winners are likely to include the United States, Canada, Australia, New Zealand, Russia, and India—although a total of 60 coastal countries could ultimately benefit to some extent from the provision.

Thus the scramble for the oceans continues, and with it the constriction of the world's high seas and open waters.

REGIONS OF THE REALM

Sail across the Pacific Ocean, and one spectacular vista follows another. Dormant and extinct volcanoes, sculpted by erosion into basalt spires draped by luxuriant tropical vegetation and encircled by reefs and lagoons, tower over azure waters. Low atolls with nearly snow-white beaches, crowned by stands of palm trees, seem to float in the water. Pacific islanders, where foreign influences have not overtaken them, appear to take life with enviable ease.

More intensive investigations reveal that such Pacific sameness is more apparent than real. Even the Pacific Realm, with its long sailing traditions, its still-diffusing populations, and its historic migrations, has a durable regional framework. Figure 12-3 outlines the three regions that constitute the Pacific Realm:

Melanesia: Papua (Indonesia), Papua New Guinea, Solomon Islands, Vanuatu, New Caledonia (France), Fiji

Micronesia: Palau, Federated States of Micronesia, Northern Mariana Islands, Republic of the Marshall Islands, Nauru, western Kiribati, Guam (United States)

Polynesia: Hawaiian Islands (United States), Samoa, American Samoa, Tuvalu, Tonga, eastern Kiribati, Cook and other New Zealand-administered islands, French Polynesia, Easter Island (Chile)

Ethnic, linguistic, and physiographic criteria are among the foundations for this regionalization of the Pacific Realm, but we should not lose sight of the dimensions. Not only is the land area small, but as we also noted earlier the 2010 population of this entire realm (including Indonesia's Papua Province) was only slightly more than 12 million—9.6 million without Papua—about the same as one very large city. Even fewer people live in this realm than in another vast area of far-flung settlements—the oases of North Africa's Sahara.

MELANESIA

The large island of New Guinea lies at the western end of a Pacific region that extends eastward to Fiji and includes the Solomon Islands, Vanuatu, and New Caledonia (Fig. 12-3). The human mosaic here is complex, both ethnically and culturally. Most of the 9.3 million people of New Guinea (including the Indonesian part—the province of Papua—and the independent state of Papua New Guinea) are Papuans, and a large minority is Melanesian. Altogether there are as many as 700 communities speaking different languages; the Papuans are most numerous in the densely forested highland interior and in the lowland south, while the Melanesians inhabit the north and east. The region as a whole has more than 8 million inhabitants, making this the most populous Pacific region by far.

With 6.8 million people today, **Papua New Guinea (PNG)** became a sovereign state in 1975 after nearly a century of British and Australian administration. Almost all of PNG's limited development is taking place along the coasts, whereas most of the interior remains hardly touched by the changes that transformed neighboring Australia. Perhaps

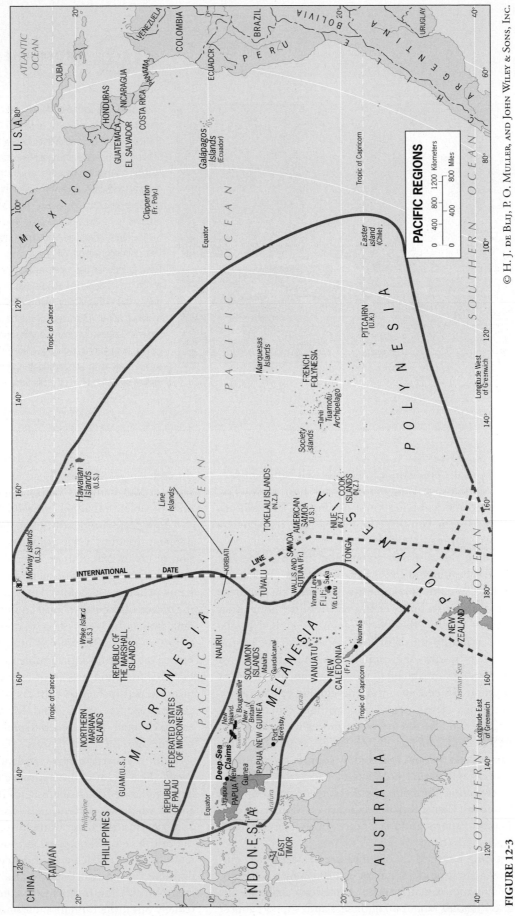

© H. J. de Blij, P. O. Muller, and John Wiley & Sons, Inc.

FIGURE 12-3

FROM THE FIELD NOTES...

© H. J. de Blij

"Arriving in the capital of New Caledonia, Nouméa, in 1996, was an experience reminiscent of French Africa 40 years earlier. The French tricolor was much in evidence, as were uniformed French soldiers. European French residents occupied hillside villas overlooking palm-lined beaches, giving the place a Mediterranean cultural landscape. And New Caledonia, like Africa, is a source of valuable minerals. It is one of the world's largest nickel producers, and from this vantage point you could see the huge treatment plants, complete with concentrate ready to be shipped (left, under conveyor). What you cannot see here is how southern New Caledonia has been ravaged by the mining operations, which have denuded whole mountainsides. Working in the mines and in this facility are the local Kanaks, Melanesians who make up about 43 percent of the population of about 240,000. Violent clashes between Kanaks and French have obstructed government efforts to change New Caledonia's political status in such a way as to accommodate pressures for independence as well as continued French administration."

www.conceptcaching.com

four-fifths of the population lives in the self-sufficient subsistence economy, growing root crops and hunting wildlife, raising pigs, and gathering forest products. Old traditions of the kind lost in Australia continue on here, protected by remoteness and the rugged terrain.

Welding this disparate population into a nation is a task hardly begun, and PNG faces numerous obstacles in addition to its cultural complexity. Not only are hundreds of languages in use, but over 40 percent of the population is illiterate. English, the official language, is used by the educated minority but is of little use beyond the coastal zone and its towns. The capital, Port Moresby, has just under 350,000 inhabitants, reflecting the low level of urbanization (13 percent) in this developing economy.

Yet Papua New Guinea is not without economic opportunities. Oil was discovered in the 1980s, and by the late 1990s crude oil was PNG's largest export by value. Gold now ranks second, followed by copper, silver, timber, and several agricultural products that include coffee and cocoa, reflecting the country's diversity of environments and resources. Pacific Rim developments have affected even PNG: most exports go to the nearest neighbor, Australia, but Japan ranks close behind.

Turning eastward, it is a measure of Melanesia's cultural fragmentation that as many as 120 languages are spoken in the approximately 1000 islands that make up the **Solomon Islands** (about 80 of these islands support almost all the people, who number about 525,000). Inter-island, historic animosities among the islanders were worsened by

the events of World War II, when U.S. forces moved thousands of Malaitans to Guadalcanal. This started a postwar cycle of violence that led Australia to intervene in 2003.

New Caledonia, still under French rule, is in a very different situation. Only around 43 percent of the population of about 250,000 are Melanesian; 37 percent are of French ancestry, many of them descended from the inhabitants of the penal colony France established here during the nineteenth century. Nickel mines, based on reserves that rank among the world's largest, dominate New Caledonia's export economy. The mining industry attracted additional French settlers, and social problems soon arose. Most of the French population lives in or near the capital city of Nouméa, steeped in French cultural landscapes, in the southeastern quadrant of the island. Melanesian demands for an end to colonial rule have led to violence, and the two communities are still in the process of coming to terms.

On its eastern margins, Melanesia includes one of the Pacific Realm's most interesting countries, **Fiji**. On two larger and over 100 smaller islands live nearly 900,000 Fijians, of whom 51 percent are Melanesians and 44 percent South Asians, the latter brought to Fiji from India during the British colonial occupation to work on the sugar plantations. When Fiji achieved independence in 1970, the indigenous Fijians owned most of the land and held political control, while the Indians were concentrated in the towns (chiefly Suva, the capital) and dominated commercial life. It was a recipe for trouble, and it was not long in coming when, in a later election, the politically active Indians out-

© H. J. de Blij

"Back in Suva, Fiji more than 20 years after I had done a study of its CBD, I noted that comparatively little had changed, although somehow the city seemed more orderly and prosperous than it was in 1978. At the Central Market I watched a Fijian woman bargain with the seller over a batch of taro, the staple starch-provider in local diets. I asked her how she would serve it. 'Well, it's like your potato,' she said. 'I can make a porridge, I can cut it up and put it in a stew, and I can even fry pieces of it and make them look like the French fries they give you at the McDonald's down the street.' Next I asked the seller where his taro came from. 'It grows over all the islands,' he said. 'Sometimes, when there's too much rain, it may rot—but this year the harvest is very good.' . . . Next I walked into the crowded Indian part of the CBD. On a side street I got a reminder of the geographic concept of agglomeration. Here was a colonial-period building, once a hotel, that had been converted into a business center. I counted 15 enterprises, ranging from a shoestore to a photographer and including a shop where rubber stamps were made and another selling diverse tobacco products. Of course (this being the Indian sector of downtown Suva) tailors outnumbered all other establishments."

www.conceptcaching.com

voted the Fijians for seats in the parliament. A coup by the Fijian military was followed by a revision of the constitution, which awarded a majority of seats to ethnic Fijians. Before long, however, the Fijian majority splintered, but a coalition government for some time proved the constitution workable—until 2000. In 1999, Fiji's first prime minister of Indian ancestry had taken office, angering certain ethnic Fijians to the point of staging another coup. The following year the prime minister and members of the government were taken hostage at the parliament building in Suva, and the perpetrators demanded that Fijians would henceforth govern the country.

This action, and Fiji's inability to counter it, had a devastating impact on the country. Foreign trading partners stopped buying Fijian products. The tourist industry suffered severely. And in the end, although the coup leaders were ousted and arrested, they secured a deal that gave ethnic Fijians the control they had sought.

The coup continues to cast a shadow across this country. The sugar industry was damaged by the nonrenewal of Indian-held leases by Fijian landowners, which also resulted in a substantial movement of Indians from the countryside to the towns, where unemployment is already high. Fiji's future remains clouded.

Melanesia, the most populous region in the Pacific Realm, is bedeviled by centrifugal forces of many kinds. No two countries present the same form of multiculturalism; each has its own challenges to confront, and some of these challenges spill over into neighboring (or more distant foreign) islands.

MICRONESIA

North of Melanesia and to the east of the Philippines lie the islands that constitute the region known as Micronesia (Fig. 12-3). The name (*micro* means small) refers to the size of the islands: the 2000-plus islands of Micronesia are not only tiny (many of them no larger than one square kilometer [half a square mile]), but they are also much lower-lying, on average, than those of Melanesia. Some are volcanic islands (**8** high islands, as the people call them), but they are outnumbered by islands composed of coral, the **9** low islands that barely protrude above sea level. Guam, with 550 square kilometers (210 sq mi), is Micronesia's largest island, and no island elevation anywhere in Micronesia reaches 1000 meters (3300 ft).

The high-island/low-island dichotomy is useful not only in Micronesia, but also throughout the realm. Both the physiographies of these islands and the economies they support differ in crucial ways. High islands wrest substantial moisture from the maritime air; they tend to be well watered and have good volcanic soils. As a result, agricultural products show some diversity, and life is reasonably secure. Populations tend to be larger on these high islands than on the low islands, where drought is the rule and fishing and the coconut palm are the mainstays of life. Small communities cluster on the low islands, and over time many of these have died out. The major migrations, which sent fleets to populate islands from Hawai'i to New Zealand, tended to originate in the high islands.

Until the mid-1980s, Micronesia was largely a United States Trust Territory (the last of the post-World War II trusteeships supervised by the United Nations), but that political status has now changed. As Figure 12-3 shows, Micronesia is now divided into countries bearing the names of independent states. Today the **Marshall Islands**, where the United States tested nuclear weapons (giving prominence to the name *Bikini*), is a republic in "free association" with the United States, having the same status as the Federated States of Micronesia and (since 1994) Palau. The **Northern Mariana Islands** are a commonwealth "in political union" with the United States. In effect, the United States provides billions of dollars in assistance to these countries, in return for which they commit themselves to avoid foreign policy actions that are contrary to American interests. There are other conditions: **Palau**, for example, granted the U.S. rights to existing military bases for 50 years following independence.

Also part of Micronesia are the U.S. territory of **Guam**, where independence is not in sight and where U.S. military installations and tourism provide the bulk of income, and

the remarkable Republic of **Nauru**. With a population of barely 15,000 and only 20 square kilometers (8 sq mi) of land, Nauru got rich by selling its phosphate deposits to Australia and New Zealand, where they are used as fertilizer. Per-capita incomes have risen above U.S. $12,000, making Nauru one of the Pacific's high-income societies. But the phosphate deposits have run out, and an island scraped bare now faces an economic crisis.

In this region of tiny islands, most people subsist on farming or fishing, and virtually all the countries need infusions of foreign aid to survive. The natural economic complementarity between the high-island farming cultures and the low-island fishing communities all too often is negated by distance, spatial as well as cultural. Life here may seem idyllic to the casual visitor, but for the Micronesians it often is a daily challenge.

POLYNESIA

To the east of Micronesia and Melanesia lies the heart of the Pacific, enclosed by a great triangle stretching from the Hawaiian Islands to Chile's Easter Island to New Zealand. This is Polynesia (Fig. 12-3), a region of numerous islands (*poly* means many), ranging from volcanic mountains rising above the Pacific's waters (Mauna Kea on Hawai'i reaches over 4200 meters [nearly 13,800 ft]), clothed by luxuriant tropical forests and drenched by well over 250 centimeters (100 in) of rainfall each year, to low coral atolls where a few palm trees form the only vegetation and where drought is a persistent problem. The Polynesians have somewhat lighter-colored skin and wavier hair than do the other peoples of the Pacific Realm; they are often also described as having an excellent physique. Anthropologists differentiate between these original Polynesians and a second group, the Neo-Hawaiians, who are a blend of Polynesian, European, and Asian ancestries. In the U.S. State of Hawai'i—actually an archipelago of more than 130 islands—Polynesian culture has been both Europeanized and Orientalized.

Its vastness and the diversity of its natural environments notwithstanding, Polynesia clearly constitutes a geographic region within the Pacific Realm. Polynesian culture, though spatially fragmented, exhibits a remarkable consistency and uniformity from one island to the next, from one end of this widely dispersed region to the other. This consistency is particularly expressed in vocabularies, technologies, housing, and art forms. The Polynesians are uniquely adapted to their maritime environment, and long before European sailing ships began to arrive in their waters, Polynesian seafarers had learned to navigate their wide expanses of ocean in huge double canoes as long as 45 meters (150 ft). They traveled hundreds of kilometers to favorite fishing zones and engaged in inter-island barter trade, using maps constructed from bamboo sticks and cowrie shells and navigating by the stars. However, modern descriptions of a

Pacific Polynesian paradise of emerald seas, lush landscapes, and gentle people distort harsh realities. Polynesian society was forced to get used to much loss of life at sea when storms claimed their boats; families were ripped apart by accident as well as migration; hunger and starvation afflicted the inhabitants of smaller islands; and the island communities were often embroiled in violent conflicts and cruel retributions.

The political geography of Polynesia is complex. In 1959, the **Hawaiian Islands** became the fiftieth State to join the United States. The State's population is now 1.4 million, with over 80 percent living on the island of Oahu. There, the superimposition of cultures is symbolized by the panorama of Honolulu's skyscrapers against the famous extinct volcano at nearby Diamond Head. The Kingdom of **Tonga** became an independent country in 1970 after seven decades as a British protectorate; the British-administered Ellice Islands were renamed **Tuvalu**, and along with the Gilbert Islands to the north (now renamed **Kiribati**), they received independence from Britain in 1978. Other islands continue under French control (including the Marquesas Islands and Tahiti), under New Zealand's administration (Rarotonga), and under British, U.S., and Chilean flags.

In the process of politico-geographical fragmentation, Polynesian culture has suffered severe blows. Land developers, hotel builders, and tourist dollars have set **Tahiti** on a course along which Hawai'i has already traveled far. The Americanization of eastern **Samoa** has created a new society different from the old. Polynesia has lost much of its ancient cultural consistency; today, the region is a patchwork of new and old—the new often bleak and barren, with the old under intensifying pressure.

The countries and cultures of the Pacific Realm lie in an ocean on whose rim a great drama of economic and political transformation will play itself out during the twenty-first century. Already, the realm's own former margins—in Hawai'i in the north and in New Zealand in the south—have been so recast by foreign intervention that little remains of the kingdoms and cultures that once prevailed. Now the Pacific Basin faces changes far greater even than those brought here by European colonizers. Once upon a time the shores and waters of the Mediterranean Sea formed an arena of regional transformation that changed the world. Then it was the Atlantic, avenue of the Industrial Revolution and stage of fateful war. Now it seems to be the turn of the Pacific as the world's largest country and next superpower (China) faces the richest and most powerful (the United States). Giants will jostle for advantage in the Pacific; how will the microstates of the Pacific Realm fare?

PARTITIONING THE ANTARCTIC

South of the Pacific Realm lies Antarctica and its encircling Southern Ocean. The combined area of these two enormous geographic expanses constitutes 40 percent of the entire planet—two-fifths of the Earth's surface containing a mere one one-thousandth of the world's population.

 FROM THE FIELD NOTES...

© H. J. de Blij

"The Antarctic Peninsula is geologically an extension of South America's Andes Mountains, and this vantage point in the Gerlach Strait leaves you in no doubt: the mountains rise straight out of the frigid waters of the Southern Ocean. We have been passing large, flat-topped icebergs, but these do not come from the peninsula's shores; rather, they form on the leading edges of ice shelves or on the margins of the mainland where continental glaciers slide into the sea. Here along the peninsula, the high-relief topography tends to produce jagged, irregular icebergs. The mountain range that forms the Antarctic Peninsula continues across Antarctica under thousands of feet of ice and is known as the Transantarctic Mountains. Looking at this place you are reminded that beneath all this ice lies an entire continent, still little-known, with fossils, minerals, even lakes yet to be discovered and studied."

www.conceptcaching.com

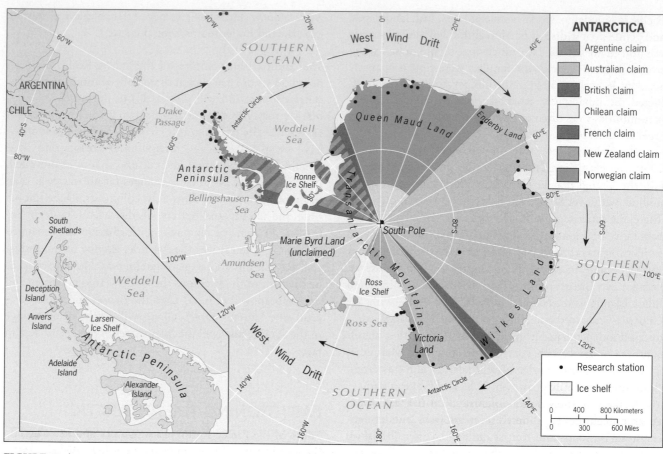

FIGURE 12-4

© H. J. de Blij, P. O. Muller, and John Wiley & Sons, Inc.

Is Antarctica a geographic realm? In physiographic terms, yes, but not on the basis of the criteria we use in this book. Antarctica is a continent, practically twice the size of Australia, but virtually all of it is covered by a dome-shaped ice sheet nearly 3.2 kilometers (2 mi) thick near its center. The continent frequently is referred to as the white desert because, despite all of its ice and snow, annual precipitation is low—less than 15 centimeters (6 in) per year. Temperatures are frigid, with winds so strong that Antarctica also is called the home of the blizzard. For all its size, no functional regions have developed here, nor have any towns or transport networks except the supply lines of research stations. Antarctica still is a frontier, even a scientific frontier still slowly giving up its secrets. Underneath all that ice lie some 70 lakes of which Lake Vostok is the largest with over 14,000 square kilometers (5400 sq mi). It may be as much as 600 meters (2000 ft) deep. No one has yet seen a sample of its water.

Like virtually all frontiers, Antarctica has always attracted pioneers and explorers. Whale and seal hunters destroyed huge populations of Southern Ocean fauna during the eighteenth and nineteenth centuries, and explorers planted the flags of their countries on Antarctic shores. Between 1895 and 1914, the quest for the South Pole became an international obsession; Roald Amundsen, the Norwegian, reached it first, just a century ago, in 1911. All this led to national claims in Antarctica during the interwar period (1918–1939).

The geographic effect was the partitioning of Antarctica into pie-shaped sectors centered on the South Pole (Fig. 12-4). In the least frigid area of the continent, the Antarctic Peninsula, British, Argentinian, and Chilean claims overlapped—and still do. One sector, Marie Byrd Land (shown in neutral beige on the main map), was never formally claimed by any country.

Why should states be interested in territorial claims in so remote and difficult an area? As we described in Chapter 2A in the section on the Arctic, these remote areas near the poles contain substantial resources and are attractive for possible future use, despite the difficulties of access. Antarctica is no different as both land and sea contain raw materials that may some day become crucial: proteins in the waters, and fuels and minerals beneath the ice. Antarctica (14.2 million square kilometers [5.5 million sq mi]) is, as already noted, almost twice as large as Australia, and the Southern Ocean is nearly as large as the North and South Atlantic. However distant actual future exploitation may be, countries want to keep their stakes here.

But the claimant states (those with territorial claims) recognize the need for cooperation. During the late 1950s, they joined in the International Geophysical Year (IGY) that launched major research programs and established a number of permanent research stations throughout the continent. This spirit of cooperation led to the 1961 signing of the **10 Antarctic Treaty**, which ensures continued scientific collaboration, prohibits military activities, protects the environment, and holds national claims in abeyance. In 1991, when the treaty was extended under the terms of the Wellington Agreement, concerns were raised that it does not do enough to control future resource exploitation.

Notice, in Figure 12-4, that the map shows no maritime claims off the pie-shaped sectors that blanket most of Antarctica. In fact, some of the claimant states did draw maps that extended their claims into the Southern Ocean, but the 1991 Wellington Agreement terminated that initiative and restricted the existing claims to the landmass. It is of course possible that those claims could be reinstated should the Wellington Agreement collapse; national self-interest has abrogated international treaties before. But delimiting maritime claims off Antarctica is particularly difficult for practical reasons. As Figure 12-4 shows, Antarctica in a number of places is flanked by *ice shelves*, permanent slabs of floating ice (such as the Ronne and the Ross). These ice shelves are replenished on the landward side and break off (*calve*) on the seaward side, where huge icebergs float away into the Southern Ocean to become part of the wide ring of *pack ice* that encircles the continent. From where would any territorial sea or EEZ be measured? From the inner edge of the ice shelf? That would put the territorial "sea" boundary on the ice shelf! From the outer edge of the ice shelf? Measuring a territorial sea or EEZ from an unstable perimeter would not work either. No UNCLOS regulations would be applicable in situations like this, so no matter what the maps showed, such claims are legally unsupportable. That is just as well. The last thing the world needs is a scramble for Antarctica.

In an age of growing national self-interest and increasing raw material consumption, the possibility exists that Antarctica and its offshore waters may yet become an arena for international rivalry. Until now, its remoteness and its forbidding environments have saved it from that fate. The entire world benefits from this because evidence is mounting that Antarctica plays a critical role in the global environmental system, so that human modifications may have worldwide (and unpredictable) consequences.

GEOPOLITICS IN THE ARCTIC

To observe how different the physiographic as well as the political situation is in the Arctic, consider the implications of Figure 12-5. Not only does the North Pole lie on the floor of a relatively small body of water grandiosely called the Arctic Ocean, but the entire Arctic is ringed by countries whose EEZs, delimited under UNCLOS rules, would allocate much of the ocean floor (the *subsoil*, to use its legal designation) to those states. And, again under UNCLOS rules, states can expand their rights to the seafloor even farther than the 200-mile EEZ if they can prove their continental shelves continue beyond that limit—in fact, up to 350 nautical miles offshore.

As the map shows, the Siberian Continental Shelf is by far the largest in the Arctic; indeed, it is the largest in the world. It extends from Russia's north coast beneath the waters of the Arctic Ocean and under the permanent ice at the ocean's center. And because several island groups off the Siberian mainland (such as Franz Josef Land, North Land, and the New Siberian Islands) belong to Russia, the Russians can claim virtually the entire Siberian Continental Shelf under existing regulations. But that is not enough: the Russians want to extend their claim all the way to the North Pole itself, where in 2007 they sent a submarine to plant a Russian flag nearly 4000 meters (13,000 ft) below the sea (see photo below).

To bolster their claim, the Russians assert that the Lomonosov Ridge, a submerged mountain range, is a "natural extension" of their Siberian Continental Shelf. Few if

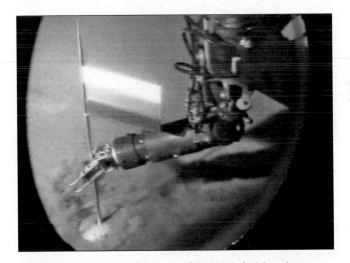

On August 3, 2007, tens of millions of Russian television viewers witnessed live the scene pictured here: the operator of a Russian mini-submarine planting a metal Russian flag at the North Pole, on the seafloor of the Arctic Ocean beneath the permanent polar ice. As Figure 12-5 shows, the North Pole lies far beyond the Russian continental shelf, but in the upcoming international negotiations relating to national claims in the Arctic region, the Russians will claim that the Lomonosov "Ridge" actually constitutes an extension of their continental shelf, thus entitling Russia to draw international boundaries up to, and even beyond, the North Pole. Other countries with actual and pending Arctic claims make less dramatic but equally assertive moves in the run-up to negotiations that have the potential to transform the map of the Arctic region. With new sea routes and newly-exploited oil and gas reserves in prospect, diplomacy will be difficult but crucial. © Visarkryeziu/AP/Wide World Photos

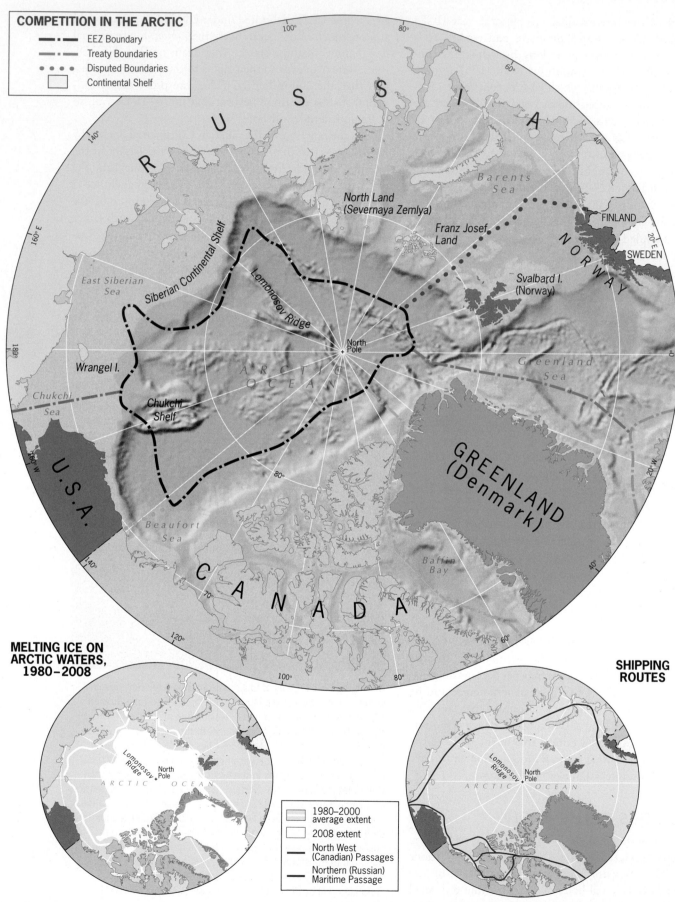

COMPETITION IN THE ARCTIC
- ▬ ·▬ · EEZ Boundary
- ▬ · ▬ · Treaty Boundaries
- • • • • Disputed Boundaries
- ▢ Continental Shelf

R U S S I A

North Land
(Severnaya Zemlya)

Franz Josef
Land

Barents
Sea

FINLAND

SWEDEN

N O R W A Y

Svalbard I.
(Norway)

Siberian Continental Shelf

East Siberian
Sea

Lomonosov Ridge

North
Pole

Greenland
Sea

Wrangel I.

Chukchi
Sea

Chukchi
Shelf

A R C T I C
O C E A N

U.S.A.

GREENLAND
(Denmark)

Beaufort
Sea

Baffin
Bay

C A N A D A

**MELTING ICE ON
ARCTIC WATERS,
1980–2008**

Lomonosov
Ridge
North
Pole

A R C T I C   O C E A N

**SHIPPING
ROUTES**

Lomonosov
Ridge
North
Pole

A R C T I C   O C E A N

- ▢ 1980–2000
  average extent
- ▢ 2008 extent
- ▬▬ North West
  (Canadian) Passages
- ▬▬ Northern (Russian)
  Maritime Passage

FIGURE 12-5

© H. J. DE BLIJ, P. O. MULLER, AND JOHN WILEY & SONS, INC.

any neutral observers would agree that this ridge, or several others rising from the deep floor of the Arctic Ocean, is a continental-shelf landform. But this is the stuff of international disputes over maritime boundaries. The legal wrangling will go on for decades.

Another look at Figure 12-5 reveals the other states with frontage on the Arctic Ocean: Norway, through its Svalbard Islands (including Spitsbergen); Denmark, because it still has ultimate authority over now-autonomous Greenland; Canada, with tens of thousands of kilometers of Arctic coastline but proportionately less continental shelf; and the United States, with Alaska's limited but important continental shelves to its north and west. The United States stands alone among these candidates by not having ratified the UNCLOS treaties, so that the Americans do not feel bound by the regulations to which everyone else adheres. In any case, you can quickly see that an Antarctic-type, pie-shaped-sector partitioning would not work here, but it is also clear that the evolving contest among claimant states is fraught with risk. And the stakes are high: estimates of the quantity of oil and natural gas reserves to be found below Arctic waters and ice run as high as 25 percent of the world's remaining total. No matter what method is used to calculate which country gets how much, the big winner will be Russia, regardless of what the resolution of its Lomonosov claim will be.

Disputation and Navigation

Even as a dispute over Arctic maritime boundaries looms, another issue is emerging, this one related to the effects of recently sustained global warming. The Greenland Ice Cap, in some ways a smaller version of the Antarctic one, is in a melting phase, and the zone of seasonally expanding and contracting Arctic sea ice has also shown significant losses in both surface extent and thickness. Researchers with the Greenland Ice Core Project reported in a 2007 issue of *Science* that during a previous warming, more than 450,000 years ago, the southernmost part of Greenland was covered by boreal forest—suggesting that the ice cap as well as the sea ice may disappear altogether if current global warming continues.

The ecological consequences would be far-reaching, threatening the habitats of polar bears, whales, walruses, seals, and other species. But the geographic implications would be extensive as well. When the sea ice melts in summer and does not recover in winter, once-blocked waterways open up and change accessibilities (Fig. 12-5, subsidiary maps). The Northwest Passage, that fabled high-latitude route between the Atlantic and Pacific oceans, has been the object of hope and despair for centuries, causing much loss of human life. But in 2007, for the first time in recorded history, it was briefly but completely ice-free. Consider what it would mean if vessels bound for East Asia from western Europe or the U.S. eastern seaboard could sail year-round from the Labrador Sea (between Greenland and Canada) to the Bering Sea (between Alaska and northeasternmost Russia) via this waterway!

But who owns the waterways vacated by the ice? Canada calls the Northwest Passage a domestic waterway, giving it the right to control shipping. The United States holds that the Northwest Passage is an international waterway, and American and Russian ships have sailed through it without Canada's permission. The Canadians have confirmed their position by starting work on a deepwater port at Nanisivik on Nunavut's Baffin Island at the eastern entrance to the Northwest Passage, and by building an army base at Resolute Bay to the west; they also announced the construction of eight Arctic patrol vessels to police the route. In the meantime, the United States is opening a new Coast Guard station to patrol the narrow Bering Strait. The lines, if not yet the battle lines, are being drawn in the warming Arctic.

Humanity's growing numbers, escalating demands, environmental impacts, and technological capacities are transforming even the most remote recesses of our resilient planet. No international resolution is more critical than the avoidance of destructive conflict in fragile polar environments.

POINTS TO PONDER

- Governments of low-elevation Pacific and Indian ocean states are demanding international help as their countries are confronted by the threat of rising sea level.

- A vast raft of garbage, much of it plastics, floats in the waters of the Pacific Ocean and continues to grow.

- Pacific-island governments are challenging the air-polluting practices of developed countries. In 2010, Micronesia sued the Czech Republic to halt the restart of a coal-fired power plant.

- Even as national claims in Antarctica are dormant, states with coastlines on Arctic waters—including Russia, Canada, and the United States—are drawing maritime boundaries and charting claims in the far north.

APPENDIX A:
METRIC (STANDARD INTERNATIONAL [SI]) AND CUSTOMARY UNITS AND THEIR CONVERSIONS

LENGTH

Metric Measure

1 kilometer (km)	= 1000 meters (m)
1 meter (m)	= 100 centimeters (cm)
1 centimeter (cm)	= 10 millimeters (mm)

Nonmetric Measure

1 mile (mi)	= 5280 feet (ft)
	= 1760 yards (yd)
1 yard (yd)	= 3 feet (ft)
1 foot (ft)	– 12 inches (in)
1 fathom (fath)	= 6 feet (ft)

Conversions

1 kilometer (km)	= 0.6214 mile (mi)
1 meter (m)	= 3.281 feet (ft)
	= 1.094 yards (yd)
1 centimeter (cm)	= 0.3937 inch (in)
1 millimeter (mm)	= 0.0394 inch (in)
1 mile (mi)	= 1.609 kilometers (km)
1 foot (ft)	= 0.3048 meter (m)
1 inch (in)	= 2.54 centimeters (cm)
	= 25.4 millimeters (mm)

AREA

Metric Measure

1 square kilometer (km^2)	= 1,000,000 square meters (m^2)
	= 100 hectares (ha)
1 square meter (m^2)	= 10,000 square centimeters (cm^2)
1 hectare (ha)	= 10,000 square meters (m^2)

Nonmetric Measure

1 square mile (mi^2)	= 640 acres (ac)
1 acre (ac)	= 4840 square yards (yd^2)
1 square foot (ft^2)	= 144 square inches (in^2)

Conversions

1 square kilometer (km^2)	= 0.386 square mile (mi^2)
1 hectare (ha)	= 2.471 acres (ac)
1 square meter (m^2)	10.764 square feet (ft^2)
	– 1.196 square yards (yd^2)
1 square centimeter (cm^2)	= 0.155 square inch (in^2)
1 square mile (mi^2)	= 2.59 square kilometers (km^2)
1 acre (ac)	= 0.4047 hectare (ha)
1 square foot (ft^2)	= 0.0929 square meter (m^2)
1 square inch (in^2)	= 6.4516 square centimeters (cm^2)

TEMPERATURE

To change from Fahrenheit (F) to Celsius (C)

$$°C = \frac{°F - 32}{1.8}$$

To change from Celsius (C) to Fahrenheit (F)

$$°F = °C \times 1.8 + 32$$

APPENDIX B: AREA AND DEMOGRAPHIC DATA FOR THE WORLD'S STATES (Categories explained on page B-8)

| | Land Area | | Population | | Population Density | | Birth Rate | Death Rate | Natural Increase % | Infant Mortality per 1,000 | Child Mortality per 1,000 | Life Expectancy | | Percent Urban Pop. | Literacy | | Corruption Index | Big Mac Price ($US) | Per Capita GNI ($US) |
	Sq km	Sq mi	2010 (Millions)	2025 (Millions)	Arithmetic	Physiologic						Male (years)	Female (years)		Male %	Female %			
WORLD	134,134,451	51,789,601	6879.9	7972.3	51	443	21	8	1.3	44.9	59	67	70	50	84.0	70.8			$9,600
REALM	5,930,511	2,289,783	594.9	597.6	26	204	11	11	0.0	6.0	7.2	72	79	71	99.4	98.4		4.38	$24,320
EUROPE																			
Albania	28,749	11,100	3.2	3.5	113	451	13	6	0.6	8.0	9	72	79	45	95.5	88.0	3.4		$6,580
Austria	83,859	32,378	8.4	8.8	100	556	9	9	0.0	3.7	5	77	83	67	100.0	100.0	8.1		$38,090
Belarus	207,598	80,154	9.6	9.0	46	150	11	14	-0.3	6.0	7	63	76	73	99.7	99.2	2.0		$10,740
Belgium	30,528	11,787	10.7	10.8	352	1408	12	10	0.2	3.7	5	77	82	97	100.0	100.0	7.3		$35,110
Bosnia	51,129	19,741	3.8	3.7	74	572	9	9	0.0	8.0	9	71	77	46	96.5	76.6	3.2		$7,280
Bulgaria	110,908	42,822	7.5	6.6	68	165	10	15	-0.5	9.2	11	69	76	71	99.1	98.0	3.6		$11,180
Croatia	56,539	21,830	4.4	4.3	77	298	9	12	-0.3	5.7	7	73	79	56	99.4	97.3	4.4		$15,050
Cyprus	9,249	3,571	1.1	1.1	120	751	12	7	0.5	6.0	8	75	80	62	98.7	95.0	6.4		$26,370
Czech Republic	78,860	30,448	10.4	10.2	132	307	11	10	0.1	3.1	4	74	80	74	100.0	100.0	5.2	3.02	$21,820
Denmark	43,090	16,637	5.5	5.6	128	229	12	10	0.2	4.0	5	76	80	72	100.0	100.0	9.3	5.07	$36,740
Estonia	45,099	17,413	1.3	1.2	29	107	12	13	-0.1	4.9	6	67	78	69	99.9	99.6	6.6		$19,680
Finland	338,149	130,560	5.3	5.6	16	225	11	9	0.2	2.7	3	76	83	63	100.0	100.0	9.0		$35,270
France	551,497	212,934	62.6	66.1	114	324	13	8	0.5	3.6	4	78	85	77	98.9	98.7	6.9		$33,470
Germany	356,978	137,830	81.9	79.6	229	655	8	10	-0.2	3.9	5	77	82	89	100.0	100.0	7.9		$33,820
Greece	131,960	50,950	11.2	11.3	85	283	10	9	0.1	3.7	4	77	81	60	98.6	96.0	4.7		$32,520
Hungary	93,030	35,919	9.9	9.6	107	194	10	13	-0.3	5.9	7	69	77	66	99.5	99.3	5.1	2.92	$17,430
Iceland	102,999	39,768	0.3	0.4	3	297	15	6	0.9	1.3	2	79	83	93	100.0	100.0	8.9		$34,060
Ireland	70,279	27,135	4.6	4.9	65	327	16	6	1.0	3.1	4	77	82	60	100.0	100.0	7.7		$37,040
Italy	301,267	116,320	59.8	62.0	198	536	9	10	-0.1	4.2	5	79	84	68	98.9	98.1	4.8		$29,900
Kosovo	10,887	4,203	2.3	2.7	208		21	7	1.4	33.0		67	71						
Latvia	64,599	24,942	2.3	2.1	35	122	10	14	-0.4	7.6	9	66	77	68	99.8	99.6	5.0		$16,890
Liechtenstein	161	62	0.1	0.1	874	3495	10	6	0.4	2.6	3	79	82	15	100.0	100.0			
Lithuania	65,200	25,174	3.4	3.1	52	112	10	14	-0.4	5.9	7	65	77	67	99.7	99.4	4.6		$17,180
Luxembourg	2,587	999	0.5	0.5	194	778	11	8	0.3	4.4	5	78	83	83	100.0	100.0	8.3		$64,400
Macedonia	25,711	9,927	2.0	2.0	78	300	11	10	0.1	13.0	15	71	76	65	94.2	83.8	3.6		$8,510
Malta	321	124	0.4	0.4	1250	3678	10	8	0.2	3.6	4	77	81	94	91.4	92.8	5.8		$20,990
Moldova	33,701	13,012	4.1	3.8	121	184	11	12	-0.1	12.0	14	65	72	41	99.6	98.3	2.9		$2,930
Montenegro	13,812	5,333	0.6	0.6	44	156	12	10	0.2	11.0	13	71	75	64	100.0	100.0	3.4		$10,290
Netherlands	40,839	15,768	16.5	16.9	404	1496	11	8	0.3	4.4	5	78	82	90	100.0	100.0	8.9		$39,500
Norway	323,878	125,050	4.8	5.6	15	497	12	9	0.3	3.1	4	78	83	79	100.0	100.0	7.9	5.79	$53,690
Poland	323,249	124,807	38.1	36.7	118	251	10	10	0.0	6.0	7	71	80	61	99.8	99.8	4.6	2.01	$15,590

| | Land Area | | Population | | Population Density | | Birth Rate | Death Rate | Natural Increase % | Infant Mortality per 1,000 | Child Mortality per 1,000 | Life Expectancy | | Percent Urban Pop. | Literacy | | Corruption Index | Big Mac Price ($US) | Per Capita GNI ($US) |
	Sq km	Sq mi	2010 (Millions)	2025 (Millions)	Arithmetic	Physiologic						Male (years)	Female (years)		Male %	Female %			
Portugal	91,981	35,514	10.6	10.5	115	397	10	10	0.0	3.5	4	75	82	55	94.8	90.0	6.1		$20,640
Romania	238,388	92,042	21.4	19.7	90	204	10	12	-0.2	12.0	14	68	75	55	99.1	97.3	3.8		$10,980
Serbia	88,357	34,115	7.3	6.7	83	219	10	14	-0.4	7.4	9	71	76	56	100.0	100.0	3.4		$10,220
Slovakia	49,010	18,923	5.4	5.2	110	324	10	10	0.0	6.1	7	70	78	56	100.0	100.0	5.0		$19,330
Slovenia	20,251	7,819	2.0	2.1	99	707	10	9	0.1	3.1	4	74	81	48	98.6	96.8	6.7		$26,640
Spain	505,988	195,363	46.7	46.2	92	237	11	9	0.2	3.7	4	77	83	77	100.0	100.0	6.5		$30,110
Sweden	449,959	173,730	9.2	9.9	21	293	12	10	0.2	2.5	3	79	83	84	100.0	100.0	9.3	4.58	$35,840
Switzerland	41,290	15,942	7.6	8.1	185	1540	10	8	0.2	4.0	5	79	84	68	99.5	97.4	9.0	5.60	$43,080
Ukraine	603,698	233,089	45.6	41.7	76	128	10	16	-0.6	11.0	14	62	74	68	100.0	100.0	2.5		$6,810
United Kingdom	244,878	94,548	61.8	68.8	252	971	13	9	0.4	4.9	6	77	81	80	97.6	89.2	7.7	3.30	$34,370
RUSSIAN REALM																			
REALM	**17,261,421**	**6,664,672**	**157.7**	**146.5**	**9**	**123**	**12**	**14**	**-0.2**	**10.0**	**11**	**61**	**73**	**71**	**99.7**	**99.1**			**$13,512**
Armenia	29,800	11,506	3.1	3.3	105	526	15	9	0.6	26.0	29	68	75	64	99.4	98.1	2.9		$5,900
Azerbaijan	86,599	33,436	8.9	9.7	103	468	18	6	1.2	12.0	14	70	75	52	98.9	95.9	1.9		$6,370
Georgia	69,699	26,911	4.6	4.2	66	441	11	10	0.1	16.0	18	70	79	53	99.7	99.4	3.9		$4,770
Russia	17,075,323	6,592,819	141.0	129.3	8	138	12	15	-0.3	9.0	10	60	73	73	99.8	99.2	2.1	1.73	$14,400
NORTH AMERICA																			
REALM	**19,599,647**	**7,567,466**	**341.9**	**393.3**	**17**	**139**	**14**	**8**	**0.6**	**6.5**	**7.9**	**76**	**81**	**79**	**97.6**	**97.4**			**$44,790**
Canada	9,970,600	3,849,670	33.6	37.6	3	67	11	7	0.4	5.4	6	78	83	81	95.7	95.3	8.7	3.36	$35,310
United States	9,629,047	3,717,796	308.2	355.7	32	160	14	8	0.6	6.6	8	75	81	79	99.0	99.0	7.3	3.54	$45,850
MIDDLE AMERICA																			
REALM	**2,714,579**	**1,048,105**	**196.3**	**225.6**	**72**	**447**	**21**	**6**	**1.6**	**24.0**	**28**	**71**	**76**	**69**	**90.3**	**87.1**			**$9,489**
Antigua and Barbuda	440	170	0.1	0.1	232	1287	17	7	1.0	20.0	23	71	75	31					$12,610
Bahamas	13,880	5,359	0.3	0.4	22	2209	17	6	1.1	14.0	15	69	75	83	95.4	96.8			
Barbados	430	166	0.3	0.3	706	1811	14	8	0.6	14.0	15	73	79	38	98.0	96.8	7.0		$10,880
Belize	22,960	8,865	0.3	0.4	14	342	27	4	2.3	18.0	21	71	74	50			2.9		$5,100
Costa Rica	51,100	19,730	4.6	5.6	90	1002	16	4	1.2	9.7	11	76	81	59	95.5	95.7	5.1		$8,340
Cuba	110,859	42,803	11.3	11.2	102	248	10	7	0.3	5.3	7	75	79	76	96.5	96.4	4.3		
Dominica	751	290	0.1	0.1	135	675	16	9	0.7	16.0	20	72	77	73			6.0		$5,650
Dominican Republic	48,731	18,815	10.3	12.1	211	679	24	6	1.8	32.0	38	69	75	67	84.0	83.7	3.0		$5,050
El Salvador	21,041	8,124	7.5	9.1	355	909	24	6	1.8	24.0	28	68	74	60	81.6	76.1	3.9		$4,840
Grenada	339	131	0.1	0.1	302	943	19	7	1.2	17.0	21	66	69	31					$6,010
Guadeloupe	1,709	660	0.4	0.5	238	1585	15	7	0.8	8.0	11	75	82	100	89.7	90.5			
Guatemala	108,888	42,042	14.5	20.0	133	739	34	6	2.8	34.0	35	66	73	47	76.2	61.1	3.1		$4,120

| | Land Area | | Population | | Population Density | | Birth Rate | Death Rate | Natural Increase % | Infant Mortality per 1,000 | Child Mortality per 1,000 | Life Expectancy | | Percent Urban Pop. | Literacy | | Corruption Index | Big Mac Price ($US) | Per Capita GNI ($US) |
	Sq km	Sq mi	2010 (Millions)	2025 (Millions)	Arithmetic	Physiologic						Male (years)	Female (years)		Male %	Female %			
Haiti	27,749	10,714	9.4	11.7	340	1030	29	11	1.8	57.0	76	56	60	43	51.0	46.5	1.4		$1,050
Honduras	112,090	43,278	7.6	9.8	68	378	27	5	2.2	23.0	27	69	74	46	72.5	72.0	2.6		$3,160
Jamaica	10,989	4,243	2.8	3.0	251	1005	17	6	1.1	21.0	25	70	75	52	82.5	90.7	3.1		$5,050
Martinique	1,101	425	0.4	0.4	368	1839	13	7	0.6	6.0		76	83	98	96.0	97.1			
Mexico	1,958,192	756,062	111.0	123.8	57	405	20	5	1.5	19.0	23	73	78	76	93.1	89.1	3.6	2.30	$12,580
Netherlands Antilles	800	309	0.2	0.2	253	2534	14	7	0.7	5.0		71	79	92	96.6	96.6			
Nicaragua	129,999	50,193	5.9	6.8	46	208	26	5	2.1	29.0	35	68	74	59	64.2	64.4	2.5		$2,080
Panama	75,519	29,158	3.5	4.2	46	516	20	4	1.6	15.0	19	73	78	64	92.6	91.3	3.4		$8,340
Puerto Rico	8,951	3,456	4.0	4.1	450	5005	12	8	0.4	9.2	15	74	82	94	93.7	94.0	5.8		$7,090
Saint Lucia	619	239	0.2	0.2	328	1172	15	7	0.8	19.4	26	71	76	28			7.1		
St. Vincent & the Grenadines	391	151	0.1	0.1	260	930	17	8	0.9	17.6		70	74	40			6.5		$5,720
Trinidad and Tobago	5,131	1,981	1.3	1.4	256	1068	14	8	0.6	24.0	27	67	71	12	99.0	97.5	3.6		$14,580
SOUTH AMERICA																			
REALM	17,867,238	6,898,579	397.5	460.9	22	356	20	6	1.4	23.0	25	69	76	81	90.1	89.0			$9,290
Argentina	2,780,388	1,073,514	40.6	46.3	15	146	19	8	1.1	13.3	15	71	79	91	96.9	96.9	2.9	3.30	$12,990
Bolivia	1,098,575	424,162	10.4	13.3	9	474	29	8	2.1	51.0	62	63	67	64	92.1	79.4	3.3		$4,140
Brazil	8,547,360	3,300,154	200.6	228.9	23	335	20	6	1.4	24.0	26	69	75	83	85.5	85.4	3.5	3.45	$9,370
Chile	756,626	292,135	17.1	19.1	23	754	14	5	0.9	8.8	10	75	81	87	95.9	95.5	6.9	2.51	$12,590
Colombia	1,138,906	439,734	45.7	53.8	40	1002	20	6	1.4	19.0	23	69	76	72	91.8	91.8	3.8		$6,640
Ecuador	283,560	109,483	14.4	17.5	51	460	26	6	2.0	25.0	28	72	78	62	93.6	90.2	2.0		$7,040
French Guiana	89,999	34,749	0.2	0.3	2	235	32	4	2.8	10.4		72	79	76	83.6	82.3			
Guyana	214,969	83,000	0.8	0.8	4	191	21	9	1.2	48.0	64	63	68	28	99.0	98.1	2.6		$2,600
Paraguay	406,747	157,046	6.5	8.0	16	265	27	6	2.1	36.0	43	69	73	57	94.4	92.2	2.4		$4,380
Peru	1,285,214	496,224	28.7	34.0	22	745	21	6	1.5	24.0	28	68	73	76	94.7	85.4	3.6	2.54	$7,240
Suriname	163,270	63,039	0.5	0.5	3	312	17	7	1.0	16.0	17	66	73	74	95.9	92.6	3.6		$6,000
Uruguay	177,409	68,498	3.3	3.5	19	268	14	9	0.5	10.5	12	72	79	94	97.4	98.2	6.9		$11,040
Venezuela	912,046	352,143	29.1	34.9	32	797	25	4	2.1	16.5	19	70	76	88	93.3	92.7	1.9		$11,920
SUBSAHARAN AFRICA																			
REALM	21,786,509	8,411,818	769.5	1106.8	32	421	40	15	2.5	87.0	121	49	51	35	73.1	59.9			$1,828
Angola	1,246,693	481,351	17.7	26.2	14	473	47	21	2.6	132.0	180	41	44	57	55.6	28.5	1.9		$4,400
Benin	112,620	43,483	9.9	14.5	88	548	42	12	3.0	98.0	155	54	57	41	47.8	23.6	3.1		$1,310
Botswana	581,727	224,606	1.8	2.2	3	316	24	14	1.0	44.0	54	50	49	57	74.4	79.8	5.8		$12,420

	Land Area		Population		Population Density		Birth Rate	Death Rate	Natural Increase %	Infant Mortality per 1,000	Child Mortality per 1,000	Life Expectancy		Percent Urban Pop.	Literacy		Corruption Index	Big Mac Price ($US)	Per Capita GNI ($US)
	Sq km	Sq mi	2010 (Millions)	2025 (Millions)	Arithmetic	Physiologic						Male (years)	Female (years)		Male %	Female %			
Burkina Faso	274,000	105,792	16.1	23.7	59	450	45	15	3.0	89.0	163	49	52	16	31.2	13.1	3.5		$1,120
Burundi	27,829	10,745	9.4	15.0	339	789	46	16	3.0	107.0	177	47	50	10	56.3	40.5	1.9		$330
Cameroon	475,439	183,568	19.4	25.5	41	255	36	13	2.3	74.0	127	51	52	57	81.8	69.2	2.3		$2,120
Cape Verde Islands	4,030	1,556	0.5	0.7	130	1185	30	5	2.5	28.0	38	68	74	59	84.3	65.3	5.1		$2,940
Central African Republic	622,978	240,533	4.6	5.5	7	244	38	19	1.9	102.0	155	43	44	38	59.6	34.5	2.0		$740
Chad	1,283,994	495,753	10.7	13.9	8	277	44	17	2.7	106.0	179	46	48	27	66.9	40.8	1.6		$1,280
Comoros	2,230	861	0.7	1.1	332	626	36	8	2.8	69.0	92	62	66	28	63.5	49.1	2.5		$1,150
Congo	341,998	132,046	4.0	5.6	12	1185	37	13	2.4	75.0	119	52	54	60	87.5	74.4	1.9		$2,750
Congo, The	2,344,848	905,351	70.7	109.7	30	754	44	13	3.1	92.0	138	49	55	33	86.6	67.7	1.7		$290
Djibouti	23,201	8,958	0.8	1.1	36	3573	30	12	1.8	67.0	101	53	55	87	65.0	38.4	3.0		$2,260
Equatorial Guinea	28,050	10,830	0.6	0.9	23	252	39	10	2.9	91.0	152	59	60	39	92.5	74.5	1.7		$21,230
Eritrea	117,598	45,405	5.3	7.7	45	1128	40	10	3.0	59.0	91	54	59	21	43.9	33.4	2.6		$400
Ethiopia	1,104,296	426,371	83.1	110.5	75	684	40	15	2.5	77.0	121	48	51	16	83.7	70.0	2.6		$780
Gabon	267,668	103,347	14	1.7	5	269	27	12	1.5	58.0	88	56	58	84	79.8	62.2	3.1		$13,080
Gambia	11,300	4,363	17	2.3	149	711	38	11	2.7	93.0	124	57	59	54	43.8	29.6	1.9		$1,140
Ghana	238,538	92,100	25.0	33.7	105	455	32	10	2.2	71.0	112	58	59	48	79.5	61.2	3.9		$1,330
Guinea	245,860	94,927	10.9	15.7	44	738	42	14	2.8	113.0	183	52	55	30	55.1	27.0	1.6		$1,120
Guinea-Bissau	36,120	13,946	1.8	2.9	50	385	50	19	3.1	117.0	196	43	47	30	53.0	21.4	1.9		$470
Ivory Coast	322,459	124,502	21.7	26.2	67	293	38	14	2.4	100.0	142	50	53	48	54.6	38.5	2.0		$1,590
Kenya	580,367	224,081	40.2	51.3	69	865	40	12	2.8	77.0	117	53	53	19	89.0	76.0	2.1		$1,540
Lesotho	30,349	11,718	1.8	1.7	60	541	27	25	0.2	91.0	112	35	36	24	73.6	93.6	3.2		$1,890
Liberia	111,369	43,000	4.2	6.8	37	932	50	18	3.2	133.0	190	45	47	58	69.9	36.8	2.4		$290
Madagascar	587,036	226,656	20.0	28.0	34	680	38	10	2.8	75.0	120	57	60	30	87.7	72.9	3.4		$920
Malawi	118,479	45,745	14.5	20.4	122	582	48	16	3.2	80.0	125	45	47	17	74.5	46.7	2.8		$750
Mali	1,240,185	478,838	13.6	20.6	11	273	48	15	3.3	96.0	161	54	59	31	47.9	33.2	3.1		$1,040
Mauritania	1,025,516	395,954	3.4	4.5	3	328	35	9	2.6	77.0	121	59	62	40	50.6	29.5	2.8		$2,010
Mauritius	2,041	788	1.3	1.4	646	1242	14	7	0.7	15.4	18	69	76	42	87.7	81.0	5.5		$11,390
Moçambique	801,586	309,494	21.3	27.5	27	663	41	20	2.1	108.0	158	42	44	29	59.9	28.4	2.6		$690
Namibia	824,287	318,259	2.1	2.3	3	260	25	15	1.0	47.0	67	48	47	35	82.9	81.2	4.5		$5,120
Niger	1,266,994	489,189	15.6	26.3	12	308	46	15	3.1	81.0	171	58	56	17	23.5	8.3	2.8		$630
Nigeria	923,766	356,668	155.6	205.4	168	495	43	18	2.5	100.0	194	46	47	47	72.3	56.2	2.7		$1,770

APPENDIX B. (Continued)

	Land Area		Population		Population Density		Birth Rate	Death Rate	Natural Increase %	Infant Mortality per 1,000	Child Mortality per 1,000	Life Expectancy		Percent Urban Pop.	Literacy		Corruption Index	Big Mac Price ($US)	Per Capita GNI ($US)
	Sq km	Sq mi	2010 (Millions)	2025 (Millions)	Arithmetic	Physiologic						Male (years)	Female (years)		Male %	Female %			
Réunion	2,510	969	0.8	1.0	328	2185	19	5	1.4	8.0		72	80	92	84.8	89.2			
Rwanda	26,340	10,170	10.1	14.6	384	915	43	16	2.7	86.0	143	47	48	18	73.7	60.6	3.0		$860
São Tomé and Príncipe	961	371	0.2	0.2	220	511	35	8	2.7	77.0	119	63	66	58	70.2	39.1	2.7		$1,630
Senegal	196,720	75,954	13.4	18.0	68	570	39	10	2.9	61.0	118	60	64	41	47.2	27.6	3.4		$1,640
Seychelles	451	174	0.1	0.1	227	1512	18	7	1.1	11.0	12	67	77	53	82.9	85.7	4.8		$8,670
Sierra Leone	71,740	27,699	5.8	7.6	81	1007	48	23	2.5	158.0	267	48	49	37	50.7	22.6	1.9		$660
Somalia	637,658	246,201	9.5	14.3	15	744	46	19	2.7	117.0	189	47	49	37	85.8	84.5	1.0		
South Africa	1,221,034	471,444	49.1	51.5	40	309	23	15	0.8	45.0	58	48	52	59	68.3	46.0	4.9	1.66	$9,560
Swaziland	17,361	6,703	1.1	1.0	63	576	31	31	0.0	85.0	116	33	34	24	80.9	78.7	3.6		$4,930
Tanzania	945,087	364,900	42.1	58.2	45	890	38	15	2.3	75.0	118	50	52	25	84.1	66.6	3.0		$1,200
Togo	56,791	21,927	7.2	9.9	127	294	38	10	2.8	91.0	140	56	60	40	72.2	42.6	2.7		$800
Uganda	241,040	93,066	31.1	56.4	129	379	48	16	3.2	76.0	121	47	48	13	77.7	57.1	2.6		$920
Zambia	752,607	290,583	12.7	15.5	17	241	43	22	2.1	100.0	164	38	37	37	85.2	71.2	2.8		$1,220
Zimbabwe	390,759	150,873	13.8	16.0	35	392	31	21	1.0	60.0	92	40	40	37	95.5	89.9	1.8		

NORTH AFRICA/SOUTHWEST ASIA

	Land Area		Population		Population Density		Birth Rate	Death Rate	Natural Increase %	Infant Mortality per 1,000	Child Mortality per 1,000	Life Expectancy		Percent Urban Pop.	Literacy		Corruption Index	Big Mac Price ($US)	Per Capita GNI ($US)
	Sq km	Sq mi	2010 (Millions)	2025 (Millions)	Arithmetic	Physiologic						Male (years)	Female (years)		Male %	Female %			
REALM	19,318,887	7,459,064	592.4	733.9	63	687	26	7	1.9	46.0	51	66	70	53	83.1	64.5			$8,309
Afghanistan	652,086	251,772	34.4	50.3	53	440	47	21	2.6	163.0	254	43	43	20	51.0	20.8	1.5		
Algeria	2,381,730	919,591	36.0	43.3	15	503	22	4	1.8	27.0	30	71	74	63	75.1	51.3	3.2		$5,490
Bahrain	689	266	0.8	1.0	1201	13,345	20	3	1.7	8.0	9	73	77	100	91.0	82.7	5.4		$34,310
Egypt	1,001,445	386,660	78.1	95.9	78	2599	27	6	2.1	33.0	40	70	74	43	66.6	43.7	2.8	2.34	$5,400
Iran	1,633,182	630,575	74.4	88.0	46	414	20	5	1.5	32.0	36	69	72	67			2.3		$10,800
Iraq	438,319	169,236	30.9	43.3	71	543	34	10	2.4	94.0	116	56	60	67	70.7	45.0	1.3		
Israel	21,059	8,131	7.7	9.3	368	1751	21	5	1.6	3.5	5	79	82	92	97.9	94.3	6.0	3.69	$25,930
Jordan	89,210	34,444	6.1	7.7	68	1363	28	4	2.4	24.0	28	71	73	83	94.9	84.4	5.1		$5,160
Kazakhstan	2,717,289	1,049,151	16.0	17.1	6	49	21	10	1.1	29.0	33	61	72	53	99.1	96.1	2.2		$9,700
Kuwait	17,819	6,880	2.8	3.6	157	15,734	21	2	1.9	8.0	9	77	79	98	84.3	79.9	4.3		$49,970
Kyrgyzstan	198,499	76,641	5.4	6.5	27	387	24	7	1.7	50.0	57	62	70	35	98.6	95.5	1.8		$1,950
Lebanon	10,399	4,015	4.1	4.6	396	1276	19	5	1.4	19.0	22	69	74	87	92.3	80.4	3.0		$10,050
Libya	1,759,532	679,359	6.6	8.1	4	373	24	4	2.0	21.0	23	71	76	77	90.9	67.6	2.6		$11,500
Morocco	446,548	172,413	32.1	36.6	72	327	21	6	1.5	43.0	46	68	72	56	61.9	36.0	3.5		$3,990
Oman	212,459	82,031	2.8	3.1	13	1325	24	3	2.1	10.0	11	73	75	71	80.4	61.7	5.5		$19,740
Palestinian Territories	6,260	2,417	4.5	6.2	716	35,797	37	4	3.3	25.0	29	72	73	72					
Qatar	11,000	4,247	0.9	1.1	84	8429	17	2	1.5	7.0	8	74	76	100	80.5	83.2	6.5		

	Land Area		Population		Population Density		Birth Rate	Death Rate	Natural Increase %	Infant Mortality per 1,000	Child Mortality per 1,030	Life Expectancy		Percent Urban Pop.	Literacy		Corruption Index	Big Mac Price ($US)	Per Capita GNI ($US)
	Sq km	Sq mi	2010 (Millions)	2025 (Millions)	Arithmetic	Physiologic						Male (years)	Female (years)		Male %	Female %			
Saudi Arabia	2,149,680	829,996	29.6	35.7	14	688	29	3	2.6	16.0	20	74	78	81	84.1	67.2	3.5	2.66	$22,910
Sudan	2,505,798	967,494	41.1	54.3	15	234	33	12	2.1	81.0	127	56	59	38	36.0	14.0	1.6		$1,880
Syria	185,179	71,498	20.9	26.8	113	376	28	4	2.4	19.0	22	71	75	50	88.3	60.4	2.1		$4,370
Tajikistan	143,099	55,251	7.6	9.5	53	76	27	5	2.2	65.0	77	64	69	26	99.6	98.9	2.0		$1,710
Tunisia	163,610	63,170	10.5	12.1	64	20	17	6	1.1	19.0	22	72	76	65	81.4	60.1	4.4		$7,130
Turkey	774,816	299,158	76.8	87.8	99	26	19	6	1.3	23.0	25	69	74	62	93.6	76.7	4.6	3.13	$12,090
Turkmenistan	488,099	188,456	5.4	6.5	11	276	24	6	1.8	74.0	82	58	67	47	98.8	96.6	1.8		$6,640
United Arab Emirates	83,600	32,278	4.6	6.2	55	5524	15	2	1.3	7.0	8	77	81	83	75.5	79.5	5.9		
Uzbekistan	447,397	172,741	28.1	33.3	63	524	24	7	1.7	48.0	55	63	70	36	98.5	96.0	1.8		$1,680
Western Sahara	252,120	97,344	0.5	0.8	2	5	28	8	2.0	53.0		62	66	81					
Yemen	527,966	203,849	23.6	35.2	45	1493	41	9	3.2	77.0	102	60	62	30	67.4	25.0	2.3		$2,200
SOUTH ASIA																			
REALM	**4,487,762**	**1,732,734**	**1569.5**	**1877.7**	**146**	**29**	**25**	**8**	**1.7**	**58.0**	**73**	**64**	**66**	**28**	**66.1**	**39.9**			**$2,576**
Bangladesh	143,998	55,598	152.4	180.1	1058	1679	24	7	1.7	52.0	67	62	64	24	51.7	29.5	2.1		$1,340
Bhutan	47,001	18,147	0.7	0.9	16	520	30	7	2.3	40.0	60	66	67	31	61.1	33.6	5.2		$4,980
India	3,287,576	1,269,340	1186.4	1407.7	361	633	24	8	1.6	57.0	75	65	66	28	68.6	42.1	3.4		$2,740
Maldives	300	116	0.3	0.4	1029	10,287	19	4	1.5	16.0	19	72	73	27	96.3	96.4	2.8		$5,040
Nepal	147,179	56,826	28.1	36.5	191	909	29	9	2.0	48.0	61	63	64	17	59.1	21.8	2.7		$1,040
Pakistan	796,098	307,375	180.8	228.9	227	783	31	8	2.3	75.0	93	62	64	35	57.6	27.8	2.5		$2,570
Sri Lanka	65,610	25,332	20.8	23.2	317	109	19	7	1.2	15.0	19	67	75	15	94.5	88.9	3.2		$4,210
EAST ASIA																			
REALM	**11,774,215**	**4,546,050**	**1573.4**	**1696.6**	**134**	**108**	**12**	**7**	**0.5**	**21.0**	**24**	**72**	**76**	**50**	**94.9**	**84.7**			**$8,380**
China	9,572,855	3,696,100	1338.0	1476.0	140	998	12	7	0.5	23.0	27	71	75	45	92.3	77.4	3.6	1.83	$5,370
Japan	377,799	145,869	127.7	119.3	338	2600	9	9	0.0	2.8	4	79	86	79	100.0	100.0	7.3	3.23	$34,600
Korea, North	120,541	46,541	23.9	25.8	198	1240	16	7	0.9	21.0	28	68	73	60	99.0	99.0			
Korea, South	99,259	38,324	49.1	49.1	495	260	10	5	0.5	4.0	4	76	82	82	99.2	96.4	5.6	2.39	$24,750
Mongolia	1,566,492	604,826	2.8	3.3	2	178	21	6	1.5	41.0	50	61	67	59	99.2	99.3	3.0		$3,160
Taiwan	36,180	13,969	23.1	23.1	640	2558	9	6	0.3	4.6		75	81	78	97.6	90.2	5.7	2.32	

	Land Area		Population		Population Density		Birth Rate	Death Rate	Natural Increase %	Infant Mortality per 1,000	Child Mortality per 1,000	Life Expectancy		Percent Urban Pop.	Literacy		Corruption Index	Big Mac Price ($US)	Per Capita GNI ($US)
	Sq km	Sq mi	2010 (Millions)	2025 (Millions)	Arithmetic	Physiologic						Male (years)	Female (years)		Male %	Female %			
SOUTHEAST ASIA																			
REALM	4,494,790	1,735,448	601.8	709.2	134	642	20	7	1.3	31.0	37	68	72	45	93.0	86.1			$4,440
Brunei	5,770	2,228	0.4	0.5	72	3578	19	3	1.6	7.0	8	72	77	72	94.7	88.2			$49,900
Cambodia	181,040	69,900	15.2	20.6	84	382	26	8	1.8	67.0	87	59	66	15	79.7	53.4	1.8		$1,690
East Timor	14,869	5,741	1.2	1.7	79	655	42	11	3.1	74.6	93	59	61	22			2.2		$3,190
Indonesia	1,904,561	735,355	247.2	291.9	130	763	21	6	1.5	34.0	43	69	72	48	91.9	82.1	2.6	1.74	$3,580
Laos	236,800	91,429	6.2	8.7	26	871	34	10	2.4	70.0	88	59	63	27	73.6	50.5	2.0		$1,940
Malaysia	329,750	127,317	28.6	34.6	87	361	21	5	1.6	9.0	11	72	76	68	91.5	83.6	5.1	1.52	$13,570
Myanmar/ Burma	676,577	261,228	50.1	55.4	74	463	19	10	0.9	70.0	98	58	64	31	89.0	80.6	1.3		
Philippines	299,998	115,830	94.3	120.2	314	953	26	5	2.1	25.0	31	66	72	63	95.5	95.2	2.3	2.07	$3,730
Singapore	619	239	4.9	5.3	7,848	392,384	11	5	0.6	2.4	3	78	83	100	96.4	88.5	9.2	2.61	$32,470
Thailand	513,118	198,116	66.8	70.2	130	325	13	8	0.5	16.0	17	68	75	36	97.2	94.0	3.5	1.77	$7,880
Vietnam	331,689	128,066	88.3	100.1	266	1210	17	5	1.2	16.0	19	71	75	27	95.7	91.0	2.7		$2,550
AUSTRAL REALM																			
REALM	8,012,942	3,093,814	26.0	29.6	3	45	14	7	0.7	4.8	5.1	79	84	87	100.0	100.0			$32,164
Australia	7,741,184	2,988,888	21.6	24.7	3	40	14	7	0.7	4.7	5	79	84	87	100.0	100.0	8.7	2.19	$33,340
New Zealand	270,529	104,452	4.4	4.9	16	135	15	7	0.8	5.0	6	78	82	86	100.0	100.0	9.3	2.48	$26,340
PACIFIC REALM																			
REALM	975,341	376,804	9.6	11.8	17	806	29	9	2.0	50.0	67	58	63	22	65.9	52.1			$1,526
Federated States of Micronesia	699	270	0.1	0.1	149	286	26	6	2.0	40.0	49	67	67	22	67.0	87.2			$3,710
Fiji	18,270	7,054	0.9	0.9	51	317	21	6	1.5	17.0	19	66	71	51	95.0	90.9			$4,370
French Polynesia	3,999	1,544	0.3	0.3	77	964	18	4	1.4	6.8		73	77	53	94.9	95.0			
Guam	549	212	0.2	0.2	375	1706	19	4	1.5	10.7		75	82	93	99.0	99.0			
Marshall Islands	179	69	0.1	0.1	596	3506	38	6	3.2	23.0	25	64	67	68	92.4	90.0			
New Caledonia	18,581	7,174	0.2	0.3	11	1105	18	5	1.3	7.0		73	80	58					
Papua New Guinea	462,839	178,703	6.8	8.6	15	1464	31	10	2.1	62.0	81	54	60	13	57.4	58.3	2.0		$1,500
Samoa	2,841	1,097	0.2	0.2	74	171	29	6	2.3	20.0	24	72	74	22	100.0	100.0	4.4		$3,570
Solomon Islands	28,899	11,158	0.5	0.7	18	607	34	8	2.6	48.0	63	62	63	17	62.4	44.9	2.9		$1,400
Vanuatu	12,191	4,707	0.2	0.4	17	172	31	6	2.5	27.0	33	66	69	21	57.3	47.8	2.9		$2,890

This Data Table is a valuable resource, and should be consulted throughout your reading. Like all else in this book, this table is subject to continuous revision and modification. Compared to previous editions, we have deleted some indices, elaborated others, and introduced new ones. For example, in a world with steadily slowing population growth, the so-called Doubling Time index—the number of years it will take for a population to double in size based on its current rate of natural increase—has lost most of its utility. On the other hand, when it comes to **Life Expectancy** and **Literacy**, general averages conceal significant differences between male and female rates, which in turn reflect conditions in individual countries, so we now report these by gender. Also in this edition, we continue to use the **Corruption Index**, not available for all countries but an important reflection of a global problem. The **Big Mac Price** index, a measure introduced by the journal *The Economist*, tells you much more than what a hamburger with all the trimmings would cost in real dollars in various countries of the world—it also reflects whether those countries' currencies are overvalued or undervalued. And the final column, in which we formerly used to reported the per-capita GNP (Gross National Product), now reveals the **GNI**, that is, the Gross National Income per person and what this would buy in each country. In the language of economic geographers, this is called the GNI-PPP, the percapita Gross National Income in terms of its Purchasing Power Parity.

Indexes that may not be immediately obvious to you include **Arithmetic Population Density**, the number of people per square kilometer in each country; **Physiologic Population Density**, the number of people per square kilometer of agriculturally productive land; **Birth** and **Death Rates** per thousand in the population, resulting in the national population's rate of **Natural Increase** (population growth measured as the excess of live births over deaths per 1000 individuals per year); a population's **Infant Mortality**, the number of deaths per thousand in the first year of life, thus reflecting largely the number of deaths at birth; **Child Mortality**, the deaths per thousand of children in their first five years; the **Corruption Index**, based on Transparency International data in which 10.0 is perfect and 0.1 is the worst; the **Big Mac Price** index, which tells you why Malaysia in 2009 was the best place to buy a hamburger in U.S. dollars; and the **Per Capita GNI ($US)**, the GNI-PPP index referred to above, which tells you how spendable income varies around the globe. For additional details on sources and data, please consult the Data Sources section of the Preface.

NOTE: Additional appendices may be found on the book's website at www.wiley.com/college/deblij

GLOSSARY*

Aboriginal land issue The legal campaign in which Australia's **indigenous peoples** have claimed title to traditional land in several parts of that country. The courts have upheld certain claims, fueling Aboriginal activism that has raised broader issues of indigenous rights.

Aboriginal population See **indigenous peoples**.

Absolute location The position or place of a certain item on the surface of the Earth as expressed in degrees, minutes, and seconds of **latitude**, 0° to 90° north or south of the equator, and **longitude**, 0° to 180° east or west of the *prime meridian* passing through Greenwich, England (a suburb of London).

Accessibility The degree of ease with which it is possible to reach a certain location from other locations. *Inaccessibility* is the opposite of this concept.

Acculturation Cultural modification resulting from intercultural borrowing. In **cultural geography**, the term refers to the change that occurs in the **culture** of **indigenous peoples** when contact is made with a society that is technologically superior.

Advantage The most meaningful distinction that can now be made to classify a country's level of economic **development**. Takes into account geographic **location, natural resources**, government, political stability, productive skills, and much more.

Agglomeration Process involving the clustering or concentrating of people or activities.

Agrarian Relating to the use of land in rural communities or to agricultural societies in general.

Agriculture The purposeful tending of crops and livestock in order to produce food and fiber.

Alluvial Referring to the mud, silt, and sand (collectively *alluvium*) deposited by rivers and streams. *Alluvial plains* adjoin many larger rivers; they consist of the renewable deposits that are laid down during floods, creating fertile and productive soils. Alluvial **deltas** mark the mouths of rivers such as the Nile (Egypt) and the Ganges (Bangladesh).

Altiplano High-elevation plateau, basin, or valley between even higher mountain ranges, especially in the Andes of South America.

Altitudinal zonation Vertical regions defined by physical-environmental zones at various elevations (see Fig. 4A-3), particularly in the highlands of South and Middle America. See *tierra caliente, tierra templada, tierra fría, tierra helada*, and *tierra nevada*.

American Manufacturing Belt North America's near-rectangular Core Region, whose corners are Boston, Milwaukee, St. Louis, and Baltimore. This region dominated the industrial geography of the United States and Canada during the industrial age; still a formidable economic powerhouse that remains the realm's geographic heart.

Animistic religion The belief that inanimate objects, such as hills, trees, rocks, rivers, and other elements of the natural landscape, possess souls and can help or hinder human efforts on Earth.

Antarctic Treaty International cooperative agreement on the use of Antarctic territory.

Antecedent boundary A political boundary that existed before the **cultural landscape** emerged and stayed in place while people moved in to occupy the surrounding area.

Anthracite coal Hardest and highest carbon-content coal, and therefore of the highest quality.

Apartheid Literally, *apartness*. The Afrikaans term for South Africa's pre-1994 policies of racial separation, a system that produced highly segregated socio-geographical patterns.

Aquaculture The use of a river segment or an artificial pond for the raising and harvesting of food products, including fish, shellfish, and even seaweed. The Japanese pioneered the practice, which is now spreading globally, particularly along heavily populated coastlines.

Aquifer An underground reservoir of water contained within a porous, water-bearing rock layer.

Arable Land fit for cultivation by one farming method or another. See also **physiologic density**.

Archipelago A set of islands grouped closely together, usually elongated into a *chain*.

Area A term that refers to a part of the Earth's surface with less specificity than **region**. For example, *urban area* alludes generally to a place where urban development has occurred, whereas *urban region* requires certain specific criteria on which such a designation is based (e.g., the spatial extent of commuting or the built townscape).

Areal functional organization A geographic principle for understanding the evolution of regional organization, whose five interrelated tenets are applied to the spatial development of Japan in Chapter 9B.

Areal interdependence A term related to **functional specialization**. When one area produces certain goods or has certain raw materials and another area has a different set of raw materials and produces different goods, their needs may be *complementary*; by exchanging raw materials and products, they can satisfy each other's requirements.

Arithmetic density A country's population, expressed as an average per unit area, without regard for its **distribution** or the limits of **arable** land. See also **physiologic density**.

Aryan From the Sanskrit *Arya* (meaning "noble"), a name applied to an ancient people who spoke an **Indo-European language** and who moved into northern India from the northwest.

Atmosphere The Earth's envelope of gases that rests on the oceans and land surface and penetrates open spaces within soils. This layer of nitrogen (78 percent), oxygen (21 percent), and traces of other gases is densest at the Earth's surface and thins with altitude.

Austral South.

Autocratic A government that holds absolute power, often ruled by one person or a small group of persons who control the country by despotic means.

Balkanization The fragmentation of a **region** into smaller, often hostile political units. Named after the historically contentious Balkan Peninsula of southeastern Europe.

Barrio Term meaning "neighborhood" in Spanish. Usually refers to an urban community in a Middle or South American city; also applied to low-income, inner-city concentrations of Hispanics in such western U.S. cities as Los Angeles.

Bauxite Aluminum ore; usually deposited at shallow depths in the wet tropics.

Biodiversity hotspot A much higher than usual, world-class geographic concentration of natural plant and/or animal species. Tropical rainforest environments have dominated, but their recent ravaging by **deforestation** has had catastrophic results. The example given in the book is the southern (Costa Rica-Panama) segment of the **land bridge** that forms Central America—over recent geologic time a biogeographic highway for evolutionary change.

Biogeography The study of *flora* (plant life) and *fauna* (animal life) in spatial perspective.

Birth rate The *crude birth rate* is expressed as the annual number of births per 1000 individuals within a given population.

Bituminous coal Softer coal of lesser quality than **anthracite**, but of higher grade than **lignite**. When heated and converted to coking coal or *coke*, it is used to make steel.

Break-of-bulk point A location along a transport route where goods must be transferred from one carrier to another. In a port, the cargoes of oceangoing ships are unloaded and put on trains, trucks, or perhaps smaller river boats for inland distribution. An *entrepôt*.

BRIC Acronym for the four biggest emerging national markets in the world today—**B**razil, **R**ussia, **I**ndia, and **C**hina.

Buffer state See **buffer zone**.

Buffer zone A country or set of countries separating ideological or political adversaries. In southern Asia, Afghanistan, Nepal, and Bhutan were parts of a buffer zone set up between British and Russian-Chinese imperial spheres. Thailand was a *buffer state* between British and French colonial domains in mainland Southeast Asia.

Caliente See *tierra caliente*.

Cartogram A specially transformed map not based on traditional representations of **scale** or area.

Cartography The art and science of making maps, including data compilation, layout, and design. Also concerned with the interpretation of mapped patterns.

Caste system The strict **social stratification** and segregation of people—specifically in India's Hindu society—on the basis of ancestry and occupation.

Cay A low-lying small island usually composed of coral and sand. Pronounced *kee* and often spelled "key."

Cell phone revolution Applied specifically to much of Subsaharan Africa and certain other parts of the developing world, the linking of farmers with centers of information for weather and market conditions via cell-phone text messaging—allowing them to negotiate higher prices for their products.

Central business district (CBD) The downtown heart of a central city; marked by high land values, a concentration of business and commerce, and the clustering of the tallest buildings.

Centrality The strength of an urban center in its capacity to attract producers and consumers to its facilities; a city's "reach" into the surrounding region.

Centrifugal forces A term employed to designate forces that tend to divide a country—such as internal religious, linguistic, ethnic, or ideological differences.

Centripetal forces Forces that unite and bind a country together—such as a strong national culture, shared ideological objectives, and a common faith.

Cerrado Regional term referring to the fertile savannas of Brazil's interior Central-West that make it one of the world's most promising agricultural frontiers (mapped in Fig. 5A-5). Soybeans are the leading crop, and other grains and cotton are expanding. Inadequate transport links to the outside world remain a problem.

Chaebol Giant corporation controlling numerous companies and benefiting from government connections as well as favors, dominant in South Korea's **economic geography**; key to that country's **development** into an **economic tiger**, but more recently a barrier to free-market growth.

Charismatic Personal qualities of certain leaders that enable them to capture and hold the popular imagination, securing the allegiance and even the devotion of the masses. Mahatma Gandhi, Mao Zedong, and Franklin D. Roosevelt were good examples in the twentieth century.

China Proper The eastern and northeastern portions of China that contain most of the country's huge population.

Choke point A narrowing of an international waterway causing marine traffic congestion, requiring reduced speeds and/or sharp turns, and increasing the risk of collision as well as vulnerability to attack. When the waterway narrows to a distance of less than 38 kilometers (24 mi), this necessitates the drawing of a **median line (maritime) boundary**. Examples are the Hormuz Strait between Oman and Iran at the entrance to the Persian Gulf, and the Strait of Malacca between Malaysia and Indonesia.

City-state An independent political entity consisting of a single city with (and sometimes without) an immediate **hinterland**. The ancient city-states of Greece have their modern equivalent in Southeast Asia's Singapore.

Climate The long-term conditions (over at least 30 years) of aggregate **weather** over a region, summarized by averages and measures of variability; a synthesis of the succession of weather events we have learned to expect at any given location.

Climate change theory An alternative to the **hydraulic civilization theory**; holds that changing **climate** (rather than a monopoly over **irrigation** methods) could have provided certain cities within the ancient Fertile Crescent with advantages over other cities.

Climate region A **formal region** characterized by the uniformity of the **climate** type within it. Figure G-7 maps the global distribution of such regions.

Climatology The geographic study of **climates**. Includes not only the classification of climates and the analysis of their regional distribution, but also broader environmental questions that concern climate change, interrelationships with soil and vegetation, and human–climate interaction.

Coal See **anthracite coal, bituminous coal, fossil fuels**, and **lignite**.

Collectivization The reorganization of a country's **agriculture** under communism that involves the expropriation of private holdings and their incorporation into relatively large-scale units, which are farmed and administered cooperatively by those who live there.

Colonialism Rule by an autonomous power over a subordinate and an alien people and place. Though often established and maintained through political structures, colonialism also creates unequal cultural and economic relations. Because of the magnitude and impact of the European colonial thrust of the last few centuries, the term is generally understood to refer to that particular colonial endeavor. Also see **imperialism**.

Command economy The tightly controlled economic system of the former Soviet Union, whereby central planners in Moscow assigned the production of particular goods to particular places, often guided more by socialist ideology than the principles of **economic geography**.

Commercial agriculture For-profit **agriculture**.

Common market A **free-trade area** that not only has created a **customs union** (a set of common tariffs on all imports from outside the area) but also has eliminated restrictions on the movement of capital, labor, and enterprise among its member countries.

Compact state A politico-geographical term to describe a state that possesses a roughly circular, oval, or rectangular territory in which the distance from the geometric center to any point on the boundary exhibits little variance. Poland and Cambodia are good examples of this shape category.

Complementarity Exists when two regions, through an exchange of raw materials and/or finished products, can specifically satisfy each other's demands. See also **areal interdependence**.

Confucianism A philosophy of ethics, education, and public service based on the writings of Confucius (*Kongfuzi*); traditionally regarded as one of the cornerstones of Chinese **culture**.

Congo Two countries in Africa have the same short-form name, *Congo*. In this book, we use *The Congo* for the larger Democratic Republic of the Congo, and *Congo* for the smaller Republic of Congo.

Coniferous forest A forest of cone-bearing, needleleaf evergreen trees with straight trunks and short branches, including spruce, fir, and pine. See also **taiga**.

Contagious diffusion The distance-controlled spreading of an idea, innovation, or some other item through a local population by contact from person to person—analogous to the communication of a contagious illness.

Conterminous United States The 48 **contiguous** or adjacent States that occupy the southern half of the North American realm. Alaska is not contiguous to these States because western Canada lies in between; neither is Hawai'i, separated from the mainland by over 3000 kilometers (2000 mi) of ocean.

Contiguous Adjoining; adjacent.

Continental drift The slow movement of continents controlled by the processes associated with **plate tectonics**.

Continental shelf Beyond the coastlines of many landmasses, the ocean floor declines very gently until the depth of about 660 feet (200 m). Beyond the

660-foot line the sea bottom usually drops off sharply, along the *continental slope*, toward the much deeper mid-oceanic basin. The submerged continental margin is called the continental shelf, and it extends from the shoreline to the upper edge of the continental slope.

Continentality The variation of the continental effect on air temperatures in the interior portions of the world's landmasses. The greater the distance from the moderating influence of an ocean, the greater the extreme in summer and winter temperatures. Continental interiors also tend to be dry when the distance from oceanic moisture sources becomes considerable.

Conurbation General term used to identify a large multimetropolitan complex formed by the coalescence of two or more major **urban areas**. The Atlantic Seaboard **Megalopolis**, extending along the northeastern U.S. coast from southern Maine to Virginia, is a classic example.

Copra The dried-out, fleshy interior of a coconut that is used to produce coconut oil.

Cordillera Mountain chain consisting of sets of parallel ranges, especially the Andes in northwestern South America.

Core See **core area; core-periphery relationships**.

Core area In geography, a term with several connotations. *Core* refers to the center, heart, or focus. The core area of a **nation-state** is constituted by the national heartland, the largest population cluster, the most productive region, and the part of the country with the greatest **centrality** and **accessibility**—probably containing the capital city as well.

Core-periphery relationships The contrasting spatial characteristics of, and linkages between, the *have* (core) and *have-not* (periphery) components of a national or regional **system**.

Corridor In general, refers to a spatial entity in which human activity is organized in a linear manner, as along a major transport route or in a valley confined by highlands. More specifically, the politico-geographical term for a land extension that connects an otherwise **landlocked state** to the sea.

Cross-border linkages The ties between two closely-connected localities or regions that face each other across an international boundary. These relationships are often longstanding, and intensify further as **supranationalism** proceeds (especially among the EU countries of western Europe). In North America, neighboring southwestern Ontario and southeastern Michigan are an example of such ties, which propel movements of people and goods across that stretch of the U.S.-Canada border.

Cultural diffusion The **process** of spreading and adopting a cultural element, from its place of origin across a wider area.

Cultural ecology The multiple interactions and relationships between a **culture** and its **natural environment**.

Cultural environment See **cultural ecology**.

Cultural geography The wide-ranging and comprehensive field of geography that studies spatial aspects of human **cultures**.

Cultural landscape The forms and artifacts sequentially placed on the **natural landscape** by the activities of various human occupants. By this progressive imprinting of the human presence, the physical (natural) landscape is modified into the cultural landscape, forming an interacting unity between the two.

Cultural pluralism See **plural(istic) society**.

Cultural revival The regeneration of a long-dormant **culture** through internal renewal and external infusion.

Culture The sum total of the knowledge, attitudes, and habitual behavior patterns shared and transmitted by the members of a society. This is anthropologist Ralph Linton's definition; hundreds of others exist.

Culture area See **culture region**.

Culture hearth Heartland, source area, innovation center; place of origin of a major **culture**.

Culture region A distinct, culturally discrete spatial unit; a **region** within which certain cultural norms prevail.

Customs union A **free-trade area** in which member countries set common tariff rates on imports from outside the area.

Death rate The *crude death rate* is expressed as the annual number of deaths per 1000 individuals within a given population.

Deciduous A deciduous tree loses its leaves at the beginning of winter or the onset of the dry season.

Definition In **political geography**, the written legal description (in a treaty-like document) of a boundary between two countries or territories. See also **delimitation**.

Deforestation The clearing and destruction of forests (especially tropical rainforests) to make way for expanding settlement frontiers and the exploitation of new economic opportunities.

Deglomeration Deconcentration.

Delimitation In **political geography**, the translation of the written terms of a boundary treaty (the **definition**) into an official cartographic representation (map).

Delta Alluvial lowland at the mouth of a river, formed when the river deposits its alluvial load on reaching the sea. Often triangular in shape, hence the use of the Greek letter whose symbol is Δ.

Demarcation In **political geography**, the actual placing of a political boundary on the **cultural landscape** by means of barriers, fences, walls, or other markers.

Demographic transition model Multi-stage **model**, based on Western Europe's experience, of changes in population growth exhibited by countries undergoing industrialization. High **birth rates** and **death rates** are followed by plunging death rates, producing a huge net population gain; birth and death rates then converge at a low overall level. See Figure 8A-10.

Demography The interdisciplinary study of population—especially **birth rates** and **death rates**, growth patterns, longevity, **migration**, and related characteristics.

***Dependencia* theory** Originating in South America during the 1960s, it was a new way of thinking about economic development and underdevelopment that explained the persistent poverty of certain countries in terms of their unequal relations with other (rich) countries.

Desert An arid expanse supporting sparse vegetation, receiving less than 25 centimeters (10 in) of precipitation per year. Usually exhibits extremes of heat and cold because the moderating influence of moisture is absent.

Desertification Process of **desert** expansion into neighboring **steppelands** as a result of human degradation of fragile semiarid environments.

Development The economic, social, and institutional growth of national **states**.

Devolution The process whereby regions within a **state** demand and gain political strength and growing autonomy at the expense of the central government.

Dhows Wooden boats with characteristic triangular sails, plying the seas between Arabian and East African coasts.

Dialect Regional or local variation in the use of a major language, such as the distinctive accents of many residents of the U.S. South.

Diffusion The spatial spreading or dissemination of a **culture** element (such as a technological innovation) or some other phenomenon (e.g., a disease outbreak). For the various channels of outward geographic spread from a source area, see **contagious, expansion, hierarchical**, and **relocation diffusion**.

Distance decay The various degenerative effects of distance on human spatial structures and interactions.

Diurnal Daily.

Divided capital In **political geography**, a country whose central administrative functions are carried on in more than one city is said to have divided capitals. The Netherlands and South Africa are examples.

Domestication The transformation of a wild animal or wild plant into a domesticated animal or a cultivated crop to gain control over food production. A necessary evolutionary step in the development of humankind: the invention of **agriculture**.

Domino theory The belief that political destabilization in one **state** can result in the collapse of order in a neighboring state, triggering a chain of events that, in turn, can affect a series of **contiguous** states.

Double cropping The planting, cultivation, and harvesting of two crops successively within a single year on the same plot of farmland.

Dry canal An overland rail and/or road **corridor** across an **isthmus** dedicated to performing the transit functions of a canalized waterway. Best adapted to the movement of containerized cargo, there must be a port at each end to handle the necessary **break-of-bulk** unloading and reloading.

Ecology The study of the many interrelationships between all forms of life and the natural environments in which they have evolved and continue to develop. The study of *ecosystems* focuses on the interactions between specific organisms and their environments. See also **cultural ecology**.

Economic geography The field of geography that focuses on the diverse ways in which people earn a living and on how the goods and services they produce are expressed and organized spatially.

Economic restructuring The transformation of China into a market-driven economy in the post-Mao era, beginning in the late 1970s.

Economic tiger One of the burgeoning beehive countries of the western **Pacific Rim**. Following Japan's route since 1945, these countries have experienced significant modernization, industrialization, and Western-style economic growth since 1980. Three leading economic tigers are South Korea, Taiwan, and Singapore. The term is increasingly used more generally to describe any fast-developing economy.

Economies of scale The savings that accrue from large-scale production wherein the unit cost of manufacturing decreases as the level of operation enlarges. Supermarkets operate on this principle and are able to charge lower prices than small grocery stores.

Ecosystem See **ecology**.

Ecumene The habitable portions of the Earth's surface where permanent human settlements have arisen.

Elite A small but influential upper-echelon social class whose power and privilege give it control over a country's political, economic, and cultural life.

El Niño-Southern Oscillation (ENSO) A periodic, large-scale, abnormal warming of the sea surface in the low latitudes of the eastern Pacific Ocean that has global implications, disturbing normal weather patterns in many parts of the world, especially South America.

Elongated state A **state** whose territory is decidedly long and narrow in that its length is at least six times greater than its average width. Chile and Vietnam are two classic examples.

Emigrant A person **migrating** away from a country or area; an out-migrant.

Empirical Relating to the real world, as opposed to theoretical abstraction.

Enclave A piece of territory that is surrounded by another political unit of which it is not a part.

Endemism Referring to a disease in a host population that affects many people in a kind of equilibrium without causing rapid and widespread deaths.

Entrepôt A place, usually a port city, where goods are imported, stored, and transshipped; a **break-of-bulk point**.

Environmental degradation The accumulated human abuse of a region's **natural landscape** that, among other things, can involve air and water pollution, threats to plant and animal ecosystems, misuse of **natural resources**, and generally upsetting the balance between people and their habitat.

Epidemic A local or regional outbreak of a disease.

Escarpment A cliff or very steep slope; often marks the edge of a plateau.

Estuary The widening mouth of a river as it reaches the sea; land subsidence or a rise in sea level has overcome the tendency to form a **delta**.

Ethanol The leading U.S. biofuel that is essentially alochol distilled from corn mash. Much of it is produced in the historic Corn Belt centered on Iowa. Many risks and problems accompany this energy source, however.

Ethnic cleansing The slaughter and/or forced removal of one **ethnic** group from its homes and lands by another, more powerful ethnic group bent on taking that territory.

Ethnicity The combination of a people's **culture** (traditions, customs, language, and religion) and racial ancestry.

European state model A **state** consisting of a legally defined territory inhabited by a population governed from a capital city by a representative government.

European Union (EU) **Supranational** organization constituted by 27 European countries to further their common economic interests. In alphabetical order, these countries are: Austria, Belgium, Bulgaria, Cyprus, the Czech Republic, Denmark, Estonia, Finland, France, Germany, Greece, Hungary, Ireland, Italy, Latvia, Lithuania, Luxembourg, Malta, the Netherlands, Poland, Portugal, Romania, Slovakia, Slovenia, Spain, Sweden, and the United Kingdom.

Exclave A bounded (non-island) piece of territory that is part of a particular **state** but lies separated from it by the territory of another state. Alaska is an exclave of the United States.

Exclusive Economic Zone (EEZ) An oceanic zone extending up to 200 **nautical miles** (370 km) from a shoreline, within which the coastal **state** can control fishing, mineral exploitation, and additional activities by all other countries.

Expansion diffusion The spreading of an innovation or an idea through a fixed population in such a way that the number of those adopting grows continuously larger, resulting in an expanding area of dissemination.

Extraterritoriality The politico-geographical concept suggesting that the property of one **state** lying within the boundaries of another actually forms an extension of the first state.

Failed state A country whose institutions have collapsed and in which anarchy prevails. Somalia is a current example.

Fatwa Literally, a legal opinion or proclamation issued by an Islamic cleric, based on the holy texts of Islam, long applicable only in the *Umma*, the realm ruled by the laws of Islam. In 1989, the Iranian Ayatollah Khomeini extended the reach of the *fatwa* by condemning to death a British citizen and author living in the United Kingdom.

Favela Shantytown on the outskirts or even well within an urban area in Brazil.

Fazenda Coffee plantation in Brazil.

Federal state A political framework wherein a central government represents the various subnational entities within a **nation-state** where they have common interests—defense, foreign affairs, and the like—yet allows these various entities to retain their own identities and to have their own laws, policies, and customs in certain spheres.

Federation See **federal state**.

Fertile Crescent Crescent-shaped zone of productive lands extending from near the southeastern Mediterranean coast through Lebanon and Syria to the **alluvial** lowlands of Mesopotamia (in Iraq). Once more fertile than today, this is one of the world's great source areas of **agricultural** and other innovations.

First Nations Name given Canada's **indigenous peoples** of American descent, whose U.S. counterparts are called Native Americans.

Fjord Narrow, steep-sided, elongated, and inundated coastal valley deepened by glacier ice that has since melted away, leaving the sea to penetrate.

Floodplain Low-lying area adjacent to a mature river, often covered by **alluvial** deposits and subject to the river's floods.

Forced migration Human **migration** flows in which the movers have no choice but to relocate.

Formal region A type of **region** marked by a certain degree of homogeneity in one or more phenomena; also called *uniform region* or *homogeneous region*.

Forward capital Capital city positioned in actually or potentially contested territory, usually near an international border; it confirms the **state's** determination to maintain its presence in the region in contention.

Fossil fuels The energy resources of **coal**, natural gas, and petroleum (oil), so named collectively because they were formed by the geologic compression and transformation of tiny plant and animal organisms.

Four Motors of Europe Rhône-Alpes (France), Baden-Württemberg (Germany), Catalonia (Spain), and Lombardy (Italy). Each is a high-technology-driven region marked by exceptional industrial vitality and economic success not only within Europe but on the global scene as well.

Fragmented state A **state** whose territory consists of several separated parts, not a **contiguous** whole. The individual parts may be isolated from each other by the land area of other states or by international waters. The United States and Indonesia are examples.

Francophone French-speaking. Quebec constitutes the heart of Francophone Canada.

Free-trade area A form of economic integration, usually consisting of two or more **states**, in which members agree to remove tariffs on trade among themselves. Usually accompanied by a **customs union** that establishes common tariffs on imports from outside the trade area, and sometimes by a **common market** that also removes internal restrictions on the movement of capital, labor, and enterprise.

Free Trade Area of the Americas (FTAA) The ultimate goal of **supranational** economic integration in North, Middle, and South America: the creation of a single-market trading bloc that would involve every country in the Western Hemisphere between the Arctic shore of Canada and Cape Horn at the southern tip of Chile.

Fria See *tierra fria*.

Frontier Zone of advance penetration, usually of contention; an area not yet fully integrated into a national **state**.

FTAA See **Free Trade Area of the Americas**.

Functional region A **region** marked less by its sameness than by its dynamic internal structure; because it usually focuses on a central node, also called *nodal region* or *focal region*.

Functional specialization The production of particular goods or services as a dominant activity in a particular location. See also **local functional specialization**.

Fundamentalism See **revivalism (religious)**.

Gentrification The upgrading of an older residential area through private reinvestment, usually in the downtown area of a central city. Frequently, this involves the displacement of established lower-income residents, who cannot afford the heightened costs of living, and conflicts are not uncommon as such neighborhood change takes place.

Geographic change Evolution of **spatial** patterns over time.

Geographic realm The basic spatial unit in our world regionalization scheme. Each realm is defined in terms of a synthesis of its total human geography—a composite of its leading cultural, economic, historical, political, and appropriate environmental features.

Geography of development The subfield of economic geography concerned with spatial aspects and regional expressions of **development**.

Geometric boundaries Political boundaries **defined** and **delimited** (and occasionally **demarcated**) as straight lines or arcs.

Geomorphology The geographic study of the configuration of the Earth's solid surface—the world's landscapes and their constituent landforms.

Ghetto An intraurban region marked by a particular **ethnic** character. Often an inner-city poverty zone, such as the black ghetto in U.S. central cities. Ghetto residents are involuntarily segregated from other income and racial groups.

Glaciation See **Pleistocene Epoch**.

Globalization The gradual reduction of regional contrasts at the world **scale**, resulting from increasing international cultural, economic, and political exchanges.

Green Revolution The successful recent development of higher-yield, fast-growing varieties of rice and other cereals in certain developing countries.

Gross domestic product (GDP) The total value of all goods and services produced in a country by that state's economy during a given year.

Gross national product (GNP) The total value of all goods and services produced in a country by that state's economy during a given year, plus all citizens' income from foreign investment and other external sources.

Growth pole An urban center with certain attributes that, if augmented by a measure of investment support, will stimulate regional economic **development** in its **hinterland**.

Growth triangle An increasingly popular economic **development** concept along the western **Pacific Rim**, especially in Southeast Asia. It involves the linking of production in growth centers of three countries to achieve benefits for all. For example, Singapore would supply capital and technical know-how to lead high-technology manufacturing projects based on raw materials and inexpensive labor supplied by adjacent areas of Malaysia (Johor State) and Indonesia (Riau Islands).

Hacienda Literally, a large estate in a Spanish-speaking country. Sometimes equated with the **plantation**, but there are important differences between these two types of agricultural enterprise.

Heartland theory The hypothesis, proposed by British geographer Halford Mackinder during the early twentieth century, that any political power based in the heart of Eurasia could gain sufficient strength to eventually dominate the world. Furthermore, since Eastern Europe controlled access to the Eurasian interior, its ruler would command the vast "heartland" to the east.

Hegemony The political dominance of a country (or even a region) by another country. The former Soviet Union's postwar grip on Eastern Europe, which lasted from 1945 to 1990, was a classic example.

Helada See *tierra helada*.

Hierarchical diffusion A form of **diffusion** in which an idea or innovation spreads by trickling down from larger to smaller adoption units. An urban **hierarchy** is usually involved, encouraging the leapfrogging of innovations over wide areas, with geographic distance a less important influence.

Hierarchy An order or gradation of phenomena, with each level or rank subordinate to the one above it and superior to the one below. The levels in a national urban hierarchy are constituted by hamlets, villages, towns, cities, and (frequently) the **primate city**.

High island Volcanic islands of the Pacific Realm that are high enough in elevation to wrest substantial moisture from the tropical ocean air (see **orographic precipitation**). They tend to be well watered, their volcanic soils enable productive agriculture, and they support larger populations than **low islands**—which possess none of these advantages and must rely on fishing and the coconut palm for survival.

High seas Areas of the oceans away from land, beyond national jurisdiction, open and free for all to use.

Highveld A term used in South Africa to identify the high, grass-covered plateau that dominates much of the country. The lowest-lying areas (mainly along the narrow coastlands) in South Africa are called *lowveld*; areas that lie at intermediate elevations are the *middleveld*.

Hinterland Literally, "country behind," a term that applies to a surrounding area served by an urban center. That center is the focus of goods and services produced for its hinterland and is its dominant urban influence as well. In the case of a port city, the hinterland also includes the inland area whose trade flows through that port.

Historical inertia A term from manufacturing geography that refers to the need to continue using the factories, machinery, and equipment of heavy industries for their full, multiple-decade lifetimes to cover major initial investments—even though these facilities may be increasingly obsolete.

Holocene The current *interglacial* epoch (the warm period of glacial contraction between the glacial expansions of an **ice age**); extends from 10,000 years ago to the present. Also known as the *Recent Epoch*.

Human evolution Long-term biological maturation of the human species. Geographically, all evidence points toward East Africa as the source of humankind. Our species, *Homo sapiens*, emigrated from this hearth to eventually populate the rest of the **ecumene**.

Hurricane Alley The most frequent pathway followed by tropical storms and hurricanes over the past 150 years in their generally westward movement across the Caribbean Basin. Historically, hurricane tracks have bundled most tightly in the center of this route, most often affecting the Lesser Antilles between Antigua and the Virgin Islands, Puerto Rico, Hispaniola (Haiti/Dominican Republic), Jamaica, Cuba, southernmost Florida, Mexico's Yucatán, and the Gulf of Mexico.

Hydraulic civilization theory The theory that cities able to control **irrigated** farming over large **hinterlands** held political power over other cities. Particularly applies to early Asian civilizations based in such river valleys as the Chang (Yangzi), the Indus, and those of Mesopotamia.

Hydrologic cycle The **system** of exchange involving water in its various forms as it continually circulates between the **atmosphere**, the oceans, and above and below the land surface.

Ice age A stretch of geologic time during which the Earth's average atmospheric temperature is lowered; causes the equatorward expansion of continental ice sheets in the higher latitudes and the growth of mountain glaciers in and around the highlands of the lower latitudes.

Immigrant A person **migrating** into a particular country or area; an in-migrant.

Immigration policies Australia's policies to regulate immigration, an issue that continues to roil that country's society.

Imperialism The drive toward the creation and expansion of a **colonial** empire and, once established, its perpetuation.

Import-substitution industries The industries local entrepreneurs establish to serve populations of remote areas when transport costs from distant sources make these goods too expensive to import.

Inaccessibility See **accessibility**.

Indentured workers Contract laborers who sell their services for a stipulated period of time.

Indigenous peoples Native or *aboriginal* peoples; often used to designate the inhabitants of areas that were conquered and subsequently colonized by the **imperial** powers of Europe.

Indo-European languages The major world language family that dominates the European **geographic realm** (Fig. 1A-7). This language family is also the most widely dispersed globally (Fig. G-9), and about half of humankind speaks one of its languages.

Industrial Revolution The term applied to the social and economic changes in agriculture, commerce, and especially manufacturing and urbanization that resulted from technological innovations and greater specialization in late-eighteenth-century Europe.

Informal sector Dominated by unlicensed sellers of homemade goods and services, the primitive form of capitalism found in many developing countries that takes place beyond the control of government.

Infrastructure The foundations of a society: urban centers, transport networks, communications, energy distribution systems, farms, factories, mines, and such facilities as schools, hospitals, postal services, and police and armed forces.

Insular Having the qualities and properties of an island. Real islands are not alone in possessing such properties of **isolation**: an **oasis** in the middle of a **desert** also has qualities of insularity.

Insurgent state Territorial embodiment of a successful guerrilla movement. The establishment by antigovernment insurgents of a territorial base in which they exercise full control; thus a **state** within a state.

Intercropping The planting of several types of crops in the same field; commonly used by **shifting cultivators**.

Interglacial See **Pleistocene Epoch**.

Intermontane Literally, between the mountains. Such a location can bestow certain qualities of natural protection or **isolation** to a community.

Internal migration Migration flow within a country, such as ongoing westward and southward movements toward the **Sunbelt** in the United States.

International migration Migration flow involving movement across an international boundary.

Intervening opportunity In trade or **migration** flows, the presence of a nearer opportunity that greatly diminishes the attractiveness of sites farther away.

Inuit Indigenous peoples of North America's Arctic zone, formerly known as Eskimos.

Irredentism A policy of cultural extension and potential political expansion by a **state** aimed at a community of its nationals living in a neighboring state.

Irrigation The artificial watering of croplands.

Islamic Front The southern border of the African Transition Zone that marks the religious **frontier** of the **Muslim** faith in its southward penetration of Subsaharan Africa (see Fig. 6B-9).

Islamization Introduction and establishment of the **Muslim** religion. A **process** still under way, most notably along the **Islamic Front**, that marks the southern border of the African Transition Zone.

Isohyet A line connecting points of equal rainfall total.

Isolated state See **von Thünen's Isolated State model**.

Isolation The condition of being geographically cut off or far removed from mainstreams of thought and action. It also denotes a lack of receptivity to outside influences, caused at least partially by poor **accessibility**.

Isotherm A line connecting points of equal temperature.

Isthmus A **land bridge**; a comparatively narrow link between larger bodies of land. Central America forms such a link between Mexico and South America.

Jakota Triangle The easternmost region of the East Asian realm, consisting of *Ja*pan, (South) *Ko*rea, and *Ta*iwan.

Juxtaposition Contrasting places in close proximity to one another.

Karst The distinctive natural landscape associated with the chemical erosion of soluble limestone rock.

Land alienation One society or culture group taking land from another. In Subsaharan Africa, for example, European **colonialists** took land from **indigenous** Africans and put it to new uses.

Land bridge A narrow **isthmian** link between two large landmasses. They are temporary features—at least in terms of geologic time—subject to appearance and disappearance as the land or sea level rises and falls.

Land Hemisphere The half of the globe containing the greatest amount of land surface, centered on western Europe (Fig. 1A-4). In **geomorphology**, this can also refer to the position of the African continent, which lies central to the world's landmasses.

Land reform The spatial reorganization of **agriculture** through the allocation of farmland (often expropriated from landlords) to **peasants** and tenants who never owned land.

Land tenure The way people own, occupy, and use land.

Landlocked An interior **state** surrounded by land. Without coasts, such a country is disadvantaged in terms of **accessibility** to international trade routes, and in the scramble for possession of areas of the **continental shelf** and control of the **exclusive economic zone** beyond.

Language family Group of languages with a shared but usually distant origin.

"Latin" American city model The Griffin-Ford model of intraurban spatial structure in the Middle American and South American realms, diagrammed in Figure 5A-7.

Latitude Lines of latitude are **parallels** that are aligned east-west across the globe, from 0° latitude at the equator to 90° North and South latitude at the poles.

Leached soil Infertile, reddish-appearing, tropical soil whose surface consists of oxides of iron and aluminum; all other soil nutrients have been dissolved and transported downward into the subsoil by percolating water associated with the heavy rainfall of moist, low-latitude climates.

Leeward The protected or downwind side of a **topographic** barrier with respect to the winds that flow across it.

Lignite Low-grade, brown-colored variety of **coal**.

Lingua franca A "common language" prevalent in a given area; a second language that can be spoken and understood by many peoples, although they speak other languages at home.

Littoral Coastal or coastland.

Llanos The interspersed **savanna** grasslands and scrub woodlands of the Orinoco River's wide basin that covers much of interior Colombia and especially Venezuela.

Local functional specialization A hallmark of Europe's **economic geography** that later spread to many other parts of the world, whereby particular people in particular places concentrate on the production of particular goods and services.

Location Position on the Earth's surface; see also **absolute location** and **relative location**.

Location theory A logical attempt to explain the locational pattern of an economic activity and the manner in which its producing areas are interrelated. The agricultural location theory that underlies the **von Thünen model** is a leading example.

Loess Deposit of very fine silt or dust that is laid down after having been windborne for a considerable distance. Notable for its fertility under **irrigation** and its ability to stand in steep vertical walls.

Longitude Angular distance (0° to 180°) east or west as measured from the *prime meridian* (0°) that passes through the Greenwich Observatory in suburban London, England. For much of its length across the mid-Pacific Ocean, the 180th meridian functions as the *international date line*.

Low island Low-lying coral islands of the Pacific Realm that—unlike **high islands**—cannot wrest sufficient moisture from the tropical maritime air to avoid chronic drought. Thus productive agriculture is impossible, and their modest populations must rely on fishing and the coconut palm for survival.

Lusitanian The Portuguese sphere, which by extension includes Brazil.

Madrassa Revivalist (**fundamentalist**) religious school in which the curriculum focuses on Islamic religion and law and requires rote memorization of the Qu'ran (Koran), Islam's holy book. Founded in former British India, these schools were most numerous in present-day Pakistan but have **diffused** as far as Turkey in the west and Indonesia in the east.

Maghreb The region occupying the northwestern corner of Africa, consisting of Morocco, Algeria, and Tunisia.

Main Street Canada's dominant **conurbation** that is home to nearly two-thirds of the country's inhabitants; extends southwestward from Quebec City in the mid-St. Lawrence Valley to Windsor on the Detroit River.

Mainland-Rimland framework The twofold regionalization of the Middle American realm based on its modern cultural history. The Euro-Amerindian *Mainland*, stretching from Mexico to Panama (minus the Caribbean coastal strip), was a self-sufficient zone dominated by **hacienda land tenure**. The Euro-African *Rimland*, consisting of that Caribbean coastal zone plus all of the Caribbean islands to the east, was the zone of the **plantation** that relied heavily on trade with Europe. See Figure 4A-6.

Maquiladora The term given to modern industrial plants in Mexico's U.S. border zone. These foreign-owned factories assemble imported components and/or raw materials, and then export finished manufactures, mainly to the United States. Import duties are disappearing under **NAFTA**, bringing jobs to Mexico and the advantages of low wage rates to the foreign entrepreneurs.

Marchland An area or **frontier** of uncertain boundaries that is subject to various national claims and an unstable political history. Refers specifically to the movement of various armies, refugees, and migrants across such zones.

Marine geography The geographic study of oceans and seas. Its practitioners investigate both the physical (e.g., coral-reef **biogeography**, ocean–**atmosphere** interactions, coastal **geomorphology**) as well as human (e.g., **maritime boundary-making**, fisheries, beachside development) aspects of oceanic environments.

Maritime boundary An international boundary that lies in the ocean. Like all boundaries, it is a vertical plane, extending from the seafloor to the upper limit of the air space in the atmosphere above the water.

Median-line boundary An international **maritime boundary** drawn where the width of a sea is less than 400 **nautical miles**. Because the **states** on either side of that sea claim **exclusive economic zones** of 200 nautical miles, it is necessary to reduce those claims to a (median) distance equidistant from each shoreline. **Delimitation** on the map almost always appears as a set of straight-line segments that reflect the configurations of the coastlines involved.

Medical geography The study of health and disease within a geographic context and from a spatial perspective. Among other things, this geographic field examines the sources, **diffusion** routes, and distributions of diseases.

Megacity Informal term referring to the world's most heavily populated cities; in this book, the term refers to a **metropolis** containing a population of greater than 10 million.

Megalopolis When spelled with a lower-case *m*, a synonym for **conurbation**, one of the large coalescing supercities forming in diverse parts of the world. When capitalized, refers specifically to the multimetropolitan (*Bosnywash*) corridor that extends along the northeastern U.S. seaboard from north of Boston to south of Washington, D.C.

Mercantilism Protectionist policy of European **states** during the sixteenth to the eighteenth centuries that promoted a **state**'s economic position in the contest with rival powers. Acquiring gold and silver and maintaining a favorable trade balance (more exports than imports) were central to the policy.

Meridian Line of **longitude**, aligned north-south across the globe, that together with **parallels** of **latitude** forms the global grid system. All meridians converge at both poles and are at their maximum distances from each other at the equator.

Mestizo Derived from the Latin word for *mixed*, refers to a person of mixed European (white) and Amerindian ancestry.

Métis Indigenous Canadian people of mixed native and European ancestry.

Metropolis Urban **agglomeration** consisting of a (central) city and its suburban ring. See also **urban (metropolitan) area**.

Metropolitan area See **urban (metropolitan) area**.

Migration A change in residence intended to be permanent. See also **forced, internal, international**, and **voluntary migration**.

Migratory movement Human relocation movement from a source to a destination without a return journey, as opposed to cyclical movement (see also **nomadism**).

Model An idealized representation of reality built to demonstrate its most important properties. A **spatial** model focuses on a geographical dimension of the real world, such as the **von Thünen model** that explains agricultural location patterns in a commercial economy.

Modernization In the eyes of the Western world, the Westernization **process** that involves the establishment of **urbanization**, a market (money) economy, improved circulation, formal schooling, adoption of foreign innovations, and the breakdown of traditional society. Non-Westerners mostly see "modernization" as an outgrowth of **colonialism** and often argue that traditional societies can be modernized without being Westernized.

Monsoon Refers to the seasonal reversal of wind and moisture flows in certain parts of the subtropics and lower-middle latitudes. The *dry monsoon* occurs during the cool season when dry offshore winds prevail. The *wet monsoon* occurs in the hot summer months, which produce onshore winds that bring large amounts of rainfall. The air-pressure differential over land and sea is the triggering mechanism, with windflows always moving from areas of relatively higher pressure toward areas of relatively lower pressure. Monsoons make their greatest regional impact in the coastal and near-coastal zones of South Asia (Fig. 8A-3), Southeast Asia, and East Asia.

Mosaic culture Emerging cultural-geographic framework of the United States, dominated by the fragmentation of specialized social groups into homogeneous communities of interest marked not only by income, race, and **ethnicity** but also by age, occupational status, and lifestyle. The result is an increasingly heterogeneous socio-spatial complex, which resembles an intricate mosaic composed of myriad uniform—but separate—tiles.

Mulatto A person of mixed African (black) and European (white) ancestry.

Multilingualism A society marked by a mosaic of local languages. Constitutes a **centrifugal force** because it impedes communication within the larger population. Often a *lingua franca* is used as a "common language," as in many countries of Subsaharan Africa.

Multinationals Internationally active corporations capable of strongly influencing the economic and political affairs of many countries in which they operate.

Muslim An adherent of the Islamic faith.

Muslim Front See **Islamic Front**.

NAFTA (North American Free Trade Agreement) The **free-trade area** launched in 1994 involving the United States, Canada, and Mexico.

Nation Legally a term encompassing all the citizens of a **state**, it also has other connotations. Most definitions now tend to refer to a group of tightly knit people possessing bonds of language, **ethnicity**, religion, and other shared **cultural** attributes. Such homogeneity actually prevails within very few states.

Nation-state A country whose population possesses a substantial degree of **cultural** homogeneity and unity. The ideal form to which most **nations** and **states** aspire—a political unit wherein the territorial state coincides with the area settled by a certain national group or people.

NATO (North Atlantic Treaty Organization) Established in 1950 at the height of the Cold War as a U.S.-led **supranational** defense pact to shield postwar Europe against the Soviet military threat. NATO is now in transition, expanding its membership while modifying its objectives in the post-Soviet era. Its 28 member-states (as of mid-2010) are: Albania, Belgium, Bulgaria, Canada, Croatia, Czech Republic, Denmark, Estonia, France, Germany, Greece, Hungary, Iceland, Italy, Latvia, Lithuania, Luxembourg, the Netherlands, Norway, Poland, Portugal, Romania, Slovakia, Slovenia, Spain, Turkey, the United Kingdom, and the United States.

Natural hazard A natural event that endangers human life and/or the contents of a **cultural landscape**.

Natural increase rate Population growth measured as the excess of live births over deaths per 1000 individuals per year. Natural increase of a population does not reflect either **emigrant** or **immigrant** movements.

Natural landscape The array of landforms that constitutes the Earth's surface (mountains, hills, plains, and plateaus) and the physical features that mark them (such as water bodies, soils, and vegetation). Each **geographic realm** has its distinctive combination of natural landscapes.

Natural resource Any valued element of (or means to an end using) the environment; includes minerals, water, vegetation, and soil.

Nautical mile By international agreement, the nautical mile—the standard measure at sea—is 6076.12 feet in length, equivalent to approximately 1.15 statute miles (1.85 km).

Near Abroad The 14 former Soviet republics that, in combination with the dominant Russian Republic, constituted the USSR. Since the 1991 breakup of the Soviet Union, Russia has asserted a sphere of influence in these now-independent countries, based on its proclaimed right to protect the interests of ethnic Russians who were settled there in substantial numbers during Soviet times.

Neocolonialism The term used by developing countries to underscore that the entrenched **colonial** system of international exchange and capital flow has not changed in the postcolonial era—thereby perpetuating the huge economic advantages of the developed world.

Network (transport) The entire regional **system** of transportation connections and nodes through which movement can occur.

Nevada See *tierra nevada*.

New World Order A description of the international system resulting from the collapse of the Soviet Union in which the balance of nuclear power theoretically no longer determines the destinies of **states**.

Nomadism Cyclical movement among a definite set of places. Nomadic peoples mostly are **pastoralists**.

North American Free Trade Agreement See **NAFTA**.

Nucleation Cluster; **agglomeration**.

Oasis An area, small or large, where the supply of water (from an **aquifer** or a major river such as the Nile) permits the transformation of the immediately surrounding **desert** into productive cropland.

Occidental Western. Also see *Oriental*.

Offshore banking Term referring to financial havens for foreign companies and individuals, who channel their earnings to accounts in such a country (usually an "offshore" island-state) to avoid paying taxes in their home countries.

Oligarchs Opportunists in post-Soviet Russia who used their ties to government to enrich themselves.

OPEC (Organization of Petroleum Exporting Countries) The international oil *cartel* or syndicate formed by a number of producing countries to promote their common economic interests through the formulation of joint pricing policies and the limitation of market options for consumers. The 12 member-states (as of late 2010) are: Algeria, Angola, Ecuador, Iran, Iraq, Kuwait, Libya, Nigeria, Qatar, Saudi Arabia, United Arab Emirates (UAE), and Venezuela.

Organic theory Friedrich Ratzel's theory of **state** development that conceptualized the state as a biological organism whose life—from birth through maturation to eventual senility and collapse—mirrors that of any living thing.

Oriental The root of the word "oriental" is from the Latin for *rise*. Thus it has to do with the direction in which one sees the sun "rise"—the east; *oriental* therefore means Eastern. *Occidental* originates from the Latin for fall, or the "setting" of the sun in the west; *occidental* therefore means Western.

Orographic precipitation Mountain-induced precipitation, especially when air masses are forced to cross **topographic** barriers. Downwind areas beyond such a mountain range experience the relative dryness known as the **rain shadow effect**.

Outback The name given by Australians to the vast, peripheral, sparsely settled interior of their country.

Outer city The non-central-city portion of the American **metropolis**; no longer "sub" to the "urb," this outer ring was transformed into a full-fledged city during the late twentieth century.

Overseas Chinese The more than 50 million ethnic Chinese who live outside China. Over half live in Southeast Asia, and many have become quite successful. A large number maintain links to China and as investors played a major economic role in stimulating the growth of **SEZs** and Open Cities in China's **Pacific Rim**.

Pacific Rim A far-flung group of countries and parts of countries (extending clockwise on the map from New Zealand to Chile) sharing the following criteria: they face the Pacific Ocean; they evince relatively high levels of economic development, industrialization, and urbanization; their imports and exports mainly move across Pacific waters.

Pacific Ring of Fire Zone of crustal instability along **tectonic plate** boundaries, marked by earthquakes and volcanic activity, that ring the Pacific Ocean Basin (see Fig. G-5).

Paddies (paddyfields) Ricefields.

Pandemic An outbreak of a disease that spreads worldwide.

Pangaea A vast, singular landmass consisting of most of the areas of the present-day continents. This supercontinent began to break up more than 200 million years ago when still-ongoing **plate** divergence and **continental drift** became dominant **processes** (see Fig. 6A-3).

Parallel An east-west line of **latitude** that is intersected at right angles by **meridians** of **longitude**.

Pastoralism A form of **agricultural** activity that involves the raising of livestock.

Peasants In a **stratified** society, peasants are the lowest class of people who depend on **agriculture** for a living. But they often own no land at all and must survive as tenants or day workers.

Peninsula A comparatively narrow, finger-like stretch of land extending from the main landmass into the sea. Florida and Korea are examples.

Peon (peone) Term used in Middle and South America to identify people who often live in serfdom to a wealthy landowner; landless **peasants** in continuous indebtedness.

Per capita Capita means *individual*. Income, production, or some other measure is often given per individual.

Perforated state A **state** whose territory completely surrounds that of another state. South Africa, which encloses Lesotho and is perforated by it, is a classic example.

Periodic market Village market that is open every third day or at some other regular interval. Part of a regional network of similar markets in a preindustrial, rural setting where goods are brought to market on foot and barter remains a leading mode of exchange.

Peripheral development Spatial pattern in which a country's or region's development (and population) is most heavily concentrated along its outer edges rather than in its interior. Australia, with its peripheral population distribution, is a classic example (note its **core area** in Fig. 11-5), and nearby New Zealand exhibits a similar configuration of people and activities.

Periphery See **core-periphery relationships**.

Permafrost Permanently frozen water in the near-surface soil and bedrock of cold environments, producing the effect of completely frozen ground. Surface can thaw during brief warm season.

Physical geography The study of the geography of the physical (natural) world. Its subfields encompass **climatology, geomorphology, biogeography, soil geography, marine geography**, and water **resources**.

Physical landscape Synonym for **natural landscape**.

Physiographic political boundaries Political boundaries that coincide with prominent physical features in the **natural landscape**—such as rivers or the crest ridges of mountain ranges.

Physiographic region (province) A **region** within which there prevails substantial **natural-landscape** homogeneity, expressed by a certain degree of uniformity in surface **relief, climate,** vegetation, and soils.

Physiography Literally means *landscape description*, but commonly refers to the total **physical geography** of a place; includes all of the natural features on the Earth's surface, including landforms, **climate,** soils, vegetation, and water bodies.

Physiologic density The number of people per unit area of **arable** land.

Pilgrimage A journey to a place of great religious significance by an individual or by a group of people (such as a pilgrimage [*hajj*] to Mecca for **Muslims**).

Plantation A large estate owned by an individual, family, or corporation and organized to produce a cash crop. Almost all plantations were established within the tropics; in recent decades, many have been divided into smaller holdings or reorganized as cooperatives.

Plate tectonics Plates are bonded portions of the Earth's mantle and crust, averaging 100 kilometers (60 mi) in thickness. More than a dozen such plates exist (see Fig. G-4), most of continental proportions, and they are in motion. Where they meet one slides under the other, crumpling the surface crust and producing significant volcanic and earthquake activity; a major mountain-building force.

Pleistocene Epoch Recent period of geologic time that spans the rise of humankind, beginning about 2 million years ago. Marked by *glaciations* (repeated advances of continental ice sheets) and more moderate *interglacials* (ice sheet contractions). Although the last 10,000 years are known as the **Holocene** Epoch, Pleistocene-like conditions seem to be continuing and we are most probably now living through another Pleistocene interglacial; thus the glaciers likely will return.

Plural(istic) society A society in which two or more population groups, each practicing its own **culture,** live adjacent to one another without mixing inside a single **state**.

Polder Land reclaimed from the sea adjacent to the shore of the Netherlands by constructing dikes and then pumping out the water trapped behind them.

Political geography The study of the interaction of geographic space and political **process**; the spatial analysis of political phenomena and processes.

Pollution The release of a substance, through human activity, which chemically, physically, or biologically alters the air or water it is discharged into. Such a discharge negatively impacts the environment, with possible harmful effects on living organisms—including humans.

Population decline A decreasing national population. Russia, which now loses about half a million people per year, is the best example. Also see **population implosion**.

Population density The number of people per unit area. Also see **arithmetic density** and **physiologic density** measures.

Population distribution The way people have arranged themselves in geographic space. One of human geography's most essential expressions because it represents the sum total of the adjustments that a population has made to its natural, cultural, and economic environments. A population distribution map is included in every chapter in this book.

Population expansion (explosion) The rapid growth of the world's human population during the past century, attended by accelerating *rates* of increase.

Population geography The field of geography that focuses on the spatial aspects of **demography** and the influences of demographic change on particular countries and regions.

Population implosion The opposite of **population explosion**; refers to the declining populations of many European countries and Russia in which the **death rate** exceeds the **birth rate** and **immigration** rate.

Population movement See **migration** and **migratory movement**.

Population projection The future population total that demographers forecast for a particular country. For example, in the Data Table in Appendix B such projections are given for all the world's countries for 2025.

Population (age-sex) structure Graphic representation (*profile*) of a population according to age and gender.

Postindustrial economy Emerging economy, in the United States and a number of other highly advanced countries, as traditional industry is increasingly eclipsed by a higher-technology productive complex dominated by services, information-related, and managerial activities.

Primary economic activity Activities engaged in the direct extraction of **natural resources** from the environment such as mining, fishing, lumbering, and especially **agriculture**.

Primate city A country's largest city—ranking atop the urban **hierarchy**—most expressive of the national culture and usually (but not in every case) the capital city as well.

Process Causal force that shapes a spatial pattern as it unfolds over time.

Productive activities The major components of the spatial economy. For individual components see: **primary economic activity, secondary economic activity, tertiary economic activity, quaternary economic activity,** and **quinary economic activity**.

Protruded state Territorial shape of a **state** that exhibits a narrow, elongated land extension (or *protrusion*) leading away from the main body of territory. Thailand is a leading example.

Push-pull concept The idea that **migration** flows are simultaneously stimulated by conditions in the source area, which tend to drive people away, and by the perceived attractiveness of the destination.

Qanat In **desert** zones, particularly in Iran and western China, an underground tunnel built to carry **irrigation** water by gravity flow from surrounding mountains (where **orographic precipitation** occurs) to the arid flatlands below.

Quaternary economic activity Activities engaged in the collection, processing, and manipulation of *information*.

Quinary economic activity Managerial or control-function activity associated with decision-making in large organizations.

Rain shadow effect The relative dryness in areas downwind of mountain ranges resulting from **orographic precipitation**, wherein moist air masses are forced to deposit most of their water content as they cross the highlands.

Rate of natural population increase See **natural increase rate**.

Realm See **geographic realm**.

Refugees People who have been dislocated involuntarily from their original place of settlement.

Region A commonly used term and a geographic concept of paramount importance. An **area** on the Earth's surface marked by specific criteria, which are discussed in the Introduction.

Regional boundary In theory, the line that circumscribes a **region**. But razor-sharp lines are seldom encountered, even in nature (e.g., a coastline constantly changes depending on the tide). In the **cultural landscape**, not only are regional boundaries rarely self-evident, but when they are ascertained by geographers they most often turn out to be **transitional** borderlands.

Regional character The personality or "atmosphere" of a **region** that makes it distinct from all other regions.

Regional complementarity See **complementarity**.

Regional concept The geographic study of **regions** and regional distinctions, as discussed in the Introduction.

Regional disparity The spatial unevenness in standard of living that occurs within a country, whose "average," overall income statistics invariably mask the differences that exist between the extremes of the wealthy **core** and the poorer, disadvantaged **periphery**.

Regional geography Approach to geographic study based on the spatial unit of the **region**. Allows for an all-encompassing view of the world, because it utilizes and integrates information from geography's topical (**systematic**) fields, which are diagrammed in Figure G-13.

Regional state A "natural economic zone" that defies political boundaries and is shaped by the global economy of which it is a part; its leaders deal directly with foreign partners and negotiate the best terms they can with the national governments under which they operate.

Regionalism The consciousness of and loyalty to a **region** considered distinct and different from the **state** as a whole by those who occupy it.

Relative location The regional position or **situation** of a place relative to the position of other places. Distance, **accessibility**, and connectivity affect relative location.

Relict boundary A political boundary that has ceased to function, but the imprint of which can still be detected on the **cultural landscape**.

Relief Vertical difference between the highest and lowest elevations within a particular area.

Religious revivalism See **revivalism (religious)**.

Relocation diffusion Sequential **diffusion process** in which the items being diffused are transmitted by their carrier agents as they relocate to new areas. The most common form of relocation diffusion involves the spreading of innovations by a **migrating** population.

Restrictive population policies Government policy designed to reduce the **rate of natural population increase**. China's one-child policy, instituted in 1979 after Mao's death, is a classic example.

Revivalism (religious) Religious movement whose objectives are to return to the foundations of that faith and to influence state policy. Often called *religious fundamentalism*; but in the case of Islam, **Muslims** prefer the term *revivalism*.

Rift valley The trough or trench that forms when a thinning strip of the Earth's crust sinks between two parallel faults (surface fractures).

Rural-to-urban migration The dominant **migration** flow from countryside to city that continues to transform the world's population, most notably in the less advantaged geographic realms.

Russification Demographic resettlement policies pursued by the central planners of the Soviet Empire, whereby ethnic Russians were encouraged to **emigrate** from the Russian Republic to the 14 non-Russian republics of the USSR.

Sahel Semiarid **steppeland** zone extending across most of Africa between the southern margins of the arid Sahara and the moister tropical **savanna** and forest zone to the south. Chronic drought, **desertification**, and overgrazing have contributed to severe famines in this area since 1970.

Savanna Tropical grassland containing widely spaced trees; also the name given to the tropical wet-and-dry climate type (*Aw*).

Scale Representation of a real-world phenomenon at a certain level of reduction or generalization. In **cartography**, the ratio of map distance to ground distance; indicated on a map as a bar graph, representative fraction, and/or verbal statement. *Macroscale* refers to a large area of national proportions; *microscale* refers to a local area no bigger than a county.

Scale economies See **economies of scale**.

Secondary economic activity Activities that process raw materials and transform them into finished industrial products; the *manufacturing* sector.

Sedentary Permanently attached to a particular area; a population fixed in its location; the opposite of **nomadic**.

Separate development The spatial expression of South Africa's "grand" **apartheid** scheme, whereby nonwhite groups were required to settle in segregated "homelands." The policy was dismantled when white-minority rule collapsed in the early 1990s.

Sequent occupance The notion that successive societies leave their cultural imprints on a place, each contributing to the cumulative **cultural landscape**.

Shantytown Unplanned slum development on the margins of cities in disadvantaged countries, dominated by crude dwellings and shelters mostly made of scrap wood and iron, and even pieces of cardboard.

Sharecropping Relationship between a large landowner and farmers on the land wherein the farmers pay rent for the land they farm by giving the landlord a share of the annual harvest.

Sharia The criminal code based in Islamic law that prescribes corporal punishment, amputations, stonings, and lashing for both major and minor offenses. Its occurrence today is associated with the spread of **religious revivalism** in **Muslim** societies.

Shatter Belt Region caught between stronger, colliding external cultural-political forces, under persistent stress, and often fragmented by aggressive rivals. Eastern Europe and mainland Southeast Asia are classic examples.

Shifting agriculture Cultivation of crops in recently cut and burned tropical-forest clearings, soon to be abandoned in favor of newly cleared nearby forest land. Also known as *slash-and-burn agriculture*.

Sinicization Giving a Chinese cultural imprint; Chinese **acculturation**.

Site The internal locational attributes of an urban center, including its local spatial organization and physical setting.

Situation The external locational attributes of an urban center; its **relative location** or regional position with reference to other non-local places.

Small-island developing economies The additional disadvantages faced by lower-income island-states because of their often small territorial size and populations as well as overland **inaccessibility**. Limited resources require expensive importing of many goods and services; the cost of government operations per capita are higher; and local production is unable to benefit from **economies of scale**. The eastern Caribbean islands of the Lesser Antilles are a good example.

Social stratification See **stratification (social)**.

Southern Ocean The ocean that surrounds Antarctica (discussed in Chapter 11).

Spatial Pertaining to space on the Earth's surface. Synonym for *geographic(al)*.

Spatial diffusion See **diffusion**.

Spatial interaction See **complementarity, intervening opportunity**, and **transferability**.

Spatial model See **model**.

Spatial process See **process**.

Spatial system The components and interactions of a **functional region**, which is defined by the areal extent of those interactions. See also **system**.

Special Administrative Region (SAR) Status accorded the former dependencies of Hong Kong and Macau that were taken over by China, respectively, from the United Kingdom in 1997 and Portugal in 1999. Both SARs received guarantees that their existing social and economic systems could continue unchanged for 50 years following their return to China.

Special Economic Zone (SEZ) Manufacturing and export center within China, created since 1980 to attract foreign investment and technology transfers. Seven SEZs—all located on China's Pacific coast—currently operate: Shenzhen, adjacent to Hong Kong; Zhuhai; Shantou; Xiamen; Hainan Island, in the far south; Pudong, across the river from Shanghai; and Binhai New Area, next to the port of Tianjin.

Squatter settlement See **shantytown**.

State A politically organized territory that is administered by a sovereign government and is recognized by a significant portion of the international community. A state must also contain a permanent resident population, an organized economy, and a functioning internal circulation system.

State boundaries The borders that surround **states** which, in effect, are derived through contracts with neighboring states negotiated by treaty. See **definition, delimitation**, and **demarcation**.

State capitalism Government-controlled corporations competing under free-market conditions, usually in a tightly regimented society. South Korea is a leading example. Also see *chaebol*.

State formation The creation of a **state** based on traditions of human **territoriality** that go back thousands of years.

State planning Involves highly centralized control of the national planning process, a hallmark of communist economic systems. Soviet central planners mainly pursued a grand political design in assigning production to particular places; their frequent disregard of the principles of economic geography contributed to the eventual collapse of the USSR.

State territorial morphology A **state's** geographical shape, which can have a decisive impact on its spatial cohesion and political viability. A **compact** shape is most desirable; among the less efficient shapes are those exhibited by **elongated, fragmented, perforated**, and **protruded** states.

Stateless nation A national group that aspires to become a **nation-state** but lacks the territorial means to do so; the Palestinians and Kurds of Southwest Asia are classic examples.

Steppe Semiarid grassland; short-grass prairie. Also the name given to the semiarid climate type (*BS*).

Stratification (social) In a layered or stratified society, the population is divided into a **hierarchy** of social classes. In an industrialized society, the working class is at the lower end; **elites** that possess capital and control the means of production are at the upper level. In the traditional **caste system** of Hindu India, the "untouchables" form the lowest class or caste, whereas the still-wealthy remnants of the princely class are at the top.

Subduction In **plate tectonics**, the **process** that occurs when an oceanic plate converges head-on with a plate carrying a continental landmass at its leading edge. The lighter continental plate overrides the denser oceanic plate and pushes it downward.

Subsequent boundary A political boundary that developed contemporaneously with the evolution of the major elements of the **cultural landscape** through which it passes.

Subsistence Existing on the minimum necessities to sustain life; spending most of one's time in pursuit of survival.

Subtropical Convergence A narrow marine **transition zone**, girdling the globe at approximately latitude 40°S, that marks the equatorward limit of the frigid **Southern Ocean** and the poleward limits of the warmer Atlantic, Pacific, and Indian oceans to the north.

Suburban downtown In the United States (and increasingly in other advantaged countries), a significant concentration of major urban activities around a highly accessible suburban location, including retailing, light industry, and a variety of leading corporate and commercial operations. The largest are now coequal to the American central city's **central business district (CBD)**.

Sunbelt The popular name given to the southern tier of the United States, which is anchored by the mega-States of California, Texas, and Florida. Its warmer climate, superior recreational opportunities, and other amenities have been attracting large numbers of relocating people and activities since the 1960s; broader definitions of the Sunbelt also include much of the western United States, particularly Colorado and the coastal Pacific Northwest.

Superimposed boundary A political boundary emplaced by powerful outsiders on a developed human landscape. Usually ignores preexisting cultural-spatial patterns, such as the border that still divides North and South Korea.

Supranational A venture involving three or more **states**—political, economic, and/or cultural cooperation to promote shared objectives. The **European Union** is one such organization.

System Any group of objects or institutions and their mutual interactions. Geography treats systems that are expressed spatially, such as in **functional regions**.

Systematic geography Topical geography: **cultural, political, economic geography**, and the like.

Taiga The subarctic, mostly **coniferous** snowforest that blankets northern Russia and Canada south of the **tundra** that lines the Arctic shore.

Takeoff Economic concept to identify a stage in a country's **development** when conditions are set for a domestic Industrial Revolution.

Taxonomy A **system** of scientific classification.

Technopole A planned techno-industrial complex (such as California's Silicon Valley) that innovates, promotes, and manufactures the products of the **postindustrial** informational economy.

Tectonics See **plate tectonics**.

Templada See *tierra templada*.

Terracing The transformation of a hillside or mountain slope into a step-like sequence of horizontal fields for intensive cultivation.

Territoriality A country's or more local community's sense of property and attachment toward its territory, as expressed by its determination to keep it inviolable and strongly defended.

Territorial sea Zone of seawater adjacent to a country's coast, held to be part of the national territory and treated as a component of the sovereign **state**.

Tertiary economic activity Activities that engage in *services*—such as transportation, banking, retailing, education, and routine office-based jobs.

Tierra caliente The lowest of the **altitudinal zones** into which the human settlement of Middle and South America is classified according to elevation. The *caliente* is the hot humid coastal plain and adjacent slopes up to 750 meters (2500 ft) above sea level. The natural vegetation is the dense and luxuriant tropical rainforest; the crops include sugar and bananas in the lower areas, and coffee, tobacco, and corn along the higher slopes.

Tierra fría Cold, high-lying **altitudinal zone** of settlement in Andean South America, extending from about 1800 meters (6000 ft) in elevation up to nearly 3600 meters (12,000 ft). **Coniferous** trees stand here; upward they change into scrub and grassland. There are also important pastures within the *fría*, and wheat can be cultivated.

Tierra helada In Andean South America, the highest-lying habitable **altitudinal zone**—ca. 3600 to 4500 meters (12,000 to 15,000 ft)—between the tree line (upper limit of the *tierra fría*) and the snow line (lower limit of the *tierra nevada*). Too cold and barren to support anything but the grazing of sheep and other hardy livestock.

Tierra nevada The highest and coldest **altitudinal zone** in Andean South America (lying above 4500 meters [15,000 ft]), an uninhabitable environment of permanent snow and ice that extends upward to the Andes' highest peaks of more than 6000 meters (20,000 ft).

Tierra templada The intermediate **altitudinal zone** of settlement in Middle and South America, lying between 750 meters (2500 ft) and 1800 meters (6000 ft) in elevation. This is the "temperate" zone, with moderate temperatures compared to the *tierra caliente* below. Crops include coffee, tobacco, corn, and some wheat.

Topography The surface configuration of any segment of **natural landscape**.

Toponym Place name.

Transculturation Cultural borrowing and two-way exchanges that occur when different **cultures** of approximately equal complexity and technological level come into close contact.

Transferability The capacity to move a good from one place to another at a bearable cost; the ease with which a commodity may be transported.

Transition zone An area of spatial change where the **peripheries** of two adjacent **realms** or **regions** join; marked by a gradual shift (rather than a sharp break) in the characteristics that distinguish these neighboring geographic entities from one another.

Transmigration The now-ended policy of the Indonesian government to induce residents of the overcrowded, **core-area** island of Jawa to move to the country's other islands.

Treaty ports Extraterritorial enclaves in China's coastal cities, established by European colonial invaders under unequal treaties enforced by gunboat diplomacy.

Triple Frontier The turbulent and chaotic area in southern South America that surrounds the convergence of Brazil, Argentina, and Paraguay. Lawlessness pervades this haven for criminal elements, which is notorious for money laundering, arms and other smuggling, drug trafficking, and links to terrorist organizations, including money flows to the Middle East.

Tropical deforestation See **deforestation**.

Tropical savanna See **savanna**.

Tsunami A seismic (earthquake-generated) sea wave that can attain gigantic proportions and cause coastal devastation. The tsunami of December 26, 2004, centered in the Indian Ocean near the Indonesian island of Sumatera (Sumatra), produced the first great natural disaster of the twenty-first century.

Tundra The treeless plain that lies along the Arctic shore in northernmost Russia and Canada, whose vegetation consists of mosses, lichens, and certain hardy grasses.

Turkestan Northeasternmost region of the North Africa/Southwest Asia realm. Known as Soviet Central Asia before 1992, its five (dominantly Islamic) former Soviet Socialist Republics have become the independent countries of Kazakhstan, Uzbekistan, Turkmenistan, Kyrgyzstan, and Tajikistan. Today Turkestan has expanded to include a sixth state, Afghanistan.

Uneven development The notion that economic development varies spatially, a central tenet of **core-periphery relationships** in realms, regions, and lesser geographic entities.

Unitary state A **nation-state** that has a centralized government and administration that exercises power equally over all parts of the **state**.

Unity of place Naturalist Alexander von Humboldt's notion that in a particular locale or region intricate connections exist among climate, geology, biology, and human cultures. This laid the foundation for modern geography as an *integrative discipline* marked by a spatial perspective.

Urbanization A term with several connotations. The proportion of a country's population living in urban places is its level of urbanization. The **process** of urbanization involves the movement to, and the clustering of, people in towns and cities—a major force in every geographic realm today. Another kind of urbanization occurs when an expanding city absorbs rural countryside and transforms it into suburbs; in the case of cities in disadvantaged countries, this also generates peripheral **shantytowns**.

Urban (metropolitan) area The entire built-up, nonrural area and its population, including the most recently constructed suburban appendages. Provides a better picture of the dimensions and population of such an area than the delimited municipality (central city) that forms its heart.

Urban realms model A spatial generalization of the contemporary large American city. It is shown to be a widely dispersed, multinodal metropolis consisting of increasingly independent zones or *urban realms*, each focused on its own **suburban downtown**; the only exception is the shrunken central realm, which is focused on the central city's **central business district** (see Fig. 3A-10).

Veld See **highveld**.

Voluntary migration Population movement in which people relocate in response to perceived opportunity, not because they are forced to **migrate**.

Von Thünen's Isolated State model Explains the location of **agricultural** activities in a commercial economy. A **process** of spatial competition allocates various farming activities into concentric rings around a central market city, with profit-earning capability the determining force in how far a crop locates from the market. The original (1826) Isolated State model now applies to the continental scale.

Wahhabism A particularly virulent form of (Sunni) **Muslim revivalism** that was made the official faith when the modern **state** of Saudi Arabia was founded in 1932. Adherents call themselves "Unitarians" to signify the strict fundamentalist nature of their beliefs.

Wallace's Line As shown in Figure 11-4, the zoogeographical boundary proposed by Alfred Russel Wallace that separates the marsupial fauna of Australia and New Guinea from the non-marsupial fauna of Indonesia.

Weather The immediate and short-term conditions of the **atmosphere** that impinge on daily human activities.

West Wind Drift The clockwise movement of water as a current that circles around Antarctica in the **Southern Ocean**.

Wet monsoon See **monsoon**.

Windward The exposed, upwind side of a **topographic** barrier that faces the winds that flow across it.

World city Either London, New York, or Tokyo. The highest-ranking urban centers of **globalization** with financial, high-technology, communications, engineering, and related industries reflecting the momentum of their long-term growth and **agglomeration**.

World geographic realm See **geographic realm**.

INDEX

Entries followed by an f *refer to figures. Any entry followed by a* t *refers to a table.*